Universitext

Universitext

Universitext is a series of textbooks that presents material from a wide variety of mathematical disciplines at master's level and beyond. The books, often well class-tested by their author, may have an informal, personal even experimental approach to their subject matter. Some of the most successful and established books in the series have evolved through several editions, always following the evolution of teaching curricula, to very polished texts.

Thus as research topics trickle down into graduate-level teaching, first textbooks written for new, cutting-edge courses may make their way into *Universitext.*

More information about this series at http://www.springer.com/series/223

Geon Ho Choe

Stochastic Analysis for Finance with Simulations

Springer

Geon Ho Choe
Department of Mathematical Sciences
and Graduate School of Finance
Korea Advanced Institute of Science
and Technology
Yuseong-gu, Daejeon
Republic of Korea

ISSN 0172-5939 ISSN 2191-6675 (electronic)
Universitext
ISBN 978-3-319-25587-3 ISBN 978-3-319-25589-7 (eBook)
DOI 10.1007/978-3-319-25589-7

Library of Congress Control Number: 2016939955

Mathematics Subject Classification: 91Gxx, 91G10, 91G20, 91G30, 91G60, 91G70, 91G80

Printed on acid-free paper

This Springer imprint is published by Springer Nature
The registered company is Springer International Publishing AG Switzerland

Preface

This book is an introduction to stochastic analysis and quantitative finance, including theoretical and computational methods, for advanced undergraduate and graduate students in mathematics and business, but not excluding practitioners in finance industry. The book is designed for readers who want to have a deeper understanding of the delicate theory of quantitative finance by doing computer simulations in addition to theoretical study. Topics include stochastic calculus, option pricing, optimal portfolio investment, and interest rate models. Also included are simulations of stochastic phenomena, numerical solutions of the Black–Scholes–Merton equation, Monte Carlo methods, and time series. Basic measure theory is used as a tool to describe probabilistic phenomena. The level of familiarity with computer programming is kept to a minimum. To make the book accessible to a wider audience, some background mathematical facts are included in the first part of the book and also in the appendices.

Financial contracts are divided into two groups: The first group consists of primary assets such as shares of stock, bonds, commodities, and foreign currencies. The second group contains financial derivatives such as options and futures on the underlying assets belonging to the first group. A financial derivative is a contract that promises payment in cash or delivery of an asset contingent on the behavior of the underlying asset in the future. The goal of this book is to present mathematical methods for finding how much one should pay for a financial derivative. To understand the option pricing theory we need ideas from various disciplines ranging over pure and applied mathematics, not to mention finance itself. We try to bridge the gap between mathematics and finance by using diagrams, graphs and simulations in addition to rigorous theoretical exposition. Simulations in this book are not only used as the computational method in quantitative finance, but can also facilitate an intuitive and deeper understanding of theoretical concepts.

Since the publications by Black, Scholes and Merton on option pricing in 1973, the theory of stochastic calculus, developed by Itô, has become the foundation for a new field called quantitative finance. In this book stochastic calculus is presented starting from the theoretical foundation. After introducing some fundamental ideas in quantitative finance in Part I, we present mathematical prerequisites such as

Lebesgue integration, basic probability theory, conditional expectation and stochastic processes in Part II. After that, fundamental properties of Brownian motion, the Girsanov theorem and the reflection principle are given in Part III. In Part IV we introduce the Itô integral and Itô's lemma, and then present the Feynman–Kac theorem. In Part V we present three methods for pricing options: the binomial tree method, the Black–Scholes–Merton partial differential equation, and the martingale method. In Part VI we analyze more examples of the martingale method, and study exotic options, American options and numeraire. In Part VII the variance minimization method for optimal portfolio investment is introduced for a discrete time model. In Part VIII some interest rate models are introduced and used in pricing bonds. In Part IX the Newton–Raphson method of finding implied volatility, time series models for estimating volatility, Monte Carlo methods for option prices, numerical solution of the Black–Scholes–Merton equation, and numerical solution of stochastic differential equations are introduced. In the appendices some mathematical prerequisites are presented which are necessary to understand the material in the main chapters, such as point set topology, linear algebra, ordinary differential equations, and partial differential equations. The graphs and diagrams in this book were plotted by the author using MATLAB and Adobe Illustrator, and the simulations were done using MATLAB.

Comments from the readers are welcome. For corrections and updates please check author's homepage http://shannon.kaist.ac.kr/choe/ or send an email.

This work was partially supported by Basic Science Research Program through the National Research Foundation of Korea (NRF) funded by the Ministry of Science, ICT and Future Planning (NRF-2015R1A2A2A01006176).

Geon H. Choe

Acknowledgements

The author wishes to thank Colin Atkinson, Suk Joon Byun, Dong-Pyo Chi, Hyeong-In Choi, Dong Myung Chung, Dan Crisan, Ole Hald, Yuji Ito, Sun Young Jang, Intae Jeon, Duk Bin Jun, Byung-Chun Kim, Dohan Kim, Hong Jong Kim, Jaewan Kim, Kyung Soo Kim, Minhyong Kim, Sol Kim, Soon Ki Kim, Tong Suk Kim, Ohsang Kwon, Yong Hoon Kwon, Jong Hoon Oh, Hyungju Park, Andrea Pascucci, Barnaby Sheppard, Sujin Shin, Ravi Shukla, Insuk Wee, Harry Zheng, and all his colleagues in the Department of Mathematical Sciences and the Graduate School of Finance at KAIST.

He wishes to thank Youngho Ahn, Soeun Choi, Hyun Jin Jang, Myeong Geun Jeong, Mihyun Kang, Bong Jo Kim, Chihurn Kim, Dong Han Kim, Kunhee Kim, Ki Hwan Koo, Soon Won Kwon, Dong Min Lee, Kyungsub Lee, Young Hoon Na, Jong Jun Park, Minseok Park, Byung Ki Seo and Seongjun Yoon. He also thanks many students who took various courses on quantitative finance taught over the years.

The author thanks the editors Joerg Sixt, Rémi Lodh and Catriona Byrne, and the staff at Springer, including Catherine Waite. He is also grateful to several anonymous reviewers who gave many helpful suggestions. The author thanks his mother and mother-in-law for their love and care. Finally, he wishes to thank his wife for her love and patience.

Contents

Part VI Examples of Option Pricing

Part VII Portfolio Management

Part VIII Interest Rate Models

List of Figures

List of Tables

List of Simulations

Acronyms

ACF	Autocorrelation function
AR	Autoregressive model
ARCH	Autoregressive conditional heteroskedasticity model
ARMA	Autoregressive moving average model
ATM	At-the-money option
BS	Black–Scholes
BSM	Black–Scholes–Merton
BTCS	Backward in time, central in space
CAPM	Capital Asset Pricing Model
cdf	Cumulative density function
CIR	Cox–Ingersoll–Ross model
DJIA	Dow Jones Industrial Average
EUR	Euro
EWMA	Exponentially weighted moving average model
FRA	Forward rate agreement
FTCS	Forward in time, central in space
FTSE	Financial Times Stock Exchange
GBP	(Great Britain) pound sterling
GARCH	Generalized ARCH model
HJB	Hamilton–Jacobi–Bellman equation
ICG	Inversive congruential generator
ITM	In-the-money
LCG	Linear congruential generator
LIBOR	London interbank offered rate
MA	Moving average model
MC	Monte Carlo method
MT19937	Mersenne twister
ODE	Ordinary differential equation
OTC	Over-the-counter
OTM	Out-of-the-money
PDE	Partial differential equation

pdf	Probability density function
SDE	Stochastic differential equation
STRIPS	Separate trading of registered interest and principal securities
T-bond	Treasury bond
USD	US dollar
YTM	Yield to maturity

List of Symbols

$\emptyset$	The empty set
∞	Infinity
$\mathbb{N}$	The set of natural numbers
$\mathbb{Z}$	The set of integers
$\mathbb{R}$	The set of real numbers
$\mathbb{C}$	The set of complex numbers
d	A metric
V	A vector space
$\|\| \cdot \|\|$	A norm
$\mathcal{F}$	A σ-algebra
$\mathcal{G}$	A sub-σ-algebra
$\{\mathcal{F}_t\}_{0 \leq t \leq T}$	A filtration
Ω	A probability space
A, B, E	Events, measurable subsets
X, Y	Random variables
$\mathbb{E}$	Expectation
$\mathrm{Var}(X)$	Variance
$\mathrm{Var}(\mathbf{X})$	Covariance matrix
Corr	Correlation
Π	A portfolio
Δ	Delta of an option
Γ	Gamma of an option
Θ	Theta of an option
B	Risk-free asset price
S	Risky asset price
r	Interest rate
$\{X_t\}$, $\{Y_t\}$	Stochastic processes
$\{W_t\}$	A Brownian motion
$\mathcal{H}_0^2$	The set of simple stochastic processes
$\mathcal{H}^2$	The closure of the set of simple stochastic processes
μ	Mean, drift coefficient

σ	Standard deviation, volatility
Σ, C	Covariance matrices
ρ	Correlation
A, B, C, M	Matrices
diag	Diagonal of a matrix
λ	An eigenvalue of a matrix
t	Index for time
K	Strike price of an option
T	Expiry date of an option, maturity date of a bond
$\mathbb{P}$	A (physical) probability
$\mathbb{Q}$	A risk-neutral probability
$N(x)$	Cdf of the standard normal distribution
$\int_\Omega X(\omega)\,\mathrm{d}\mathbb{P}$	Lebesgue integral
$\int_0^T f(t, W_t)\,\mathrm{d}t$	Itô integral
$\mathscr{L}$	Laplace transform
$\mathrm{erf}(x)$	The error function
$\mathbb{Z}_p$	The field of integers modulo p for a prime number p
$\mathbb{Z}_n$	The ring of integers modulo n
B	The backshift operator
$f * g$	Convolution
$C(t_1, t_2; I_1, I_2)$	A cylinder subset of Brownian sample paths
d_1	$\{\log(S/K) + (r + \frac{1}{2}\sigma^2)T\}/(\sigma\sqrt{T})$
d_2	$\{\log(S/K) + (r - \frac{1}{2}\sigma^2)T\}/(\sigma\sqrt{T})$
U	A uniformly distributed random variable
Z	A standard normal variable
$V_a^b(f)$	Variation of f on $[a, b]$

Part I
Introduction to Financial Mathematics

Chapter 1
Fundamental Concepts

A stock exchange is an organization of brokers and financial companies which has the purpose of providing the facilities for trade of company stocks and other financial instruments. Trade on an exchange is by members only. In Europe, stock exchanges are often called bourses. The trading of stocks on stock exchanges, physical or electronic, is called the stock market.

A *portfolio* is a collection of financial assets such as government bonds, stocks, commodities and financial derivatives. Mathematically speaking, an asset is a time-indexed sequence of random variables, and a portfolio is a linear combination of assets. For example, consider a bank deposit with value 1 at time $t = 0$ and B_t at time t and a stock with price S_t at time t per share. Then a portfolio consisting of a units of bond with value B_t at t and b shares of stock with value S_t at t has value $aB_t + bS_t$ at t if there is no trading before or at t. The coefficients a and b can be negative, which means borrowing bonds or stocks from someone else. A *contingent claim* is a claim that can be made depending on whether one or more specified outcomes occur.

In this chapter we briefly introduce three fundamental concepts in financial mathematics: risk, time value of money, and the no arbitrage principle.

1.1 Risk

Risk is the possibility of exposure to uncertain losses in the future. It is different from danger. Investment in stock is said to be risky not dangerous. Investors prefer less risky investment if all other conditions are identical. The basic principle of investment is that the payoff should be large if the risk level is high. To reduce risk in investment, it is better to diversify: Don't put all your eggs in one basket, as the proverb says. For that purpose, we usually construct a portfolio consisting of assets which are not strongly correlated or even negatively correlated.

G.H. Choe, *Stochastic Analysis for Finance with Simulations*, Universitext,
DOI 10.1007/978-3-319-25589-7_1

There are several different types of risk. First, credit risk is the risk of not receiving back loan from the borrower. Second, market risk is the risk caused by unexpected asset price movement. Finally, operational risk means the risks from malfunctioning of a computer system, or fraudulent activity of employees, etc.

Risk management refers to management activities to protect asset values from the risks in investment. Regulating agencies are concerned with risk management of the whole financial system as well as individual financial companies. If financial derivatives are used as tools for excessive speculation, financial crisis can occur.

When we completely eliminate the risk in an investment, we call it *hedging*. Consult [5] for the viewpoint that probability is a tool for understanding risks in investment. For general risks caused by natural catastrophes consult [105].

1.2 Time Value of Money

A dollar today is usually worth more than a dollar in the future due to the opportunity to lend it and receive interest or to invest in a business and make profit. When there is inflation, a dollar in the future has less purchasing power than a dollar today. Furthermore, there is credit risk: receiving money in the future rather than now has uncertainty in recovering the money. Time value of money is expressed the in terms of interest rate. For example, \$100 today, invested for one year at a 3% interest rate, will be \$103 one year from now. Interest rates are usually positive, however, zero or negative interest rates are possible, especially in a time of deflation or economic recession. As an economic policy, the central bank in a country may lower its interest rate to stimulate the national economy. Throughout the book the risk-free interest rate is assumed to be positive unless stated otherwise.

1.3 No Arbitrage Principle

An opportunity to make a profit without risking any future loss is called an *arbitrage*, which exists as a result of market inefficiencies. An *arbitrage trade* means a transaction generating risk-free profit by buying and selling related assets with no net investment of cash. For example, if the price of a product is \$10 in New York, but in Los Angeles, the same product is selling for \$15, and if someone buys the merchandise in New York and sells it in Los Angeles, he/she can profit from the difference without any risk. When expenses such as transportation cost and taxes are considered there might be no overall profit to the arbitrager, and no arbitrage exists in this case. If the market functions perfectly, there would be no arbitrage opportunities. All the participants in the financial market look for any possibility of arbitrage and make riskless profit if there exists one. Arbitrage chances only last for a very short period of time. Once an arbitrage opportunity is exploited, there would be no remaining arbitrage opportunity for others.

The concept of arbitrage may be defined in a formal and precise way in the following simple model of the financial market in which we consider only two time points $t = 0$ and $t = T > 0$. The values of a portfolio V at $t = 0$ and $t = T$ are denoted by V_0 and V_T, respectively. At $t = 0$, which represents the present time, all asset values and portfolio values are known and there is no uncertainty. The asset prices other than the risk-free bank deposit at time T are unknown and random, and hence a portfolio value V_T containing a risky asset is a random variable.

Definition 1.1 (Arbitrage) We say that there exists an *arbitrage* if there is a portfolio V satisfying one of the following conditions:

(i) $V_0 = 0$, and $V_T \geq 0$ with probability one and $V_T > 0$ with positive probability,
or

(ii) $V_0 < 0$, and $V_T \geq 0$ with probability one.

See also Definition 1.2. In more general cases, we assume that there exists a generally accepted concept of arbitrage even when it is not stated explicitly. The *no arbitrage principle* means that one cannot create positive value out of nothing, i.e., there is no free lunch.

1.4 Arbitrage Free Market

Consider a one-period model of a financial market which is observed at times $t = 0$ and $t = T$. Assume that there are N assets with prices $S_t^1, \ldots, S_t^N$ at $t = 0, T$. All the prices of N assets are already known at $t = 0$, and $S_T^1, \ldots, S_T^N$ are random. Uncertainty in the financial market at T is modelled using a finite number of possible future states. Suppose that there are M states at time T, which is represented by a sample space $\Omega = \{\omega_1, \ldots, \omega_M\}$. The probability that the future state is ω_j is equal to $p_j > 0$, $1 \leq i \leq M$, $p_1 + \cdots + p_M = 1$, and S_T^i takes one of the values $S_T^i(\omega_1), \ldots, S_T^i(\omega_M)$ depending on the future state. Let

$$\mathbf{S}_t = \begin{bmatrix} S_t^1 \\ \vdots \\ S_t^N \end{bmatrix}$$

and define the *matrix of securities* by

$$D = \begin{bmatrix} S_T^1(\omega_1) & \cdots & S_T^1(\omega_M) \\ \vdots & & \vdots \\ S_T^N(\omega_1) & \cdots & S_T^N(\omega_M) \end{bmatrix} .$$

An interpretation of D is given by the diagram

$$\begin{array}{c} \\ i\text{th asset} \rightarrow \end{array} \begin{array}{c} j\text{th state} \\ \downarrow \\ \left[\quad D_{ij} \quad \right] \end{array}$$

i.e., the entry D_{ij} represents the value (including dividend payments) of the ith asset in the jth state at time T. Let $\mathbf{v} \cdot \mathbf{w}$ denote the scalar product of $\mathbf{v}$, $\mathbf{w}$. A *portfolio* $V_t^{\boldsymbol{\pi}}$ of securities is defined by $\boldsymbol{\pi} \cdot \mathbf{S}_t$ where

$$\boldsymbol{\pi} = \begin{bmatrix} \pi_1 \\ \vdots \\ \pi_N \end{bmatrix} \in \mathbb{R}^N$$

and π_i is the number of units of the ith asset held in the portfolio. It is assumed to be constant over the time interval $[0, T]$. If $\pi_i > 0$, then the investor has a *long position* in the ith asset and will receive cash flow $\pi_i D_{ij}$ at time T depending on the outcome. If $\pi_i < 0$ for some i, then we have a *short position* in the ith asset, i.e., we borrow π_i unit of S^i at $t = 0$ and have the obligation to return it at time T. We assume that any investor can take short and long positions in arbitrary amounts of securities. That is, even though in the real financial market the values π_i are integers, we assume that π_i are real numbers in our mathematical model. With abuse of language, $\boldsymbol{\pi}$ is also called a portfolio since there is a one-to-one correspondence between the collections of $V_t^{\boldsymbol{\pi}}$ and $\boldsymbol{\pi}$. The total value of the portfolio defined by $\boldsymbol{\pi}$ is equal to

$$V_t^{\boldsymbol{\pi}} = \sum_{i=1}^{N} \pi_i S_t^i = \boldsymbol{\pi} \cdot \mathbf{S}_t$$

for $t = 0, T$. In particular, for $t = T$, the cash flow of the portfolio in the jth state is

$$V_T^{\boldsymbol{\pi}}(\omega_j) = \sum_{i=1}^{N} \pi_i S_T^i(\omega_j) = \sum_{i=1}^{N} \pi_i D_{ij} .$$

Now the definition of arbitrage given in Definition 1.1 can be restated as follows:

Definition 1.2 (Arbitrage Portfolio) Let $\mathbf{d}_1, \ldots, \mathbf{d}_M$ denote the columns of D. An *arbitrage portfolio* is a portfolio $\boldsymbol{\pi}$ satisfying one of the following conditions:

(i) $\boldsymbol{\pi} \cdot \mathbf{S}_0 = 0$ (zero cost), and $\boldsymbol{\pi} \cdot \mathbf{d}_j \geq 0$ for all j (no loss) and $\boldsymbol{\pi} \cdot \mathbf{d}_k > 0$ for some k (possibility of profit),

or

(ii) $\boldsymbol{\pi} \cdot \mathbf{S}_0 < 0$, and $\boldsymbol{\pi} \cdot \mathbf{d}_j \geq 0$ for all j.

If there does not exist any arbitrage portfolio, we say that the market is *arbitrage free*.

The second case can be reduced to the first one if an investor can buy risk-free bonds and lock in a profit in the following way: Buy risk-free bonds using the borrowed money $-\boldsymbol{\pi} \cdot \mathbf{S}_0 > 0$ and receive $e^{rT}(-\boldsymbol{\pi} \cdot \mathbf{S}_0)$ at time T, making a risk-free profit of

$$e^{rT}(-\boldsymbol{\pi} \cdot \mathbf{S}_0) - (-\boldsymbol{\pi} \cdot \mathbf{S}_0) > 0$$

where $r > 0$ is the risk-free interest rate. An efficient market does not allow arbitrage opportunities to last for long.

Definition 1.3 (Arrow–Debreu Security) If a security θ_t^i, $1 \le i \le M$, satisfies the condition that $\theta_T^i(\omega_i) = 1$ and $\theta_T^i(\omega_j) = 0$ for $j \ne i$, then it is called an *Arrow–Debreu security*.

A contingent claim at time T can be expressed as a linear combination of the Arrow–Debreu securities if they exist. A portfolio $\boldsymbol{\pi} = (1, \dots, 1)$ consisting of one unit of each Arrow–Debreu security is a risk-free asset that pays 1 in any future state. It may be regarded as a bond. The difference of prices of $\boldsymbol{\pi}$ at times 0 and T is regarded as interest payment.

Definition 1.4 (State Price Vector) A column vector $\boldsymbol{\psi} = (\psi_1, \dots, \psi_M)^t$ is called a *state price vector* if $\psi_i > 0$ for every i, $1 \le i \le M$, and

$$\mathbf{S}_0 = D\,\boldsymbol{\psi}\ . \tag{1.1}$$

Equivalently, there exist positive constants $\psi_1, \dots, \psi_M$ such that

$$S_0^i = \psi_1 S_T^i(\omega_1) + \cdots + \psi_M S_T^i(\omega_M) \tag{1.2}$$

for every $1 \le i \le N$, or

$$\mathbf{S}_0 = \psi_1 \mathbf{d}_1 + \cdots + \psi_M \mathbf{d}_M\ . \tag{1.3}$$

Remark 1.1 Note that (1.2) implies that the asset price today is equal to a weighted average of its future values, with weights given by the state price vector $\boldsymbol{\psi}$. The weights ψ_i, $1 \le i \le M$, are called the state prices, and ψ_i is the value of one unit of the Arrow–Debreu security θ^i at time 0, i.e., $\psi_i = \theta_0^i$, $1 \le i \le M$.

Throughout the rest of the section we use the notation $(x_1, \dots, x_n) \ge 0$ to mean that $x_i \ge 0$ for every i, $1 \le i \le n$. Similarly, $(x_1, \dots, x_n) > 0$ means that $x_i > 0$ for every i, $1 \le i \le n$. The following fact implies that asset prices and cash-flows satisfy certain relations in a no-arbitrage financial world.

Theorem 1.1 *A market is arbitrage free if and only if there exists a state price vector.*

Proof ($\Leftarrow$) Suppose that there exists an arbitrage portfolio $\boldsymbol{\pi}$. Then either it satisfies (i) or (ii) in Definition 1.2. In the first case the condition that $\boldsymbol{\pi} \cdot \mathbf{d}_j \geq 0$ for all j is equivalent to the condition that $\boldsymbol{\pi}^t D \geq 0$ with at least one of its entries being positive. Since $\boldsymbol{\psi} > 0$, we have

$$0 = \boldsymbol{\pi} \cdot \mathbf{S}_0 = \boldsymbol{\pi} \cdot (D\,\boldsymbol{\psi}) = \boldsymbol{\pi}^t (D\,\boldsymbol{\psi}) = (\boldsymbol{\pi}^t D)\boldsymbol{\psi} > 0 \;,$$

and thus we have a contradiction. Now we consider the second case. The condition that $\boldsymbol{\pi} \cdot \mathbf{d}_j \geq 0$ for all j is equivalent to the condition that $\boldsymbol{\pi}^t D \geq 0$. Since $\boldsymbol{\psi} > 0$, we have

$$0 > \boldsymbol{\pi} \cdot \mathbf{S}_0 = \boldsymbol{\pi} \cdot (D\,\boldsymbol{\psi}) = \boldsymbol{\pi}^t (D\,\boldsymbol{\psi}) = (\boldsymbol{\pi}^t D)\boldsymbol{\psi} \geq 0 \;,$$

and thus we have a contradiction as well.

($\Rightarrow$) Let

$$\mathbb{R}^{M+1}_+ = \{\mathbf{x} = (x_0, x_1, \ldots, x_n) \in \mathbb{R}^{M+1} : \mathbf{x} \geq 0\} \;,$$

and let

$$W = \{(-\boldsymbol{\pi} \cdot \mathbf{S}_0, \boldsymbol{\pi} \cdot \mathbf{d}_1, \ldots, \boldsymbol{\pi} \cdot \mathbf{d}_M) \in \mathbb{R}^{M+1} : \boldsymbol{\pi} \in \mathbb{R}^N\} \;.$$

Let $\widetilde{D}$ be an $N \times (M+1)$ matrix with its columns given by $-\mathbf{S}_0, \mathbf{d}_1, \ldots, \mathbf{d}_M$. Then $W = \{\boldsymbol{\pi}^t \widetilde{D} : \boldsymbol{\pi} \in \mathbb{R}^N\}$. Since an arbitrage portfolio does not exist by the assumption, the subspace W and the cone $\mathbb{R}^{M+1}_+$ intersect only at the origin $(0, \ldots, 0)$. For, if there exists (a portfolio) $\boldsymbol{\pi} \in \mathbb{R}^N$ such that $\boldsymbol{\pi}^t \widetilde{D} \geq 0$ and $\boldsymbol{\pi}^t \widetilde{D} \neq \mathbf{0}$, then either (i) $\boldsymbol{\pi} \cdot \mathbf{S}_0 \leq 0$, $\boldsymbol{\pi} \cdot \mathbf{d}_j \geq 0$ for every j, and there exists $1 \leq j \leq M$ such that $\boldsymbol{\pi} \cdot \mathbf{d}_j > 0$, or (ii) $\boldsymbol{\pi} \cdot \mathbf{S}_0 < 0$, $\boldsymbol{\pi} \cdot \mathbf{d}_j = 0$ for every j, both of which are impossible.

Hence there exists $\boldsymbol{\lambda} = (\lambda_0, \lambda_1, \ldots, \lambda_M) \in \mathbb{R}^{M+1}$ and a hyperplane

$$H = \{\mathbf{x} \in \mathbb{R}^{M+1} : \boldsymbol{\lambda} \cdot \mathbf{x} = 0\}$$

that separates $\mathbb{R}^{M+1}_+ \setminus \{\mathbf{0}\}$ and W in such a way that $W \subset H$ and $\boldsymbol{\lambda} \cdot \mathbf{x} > 0$ for every $\mathbf{x} \in \mathbb{R}^{M+1}_+ \setminus \{\mathbf{0}\}$, which holds if and only if $\lambda_j > 0$ for every j.

Since W is contained in H, for every $\boldsymbol{\pi} \in \mathbb{R}^N$ we have

$$-\lambda_0 \boldsymbol{\pi} \cdot \mathbf{S}_0 + \sum_{j=1}^{M} \lambda_j \boldsymbol{\pi} \cdot \mathbf{d}_j = 0 \;,$$

which implies that

$$-\lambda_0 \mathbf{S}_0 + \sum_{j=1}^{M} \lambda_j \mathbf{d}_j = \mathbf{0} \;,$$

or

$$\mathbf{S}_0 = \sum_{j=1}^{M} \frac{\lambda_j}{\lambda_0} \mathbf{d}_j = \sum_{j=1}^{M} \psi_j \mathbf{d}_j$$

where $\psi_j = \lambda_j/\lambda_0 > 0$, $1 \le j \le M$. □

For more details of the proof, consult [1, 29]. For an equivalent version, see Corollary 1.1.

Definition 1.5 (Complete Market) If a contingent claim at time T is expressed as a linear combination of the assets traded in the market, we say that the claim is *replicated*, or *hedged*. If any contingent claim at T can be replicated, then we say that the market is *complete*.

Theorem 1.2 *The market is complete if and only if the rank of the matrix D of securities is equal to M.*

Proof (⇒) Let θ_t^i, $1 \le i \le M$, be the Arrow–Debreu securities. Since the market is complete, there exist a_{ij}, $1 \le i \le N$, $1 \le j \le M$, such that

$$\begin{cases} \theta_T^1 &= a_{11} S_T^1 + \cdots + a_{N1} S_T^N \\ \vdots & \quad \vdots \qquad\qquad\quad \vdots \\ \theta_T^M &= a_{1M} S_T^1 + \cdots + a_{NM} S_T^N \end{cases}$$

Let $A = [a_{ij}]_{1 \le i \le N, 1 \le j \le M}$. Thus, if the future is in state ω_j, then

$$\begin{bmatrix} \theta_T^1(\omega_j) \\ \vdots \\ \theta_T^M(\omega_j) \end{bmatrix} = A^t \begin{bmatrix} S_T^1(\omega_j) \\ \vdots \\ S_T^N(\omega_j) \end{bmatrix},$$

and hence

$$\begin{bmatrix} \theta_T^1(\omega_1) & \cdots & \theta_T^1(\omega_M) \\ \vdots & & \vdots \\ \theta_T^M(\omega_1) & \cdots & \theta_T^M(\omega_M) \end{bmatrix} = A^t \begin{bmatrix} S_T^1(\omega_1) & \cdots & S_T^1(\omega_M) \\ \vdots & & \vdots \\ S_T^N(\omega_1) & \cdots & S_T^N(\omega_M) \end{bmatrix}.$$

Since $\theta_T^i(\omega_j) = 0$ or 1 depending on whether $i = j$ or not, we have

$$I = A^t D \tag{1.4}$$

where I denotes the $M \times M$ identity matrix. Hence

$$M = \text{rank}(I) \le \min\{\text{rank}(A^t), \text{rank}(D)\} \le \min\{N, M\},$$

and thus $M \le N$. Since rank$(D) \ge$ rank$(I) = M$ from (1.4), and since D is an $N \times M$ matrix, we conclude that rank$(D) = M$.

($\Leftarrow$) To prove that the market is complete, it suffices to find a matrix $A = [a_{ij}]$ satisfying (1.4). Taking the transposes of (1.4), we have $I = D^t A$. □

Note that if the market is complete then $N \ge M$. For more information, consult [1, 2, 9, 24, 29]. For continuous time market models the conditions for no-arbitrage and completeness are essentially the same, even though the necessary mathematical techniques are sophisticated. Interested readers are referred to [8].

1.5 Risk-Neutral Pricing and Martingale Measures

Theorem 1.3 *Let* $\boldsymbol{\psi} = (\psi_1, \ldots, \psi_M)$ *be a state price vector. Assume that there exists a risk-free security, i.e., a risk-free bond,* S^0 *such that* $S^0_0 = 1$ *and* $S^0_T = 1 + r$ *regardless of future state, where* $r > 0$ *is a constant that is regarded as the risk-free interest rate. Then*

$$\sum_{j=1}^{M} \psi_j = \frac{1}{1+r} \, .$$

Proof Since

$$S^0_0 = \psi_1 S^0_T(\omega_1) + \cdots + \psi_M S^0_T(\omega_M) \, ,$$

we have

$$1 = \psi_1(1+r) + \cdots + \psi_M(1+r) \, ,$$

and the proof is complete. □

Definition 1.6 (Risk-Neutral Probability or Martingale Measure) Let

$$p^*_j = \frac{\psi_j}{\sum_{k=1}^{M} \psi_k} = (1+r)\psi_j > 0 \, , \quad 1 \le j \le M \, .$$

Then the positive constants $p^*_1, \ldots, p^*_M$ add up to 1, and define a probability on the future states $\omega_1, \ldots, \omega_M$, which is called a *risk-neutral probability* or a *martingale measure*. It is not a real probability, but a formal probability representing the coefficients of a convex linear combination of given values.

We can rewrite (1.2) as

$$S_0^i = \sum_{k=1}^{M} p_k^* \frac{S_T^i(\omega_k)}{1+r} = \frac{1}{1+r} \mathbb{E}^* [S_T^i] \tag{1.5}$$

where $\mathbb{E}^*$ denotes the expectation under the martingale measure $(p_1^*, \ldots, p_M^*)$. The idea given in (1.5) is called the *risk-neutral pricing* of financial claims, and is the central theme for the rest of the book.

Remark 1.2 The state prices ψ_i, $1 \le i \le M$, are the prices at time $t = 0$ of the Arrow–Debreu securities θ^j, $1 \le j \le M$. To see why, note that

$$\theta_0^j = \mathbb{E}^* \left[\frac{1}{1+r} \theta_T^j \right] = \frac{1}{1+r} \sum_{k=1}^{M} p_k^* \, \theta_T^j(\omega_k) = \frac{1}{1+r} p_j^* = \psi_j \, .$$

This is why $\boldsymbol{\psi}$ is called the state price vector.

Corollary 1.1 *The following statements are equivalent:*

(i) There exists a state price vector.
(ii) There exists a martingale measure.
(iii) The market is arbitrage-free.

Proof Use Theorem 1.1. □

Theorem 1.4 (Uniqueness of Martingale Measure) *Suppose that the market is arbitrage-free. Then the market is complete if and only if there exists a unique martingale measure.*

Proof ($\Rightarrow$) Recall that $M \le N$ and $\text{rank}(D) = M$. Since D is an $N \times M$ matrix, and since $\text{rank}(D) + \text{nullity}(D) = M$, we have $\text{nullity}(D) = 0$. Suppose that there exist two distinct state price vectors $\boldsymbol{\psi}$ and $\boldsymbol{\phi}$. Since they are solutions of (1.1), we have $D(\boldsymbol{\psi} - \boldsymbol{\phi}) = \mathbf{0}$, and hence $D\mathbf{x} = \mathbf{0}$ for some $\mathbf{x} \neq \mathbf{0}$, which is a contradiction.

($\Leftarrow$) Let $\boldsymbol{\psi} > 0$ be the unique price. If the market were not complete, then $\text{rank}(D) < M$, and hence there exists $\mathbf{0} \neq \mathbf{v} \in \mathbb{R}^M$ such that $D\mathbf{v} = \mathbf{0}$. Hence $D(\boldsymbol{\psi} + \alpha\mathbf{v}) = D\boldsymbol{\psi} + \alpha D\mathbf{v} = \mathbf{S}_0$ and $\boldsymbol{\psi} + \alpha\mathbf{v} > 0$ for $|\alpha| < \delta$ and $\delta > 0$ sufficiently small. If we choose $\alpha \neq 0$, then $\boldsymbol{\psi} + \alpha\mathbf{v}$ is another state price vector, which contradicts the assumption. □

1.6 The One Period Binomial Tree Model

Consider the case when $N = 2$ and $D = 2$. There are two future states: up state ω_u and down state ω_d. Let C be any contingent claim with cash flows C^u in state ω_u and C^d in state ω_d at time T. (See Fig. 1.1.) If there are no arbitrage opportunities,

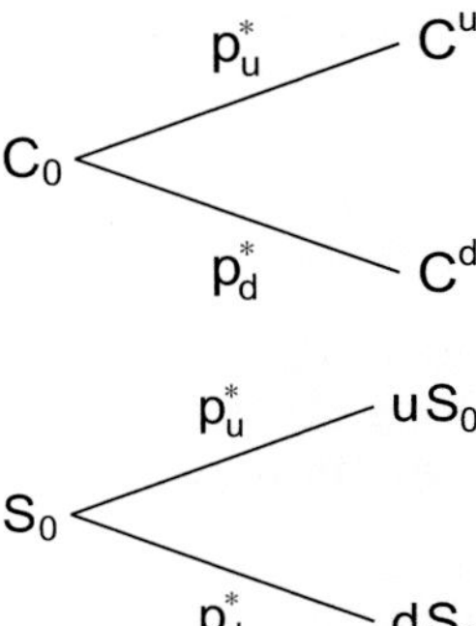

Fig. 1.1 The one period binomial tree and the risk-neutral probabilities

Fig. 1.2 The one period binomial tree and the risk-neutral probabilities

then there exists a martingale measure (p_u^*, p_d^*) such that the price C_0 at time 0 is given by

$$C_0 = \frac{1}{1+r}\mathbb{E}^*[C_T] = \frac{1}{1+r}\left(p_u^* C^u + p_d^* C^d\right) . \tag{1.6}$$

To see why, note that C can be replicated by a linear combination of two given assets, each of which has price given by (1.5). (See Exercise 1.2(i), which shows that the market is complete.)

We assume that the market consists of a risky asset S and a risk-free asset B such as a bond. Furthermore, S_T has values uS_0 and dS_0 where $0 < d < u$ depending on whether the future state is up or down. (See Fig. 1.2.)

Then, by (1.5) or (1.6) we have

$$S_0 = \frac{1}{1+r}\mathbb{E}^*[S_T] = \frac{1}{1+r}\left(p_u^* uS_0 + p_d^* dS_0\right) \tag{1.7}$$

and

$$B_0 = \frac{1}{1+r}\mathbb{E}^*[C_T] = \frac{1}{1+r}\left(p_u^*(1+r)B_0 + p_d^*(1+r)B_0\right) = B_0 . \tag{1.8}$$

We obtain no new information from (1.8). From (1.7), we have

$$1 + r = p_u^* u + p_d^* d . \tag{1.9}$$

Since $p_u^* + p_d^* = 1$, we finally obtain

$$p_u^* = \frac{1+r-d}{u-d} \quad \text{and} \quad p_u^* = \frac{u-(1+r)}{u-d} .$$

Thus $p_u^* > 0$ and $p_d^* > 0$ if and only if $d < 1+r < u$. Note that $p_u^* > 0$ and $p_d^* > 0$ is the condition for the existence of a martingale measure. Also note that the inequality

$d < 1 + r < u$ is the no arbitrage condition since if $d < u \le 1 + r$, for example, then every investor would buy the risk-free asset B making a risk-free profit that is at least the maximum possible profit from investment in the risky asset S. For more information, consult [77]. The one period binomial tree model is extended to the multiperiod binomial tree model in Chap. 14.

1.7 Models in Finance

Why do we need new financial models in addition to existing ones? A model reflects only some aspects of reality. Information obtained from a model is a shadow on the cave wall. The real financial market represented by a model can change over time, and models should be updated constantly. Technological advances allow better models, and increased computing power enables us to develop more computationally intensive models. New models are needed constantly, and are invented for new problems and techniques.

The more the better. A model is an approximation, and parameters in the model are not precisely known. Some models are practically oriented, while others give insights. A model should be stable under small perturbations. A slightly modified model of a classical model (e.g. the Black–Scholes–Merton differential equation) should produce a result that is close to a conventional solution. Using a modification of an existing classical model, we can test the existing model. Different models provide different viewpoints. A practitioner chooses a useful and convenient model.

Exercises

1.1 (Farkas' Lemma) We are given $M + 1$ vectors $\mathbf{d}_0, \mathbf{d}_1, \ldots, \mathbf{d}_M$ in $\mathbb{R}^N$. Then exactly one of following two statements is true:

(i) There exist nonnegative constants $\lambda_1, \ldots, \lambda_M$ such that

$$\mathbf{d}_0 = \lambda_1 \mathbf{d}_1 + \cdots + \lambda_M \mathbf{d}_M \ .$$

(ii) There exists a $\boldsymbol{\pi} \in \mathbb{R}^N$ such that

$$\boldsymbol{\pi} \cdot \mathbf{d}_0 < 0 \quad \text{and} \quad \boldsymbol{\pi} \cdot \mathbf{d}_j \ge 0$$

for every $1 \le j \le M$.

1.2 Consider the binomial tree model in Sect. 1.6.

(i) Find a condition under which there exists a unique solution (a, b) such that

$$\begin{cases} C^u = aS_T^1(\omega_u) + bS_T^2(\omega_u) \\ C^d = aS_T^1(\omega_d) + bS_T^2(\omega_d) \end{cases} .$$

(ii) Let B denote the risk-free bond. For $d < u$, show that there exists a unique solution (a, b) such that

$$\begin{cases} C^u = aS_T(\omega_u) + bB_T(\omega_u) \\ C^d = aS_T(\omega_d) + bB_T(\omega_d) \end{cases} .$$

Chapter 2
Financial Derivatives

Financial assets are divided into two categories: The first group consists of primary assets including shares of a stock company, bonds issued by companies or governments, foreign currencies, and commodities such as crude oil, metal and agricultural products. The second group consists of financial contracts that promise some future payment of cash or future delivery of the primary assets contingent on an event in the future date. The event specified in the financial contract is usually defined in terms of the behavior of an asset belonging to the first category, and such an asset is called an *underlying asset* or simply an *underlying*. The financial assets belonging to the second category are called derivatives since their values are derived from the values of the underlying asset belonging to the first category. *Securities* are tradable financial instruments such as stocks and bonds.

How much should one pay for a financial derivative? The price of a financial derivative depends upon the underlying asset, and that is why it is called a derivative. The buyer of a financial contract is said to hold the *long* position, and the seller the *short* position.

Short selling means selling securities one does not own. A trader borrows the securities from someone else and sells them in the market. At some stage one must buy back the securities so they can be returned to the lender. The short-seller must pay dividends and other benefits the owner of the securities would normally receive.

2.1 Forward Contracts and Futures

A *forward* contract is an agreement to buy or sell an asset on a specified future date, called the maturity date, for the forward price specified at the time the contract is initiated. There is no initial payment at the time the contract is signed, and there is no daily settlement during the lifetime of the contract except at the end of the life of the contract. At the maturity date, one party buys the asset for the agreed price from

G.H. Choe, *Stochastic Analysis for Finance with Simulations*, Universitext,
DOI 10.1007/978-3-319-25589-7_2

Table 2.1 Comparison of forward and futures

Forward	Futures
Private contract between two parties	Traded on an exchange
Not standardized	Standardized
Settled at end of contract	Settled daily
Delivery or final settlement	Usually closed out prior to maturity
Credit risk exists	Practically no credit risk

the other party. The buyer expects that the asset price will increase, while the seller hopes that it will decrease in the future.

A *futures* contract is the same as a forward contract except that futures are traded on exchanges and the exchange specifies certain standard features of the contract and a particular form of settlement. No money changes hands initially. However, futures contracts are settled daily, and on each day the difference between the price of the previous day and the present day is calculated and given to the appropriate party. Futures are available on a wide range of underlyings, and they are traded on an exchange. We need to specify what can be delivered, where it can be delivered, and when it can be delivered. A *margin* is cash or marketable securities deposited as collateral by an investor with his or her broker. The balance in the margin account is adjusted to reflect daily settlement. Margins minimize the risk of a default on a contract either by a buyer when the price falls or a seller when the price rises. Closing out a futures position involves entering into an offsetting trade and most futures contracts are closed out before maturity. If a futures contract is not closed out before maturity, it is usually settled by delivering the underlying asset. See Table 2.1. Forward and futures prices are equal if the interest rate is constant. For the proof, consult [41].

2.2 Options

An option is a financial contact that allows the option holder (or a buyer) a right to ask the issuer (or writer or seller) to do what is stated in the contract. The issuer has the obligation to do the work. For example, there are options on the right to sell or buy a stock at a fixed price, called an exercise price or a strike price, at a fixed date, called the expiry date or maturity date. A relatively recent type of option is on credit risk for a bond issued by a company. If the company is in good shape and the default risk level is low, then the issuer of the option on credit risk can make a substantial profit. On the other hand, if default risk level is high, then the issuer of the option is exposed to a high level of loss.

Options have financial values derived from underlying assets such as stocks, bonds and foreign exchange, and hence they are called *derivatives*. European options can be exercised only at expiry date, while American options can be exercised before

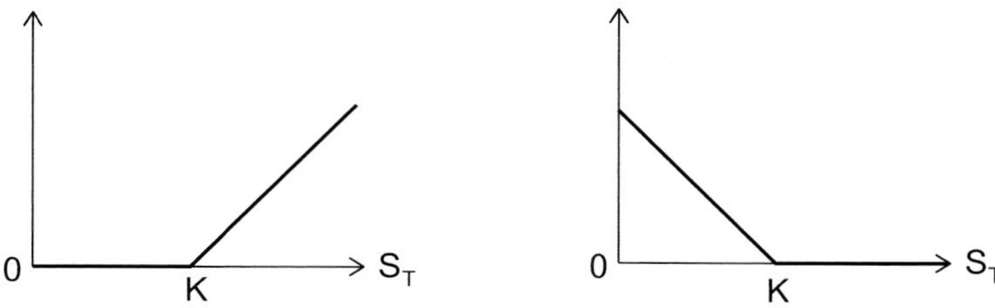

Fig. 2.1 Payoffs of a European call (*left*) and a European put (*right*)

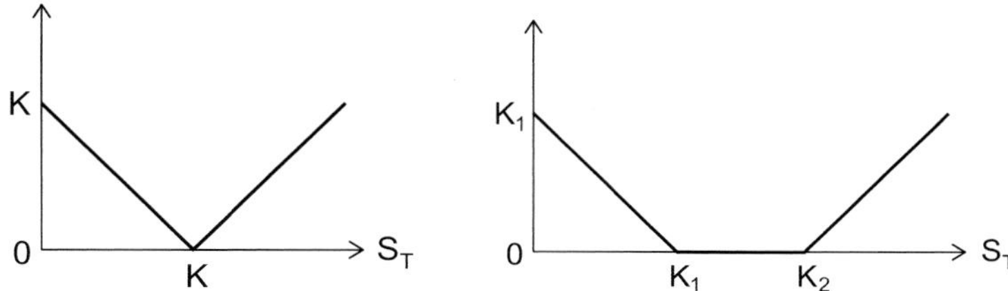

Fig. 2.2 Payoffs of a European straddle (*left*) and a European strangle (*right*)

or at expiry date. The value of an option if it can be exercised immediately is called its *intrinsic value*. Standardized options are traded at a regulated exchange, while over-the-counter (OTC) options are customized, and not traded on exchanges.

A call option gives the holder the right to buy an underlying asset at a strike price (or exercise price), and a put option gives the holder the right to sell. In Fig. 2.1 payoff functions at the expiry date T of a European call option and a European put option are plotted as functions of asset price S_T. A call option is insurance for an asset we plan to buy in the future, and a put option is insurance for an asset we already own.

European call options are bought when at expiry date the underlying asset price is expected to rise above the exercise price, and European put options are bought when underlying asset price is expected to fall below the exercise price. If we expect that the asset price will rise or fall above or below the exercise price and buy a call option and a put option of the same expiry date T and exercise price K, then that is equivalent to buying one option, called a straddle, with a payoff given on the left in Fig. 2.2. Therefore, its price is given by the sum of the prices of a European call option and a European put option. Since the set of all payoffs of European options of the same expiry date is a vector space, by taking linear combinations of several payoffs we can construct a new option as seen in the previous example.

If we expect that the asset price will move sufficiently far away from the strike price at expiry date, it is preferable to buy an option with a payoff, called a strangle given on the right in Fig. 2.2. In this case, the payoff is zero if the asset price at expiry date is between K_1 and K_2. Hence its price is the sum of prices of a put option with strike price K_1 and a call option with strike price K_2.

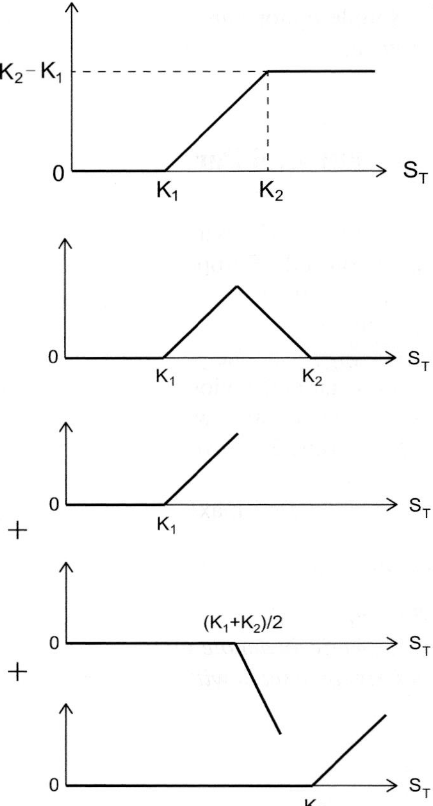

Fig. 2.3 Payoff of a European spread option

Fig. 2.4 Payoff of a butterfly

Fig. 2.5 A butterfly as a sum of two long calls and two contracts of a short call

If we expect that at expiry date the asset price will rise above the exercise price but not too much, then it is preferable to buy an option, called a *spread*, which is a combination of two or more options (both calls or both puts) on the same underlying asset with different strikes. Buying a spread option is equivalent to buying a call option of strike K_1 and selling a call option of strike K_2. See Fig. 2.3. Therefore, the price of a spread is the difference between the prices of a European call option with exercise price K_1 and a European call option with exercise price K_2. From the viewpoint of the seller of the option, the maximum of possible loss is bounded.

A European option called a *butterfly* has a payoff given in Fig. 2.4. To find its price, we first express the given payoff as a linear combination of payoffs of other options with their prices known. For example, we take three European call options with strike prices $K_1 < \frac{1}{2}(K_1 + K_2) < K_2$. We buy two call options, one with strike K_1 and one with strike K_2, and sell two contracts of a call option with strike price $\frac{1}{2}(K_1 + K_2)$. In other words, long two calls with strike prices K_1 and K_2, respectively, and short two contracts of a call with strike price $\frac{1}{2}(K_1 + K_2)$. See Fig. 2.5 where a butterfly is decomposed into a sum of long and short positions.

Simple options are called *vanilla*[1] options, and complicated options are called *exotic* options. There are many exotic options, and new varieties are being invented.

2.3 Put-Call Parity

The *put-call parity* is an equality which describes a relationship between a European put option and a European call option of the same expiry date and strike price. Thus, if we know the price of a call, we can find the price of a put with the same expiry date and exercise price, and vice versa.

Let $\{S_t\}_{t\geq 0}$ be the price process of the underlying asset S. Suppose that we buy a European call option on S with expiry date T and exercise price K and sell a European put option with the same expiry date and exercise price. Then the overall payoff at time T is given by

$$\max\{S_T - K, 0\} - \max\{K - S_T, 0\} = S_T - K .$$

See the graph in Fig. 2.6 and also Fig. 2.1.

Theorem 2.1 (Put-Call Parity) *Let $r \geq 0$ be a risk-free interest rate. Let C_t and P_t respectively denote the prices of European call and put options at time t on the underlying asset S_t with expiry date T and exercise price K. Then*

$$C_0 - P_0 = S_0 - K\mathrm{e}^{-rT} .$$

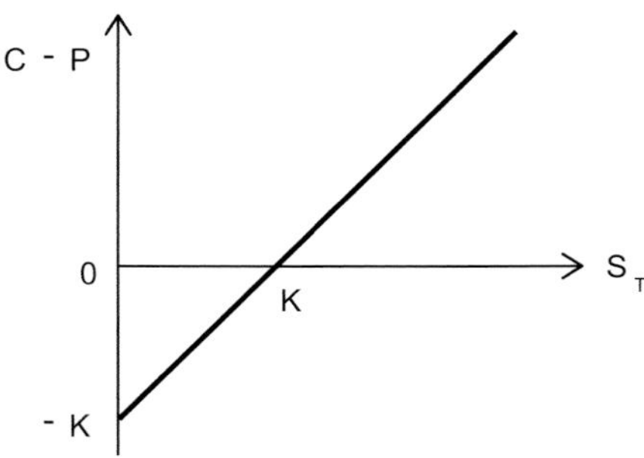

Fig. 2.6 Payoff when we buy a European call option and sell a put option

[1] Vanilla is an orchid with runners climbing up a tree. Its pods contain fragrant material. Although natural vanilla is expensive, synthesized vanilla fragrance is widely used, and thus the word 'vanilla' is a synonym for 'common'.

Table 2.2 Comparison of two portfolios

Time	0	T
Π^1	Call option and bank deposit Ke^{-rT}	$\max\{S_T - K, 0\} + K$
Π^2	Put option and a share of stock S_0	$\max\{K - S_T, 0\} + S_T$

Proof We will construct portfolios Π_t^1 and Π_t^2 with equal payoff at maturity. The portfolio Π^1 consists of a call and risk-free deposit Ke^{-rT} at $t = 0$. Its payoff at T is equal to

$$\Pi_T^1 = \max\{S_T - K, 0\} + K = \max\{S_T, K\} .$$

The portfolio Π^2 consists of a put and a share of the stock. Its payoff is equal to

$$\Pi_T^2 = \max\{K - S_T, 0\} + S_T = \max\{K, S_T\} .$$

See Table 2.2. Define a new portfolio $\Pi_t^3 = \Pi_t^1 - \Pi_t^2$. Since $\Pi_T^3 = \Pi_T^1 - \Pi_T^2 = 0$ with probability 1, the no arbitrage principle implies that $\Pi_0^3 = 0$, and hence $\Pi_0^1 - \Pi_0^2 = 0$. Hence $C_0 + Ke^{-rT} = P_0 + S_0$. □

Remark 2.1 The put-call parity makes sense even in the following extreme cases:

(i) If $K = 0$, then holding a call option is equivalent to owning a share of the stock, and hence $C_0 = S_0$. Since it is impossible for the asset price to fall below 0, the put option has no value. Hence $P_0 = 0$.
(ii) If $T = 0$, i.e., the option is exercised immediately, then we have $C_0 = \max\{S_0 - K, 0\}$ and $P_0 = \max\{K - S_0, 0\}$.
(iii) If C_t and P_t denote the European call and put option prices at time $0 < t < T$, then we have

$$C_t - P_t = S_t - Ke^{-r(T-t)} ,$$

where $T - t$ is called the *time to expiry*.

2.4 Relations Among Option Pricing Methods

In the rest of book we study several different methods for option pricing: First, the binomial tree method for discrete time models where the underlying asset can move only upward and downward, second, the partial differential equation approach introduced by F. Black, M. Scholes and R. Merton, and third, the martingale

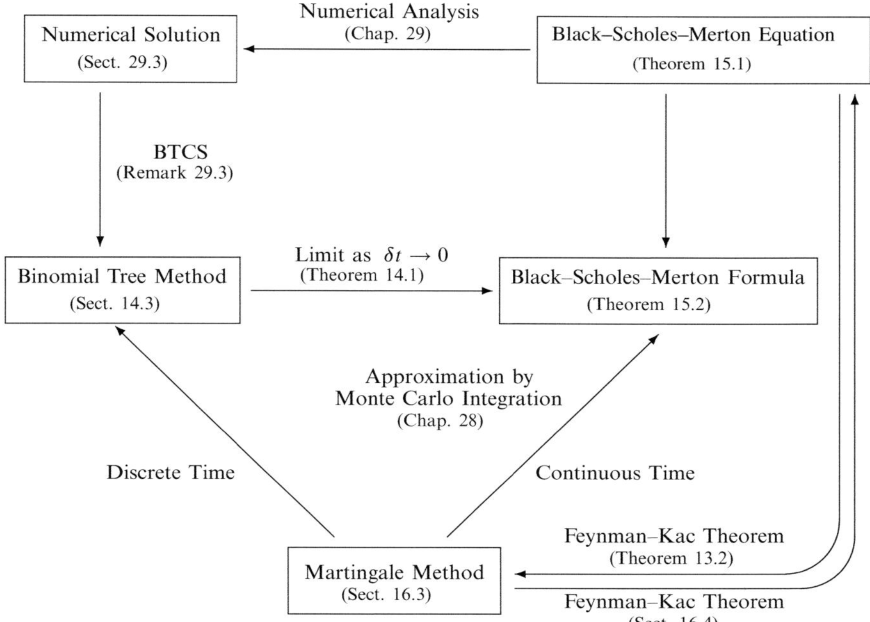

Fig. 2.7 Logical relations among option pricing methods

method that is most sophisticated mathematically and is the theoretical basis for the Monte Carlo method. The Girsanov theorem is at the center of the method. The logical relations among them are presented in Fig. 2.7. The N-period binomial tree method has a solution which converges to the Black–Scholes–Merton formula as $N \to \infty$ for some special choice of parameters as mentioned in Sect. 14.4. The partial differential equation approach is combined with numerical methods when it is hard to find closed form solutions. The Feynman–Kac theorem provides the bridge between the partial differential equations approach and the martingale method.

Exercises

2.1 Write down the formula for a payoff of a portfolio consisting of a short position on an asset and a call option on the same asset. (Hint: Consult the graph for a payoff in Fig. 2.8.)

2.2 Show that a call option is insurance for a short position while a put option is insurance for a long position.

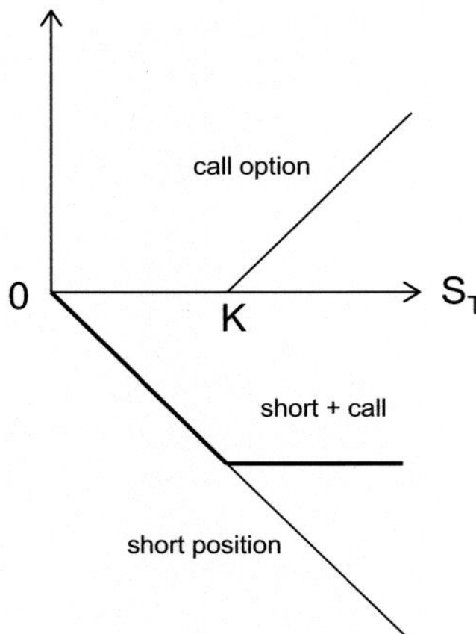

Fig. 2.8 Sum of a short position on an asset and a call option

2.3 Express a strangle as a linear combination of other options.

2.4 Using Fig. 2.5 as a hint, express the price of the given butterfly option as a sum of prices of calls and a put.

Part II
Probability Theory

Chapter 3
The Lebesgue Integral

Given an abstract set Ω, how do we measure the size of one of its subsets A? When Ω has finite or countably infinite elements, it is natural to count the number of elements in A. However, if A is an uncountable set, we need a rigorous and systematic method. In many interesting cases there is no logical way to measure the sizes of *all* the subsets of Ω, however, we define the concept of size for sufficiently many subsets. Subsets whose sizes can be determined are called measurable subsets, and the collection of these measurable subsets is called a σ-algebra where the set operations such as union and intersection resemble operations on numbers such as addition and multiplication. After measures are introduced we define the Lebesgue integral.

3.1 Measures

When a set is endowed with two operations which resemble the addition and the multiplication of numbers, we call the set an algebra or a field. If the set under consideration is given by a collection of subsets of a given set, then the operations are usually given by set union and set intersection.

Definition 3.1 (Algebra) A collection $\mathcal{F}_0$ of subsets of a set Ω is called an *algebra* on Ω if it satisfies the following conditions:

(i) $\emptyset, \Omega \in \mathcal{F}_0$.
(ii) If $A \in \mathcal{F}_0$, then $X \setminus A \in \mathcal{F}_0$.
(iii) If $A, B \in \mathcal{F}_0$, then $A \cup B \in \mathcal{F}_0$.

G.H. Choe, *Stochastic Analysis for Finance with Simulations*, Universitext,
DOI 10.1007/978-3-319-25589-7_3

Definition 3.2 (σ-Algebra) A collection $\mathcal{F}$ of subsets of a set Ω is called a *σ-algebra* on Ω if it satisfies the following conditions:

(i) $\emptyset, \Omega \in \mathcal{F}$.
(ii) If $A \in \mathcal{F}$, then $X \setminus A \in \mathcal{F}$.
(iii) If $A_1, A_2, A_3, \ldots \in \mathcal{F}$, then $\bigcup_{n=1}^{\infty} A_n \in \mathcal{F}$. A subset belonging to $\mathcal{F}$ is called an *$\mathcal{F}$-measurable* subset.

Remark 3.1

(i) A σ-algebra is an algebra.
(ii) A σ-algebra is not necessarily closed under uncountable unions. The prefix σ indicates 'countable', just as Σ represents countable summation.
(iii) If $\mathcal{P} = \{B_1, \ldots, B_k\}$ is a finite partition of Ω, then there exists an associated σ-algebra $\mathcal{F}$ whose measurable subsets are unions of some subsets in $\mathcal{P}$. More precisely, if $A \in \mathcal{F}$ then $A = C_1 \cup \cdots \cup C_k$ where $C_i = B_i$ or $C_i = \emptyset$ for $1 \leq i \leq k$.
(iv) If $\mathcal{F}_\gamma$ is a σ-algebra for every $\gamma \in \Gamma$, then $\bigcap_{\gamma \in \Gamma} \mathcal{F}_\gamma$ is also a σ-algebra.
(v) For a collection $\mathcal{S}$ of subsets of a given set Ω there exists the smallest σ-algebra among all the sub-σ-algebras containing $\mathcal{S}$, which is denoted by $\sigma(\mathcal{S})$ and called the σ-algebra generated by $\mathcal{S}$.
(vi) When a σ-algebra $\mathcal{F}$ is given on Ω, we call it a measurable space and write $(\Omega, \mathcal{F})$.More than one σ-algebra may be defined on the same set Ω.
(v) Let $X : \Omega \to \mathbb{R}^1$ be a function on a measurable space $(\Omega, \mathcal{F})$. Then the smallest sub-σ-algebra generated by subsets of the form $X^{-1}(B)$, where $B \subset \mathbb{R}^1$ are Borel subsets, is called the sub-σ-algebra generated by X and denoted by $\sigma(X)$.

Definition 3.3 (Measure) A *measure* is a rule which assigns a nonnegative number or $+\infty$ to each measurable subset. More precisely, a measure $\mathbb{P}$ on $(\Omega, \mathcal{F})$ is a function

$$\mathbb{P} : \mathcal{F} \to [0, +\infty) \cup \{+\infty\}$$

satisfying the following conditions:

(i) $\mathbb{P}(\emptyset) = 0$.
(ii) If $A_1, A_2, A_3, \ldots \in \mathcal{F}$ are pairwise disjoint, then

$$\mathbb{P}\left(\bigcup_{n=1}^{\infty} A_n\right) = \sum_{n=1}^{\infty} \mathbb{P}(A_n) .$$

When $\mathbb{P}(\Omega) = 1$, then $\mathbb{P}$ is called a probability measure.

When a measure μ is defined on $(\Omega, \mathcal{F})$, we call Ω a measure space, and write $(\Omega, \mathcal{F}, \mathbb{P})$ or $(\Omega, \mathbb{P})$. If $\mathbb{P}$ is a probability measure, then $(\Omega, \mathcal{F}, \mathbb{P})$ is called a probability measure space, or a probability space.

Example 3.1 In $\mathbb{R}^n$ a set of the form $[a_1, b_1]\times\cdots\times[a_n, b_n]$ is called an n-dimensional rectangle. Consider the collection $\mathcal{R}$ of n-dimensional rectangles. Using the concept of volume, define $\mu_{\mathcal{R}} : \mathcal{R} \to [0, +\infty) \cup \{+\infty\}$ by

$$\mu_{\mathcal{R}}([a_1, b_1] \times \cdots \times [a_n, b_n]) = (b_1 - a_1) \times \cdots \times (b_n - a_n) .$$

Let $\mathcal{B}_0$ be the σ-algebra generated by $\mathcal{R}$. Then we can extend $\mu_{\mathcal{R}}$ to a measure μ_0 defined on $\mathcal{B}_0$. (We call $\mathcal{B}_0$ the Borel σ-algebra, and a measurable set belonging to $\mathcal{B}_0$ a Borel set.) Next, extend the σ-algebra $\mathcal{B}_0$ by including all the subsets $N \subset A$ such that $\mu_0(A) = 0$, and form a new σ-algebra $\mathcal{B}$. Also extend μ_0 to $\mathcal{B}$ as a measure on $\mathcal{B}$, which is denoted by μ. The extended measure μ is called Lebesgue measure. If $A \in \mathcal{B}$ and $\mu(A) = 0$ and if $B \subset A$, then $B \in \mathcal{B}$ and $\mu(B) = 0$. Such a measure is said to be complete.

Fact 3.1 (Monotonicity of Measure) *Consider a sequence of measurable subsets $A_1, A_2, A_3, \cdots$ in a probability measure space $(\Omega, \mathcal{F}, \mathbb{P})$.*

(i) If $A_1 \supset A_2 \supset A_3 \supset \cdots$, then

$$\lim_{n\to\infty} \mathbb{P}(A_n) = \mathbb{P}\Big(\bigcap_{n=1}^{\infty} A_n\Big) .$$

(ii) If $A_1 \subset A_2 \subset A_3 \subset \cdots$, then

$$\lim_{n\to\infty} \mathbb{P}(A_n) = \mathbb{P}\Big(\bigcup_{n=1}^{\infty} A_n\Big) .$$

Example 3.2 (Binary Expansion) Consider an infinite product $\Omega = \prod_1^{\infty}\{0, 1\}$. Elements of Ω are infinitely long binary sequences $\omega = (\omega_1, \omega_2, \omega_3 \ldots)$ where $\omega_i = 0$ or 1. By the binary expansion a real number belonging to the unit interval can be written as $\sum_{i=1}^{\infty} a_i 2^{-i}$, $a_i \in \{0, 1\}$, which is identified with a sequence $(a_1, a_2, a_3, \ldots) \in \Omega$. Define a cylinder subset of length n by

$$[a_1, \ldots, a_n] = \{\omega \in \Omega : \omega_1 = a_1, \ldots, \omega_n = a_n\} .$$

The collection of all cylinder sets is denoted by $\mathcal{R}$. For $0 \le p \le 1$ define the set function $\mu_p : \mathcal{R} \to [0, \infty)$ by

$$\mu_p([a_1, \ldots, a_n]) = p^k (1 - p)^{n-k}$$

where k is the number of times that the symbol '0' appears in the string $a_1, \ldots, a_n$. Then the Kolmogorov Extension Theorem implies that μ_p extends to the σ-algebra generated by $\mathcal{R}$, which is still denoted by μ_p.

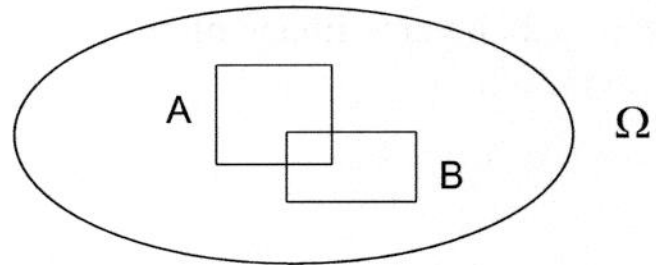

Fig. 3.1 The intersection of A and B

Fact 3.2 (Kolmogorov Extension Theorem) *If μ is a function defined on the collection of all cylinder subsets satisfying the consistency conditions such as $\mu([a_1, \ldots, a_{n-1}]) = \sum_a \mu([a_1, \ldots, a_{n-1}, a])$, then it can be extended to a measure on the σ-algebra generated by the cylinder subsets.*

Definition 3.4 (Conditional Measure) Given a measurable subset A of positive measure on a probability space $(\Omega, \mathcal{F}, \mathbb{P})$, define a new measure $\mathbb{P}_A$ by

$$\mathbb{P}_A(B) = \frac{\mathbb{P}(B \cap A)}{\mathbb{P}(A)} .$$

Then $\mathbb{P}_A$ is called a *conditional measure*, usually denoted by $\mathbb{P}(\cdot\,|A)$, i.e., $\mathbb{P}(B|A) = \mathbb{P}_A(B)$. See Fig. 3.1. Or, we may define a σ-algebra on A by $\mathcal{F}_A = \{B \in \mathcal{F} : B \subset A\}$, and define a probability measure $\mathbb{P}_A$ on $(A, \mathcal{F}_A)$ by $\mathbb{P}_A(B) = \mathbb{P}(B)/\mathbb{P}(A)$.

Definition 3.5 Let $(\Omega, \mathbb{P})$ be a probability space. Suppose that a statement H depends on $\omega \in \Omega$, i.e., H is true or false depending ω. If there exists an A such that $\mathbb{P}(\Omega \setminus A) = 0$ and if H holds for every $\omega \in A$, then we say that H holds almost everywhere or for almost every ω with respect to $\mathbb{P}$. In probability theory we say that H holds almost surely, or with probability 1.

3.2 Simple Functions

Definition 3.6 (Measurable Function) A function $X : (\Omega, \mathcal{F}) \to \mathbb{R}$ is said to be *measurable* if

$$X^{-1}((a, b)) \in \mathcal{F} \quad \text{for every } a < b .$$

Some other equivalent conditions are

$$X^{-1}((-\infty, a]) \in \mathcal{F} \ \text{ for every } a ,$$

$$X^{-1}((-\infty, a)) \in \mathcal{F} \ \text{ for every } a ,$$

$$X^{-1}((a, +\infty)) \in \mathcal{F} \ \text{ for every } a ,$$

$$X^{-1}([a, +\infty)) \in \mathcal{F} \ \text{ for every } a .$$

(See Fig. 3.2.) In probability theory a measurable function is called a *random variable.*

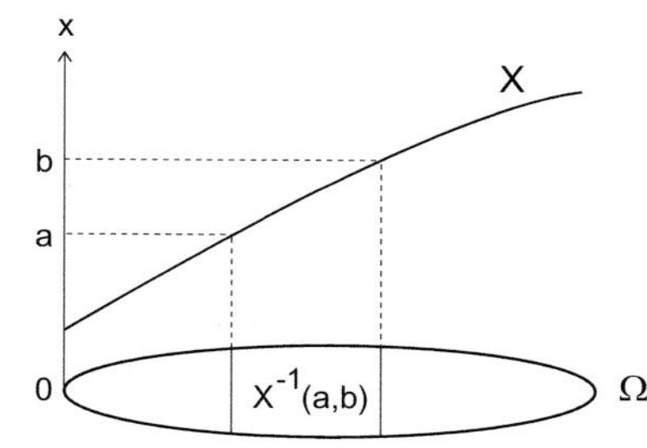

Fig. 3.2 Inverse image of an interval (a, b) under X

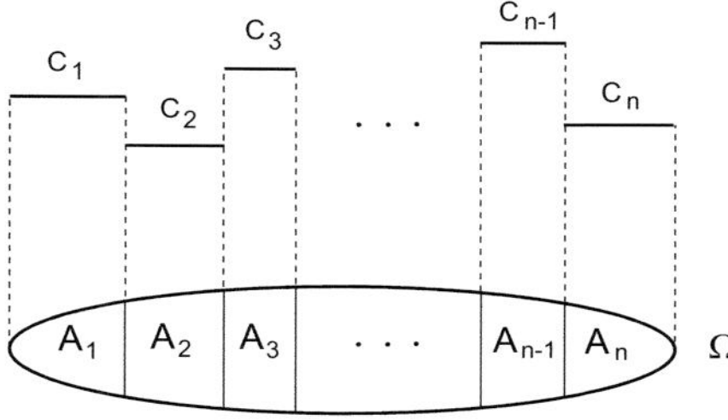

Fig. 3.3 A simple function on a measurable space

Definition 3.7 (Indicator Function) Let $A \subset \Omega$. A function $\mathbf{1}_A : \Omega \to \mathbb{R}^1$ defined by $\mathbf{1}_A(\omega) = 1$ for $\omega \in A$ and $\mathbf{1}_A(\omega) = 0$ for $\omega \notin A$ is called an *indicator function*. If A is a measurable subset, then $\mathbf{1}_A$ is a measurable function.

Remark 3.2 If σ-algebras $\mathcal{F}$ and $\mathcal{G}$ satisfy $\mathcal{G} \subset \mathcal{F}$, i.e., there are additional measurable subsets in $\mathcal{F}$, then a $\mathcal{G}$-measurable function is also $\mathcal{F}$-measurable. A function is more likely to be measurable with respect to a larger σ-algebra.

Definition 3.8 (Simple Function) A measurable function $s : \Omega \to \mathbb{R}$ is called a *simple function* if it takes finitely many values, i.e., it is of the form

$$s(\omega) = \sum_{i=1}^{n} c_i \mathbf{1}_{A_i}(\omega)$$

where A_i is a measurable set and c_i is constant for $1 \leq i \leq n$. We may assume that A_i are pairwise disjoint. See Fig. 3.3. Note that a simple function is a linear combination of indicator functions.

3.3 The Lebesgue Integral

H. Lebesgue in the early 1900s invented a new integration technique, which extended the classical Riemann integral in the sense that the domain of integrand for the new method can be an arbitrary set, not necessarily a subset of a Euclidean space.

In contrast to the partitioning of the x-axis for the Riemann integral, the Lebesgue integral partitions the y-axis. In the Lebesgue integral the concept corresponding to length, area, or volume in Euclidean spaces is measure. One of the advantages of the Lebesgue integral is that the operations of taking integral and taking limit can be done interchangeably.

When the Lebesgue integral was first invented, it was regarded as an unnecessarily abstract mathematical concept compared to Riemann integral, but in 1933 A.N. Kolmogorov used it in his book [52] to build an axiomatic foundation for probability theory, and the Lebesgue integral has become the language of probability theory. However, there exists a linguistic gap between real analysis, which is the area built upon the Lebesgue integral, and probability theory, and both disciplines still have their own terminology. For example, a measurable function in real analysis corresponds to a random variable in probability theory. Consult Table 4.1.

Now we find the average of a function defined on Ω with respect to a measure $\mathbb{P}$. We consider only measurable functions to define Lebesgue integration. Let $X \geq 0$ be a bounded measurable function on X, in other words, there exists a constant $M > 0$ such that $0 \leq X(\omega) < M$ for all $\omega \in \Omega$. Put

$$A_{n,k} = X^{-1}\left(\left[\frac{k-1}{2^n}M, \frac{k}{2^n}M\right)\right) .$$

From the fact that X is a measurable function, it can be easily seen that $A_{n,k}$ is a measurable set, whose size can be measured by a measure $\mathbb{P}$. Note that the sequence of simple functions

$$\sum_{k=1}^{2^n} \frac{k}{2^n} M \mathbf{1}_{A_{n,k}}$$

converges to X everywhere uniformly as $n \to \infty$. The limit of

$$\sum_{k=1}^{2^n} \frac{k}{2^n} M \times \mathbb{P}(A_{n,k})$$

is called the Lebesgue integral of X on $(\Omega, \mathbb{P})$, and is denoted by

$$\int_{\Omega} X \, \mathrm{d}\mathbb{P} .$$

(See Fig. 3.4.) In probability theory this Lebesgue integral is called the expectation, and denoted by $\mathbb{E}[X]$ or $\mathbb{E}^{\mathbb{P}}[X]$. For more details see Sect. 4.1.

In defining a Riemann integral of a continuous function the domain is a subset of a Euclidean space such as an interval or a rectangle, and we partition the domain which is usually represented by a horizontal axis or a horizontal plane. On the other hand, in Lebesgue theory the domain of a function under consideration is

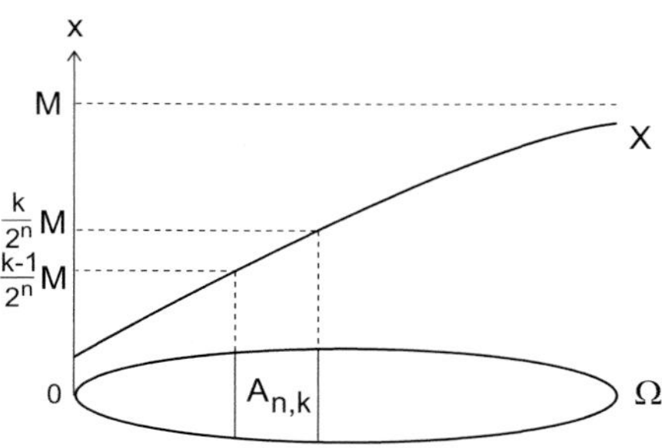

Fig. 3.4 Partition of a domain for a Lebesgue integral

not necessarily a subset of a Euclidean space and it is not possible to partition the domain in a natural way, and hence the vertical real axis representing the function values is partitioned. That is, the vertical axis is partitioned into sufficiently short intervals, and their inverse images are used to partition the domain of the function. Since the function is measurable, the inverse images are measurable subsets, and their sizes are measured by a given measure defined on the domain.

A real-valued function X can be written as

$$X = X^+ - X^-$$

where $X^+ = \max(X, 0) \geq 0$ and $X^- = \max(-X, 0) \geq 0$, and we define the integral of X by

$$\int_\Omega X \, d\mathbb{P} = \int_\Omega X^+ \, d\mathbb{P} - \int_\Omega X^- \, d\mathbb{P} \, .$$

Let $\mathbf{1}_A$ denote the indicator function of a measurable subset A. Then $\mathbf{1}_A$ is a measurable function, and for an arbitrary measure $\mathbb{P}$ we have

$$\int_\Omega \mathbf{1}_A \, d\mathbb{P} = \mathbb{P}(A) \, .$$

If two measurable sets A_1 and A_2 satisfy $\mathbb{P}(A_1 \triangle A_2) = 0$, then from the viewpoint of measure theory they may be regarded as the same sets, and we say that $A_1 = A_2$ modulo measure zero sets. When the integral of a measurable function X is finite, then X is said to be integrable. If the values of two measurable functions are different only on a subset of measure 0, then two functions have the same integral.

Example 3.3 For a finite or countably infinite set $\Omega = \{\omega_1, \ldots, \omega_n, \ldots\}$ we define its σ-algebra as the collection of all subsets of Ω. Note that every function is measurable on Ω. If a measure $\mathbb{P}$ on Ω is given by $\mathbb{P}(\{\omega_i\}) = p_i$ for every i, then

$$\int_\Omega X \, d\mathbb{P} = \sum_i p_i X(\omega_i) \, .$$

Such a measure is said to be *discrete*. It is a probability measure if and only if $\sum_i p_i = 1$.

The smallest σ-algebra on a Euclidean space generated by open subsets is called the Borel σ-algebra, on which Borel measure is defined. If we extend the Borel σ-algebra by including all the subsets of Borel measure zero subsets, then it is called a completion, and the corresponding extension of Borel measure is called Lebesgue measure. From the practical point of view, Lebesgue measure is an extension of the concept of volume in Euclidean space to subsets of irregular shape. On a finite closed interval the Riemann integral of a continuous function is equal to the Lebesgue integral.

Remark 3.3 Consider a measure μ on the real line $\mathbb{R}$. If a measurable function $\alpha(x) \geq 0$ satisfies

$$\mu(A) = \int_A \alpha(x)\,\mathrm{d}x$$

for every measurable set $A \subset \mathbb{R}$ where $\mathrm{d}x$ denotes Lebesgue measure, then μ is called an *absolutely continuous* measure, and we write $\mathrm{d}\mu = \alpha\,\mathrm{d}x$. In this case, for any measurable function $f : \mathbb{R} \to \mathbb{R}$ we have

$$\int_{\mathbb{R}} f(x)\,\mathrm{d}\mu = \int_{\mathbb{R}} f(x)\alpha(x)\,\mathrm{d}x\,.$$

Here some of important properties of the Lebesgue integral.

Fact 3.3 (Monotone Convergence Theorem) *Let $0 \leq X_1 \leq X_2 \leq \cdots$ be a monotonically increasing sequence of measurable functions on a measure space $(\Omega, \mathbb{P})$. Then*

$$\int_\Omega \lim_{n\to\infty} X_n\,\mathrm{d}\mathbb{P} = \lim_{n\to\infty} \int_\Omega X_n\,\mathrm{d}\mathbb{P}\,.$$

Fact 3.4 (Fatou's Lemma) *Let $X_n \geq 0$, $n \geq 1$, be a sequence of measurable functions on a measure space $(\Omega, \mathbb{P})$. Then*

$$\int_\Omega \liminf_{n\to\infty} X_n\,\mathrm{d}\mathbb{P} \leq \liminf_{n\to\infty} \int_\Omega X_n\,\mathrm{d}\mathbb{P}\,.$$

Fact 3.5 (Lebesgue Dominated Convergence Theorem) *Let X_n, $n \geq 1$, be a sequence of measurable functions on a measure space $(\Omega, \mathbb{P})$. Suppose that $\lim_{n\to\infty} X_n(\omega)$ exists at every $\omega \in \Omega$, and that there exists an integrable function $Y \geq 0$ such that $|X_n(\omega)| \leq Y(\omega)$ for every n. Then*

$$\int_\Omega \lim_{n\to\infty} X_n\,\mathrm{d}\mathbb{P} = \lim_{n\to\infty} \int_\Omega X_n\,\mathrm{d}\mathbb{P}\,.$$

Definition 3.9 (Lebesgue Space)

(i) For $1 \leq p < \infty$ define the L^p-norm $||\cdot||_p$ and the Lebesgue space L^p by

$$||X||_p = \left(\int_\Omega |X|^p \,\mathrm{d}\mathbb{P} \right)^{1/p} ,$$

and

$$L^p(\Omega, \mathbb{P}) = \left\{ X : ||X||_p < \infty \right\} .$$

Then $L^p(\Omega, \mathbb{P})$ is a vector space, and $||\cdot||_p$ is a norm. If $\mathbb{P}(\Omega) < \infty$ and $1 \leq r < p$, then $L^p(\Omega) \subset L^r(\Omega)$, and hence $L^p(\Omega) \subset L^1(\Omega)$ for $p \geq 1$. For $p = \infty$ define

$$||X||_\infty = \min\{0 \leq K < \infty : |X(\omega)| \leq K \text{ for almost every } \omega\}$$

$$L^\infty(X, \mathbb{P}) = \{X : ||X||_\infty < \infty\} .$$

For every $1 \leq p \leq \infty$ the normed space L^p is complete.

Definition 3.10 (Convergence)

(i) A sequence of functions $X_n \in L^p$ is said to *converge in* L^p to some $X \in L^p$ if

$$\lim_{n\to\infty} ||X_n - X||_p = 0 .$$

(ii) A sequence of measurable functions Y_n *converges in probability* (or, *in measure*) to Y if for every $\varepsilon > 0$ we have

$$\lim_{n\to\infty} \mathbb{P}(\{\omega : |Y_n(\omega) - Y(\omega)| > \varepsilon\}) = 0 .$$

Example 3.4 Take $\Omega = (0, 1]$. Let $A_n = (0, \frac{1}{n}]$, $n \geq 1$. Define a sequence of indicator functions $X_n = n \times \mathbf{1}_{A_n}$, $n \geq 1$. Then X_n converges to 0 as $n \to \infty$ at every point, but $||X_n - 0||_1 = 1$ and $||X_n - 0||_2 = \sqrt{n} \to \infty$.

Now we compare various modes of convergence.

Fact 3.6 *For* $1 \leq p < \infty$ *the following statements hold:*

(i) If X_n converges to X in L^p, then X_n converges to X in probability.
(ii) If X_n converges to X in probability, and if $|X_n| \leq Y$ for every n for some $Y \in L^p$, then X_n converges to X in L^p.
(iii) If $X_n, X \in L^p$ and if X_n converges to X almost everywhere, then the L^p-convergence of X_n to X is equivalent to $||X_n||_p \to ||X||_p$.

Fact 3.7 *Assume that* $\mathbb{P}(\Omega) < \infty$.

(i) *For* $1 \leq p \leq \infty$, *if* X_n *converges to* X *in* L^p, *then* X_n *converges to* X *in probability.*
(ii) *If* $X_n(\omega)$ *converges to* $X(\omega)$ *at almost every* ω, *then* X_n *converges to* X *in probability.*

Fact 3.8 *Let* $1 \leq p \leq \infty$. *If* X_n *converges to* X *in* L^p, *then there exists an increasing sequence of natural numbers* $\{n_k\}_{k=1}^{\infty}$ *such that* $X_{n_k}(\omega)$ *converges to* $X(\omega)$ *for almost every* ω *as* $k \to \infty$.

3.4 Inequalities

Definition 3.11 (Convex Function) A function $\phi : \mathbb{R} \to \mathbb{R}$ is said to be *convex* if

$$\phi\Big(\sum_{i=1}^{n} \lambda_i x_i\Big) \leq \sum_{i=1}^{n} \lambda_i \phi(x_i)$$

for $\lambda_1, \ldots, \lambda_n \geq 0$ such that $\sum_{i=1}^{n} \lambda_i = 1$ and for $x_1, \ldots, x_n \in \mathbb{R}$. If the inequality is in the opposite direction, then it is called a concave function. If ϕ is convex, then $-\phi$ is concave. Convex and concave functions are continuous. A linear combination of the form $\sum_{i=1}^{n} \lambda_i x_i$, $\sum_{i=1}^{n} \lambda_i = 1$, $\lambda_i \geq 0$, is called a convex linear combination of x_i.

Fact 3.9 (Jensen's Inequality) *Let* $X : \Omega \to \mathbb{R}$ *be a measurable function on a probability measure space* $(\Omega, \mathbb{P})$. *If* $\phi : \mathbb{R} \to \mathbb{R}$ *is convex and* $\phi \circ X$ *is integrable, then*

$$\phi\Big(\int_{\Omega} X \, d\mathbb{P}\Big) \leq \int_{\Omega} \phi(X(\omega)) \, d\mathbb{P} .$$

(See Fig. 3.5.) For a concave function the inequality is in the opposite direction.

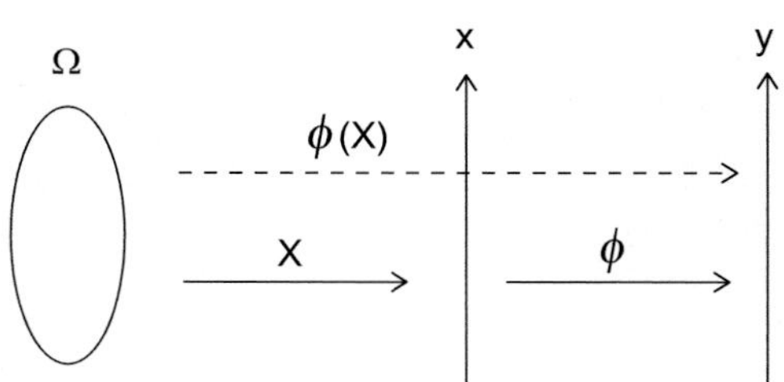

Fig. 3.5 Composition of a random variable X and a function ϕ

Example 3.5 Since $\phi(x) = |x|$ is a convex function, we have

$$\left| \int_\Omega X \, \mathrm{d}\mathbb{P} \right| \leq \int_\Omega |X(\omega)| \, \mathrm{d}\mathbb{P} \, .$$

Example 3.6 Since $\phi(x) = \dfrac{1}{x}$ is a convex function on $(0, \infty)$, we have

$$\frac{1}{\mathbb{E}[X]} \leq \mathbb{E}\left[\frac{1}{X}\right]$$

for $X > 0$.

Example 3.7 Since $\phi(x) = \log x$ is a concave function on $(0, \infty)$, we have

$$\mathbb{E}\left[\log X\right] \leq \log \mathbb{E}[X]$$

for $X > 0$. Now we consider a special case. In Example 3.3 we choose $X(\omega_i) = 1/p_i$ for every $1 \leq i \leq n$, and hence $\mathbb{E}[X] = n$ and

$$-\sum_{i=1}^{n} p_i \log p_i \leq \log n$$

by Jensen's inequality.

Example 3.8 Take a convex function $\phi(x) = \max\{x - K, 0\}$ for a constant K, then we have

$$\max\{\mathbb{E}[X] - K, 0\} \leq \mathbb{E}[\max\{X - K, 0\}]$$

for a random variable X. In option pricing theory, X is the asset price at expiry date and ϕ is the payoff of a European call option with strike price K.

Fact 3.10 (Hölder's Inequality) *Let X, Y be measurable functions on a measure space $(\Omega, \mathbb{P})$ such that $X \in L^p$, $Y \in L^q$, $XY \in L^1$. If p, q satisfy $1 \leq p \leq \infty$, $1 \leq q \leq \infty$, $\frac{1}{p} + \frac{1}{q} = 1$, then*

$$||XY||_1 \leq ||X||_p \, ||Y||_q \, .$$

Equality holds if and only if there exist two constants $C_1 \geq 0$ and $C_2 \geq 0$, not both zero, such that for almost every $\omega \in \Omega$

$$C_1 |X(\omega)|^p = C_2 |Y(\omega)|^q \, .$$

Hölder's inequality for $p = 2$ is called the Cauchy–Schwarz inequality.

Fact 3.11 (Minkowski's Inequality) *For $p \geq 1$ we have*

$$||X + Y||_p \leq ||X||_p + ||Y||_p \ .$$

For $1 < p < \infty$ a necessary and sufficient condition for equality is that there exist constants $C_1 \geq 0$ and $C_2 \geq 0$, not both 0, such that for almost every ω we have $C_1 X(\omega) = C_2 Y(\omega)$. For $p = 1$ a necessary and sufficient condition for equality is that there exists a measurable function $Z \geq 0$ such that $X(\omega)Z(\omega) = Y(\omega)$ for almost every ω satisfying $X(\omega)Y(\omega) \neq 0$.

Minkowski's inequality implies that an L^p space is a normed space.

Fact 3.12 (Chebyshev's Inequality) *If $p \geq 1$ and $X \in L^p(\Omega, \mathbb{P})$, then for every $\varepsilon > 0$ we have*

$$\mathbb{P}(\{\omega : |X(\omega)| > \varepsilon\}) \leq \frac{1}{\varepsilon^p} \int_\Omega |X|^p \, \mathrm{d}\mathbb{P} \ .$$

3.5 The Radon–Nikodym Theorem

Definition 3.12 (i) A measure $\mathbb{P}$ on Ω is *discrete* if there is a countable set $B \subset \Omega$ such that $\mathbb{P}(B^c) = 0$. (ii) A measure $\mathbb{P}$ on Ω is *continuous* if $\mathbb{P}(\{\omega\}) = 0$ for any single element $\omega \in \Omega$. Clearly, if $\mathbb{P}$ is continuous, then a countable set A satisfies $\mathbb{P}(A) = 0$.

Let $X \geq 0$ be an integrable function on a probability measure space $(\Omega, \mathcal{F}, \mathbb{P})$. For arbitrary $A \in \mathcal{F}$ define

$$\mathbb{Q}(A) = \int_A X \, \mathrm{d}\mathbb{P} = \mathbb{E}[\mathbf{1}_A X] \ . \tag{3.1}$$

Then $\mathbb{Q}$ is a measure on $(\Omega, \mathcal{F})$. If $\mathbb{E}[X] = 1$, then $\mathbb{Q}$ is a probability measure. If $\mathbb{P}(A) = 0$, then $\mathbb{Q}(A) = 0$.

Definition 3.13 (Absolute Continuity)

(i) Given two measures $\mathbb{P}$ and $\mathbb{Q}$ on a measurable space $(\Omega, \mathcal{F})$, we say that $\mathbb{Q}$ is *absolutely continuous* with respect to $\mathbb{P}$ if $\mathbb{P}(A) = 0$ implies $\mathbb{Q}(A) = 0$, and write $\mathbb{Q} \ll \mathbb{P}$. Here is an equivalent condition for absolute continuity: for every $\varepsilon > 0$ there exists a $\delta > 0$ such that $\mathbb{P}(A) < \delta$ implies $\mathbb{Q}(A) < \varepsilon$. If $\mathbb{Q}$ by $\mathbb{Q}(A) = \mathbb{E}[\mathbf{1}_A X]$ as in (3.1), then $\mathbb{Q} \ll \mathbb{P}$. Theorem 3.13 states that the converse is also true.

(ii) If $\mathbb{Q} \ll \mathbb{P}$ and also if $\mathbb{P} \ll \mathbb{Q}$, then we say that $\mathbb{P}$ and $\mathbb{Q}$ are *equivalent*, and write $\mathbb{Q} \approx \mathbb{P}$.

A measure ν on $\mathbb{R}$ is absolutely continuous (with respect to Lebesgue measure) if there exists an integrable function $f(x) \geq 0$ such that $\mathrm{d}\nu = f \, \mathrm{d}x$ where $\mathrm{d}x$ is

Lebesgue measure. Note that f is the Radon–Nikodym derivative of ν with respect to Lebesgue measure. An absolutely continuous measure is continuous. A measure ν on $\mathbb{R}$ is *singular continuous* (with respect to Lebesgue measure) if ν is continuous and there is a set A of Lebesgue measure zero such that $\nu(A^c) = 0$. A measure ν can be decomposed into a sum $\nu = \nu_{ac} + \nu_{\mathrm{sc}} + \nu_{\mathrm{d}}$, the components respectively representing the absolutely continuous part, singular continuous part and discrete part of ν. If ν is absolutely continuous, then $\nu = \nu_{ac}$, i.e., $\nu_{\mathrm{sc}} = 0$ and $\nu_{\mathrm{d}} = 0$. For an example of an uncountable subset of Lebesgue measure zero, see Exercise 3.8.

Theorem 3.13 (Radon–Nikodym) *Let $\mathbb{P}$ and $\mathbb{Q}$ be two finite measures on a measurable space $(\Omega, \mathcal{F})$. If $\mathbb{Q} \ll \mathbb{P}$, then there exists a nonnegative measurable function $X : \Omega \to \mathbb{R}^1$ such that for arbitrary $A \in \mathcal{F}$ we have*

$$\mathbb{Q}(A) = \int_A X \, \mathrm{d}\mathbb{P} \, .$$

We write $X = \frac{\mathrm{d}\mathbb{Q}}{\mathrm{d}\mathbb{P}}$ and call it the Radon–Nikodym derivative of $\mathbb{Q}$ with respect to $\mathbb{P}$.

Remark 3.4

(i) If $\mathbb{P}$ and $\mathbb{Q}$ are probability measures, then for any measurable function $X : \Omega \to \mathbb{R}^1$ we have

$$\mathbb{E}^{\mathbb{Q}}[X] = \mathbb{E}^{\mathbb{P}}\left[X \frac{\mathrm{d}\mathbb{Q}}{\mathrm{d}\mathbb{P}}\right]$$

where $\mathbb{E}^{\mathbb{P}}$ and $\mathbb{E}^{\mathbb{Q}}$ denote integrals with respect to $\mathbb{P}$ and $\mathbb{Q}$, respectively.

(ii) We say that two measures are equivalent when $\mathbb{P}(A) = 0$ if and only if $\mathbb{Q}(A) = 0$. In this case we have

$$\frac{\mathrm{d}\mathbb{P}}{\mathrm{d}\mathbb{Q}} = \left(\frac{\mathrm{d}\mathbb{Q}}{\mathrm{d}\mathbb{P}}\right)^{-1} .$$

(iv) The Radon–Nikodym Theorem is an essential concept in the martingale method for option pricing. Instead of the real world probability distribution $\mathbb{P}$ for the stock price movement we find an equivalent probability distribution $\mathbb{Q}$ and compute the expectation.

3.6 Computer Experiments

Simulation 3.1 (Cantor Set)

We try to visualize the Cantor set in Exercise 3.8, and see why its Lebesgue measure is equal to 0. In the following we generate random sequences of 0's and 2's in the ternary expansion of the real numbers belonging to the unit interval, and plot

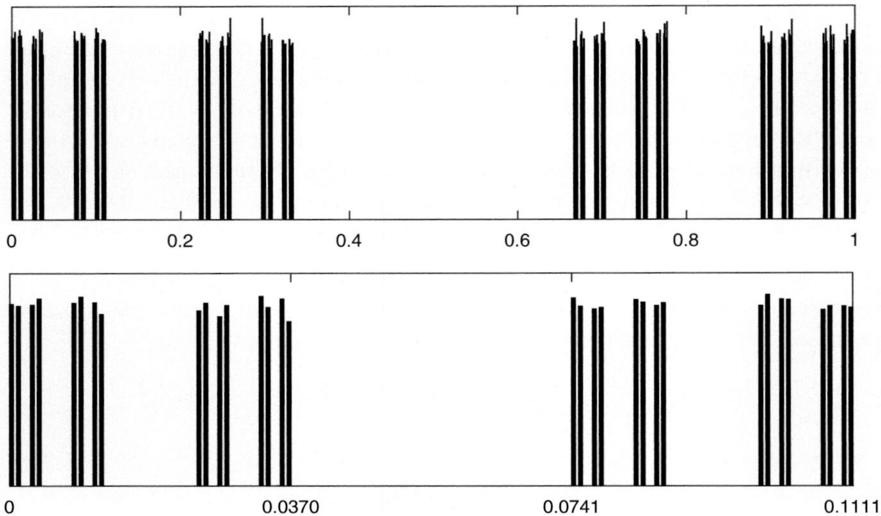

Fig. 3.6 The probability distribution supported by the Cantor set

the distribution that is supported by the Cantor set. See Fig. 3.6 where the bottom graph is the nine times magnification of a part of the top one. Observe the fractal nature of the Cantor set.

```
digits         % Check the number of decimal digits in computation.
k = 20;        % number of ternary digits in the ternary expansion
M = 100000;    % number of sample points for the histogram
Cantor = zeros(M,1);
for i=1:M
    for j=1:k
        Cantor(i) = (Cantor(i) + 2.0*randi([0,1]))/3.0;
    end
end

histogram(Cantor,3^7); % Choose the number of bins.
```

Remark 3.5 Sample points with many significant digits will result in a good resolution when plotting the histogram representing the Cantor set, even in the case when there are many bins. However, too much precision in numerical experiments slows down the computation. If we construct numerically a point $x_0 = \sum_{i=1}^{k} a_i 3^{-i}$ belonging to the Cantor set within an error bound 10^{-D} by specifying the first k ternary digits, we have $10^{-D} \geq 3^{-k}$, i.e., $D \leq \log_{10} 3 \times k$. If there are N bins, then $10^{-D} \leq N^{-1}$, i.e., $D \geq \log_{10} N$. In Simulation 3.1 we take $D = 5$, $N = 3^7$, $k = 20$. By default, the number of decimal digits in numerical computation is thirty two. (For numerical experiments with singular continuous measures, consult [21].)

Exercises

3.1 Let $X : \Omega \to \mathbb{R}^1$ be a function defined on a measurable space $(\Omega, \mathcal{F})$.

(i) Show that $\sigma(X + \alpha) = \sigma(X)$ for any constant α.
(ii) Show that $\sigma(\beta X) = \sigma(X)$ for any constant $\beta \neq 0$.
(iii) Let $\phi : \mathbb{R}^1 \to \mathbb{R}^1$ be a continuous function. Show that $\sigma(\phi(X))$ is a sub-σ-algebra of $\sigma(X)$. For example, $\sigma(X^2)$ is a sub-σ-algebra of $\sigma(X)$.

3.2 Let $\Omega = \{a, b, c\}$ and consider a σ-algebra $\mathcal{F} = \{\{a, b\}, \{c\}, \emptyset, \Omega\}$. Prove that a function $X : \Omega \to \mathbb{R}^1$ is $\mathcal{F}$-measurable if and only if $X(a) = X(b)$.

3.3 Let Ω be a measurable space. Let $f, g : \Omega \to \mathbb{R}$ be measurable functions. Show that for every measurable subset A the function $h(x)$ defined by $h(x) = f(x)$ for $x \in A$, and $h(x) = g(x)$ for $x \notin A$, is measurable.

3.4 Let $(\Omega, \mathcal{A})$ be a measurable space and let $\mathcal{B}$ be a sub-σ-algebra of $\mathcal{A}$. Show that if X is measurable with respect to $\mathcal{B}$, then it is also measurable with respect to $\mathcal{A}$.

3.5 Let $(\Omega, \mathcal{F}, \mathbb{P})$ be a probability measure space and let $A_1, A_2, A_3, \ldots$ be a sequence of measurable subsets such that $\sum_{n=1}^{\infty} P(A_n) < \infty$ and let $B_n = \bigcup_{k=n}^{\infty} A_k$. Prove that $\mathbb{P}\left(\bigcap_{n=1}^{\infty} B_n\right) = 0$.

3.6 Let $(\Omega, \mathcal{F}, \mathbb{P})$ be a probability measure space and let $A_1, A_2, A_3, \ldots$ be a sequence of measurable subsets with $\mathbb{P}(A_n) = 1$ for all $n \geq 1$. Show that $\mathbb{P}\left(\bigcap_{n=1}^{\infty} A_n\right) = 1$.

3.7 Let $A \subset [0, 1]$ be a set defined by

$$A = \left\{ \sum_{n=1}^{\infty} \frac{a_n}{2^n} \,\middle|\, a_n = 0, 1,\ n \geq 1,\ \text{and } a_n = 0 \text{ except for finitely many } n \right\}.$$

Find the Lebesgue measure of A.

3.8 Define the Cantor set $A \subset [0, 1]$ by

$$A = \left\{ \sum_{n=1}^{\infty} \frac{a_n}{3^n} = (0.a_1 a_2 a_3 \ldots)_3 \,\middle|\, a_n = 0, 2,\ n \geq 1 \right\}.$$

(See Simulation 3.1.)

(i) Show that $\frac{1}{4} \in A$. (Hint: Find the ternary expansion of $\frac{1}{4}$.)
(ii) Find the Lebesgue measure of A.
(iii) Show that A has uncountably many points.
(iii) Explain how to construct a singular continuous measure using the one-to-one correspondence $\phi : A \to [0, 1]$ defined by

$$\phi((0.a_1 a_2 a_3 \ldots)_3) = (0.b_1 b_2 b_3 \ldots)_2$$

where $(0.b_1b_2b_3\ldots)_2 = \sum_{n=1}^{\infty} b_n 2^{-n}$, $b_i = \dfrac{a_i}{2} \in \{0,1\}$, $i \geq 1$. (We ignore the set of all numbers with multiple representations, which has measure zero.)

3.9 Prove that for any measurable function $X \geq 0$ defined on a measurable space Ω there exists a monotonically increasing sequence of simple functions $s_n \geq 0$ such that $\lim_{n\to\infty} s_n(\omega) = X(\omega)$ for every $\omega \in \Omega$.

3.10 Let $X_n \geq 0$, $n \geq 1$, be a sequence of random variables on a probability space $(\Omega, \mathbb{P})$ such that

$$\int_\Omega X_n \, \mathrm{d}\mathbb{P} < \frac{1}{2^n}$$

for $n \geq 1$. Show that X_n converges to 0 almost everywhere.

3.11 Compute the limit of

$$\int_0^n \left(1 - \frac{x}{n}\right)^n \mathrm{e}^{x/2} \, \mathrm{d}x \, .$$

3.12 Consider a sequence of functions on [0, 1] defined by

$$f_n = \mathbf{1}_{[j2^{-k},(j+1)2^{-k}]} \, , \quad n = j + 2^k \, , \quad 0 \leq j < 2^k \, .$$

Show that f_n converges to 0 in probability, but not pointwise.

3.13 Let $(\Omega, \mathbb{P})$ be a probability measure space. Suppose that $X \geq 0$ and $Y \geq 0$ are measurable functions on Ω such that $XY \geq 1$. Show that

$$\int_\Omega X \, \mathrm{d}\mathbb{P} \int_\Omega Y \, \mathrm{d}\mathbb{P} \geq 1 \, .$$

(For an example in finance, see Example 4.4.)

3.14 Suppose that $0 \leq X \in L^1(\Omega, \mathbb{P})$. Prove that for every $\varepsilon > 0$ there exists a $\delta > 0$ such that $\int_E X \, \mathrm{d}\mu < \varepsilon$ whenever $\mathbb{P}(E) < \delta$.

Chapter 4
Basic Probability Theory

When H. Lebesgue invented the Lebesgue integral, it was regarded as an abstract concept without applications. It was A.N. Kolmogorov [52] who first showed how to formulate rigorous axiomatic probability theory based on Lebesgue integration.

4.1 Measure and Probability

There is a linguistic gap between terminologies in Lebesgue integral theory and probability theory. To compare the definitions, a measure in Lebesgue integral theory is called probability, a measure space is a sample space, a measurable subset is an event, a σ-algebra means a σ-field, a measurable function corresponds to a random variable, and a Lebesgue integral is called an expectation. The characteristic function χ_E of a subset E is called an indicator function, and denoted by $\mathbf{1}_E$. In set theory, the domain of a function is usually denoted by X, and a variable denoted by x, however, in probability theory a probability measure space is denoted by Ω and the values of a random variable X denoted by x.[1] If we may repeat, a random variable is a function.

Once we get used to two different sets of notations, we find them convenient in practice. While in real analysis such expressions as 'almost everywhere' or 'for almost every point' are used, in probability theory more intuitive expressions such as 'almost surely' or 'with probability 1' are used. Consult Table 4.1.

More confusing is the fact that in probability theory a Fourier transform is called a characteristic function. In this case, the Fourier transform of a random variable

[1] In probability theory a random variable is the most important object of study, and its values are regarded as unknown, so the most prestigious and mysterious symbol 'X' is reserved to denote a random variable.

G.H. Choe, *Stochastic Analysis for Finance with Simulations*, Universitext,
DOI 10.1007/978-3-319-25589-7_4

Table 4.1 Comparison of terminology and notation

Probability theory	Lebesgue integral
Sample space Ω	Measure space X
$\omega \in \Omega$	$x \in X$
σ-field $\mathcal{F}$	σ-algebra $\mathcal{A}$
Event A	Measurable set A
Random variable X	Measurable function f
$\{X \leq x\}$	$f^{-1}((-\infty, \alpha])$
Probability $\mathbb{P}$	Measure μ
Indicator function $\mathbf{1}_A$	Characteristic function χ_A
Expectation $\mathbb{E}[X]$	Lebesgue integral $\int_X f \, \mathrm{d}\mu$
Almost surely, with probability 1	For almost every point
Conditional probability $\mathbb{P}(\cdot\|A)$	Conditional measure μ_A

$X : \Omega \to \mathbb{R}$ is defined by

$$\mathbb{E}[\mathrm{e}^{\mathrm{i}tX}] = \int_\Omega \mathrm{e}^{\mathrm{i}tX(\omega)} \, \mathrm{d}\mathbb{P}(\omega) \,, \quad -\infty < t < \infty$$

where i denotes a complex number satisfying $\mathrm{i}^2 = -1$. (We may define the Fourier transform by using $\mathrm{e}^{-\mathrm{i}tX(\omega)}$.)

For an event A of positive probability, we define the conditional probability $\mathbb{P}(B|A)$ that the event B occurs on the condition that the event A occurs by

$$\mathbb{P}(B|A) = \frac{\mathbb{P}(A \cap B)}{\mathbb{P}(A)} \,.$$

The conditional probability measure $\mathbb{P}_A$ is defined by $\mathbb{P}_A(B) = \mathbb{P}(B|A)$.

Definition 4.1 Given a probability measure space $(\Omega, \mathcal{F}, \mathbb{P})$, a measurable function $X : (\Omega, \mathcal{F}) \to \mathbb{R}$ is called a *random variable*. We define a probability measure μ_X on $\mathbb{R}$ by

$$\mu_X(A) = \mathbb{P}(X^{-1}(A))$$

where A is a Borel measurable subset of $\mathbb{R}$. If there exists an $f_X \geq 0$ on $\mathbb{R}$ such that

$$\int_A f_X(x) \, \mathrm{d}x = \mu_X(A)$$

for every A, we call f_X the *probability density function*, or pdf for short, of X.

Note that $\mu_X(A)$ is the probability that a value of X belongs to A, i.e., $\mu_X(A) = \Pr(X \in A)$. The probability measure μ_X has all the information on the distribution of the values of X, and we focus our attention on μ_X instead of Ω.

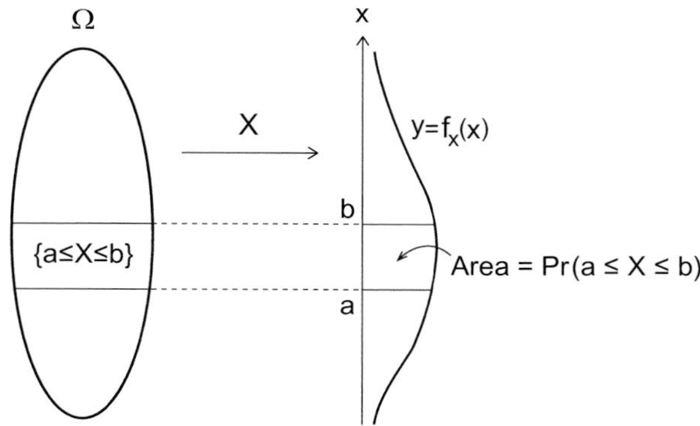

Fig. 4.1 An abstract model of a probability density function

Example 4.1 (Coin Tossing) To solve a given problem we may choose any convenient probability space. For example, when we consider the problem of tossing a coin once, if we let H denote heads, T tails, then we may take $\Omega = \Omega_1 = \{H, T\}$. However, in the problem of tossing a coin twice or tossing two coins once, we take $\Omega = \Omega_2 = \{HH, HT, TH, TT\}$. Even in the first problem of tossing a coin once, it is possible to use Ω_2, but it is simpler and more convenient to use Ω_1. For example, suppose that a random variable X takes the value $X = 1$ if a coin turns up heads and takes the value $X = -1$ if a coin turn up tails. In the case of employing Ω_2 we define $X(HH) = X(HT) = 1, X(TH) = X(TT) = -1$, which is more complicated than the case of Ω_1 with $X(H) = 1, X(T) = -1$.

In Fig. 4.1 the probability of a subset of Ω on the left is equal to the integral of the probability density function on an interval $[a, b]$ on the right. The mapping X is a probability preserving transformation. In the diagram it is intuitive to regard the random variable X as a function that measures the level of the domain Ω.

For a measurable function $h : \mathbb{R} \to \mathbb{R}$ we have

$$\int_{\Omega} h(X(\omega))\, \mathrm{d}\mathbb{P}(\omega) = \int_{-\infty}^{\infty} h(x)\, \mathrm{d}\mu_X(x) .$$

If we take $h(x) = |x|^p$, then

$$\mathbb{E}[\,|X|^p] = \int_{\omega} |X(\omega)|^p\, \mathrm{d}\mathbb{P}(\omega) = \int_{-\infty}^{\infty} |x|^p\, \mathrm{d}\mu_X(x) .$$

The *expectation* or *mean* of X is defined by

$$\mathbb{E}[X] = \int_{-\infty}^{\infty} x\, \mathrm{d}\mu_X(x) ,$$

and its *variance* is defined by

$$\sigma^2(X) = \mathbb{E}[(X - \mathbb{E}[X])^2] = \int_{-\infty}^{\infty} (x - \mathbb{E}[X])^2 \, \mathrm{d}\mu_X(x) .$$

The square root of the variance is called the *standard deviation*, and denoted by σ. If two random variables X and Y have the same distribution, in other words, if μ_X and μ_Y are equal on $\mathbb{R}$, then they have the same probabilistic properties. For example, their expectations are equal.

Example 4.2 (Coin Tossing) To model coin tossing we take $\Omega = \{H, T\}$ and define a σ-algebra $\mathcal{F} = \{\emptyset, \Omega, \{H\}, \{T\}\}$, and a probability measure $\mathbb{P}$ by $\mathbb{P}(\{H\}) = p$, $\mathbb{P}(\{T\}) = 1-p$ where the probability of obtaining heads and tails are equal to p and $1-p$, respectively. Define a random variable $X : \Omega \to \mathbb{R}$ by $X(H) = 0$, $X(T) = 1$. Then $\mathbb{E}[X] = \mathbb{E}[X^2] = p \times 0 + (1-p) \times 1 = 1-p$ and $\mathrm{Var}(X) = \mathbb{E}[X^2] - \mathbb{E}[X]^2 = 1 - p - (1-p)^2 = p - p^2$.

Example 4.3 If the values of a random variable $X : \Omega \to \mathbb{R}$ are uniformly distributed in the interval $[a, b]$, then the probability density function of X is given by

$$f_X(x) = \frac{1}{b-a}\mathbf{1}_{[a,b]}(x) = \begin{cases} \dfrac{1}{b-a}, & x \in [a, b] , \\ 0, & x \notin [a, b] . \end{cases}$$

See Fig. 4.2. Note that $\mathbb{E}[X] = \int_0^1 x \, \mathrm{d}x = \frac{1}{2}$ and $\mathrm{Var}(X) = \mathbb{E}[X^2] - \mathbb{E}[X]^2 = \int_0^1 x^2 \mathrm{d}x - (\frac{1}{2})^2 = \frac{1}{3} - \frac{1}{4} = \frac{1}{12}$.

Example 4.4 (Siegel's Paradox) We consider a seemingly contradictory fact on exchange rate between two currencies. Let X represent the future exchange rate between US dollar and euro, i.e., if a dollar will be equal to X euro in the future then Jensen's inequality in Theorem 3.9 implies $\frac{1}{\mathbb{E}[X]} \leq \mathbb{E}\left[\frac{1}{X}\right]$ where $1/X$ represents the

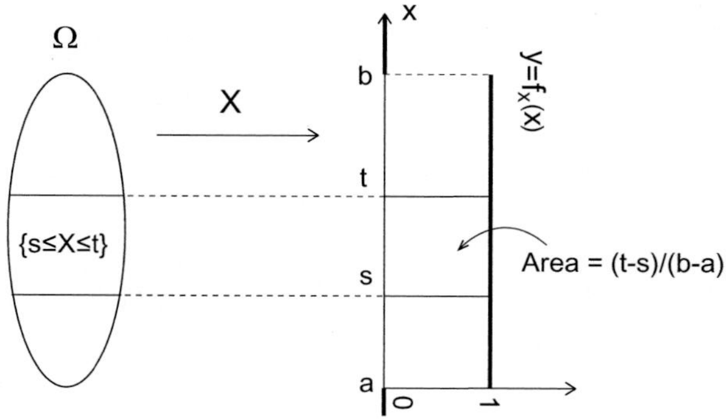

Fig. 4.2 A model of a uniformly distributed random variable in the interval $[a, b]$

exchange rate for converting one euro into dollars. Note that $1/\mathbb{E}[X] \neq \mathbb{E}[1/X]$ in general.

For a concrete example, consider a case when the present exchange rate 1.00 (EUR/USD) changes to either $X = 2.00$ (EUR/USD) with probability $\frac{1}{2}$ or $X = 0.50$ (EUR/USD) with probability $\frac{1}{2}$ in a year. The opposite exchange rate $1/X$ is currently 1.00 (USD/EUR) and in a year it will change to 0.50 (USD/EUR) or 2.00 (USD/EUR) with probability $\frac{1}{2}$. Hence

$$\mathbb{E}[X] = \frac{1}{2} \times 2.00 + \frac{1}{2} \times 0.50 = 1.25 \text{ (EUR/USD)} ,$$

$$\mathbb{E}\left[\frac{1}{X}\right] = \frac{1}{2} \times 0.50 + \frac{1}{2} \times 2.00 = 1.25 \text{ (USD/EUR)} .$$

The exchange rate will rise 1.25 times in a year whether we exchange euro with USD or do the opposite. In other words, we have the same expected return in any case, which is called Siegel's paradox. It is not a paradox, but seems to be counterintuitive at first glance. Of course, if the exchange rate at future date is determined, one side gains profit while the other side suffers loss.

Example 4.5 (Payoff of an Option) Figure 4.3 illustrates the payoff $(S_T - K)^+$ of a European call option as a composite of a random variable $S_T : \Omega \to [0, \infty)$ representing the stock price at time T and a function

$$(x - K)^+ = \max\{x - K, 0\} : [0, \infty) \to [0, \infty) .$$

The new composite function is again a random variable.

Example 4.6 If X is a random variable with normal probability distribution with mean 0 and variance σ^2, then

$$\mathbb{E}[X^{2n}] = \frac{(2n)!}{2^n n!} \sigma^{2n}$$

and $\mathbb{E}[X^{2n+1}] = 0$ for every $n \geq 0$.

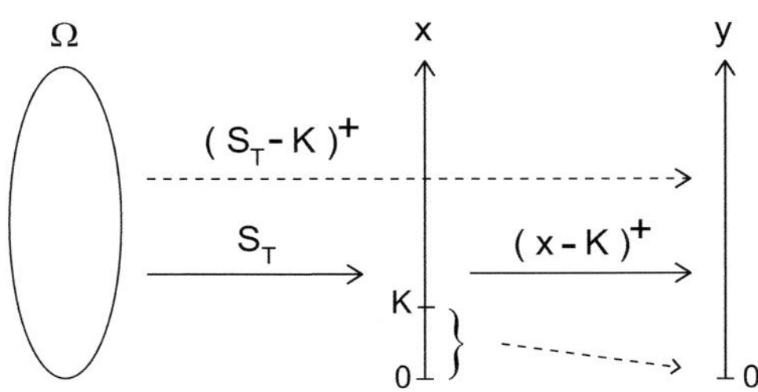

Fig. 4.3 Payoff of a European call option, $(S_T - K)^+$, as a composite function

4.2 Characteristic Functions

Definition 4.2 (Cumulative Distribution Function) Given a random variable X, its *cumulative distribution function* (cdf for short) is defined by

$$F(x) = \Pr(X \le x) , \quad -\infty < x < \infty .$$

If a cumulative distribution function satisfies

$$F(x) = \int_{-\infty}^{x} f(t)\,\mathrm{d}t , \quad -\infty < x < \infty ,$$

for some integrable function $f \ge 0$, then $f(x)$ is called a *probability density function* (pdf for short) for X.

If a probability density function $f(x)$ is continuous at x, then $F'(x) = f(x)$ since $\mathbb{P}\{X < x + \delta x\} - \mathbb{P}\{X < x\} \approx f_X(x)\,\delta x$ for small $\delta x > 0$. A cumulative distribution function is monotonically increasing, and satisfies $\lim_{x\to-\infty} F(x) = 0$ and $\lim_{x\to\infty} F(x) = 1$. A cumulative distribution function F is continuous from the right, i.e., $F(x) = \lim_{h\to 0+} F(x+h)$. Furthermore, we have $\mathbb{P}(X < c) = \lim_{x\to c-} F(x)$.

Example 4.7 The probability density function $f(x)$ of the standard normal distribution is given by

$$f(x) = \frac{1}{\sqrt{2\pi}} \mathrm{e}^{-x^2/2}$$

and a standard normal random variable X has the cumulative distribution function

$$N(x) = \mathbb{P}(X \le x) = \int_{-\infty}^{x} \frac{1}{\sqrt{2\pi}} \mathrm{e}^{-t^2/2}\,\mathrm{d}t .$$

(See Figs. 4.4 and 4.5.)

Definition 4.3 (Fourier Transform) For an integrable function $f(x)$ on $\mathbb{R}$, its *Fourier transform* $\widehat{f}$ is defined by $\widehat{f}(t) = \int_{-\infty}^{\infty} f(x)\mathrm{e}^{-\mathrm{i}tx}\,\mathrm{d}x$, $t \in \mathbb{R}$, where $\mathrm{i}^2 = -1$. It is continuous on the real line.

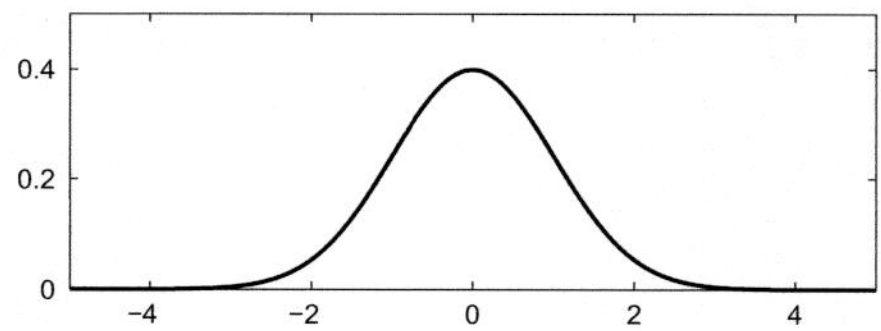

Fig. 4.4 Probability density function of the standard normal distribution

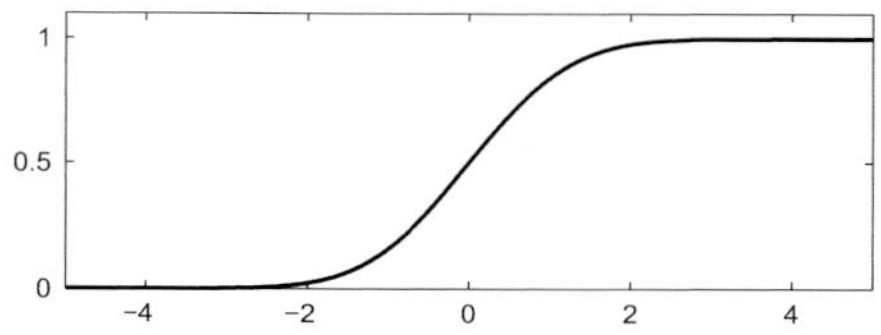

Fig. 4.5 Cumulative distribution function of the standard normal distribution

Definition 4.4 (Characteristic Function) The *characteristic function* ϕ_X of a random variable X is defined by

$$\phi_X(t) = \mathbb{E}[e^{itX}], \quad -\infty < t < \infty ,$$

where $i = \sqrt{-1}$. Note that X need not be continuous for the expectation to exist. For example, the characteristic function of a binomial distribution B(n,k) defined by $\Pr(X = k) = \binom{n}{k} p^k (1-p)^k$ is given by $\phi(t) = (pe^{it} + (1-p))^n$. For a continuous random variable X with its pdf f_X the characteristic function is the Fourier transform of f_X, i.e.,

$$\phi_X(t) = \int_{-\infty}^{\infty} e^{itx} f_X(x)\, dx , \quad -\infty < t < \infty .$$

(In probability theory it is more common to use e^{itx} instead of e^{-itx}.)

Theorem 4.1 *The following facts are known:*

(i) $\phi_X(0) = 1$,
(ii) $|\phi_X(t)| \leq 1$,
(iii) $\phi_X(t)$ *is continuous,*
(iv) $\phi_{a+bX}(t) = e^{iat}\phi_X(bt)$,
(v) $\phi'_X(0) = i\mu$,
(vi) $\phi''_X(0) = -\mathbb{E}[X^2] = -\mu^2 - \sigma^2$, *and*
(vii) *the Fourier inversion formula holds, i.e.,*

$$f_X(x) = \frac{1}{2\pi} \int_{-\infty}^{\infty} e^{-ixt} \phi_X(t)\, dt .$$

Remark 4.1

(i) The power series expansion of the characteristic function is given by

$$\phi_X(t) = \sum_{n=0}^{\infty} \frac{i^n \mathbb{E}[X^n]}{n!} t^n .$$

(ii) The characteristic function of a random variable completely determines the distribution of X. More precisely, if X and Y have the same characteristic function, then their probability distributions are identical due to the Fourier inversion formula.

Example 4.8 If X is normally distributed with mean 0 and variance σ^2, then

$$\mathbb{E}[X^{2k}] = \frac{\sigma^{2k}(2k)!}{2^k k!} ,$$

and hence

$$\phi_X(t) = \sum_{k=0}^{\infty} \frac{\mathrm{i}^{2k}\mathbb{E}[X^{2k}]}{(2k)!} t^{2k} = \sum_{k=0}^{\infty} \frac{\mathrm{i}^{2k}\sigma^{2k}(2k)!}{(2k)!2^k k!} t^{2k} = \sum_{k=0}^{\infty} \frac{(-\sigma^2)^k}{2^k k!} t^{2k} = \mathrm{e}^{-\sigma^2 t^2/2} .$$

(If X has mean μ, then $\phi_X(t) = \mathrm{e}^{\mathrm{i}t\mu}\mathrm{e}^{-\sigma^2 t^2/2}$.) Thus

$$\int_{-\infty}^{\infty} \mathrm{e}^{\mathrm{i}tx} \frac{1}{\sqrt{2\pi}} \mathrm{e}^{-x^2/2} \mathrm{d}x = \mathrm{e}^{-t^2/2}$$

for $t \in \mathbb{R}$ where the left-hand side is equal to the characteristic function of the standard normal distribution. Since the both sides are real, by taking complex conjugates, we also have

$$\int_{-\infty}^{\infty} \mathrm{e}^{-\mathrm{i}tx} \frac{1}{\sqrt{2\pi}} \mathrm{e}^{-x^2/2} \mathrm{d}x = \mathrm{e}^{-t^2/2} .$$

Remark 4.2 Let X_n, $n \geq 1$, and X be random variables with their cumulative distribution functions F_{X_n} and F_X satisfying

$$\lim_{n\to\infty} \phi_{X_n}(t) = \phi_X(t) , \quad -\infty < t < \infty .$$

Then $\lim_{n\to\infty} F_{X_n}(x) = F_X(x)$ at the points x where F_X is continuous.

Remark 4.3 Define $\log z = \log |z| + \mathrm{i}\mathrm{Arg}(z)$, $-\pi < \mathrm{Arg}(z) < \pi$. Then for $|z-1| < 1$ we have the power series expansion $\log z = (z-1) - \frac{1}{2}(z-1)^2 + \cdots$. Since

$$\begin{aligned} \phi_X(t) &= \phi_X(0) + \phi_X'(0)t + \frac{\phi_X''(0)}{2}t^2 + \cdots \\ &= 1 + \mathrm{i}\mu t - \frac{\mu^2+\sigma^2}{2}t^2 + \cdots , \end{aligned}$$

we have

$$\log \phi_X(t) = \left(\mathrm{i}\mu t - \frac{\mu^2 + \sigma^2}{2} t^2 + \cdots \right) - \frac{1}{2} \left(\mathrm{i}\mu t - \frac{\mu^2 + \sigma^2}{2} t^2 + \cdots \right)^2 + \cdots$$

$$= \mathrm{i}\mu t - \frac{\sigma^2}{2} t^2 + \cdots ,$$

and hence

$$\frac{\log \phi_X(t) - \mathrm{i}\mu t}{t^2} = -\frac{\sigma^2}{2} + \cdots .$$

Theorem 4.2 *If X_n, $n \geq 1$, are normal random variables with mean μ_n and variance σ_n^2 and if X_n converges to X in L^2, then X is also normally distributed with mean $\mu = \lim_{n\to\infty} \mu_n$ and variance $\sigma^2 = \lim_{n\to\infty} \sigma_n^2$ under the assumption that the limits exist.*

Proof Since the characteristic functions of X_n converge everywhere to a limit that is also a characteristic function of a normal variable, the sequence X_n itself converges to a normal variable. □

Definition 4.5 (Moment Generating Function) The *moment generating function* of a random variable X is defined for u in a neighborhood of 0 by

$$M_X(u) = \mathbb{E}[\mathrm{e}^{uX}]$$

if the expectation exists. Depending on X, the expectation may not exist for large real values of u. Note that

$$M_X(0) = 1$$

and

$$M_X(u) = 1 + m_1 u + \frac{m_2}{2!} u^2 + \frac{m_3}{3!} u^3 + \cdots + \frac{m_k}{k!} u^k + \cdots$$

where $m_k = \mathbb{E}[X^k]$ and the power series converges. If the kth moment m_k grows too fast then the power series may not converge to a finite limit. If $X \geq 0$, the moment generating function is essentially the Laplace transform of the probability density function. Note that u need not be real. If u is purely imaginary, then we obtain the characteristic function in Definition 4.4.

Example 4.9 Let Z be a standard normal variable and take $u = a+\mathrm{i}b \in \mathbb{C}, a, b \in \mathbb{R}$. Then

$$\begin{aligned}
\mathbb{E}[\mathrm{e}^{uZ}] &= \int_{-\infty}^{\infty} \mathrm{e}^{uz} \frac{1}{\sqrt{2\pi}} \mathrm{e}^{-z^2/2} \mathrm{d}z \\
&= \mathrm{e}^{a^2/2} \int_{-\infty}^{\infty} \mathrm{e}^{\mathrm{i}bz} \frac{1}{\sqrt{2\pi}} \mathrm{e}^{-(z-a)^2/2} \mathrm{d}z \qquad \text{(take } y = z - a\text{)} \\
&= \mathrm{e}^{a^2/2} \mathrm{e}^{\mathrm{i}ab} \int_{-\infty}^{\infty} \mathrm{e}^{\mathrm{i}by} \frac{1}{\sqrt{2\pi}} \mathrm{e}^{-y^2/2} \mathrm{d}y \\
&= \mathrm{e}^{a^2/2} \mathrm{e}^{\mathrm{i}ab} \mathrm{e}^{-b^2/2} \\
&= \mathrm{e}^{u^2/2} \ .
\end{aligned}$$

In general, for $X \sim N(\mu, \sigma^2)$ we have

$$\mathbb{E}[\mathrm{e}^{uX}] = \mathrm{e}^{\mu u + \frac{1}{2}\sigma^2 u^2} \ .$$

4.3 Independent Random Variables

Definition 4.6 (Independence of Events) If measurable sets (or events) $A_1, \ldots, A_n$ satisfy

$$\mathbb{P}(A_1 \cap \cdots \cap A_n) = \mathbb{P}(A_1) \cdots \mathbb{P}(A_n) \ ,$$

then $A_1, \ldots A_n$ are said to be *independent*. A collection of infinitely many measurable subsets $\{A_\gamma\}_{\gamma \in \Gamma}$ is said to be independent if any finitely many subsets $A_{\gamma_1}, \ldots, A_{\gamma_n}$ are independent.

If two events A_1 and A_2 are independent, then the pairs A_1 and A_2^c, A_1^c and A_2, A_1^c and A_2^c are independent where A_i^c denotes the complement of A_i, $i = 1, 2$.

Definition 4.7 (Independence of Random Variables) Let $X_i : \Omega \to \mathbb{R}$, $i \geq 1$, be random variables. If $\{X_i^{-1}(B_i)\}_{i=1}^{\infty}$ are independent for arbitrary Borel measurable subsets $B_i \subset \mathbb{R}$, then X_i are said to be *independent*.

Since the collection of all intervals generates the Borel measurable subsets of $\mathbb{R}$, it is sufficient to apply the above criterion only with intervals for B_i. The same comment holds in similar situations.

Definition 4.8 (Independence of σ-Algebras)

(i) Let $\mathcal{F}_i$, $1 \leq i \leq n$, be sub-σ-algebras. If $A_1, \ldots, A_n$ are independent for any choice of $A_1 \in \mathcal{F}_1, \ldots, A_n \in \mathcal{F}_n$, then we say that $\mathcal{F}_1, \ldots, \mathcal{F}_n$ are *independent*.

(ii) Given a random variable X and a sub-σ-algebra $\mathcal{F}$, if $X^{-1}(B)$ and A are independent for arbitrary Borel subset $B \subset \mathbb{R}$ and $A \in \mathcal{F}$, then X and $\mathcal{F}$ are said to be *independent*.

Let $\sigma(X)$ denote the smallest σ-algebra containing all the subsets of the form $X^{-1}(B)$ where $B \subset \mathbb{R}$ is an arbitrary Borel set. Then the independence of X and $\mathcal{F}$ is equivalent to the independence of $\sigma(X)$ and $\mathcal{F}$. Similarly, for random variables X_α, $\alpha \in A$, let $\sigma(\{X_\alpha : \alpha \in A\})$ denote the smallest σ-algebra containing all the σ-algebras $\sigma(X_\alpha)$.

Lemma 4.1 *Suppose that random variables X and Y are independent. For arbitrary Borel measurable functions $f : \mathbb{R} \to \mathbb{R}$ and $g : \mathbb{R} \to \mathbb{R}$, the random variables $f(X)$ and $g(Y)$ are independent.*

Proof Use $\sigma(f(X)) \subset \sigma(X)$ and $\sigma(g(Y)) \subset \sigma(Y)$. □

Example 4.10 Suppose that random variables X and Y are independent.

(i) For constants α and β we take $f(x) = \alpha x + \beta$. Then we see that $\alpha X + \beta$ and Y are independent.
(ii) $X - \mathbb{E}[X]$ and $Y - \mathbb{E}[Y]$ are independent.
(iii) Take $f(x) = \max\{x, 0\}$. Then $X^+ = \max\{X, 0\}$ and Y are independent.

Theorem 4.3 *Suppose that $X_1, \ldots, X_n$ are independent random variables.*

(i) For an arbitrary Borel measurable $B_i \subset \mathbb{R}$ we have

$$\mathbb{P}(X_1 \in B_1, \ldots, X_n \in B_n) = \mathbb{P}(X_1 \in B_1) \times \cdots \times \mathbb{P}(X_n \in B_n) .$$

(ii) If $-\infty < \mathbb{E}[X_i] < \infty$ for every i, then

$$\mathbb{E}[X_1 \cdots X_n] = \mathbb{E}[X_1] \times \cdots \times \mathbb{E}[X_n] .$$

(iii) If $\mathbb{E}[X_i^2] < \infty$ for every i, then

$$\mathrm{Var}(X_1 + \cdots + X_n) = \mathrm{Var}(X_1) + \cdots + \mathrm{Var}(X_n)$$

where Var *denotes variance.*

Proof For the sake of notational convenience we prove only the case $n = 2$.

(i) Use

$$\mathbb{P}(X_1 \in B_1, X_2 \in B_2) = \mathbb{P}(X_1^{-1}(B_1) \cap X_2^{-1}(B_2)) = \mathbb{P}(X_1^{-1}(B_1))\, \mathbb{P}(X_2^{-1}(B_2)) .$$

(ii) In part (i) we take $X_1 = \sum_i \alpha_i \mathbf{1}_{A_i}$, $X_2 = \sum_i \beta_i \mathbf{1}_{B_j}$, and obtain

$$\mathbb{E}[X_1 X_2] = \sum_{i,j} \alpha_i \beta_j \mathbb{P}(A_i \cap B_j) = \sum_i \alpha_i \mathbb{P}(A_i) \sum_j \beta_j \mathbb{P}(B_j) = \mathbb{E}[X_1]\, \mathbb{E}[X_2] .$$

For the general case, approximate the given random variables by sequences of simple functions and take the limits.

(iii) Note that

$$\begin{aligned}&\mathrm{Var}(X_1+X_2)\\&=\mathbb{E}[(X_1-\mathbb{E}[X_1]+X_2-\mathbb{E}[X_2])^2]\\&=\mathbb{E}[(X_1-\mathbb{E}[X_1])^2]+\mathbb{E}[(X_2-\mathbb{E}[X_2])^2]+2\mathbb{E}[(X_1-\mathbb{E}[X_1])(X_2-\mathbb{E}[X_2])]\,.\end{aligned}$$

By Example 4.10(ii) and the part (ii) of the above theorem,

$$\mathbb{E}[(X_1-\mathbb{E}[X_1])(X_2-\mathbb{E}[X_2])]=\mathbb{E}[X_1-\mathbb{E}[X_1]]\,\mathbb{E}[X_2-\mathbb{E}[X_2]]=0\,.$$

Thus $\mathrm{Var}(X_1+X_2)=\mathbb{E}[(X_1-\mathbb{E}[X_1])^2]+\mathbb{E}[(X_2-\mathbb{E}[X_2])^2]$. □

Definition 4.9 (Joint Probability) For random variables X_i, $1\le i\le n$, on a probability space Ω, we define a *random vector* $(X_1,\dots,X_n):\Omega\to\mathbb{R}^n$ by $(X_1,\dots,X_n)(\omega)=(X_1(\omega),\dots,X_n(\omega))$, $\omega\in\Omega$. The corresponding joint probability density function $f_{X_1,\dots,X_n}:\mathbb{R}^n\to\mathbb{R}$ is given by

$$\int\cdots\int_A f_{X_1,\dots,X_n}(x_1,\dots,x_n)\,\mathrm{d}x_1\cdots\mathrm{d}x_n=\mathbb{P}((X_1,\dots,X_n)^{-1}(A))\,.$$

The characteristic function for a random vector $\mathbf{X}=(X_1,\dots,X_n)$ is defined by the n-dimensional Fourier transform

$$\phi_{\mathbf{X}}(t)=\mathbb{E}\left[\mathrm{e}^{\mathrm{i}\mathbf{t}\cdot\mathbf{X}}\right]=\int_{-\infty}^{\infty}\cdots\int_{-\infty}^{\infty}\mathrm{e}^{\mathrm{i}\mathbf{t}\cdot\mathbf{x}}f_{X_1,\dots,X_n}(x_1,\dots,x_n)\,\mathrm{d}x_1\cdots\mathrm{d}x_n$$

where $\mathbf{t}\cdot\mathbf{x}=t_1x_1+\cdots+t_nx_n$ and $\mathbf{t}\cdot\mathbf{X}=t_1X_1+\cdots+t_nX_n$.

Definition 4.10 (Jointly Normal Distribution) The random variables $X_1,\dots,X_n$ are *jointly normal* with a *covariance matrix* Σ if their joint pdf is given by

$$f(\mathbf{x})=\frac{1}{(2\pi)^{n/2}(\det\Sigma)^{1/2}}\exp\Big(-\frac{1}{2}(\mathbf{x}-\boldsymbol{\mu})^t\Sigma^{-1}(\mathbf{x}-\boldsymbol{\mu})\Big)$$

where

$$\begin{aligned}\Sigma&=\mathbb{E}[(\mathbf{X}-\boldsymbol{\mu})^t(\mathbf{X}-\boldsymbol{\mu})]\\&=\begin{bmatrix}\mathrm{Cov}(X_1,X_1)&\cdots&\mathrm{Cov}(X_1,X_n)\\\vdots&&\vdots\\\mathrm{Cov}(X_n,X_1)&\cdots&\mathrm{Cov}(X_n,X_n)\end{bmatrix}\end{aligned}$$

and

$$\mathbf{x} = (x_1, \ldots, x_n)\ ,\ \boldsymbol{\mu} = (\mu_1, \ldots, \mu_n) \text{ and } \mathbf{X} = (X_1, \ldots, X_n)\ .$$

In this case, we write $\mathbf{X} \sim N(\boldsymbol{\mu}, \Sigma)$. Note that $\mathbb{E}[X_i] = \mu_i$, $1 \le i \le n$, and that Σ is symmetric and nonnegative definite.

Remark 4.4

(i) Here is a generalization of the jointly normal distribution. The variables $X_1, \ldots, X_n$ are said to be jointly normal in the extended sense if there exists a (not necessarily invertible) symmetric and nonnegative definite $n \times n$ matrix C and a vector $\boldsymbol{\mu} = (\mu_1, \ldots, \mu_n)$ satisfying

$$\phi_{\mathbf{X}}(x_1, \ldots, x_n) = \exp\Big(-\frac{1}{2} \sum_{j,k=1}^{n} c_{jk} x_j x_k + \mathrm{i} \sum_{j=1}^{n} \mu_j x_j \Big)\ . \tag{4.1}$$

This condition (4.1) is satisfied if we choose $C = \Sigma^{-1}$ in Definition 4.10. It is equivalent to the statement that any linear combination of $X_1, \ldots, X_n$ is normal. When $X_1, \ldots, X_n$ are jointly normal, and X_1 and X_j are uncorrelated for $2 \le j \le n$, then X_1 is independent of $X_2, \ldots, X_n$.

(ii) Suppose that X_n is normal for $n \ge 1$ and that X_n converges to X in L^2. Then X is normal. For the proof, consult [74].

Theorem 4.4 (Independent Random Variables) *Let X and Y be random variables with their probability density functions f_X and f_Y, respectively. If $f_{X,Y}(x, y) = f_X(x) f_Y(y)$, then X and Y are independent. Conversely, if X, Y are independent, then the probability density function of the two-dimensional random variable $(X, Y) : \Omega \to \mathbb{R}^2$, denoted by $f_{X,Y}$, is given by $f_{X,Y}(x, y) = f_X(x) f_Y(y)$.*

Proof First, if $f_{X,Y}(x, y) = f_X(x) f_Y(y)$, then for any Borel subsets B_1, B_2 we have

$$X^{-1}(B_1) \cap Y^{-1}(B_2) = (X, Y)^{-1}(B_1 \times B_2)\ .$$

Hence

$$\begin{aligned}
\mathbb{P}(X^{-1}(B_1) \cap Y^{-1}(B_2)) &= \iint_{B_1 \times B_2} f_{X,Y}(x, y)\, \mathrm{d}x\, \mathrm{d}y \\
&= \int_{B_2} \Big(\int_{B_1} f_X(x)\, \mathrm{d}x \Big) f_Y(y)\, \mathrm{d}y \\
&= \mathbb{P}(X^{-1}(B_1)) \mathbb{P}(Y^{-1}(B_1))\ .
\end{aligned}$$

Next, for any pair of real numbers a, b, let $A = (-\infty, a) \times (-\infty, b)$. Then, by definition,

$$\mathbb{P}((X, Y) \in A) = \iint_A f_{X,Y}(x, y)\, \mathrm{d}x\, \mathrm{d}y .$$

By the independence,

$$\begin{aligned}
\mathbb{P}((X, Y) \in A) &= \mathbb{P}(X \in (-\infty, a))\, \mathbb{P}(Y \in (-\infty, b)) \\
&= \int_{-\infty}^{a} f_X(x)\mathrm{d}x \int_{-\infty}^{b} f_Y(y)\mathrm{d}y \\
&= \int_{-\infty}^{a} \int_{-\infty}^{b} f_X(x) f_Y(y)\, \mathrm{d}x\mathrm{d}y .
\end{aligned}$$

Hence $f_{X,Y}(x, y) = f_X(x) f_Y(y)$. □

Definition 4.11 (Convolution) Let f and g be two integrable functions on $\mathbb{R}$. Then their *convolution* $f * g$ is defined by

$$(f * g)(x) = \int_{-\infty}^{\infty} f(x - y) g(y) \mathrm{d}y .$$

It can be shown that

$$f * g = g * f$$

and

$$f * (g * h) = (f * g) * h .$$

Theorem 4.5 (Sum of Random Variables) *Let X and Y be independent random variables with probability density functions f_X and f_Y, respectively. Then the probability density function of $X + Y$, denoted by f_{X+Y}, is given by the convolution, i.e., $f_{X+Y} = f_X * f_Y$.*

Proof First, we present a heuristic proof. Let $F_X(x) = \mathbb{P}(X \le x)$ and let $-\infty < \cdots < y_i < y_{i+1} < \cdots < \infty$ be a partition of $(-\infty, \infty)$ such that $\sup_i \delta_i$ is sufficiently small where $\delta_i = y_{i+1} - y_i > 0$. Then

$$\begin{aligned}
\mathbb{P}(X + Y \le a) &\approx \sum_i \mathbb{P}(X \le a - y_i | y_i \le Y < y_{i+1}) \times \mathbb{P}(y_i \le Y < y_{i+1}) \\
&= \sum_i \mathbb{P}(X \le a - y_i) \times \mathbb{P}(y_i \le Y < y_{i+1}) \\
&\approx \sum_i F_X(a - y_i) f_Y(y_i) \delta y_i .
\end{aligned}$$

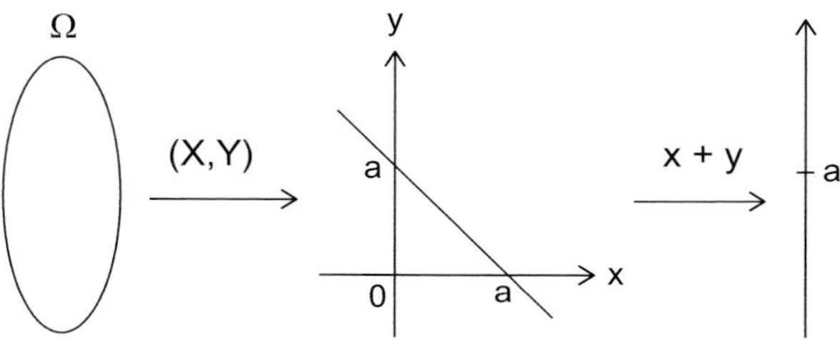

Fig. 4.6 Sum of two independent random variables

(See Fig. 4.6.) By letting $\sup_i \delta_i \to 0$, we obtain

$$\mathbb{P}(X + Y \leq a) = \int_{-\infty}^{\infty} F_X(a - y) f_Y(y)\, \mathrm{d}y\, .$$

Taking the derivatives of both sides with respect to a, we have

$$f_{X+Y}(a) = \int_{-\infty}^{\infty} f_X(a - y) f_Y(y)\, \mathrm{d}y = f_X * f_Y(a)\, .$$

We used the fact that F_X is sufficiently smooth.

A more mathematically rigorous proof goes as follows: Since the joint probability density function of (X, Y) is equal to $f_X(x)f_Y(y)$, we have

$$\mathbb{P}(X + Y \leq a) = \int_{-\infty}^{\infty} \int_{-\infty}^{a-y} f_X(x) f_Y(y)\, \mathrm{d}x\mathrm{d}y\, .$$

Differentiating both sides with respect to a, we obtain

$$f_{X+Y}(a) = \int_{-\infty}^{\infty} f_X(a - y) f_Y(y)\, \mathrm{d}y\, .$$

□

Remark 4.5 Let ϕ_X, ϕ_Y and ϕ_{X+Y} denote the characteristic function of X, Y and $X + Y$, respectively. If X and Y are independent, then $\phi_{X+Y}(t) = \phi_X(t)\, \phi_Y(t)$. For, if we let f_X, f_Y and f_{X+Y} denote the probability density functions of X, Y and $X + Y$, respectively, then

$$\phi_{X+Y}(t) = \widehat{f_{X+Y}} = \widehat{f_X * f_Y} = \widehat{f_X}\,\widehat{f_Y} = \phi_X(t)\, \phi_Y(t)$$

by Theorem 4.5. This can be shown directly since

$$\phi_{X+Y}(t) = \mathbb{E}[\mathrm{e}^{\mathrm{i}t(X+Y)}] = \mathbb{E}[\mathrm{e}^{\mathrm{i}tX}]\, \mathbb{E}[\mathrm{e}^{\mathrm{i}tY}] = \phi_X(t)\, \phi_Y(t)\, .$$

Example 4.11 Let U_1, U_2 be independent and uniformly distributed in $[0, 1]$. Then the pdf f_V of $V = U_1 + U_2$ is given by

$$f_V(x) = \begin{cases} x, & x \in [0, 1] , \\ 2 - x, & x \in (1, 2] , \\ 0, & x \notin [0, 2] . \end{cases}$$

We can prove this fact by computing the convolution $f * f$ where $f(x) = 1$ on $[0, 1]$ and $f(x) = 0$ elsewhere. See also Exercise 4.20.

Example 4.12 Let X_1, X_2 be two independent normally distributed random variables with means μ_1 and μ_2 and variances σ_1 and σ_2, respectively. Then $X_1 + X_2$ is also normally distributed with mean $\mu_1 + \mu_2$ and variance $\sigma_1 + \sigma_2$. To prove the fact it suffices to compute the convolution of two normal density functions f_{X_1} and f_{X_2}. For notational simplicity, consider the case $\mu_1 = \mu_2 = 0$ and $\sigma_1 = \sigma_2 = 1$. Then

$$\begin{aligned} (f_{X_1} * f_{X_2})(z) &= \frac{1}{2\pi} \int_{-\infty}^{\infty} \mathrm{e}^{-(z-t)^2/2} \mathrm{e}^{-t^2/2} \mathrm{d}t \\ &= \frac{1}{2\pi} \mathrm{e}^{-z^2/4} \int_{-\infty}^{\infty} \mathrm{e}^{-(t-z/2)^2} \mathrm{d}t \\ &= \frac{1}{2\pi} \mathrm{e}^{-z^2/4} \int_{-\infty}^{\infty} \mathrm{e}^{-t^2} \mathrm{d}t \\ &= \frac{1}{2\pi} \mathrm{e}^{-z^2/4} \sqrt{\pi} \\ &= \frac{1}{\sqrt{2\pi \times 2}} \mathrm{e}^{-z^2/(2\times 2)} , \end{aligned}$$

which implies that $X_1 + X_2$ is also normal with mean 0 and variance 2. In general, if X_i are independent normal variables with mean μ_i and variance σ_i^2 for $1 \le i \le n$, then $S_n = X_1 + \cdots + X_n$ is normally distributed with mean $\mu_1 + \cdots + \mu_n$ and variance $\sigma_1^2 + \cdots + \sigma_n^2$.

4.4 Change of Variables

Theorem 4.6 (Composition) *If $h(x)$ is a differentiable and monotone function, and if a random variable X has a continuous probability density function $f_X(x)$, then the probability density function of $Y = h(X)$ is given by*

$$f_Y(y) = \frac{1}{|h'(h^{-1}(y))|} f_X(h^{-1}(y))$$

where h' denotes the derivative of h. See Fig. 4.7.

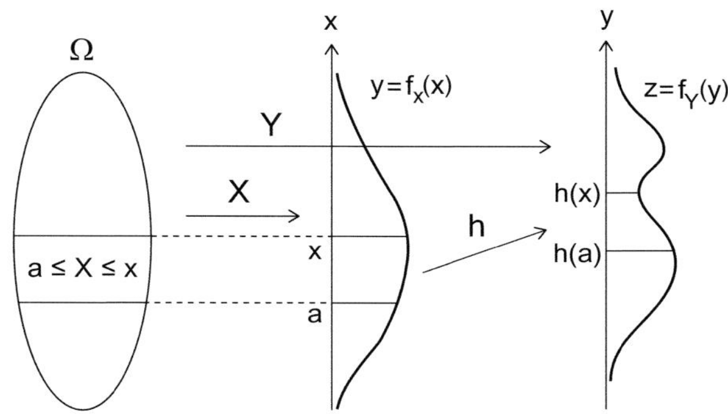

Fig. 4.7 A new probability density function obtained by change of variables

Proof First, consider the case when h is monotonically increasing. Since

$$\Pr(a \le X \le x) = \Pr(h(a) \le Y \le h(x)) ,$$

we have

$$\int_a^x f_X(x)\,\mathrm{d}x = \int_{h(a)}^{h(x)} f_Y(y)\,\mathrm{d}y .$$

By differentiating with respect to x, we have

$$f_X(x) = f_Y(h(x))h'(x) .$$

Now substitute $x = h^{-1}(y)$. For the case when h is decreasing, note that

$$\int_a^x f_X(x)\,\mathrm{d}x = \int_{h(x)}^{h(a)} f_Y(y)\,\mathrm{d}y = -\int_{h(a)}^{h(x)} f_Y(y)\,\mathrm{d}y .$$

□

Example 4.13 If a random variable X has a probability density function $f_X(x)$, let us find the probability density function $f_Y(y)$ of $Y = aX + b$ where $a > 0$. Note that for arbitrary y we have

$$\int_{-\infty}^{y} f_Y(y)\,\mathrm{d}y = \mathbb{P}(aX + b \le y) = \mathbb{P}\left(X \le \frac{y-b}{a}\right) = \int_{-\infty}^{(y-b)/a} f_X(x)\,\mathrm{d}x .$$

By differentiating with respect to y, we obtain $f_Y(y) = \frac{1}{a} f_X(\frac{y-b}{a})$. For a standard normal variable X, the pdf of $Y = aX + b$ is given by

$$f_Y(y) = \frac{1}{\sigma} f_X\Big(\frac{y-\mu}{\sigma}\Big) = \frac{1}{\sqrt{2\pi}\sigma} \exp\Big(-\frac{(x-\mu)^2}{2\sigma^2}\Big) .$$

Example 4.14 Let U be uniformly distributed in $[0, 1]$, i.e.,

$$f_U(x) = \begin{cases} 1, & u \in (0, 1] , \\ 0, & u \notin (0, 1] . \end{cases}$$

Consider $Y = \sqrt{U}$, i.e., $h(x) = \sqrt{x}$ in Theorem 4.6. Then

$$f_Y(y) = 2y , \quad 0 \leq y \leq 1 .$$

Example 4.15 Let X be uniformly distributed in $(0, 1]$, and define $Y = -\log X$ by choosing $h(x) = -\ln x$. Since $h'(x) = -\frac{1}{x}$ and $x = \mathrm{e}^{-y}$,

$$f_Y(y) = \mathrm{e}^{-y}, \quad y \geq 0 .$$

Example 4.16 (Lognormal Distribution) If X is normally distributed with mean μ and variance σ^2, then $Y = \mathrm{e}^X$ has the probability density function given by

$$f_Y(y) = \frac{1}{\sigma y \sqrt{2\pi}} \exp\left(-\frac{(\log y - \mu)^2}{2\sigma^2}\right) , \quad y > 0 .$$

It can be shown that $\mathbb{E}[Y] = \mathrm{e}^{\mu+\sigma^2/2}$ and $\mathrm{Var}(Y) = \mathrm{e}^{2\mu+\sigma^2}(\mathrm{e}^{\sigma^2} - 1)$. (For the proof see Exercise 4.8.) See Fig. 4.8 for the graph for $\mu = 0, \sigma = 1$.

Example 4.17 (Geometric Brownian Motion) This example is a special case of Example 4.16. Suppose that an asset price S_t at time $t \geq 0$ is given by

$$S_t = S_0 \mathrm{e}^{(\mu - \frac{1}{2}\sigma^2)t + \sigma W_t}$$

where a Brownian motion W_t, $t \geq 0$, is normally distributed with mean 0 and variance t. (See Chap. 7 for the definition of a Brownian motion.) Put $Y = S_t$. Since $\log Y = \log S_0 + (\mu - \frac{1}{2}\sigma^2)t + \sigma W_t$, we see that $X = \log Y$ is normally distributed with mean $\log S_0 + (\mu - \frac{1}{2}\sigma^2)t$ and variance $\sigma^2 t$. Hence the pdf of $Y = \mathrm{e}^X$ is given by

$$f_Y(y) = \frac{1}{\sigma y \sqrt{2\pi t}} \exp\left(- \frac{(\log y - \log S_0 - (\mu - \frac{1}{2}\sigma^2)t)^2}{2\sigma^2 t} \right) ,$$

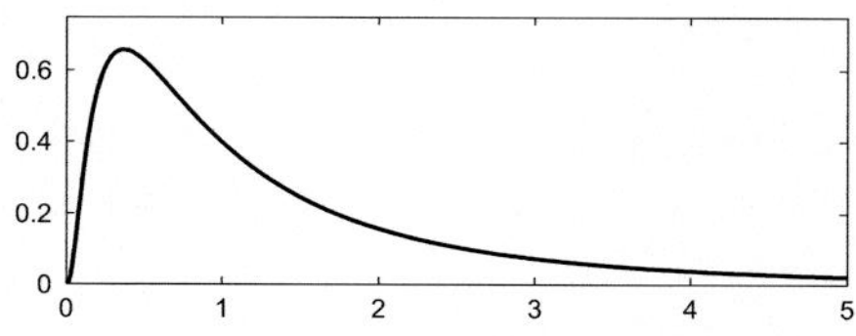

Fig. 4.8 The probability density function of $\exp X$ when $X \sim N(0, 1)$

and hence

$$\mathbb{E}[S_t] = e^{\log S_0 + (\mu - \frac{1}{2}\sigma^2)t + \frac{1}{2}\sigma^2 t} = S_0 e^{\mu t} ,$$

$$\mathrm{Var}(S_t) = S_0^2 e^{2\mu t}(e^{\sigma^2 t} - 1)$$

by Example 4.16. This result will be used in Sect. 16.2.

Remark 4.6

(i) If X has a pdf $f_X(x)$, and if $y = h(x)$ is differentiable but not necessarily one-to-one, then $Y = h(X)$ has its pdf given by

$$f_Y(y) = \sum_{h(z)=y} \frac{1}{|h'(z)|} f_X(z)$$

where the symbol $\sum$ represents the summation over z satisfying $h(z) = y$.

(ii) For two-dimensional examples and applications, see Exercises 4.15, 4.16 and Theorems 27.2, 27.3.

Example 4.18 (Square of the Standard Normal Variable) Suppose that X is the standard normal variable, i.e.,

$$f_X(x) = \frac{1}{\sqrt{2\pi}} e^{-x^2/2} .$$

Let us find the pdf of $Y = X^2$ on $[0, \infty)$. Since $h(x) = x^2$, we have $h^{-1}(y) = \pm\sqrt{y}$ for $y \geq 0$ and

$$f_Y(y) = \frac{1}{|-2\sqrt{y}|} f_X(-\sqrt{y}) + \frac{1}{|2\sqrt{y}|} f_X(\sqrt{y}) = \frac{1}{\sqrt{2\pi y}} e^{-y/2} .$$

For $y < 0$, clearly $f_Y(y) = 0$.

Theorem 4.7 (Uniform Distribution) *Let X be a random variable with its cumulative distribution function F. Assume that F is invertible. Then $F(X)$ is uniformly distributed in* $[0, 1]$. *(See Fig. 4.9.)*

Proof Since

$$\Pr(F(X) \leq x) = \Pr(X \leq F^{-1}(x)) = F(F^{-1}(x)) = x ,$$

we have

$$\Pr(a \leq F(X) \leq b) = \Pr(F(X) \leq b) - \Pr(F(X) \leq a) = b - a$$

and $F(X)$ is uniformly distributed in $[0, 1]$. □

For a computer experiment, see Simulations 4.1 and 4.2.

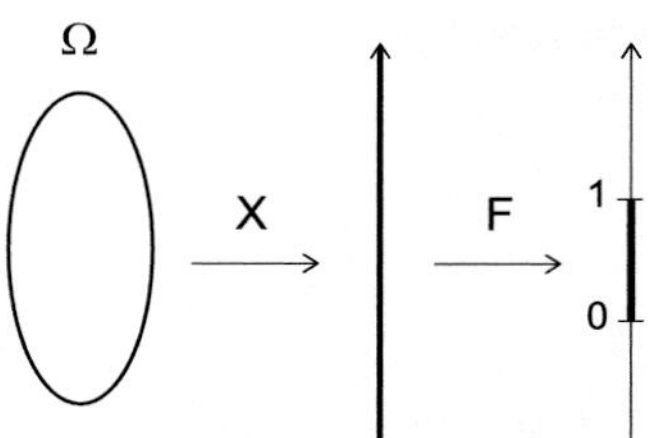

Fig. 4.9 The composite random variable $F(X)$

4.5 The Law of Large Numbers

In this section the symbol μ denotes an average not a measure.

Theorem 4.8 (Weak Law of Large Numbers) *Let X_n, $n \geq 1$, be a sequence of independent and identically distributed random variables with $\mathbb{E}[X_n] = \mu$ and $\mathrm{Var}(X_n) = \sigma^2$. Put $S_n = X_1 + \cdots + X_n$. Then $\frac{1}{n}S_n$ converges to μ in probability as $n \to \infty$. In other words, for every $\varepsilon > 0$ we have*

$$\lim_{n\to\infty} \mathbb{P}\left(\left| \frac{S_n}{n} - \mu \right| > \varepsilon \right) = 0 \,.$$

Proof Since $\frac{S_n}{n} - \mu = \frac{1}{n}(X_1 + \cdots + X_n) - \mu$, the characteristic function is equal to $e^{-i\mu t}\left(\phi_{X_1}(\frac{t}{n})\right)^n$, and its logarithm is equal to $t(\log \phi_{X_1}(\frac{t}{n}) - i\mu\frac{t}{n})/(\frac{t}{n})$, which converges to 0 as $n \to \infty$. Hence the characteristic function of $\frac{S_n}{n} - \mu$ converges to 1, which is the characteristic function of the constant random variable $X = 0$. The cdf of X is given by

$$F_X(x) = \begin{cases} 1, & x \geq 0 \,, \\ 0, & x < 0 \,. \end{cases}$$

Choose $\varepsilon > 0$. By continuity, we have

$$\lim_{n\to\infty} \Pr\left(\frac{S_n}{n} - \mu \leq -\varepsilon \right) = F_X(-\varepsilon) = 0$$

and

$$\lim_{n\to\infty} \Pr\left(\frac{S_n}{n} - \mu \leq \varepsilon \right) = F_X(\varepsilon) = 1 \,.$$

Hence the proof is complete. □

Remark 4.7 With an additional condition that X_n are square-integrable, the proof of Theorem 4.8 is simple. Using Chebyshev's inequality for $p = 2$, we obtain

$$\mathbb{P}\left(\left| \frac{1}{n} \sum_{j=1}^{n} (X_j - \mu) \right| > \varepsilon \right) \le \frac{1}{n^2 \varepsilon^2} \sum_{j=1}^{n} \mathbb{E}[(X_j - \mu)^2] = \frac{\sigma^2}{n\varepsilon^2}$$

and let $n \to \infty$, completing the proof.

Theorem 4.9 (Strong Law of Large Numbers) *Let X_n, $n \ge 1$, be a sequence of independent and identically distributed random variables with $\mathbb{E}[X_n] = \mu$. Put $S_n = X_1 + \cdots + X_n$. Then $\frac{1}{n} S_n$ converges to μ almost surely as $n \to \infty$.*

Proof Consult [31]. □

Let $A_1, A_2, A_3, \ldots$ be a sequence of events in a probability space $(\Omega, \mathbb{P})$. Let $\limsup A_n = \bigcap_{k=1}^{\infty} \bigcup_{n=k}^{\infty} A_n$. Note that $\omega \in \limsup A_n$ if and only if $\omega \in A_n$ for infinitely many n.

Theorem 4.10 (Borel–Cantelli Lemma) *Let A_n, $n \ge 1$, be a sequence of events.*

(i) If $\sum_{n=1}^{\infty} \mathbb{P}(A_n) < \infty$, then $\mathbb{P}(\limsup A_n) = 0$.
(ii) If A_n are independent and if $\sum_{n=1}^{\infty} \mathbb{P}(A_n) = \infty$, then $\mathbb{P}(\limsup A_n) = 1$.

Proof

(i) For each $k \ge 1$, we have

$$\mathbb{P}(\limsup A_n) \le \mathbb{P}\Big(\bigcup_{n \ge k} A_n \Big) \le \sum_{n=k}^{\infty} \mathbb{P}(A_n) \to 0 \,.$$

(ii) If $\mathbb{P}(A_n) = 1$ for infinitely many n, then it is obvious that the statement is true. Assume that $\mathbb{P}(A_n) < 1$ for every $n \ge N$ for some N. Recall the fact that $\prod_n (1 - a_n) = 0$ if and only if $\sum_n a_n = \infty$ where $0 \le a_n < 1$ for every n. For each k, we have

$$\mathbb{P}\Big(\Omega \setminus \bigcup_{n=k}^{\infty} A_n \Big) = \mathbb{P}\Big(\bigcap_{n=k}^{\infty} (\Omega \setminus A_n) \Big) = \prod_{n=k}^{\infty} (1 - \mathbb{P}(A_n)) = 0 \,.$$

We use the independence of the complements of $A_1, A_2, A_3, \ldots$ for the second equality. Hence $\mathbb{P}(\bigcup_{n=k}^{\infty} A_n) = 1$ for every k. □

For example, if a fair coin is tossed n times, then the probability of the event A_n that heads comes up every time is 2^{-n}. Since $\sum_1^{\infty} \mathbb{P}(A_n) < \infty$, we conclude that tails will eventually come up with probability 1.

4.6 The Central Limit Theorem

Definition 4.12 (Convergence in Distribution) A sequence of random variables $\{X_n\}_{n=1}^{\infty}$ *converges in distribution* to a random variable X if

$$\lim_{n\to\infty} \mathbb{P}\{X_n \le x\} = \mathbb{P}\{X \le x\}$$

at all points where the cumulative distribution function of X is continuous. If X_n converges to X in distribution, we write $X_n \xrightarrow{D} X$. It is known that if $\{X_n\}_{n=1}^{\infty}$ converges almost surely to X as $n \to \infty$, then $X_n \xrightarrow{D} X$.

Theorem 4.11 (Central Limit Theorem) *Let $\{X_n\}_{n=1}^{\infty}$ be independent, identically distributed L^2-random variables with $\mathbb{E}[X_n] = \mu$ and $\mathrm{Var}(X_n) = \sigma^2$ for $n \ge 1$. Put $S_n = X_1 + \cdots + X_n$. Then*

$$\lim_{n\to\infty} \mathbb{P}\left(\frac{S_n - n\mu}{\sigma\sqrt{n}} \le x\right) = N(x) \ , \quad -\infty < x < \infty$$

where $N(x)$ is the cumulative distribution function for the normal distribution defined in Example 4.7. In other words, $\frac{S_n - n\mu}{\sigma\sqrt{n}}$ converges in distribution to a standard normal variable Z.

Proof Here is a sketch of the proof. Put $S_n^* = (S_n - n\mu)/(\sigma\sqrt{n})$. Then

$$\phi_{S_n^*}(t) = \exp\left(\frac{-in\mu t}{\sigma\sqrt{n}}\right)\phi_{S_n}\left(\frac{t}{\sigma\sqrt{n}}\right) = \exp\left(\frac{-in\mu t}{\sigma\sqrt{n}}\right)\left(\phi_{X_1}\left(\frac{t}{\sigma\sqrt{n}}\right)\right)^n .$$

The logarithm of the leftmost side is

$$n\left(\log\phi_{X_1}\left(\frac{t}{\sigma\sqrt{n}}\right) - \mathrm{i}\mu\frac{t}{\sigma\sqrt{n}}\right) = \frac{t^2}{\sigma^2}\frac{\log\phi_{X_1}(\frac{t}{\sigma\sqrt{n}}) - \mathrm{i}\mu\frac{t}{\sigma\sqrt{n}}}{(\frac{t}{\sigma\sqrt{n}})^2} ,$$

which converges to $\frac{t^2}{\sigma^2}(-\frac{\sigma^2}{2}) = -\frac{t^2}{2}$ by the power series expansion given in Remark 4.3. Hence

$$\lim_{n\to\infty} \phi_{S_n^*} = \mathrm{e}^{-t^2/2} ,$$

which is the characteristic function of the standard normal distribution. (Consult Example 4.8.) □

The Central Limit Theorem states that if we take n samples, not necessarily normally distributed, with average μ and standard deviation σ, then the average of sample S_n/n is approximately normally distributed with average μ and standard deviation $\sigma/\sqrt{n}$ for sufficiently large n. This idea is behind the Monte Carlo integration method for option pricing.

4.7 Statistical Ideas

Definition 4.13 (Covariance) The *covariance* $\mathrm{Cov}(X,Y)$ between two random variables X and Y is defined by

$$\mathrm{Cov}(X,Y) = \mathbb{E}\left[\,(X-\mathbb{E}[X])\,(Y-\mathbb{E}[Y])\,\right] .$$

Theorem 4.12 *For random variables X,Y,Z,X_i,Y_j the following facts hold:*

(i) $\mathrm{Cov}(X,Y) = \mathrm{Cov}(Y,X)$.
(ii) $\mathrm{Cov}(X,X) = \mathrm{Var}(X)$.
(iii) $\mathrm{Cov}(X,Y) = \mathbb{E}[XY] - \mathbb{E}[X]\,\mathbb{E}[Y]$.
(iv) $\mathrm{Cov}(X+Y,Z) = \mathrm{Cov}(X,Z) + \mathrm{Cov}(Y,Z)$.
(v) $\mathrm{Cov}(\sum_i X_i, \sum_j Y_j) = \sum_i \sum_j \mathrm{Cov}(X_i, Y_j)$.
(vi) *For a constant a,* $\mathrm{Cov}(X,aY) = a\,\mathrm{Cov}(X,Y)$.
(vii) *For a constant b,* $\mathrm{Cov}(X,b) = 0$.
(viii) *If X and Y are independent then* $\mathrm{Cov}(X,Y) = 0$ *since* $\mathrm{Cov}(X,Y) = \mathbb{E}[XY] - \mathbb{E}[X]\,\mathbb{E}[Y]$.

Definition 4.14 (Correlation) The *correlation coefficient* $\rho(X,Y)$ for random variables X and Y with $\mathrm{Var}(X) > 0$ and $\mathrm{Var}(Y) > 0$ is defined by

$$\rho(X,Y) = \frac{\mathrm{Cov}(X,Y)}{\sqrt{\mathrm{Var}(X)\mathrm{Var}(Y)}} .$$

By the Cauchy–Schwarz inequality we have $-1 \le \rho(X,Y) \le 1$.

Theorem 4.13 *Given a random variable X with* $\mathrm{Var}(X) > 0$, *and constants $a \neq 0$, b, we define $Y = aX + b$. Then*

$$\rho(X,Y) = \begin{cases} 1, & a > 0 \,, \\ -1, & a < 0 \,. \end{cases}$$

Proof Note that

$$\mathrm{Cov}(X,Y) = a\mathrm{Cov}(X,X) + \mathrm{Cov}(X,b) = a\mathrm{Var}(X)$$

and

$$\mathrm{Var}(Y) = \mathrm{Var}(aX+b) = \mathrm{Var}(aX) = a^2\mathrm{Var}(X) .$$

Now use the fact $\mathrm{Var}(X) >$ and $\mathrm{Var}(Y) > 0$ to find $\rho(X,Y)$. □

Example 4.19

(i) Let U be a random variable uniformly distributed in $[0, 1]$. Then U and $1 - U$ have negative covariance since

$$\mathrm{Cov}(U, 1-U) = \mathrm{Cov}(U, 1) - \mathrm{Cov}(U, U) = 0 - \mathrm{Var}(U) = -\frac{1}{12}$$

by Example 4.3.

(ii) If Z is a standard normal variable, then $\mathrm{Cov}(Z, -Z) = -\mathrm{Var}(Z) = -1$.

Remark 4.8

(i) For $X_1, X_2 \in L^2(\Omega, \mathbb{P})$ on a probability space $(\Omega, \mathbb{P})$, note that $\mathrm{Cov}(X_1, X_2) = 0$ if and only if $X_1 - \mu_1 \mathbf{1}$ and $X_2 - \mu_2 \mathbf{1}$ are orthogonal where $\mu_i = \mathbb{E}[X_i]$, $i = 1, 2$, and $\mathbf{1}$ is the constant function equal to 1 on Ω.

(ii) Define

$$Y = X_2 - \frac{\mathbb{E}[X_2 X_1]}{\mathbb{E}[X_1^2]} X_1 \,.$$

Then $\mathbb{E}[Y X_1] = 0$ since $\frac{\mathbb{E}[X_2 X_1]}{\mathbb{E}[X_1^2]} X_1$ is the component of X_2 in the direction of X_1.

(iii) Let X_1, X_2 be random variables with mean μ_i and variance $\sigma_i^2 > 0$, $i = 1, 2$, and the correlation coefficient ρ. Define

$$Y = X_2 - \rho \frac{\sigma_2}{\sigma_1} X_1 \,.$$

Then $\mathrm{Cov}(Y, X_1) = 0$ since $\mathbb{E}[Y] = \mu_2 - \rho \frac{\sigma_2}{\sigma_1} \mu_1$ and since

$$\begin{aligned}
\mathrm{Cov}(Y, X_1) &= \mathbb{E}[(X_2 - \rho \frac{\sigma_2}{\sigma_1} X_1) X_1] - (\mu_2 - \rho \frac{\sigma_2}{\sigma_1} \mu_1) \mu_1 \\
&= \mathbb{E}[X_1 X_2] - \rho \frac{\sigma_2}{\sigma_1} \mathbb{E}[X_1^2] - (\mu_2 - \rho \frac{\sigma_2}{\sigma_1} \mu_1) \mu_1 \\
&= (\rho \sigma_1 \sigma_2 + \mu_1 \mu_2) - \rho \frac{\sigma_2}{\sigma_1} (\sigma_1^2 + \mu_1^2) - (\mu_2 - \rho \frac{\sigma_2}{\sigma_1} \mu_1) \mu_1 \\
&= 0 \,.
\end{aligned}$$

Remark 4.9 If jointly normal variables $X_1, \ldots, X_n$ are uncorrelated, i.e., $\mathrm{Cov}(X_i, X_j) = 0$ for $i \neq j$, then $\Sigma = \mathrm{diag}(\sigma_1^2, \ldots, \sigma_n^2)$ is a diagonal matrix where $\sigma_i^2 = \mathrm{Var}(X_i)$, and

$$f(x_1, \ldots, x_n) = \prod_{i=1}^{n} \frac{1}{\sqrt{2\pi}\sigma_i} \exp\left(-\frac{1}{2\sigma_i^2} (x_i - \mu_i)^2 \right) \,.$$

Since the joint pdf is the product of individual densities for X_i, we conclude that $X_1, \ldots, X_n$ are independent. (See Theorem 4.4.)

Remark 4.10 For jointly normal variables $X_1, \ldots, X_n$ and an $m\times n$ matrix $A = (a_{ij})$, it can be shown that the linear combinations $Y_i = \sum_{j=1}^{n} a_{ij}X_j$, $1 \le i \le m$, are jointly normal. Since independent normal variables are jointly normal, their linear combinations are jointly normal.

Theorem 4.14 *Using uncorrelated standard normal variables Z_1 and Z_2, we can construct a pair of correlated normal variables X_1 and X_2 with a given correlation ρ through the linear transformation given by*

$$\begin{cases} X_1 = & Z_1 \\ X_2 = \rho Z_1 + \sqrt{1-\rho^2}Z_2 \end{cases}$$

Proof Note that

$$\mathrm{Var}(X_2) = \rho^2\mathrm{Var}(Z_1) + (1-\rho^2)\mathrm{Var}(Z_2) + 2\rho\sqrt{1-\rho^2}\,\mathrm{Cov}(Z_1, Z_2) = 1 \ ,$$

and the correlation $\mathrm{Corr}(X_1, X_2)$ is equal to

$$\mathrm{Corr}(Z_1, \rho Z_1 + \sqrt{1-\rho^2}Z_2) = \rho\,\mathrm{Corr}(Z_1, Z_1) + \sqrt{1-\rho^2}\,\mathrm{Corr}(Z_1, Z_2)) = \rho \ .$$

Now use Remark 4.10 to prove that X_1 and X_2 are normal variables. □

Example 4.20 Using a pair of independent standard normal variables Z_1 and Z_2, two correlated normal variables X_1 and X_2 with $\rho(X_1, X_2) = -0.5$ are constructed. Fig. 4.10 shows (Z_1, Z_2) and (X_1, X_2) in the left and right panels, respectively. Consult Simulation 4.4.

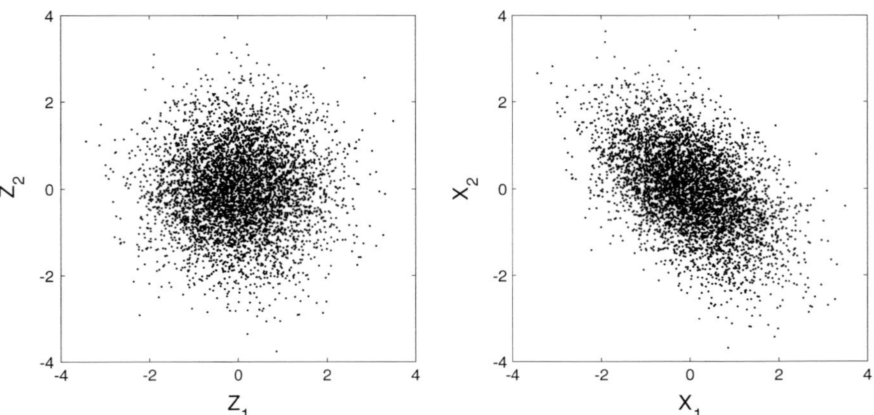

Fig. 4.10 Scatter plots for independent standard normal variables Z_1 and Z_2 (*left*), and negatively correlated normal variables X_1 and X_2 (*right*)

Definition 4.15 (Skewness and Kurtosis) Let X be a random variable with mean μ and variance σ^2.

(i) The *skewness* of X is defined by $\mathbb{E}\left[\left(\frac{X-\mu}{\sigma}\right)^3\right]$.

(ii) The *kurtosis* of X is defined by $\mathbb{E}\left[\left(\frac{X-\mu}{\sigma}\right)^4\right]$.

Example 4.21 A normal variable with mean μ and variance σ^2 has skewness 0 and kurtosis 3.

Definition 4.16 (χ^2-Distribution) For independent standard normal variables $Z_1, \ldots, Z_n$, the sum $X = Z_1^2 + \cdots + Z_n^2$ is said to have the *chi-squared distribution* denoted by $X \sim \chi^2(n)$. Its probability density function is given by

$$f(x) = \frac{1}{2^{n/2}\Gamma(\frac{n}{2})} e^{-x/2} x^{n/2-1}, \quad x > 0,$$

where Γ denotes the gamma function. The mean and variance of the $\chi^2(n)$-distribution are k and $2k$, respectively. For $n = 1$ see Example 4.18, and for $n = 2$ see Exercise 4.24. For the graph of $f(x)$ for $1 \leq n \leq 6$, see Fig. 4.11.

Definition 4.17 (Noncentral χ^2-Distribution) For independent normal variables $Y_1, \ldots, Y_n$ with mean μ_i and variance 1, $1 \leq i \leq n$, the sum $X = Y_1^2 + \cdots + Y_n^2$ is said to have the *noncentral chi-squared distribution* with noncentrality parameter $\lambda = \mu_1^2 + \cdots + \mu_n^2$. Its probability density function is given by

$$f(x) = e^{-\lambda/2} {}_0F_1\left(;\frac{n}{2}; \frac{\lambda x}{4}\right) \frac{1}{2^{n/2}\Gamma(\frac{n}{2})} e^{-x/2} x^{n/2-1}, \quad x > 0,$$

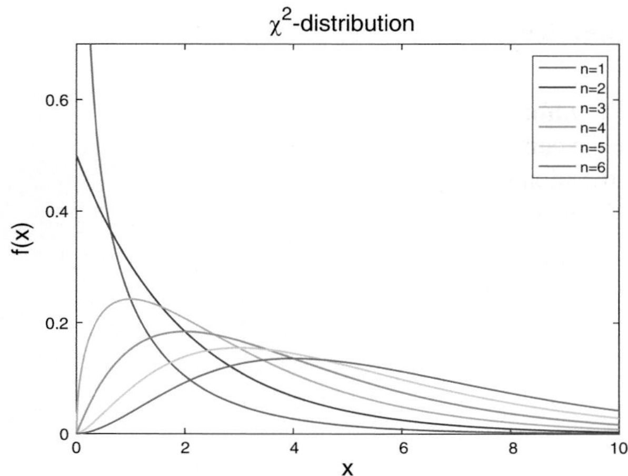

Fig. 4.11 Probability density functions of $\chi^2(n)$ for $1 \leq n \leq 6$

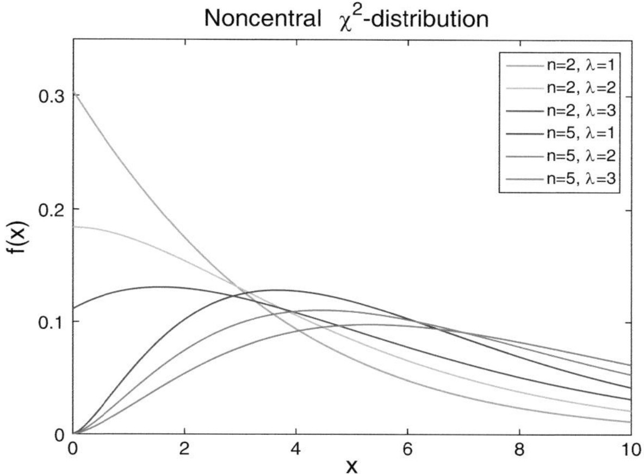

Fig. 4.12 Probability density functions of noncentral $\chi^2(n)$ for $n = 2, 5$ and $\lambda = 1, 2, 3$

where ${}_pF_q$ is the hypergeometric function defined by

$$ {}_pF_q(a_1, \ldots, a_p; b_1, \ldots, b_q; x) = \sum_{k=0}^{\infty} \frac{(a_1)_k \cdots (a_p)_k}{(b_1)_k \cdots (b_q)_k} \frac{x^k}{k!} $$

where $(a)_k = a(a+1)\cdots(a+k-1)$. The mean and variance are given by $n + \lambda$ and $2n + 4\lambda$. Consult [72]. For the graph of $f(x)$, see Fig. 4.12 and Simulation 4.5. For an example in interest rate modeling see Theorem 23.4.

4.8 Computer Experiments

Simulation 4.1 (Uniform Distribution)

Let $F_X(x)$ be the cumulative distribution function of X. We check that $F_X(X)$ has the uniform distribution by plotting its probability density function. In Fig. 4.13 we choose X with the standard normal distribution. (See Theorem 4.7.)

```
N = 20;  % number of bins
Sample=10^4; % number of samples
width = 1/N
bin = zeros(1,N);  % a zero matrix of size 1 x N

for i=1:Sample
j_bin=ceil(normcdf(randn)/width);
bin(j_bin)=bin(j_bin)+1;
end

x = 0:width:1;
```

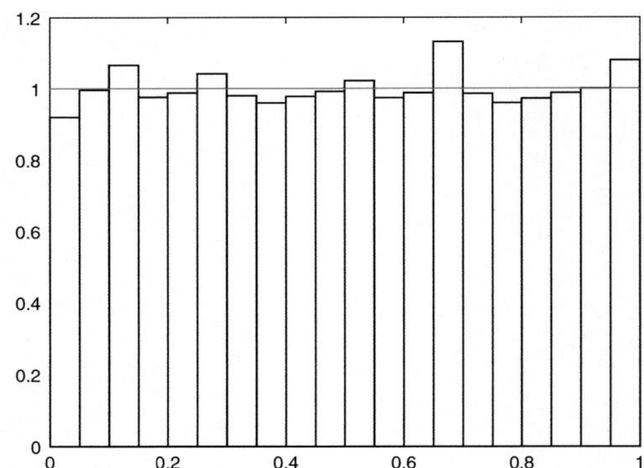

Fig. 4.13 Uniform distribution of $F_X(X)$

```
y = ones(1,N+1);
mid_points = 0+width/2:width:1; % middle points of bins
bar(mid_points,bin/Sample/width, 1, 'w')
hold on
plot(x,y)
```

Simulation 4.2 (Equally Spaced Points in the Unit Interval)

Let $F_X(x)$ be the cdf of X. Generate N random values $x_1, \ldots, x_N$ of X. Sort them in ascending order, which are written as $y_1 < y_2 < \cdots < y_N$. Since $F_X(x_1), \ldots, F_X(x_N)$ are approximately uniformly distributed for large N by Theorem 4.7, so are $F_X(y_1), \ldots, F_X(y_N)$ since they are a rearrangement of $F_X(x_1), \ldots, F_X(x_N)$. Note that

$$F_X(y_1) < F_X(y_2) < \cdots < F_X(y_N)$$

since F_X is monotonically increasing. Hence the sequence is close to the equally spaced points $\frac{1}{N+1} < \cdots < \frac{N}{N+1}$, and hence $F_X(y_i)$ is close to $\frac{i}{N+1}$. Therefore, if we plot the points $(F_X(y_i), \frac{i}{N+1})$ they are concentrated along the diagonal of the unit square. For a scatter plot of one hundred points for the standard normal distribution, see Fig. 4.14.

```
N = 100; % number of points
X = randn(N,1);
Y = sort(X);

U = 1/(N+1):1/(N+1):N/(N+1);
V = normcdf(Y); % the normal cdf

x = 0:0.01:1;
plot(x,x)
hold on
plot(V,U,'o')
```

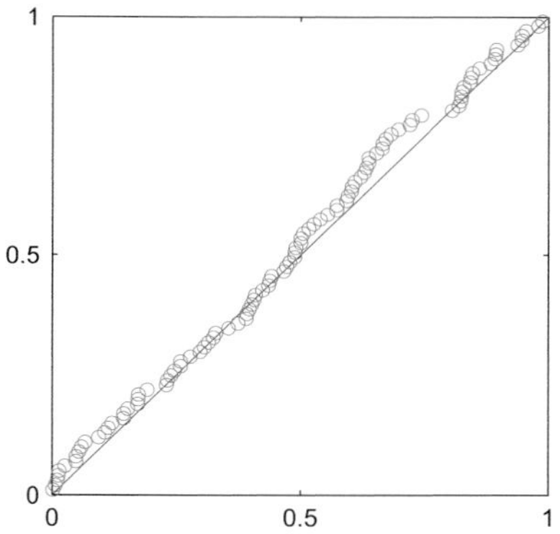

Fig. 4.14 Equally spaced points along the diagonal

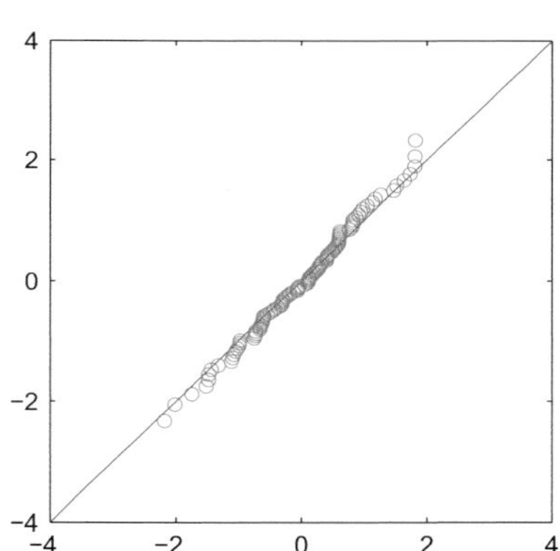

Fig. 4.15 A quantile-quantile plot for the standard normal distribution

Simulation 4.3 (Quantile-Quantile Plot)

Let $F_X(x)$ be the cumulative distribution function of X. Generate N values of X, and sort them in the increasing order. Sort them into an increasing sequence $y_1 < y_2 < \cdots < y_N$. Since $F_X(y_1) < F_X(y_2) < \cdots < F_X(y_N)$ are more or less uniformly distributed, they are close to $\frac{1}{N+1}, \frac{2}{N+1}, \ldots, \frac{N}{N+1}$, respectively, and hence $F_X(y_i)$ is close to $\frac{i}{N+1}$. Therefore, if we plot the points $\left(y_i, F_X^{-1}\left(\frac{i}{N+1}\right)\right)$, then they are scattered around the straight line $y = x$, $-\infty < x < \infty$. For a scatter plot of one hundred points for the standard normal distribution, see Fig. 4.15.

```
N = 100; % number of points
X = randn(N,1);
Y = sort(X);

U = 1/(N+1):1/(N+1):N/(N+1);
V = norminv(U); % inverse of the normal cdf

x = -4:0.1:4;
plot(x,x)
hold on
plot(Y,V,'o')
```

Simulation 4.4 (Correlated Normal Variables)
We plot the ordered pairs (Z_1, Z_2) and (X_1, X_2) as in Example 4.20.

```
N = 5000; % number of points
Z1 = randn(N,1);
Z2 = randn(N,1);
CorrelationZ1Z2 = corr(Z1,Z2)
rho = -0.5;
X1 = Z1;
X2 = rho*Z1 + sqrt(1-rho^2)*Z2;
CorrelationX1X2 = corr(X1,X2)

figure(1)
plot(Z1,Z2,'.');
hold off;

figure(2)
plot(X1,X2,'.')
```

We have the following output:

```
CorrelationZ1Z2 = 0.0088

CorrelationX1X2 = -0.4897
```

Simulation 4.5 (Noncentral χ^2-Distribution)
We plot the pdf of the noncentral χ^2-distribution for various values of k and λ. For the output see Fig. 4.12.

```
for k = 2:3:5
    for lambda = 1:1:3
x = 0:0.1:10;
y = ncx2pdf(x,k,lambda);
plot(x,y,'color',hsv2rgb([(k+lambda)/7 1 1]))
hold on;
    end
end
```

Exercises

4.1 Let $(\Omega, \mathbb{P})$ be a probability space, and let $X \geq 0$ be a random variable. Show that $\mathbb{E}[X] < \infty$ if and only if $\sum_{j=0}^{\infty} \mathbb{P}(\{X \geq j\}) < \infty$.

4.2 Let X be a random variable taking values ± 1 with probability $\frac{1}{2}$ for each value. Show that if we choose $\phi(x) = x^2$ in Jensen's inequality then we have a strict inequality.

4.3 For a random variable X satisfying $\mathbb{P}(X = a) = 1$ find the cumulative distribution function $F(x)$.

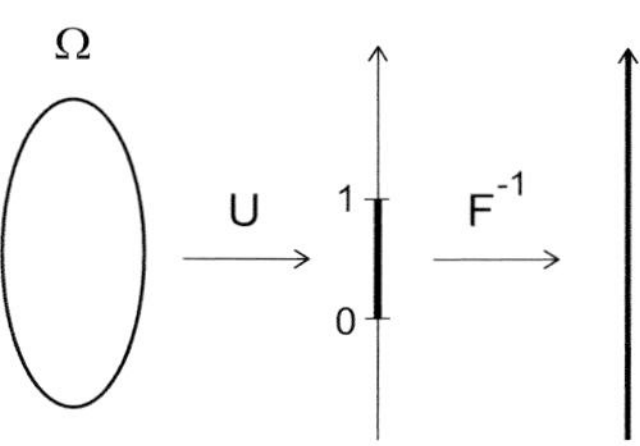

Fig. 4.16 The composite random variable $F^{-1}(U)$

4.4 Let U be a random variable uniformly distributed in $[0, 1]$. Show that $\mathbb{E}[e^{\sqrt{U}}] = 2$.

4.5 Let N denote the cdf of the standard normal distribution. Prove that $N(\alpha) + N(-\alpha) = 1$.

4.6 Let X be a random variable with its cumulative distribution function $F(x)$. Assume that F is invertible. Show that if U is uniformly distributed in $[0, 1]$, then the cumulative distribution function of $F^{-1}(U)$ is F. In other words, $F^{-1}(U)$ and X are identically distributed. (See Fig. 4.16.) (For an application, see Exercise 27.2. See also Exercise 28.1.)

4.7

(i) Let $t > 0$ and $f(x) = \frac{1}{\sqrt{2\pi t}} \exp\left(-\frac{x^2}{2t}\right)$. Show $\int_{-\infty}^{\infty} x^4 f(x)\, dx = 3t^2$ by direct computation.
(ii) Suppose that X is normally distributed with mean μ and variance σ^2. Find the moment generating function $\mathbb{E}[e^{\theta X}]$, and thus evaluate $\mathbb{E}[X^4]$. What is $\mathbb{E}[X^{2k}]$?

4.8 Let X be a normal variable with mean μ and variance σ^2. Show that the mean and the variance of e^X are given by $e^{\mu+\sigma^2/2}$ and $e^{2\mu+\sigma^2}(e^{\sigma^2} - 1)$, respectively. For example, for a standard normal variable Z we have $\mathbb{E}[e^Z] = \sqrt{e}$.

4.9 Let Z denote a standard normal variable. For a real constant α, show that $e^{\alpha Z} \in L^p$ for every $1 \le p < \infty$.

4.10 Let Z denote a standard normal variable. For $0 < \alpha < \frac{1}{2}$, show that

$$\mathbb{E}\left[e^{\alpha Z^2}\right] = \frac{1}{\sqrt{1 - 2\alpha}} .$$

4.11 Calculate the Fourier transform of the step function $\mathbf{1}_{[-n,n]}$. Discuss its behavior as $n \to \infty$.

4.12 Let $U_1, \ldots, U_{12}$ be independent random variables uniformly distributed in the unit interval. Compute the kurtosis of $Y = U_1 + \cdots + U_{12} - 6$.

4.13 Suppose that a random variable X is uniformly distributed in $[-1, 1]$. Find the probability distribution of $Y = X^2$ on $[0, 1]$. (Hint: Note that $f_X = \frac{1}{2} \times \mathbf{1}_{[-1,1]}$. For $y \notin (0, 1], f_Y(y) = 0$.)

4.14 Let X and Y be independent standard normal variables. Show that $\tan^{-1}\left(\frac{Y}{X}\right)$ is uniformly distributed in $(-\frac{\pi}{2}, \frac{\pi}{2})$.

4.15 Suppose that (X, Y) is a continuous two-dimensional random vector with probability density function $f_{X,Y}(x, y)$. Prove that, if $(V, W) = \Phi(X, Y)$ for some bijective differentiable mapping $\Phi : \mathbb{R}^2 \to \mathbb{R}^2$, then

$$f_{V,W}(v, w) = f_{X,Y}(\Phi^{-1}(v, w)) \frac{1}{|J_\Phi(\Phi^{-1}(v, w))|}$$

where J_Φ denotes the Jacobian of Φ.

4.16 Suppose that a two-dimensional random vector (X, Y) is uniformly distributed on the disk $D = \{(x, y) : x^2 + y^2 \leq a^2\}$. Let (R, Θ) be the random vector given by the polar coordinates of (X, Y). Show that

(i) R and Θ are independent,
(ii) R has the pdf given by $f_R(r) = 2r/a^2, 0 < r < a$, and
(iii) Θ is uniformly distributed in $(0, 2\pi)$.

(Hint: Modify the proof of Theorem 27.3.)

4.17 Find a pair of random variables X, Y which are not independent but $\mathrm{Cov}(X, Y) = 0$.

4.18 Find a pair of standard normal variables with correlation $-1 \leq \rho \leq 1$.
(Hint: First, generate a pair of independent standard normal variables Z_1, Z_2. Define X_1, X_2 by linear combinations of Z_1, Z_2.)

4.19 Let $X \geq 0$ be a square-integrable random variable.

(i) Show that if X has continuous distribution, then

$$\mathbb{E}[X^2] = 2 \int_0^\infty t\, \mathbb{P}(X > t)\, \mathrm{d}t\ .$$

(ii) Show that if X has discrete distribution, then

$$\mathbb{E}[X^2] = \sum_{k=1}^{\infty} (2k + 1)\mathbb{P}(X > k)\ .$$

4.20 Let $U_1, \ldots, U_n$ be independent random variables uniformly distributed in $[0, 1]$, and let $S_n = U_1 + \cdots + U_n$. Show that the pdf of S_n is given by

$$f_{S_n}(x) = f^{*n}(x) = \frac{1}{(n-1)!} \sum_{0 \leq k \leq x} (-1)^k C(n,k)(x-k)^{n-1}$$

for $0 < x < n$, and $f_{S_n}(x) = 0$ elsewhere. Plot the graph of $f_{S_n}(x)$.

4.21 (Product of Random Variables) Let X, Y be independent random variables with their probability density functions f_X and f_Y, respectively. Prove that the probability density function of XY is given by

$$f_{XY}(a) = \int_{-\infty}^{\infty} \frac{1}{|y|} f_X\left(\frac{a}{y}\right) f_Y(y)\, dy .$$

(Hint: See Fig. 4.17.)

4.22 Suppose that U, V are independent and uniformly distributed in $[0, 1]$.

(i) Prove that $f_{UV}(x) = -\log x$ for $0 \leq a \leq 1$, and $f_{UV}(x) = 0$ elsewhere.
(ii) Find the cumulative distribution function for $X = UV$. (Hint: See Fig. 4.18.)

4.23 Let Σ be a nonnegative-definite symmetric matrix with an eigenvalue decomposition $\Sigma = PDP^t$ where $D = \mathrm{diag}(\lambda_1, \ldots, \lambda_n)$ is a diagonal matrix whose diagonal entries are eigenvalues of Σ arranged in decreasing order and the columns of P are given by the eigenvectors $\mathbf{v}_1, \ldots, \mathbf{v}_n$ of Σ corresponding to $\lambda_1, \ldots, \lambda_n$, respectively. Let $\sqrt{D} = \mathrm{diag}(\sqrt{\lambda_1}, \ldots, \sqrt{\lambda_n})$ and $\mathbf{Z} \sim N(\mathbf{0}, I)$.

(i) Let $\mathbf{X} = \boldsymbol{\mu} + P\sqrt{D}\mathbf{Z}$. Show that $\mathbf{X} \sim N(\boldsymbol{\mu}, \Sigma)$.
(ii) Let $\mathbf{X} = P\sqrt{D}\mathbf{Z}$. Show that $\mathbf{X} = \sqrt{\lambda_1}\mathbf{v}_1 Z_1 + \cdots + \sqrt{\lambda_n}\mathbf{v}_n Z_n$.

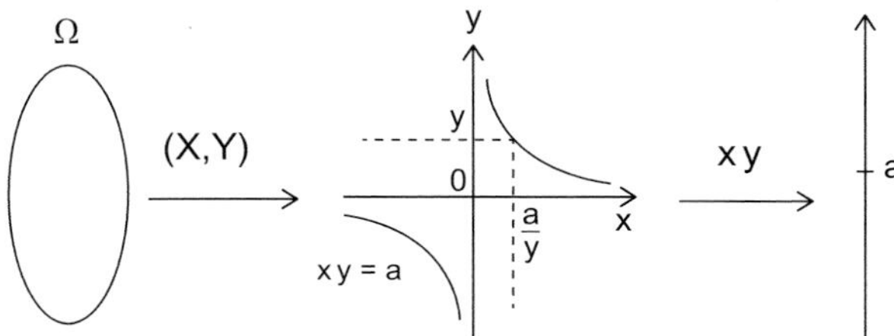

Fig. 4.17 The product of two independent random variables X and Y

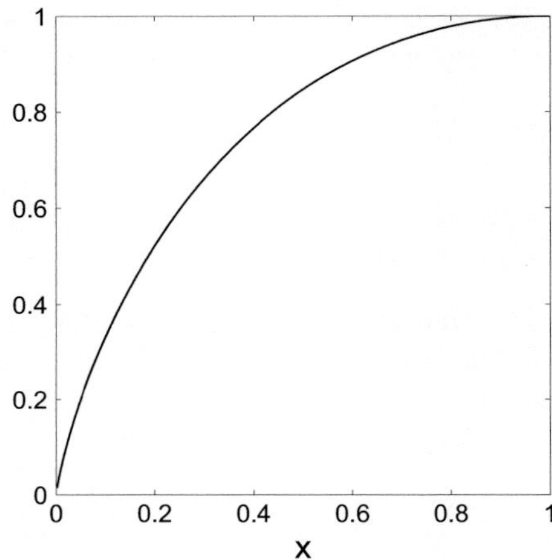

Fig. 4.18 The cumulative distribution function for the product of two independent uniformly distributed random variables

4.24

(i) Show that the pdf of the chi-squared distribution for $k = 2$ is given by $f(x) = \frac{1}{2}e^{-x/2}, x \geq 0$.
(ii) Find the pdf of the chi-squared distribution for $k = 3$.

Chapter 5
Conditional Expectation

The concept of conditional expectation will be developed in three stages. First, we define the conditional expectation on a given event, $\mathbb{E}[X|A]$, and next define the conditional expectation with respect to a sub-σ-algebra, $\mathbb{E}[X|\mathcal{G}]$, then finally the conditional expectation with respect to a random variable, $\mathbb{E}[X|Y]$. Recall that a sub-σ-algebra represents a collection of currently available information.

5.1 Conditional Expectation Given an Event

For measurable subsets A and B, $\mathbb{P}(A) > 0$, define the *conditional probability* of B given A by

$$\mathbb{P}(B|A) = \frac{\mathbb{P}(A \cap B)}{\mathbb{P}(A)} .$$

(See also Definition 3.4.)

Definition 5.1 (Conditional Expectation) Consider a random variable X on a probability space $(\Omega, \mathcal{F}, \mathbb{P})$ and an event $A \in \mathcal{F}$ of positive measure. The *conditional expectation* of X given A, denoted by $\mathbb{E}[X|A]$, is defined by

$$\mathbb{E}[X|A] = \frac{\int_A X \, d\mathbb{P}}{\mathbb{P}(A)} = \frac{\mathbb{E}[\mathbf{1}_A X]}{\mathbb{P}(A)} .$$

G.H. Choe, *Stochastic Analysis for Finance with Simulations*, Universitext,
DOI 10.1007/978-3-319-25589-7_5

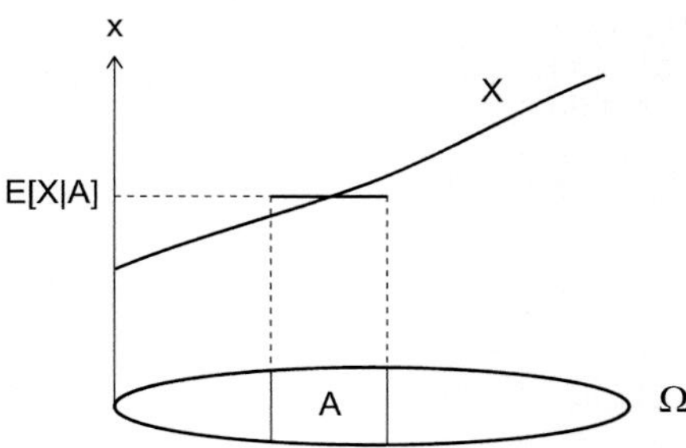

Fig. 5.1 Conditional expectation of X given an event A

In other words, $\mathbb{E}[X|A]$ is the average of X on a set A. (See Fig. 5.1.) If a random variable X is given by $X = \mathbf{1}_B$ for a measurable set B, we have

$$\mathbb{E}[X|A] = \frac{\mathbb{P}(A \cap B)}{\mathbb{P}(A)} = \mathbb{P}(B|A) , \tag{5.1}$$

which is an extension of the concept of conditional probability.

5.2 Conditional Expectation with Respect to a σ-Algebra

As a motivation, we first consider a sample space Ω which is a disjoint union $\Omega = A_1 \cup \cdots \cup A_n$ of measurable subsets $A_1, \ldots, A_n$ of positive measure. Let $\mathcal{G}$ be the σ-algebra generated by $A_1, \ldots, A_n$, then $\mathcal{G}$ is a sub-σ-algebra of $\mathcal{F}$. Note that if $B \in \mathcal{G}$ and $\mathbb{P}(B) > 0$, then B is a union of at least one subset from the collection $A_1, \ldots, A_n$. If a subset B of A_i has positive measure, then $B = A_i$. Sometimes $\mathcal{G}$ is denoted by $\sigma(\{A_1, \ldots, A_n\})$. In this case, a new random variable $E[X|\mathcal{G}]$ is defined by

$$\mathbb{E}[X|\mathcal{G}](\omega) = \mathbb{E}[X|A_i] , \quad \omega \in A_i .$$

In other words,

$$\mathbb{E}[X|\mathcal{G}] = \sum_{i=1}^{n} \mathbb{E}[X|A_i] \, \mathbf{1}_{A_i} .$$

(See Fig. 5.2.) Note that $E[X|\mathcal{G}]$ is $\mathcal{G}$-measurable and for $B \in \mathcal{G}$ we have

$$\int_B \mathbb{E}[X|\mathcal{G}] \, d\mathbb{P} = \int_B X \, d\mathbb{P} .$$

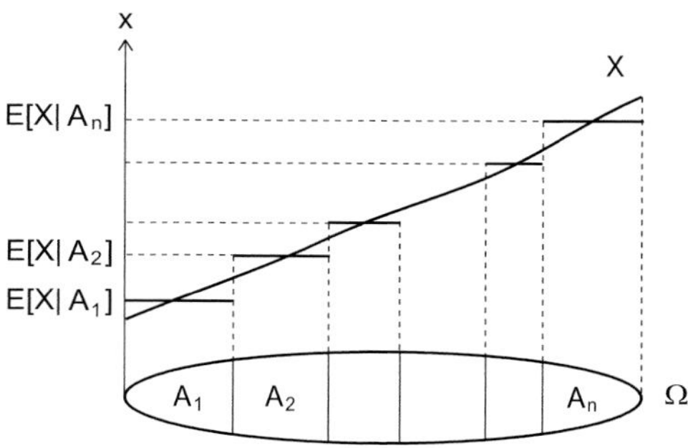

Fig. 5.2 Conditional expectation when a probability space is decomposed into subsets of positive measure

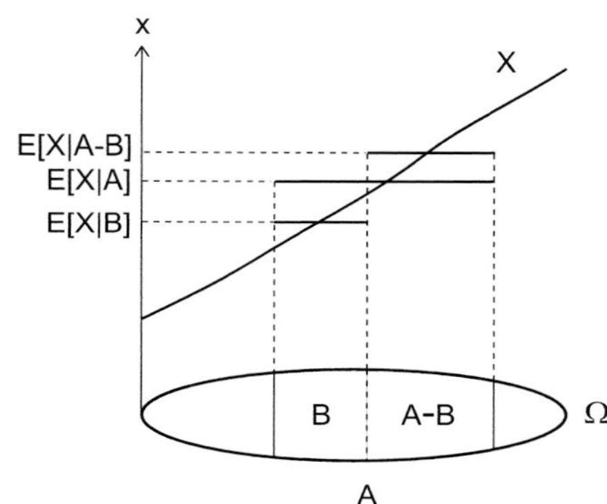

Fig. 5.3 Comparison of $\mathbb{E}[X|A]$, $\mathbb{E}[X|B]$, $\mathbb{E}[X|A-B]$ where $B \subset A$

The idea can be easily extended to the case when the sample space is a countable union of pairwise disjoint subsets.

Now consider a general sub-σ-algebra $\mathcal{G}$ where the probability space need not be a disjoint union of countably many subsets. In this case, even if we define the value of the function $\mathbb{E}[X|\mathcal{G}]$ on a sufficiently small size subset $A \in \mathcal{G}$ it is necessary that we define on a further smaller size subset $B \in \mathcal{G}$ such that $B \subset A$, and with the property that

$$\frac{1}{\mathbb{P}(B)} \int_B \mathbb{E}[X|\mathcal{G}] \, \mathrm{d}\mathbb{P} = \frac{1}{\mathbb{P}(B)} \int_B X \, \mathrm{d}\mathbb{P} = \mathbb{E}[X|B] \, .$$

More precisely, if $A = B \cup (A - B)$, then the average of the averages $\mathbb{E}[X|B]$ and $\mathbb{E}[X|A-B]$ on B and $A - B$, respectively, should be equal to $\mathbb{E}[X|A]$. (See Fig. 5.3.) If there exists a $C \subset B$ such that $C \in \mathcal{G}$ and $\mathbb{P}(C) > 0$, then this procedure should be repeated. In general, such a procedure continues forever and does not come to an end, and hence we have to define $\mathbb{E}[X|\mathcal{G}]$ in an abstract way in what follows.

Definition 5.2 (Conditional Expectation) Let $\mathcal{F}$ be a σ-algebra on Ω, and $\mathcal{G}$ a sub-σ-algebra of $\mathcal{F}$. If X is an $\mathcal{F}$-measurable random variable on a probability space $(\Omega, \mathcal{F}, \mathbb{P})$, and if there exists a $\mathcal{G}$-measurable random variable Y such that

$$\int_A Y \, \mathrm{d}\mathbb{P} = \int_A X \, \mathrm{d}\mathbb{P}$$

for every $A \in \mathcal{G}$, then Y is called the *conditional expectation* of X with respect to $\mathcal{G}$ and denoted by $\mathbb{E}[X|\mathcal{G}]$.

Theorem 5.1 *The conditional expectation Y in Definition 5.2 exists and is unique.*

Proof Given a random variable X, let $X^+ = \max\{X, 0\}$, $X^- = \max\{-X, 0\}$, then $X = X^+ - X^-$. For a sub-σ-algebra $\mathcal{G}$ define $\mathbb{Q} : \mathcal{G} \to [0, \infty)$ by

$$\mathbb{Q}(A) = \int_A X^+ \, \mathrm{d}\mathbb{P} \,, \quad A \in \mathcal{G} \,.$$

Then $\mathbb{Q}$ is a measure on $\mathcal{G}$. Since $\mathbb{Q}$ is absolutely continuous with respect to the measure $\mathbb{P}|_{\mathcal{G}}$, the restriction of $\mathbb{P} : \mathcal{F} \to [0, \infty)$ to $\mathcal{G}$, by the Radon–Nikodym theorem there exists a $\mathcal{G}$-measurable random variable Y^+ satisfying

$$\int_A X^+ \, \mathrm{d}\mathbb{P} = \mathbb{Q}(A) = \int_A Y^+ \, \mathrm{d}\mathbb{P} \,, \quad A \in \mathcal{G} \,.$$

Similarly for X^- there exists a $\mathcal{G}$-measurable random variable Y^- satisfying

$$\int_A X^- \, \mathrm{d}\mathbb{P} = \int_A Y^- \, \mathrm{d}\mathbb{P} \,, \quad A \in \mathcal{G} \,.$$

Now let $Y = Y^+ - Y^-$, then Y is $\mathcal{G}$-measurable and satisfies

$$\int_A X \, \mathrm{d}\mathbb{P} = \int_A Y \, \mathrm{d}\mathbb{P} \,, \quad A \in \mathcal{G} \,.$$

To show the uniqueness, suppose that there exist $\mathcal{G}$-measurable random variables Y_1 and Y_2 satisfying

$$\int_A X \, \mathrm{d}\mathbb{P} = \int_A Y_1 \, \mathrm{d}\mathbb{P} = \int_A Y_2 \, \mathrm{d}\mathbb{P}$$

for every $A \in \mathcal{G}$. Hence $\int_A (Y_1 - Y_2) \, \mathrm{d}\mathbb{P} = 0$ for every $A \in \mathcal{G}$, thus $Y_1 - Y_2 = 0$, i.e., $Y_1 = Y_2$ modulo $\mathbb{P}$-measure zero subsets. □

Example 5.1 (Binary Expansion) Let $\Omega = \prod_1^\infty \{0, 1\}$, a convenient choice for the fair coin tossing problem, identified with the unit interval via the binary expansion of real numbers. (See also Example 3.2.) For a *cylinder subset*

$$[a_1, \ldots, a_n] = \{(x_1 x_2 x_3 \ldots) : x_i = a_i, 1 \le i \le n\}$$

define its size by $\mu([a_1, \ldots, a_n]) = 2^{-n}$. Then the set function μ is extended to a probability measure, also denoted by μ, defined on the σ-algebra generated by the cylinder sets. (If we regard μ as a measure on the unit interval, it is Lebesgue measure.)

Now let $\mathcal{F}_n$ be the sub-σ-algebra generated by cylinder sets $[a_1, \ldots, a_n]$. (From the perspective of the unit interval, $\mathcal{F}_n$ is the sub-σ-algebra generated by the intervals $[(i-1) \times 2^{-n}, i \times 2^{-n})$, $1 \le i \le 2^n$.) An $\mathcal{F}$-measurable function is of the form $\xi(x_1, x_2, x_3, \ldots)$, $x_i \in \{0, 1\}$, $i \ge 1$, and an $\mathcal{F}_n$-measurable function is of the form $\eta(x_1, \ldots, x_n)$, $x_i \in \{0, 1\}$, which does not depend on the variables $x_{n+1}, x_{n+2}, \ldots$. Hence $\mathbb{E}[\xi|\mathcal{F}_n] = \eta(x_1, \ldots, x_n)$ for some function $\eta : \prod_1^n \{0, 1\} \to \mathbb{R}$. With abuse of notation, we may write

$$\mathbb{E}[\xi|\mathcal{F}_n](x_1, \ldots, x_n) = \int \cdots \int \xi(x_1, \ldots, x_n, x_{n+1}, \ldots)\, \mathrm{d}x_{n+1} \cdots \mathrm{d}x_\infty$$

where integration is done over $\prod_{n+1}^\infty \{0, 1\}$.

Theorem 5.2 (Conditional Expectation) *For a given measure space $(\Omega, \mathcal{F}, \mathbb{P})$, let $\mathcal{G}$ and $\mathcal{H}$ be sub-σ-algebras of a σ-algebra $\mathcal{F}$. Then the following facts hold:*

(i) (Linearity) $\mathbb{E}[aX + bY|\mathcal{G}] = a\,\mathbb{E}[X|\mathcal{G}] + b\,\mathbb{E}[Y|\mathcal{G}]$ for any constants a, b.
(ii) (Average of average) $\mathbb{E}[\mathbb{E}[X|\mathcal{G}]] = \mathbb{E}[X]$.
(iii) (Taking out what is known) If X is $\mathcal{G}$-measurable and XY is integrable, then $\mathbb{E}[XY|\mathcal{G}] = X\,\mathbb{E}[Y|\mathcal{G}]$.
(iv) (Independence) If X and $\mathcal{G}$ are independent, then $\mathbb{E}[X|\mathcal{G}] = \mathbb{E}[X]$.
(v) (Tower property) If $\mathcal{H} \subset \mathcal{G}$, then $\mathbb{E}[\mathbb{E}[X|\mathcal{G}]\,|\mathcal{H}] = \mathbb{E}[X|\mathcal{H}]$. Hence, if $\mathcal{H} = \mathcal{G}$, the conditional expectation is a projection, i.e., $\mathbb{E}[\mathbb{E}[X|\mathcal{G}]\,|\mathcal{G}] = \mathbb{E}[X|\mathcal{G}]$.
(vi) (Positivity) If $X \ge 0$, then $\mathbb{E}[X|\mathcal{G}] \ge 0$.

Proof

(i) Use the fact that for every $A \in \mathcal{G}$

$$\begin{aligned}\int_A (a\,\mathbb{E}[X|\mathcal{G}] + b\,\mathbb{E}[Y|\mathcal{G}])\mathrm{d}\mathbb{P} &= a\int_A \mathbb{E}[X|\mathcal{G}]\mathrm{d}\mathbb{P} + b\int_A \mathbb{E}[Y|\mathcal{G}]\mathrm{d}\mathbb{P} \\ &= a\int_A X\mathrm{d}\mathbb{P} + b\int_A Y\mathrm{d}\mathbb{P} \\ &= \int_A (aX + bY)\mathrm{d}\mathbb{P}\,.\end{aligned}$$

(ii) For a proof choose $A = \Omega$ in the definition of conditional expectation. For another proof take $\mathcal{H} = \{\emptyset, \Omega\}$ in item (v) and use $\mathbb{E}[\,\cdot\,|\mathcal{H}] = \mathbb{E}[\,\cdot\,]$.

(iii) First, consider the case when $X = \mathbf{1}_A$ for $A \in \mathcal{G}$. For any $B \in \mathcal{G}$ we have

$$\int_B \mathbf{1}_A \mathbb{E}[Y|\mathcal{G}] \mathrm{d}\mathbb{P} = \int_{A\cap B} \mathbb{E}[Y|\mathcal{G}] \mathrm{d}\mathbb{P} = \int_{A\cap B} Y \mathrm{d}\mathbb{P} = \int_B \mathbf{1}_A Y \mathrm{d}\mathbb{P} ,$$

and hence

$$\mathbf{1}_A \mathbb{E}[Y|\mathcal{G}] = \mathbb{E}[\mathbf{1}_A Y|\mathcal{G}] .$$

This holds for simple functions which are linear combinations of indicator functions by item (i). Next, use the fact that a measurable function can be approximated by a convergent sequence of simple functions.

(iv) Since X and $\mathcal{G}$ are independent, if we let $0 = a_0 < a_1 < a_2 < \cdots$ be a partition of the real line that is the range of X, then for any $A \in \mathcal{G}$ we have

$$\mathbb{P}(X^{-1}([a_i, a_{i+1}]) \cap A) = \mathbb{P}(X^{-1}([a_i, a_{i+1}]))\, \mathbb{P}(A) .$$

Hence

$$\begin{aligned} \int_A X \mathrm{d}\mathbb{P} &= \lim \sum_i a_i \mathbb{P}(X^{-1}([a_i, a_{i+1}]) \cap A) \\ &= \lim \sum_i a_i \mathbb{P}(X^{-1}([a_i, a_{i+1}]))\, \mathbb{P}(A) \\ &= \mathbb{E}[X]\, \mathbb{P}(A) \\ &= \int_A \mathbb{E}[X]\, \mathrm{d}\mathbb{P} . \end{aligned}$$

(v) Note that for any $B \in \mathcal{G}$ we have

$$\int_B \mathbb{E}[X|\mathcal{G}]\, \mathrm{d}\mathbb{P} = \int_B X\, \mathrm{d}\mathbb{P}$$

and that for any $B \in \mathcal{H}$ we have

$$\int_B \mathbb{E}[X|\mathcal{H}]\, \mathrm{d}\mathbb{P} = \int_B X\, \mathrm{d}\mathbb{P} .$$

Thus if $\mathcal{H} \subset \mathcal{G}$ then for any $B \in \mathcal{H}$ we have

$$\int_B \mathbb{E}[X|\mathcal{G}]\, \mathrm{d}\mathbb{P} = \int_B \mathbb{E}[X|\mathcal{H}]\, \mathrm{d}\mathbb{P} .$$

By the definition of conditional expectation we conclude

$$\mathbb{E}[\mathbb{E}[X|\mathcal{G}]|\mathcal{H}] = \mathbb{E}[X|\mathcal{H}] .$$

(vi) For each $n \geq 1$, let $A_n = \{\omega : \mathbb{E}[X|\mathcal{G}](\omega) \leq -\frac{1}{n}\}$. If $X \geq 0$, then $A_n \in \mathcal{G}$. Put $A = \bigcup_{n=1}^{\infty} A_n$. Then $A = \{\mathbb{E}[X|\mathcal{G}] < 0\}$ and

$$0 \leq \int_{A_n} X \, \mathrm{d}\mathbb{P} = \int_{A_n} \mathbb{E}[X|\mathcal{G}] \, \mathbb{P} \leq -\frac{1}{n} \mathbb{P}(A_n)$$

and $\mathbb{P}(A_n) = 0$. By Fact 3.1(ii) we conclude that $\mathbb{P}(A) = 0$. □

The following is Jensen's inequality for conditional expectation.

Theorem 5.3 *Let X be a random variable on a probability space $(\Omega, \mathcal{F}, \mathbb{P})$. Let $\mathcal{G}$ be a sub-σ-algebra of $\mathcal{F}$, and let $\phi : \mathbb{R} \to \mathbb{R}$ be convex. If X and $\phi \circ X$ are integrable, then*

$$\phi\left(\mathbb{E}[X|\mathcal{G}]\right) \leq \mathbb{E}[\phi(X)|\mathcal{G}] .$$

For a concave function the inequality is in the opposite direction.

Proof Note that

$$\phi(x_0) = \sup\{ax_0 + b : ay + b \leq \phi(y) \text{ for all } y \in \mathbb{R}\}$$

for each x_0. Since $aX + b \leq \phi(X)$ for a, b such that $ay + b \leq \phi(y)$ for $y \in \mathbb{R}$, we have

$$a\mathbb{E}[X|\mathcal{G}] + b = \mathbb{E}[aX + b|\mathcal{G}] \leq \mathbb{E}[\phi(X)|\mathcal{G}] .$$

Now use the fact that the supremum of the left-hand side is equal to $\phi\left(\mathbb{E}[X|\mathcal{G}]\right)$. □

Corollary 5.1 *Let X be a random variable on a probability space $(\Omega, \mathcal{F}, \mathbb{P})$. Let $\mathcal{G}$ be a sub-σ-algebra of $\mathcal{F}$. If $|X|^p$ is integrable for $1 \leq p < \infty$, then*

$$|| \, \mathbb{E}[X|\mathcal{G}] \, ||_p \leq ||X||_p .$$

In other words, the linear transformation

$$\mathbb{E}[\,\cdot\,|\mathcal{G}] : L^p(\Omega, \mathcal{F}) \to L^p(\Omega, \mathcal{G}) \subset L^p(\Omega, \mathcal{F})$$

has norm 1, and hence is continuous.

Proof Take a convex function $\phi(x) = |x|^p$. By Jensen's inequality we obtain

$$|\mathbb{E}[X|\mathcal{G}]|^p \leq \mathbb{E}[\,|X|^p|\mathcal{G}]$$

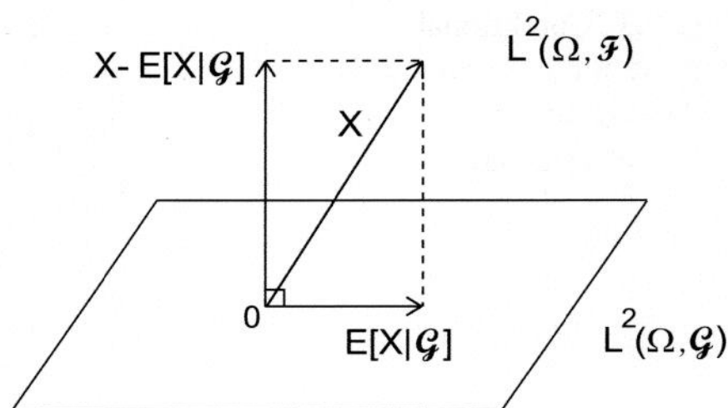

Fig. 5.4 Conditional expectation as an orthogonal projection

and

$$\mathbb{E}[|\mathbb{E}[X|\mathcal{G}]|^p] \leq \mathbb{E}[\mathbb{E}[\,|X|^p|\mathcal{G}]] = \mathbb{E}[\,|X|^p]\,.$$

Now take the pth root of the both sides. It is clear that the norm is less than or equal to 1. Since $\mathbb{E}[\,1\,|\mathcal{G}] = 1$ we conclude that the norm is equal to 1. □

Recall that if $\mathcal{G}$ is a sub-σ-algebra of $\mathcal{F}$, then $L^2(\Omega,\mathcal{G})$ is a subspace of $L^2(\Omega,\mathcal{F})$. Define an inner product for $X, Y \in L^2(\Omega,\mathcal{F})$ by $\mathbb{E}[XY]$, which is equal to $\int_\Omega XY\mathrm{d}\mathbb{P}$.

Theorem 5.4 *Let $\mathcal{G}$ be a sub-σ-algebra of $\mathcal{F}$. Then the linear transformation*

$$\mathbb{E}[\cdot\,|\mathcal{G}] : L^2(\Omega,\mathcal{F}) \to L^2(\Omega,\mathcal{G}) \subset L^2(\Omega,\mathcal{F})$$

is an orthogonal projection onto $L^2(\Omega,\mathcal{G})$. (See Fig. 5.4.)

Proof To show that $\mathbb{E}[X|\mathcal{G}]$ and $X - \mathbb{E}[X|\mathcal{G}]$ are orthogonal, we use

$$\begin{aligned}\mathbb{E}[\mathbb{E}[X|\mathcal{G}](X - \mathbb{E}[X|\mathcal{G}])] &= \mathbb{E}[\mathbb{E}[X|\mathcal{G}]X] - \mathbb{E}[\mathbb{E}[X|\mathcal{G}]\mathbb{E}[X|\mathcal{G}]] \\ &= \mathbb{E}[\mathbb{E}[X|\mathcal{G}]X] - \mathbb{E}[\mathbb{E}[\mathbb{E}[X|\mathcal{G}]X|\mathcal{G}]] \\ &= \mathbb{E}[\mathbb{E}[X|\mathcal{G}]X] - \mathbb{E}[\mathbb{E}[X|\mathcal{G}]X] \\ &= 0\,.\end{aligned}$$

Since $\mathbb{E}[X|\mathcal{G}]$ is measurable with respect to $\mathcal{G}$, Theorem 5.2(iii) implies that

$$\mathbb{E}[X|\mathcal{G}]\,\mathbb{E}[X|\mathcal{G}] = \mathbb{E}[\mathbb{E}[X|\mathcal{G}]X|\mathcal{G}]\,,$$

from which we obtain the second equality in the above. To prove the third equality we use Theorem 5.2(ii). □

Corollary 5.2 *Let $\mathcal{G}$ be a sub-σ-algebra of $\mathcal{F}$. Let X be an $\mathcal{F}$-measurable random variable such that $\mathbb{E}[X^2] < \infty$. Then*

$$\mathbb{E}[\mathbb{E}[X|\mathcal{G}]^2] \leq \mathbb{E}[X^2]\,.$$

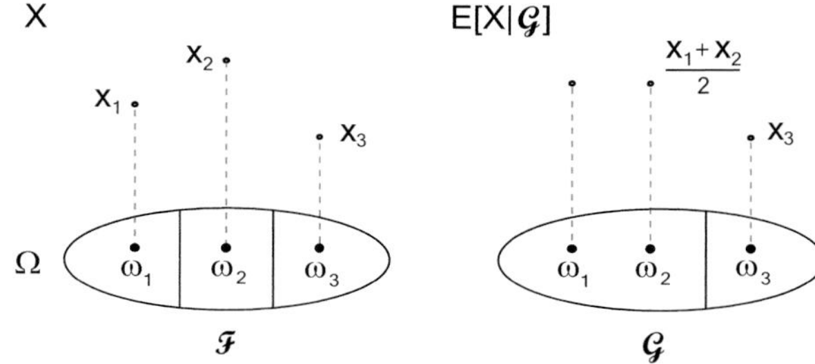

Fig. 5.5 Conditional expectation with respect to a sub-σ-algebra $\mathcal{G}$ on $\Omega = \{\omega_1, \omega_2, \omega_3\}$

Proof Either we use the preceding theorem, or apply Jensen's inequality $\mathbb{E}[X|\mathcal{G}]^2 \leq \mathbb{E}[X^2|\mathcal{G}]$, which implies $\mathbb{E}[\mathbb{E}[X|\mathcal{G}]^2] \leq \mathbb{E}[\mathbb{E}[X^2|\mathcal{G}]] = \mathbb{E}[X^2]$. □

Example 5.2 A probability measure space $(\Omega, \mathcal{F}, \mathbb{P})$ is given by

$$\Omega = \{\omega_1, \omega_2, \omega_3\} , \quad \mathcal{F} = \sigma(\{\omega_1\}, \{\omega_2\}, \{\omega_3\})$$

with a probability measure $\mathbb{P}$ defined by $\mathbb{P}(\{\omega_i\}) = \frac{1}{3}$, $1 \leq i \leq 3$. If a random variable X is defined by

$$X(\omega_1) = x_1 , \quad X(\omega_2) = x_2 , \quad X(\omega_3) = x_3$$

and if a sub-σ-algebra $\mathcal{G}$ is given by

$$\mathcal{G} = \sigma(\{\omega_1, \omega_2\}, \{\omega_3\}) = \{\emptyset, \Omega, \{\omega_1, \omega_2\}, \{\omega_3\}\} ,$$

then

$$\mathbb{E}[X|\mathcal{G}](\omega) = \begin{cases} \frac{1}{2}(x_1 + x_2), & \omega \in \{\omega_1, \omega_2\} \\ x_3, & \omega \in \{\omega_3\} . \end{cases}$$

(See Fig. 5.5. Note that if a random variable Y is measurable with respect to $\mathcal{G}$, then $Y(\omega_1) = Y(\omega_2)$.) Since

$$X(\omega) - \mathbb{E}[X|\mathcal{G}](\omega) = \begin{cases} \frac{1}{2}(x_1 - x_2), & \omega = \omega_1 \\ \frac{1}{2}(-x_1 + x_2), & \omega = \omega_2 \\ 0, & \omega = \omega_3 , \end{cases}$$

the expectation of the product of $\mathbb{E}[X|\mathcal{G}]$ and $X - \mathbb{E}[X|\mathcal{G}]$ is zero, and they are orthogonal to each other.

Remark 5.1 If $\mathbb{E}[X^2] < \infty$, then

$$\mathbb{E}[X - \mathbb{E}[X|\mathcal{G}]]^2 = \min_{Y \in L^2(\Omega, \mathcal{G}, \mathbb{P})} \mathbb{E}[X - Y]^2 .$$

Hence $\mathbb{E}[X|\mathcal{G}]$ is the best approximation based on the information contained in $\mathcal{G}$ from the viewpoint of the least squares method in $L^2(\Omega, \mathcal{F})$.

5.3 Conditional Expectation with Respect to a Random Variable

Definition 5.3 (Conditional Expectation) On a probability measure space $(\Omega, \mathcal{F}, \mathbb{P})$ we are given $\mathcal{F}$-measurable random variables X and Y. We assume that X is integrable. Define the conditional expectation of X with respect to Y, denoted by $\mathbb{E}[X|Y]$, by

$$\mathbb{E}[X|Y] = \mathbb{E}[X|\sigma(Y)] \ .$$

Then $\mathbb{E}[X|Y]$ is measurable with respect to $\sigma(Y)$, and satisfies

$$\int_{\{a \le Y \le b\}} \mathbb{E}[X|Y] \, \mathrm{d}\mathbb{P} = \int_{\{a \le Y \le b\}} X \, \mathrm{d}\mathbb{P}$$

for an arbitrary interval $[a, b] \subset \mathbb{R}$.

Remark 5.2 The conditional expectation of a random variable given an event is a constant. However, the conditional expectation with respect to a sub-σ-algebra or a random variable is again a random variable.

Theorem 5.5 *If the conditions given in Definition 5.3 are satisfied, then* $\mathbb{E}[X|Y]$ *is a function of* Y*. More precisely, there exists a measurable function* $f : \mathbb{R}^1 \to \mathbb{R}^1$ *such that*

$$\mathbb{E}[X|Y] = f \circ Y \ .$$

Proof We sketch the proof. First, consider the case when Y is a simple function of the form

$$Y = \sum_{i=1}^{n} y_i \times \mathbf{1}_{A_i}$$

where y_i are distinct and the subsets $A_i = Y^{-1}(y_i)$ are pairwise disjoint. Then for $\omega \in A_i$ we have $\mathbb{E}[X|Y](\omega) = \mathbb{E}[X|A_i]$. Hence

$$\mathbb{E}[X|Y] = \sum_{i=1}^{n} \mathbb{E}[X|A_i] \times \mathbf{1}_{A_i} \ .$$

Therefore if we take a measurable function f satisfying $f(y_i) = \mathbb{E}[X|A_i]$ for every i, then $\mathbb{E}[X|Y] = f \circ Y$.

For the general case when Y is continuous, choose a sequence of partitions

$$-\infty < \cdots < y_{i-1}^{(n)} < y_i^{(n)} < y_{i+1}^{(n)} < \cdots < +\infty$$

such that $\delta y_i^{(n)} = y_{i+1}^{(n)} - y_i^{(n)} \to 0$ as $n \to \infty$, and approximate Y by a sequence of simple functions

$$Y_n = \sum_{i=1}^{n} y_i^{(n)} \times \mathbf{1}_{A_i^{(n)}}$$

where

$$A_i^{(n)} = Y^{-1}([y_i^{(n)}, y_{i+1}^{(n)})) ,$$

and find $\mathbb{E}[X|Y_n] = f_n \circ Y_n$, and finally take the limit of f_n as $n \to \infty$. □

Remark 5.3 In Fig. 5.6 we are given two random variables X and Y. We assume that the subset $A = \{y \leq Y \leq y + \delta y\}$ has sufficiently small measure, in other words, $\delta y \approx 0$. If $\omega \in A$, then $\mathbb{E}[X|Y](\omega)$ is approximated by $\mathbb{E}[X|A]$. The inverse images under Y of short intervals are represented by horizontal strips in the sample space Ω, and they generate the σ-algebra $\sigma(Y)$. In the limiting case we regard the widths of these strips as being equal to 0, and the strips are treated as line segments. Since $\mathbb{E}[X|Y]$ is measurable with respect to $\sigma(Y)$, $\mathbb{E}[X|Y]$ is constant on each strip. Since each strip is identified with a point y on the real line via Y, we can define f naturally. Usually, instead of $f(y)$ we use the notation $\mathbb{E}[X|Y = y]$ and regard it as a function of y. Strictly speaking, $\mathbb{E}[X|Y = y]$ is a function defined on Ω satisfying

$$\mathbb{E}[X|Y = y] = \mathbb{E}[X|Y]\big|_{\{\omega: Y(\omega)=y\}} .$$

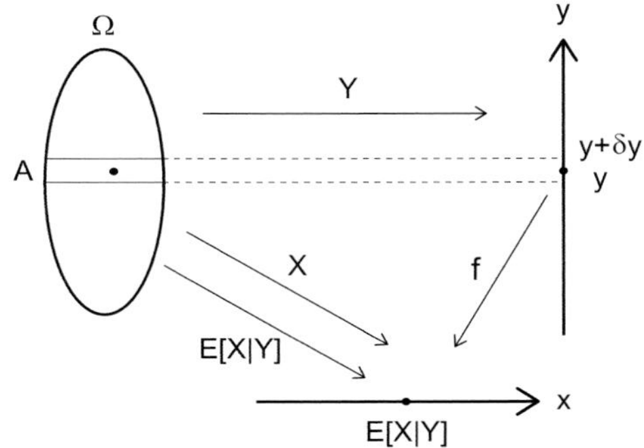

Fig. 5.6 Conditional expectation of X with respect to Y

5.4 Computer Experiments

Simulation 5.1 (Monty Hall Problem)

In a game show there are three closed doors: only one door has a car behind it, and behind the others, goats. A guest chooses one door and can have the car if the present exists when the door is opened. Suppose that the guest picks a door, say Door 1. Before checking what is behind Door 1, the host opens another door, say Door 3, and shows there is a goat and asks if the guest wants to switch the doors, i.e., to choose Door 2 instead of Door 1. Does it make sense for the guest to choose Door 2? The host knows what is behind the doors, and opens a door with a goat. Prove that the guest should switch the doors to have a better chance of winning the prize.

If the guest sticks with the first choice, then the probability of having chosen the door with a car is $\frac{1}{3}$. On the other hand, if the door is switched then the probability increases to $\frac{2}{3}$. Here is why: Let C denote the door behind which there is a car, X the first choice made by the guest, H the door opened by the host. Since

$$\begin{aligned}&\mathbb{P}(C=2,X=1,H=3)\\&=\mathbb{P}(C=2|X=1,H=3)\mathbb{P}(H=3|X=1)\mathbb{P}(X=1)\ ,\end{aligned}$$

and since

$$\begin{aligned}&\mathbb{P}(C=2,X=1,H=3)\\&=\mathbb{P}(H=3|C=2,X=1)\mathbb{P}(C=2|X=1)\mathbb{P}(X=1)\ ,\end{aligned}$$

we have

$$\begin{aligned}&\mathbb{P}(C=2|X=1,H=3)\mathbb{P}(H=3|X=1)\\&=\mathbb{P}(H=3|C=2,X=1)\mathbb{P}(C=2|X=1)\ ,\end{aligned}$$

and finally

$$\begin{aligned}&\mathbb{P}(C=2|X=1,H=3)\\&=\frac{\mathbb{P}(H=3|C=2,X=1)\mathbb{P}(C=2|X=1)}{\mathbb{P}(H=3|X=1)}\\&=\frac{1\times\frac{1}{3}}{\frac{1}{2}\times\frac{1}{3}+1\times\frac{1}{3}+0\times\frac{1}{3}}=\frac{2}{3}\ .\end{aligned}$$

```
N = 1000; % number of trials
prize = randi([0,2],N,1);
% N random integers from the set {0,1,2} for the location of the car
% Doors are numbered 0,1,2.

first = randi([0,2],N,1);
% N random integers from the set {0,1,2} representing the first choice made

second = zeros(N,1);
% the second choice to be made by switching

for i = 1:N
    if first(i) == prize(i)
        k=randi([1,2],1,1); % Randomly choose 1 or 2.
        % The host opens the door with number mod((3-k) + first(i),3)
        % out of two remaining doors without prize.
        second(i) = mod( k + first(i),3); % The guest switches the doors.
    else
        second(i) = prize(i);
        % Switching the doors always wins a prize in this case.
    end
end
n = 0; % number of right answers;
for i = 1:N
    if prize(i) == second(i)
        n = n + 1;
    end
end
fprintf('Probability of winning if door is switched = %f\n',n/N)
```

Exercises

5.1 Let $\Omega = \{a, b, c\}$ and let $X : \Omega \to \mathbb{R}^1$ be a random variable defined by $X(a) = 0$, $X(b) = X(c) = 1$.

(i) What is the σ-algebra generated by X?
(ii) If Y is defined by $Y(a) = 0$, $Y(b) = 1$, $Y(c) = 2$, then what is $\mathbb{E}[X|Y]$?
(iii) Prove or disprove $\mathbb{E}[X] = \mathbb{E}[\,\mathbb{E}[X|Y]\,]$.

5.2 Let $\Omega = \{a, b, c, d\}$ and let $\mathcal{F}$ be the σ-algebra consisting of all the subsets of Ω. And let $\mathcal{G}$ be the sub-σ-algebra generated by $\{a, b\}$ and $\{c, d\}$.

(i) List all of the measurable subsets belonging to $\mathcal{G}$.
(ii) Let $Z : \Omega \to \mathbb{R}^1$ a function defined by $Z(a) = Z(b) = Z(c) = Z(d) = 1$. Is Z measurable with respect to $\mathcal{G}$?
(iii) Suppose that $Y : \Omega \to \mathbb{R}^1$ is a $\mathcal{G}$-measurable function such that $Y(a) = 5$. What are the possible values of $Y(b)$?

(iv) Let $X : \Omega \to \mathbb{R}^1$ be a random variable defined by $X(a) = 0$, $X(b) = X(c) = 3$ and $X(d) = 1$. Let $\mathcal{H}$ be the sub-σ-algebra generated by X. List all the elements of $\mathcal{H}$.

(v) Let X be the random variable given in (iv). Let $W : \Omega \to \mathbb{R}^1$ be a random variable defined by $W(a) = 10$, $W(b) = W(c) = W(d) = 20$. Find $\mathbb{E}[W|X]$.

5.3 Let $\Omega = \{uu, ud, du, dd\}$ and let $\mathcal{F}$ be the σ-algebra consisting of all the subsets of Ω. Let $X : \Omega \to \mathbb{R}^1$ be a random variable defined by $X(uu) = 5$, $X(ud) = X(du) = 3$ and $X(dd) = 1$.

(i) What is the sub-σ-algebra generated by X?

(ii) Let $Y : \Omega \to \mathbb{R}^1$ be a random variable defined by $Y(uu) = 1$, $Y(uu) = Y(du) = Y(ud) = -1$. Find $\mathbb{E}[Y|X]$.

(iii) Prove or disprove $(\mathbb{E}[Y|X])^2 = \mathbb{E}[Y^2|X]$.

5.4 Let $\Omega = [0, 1]$ with $\mathbb{P}$ the Lebesgue measure on $[0, 1]$. Suppose that X and Y are random variables on $(\Omega, \mathbb{P})$, and $Y(\omega) = \omega(1 - \omega)$. Find $\mathbb{E}[X|Y]$. (We assume that all the integrals under consideration exist.)

5.5 Show that Example 5.2 can be interpreted as a linear algebra problem. (Hint: First, note that $L^2(\Omega, \mathcal{F}, \mathbb{P})$ is identified with $\mathbb{R}^3$, and $L^2(\Omega, \mathcal{G}, \mathbb{P})$ with $\{(a, a, b) : a, b \in \mathbb{R}\}$. Let $T : \mathbb{R}^3 \to \mathbb{R}^3$ be a linear transformation defined by $T(x_1, x_2, x_3) = (\frac{1}{2}(x_1 + x_2), \frac{1}{2}(x_1 + x_2), x_3)$. Its associated matrix is given by

$$M = \begin{bmatrix} \frac{1}{2} & \frac{1}{2} & 0 \\ \frac{1}{2} & \frac{1}{2} & 0 \\ 0 & 0 & 1 \end{bmatrix}$$

Note that the conditional expectation operator can be identified with T. Since M is real symmetric, it has real eigenvectors which are orthogonal. More precisely, there exist eigenvalues 0 and 1 with corresponding eigenspaces H_0 and H_1 respectively, where H_0 is spanned by $(1, -1, 0)$, and H_1 is spanned by $(1, 1, 0)$ and $(0, 0, 1)$. The conditional expectation $\mathbb{E}[\,\cdot\,|\mathcal{G}]$ can be identified with T which is the orthogonal projection to H_1.)

5.6 Suppose that X and Y have the joint density $f_{X,Y}(x, y) = x + y$ for any $x, y \in [0, 1]$ and $f_{X,Y}(x, y) = 0$ otherwise. Find $\mathbb{E}[X|Y]$.

5.7 For two random variables X and Y, define the conditional variance of X given Y by

$$\mathrm{Var}(X|Y) = \mathbb{E}[(X - \mathbb{E}[X|Y])^2|Y]\ .$$

Show that

$$\mathrm{Var}(X|Y) = \mathbb{E}[X^2|Y] - \mathbb{E}[X|Y]^2$$

and

$$\mathrm{Var}(X) = \mathbb{E}[\mathrm{Var}(X|Y)] + \mathrm{Var}(\mathbb{E}[X|Y]) \ .$$

5.8 (Bayes' Rule)

(i) Let $B_1, \ldots, B_n$ be a partition of a probability space $(\Omega, \mathbb{P})$ such that $\mathbb{P}(B_i) > 0$, $1 \leq i \leq n$, and take $A \subset \Omega$, $\mathbb{P}(A) > 0$. Prove that

$$\mathbb{P}(B_i|A) = \frac{\mathbb{P}(A|B_i)\mathbb{P}(B_i)}{\sum_{k=1}^{n} \mathbb{P}(A|B_k)\mathbb{P}(B_k)} \qquad \text{for } 1 \leq i \leq n \ .$$

(ii) Consider a test of a certain disease caused by a virus which infects one out of a thousand people on average. The test method is effective, but not perfect. For a person who has already been infected there is 95% chance of detection, and for a healthy person there is 1% probability of false alarm. Suppose that a person, who is not yet known to be infected or not, has a positive test result. What is the probability of his being truly infected?

Chapter 6
Stochastic Processes

We collect and analyze sequential data from nature or society in the form of numerical sequences which are indexed by the passage of time, and try to predict what will happen next. Due to uncertainty of payoff in financial investment before maturity date, the theory of stochastic processes, a mathematical discipline which studies a sequence of random variables, has become the language of mathematical finance.

6.1 Stochastic Processes

Definition 6.1 (Filtration) Let Ω be a measurable space with a σ-algebra $\mathcal{F}$. Consider a collection of sub-σ-algebras $\{\mathcal{F}_t\}_{t\in I}$ of $\mathcal{F}$, indexed by $I \subset \mathbb{R}$. (For example, $I = \{0, 1, \ldots, n\}$, $I = \mathbb{N}$, $I = [0, T]$ and $I = \mathbb{R}$. The parameter or index t represents time in general.) If $\mathcal{F}_s \subset \mathcal{F}_t \subset \mathcal{F}$ for $s, t \in I$ such that $s \leq t$, then $\{\mathcal{F}_t\}_{t\in I}$ is called a *filtration*. Unless stated otherwise, we assume that 0 is the smallest element in I and $\mathcal{F}_0 = \{\emptyset, \Omega\}$.

A filtration is interpreted as a monotone increment of information as time passes by. As an example, consider the game of twenty questions. One of the players of the game is chosen as 'answerer', and the remaining players ask him/her up to twenty questions. The answerer chooses an object that is not revealed to other players until they correctly guess the answer. The answerer can answer 'Yes' or 'No', and after twenty questions are asked, the players present their guess. To model the game mathematically, we define the sample space Ω as the set of all objects in the world and after each question is asked we increase the information on the object $\omega \in \Omega$. Each question X_n is something like 'Is ω alive?', or 'Is ω visible?', and it may be regarded as a random variable $X_n : \Omega \to \{0, 1\}$ depending on the correctness of the guess. Depending on whether the answer to the question is correct or incorrect, we have $X_n = 0$ or 1, respectively. A wise player would ask questions in such a

G.H. Choe, *Stochastic Analysis for Finance with Simulations*, Universitext,
DOI 10.1007/978-3-319-25589-7_6

way that $\{\omega : X_n(\omega) = 0\}$ and $\{\omega : X_n(\omega) = 1\}$ have comparable probabilities. After n questions we obtain information $X_i = 0$ or 1, $1 \le i \le n$, which defines a partition of Ω, or equivalently a σ-algebra $\mathcal{F}_n = \sigma(\{X_1, \ldots, X_n\})$. As n increases, $\mathcal{F}_n$ also increases, and we see that $\mathcal{F}_n$, $1 \le n \le 20$, is a filtration. If we don't ask a wise question, i.e., if we don't define X_n wisely, then some subsets among 2^n subsets are empty and some others are too big, containing many objects, thus making the game difficult to win. If we ask twenty reasonable questions, the sample space is partitioned into $2^{20} \approx 1,000,000$ subsets, and each subset contains one and only one object in the world, corresponding to almost all the words in a dictionary, and we can find the correct answer.

Definition 6.2 (Stochastic Process)

(i) A *stochastic process* is a sequence of random variables $X_t : \Omega \to \mathbb{R}$ parameterized by time t belonging to an index set $I \subset \mathbb{R}$. In other words, when a stochastic process $X : I \times \Omega \to \mathbb{R}$ is given, $X(t, \cdot) = X_t : \Omega \to \mathbb{R}$ is a measurable mapping for each $t \in I$.
(ii) If I is a discrete set, then a process $\{X_t\}_{t\in I}$ is called a discrete time stochastic process, and if I is an interval then $\{X_t\}_{t\in I}$ is called a continuous time stochastic process.
(iii) For each $\omega \in \Omega$ the mapping $t \mapsto X_t(\omega)$ is called a *sample path*.
(iv) If almost all sample paths of a continuous time process are continuous, then we call the process a *continuous process*.
(v) The filtration $\mathcal{F}_t$ generated by a process X_t, i.e., $\mathcal{F}_t = \sigma(\{X_s : 0 \le s \le t\})$ is called a *natural filtration* for X_t.

Definition 6.3 (Adapted Process) Consider a filtration $\{\mathcal{F}_t\}_{t\in I}$ and a stochastic process $\{X_t\}_{t\in A}$. If X_t is measurable with respect to $\mathcal{F}_t$ for every t, then $\{X_t\}_{t\in I}$ is said to be *adapted* to the filtration $\{\mathcal{F}_t\}_{t\in I}$.

Definition 6.4 (Markov Property) Let $\{X_t\}$ be a stochastic process and let $\mathcal{F}_t = \sigma(\{X_u : 0 \le u \le t\})$ be the sub-σ-algebra determined by the history of the process up to time t, i.e., $\mathcal{F}_t$ is generated by $X_u^{-1}(B)$ where $0 \le u \le t$ and B is an arbitrary Borel subset of $\mathbb{R}$. We say that $\{X_t\}$ has the *Markov property* if for every $0 \le s \le t$ the conditional probability of X_t given $\mathcal{F}_s$ is the same as the conditional probability of X_t given X_s, i.e.,

$$\mathbb{P}(X_t \le y|\mathcal{F}_s) = \mathbb{P}(X_t \le y|X_s)$$

with probability 1. In other words, if the present state of the process, X_s, is known, then its future movement is independent of its past history $\mathcal{F}_s$. A stochastic process with the Markov property is called a *Markov process*. The *transition probability* of a Markov process $\{X_t\}$ is defined by

$$\mathbb{P}(X_t \in (y, y + \mathrm{d}y)|X_s = x)$$

which is the conditional probability of the process at time t to be found in a very short interval $(y, y + dy)$ given the condition that $X_s = x$.

Example 6.1 (Random Walk) Let Z_n, $n \geq 1$, be a sequence of independent and identically distributed random variables defined on a probability space $(\Omega, \mathcal{F}, \mathbb{P})$ such that $\mathbb{P}(Z_n = 1) = p$ and $\mathbb{P}(Z_n = -1) = 1 - p$ for some $0 \leq p \leq 1$. A one-dimensional *random walk* $X_0, X_1, X_2, \ldots$ is defined by $X_0 = 0$, $X_n = Z_1 + \cdots + Z_n$, $n \geq 1$. Let $\mathcal{F}_0$ be the trivial σ-algebra $\{\emptyset, \Omega\}$ and let $\mathcal{F}_n = \sigma(Z_1, \ldots, Z_n)$ be the sub-σ-algebra generated by $Z_1, \ldots, Z_n$. If $p = \frac{1}{2}$, the process $\{X_n\}_{n\geq 0}$ is called a *symmetric* random walk.

Example 6.2 (Coin Tossing) Take

$$\Omega = \prod_{n=1}^{\infty} \{0, 1\} = \{\omega = \omega_1\omega_2\omega_3 \cdots \mid \omega_n = 0, 1\}$$

which is the set of all possible outcomes of coin tossing with the symbols 0 and 1 representing heads and tails, respectively, and may be identified with the unit interval via the binary representation $\omega = \sum_{n=1}^{\infty} \omega_n 2^{-n}$. Define $Z_n(\omega) = (-1)^{\omega_n}$. Then $\mathcal{F}_n$ is the σ-algebra generated by the blocks or the cylinder sets of length n defined by

$$[a_1, \ldots, a_n] = \{\omega : \omega_1 = a_1, \ldots, \omega_n = a_n\}$$

whose probability is equal to $p^k(1 - p)^{n-k}$ where k is the number of 0's among $a_1, \ldots, a_n$. Then by the Kolmogorov Extension Theorem we obtain a probability measure on arbitrary measurable subsets of Ω. If we regard Ω as the unit interval, then $\mathcal{F}_n$ is generated by subintervals of the form $[\frac{i-1}{2^n}, \frac{i}{2^n})$, $1 \leq i \leq 2^n$. If $p = \frac{1}{2}$, we have a fair coin tossing and the probability measure on $[0, 1]$ is Lebesgue measure. If $p \neq \frac{1}{2}$, the coin is biased and the corresponding measure on $[0, 1]$ is singular. For more information on singular measures, consult [21].

6.2 Predictable Processes

We cannot receive information that influences the financial market and act on it simultaneously. There should be some time lag, however short, before we decide on an investment strategy. This idea is reflected in the definition of predictability or previsibility of a stochastic process.

Definition 6.5 (Predictable Process)

(i) (Discrete time) If a stochastic process $\{X_n\}_{n=0}^{\infty}$ is adapted to a filtration $\{\mathcal{F}_n\}_{n=0}^{\infty}$, and if X_n is measurable with respect to $\mathcal{F}_{n-1}$ for every $n \geq 1$, then $\{X_n\}_{n=0}^{\infty}$ is said to be *predictable*.

(ii) (Continuous time) If a process $\{X_t\}_{t\geq 0}$ is adapted to a filtration $\{\mathcal{F}_t\}_{t\geq 0}$, and if X_t is measurable with respect to a sub-σ-algebra $\mathcal{F}_{t-}$ defined by

$$\mathcal{F}_{t-} = \sigma\Big(\bigcup_{0\leq s<t} \mathcal{F}_s\Big)$$

for every t, then $\{X_t\}_{t\geq 0}$ is said to be *predictable*.

Example 6.3

(i) (Discrete time) Let $\mathcal{S} = \{s_1, s_2, \ldots\}$ be a set of symbols and let $\Omega = \prod_{n=1}^{\infty} \mathcal{S}$ be the set of all infinite sequences of the symbols in $\mathcal{S}$. Let $\mathcal{F}_n$ be a σ-algebra on Ω generated by the cylinder sets of the form

$$[a_1, \ldots, a_n] = \{(s_{i_1}, s_{i_2}, \ldots) \in \Omega : s_{i_k} = a_k, 1 \leq k \leq n\} .$$

Note that $\{X_n\}$ is adapted to $\{\mathcal{F}_n\}$ if and only if X_n is of the form

$$X_n(s_{i_1}, \ldots, s_{i_n}, \ldots) = X_n(s_{i_1}, \ldots, s_{i_n})$$

as a function defined on $\prod_{i=1}^{n} \mathcal{S}$, i.e., its values depend only on the first n coordinates. Also note that X_n is predictable if and only if X_n is of the form

$$X_n(s_{i_1}, \ldots, s_{i_{n-1}}, \ldots) = X_n(s_{i_1}, \ldots, s_{i_{n-1}})$$

as a function defined on $\prod_{i=1}^{n-1} \mathcal{S}$, i.e., its values depend only on the first $n-1$ coordinates.

(ii) (Continuous time) It is known that a continuous process is predictable. For example, a Brownian motion is predictable.

Example 6.4 Take $\Omega = \{\omega_1, \omega_2\}$ where $\omega_i : [0,\infty) \to \mathbb{R}$ is a function to be specified later, and define $X_t : \Omega \to \mathbb{R}$ by $X_t(\omega_i) = \omega_i(t)$ for $t \geq 0$. Let $\mathcal{F}_t$ denote the σ-algebra $\sigma(\{X_s\}_{0\leq s\leq t})$ generated by $\{X_s : 0 \leq s \leq t\}$. Then $\{\mathcal{F}_t\}_{t\geq 0}$ is a filtration and X_t is a stochastic process adapted to $\mathcal{F}_t$. (See Fig. 6.1.) In the following examples we choose constants $a < b$ and fix $t_0 > 0$.

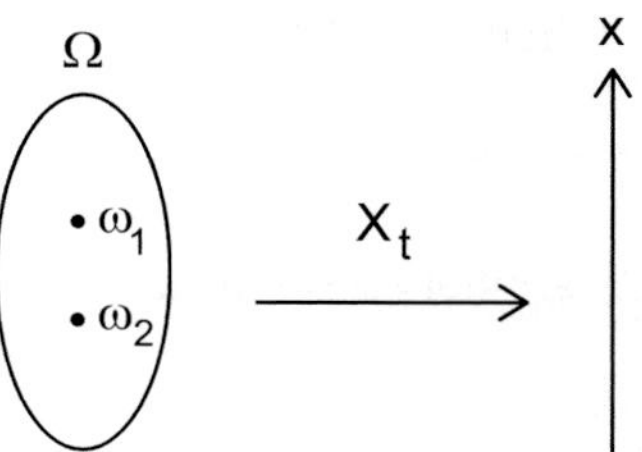

Fig. 6.1 Predictability

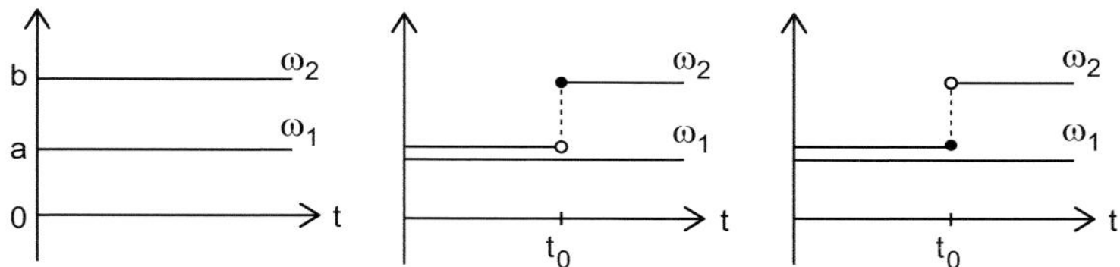

Fig. 6.2 Predictable, not predictable, and predictable processes (left to right)

(i) Let $\omega_1(t) = a$ and $\omega_2(t) = b$ for $t \geq 0$. Note that $\mathcal{F}_0 = \{\emptyset, \Omega\}$, $\mathcal{F}_t = \{\emptyset, \Omega, \{\omega_1\}, \{\omega_2\}\}$ for $t > 0$. To check whether $\{X_t\}_{t\geq 0}$ is predictable, it suffices to observe that X_t is measurable with respect to $\mathcal{F}_{t-}$ for $t > 0$. This holds true since for $t > 0$ we have $\mathcal{F}_{t-} = \mathcal{F}_t = \{\emptyset, \Omega, \{\omega_1\}, \{\omega_2\}\}$.

(ii) Let $\omega_1(t) = a$ for $t \geq 0$ and,

$$\omega_2(t) = \begin{cases} a, \ 0 \leq t < t_0 \ , \\ b, \ t \geq t_0 \ . \end{cases}$$

(See the middle graph in Fig. 6.2.) Note that $\mathcal{F}_t = \{\emptyset, \Omega\}$ for $0 \leq t < t_0$, and $\mathcal{F}_t = \{\emptyset, \Omega, \{\omega_1\}, \{\omega_2\}\}$ for $t \geq t_0$. Since $\mathcal{F}_{t_0-} = \{\emptyset, \Omega\}$, X_{t_0} is not measurable with respect to $\mathcal{F}_{t_0-}$. Thus $\{X_t\}_{t\geq 0}$ is not predictable.

(iii) Let $\omega_1(t) = a$ for $t \geq 0$ and

$$\omega_2(t) = \begin{cases} a, \ 0 \leq t \leq t_0 \ , \\ b, \ t > t_0 \ . \end{cases}$$

(See the right graph in Fig. 6.2.) Note that $\mathcal{F}_t = \{\emptyset, \Omega\}$ for $0 \leq t \leq t_0$ and $\mathcal{F}_t = \{\emptyset, \Omega, \{\omega_1\}, \{\omega_2\}\}$ for $t > t_0$. Hence $\mathcal{F}_{t-} = \mathcal{F}_t$ for $t \geq 0$, and X_t is predictable.

6.3 Martingales

Definition 6.6 (Martingale) Suppose that a stochastic process $\{X_t\}_{t\in I}$ is adapted to a filtration $\{\mathcal{F}_t\}_{t\in I}$, and that X_t is integrable for every t, i.e., $\mathbb{E}[\,|X_t|\,] < \infty$. If

$$X_s = \mathbb{E}[X_t|\mathcal{F}_s]$$

for arbitrary $s \leq t$, then $\{X_t\}_{t\in I}$ is called a *martingale* with respect to $\{\mathcal{F}_t\}_{t\in I}$. If

$$X_s \leq \mathbb{E}[X_t|\mathcal{F}_s]$$

for $s \le t$, then it is called a *submartingale*, and if

$$X_s \ge \mathbb{E}[X_t|\mathcal{F}_s]$$

for $s \le t$, then it is a *supermartingale*..

A martingale is both a submartingale and a supermartingale. A stochastic process that is both a submartingale and a supermartingale is a martingale.

Remark 6.1 If a martingale $\{X_n\}_{n\ge0}$ is *increasing*, i.e., $X_n \le X_{n+1}$ for $n \ge 0$, then it is constant. For the proof, first note that

$$\mathbb{E}[X_n - X_{n-1}|\mathcal{F}_{n-1}] = \mathbb{E}[X_n|\mathcal{F}_{n-1}] - \mathbb{E}[X_{n-1}|\mathcal{F}_{n-1}] = X_{n-1} - X_{n-1} = 0\ .$$

Since $X_n - X_{n-1} \ge 0$, we have $X_n - X_{n-1} = 0$, and hence

$$X_n = X_{n-1} = \cdots = X_1 = X_0\ .$$

Since $X_0 = \mathbb{E}[X_1|\mathcal{F}_0]$, X_n is the constant X_0 for $n \ge 0$.

Theorem 6.1 *If $\{X_t\}_{t\ge0}$ is a martingale with respect to a filtration $\{\mathcal{F}_t\}_{t\ge0}$, $\mathcal{F}_0 = \{\emptyset, \Omega\}$, then $\mathbb{E}[X_t] = \mathbb{E}[X_0]$ for every t.*

Proof For $0 \le s \le t$, we have

$$\mathbb{E}[X_t] = \mathbb{E}[X_t|\mathcal{F}_0] = \mathbb{E}[\mathbb{E}[X_t|\mathcal{F}_s]|\mathcal{F}_0] = \mathbb{E}[X_s|\mathcal{F}_0] = \mathbb{E}[X_s]\ .$$

Now take $s = 0$. □

Example 6.5 Let $X_n, n \ge 0$, be the symmetric random walk in Example 6.1. Since, for $j = k \pm 1$,

$$\begin{aligned}\mathbb{P}(X_{n+1} = j|X_n = k) &= \mathbb{P}(k + Z_{n+1} = j|X_n = k) \\ &= \mathbb{P}(Z_{n+1} = j - k|X_n = k) \\ &= \mathbb{P}(Z_{n+1} = j - k) = \frac{1}{2}\ ,\end{aligned}$$

we have

$$\mathbb{E}[X_{n+1}|X_n = k] = \mathbb{E}[k + Z_{n+1}|X_n = k] = k + \mathbb{E}[Z_{n+1}|X_n = k] = k\ .$$

Hence $\mathbb{E}[X_{n+1}|X_n = k] = X_n$ on the subset $\{\omega \in \Omega : X_n = k\}$ for every k. Thus $\mathbb{E}[X_{n+1}|\mathcal{F}_n] = X_n$ and $\{X_n\}_{n\ge0}$ is a martingale. See Fig. 6.3 where the probability space Ω is partitioned into the subsets $\{\omega \in \Omega : X_n = k\}$, $-n \le k \le n$, for $n \ge 1$. Note that the averages of X_{n+1} over those subsets are equal to the values of X_n.

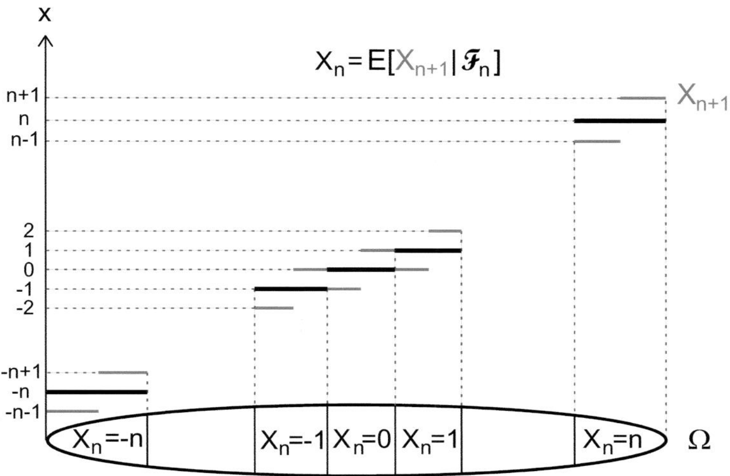

Fig. 6.3 Martingale property of the symmetric random walk

Remark 6.2 The name 'martingale' came from an old French gambling game called 'martingale'. We bet one dollar that the first flip of a coin comes up heads. If it does not come up with heads then we stop, having won one dollar. Otherwise, we bet two dollars that the second toss comes up heads. If it does, then the net gain is one dollar, and we stop. Otherwise, we have lost three dollars and this time we bet $2^2 = 4$ dollars that the third flip is a head. And so on. If the first $n-1$ flips all come up tails, then we have lost $\sum_{j=0}^{n-1} 2^j = (2^n - 1)$ dollars and we bet 2^n dollars on the nth toss. Since the coin will eventually come up heads with probability 1, we are guaranteed to win one dollar. This works if we have an infinite amount of funds. See also Example 6.13.

Theorem 6.2 *If X_t is a martingale and ϕ is a convex function, then $\phi(X)$ is a submartingale if $\phi(X_t)$ is integrable for every t.*

Proof By Theorem 5.3, $\mathbb{E}[\phi(X_t)|\mathcal{F}_s] \geq \phi(\mathbb{E}[X_t|\mathcal{F}_s])$, $s < t$, almost surely for a stochastic process X_t. If X_t is a martingale, then $\phi(\mathbb{E}[X_t|\mathcal{F}_s]) = \phi(X_s)$, $s < t$, almost surely. □

Example 6.6 Let X_t be a martingale. Then $|X_t|$, X_t^2, $X^+ = \max\{X, 0\}$ and e^X are submartingales since $|x|$, x^2, $\max\{x, 0\}$ and e^x are convex functions. (We assume that all the integrals under consideration exist.)

The following fact for discrete time is due to Joseph L. Doob. See [26].

Theorem 6.3 (Doob Decomposition) *If $\{X_n\}_{n\geq 0}$ is a submartingale with respect to a filtration $\{\mathcal{F}_n\}_{n\geq 0}$. Define a stochastic process $\{A_n\}_{n\geq 0}$, called the* compensator*,* *by* $A_0 = 0$ *and*

$$A_n = \sum_{i=1}^{n} (\mathbb{E}[X_i|\mathcal{F}_{i-1}] - X_{i-1}) , \quad n \geq 0 .$$

Then A_n *is* $\mathcal{F}_{n-1}$*-measurable, and* $\{A_n\}$ *is increasing almost surely. Let*

$$M_n = X_n - A_n .$$

Then $\{M_n\}_{n\geq 0}$ *is a martingale. If* $A_0 = 0$ *and* A_n *is* $\mathcal{F}_{n-1}$*-measurable, then the decomposition*

$$X_n = M_n + A_n , \quad n \geq 0 ,$$

is unique.

Proof Note that $\mathbb{E}[X_i|\mathcal{F}_{i-1}] - X_{i-1} \geq 0$ since $\{X_n\}_{n\geq 0}$ is a submartingale, and hence $\{A_n\}$ is an increasing sequence. □

Example 6.7 Let X_n be the symmetric random walk in Example 6.1. Then the process $X_n^2 - n$ is a martingale. For the proof, note that

$$\begin{aligned}
\mathbb{E}[X_{n+1}^2 - (n+1)|\mathcal{F}_n] &= \mathbb{E}[(X_n + Z_{n+1})^2 - (n+1)|\mathcal{F}_n] \\
&= \mathbb{E}[X_n^2 + 2X_nZ_{n+1} + Z_{n+1}^2 - (n+1)|\mathcal{F}_n] \\
&= \mathbb{E}[X_n^2 + 2X_nZ_{n+1} + 1 - (n+1)|\mathcal{F}_n] \\
&= \mathbb{E}[X_n^2 + 2X_nZ_{n+1} - n|\mathcal{F}_n] \\
&= \mathbb{E}[X_n^2|\mathcal{F}_n] + \mathbb{E}[2X_nZ_{n+1}|\mathcal{F}_n] - \mathbb{E}[n|\mathcal{F}_n] \\
&= X_n^2 + 2X_n\mathbb{E}[Z_{n+1}|\mathcal{F}_n] - n \\
&= X_n^2 - n .
\end{aligned}$$

For the equality just before the last one, we take out what is known at time n, and for the last equality we use $\mathbb{E}[Z_{n+1}|\mathcal{F}_n] = \mathbb{E}[Z_{n+1}] = 0$ since Z_{n+1} and $\mathcal{F}_n$ are independent.

Remark 6.3 The continuous limit of a symmetric random walk is Brownian motion W_t. Hence the fact that $X_n^2 - n$ is a martingale corresponds to the fact that $W_t^2 - t$ is a martingale. See Theorem 7.10.

The following fact for continuous time is due to Paul-André Meyer. See [67, 68].

Theorem 6.4 (Doob–Meyer Decomposition) *Let* $\{Y_t\}_{0\leq t\leq T}$ *be a submartingale. Under certain conditions for* Y_t*, we have a unique decomposition given by*

$$Y_t = L_t + C_t , \quad 0 \leq t \leq T ,$$

where L_t is a martingale, and C_t is predictable and increasing almost surely with $C_0 = 0$ and $\mathbb{E}[C_t] < \infty$, $0 \le t \le T$.

Definition 6.7 (Compensator) The process C_t given in Theorem 6.4 is called the *compensator* of Y_t. Let $\{X_t\}_{0\le t\le T}$ be a continuous, square-integrable martingale. Since $Y_t = X_t^2$ is a submartingale, by the Doob–Meyer theorem we have the decomposition

$$X_t^2 = L_t + C_t \ , \quad 0 \le t \le T \ .$$

The *compensator* C_t of X_t^2 is denoted by $\langle X, X\rangle_t$, or simply $\langle X\rangle_t$. (For the details, consult [58].)

Definition 6.8 (Quadratic Variation) Let $\{X_t\}_{0\le t\le T}$ be a continuous martingale. The *quadratic variation* process of X_t, denoted by $[X,X]_t$ or $[X]_t$, is defined by

$$[X,X]_t = \lim_{n\to\infty} \sum_{j=1}^{n} |X_{t_j} - X_{t_{j-1}}|^2$$

where the convergence is in probability as $\max_{1\le j\le n} |t_j - t_{j-1}| \to 0$ as $n \to \infty$.

Using the idea from the identity

$$||\mathbf{v} + \mathbf{w}||^2 = ||\mathbf{v}||^2 + ||\mathbf{w}||^2 + 2\mathbf{v}\cdot\mathbf{w}$$

for $\mathbf{v}, \mathbf{w} \in \mathbb{R}^k$, we define the *quadratic covariation* process $[X, Y]_t$ of two continuous square-integrable martingales X_t and Y_t by

$$[X, Y]_t = \frac{1}{2}\left([X + Y, X + Y]_t - [X,X]_t - [Y, Y]_t\right) \ .$$

Remark 6.4

(i) It is known that $X_t^2 - [X]_t$ is a martingale.
(ii) Since $X_t^2 - \langle X\rangle_t$ and $X_t^2 - [X]_t$ are martingales, $[X]_t - \langle M\rangle_t$ is also a martingale.
(iii) If X_t is a continuous martingale, then $[X,X]_t = \langle X,X\rangle_t$.

Example 6.8 For a Brownian motion W_t, we have

$$\langle W, W\rangle_t = [W, W]_t = t \ .$$

Definition 6.9 (Integrability)

(i) A random variable X is *integrable* if $\mathbb{E}[|X|] < \infty$, which holds if and only if

$$\lim_{n\to\infty} \mathbb{E}[\mathbf{1}_{\{|X|>n\}}|X|] = 0 \ .$$

(ii) A stochastic process $\{X_t\}_{0\le t<\infty}$ is *integrable* if

$$\sup_{0\le t<\infty} \mathbb{E}[|X_t|] < \infty .$$

(iii) A stochastic process $\{X_t\}_{0\le t<\infty}$ is *uniformly integrable* if, as $n \to \infty$,

$$\sup_{0\le t<\infty} \mathbb{E}[\mathbf{1}_{\{|X_t|>n\}}|X_t|] = \sup_{0\le t<\infty} \int_{\{|X_t|>n\}} |X_t| \, \mathrm{d}\mathbb{P} \to 0 .$$

Remark 6.5

(i) The condition that $\sup_{0\le t<\infty} \mathbb{E}[|X_t|] < \infty$ is equivalent to $\lim_{t\to\infty} \mathbb{E}[|X_t|] < \infty$ since $|X_t|$ is a submartingale (see Example 6.6) and since the expectation of a submartingale is an increasing function of t.
(ii) If $\{X_t\}$ is uniformly integrable, then it is integrable since

$$\sup_{0\le t<\infty} \mathbb{E}[|X_t|] < \sup_{0\le t<\infty} \mathbb{E}[\mathbf{1}_{\{|X_t|>n\}}|X_t|] + n < \infty .$$

Example 6.9 The sequence $f_k = k\,\mathbf{1}_{[0,1/k]}$, $k \ge 1$, is bounded in $L^1([0,1], \mathrm{d}x)$, i.e., $||f_k||_1 = 1 < \infty$, but not uniformly integrable for Lebesgue measure $\mathrm{d}x$.

The following fact is due to Joseph L. Doob. Consult [26, 81, 103] for the details.

Theorem 6.5 (Martingale Convergence Theorem) *If a martingale $\{X_t\}_{0\le t<\infty}$ is integrable, then there exists an $X_\infty \in L^1$ such that X_t converges to X_∞ almost surely as $t \to \infty$.*

Corollary 6.1

(i) *Uniformly integrable martingales converge almost surely.*
(ii) *Square-integrable martingales converge almost surely.*
(iii) *Positive martingales converge almost surely.*

Proof

(i) Note that uniformly integrable martingales are integrable.
(ii) Note that square-integrable martingales are uniformly integrable.
(iii) If a martingale satisfies $X_t \ge 0$, then $\mathbb{E}[|X_t|] = \mathbb{E}[X_t] = \mathbb{E}[X_0] < \infty$. □

The following fact shows a typical method for constructing a uniformly integrable martingale.

Theorem 6.6 *On a probability space $(\Omega, \mathcal{F}, \{\mathcal{F}_t\}_{t\ge0}, \mathbb{P})$ with a filtration $\{\mathcal{F}_t\}_{t\ge0}$ we are given an $\mathcal{F}$-measurable and integrable random variable X. Define $X_t = \mathbb{E}[X|\mathcal{F}_t]$. Then $\{X_t\}$ is a uniformly integrable martingale.*

Proof By Jensen's inequality for conditional expectation in Theorem 5.3 we have $|X_t| = |\mathbb{E}[X|\mathcal{F}_t]| \leq \mathbb{E}[|X||\mathcal{F}_t]$, and hence

$$\mathbb{E}[|X_t|] \leq \mathbb{E}[\mathbb{E}[|X||\mathcal{F}_t]] = \mathbb{E}[|X|] < \infty \, .$$

Using the tower property, we have

$$\mathbb{E}[X_t|\mathcal{F}_s] = \mathbb{E}[\mathbb{E}[X|\mathcal{F}_t]|\mathcal{F}_s] = \mathbb{E}[X|\mathcal{F}_s] = X_s$$

for $s < t$, which proves that X_t is a martingale.

As mentioned in Definition 3.13, due to absolute continuity, for every $\varepsilon > 0$ there exists a $\delta > 0$ such that $\mathbb{P}(A) < \delta$ implies $\mathbb{E}[\mathbf{1}_A|X|] < \varepsilon$. By Jensen's inequality we have $|X_t| \leq \mathbb{E}[|X||\mathcal{F}_t]$, and hence

$$\mathbb{E}[|X|] \geq \mathbb{E}[|X_t|] \geq \mathbb{E}[\mathbf{1}_{\{|X_t| \geq K\}}|X_t|] \geq K\mathbb{P}(\{|X_t| \geq K\})$$

for every $K > 0$. Choose $K > \mathbb{E}[|X|]/\delta$, then $\mathbb{P}(\{X_t| > K\}) < \delta$. Then

$$\begin{aligned}
\mathbb{E}[\mathbf{1}_{\{|X_t|>K\}}|X_t|] &\leq \mathbb{E}[\mathbf{1}_{\{|X_t|>K\}}\mathbb{E}[|X||\mathcal{F}_t]] \quad \text{(Jensen's inequality)} \\
&= \mathbb{E}[\mathbb{E}[\mathbf{1}_{\{|X_t|>K\}}|X||\mathcal{F}_t]] \quad (\mathbf{1}_{\{|X_t|>K\}} \text{ is } \mathcal{F}_t\text{-measurable}) \\
&= \mathbb{E}[\mathbf{1}_{\{|X_t|>K\}}|X|] \qquad \text{(the tower property)} \\
&< \varepsilon \qquad \text{(absolute continuity)}
\end{aligned}$$

for every t. □

Remark 6.6 The above fact is essential in the martingale method for option pricing, where T is the expiry date of an option, X is a discounted payoff function, and $\mathbb{E}[X|\mathcal{F}_t]$ is the discounted option price at time $0 \leq t \leq T$. For more information see Chap. 16.

Theorem 6.6 states that the conditional expectation with respect to a filtration is a uniformly integrable martingale. Hence, by Corollary 6.1, it converges almost surely. The following shows when its converse holds.

Theorem 6.7 (Uniformly Integrable Martingale) *Let $\{X_t\}_{t\geq 0}$ be a martingale defined on a probability space with a filtration $(\Omega, \mathcal{F}, \{\mathcal{F}_t\}_{t\geq 0}, \mathbb{P})$. Then the following statements are equivalent:*

(i) *X_t converges in L^1.*
(ii) *$\{X_t\}$ is integrable, X_t converges to $X_\infty \in L^1$ almost surely, and $X_t = \mathbb{E}[X_\infty|\mathcal{F}_t]$, $t \geq 0$.*
(iii) *$\{X_t\}$ is uniformly integrable.*

Proof

(i) $\Rightarrow$ (ii). Since X_t converges in L^1, $\mathbb{E}[|X_t|]$ also converges and is bounded. Hence $\{X_t\}$ is integrable. By Theorem 6.5 X_t converges to $X_\infty \in L^1$ almost surely. Since $X_t = \mathbb{E}[X_u|\mathcal{F}_t]$ for $u \geq t$ and since $\mathbb{E}[X_u|\mathcal{F}_t]$ converges to $\mathbb{E}[X|\mathcal{F}_t]$ in L^1 as $u \to \infty$, $X_t = \mathbb{E}[X|\mathcal{F}_t]$. It remains to show that $X \in L^1$. Fatou's inequality implies that

$$\mathbb{E}[|X|] = \mathbb{E}[\lim_{t\to\infty} |X_t|] = \mathbb{E}[\liminf_{t\to\infty} |X_t|] \leq \liminf_{t\to\infty} \mathbb{E}[|X_t|]$$
$$\leq \sup_{0\leq t\leq\infty} \mathbb{E}[|X_t|] < \infty .$$

(ii) $\Rightarrow$ (iii). This is Theorem 6.6.

(iii) $\Rightarrow$ (i). If $\{X_t\}$ is uniformly integrable, it is bounded in L^1 by Remark 6.5. Thus by Theorem 6.5 there exists an X such that X_t converges to X almost surely. Since almost sure convergence implies convergence in probability by Fact 3.7, combining uniform integrability and almost sure convergence we have L^1-convergence. □

6.4 Stopping Time

Consider the situation when an investor wants to reduce the risk of stock market crash by selling some shares of a stock, whose price at time t is denoted by S_t, when its value falls below a level b, $b < S_0$, i.e., the shares will be sold at time $\tau = \min\{n > 0 : S_{t_n} \leq b\}$ if the stock price is observed at discrete times $0 < t_1 < t_2 < \cdots$. If $\{n \geq 1 : S_{t_n} \leq b\} = \emptyset$, then $\tau = +\infty$ by convention. Since the decision to trade depends on information obtained by the time t_n, which is represented by $\mathcal{F}_n = \sigma(\{S_{t_k} : 0 \leq k \leq n\})$, we require the condition that $\{\tau = n\}$ is $\mathcal{F}_n$-measurable.

Definition 6.10 (Stopping Time)

(i) (Discrete time) Let $\{\mathcal{F}_n\}_{n\geq 0}$ be a filtration on a probability space $(\Omega, \mathcal{F})$. A random variable

$$\tau : \Omega \to \{0, 1, 2, \ldots\} \cup \{+\infty\}$$

satisfying

$$\{\omega \in \Omega : \tau(\omega) \leq n\} \in \mathcal{F}_n , \quad n \geq 1 ,$$

is called a *stopping time*. This condition is equivalent to $\{\tau = n\} \in \mathcal{F}_n$ for every $n \geq 1$ since

$$\{\tau = n\} = \{\tau \leq n\} \cap \{\tau \leq n-1\}^c \in \mathcal{F}_n$$

and

$$\{\tau \le n\} = \{\tau = 1\} \cup \cdots \cup \{\tau = n\} \in \mathcal{F}_n .$$

(ii) (Continuous time) Let $\{\mathcal{F}_t\}_{t\ge0}$ be a filtration on a probability space $(\Omega, \mathcal{F})$. A random variable $\tau : \Omega \to [0, \infty]$ satisfying

$$\{\omega \in \Omega : \tau(\omega) \le t\} \in \mathcal{F}_t \qquad \text{for every } t \ge 0$$

is called a *stopping time*. (If we consider a finite maturity date T in financial applications, the filtration in the above condition may be defined only on $0 \le t \le T$.)

Remark 6.7 (Stopped Process) Let τ be a stopping time, and let $\{X_n\}_{n\ge1}$ be a stochastic process. Here are some standard notations: $a \wedge b = \min\{a, b\}$, $(\tau \wedge n)(\omega) = \tau(\omega) \wedge n$ and $X_{\tau\wedge n} = X_n \mathbf{1}_{\{n<\tau\}} + X_\tau \mathbf{1}_{\{n\ge\tau\}}$. If X_n is adapted to a filtration $\mathcal{F}_n$, then $X_{\tau\wedge n}$ is also adapted to $\mathcal{F}_n$ since for a Borel subset $B \subset \mathbb{R}$ we have

$$\{X_{\tau\wedge n} \in B\} = \{X_n \in B, \tau > n\} \cup \left(\bigcup_{k=1}^{n} \{X_k \in B, \tau = k\} \right) .$$

Example 6.10 (Binary Expansion) Let $\Omega = [0, 1)$ be the unit interval with its filtration $\{\mathcal{F}_n\}_{n\ge0}$ defined by

$$\mathcal{F}_n = \sigma\left(\left\{ \left[\frac{i}{2^n}, \frac{i+1}{2^n} \right) : 0 \le i \le 2^n - 1 \right\} \right) .$$

Note that $\mathcal{F}_n = \sigma(X_1, \ldots, X_n)$. Define $T : \Omega \to \{0, 1, 2, \ldots\}$ by $T(\omega) = k$ if $\omega = \omega_1\omega_2\omega_3\cdots$ and $k = \min\{i \ge 1 : \omega_i = 0\}$. In other words, T is the first time that the digit in the binary expansion of ω is equal to 0 for the first time. For example, $T = 1$ on $[0, \frac{1}{2})$. See Fig. 6.4. There are points with two different binary expansions such as $\omega = 0111\cdots = 1000\cdots$, but those points form a subset of

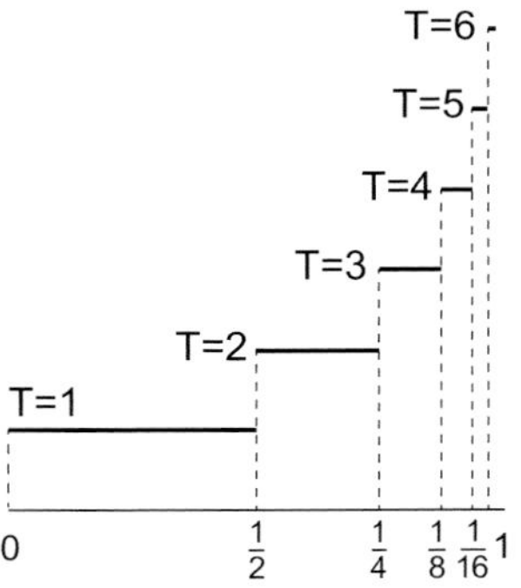

Fig. 6.4 The first time that the digit '0' appears in the binary expansion

probability zero. Except at such points, T is uniquely defined, and hence it is well-defined at almost every point. Since $\{\omega : T \le n\}$ is $\mathcal{F}_n$-measurable, T is a stopping time.

Definition 6.11 (The First Hitting Time) *The first hitting time* τ_B of a Borel set $B \subset \mathbb{R}$ by an adapted process $\{X_n\}_{n\ge 0}$ is defined by

$$\tau_B(\omega) = \min\{n \ge 0 : X_n(\omega) \in B\} \,.$$

It is a stopping time since for every $n \ge 0$ we have

$$\{\omega : \tau_B(\omega) \le n\} = \bigcup_{k\le n}\{\omega : X_k(\omega) \in B\} \in \mathcal{F}_n \,.$$

For the continuous time case, we define the first hitting time τ_B by

$$\tau_B(\omega) = \inf\{t \ge 0 : X_t(\omega) \in B\} \,.$$

Example 6.11 Consider a random walker starting from $0 \in \mathbb{Z}$. Toss a coin, and he moves to the right or left by the distance 1 depending on the outcome 0 and 1 (for heads and tails) until he arrives at $-5 \in \mathbb{Z}$ or $5 \in \mathbb{Z}$. Let X_n denote the position of the random walker at time n. Note that $X_0 = 0$. Define

$$\tau(\omega) = \min\{n : X_n(\omega) = -5 \text{ or } 5\}$$

where $\omega \in \Omega = \prod_{n=1}^{\infty}\{0, 1\}$, which is identified with $[0, 1]$ as in Example 6.10. As before, the filtration is given by $\mathcal{F}_n = \sigma(X_1, \ldots, X_n)$. Note that

$$\{\tau = 1\} = \{\tau = 2\} = \{\tau = 3\} = \{\tau = 4\} = \emptyset$$

and $\tau(\omega) = 5$ if and only if $\omega = 00000\cdots$ or $\omega = 11111\cdots$, i.e.,

$$\omega \in \left[0, \frac{1}{2^5}\right] \cup \left[1 - \frac{1}{2^5}, 1\right] .$$

Then τ is a stopping time. To see why, note that

$$\{\tau = n\} = \left(\bigcap_{k=1}^{n-1}\{-5 < X_k < 5\}\right) \cap \{X_n = \pm 5\}$$

which is an intersection of two subsets belonging to $\mathcal{F}_{n-1}$ and $\mathcal{F}_n$.

Theorem 6.8 (Stopped Martingale) *Let τ be a stopping time, and let $\{X_t\}$ be a martingale. Then $\{X_{t\wedge\tau}\}$ is also a martingale.*

Proof Consult [85]. □

Example 6.12 As a corollary of Theorem 6.8, we have $\mathbb{E}[X_{\tau\wedge n}] = \mathbb{E}[X_0]$. Recall that the symmetric random walk $\{X_n\}$ on $\mathbb{Z}$ starting from 0 moving to the next position by the distance ± 1 is a martingale. Let τ be the stopping time defined by $\tau = \inf\{n : X_n = 1\}$. Then

$$\mathbb{E}[X_{\tau\wedge n}] = \mathbb{E}[X_0] = 0 < 1 = \mathbb{E}[X_\tau]$$

since $X_\tau = 1$.

Example 6.13 (Martingale) A fair coin is being tossed repeatedly, and the outcome is given by $Z_i = \pm 1$, $i \geq 1$, representing a win or a loss. A strategy, or a betting system, is defined by $\beta_n = 2^{n-1}$ dollars if $Z_1 = \cdots = Z_{n-1} = -1$ ($n-1$ consecutive losses), and $\beta_n = 0$, otherwise. Note that the nth bet is placed after the $n-1$ observations. Let

$$Y_n = Z_1 + 2Z_2 + \cdots + 2^{n-1}Z_n \ .$$

Define a stopping time $\tau = \min\{n : Y_n = 1\}$. Then $\tau < \infty$ almost surely, and τ is the total number of wins if we play until we collect one dollar, and $Y_{\tau\wedge n}$ is the total amount of winnings after n plays. It is a martingale. Note that for $n-1$ consecutive losses from the beginning, we have

$$Y_\tau(\omega) = -1 - 2 - \cdots - 2^{n-2} + 2^{n-1} = 1 \ ,$$

and hence $Y_\tau = 1$ almost surely. This game was called 'martingale', and is the origin of the mathematical term.

Definition 6.12 (Local Martingale) Let $\{X_t\}_{0\leq t<\infty}$ be a (continuous) stochastic process. Suppose that there exists a nondecreasing sequence of stopping times τ_n, $n \geq 1$, such that $\{X_{t\wedge\tau_n}\}_{0\leq t<\infty}$ is a martingale for every $n \geq 1$ and $\lim_{n\to\infty} \tau_n = \infty$ almost surely. Then we call $\{X_t\}_{0\leq t<\infty}$ a (continuous) *local martingale*.

Remark 6.8

(i) A martingale $\{M_t\}_{0\leq t<\infty}$ is a local martingale. For, if we take a localizing sequence $\tau_n = n$, $n \geq 1$, then $\{M_{t\wedge\tau_n}\}_{0\leq t<\infty}$ is a martingale for every $n \geq 1$ by Theorem 6.8.

(ii) $\mathbb{E}[X_t]$ need not exist for a local martingale X_t.

6.5 Computer Experiments

Simulation 6.1 (Estimation of Exponential)

Let $N = \min\{n \geq 2 : U_1 \leq U_2 \leq \cdots \leq U_{n-1} > U_n\}$ where U_i, $i \geq 1$, are independent and uniformly distributed in $[0, 1]$. We check the fact that the expectation of N equals $\mathrm{e} = 2.7183\ldots$ (For a proof, consult Exercise 6.6.)

```
num = 10^7; % number of iterations
N = ones(num,1);
for j=1:num
    count = 2;
    U1 = rand;
    U2 = rand;
    while (U1 <= U2)
        U1 = U2;
        U2 = rand;
        count = count + 1;
    end
    N(j) = count;
end
ave = mean(N)
```

Simulation 6.2 (Estimation of Exponential)

Let $N = \min\{n : U_1 + \cdots + U_n > 1\}$ where U_i, $i \geq 1$, are independent and uniformly distributed in $[0, 1]$. We check the fact that the expectation of N equals $\mathrm{e} = 2.7183\ldots$ (For a proof, consult Exercise 6.7.)

```
num = 10^7; % number of iterations
N = zeros(num,1);
for j=1:num
    count = 0;
    total = 0;
    while(total <= 1)
        total = total + rand;
        count = count + 1;
    end
    N(j) = count;
end
ave = mean(N)
```

Exercises

6.1 Let $s < t < u$. Show that $P(X_u \leq y \,|\, X_s, X_t) = P(X_u \leq y \,|\, X_t)$ for a Markov process $\{X_t\}$.

6.2 Let $Z_1, Z_2, Z_3, \ldots$ be independent and identically distributed random variables such that $\Pr(Z_n = 1) = p > \frac{1}{2}$ and $\Pr(Z_n = -1) = q = 1 - p$. Using the notations in Example 6.1, let $X_0 = 0$, $X_n = Z_1 + \cdots + Z_n$, $n \geq 1$, be a random walk. Prove that X_n is not a martingale.

6.3 Let $M_0, M_1, M_2, \ldots$ be a martingale. Show that $\exp(M_0), \exp(M_1), \exp(M_2), \ldots$ is a submartingale. (We assume that all the integrals under consideration exist.)

6.4 Show that a discrete time previsible martingale is constant.

6.5 Suppose that $|X_t| \leq Y$ for every t where $Y \geq 0$ and $\mathbb{E}[Y] < \infty$. Show that $\{X_t\}_t$ be uniformly integrable.

6.6 Let $N = \min\{n \geq 2 : U_1 \leq U_2 \leq \cdots \leq U_{n-1} > U_n\}$ where U_i, $i \geq 1$, are independent and uniformly distributed in $[0, 1]$.

(i) Prove that $\mathbb{P}\{N > n\} = \frac{1}{n!}$, $n \geq 1$.
(ii) Show that $\mathbb{E}[N] = \mathrm{e}$.

6.7 Let U_i, $i \geq 1$, be independent and uniformly distributed in $[0, 1]$. Define $N = \min\{n : U_1 + \cdots + U_n > 1\}$. Show that $\mathbb{E}[N] = \mathrm{e}$.

6.8 Using the same notations as in Example 6.1, define the first hitting time $\tau = \inf\{k \geq 0 : X_k = 1\}$.

(i) Prove that τ is a stopping time.
(ii) Define $T = \sup\{k \geq 0 : X_k = 1\}$ and show that T is not a stopping time. (Hint: $\{\tau = 1\} = \bigcup_{j=1}^{\infty} B_{2j+1}$, $B_{2j+1} \in \mathcal{F}_{2j+1}$.)
(iii) Taking $\Omega = [0, 1]$ and using the binary expansion representation, plot the graphs of τ and T.

6.9 Given a stochastic process $\{X_t\}_{t \geq 0}$, let $\mathcal{G}_t$ be the σ-algebra generated by $\{X_s : 0 \leq s \leq t\}$. Prove that if X_t is a martingale with respect to some filtration $\{\mathcal{H}_t\}_{t \geq 0}$, then X_t is also a martingale with respect to the filtration $\{\mathcal{G}_t\}_{t \geq 0}$.

6.10 Assume that $\mathcal{F}_0 = \{\emptyset, \Omega\}$. Show that if a discrete time martingale $\{X_n\}$ is predictable, then it is constant.

6.11 Let X_t be a submartingale and ϕ be a nondecreasing convex function. Show that $\phi(X)$ is a submartingale if $\phi(X_t)$ is integrable for every t.

6.12 Define the last hitting time $\lambda_B : \Omega \to \{0, 1, 2, \ldots\} \cup \{+\infty\}$ by

$$\lambda_B(\omega) = \sup\{n \geq 0 : X_n(\omega) \in B\} .$$

Show that λ_B is not a stopping time.

6.13 If τ_1 and τ_2 are stopping times with respect to a filtration $\{\mathcal{F}_n\}_{n \geq 0}$, then we define $(\tau_1 \vee \tau_2)(\omega) = \max\{\tau_1(\omega), \tau_2(\omega)\}$. Show that $\tau_1 \vee \tau_2$ is also a stopping time.

Part III
Brownian Motion

Chapter 7
Brownian Motion

After the botanist Robert Brown discovered Brownian motion under a microscope in 1827, it was studied by Louis Bachelier in 1900 to study option price, and Albert Einstein did research on Brownian motion in 1905. Later, Norbert Wiener gave a rigorous framework. While the Brownian motions found in nature should have bounded speed, the mathematical model allows unbounded speed of transition from one location to another with small but positive probability. Therefore, when we need to distinguish the mathematical definition of Brownian motion from physical Brownian motion, we use the terminology Wiener process instead of Brownian motion. In this book we use the terminology Brownian motion on most occasions. In this chapter we introduce the basic properties of Brownian motion needed in finance. To denote a Brownian motion we use two notations $W(t)$ and W_t interchangeably.

7.1 Brownian Motion as a Stochastic Process

We consider only one-dimensional Brownian motion in this chapter. Brownian motion as an idealized physical phenomena can be described as an axiomatic mathematical system. Assume that at time $t = 0$ a particle is located at $0 \in \mathbb{R}^1$, and as time passes by it continuously moves in the positive or negative direction at random. The probability that the particle is found in the interval $[a, b]$ at time $t > 0$ is given by

$$\int_a^b \frac{1}{\sqrt{2\pi t}} \mathrm{e}^{-x^2/(2t)} \,\mathrm{d}x \,.$$

Now, we present an axiomatic definition of a Brownian motion, and in the next section we introduce a sample space consisting of continuous curves $\omega(t)$ defined on

G.H. Choe, *Stochastic Analysis for Finance with Simulations*, Universitext,
DOI 10.1007/978-3-319-25589-7_7

$[0, \infty)$ such that $\omega(0) = 0$, and call the associated stochastic process W_t, $W_t(\omega) = \omega(t)$ for $t \geq 0$, a Brownian motion.

Definition 7.1 (Brownian Motion) A stochastic process W_t, $t \geq 0$, is called a Brownian motion if it has the following properties:

(i) $W_0 = 0$ and $t \mapsto W_t$, $t \geq 0$, is continuous with probability 1.
(ii) For $0 \leq s \leq t$ the increment $W_t - W_s$ has normal distribution with mean 0 and variance $t - s$.
(iii) For $0 \leq t_1 < t_2 \leq t_3 < t_4 \leq \cdots \leq t_{2n-1} < t_{2n}$ the increments

$$W(t_2) - W(t_1), \ldots\ldots, W(t_{2n}) - W(t_{2n-1})$$

are independent.

Remark 7.1 The transition probability density of moving from a fixed point x at time s to y at time t, $s \leq t$, is given by

$$p(t - s; x, y) = \frac{1}{\sqrt{2\pi(t - s)}} \exp\left(-\frac{(x - y)^2}{2(t - s)}\right), \tag{7.1}$$

which has variance $t - s$ and mean x.

In Fig. 7.1 thirty sample paths of Brownian motion are given together with the parabola $t = W^2$ (left) and the probability density functions for W_t, $t = 1, \ldots, 10$.

In Fig. 7.2 sample paths of Brownian motion are given for $0 \leq t \leq 10^{-8}$ (left) and for $0 \leq t \leq 10^8$ (left), respectively. Compare the scalings.

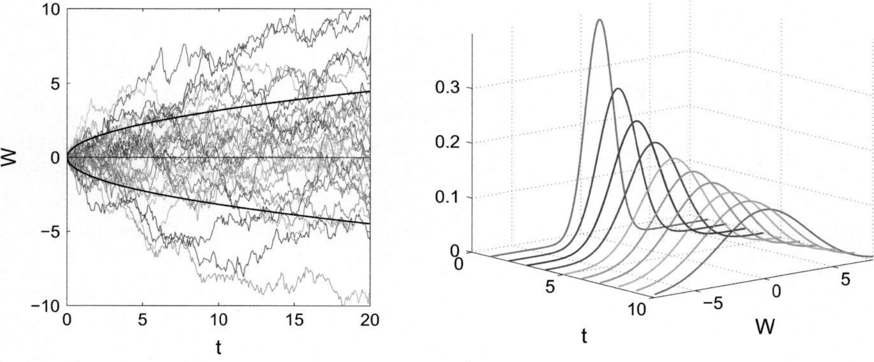

Fig. 7.1 Sample paths of Brownian motion together with the parabola $t = W^2$ (*left*) and the probability density functions for W_t as t increases (*right*)

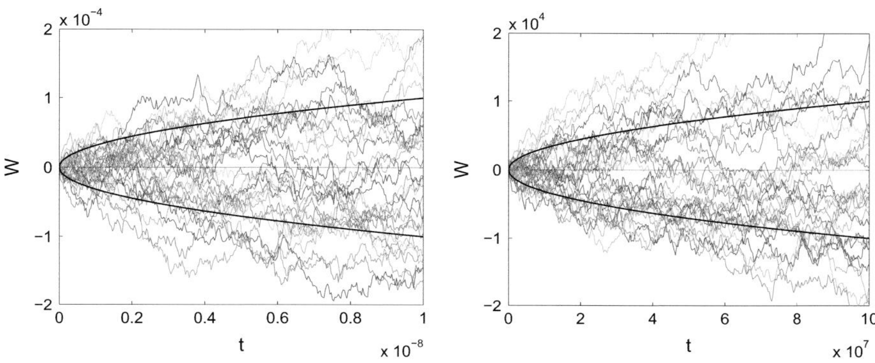

Fig. 7.2 Sample paths of Brownian motion for $0 \le t \le 10^{-8}$ and for $0 \le t \le 10^{8}$

Theorem 7.1 *For* $0 = t_0 < t_1 < \cdots < t_n$ *and for intervals* $I_1, \ldots, I_n \subset \mathbb{R}$, *put* $\delta t_j = t_j - t_{j-1}$. *Then*

$$\begin{aligned}&\mathbb{P}(W_{t_1} \in I_1, \ldots, W_{t_n} \in I_n)\\ &= \int_{I_1} \cdots \int_{I_n} p(\delta t_1; 0, x_1)\, p(\delta t_2; x_1, x_2) \cdots p(\delta t_n; x_{n-1}, x_n)\, \mathrm{d}x_n \cdots \mathrm{d}x_1 \,.\end{aligned}$$

Theorem 7.2 (Invariance Under Time Translation) *Given a Brownian motion* $\{W_t\}_{t\ge 0}$, *for an arbitrary constant* $a \ge 0$ *put* $B_t = W_{t+a} - W_a$. *Then* $\{B_t\}_{t\ge 0}$ *is also a Brownian motion.*

Proof Put $\delta t_k = (t_k + a) - (t_{k-1} + a) = t_k - t_{k-1}$ and $J_k = I_k + W_a$. Then

$$\begin{aligned}&\mathbb{P}(B_{t_1} \in I_1, \ldots, B_{t_n} \in I_n)\\ &= \mathbb{P}(W_a \in \mathbb{R}, W_{t_1+a} \in J_1, \ldots, W_{t_n+a} \in J_n)\\ &= \int_{J_1} \cdots \int_{J_n} p(\delta t_1; W_a, x_1)\, p(\delta t_2; x_1, x_2) \cdots p(\delta t_n; x_{n-1}, x_n)\, \mathrm{d}x_n \cdots \mathrm{d}x_1\\ &= \int_{J_1} \cdots \int_{J_n} p(\delta t_1; 0, x_1 - W_a) \cdots p(\delta t_n; x_{n-1} - W_a, x_n - W_a)\, \mathrm{d}x_n \cdots \mathrm{d}x_1\\ &= \int_{I_1} \cdots \int_{I_n} p(\delta t_1; 0, y_1)\, p(\delta t_2; y_1, y_2) \cdots p(\delta t_n; y_{n-1}, x_n)\, \mathrm{d}y_n \cdots \mathrm{d}y_1\end{aligned}$$

which coincides with the probability of Brownian motion. □

Theorem 7.3

(i) $\mathbb{E}[W_t W_s] = \min\{t, s\}$. *As a corollary,* $\mathbb{E}\left[W_t^2\right] = t$.
(ii) $\mathbb{E}\left[(W_t - W_s)^2\right] = |t - s|$.

Proof

(i) Assume that $s \le t$. Then we have

$$\begin{aligned}\mathbb{E}\left[W_t W_s\right] &= \int_{-\infty}^{\infty}\int_{-\infty}^{\infty} x\,y\,p(s,0,x)\,p(t-s,x,y)\,\mathrm{d}x\,\mathrm{d}y \\ &= \int_{-\infty}^{\infty} x\,p(s,0,x)\left(\int_{-\infty}^{\infty} y\,p(t-s,x,y)\,\mathrm{d}y\right)\mathrm{d}x \\ &= \int_{-\infty}^{\infty} x\,p(s,0,x)\,x\,\mathrm{d}x = \int_{-\infty}^{\infty} x^2\,p(s,0,x)\,\mathrm{d}x = s\,.\end{aligned}$$

(ii) Note that $\mathbb{E}\left[W_t^2 - 2W_tW_s + W_s^2\right] = t - 2\min\{t,s\} + s = |t-s|$. Or, we may apply Theorem 7.2: Since $W_t - W_s$ and W_{t-s} are identically distributed for $s \le t$, Part (i) implies that $\mathbb{E}\left[(W_t - W_s)^2\right] = \mathbb{E}\left[W_{t-s}^2\right] = t - s$. □

Remark 7.2 Either by direct computation, or by comparing the coefficients of Taylor expansions of the equality in Lemma 7.1 we have $\mathbb{E}[W_t^n] = 0$ for n odd, $\mathbb{E}[W_t^2] = t$ and $\mathbb{E}[W_t^4] = 3t^2$.

Definition 7.2 (Quadratic Variation) Let $0 = t_0 < t_1 < \cdots < t_n = T$ be a partition of the interval $[0,T]$ such that $\lim_{n\to\infty}\max_i \delta_i t = 0$ where $\delta_i t = t_{i+1} - t_i$. The *quadratic variation* of Brownian motion is defined by

$$\lim_{n\to\infty}\sum_{i=0}^{n-1}|W(t_{i+1}) - W(t_i)|^2$$

where the convergence is in the L^2-sense. Put $\delta_i W = W_{t_{i+1}} - W_{t_i}$. Note that $\mathbb{E}[\sum_i |\delta_i W|^2] = \sum_i \delta_i t = T$.

Theorem 7.4 *The quadratic variation of Brownian motion is given by*

$$\lim_{n\to\infty}\sum_{i=0}^{n-1}\left|W_{t_{i+1}} - W_{t_i}\right|^2 = T$$

where the convergence is in the L^2-sense.

Proof Since $(\delta_i W)^2$ and $(\delta_j W)^2$ are independent for $i \ne j$, we have

$$\mathbb{E}[(\delta_i W)^2(\delta_j W)^2] = \mathbb{E}[(\delta_i W)^2]\,\mathbb{E}[(\delta_j W)^2]\,.$$

By Remark 7.2, $\mathbb{E}[(\delta_i W)^4] = 3(\delta_i t)^2$ and $\mathbb{E}[(\delta_i W)^2] = \delta_i t$. As $n \to \infty$,

$$\begin{aligned}
&\mathbb{E}\left[\left|\sum_{i=0}^{n-1}(\delta_i W)^2 - T\right|^2\right] \\
&= \sum_{i=0}^{n-1}\sum_{j=0}^{n-1}\mathbb{E}\left[(\delta_i W)^2(\delta_j W)^2\right] - 2T\sum_{i=0}^{n-1}\mathbb{E}\left[(\delta_i W)^2\right] + T^2 \\
&= \sum_{i=0}^{n-1}\mathbb{E}\left[(\delta_i W)^4\right] + 2\sum_{i<j}\mathbb{E}\left[(\delta_i W)^2(\delta_j W)^2\right] - 2T\sum_{i=0}^{n-1}\mathbb{E}\left[(\delta_i W)^2\right] + T^2 \\
&= \sum_{i=0}^{n-1}3(\delta_i t)^2 + 2\sum_{i<j}(\delta_i t)\,(\delta_j t) - 2T\sum_{i=0}^{n-1}\delta_i t + T^2 \\
&= 2\sum_{i=0}^{n-1}(\delta_i t)^2 + \left(\sum_{i=0}^{n-1}\delta_i t\right)^2 - 2T\sum_{i=0}^{n-1}\delta_i t + T^2 \\
&= 2\sum_{i=0}^{n-1}(\delta_i t)^2 + T^2 - 2T^2 + T^2 \le 2\max_i \delta_i t\sum_{i=0}^{n-1}\delta_i t = 2\max_i \delta_i t \times T \to 0
\end{aligned}$$

and the proof is complete. □

Corollary 7.1 (The First Order Variation) *The first order variation of Brownian motion diverges. In other words, with probability 1 we have*

$$\lim_{n\to\infty}\sum_{i=0}^{n-1}\left|W_{t_{i+1}} - W_{t_i}\right| = \infty\,.$$

Proof For a continuous sample path ω belonging to a set Ω_1 of measure 1 we put $C_n = \max_{0\le i\le n-1}\left|W_{t_{i+1}} - W_{t_i}\right|$. Since ω is a uniformly continuous function defined on a compact set $[0,T]$, we have $\lim_{n\to\infty} C_n = 0$ and

$$\sum_{i=0}^{n-1}\left|W_{t_{i+1}} - W_{t_i}\right|^2 \le C_n\sum_{i=0}^{n-1}\left|W_{t_{i+1}} - W_{t_i}\right|\,.$$

The left-hand side converges to T for almost every ω belonging to a set Ω_2 of measure 1 along a subsequence $\{n_k\}_{k=1}^{\infty}$ by Theorem 7.4. Hence $\sum_{i=0}^{n-1}\left|W_{t_{i+1}} - W_{t_i}\right|$ increases to infinity with probability 1. □

Lemma 7.1 *For a real constant θ we have*

$$\mathbb{E}\left[e^{\theta W_t}\right] = e^{\frac{1}{2}\theta^2 t}\,, \quad t \ge 0\,.$$

Proof Note that

$$\begin{aligned}\mathbb{E}\left[e^{W_t}\right] &= \frac{1}{\sqrt{2\pi t}} \int_{-\infty}^{\infty} e^x e^{-\frac{x^2}{2t}}\, dx \\ &= \frac{1}{\sqrt{2\pi t}} \int_{-\infty}^{\infty} e^{-\frac{(x-t)^2-t^2}{2t}}\, dx \\ &= e^{\frac{t}{2}} \frac{1}{\sqrt{2\pi t}} \int_{-\infty}^{\infty} e^{-\frac{y^2}{2t}}\, dy = e^{\frac{1}{2}t}\,.\end{aligned}$$

Since θW_t and $W_{\theta^2 t}$ have the same distribution, we have $\mathbb{E}\left[e^{\theta W_t}\right] = \mathbb{E}\left[e^{W_{\theta^2 t}}\right] = e^{\frac{1}{2}\theta^2 t}$. □

Remark 7.3 By replacing θ by $-\theta$ in Lemma 7.1, we have $\mathbb{E}\left[e^{-\theta W_t}\right] = e^{\frac{1}{2}\theta^2 t}$, which can be proved also by the fact that θW_t and $-\theta W_t$ are identically distributed.

Example 7.1 In the geometric Brownian motion model the stock price S_t satisfies

$$S_t = S_0 e^{(\mu - \frac{1}{2}\sigma^2)t + \sigma W_t}\,.$$

Since $\mathbb{E}[e^{-\frac{1}{2}\theta^2 t + \theta W_t}] = e^{-\frac{1}{2}\theta^2 t}\mathbb{E}\left[e^{\theta W_t}\right] = 1$, we have $\mathbb{E}[S_t] = S_0 e^{\mu t}$, i.e., the stock price increases exponentially on average regardless of the volatility σ as time increases. In Fig. 7.3 sample paths of S_t are plotted together with the curve $y = S_0 e^{\mu t}$ for $\mu = 0.25$ and $\sigma = 0.3$. Consult Simulation 7.2. See also Remark 7.8 for a less intuitive result.

Lemma 7.2 *If* $i = \sqrt{-1}$, *then* $\mathbb{E}\left[e^{i\theta W_t}\right] = e^{-\frac{1}{2}\theta^2 t}$ *for any real constant* θ.

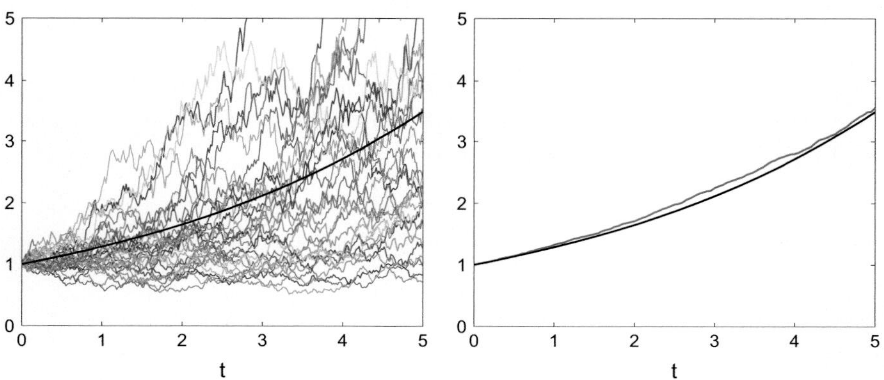

Fig. 7.3 Exponential growth of the geometric Brownian motion: The sample average of S_t is close to the exponential curve given in the right

Proof Recall that the Fourier transform of the probability density function of the standard normal variable is the probability density function itself up to a constant multiple. (See Example 4.8.) Hence

$$\begin{aligned}\mathbb{E}\left[e^{iW_t}\right] &= \int_{-\infty}^{\infty} \frac{1}{\sqrt{2\pi t}} e^{-x^2/(2t)} e^{ix} dx \\ &= \int_{-\infty}^{\infty} \frac{1}{\sqrt{2\pi}} e^{-y^2/2} e^{i\sqrt{t}y} dy \\ &= e^{-(\sqrt{t})^2/2} = e^{-t/2} .\end{aligned}$$

Since θW_t and $W_{\theta^2 t}$ are identically distributed, we have $\mathbb{E}[e^{i\theta W_t}] = \mathbb{E}[e^{iW_{\theta^2 t}}] = e^{-\frac{1}{2}\theta^2 t}$. □

Remark 7.4 The probability distribution of W_t is symmetric around 0 and hence $\mathbb{E}[\sin W_t] = 0$. Thus

$$\mathbb{E}\left[e^{iW_t}\right] = \mathbb{E}\left[\cos W_t\right] + i\,\mathbb{E}\left[\sin W_t\right] = \mathbb{E}\left[\cos W_t\right]$$

and we obtain a real number even when the function under integration is complex-valued in the above. As a by-product we have $\mathbb{E}\left[\cos W_t\right] = e^{-t/2}$. For $t = 1$ we have

$$\mathbb{E}\left[\cos Z\right] = \frac{1}{\sqrt{e}}$$

where Z is a standard normal variable.

Theorem 7.5 *Let $0 \leq s < t$. Under the conditions that $W_s = x$ and $W_t = z$ the distribution of $W_{(s+t)/2}$ is normal with expectation $\frac{1}{2}(x+z)$ and variance $\frac{1}{4}(t-s)$. That is, the conditional probability density function $f(y)$ is given by*

$$f(y) = \frac{1}{\sqrt{2\pi\frac{t-s}{4}}} \exp\left(-\frac{(y-\frac{x+z}{2})^2}{2\frac{t-s}{4}}\right) .$$

Proof For the sake of notational simplicity, we prove the statement for $s = 0$, in which case we have $x = 0$. To find the conditional density function, we compute first the conditional expectation $\mathbb{E}[\phi(W_{t/2})|W_t = z]$ for an arbitrary function $\phi : \mathbb{R} \to \mathbb{R}$. For $\delta z > 0$ let

$$A = \{\omega : W_t(\omega) \in [z, z+\delta z]\} .$$

Then

$$\begin{aligned}
&\frac{1}{\mathbb{P}(A)}\int_A \phi(W_{\frac{t}{2}})\mathrm{d}\mathbb{P} \\
&= \frac{\int_z^{z+\delta z}\int_{-\infty}^{\infty}\phi(y)\,p(\frac{t}{2};0,y)\,p(\frac{t}{2};y,u)\,\mathrm{d}y\,\mathrm{d}u}{\int_z^{z+\delta y} p(t;0,u)\,\mathrm{d}u} \\
&\approx \frac{\int_{-\infty}^{\infty}\phi(y)\,p(\frac{t}{2};0,y)\,p(\frac{t}{2};y,z)\,\mathrm{d}y\,\delta z}{p(t;0,z)\,\delta z} \\
&= \int_{-\infty}^{\infty}\phi(y)\,\frac{p(\frac{t}{2};0,y)\,p(\frac{t}{2};y,z)}{p(t;0,z)}\,\mathrm{d}y\,.
\end{aligned}$$

Thus the conditional probability density function of $W_{\frac{t}{2}}$ given $W_t = z$ is obtained by

$$\begin{aligned}
&\frac{p(\frac{t}{2};0,y)\,p(\frac{t}{2};y,z)}{p(t;0,z)} \\
&= \frac{1}{\sqrt{2\pi\frac{t}{2}}}\exp\left(-\frac{y^2}{2\frac{t}{2}}\right)\frac{1}{\sqrt{2\pi\frac{t}{2}}}\exp\left(-\frac{(z-y)^2}{2\frac{t}{2}}\right)\sqrt{2\pi t}\exp\left(\frac{z^2}{2t}\right) \\
&= \frac{\sqrt{2}}{\sqrt{\pi t}}\exp\left(-\frac{2(y-\frac{1}{2}z)^2}{t}\right)\,.
\end{aligned}$$

□

In Theorem 7.6 presented in the next paragraph, we construct a sample Brownian path satisfying $W_T = b$ using the above result. First, we find the value at $t = \frac{1}{2}T$, then find the values at $t = \frac{1}{4}T, \frac{3}{4}T$, and so on by adding midpoints. We let $\delta t = \frac{1}{2^n}T$ for some n. Connect the points

$$(0,0), (\delta t, W(\delta t)), \ldots, (k\delta t, W(k\delta t)), \ldots, (T, W(T))$$

by line segments, and obtain a piecewise curve which converges to a Brownian sample path. See also Sect. 12.3.

Theorem 7.6 (Brownian Motion Conditional on $W_T = b$**)** *By using successive approximation we construct a sample path of Brownian motion over a finite interval* $[0, T]$. *For the sake of notational simplicity we consider the case* $T = 1$. *We will define a sequence of piecewise linear functions* $W^{(n)}$, $n \geq 1$, *recursively. Given* $n \geq 1$, *consider the points* $\frac{k}{2^n}$, $0 \leq k \leq 2^n$, *and define* $W^{(n)}$ *using* $W^{(n-1)}$ *at those points depending on whether k is even or odd.*

(i) If k is even, then $k = 2i$, *and define*

$$W^{(n)}\left(\frac{k}{2^n}\right) = W^{(n)}\left(\frac{i}{2^{n-1}}\right) = W^{(n-1)}\left(\frac{i}{2^{n-1}}\right)\,.$$

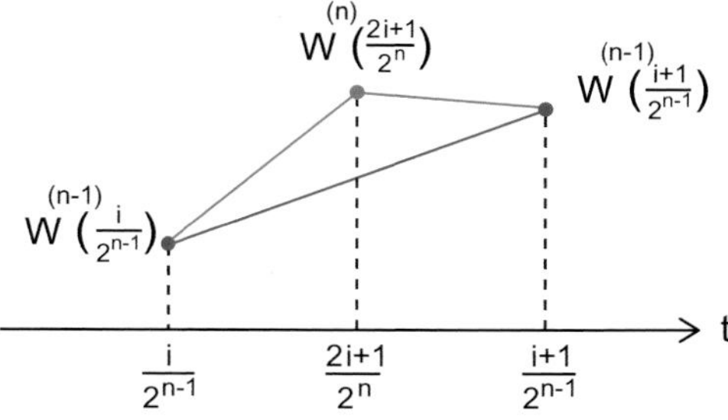

Fig. 7.4 Construction of a sample path of Brownian motion

(ii) If k is odd, then $k = 2i + 1$, and define

$$W^{(n)}\left(\frac{k}{2^n}\right) = \frac{1}{2}\left[W^{(n)}\left(\frac{2i}{2^n}\right) + W^{(n)}\left(\frac{2i+2}{2^n}\right)\right] + 2^{-\frac{n+1}{2}} Z^{(n)}_{2i+1}$$

where $Z^{(n)}_{2i+1}$, $1 \le 2i + 1 \le 2^n$, are independent standard normal variables. In other words,

$$W^{(n)}\left(\frac{2i+1}{2^n}\right) = \frac{1}{2}\left[W^{(n-1)}\left(\frac{i}{2^n}\right) + W^{(n-1)}\left(\frac{i+1}{2^n}\right)\right] + 2^{-\frac{n+1}{2}} Z^{(n)}_{2i+1} \ .$$

(See Fig. 7.4.) Finally, we connect the points $(\frac{k}{2^n}, W^{(n)}(\frac{k}{2^n}))$ to obtain a piecewise linear continuous function $W^{(n)}$ on the whole interval. Then $W^{(n)}$, $n \ge 1$, converges uniformly to a continuous function with probability 1.

Proof Let $(\Omega, \mathbb{P})$ denote the probability space under consideration, and put

$$M_n(\omega) = \max\{|Z^{(n)}_k(\omega)| : 0 < k < 2^n, k \text{ is odd}\}$$

and $A_n = \{M_n > n\}$. Take $\alpha > 0$. Since

$$\mathbb{P}(|Z^{(n)}_k| > \alpha) = 2\frac{1}{\sqrt{2\pi}} \int_\alpha^\infty \mathrm{e}^{-z^2/2}\mathrm{d}z < \sqrt{\frac{2}{\pi}} \int_\alpha^\infty \frac{\alpha}{z}\mathrm{e}^{-z^2/2}\mathrm{d}z = \sqrt{\frac{2}{\pi}}\frac{1}{\alpha}\mathrm{e}^{-\alpha^2/2} \ ,$$

we have

$$\mathbb{P}(A_n) = \mathbb{P}\left(\bigcup_k \left\{|Z^{(n)}_k| > n\right\}\right) \le 2^{n-1}\mathbb{P}(|Z^{(n)}_1| > n) \le \sqrt{\frac{2}{\pi}}\frac{2^{n-1}}{n}\mathrm{e}^{-n^2/2} \ ,$$

and

$$\sum_{n=1}^\infty \mathbb{P}(A_n) < \infty \ .$$

By the Borel–Cantelli lemma (Theorem 4.10) there exists a set Ω_1 such that $\mathbb{P}(\Omega_1) = 1$ and for $\omega \in \Omega_1$ there exists $N = N(\omega)$ satisfying $M_n(\omega) \le n$ for $n \ge N$. Then, for $\omega \in \Omega_1$ and $n \ge N$, we have

$$\sup_{0 \le t \le 1} |W^{(n+1)}(t) - W^{(n)}(t)| = 2^{-\frac{n+1}{2}} M_n \le 2^{-\frac{n+1}{2}} n\,,$$

and hence, for $n, m \ge N$,

$$\sup_{0 \le t \le 1} |W^{(m)}(t) - W^{(n)}(t)| = \sum_{j=n}^{\infty} 2^{-\frac{j+1}{2}} j\,.$$

Since the infinite sum in the above converges monotonically to 0 as $n \to \infty$, we conclude that $W^{(n)}(t)$ converges uniformly on $[0, 1]$. □

Fact 7.7 *Let $W(t)$ be the limit of $W^{(n)}(t)$ in the preceding construction.*

(i) The increments

$$W^{(n)}\left(\frac{k}{2^n}\right) - W^{(n)}\left(\frac{k-1}{2^n}\right)$$

are independent and normally distributed with mean 0 and variance $\frac{1}{2^n}$.

(ii) If $0 = t_0 < t_1 < \cdots < t_n \le 1$, then the increments $W(t_{i+1}) - W(t_i)$ are independent, normally distributed, with mean 0 and variance $t_{i+1} - t_i$. Thus $W(t)$ is a Brownian motion. (See Fig. 7.5.)

Proof Consult [45]. □

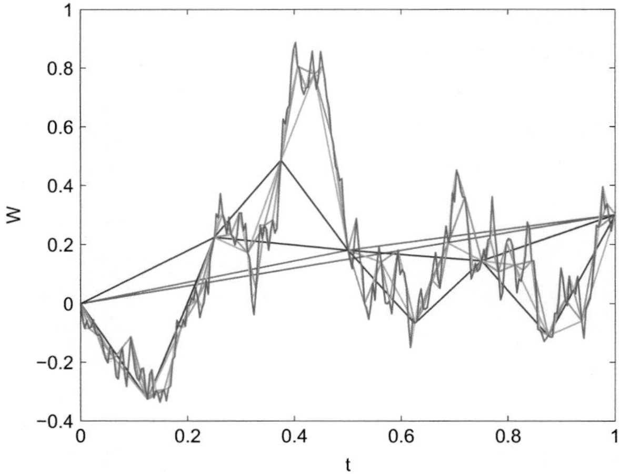

Fig. 7.5 A sample path of Brownian motion constructed by the bisection method

Remark 7.5 Sometimes a generalized Brownian motion $a + W_t$ is called a Brownian motion conditional on $W_0 = a$. (See also Theorem 7.2.)

7.2 Sample Paths of Brownian Motion

In this section we introduce a concrete example of a Brownian motion which will be useful in computer simulations.

Definition 7.3 (Cylinder Subset) Let Ω be the set of functions $\omega : [0, \infty) \to \mathbb{R}^1$, called sample paths, satisfying the following conditions:

(i) $\omega(0) = 0$,
(ii) $\omega(t)$ is a continuous function of t.

To endow a measurable structure on Ω we consider cylinder subsets which are building blocks of all measurable subsets: For arbitrary time points $0 = t_0 < t_1 < \cdots < t_n$ and arbitrary intervals $I_1, \ldots, I_n \subset \mathbb{R}$, we define a *cylinder subset* by

$$C(t_1, \ldots, t_n; I_1, \ldots, I_n) = \{\omega \in \Omega \mid \omega(t_1) \in I_1, \ldots, \omega(t_n) \in I_n\} .$$

(See Fig. 7.6 for a graphical representation of a cylinder subset where a sample path passes through the gates represented by the intervals.) The σ-algebra generated by the cylinder subsets $C(t_1, \ldots, t_n; I_1, \ldots, I_n)$ for $n \geq 1$ and $0 = t_0 < t_1 < \cdots < t_n \leq t$, is denoted by $\mathcal{F}_t$. Note that $\{\mathcal{F}_t\}_{t \geq 0}$ is a filtration and that a σ-algebra $\mathcal{F}$ is generated by $\bigcup_{t \geq 0} \mathcal{F}_t$. Then we obtain a filtered measurable space $(\Omega, \{\mathcal{F}_t\}_{t \geq 0}, \mathcal{F})$, which is called a sample space of the Brownian motion.

An infinitely long cylinder $\{(x, y) | x^2 + y^2 = 1\} \times \mathbb{R}^1 \subset \mathbb{R}^3$ is a set of the points whose third coordinates are arbitrary. A cylinder subset for a Brownian motion is regarded as a subset of the infinite product $\prod_{t \geq 0} \mathbb{R}^1_t$ of the real line $\mathbb{R}^1_t = \mathbb{R}^1$, $t \geq 0$, whose points are arbitrary except at finitely many values $0 = t_0 < t_1 < \cdots < t_n$.

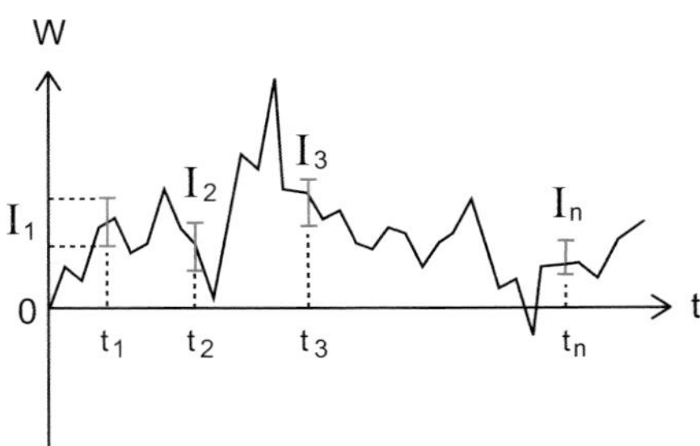

Fig. 7.6 A cylinder subset for Brownian motion

Remark 7.6 The following are basic properties of the cylinder subsets.

(i) $C(t; I_1) \cup C(t; I_2) = C(t; I_1 \cup I_2)$.
(ii) $C(t; I_1) \cap C(t; I_2) = C(t; I_1 \cap I_2)$.
(iii) For $t_1 < t_2$ we have $C(t_1; I_1) \cap C(t_2; I_2) = C(t_1, t_2; I_1, I_2)$.
(iv) If $s < t$, then W_t is not $\mathcal{F}_s$-measurable.

Now we define a measure on Ω using the transition probability of Brownian motion.

Definition 7.4 (Probability of Brownian Motion) Put $\delta t_j = t_j - t_{j-1}$ and define the size $\mathbb{P}_0$ of the cylinder subset $C(t_1, \ldots, t_n; I_1, \ldots, I_n)$ of Ω by

$$\begin{aligned}&\mathbb{P}_0(C(t_1, \ldots, t_n; I_1, \ldots, I_n)) \\ &= \int_{I_1} \cdots \int_{I_n} p(\delta t_1; 0, x_1)\, p(\delta t_2; x_1, x_2) \cdots p(\delta t_n; x_{n-1}, x_n)\, \mathrm{d}x_n \cdots \mathrm{d}x_1\end{aligned}$$

where $p(\delta t; x, y)$ is the transition probability density of the normal distribution defined in (7.1). By the Kolmogorov Extension Theorem, $\mathbb{P}_0$ is extended to a probability measure $\mathbb{P}$ defined on $\mathcal{F}$. To be more precise, we first consider the algebra $\mathcal{F}_0$ generated by finite unions of cylinder subsets and define an additive set function $\mathbb{P}_0 : \mathcal{F}_0 \to [0, 1]$, and next we extend the domain of $\mathbb{P}_0$ to the whole σ-algebra $\mathcal{F}$ and obtain a countably additive set function, i.e., a measure, $\mathbb{P} : \mathcal{F} \to [0, 1]$.

Definition 7.5 (Brownian Motion as a Stochastic Process) Using the notations given in Definition 7.4, we define a Brownian motion as a sequence of random variables W_t, $t \geq 0$, defined on $(\Omega, \mathcal{F}, \mathbb{P})$ by $W_t(\omega) = \omega(t)$. Since $W_t^{-1}(I) = C(t; I)$ for a Borel subset $I \subset \mathbb{R}$, we have $\sigma(W_t) \subset \mathcal{F}_t$. In other words, $\{W_t\}_{t\geq 0}$ is adapted to $\{\mathcal{F}_t\}_{t\geq 0}$ (Fig. 7.7).

The following theorem shows that the concrete Brownian motion constructed in Definition 7.5 satisfies all the axioms for the abstract Brownian motion given in Definition 7.1. Hence the Brownian motion in Definition 7.5 is a Brownian motion in the original sense.

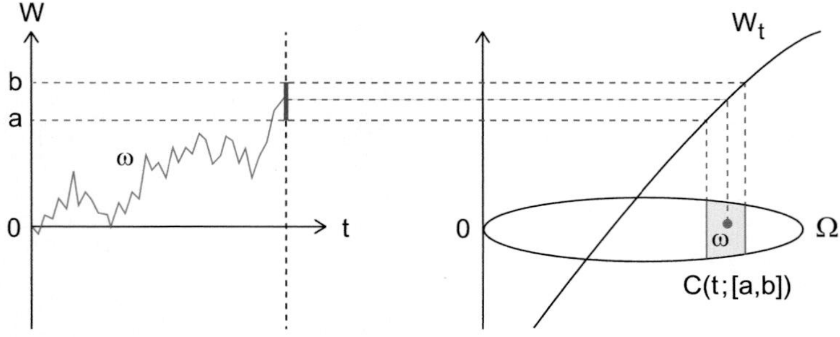

Fig. 7.7 Brownian motion and a cylinder subset

Theorem 7.8 (Brownian Motion) *Let W_t be the Brownian motion in Definition 7.5. Then we have the following:*

(i) For $0 \le s \le t$ the increment $W_t - W_s$ has normal distribution with expectation 0 and variance $t - s$.
(ii) For $0 \le t_1 < t_2 \le t_3 < t_4 \le \cdots \le t_{2n-1} < t_{2n}$, the increments

$$W_{t_2} - W_{t_1}\,,\ W_{t_4} - W_{t_3}\,,\ \ldots\,,\ W_{t_{2n}} - W_{t_{2n-1}}$$

are independent.

Proof

(i) Note that for an arbitrary Borel set $I \subset \mathbb{R}$ we have

$$\begin{aligned}
\mathbb{P}(W_t - W_s \in I) &= \iint_{\{(x,y):y-x\in I\}} p(s;0,x)\,p(t-s;x,y)\,\mathrm{d}x\,\mathrm{d}y \\
&= \int_{-\infty}^{\infty} p(s;0,x)\left(\int_{\{y:y-x\in B\}} p(t-s;x,y)\,\mathrm{d}y\right)\mathrm{d}x \\
&= \int_{-\infty}^{\infty} p(s;0,x)\left(\int_B p(t-s;x,x+z)\,\mathrm{d}z\right)\mathrm{d}x \\
&= \int_{-\infty}^{\infty} p(s;0,x)\left(\int_B p(t-s;0,z)\,\mathrm{d}z\right)\mathrm{d}x \\
&= \int_I p(t-s;0,z)\,\mathrm{d}z \int_{-\infty}^{\infty} p(s;0,x)\,\mathrm{d}x \\
&= \int_I p(t-s;0,z)\,\mathrm{d}z\,.
\end{aligned}$$

Hence $W_t - W_s$ is normally distributed with mean 0 and variance $t - s$.

(ii) For notational convenience we consider the case $n = 2$. For arbitrary Borel sets $A, B \subset \mathbb{R}$ we put

$$D_1 = \{(x_1, x_2, x_3, x_4) : x_2 - x_1 \in A, x_4 - x_3 \in B\}$$

and

$$D_2 = \{(x_3, x_4) : x_4 - x_3 \in B\}$$

and obtain

$$\begin{aligned}
&\mathbb{P}\left(W_{t_2} - W_{t_1} \in A, W_{t_4} - W_{t_3} \in B\right) \\
&= \iiiint_{D_1} p(t_1;0,x_1)\,p(t_2-t_1,x_1,x_2)\,p(t_3-t_2,x_2,x_3) \\
&\qquad \times p(t_4-t_3,x_3,x_4)\,\mathrm{d}x_1\,\mathrm{d}x_2\,\mathrm{d}x_3\,\mathrm{d}x_4
\end{aligned}$$

$$= \int_{-\infty}^{\infty} p(t_1; 0, x_1) \Big[\int_{\{x_2 : x_2 - x_1 \in A\}} \iint_{D_2} p(t_2 - t_1, x_1, x_2)\, p(t_3 - t_2, x_2, x_3)$$

$$\times p(t_4 - t_3, x_3, x_4)\, \mathrm{d}x_2\, \mathrm{d}x_3\, \mathrm{d}x_4 \Big] \mathrm{d}x_1 \quad (\text{Put } x_2 - x_1 = u.)$$

$$= \int_{-\infty}^{\infty} p(t_1; 0, x_1) \left(\int_A p(t_2 - t_1, x_1, u + x_1) I_2(x_3, x_4) \mathrm{d}u \right) \mathrm{d}x_1$$

$$= \int_{-\infty}^{\infty} p(t_1; 0, x_1) \left(\int_A p(t_2 - t_1, 0, u) I_2(x_3, x_4) \mathrm{d}u \right) \mathrm{d}x_1$$

$$= \int_A p(t_2 - t_1, 0, u) I_2(x_3, x_4) \mathrm{d}u$$

where

$$I_2(x_3, x_4) = \iint_{D_2} p(t_3 - t_2, u + x_1, x_3) p(t_4 - t_3, x_3, x_4)\, \mathrm{d}x_3\, \mathrm{d}x_4 \,.$$

Now we put $x_4 - x_3 = v$ and obtain

$$\begin{aligned} I_2(x_3, x_4) &= \int_{-\infty}^{\infty} \int_B p(t_3 - t_2, u + x_1, x_3) p(t_4 - t_3, x_3, x_3 + v)\, \mathrm{d}v\, \mathrm{d}x_3 \\ &= \int_{-\infty}^{\infty} p(t_3 - t_2, u + x_1, x_3) \left(\int_B p(t_4 - t_3, 0, v)\, \mathrm{d}v \right) \mathrm{d}x_3 \\ &= \int_B p(t_4 - t_3, 0, v)\, \mathrm{d}v \,. \end{aligned}$$

Hence

$$\begin{aligned} &\mathbb{P}(W_{t_2} - W_{t_1} \in A, W_{t_4} - W_{t_3} \in B) \\ &= \int_A p(t_2 - t_1, 0, u) \left(\int_B p(t_4 - t_3, 0, v)\, \mathrm{d}v \right) \mathrm{d}u \\ &= \int_A p(t_2 - t_1, 0, u)\, \mathrm{d}u \int_B p(t_4 - t_3, 0, v)\, \mathrm{d}v \\ &= \mathbb{P}(W_{t_2} - W_{t_1} \in A)\, \mathbb{P}(W_{t_4} - W_{t_3} \in B) \,. \end{aligned}$$

For the last equality we use the result from the proof of Theorem 7.8(i). □

Remark 7.7 (Nondifferentiability of Brownian Motion) From the equation

$$\frac{\delta W_t}{\delta t} = \frac{W_{t+\delta t} - W_t}{\delta t} \sim \frac{1}{\sqrt{\delta t}} N(0, 1)$$

we observe that as the increment $\delta t > 0$ converges to 0 the increment of Brownian motion can be arbitrarily large. This is the reason why almost every sample path of Brownian motion is not differentiable at every t. Therefore we use short line segments in plotting sample paths of Brownian motion, and should not use any smoothing technique.

Theorem 7.9 (Time Inversion) *Let W_t, $t \geq 0$, be a Brownian motion. The process X_t defined by*

$$X_t = \begin{cases} 0, & t = 0 \\ tW_{1/t}, & t > 0 \end{cases}$$

is also a Brownian motion.

Proof Consult [71]. □

Corollary 7.2 (Law of Large Numbers for Brownian Motion) *Let W_t, $t \geq 0$, be a Brownian motion. Then*

$$\lim_{t \to \infty} \frac{W_t}{t} = 0$$

almost surely.

Proof Note that

$$\lim_{t \to \infty} \frac{W_t}{t} = \lim_{u \to 0+} uW_{1/u} = \lim_{u \to 0+} X_u = 0$$

almost surely since X_u, $u \geq 0$, is a Brownian motion by Theorem 7.9. □

For a simulation of Corollary 7.2, see Simulation 7.4 and Fig. 7.8. For a related result, see Exercise 8.7.

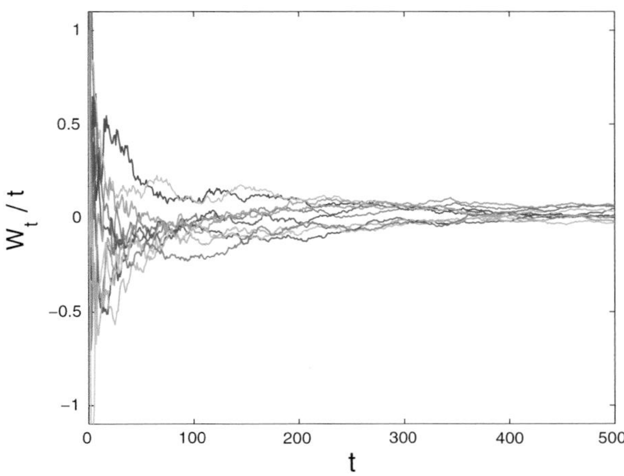

Fig. 7.8 The Law of Large Numbers for Brownian motion

Remark 7.8 In Example 7.1, for any constants μ and σ we showed that

$$\mathbb{E}[S_t] = \mathbb{E}[S_0 e^{(\mu-\frac{1}{2}\sigma^2)t+\sigma W_t}] = S_0 e^{\mu t}, \quad t \geq 0.$$

However, if $\mu - \frac{1}{2}\sigma^2 < 0$, then

$$\lim_{t\to\infty} S_0 e^{(\mu-\frac{1}{2}\sigma^2)t+\sigma W_t} = \lim_{t\to\infty} S_0 e^{t[(\mu-\frac{1}{2}\sigma^2)+\sigma\frac{1}{t}W_t]} = \lim_{t\to\infty} S_0 e^{t(\mu-\frac{1}{2}\sigma^2)} = 0$$

almost surely by the Law of Large Numbers. Since

$$\lim_{t\to\infty} \mathbb{E}[S_t] = S_0 e^{\mu t} = \begin{cases} \infty, & \mu > 0 \\ S_0, & \mu = 0 \\ 0, & \mu < 0 \end{cases}$$

we have

$$0 = \mathbb{E}\left[\lim_{t\to\infty} S_t\right] \neq \lim_{t\to\infty} \mathbb{E}[S_t]$$

for $\mu \geq 0$. If $\mu = 0$ then S_t is a martingale and $\mathbb{E}[S_t] = S_0$ for every $t \geq 0$ by Theorem 6.1, which is in agreement with the above result. Note that even in this case we have $\lim_{t\to\infty} S_t = 0$.

See Fig. 7.3 for the simulations for $\mu > 0$ and $\mu - \frac{1}{2}\sigma^2 > 0$ where individual sample paths tend to increase as $t \to \infty$ and the average of sample paths grows exponentially. See also Fig. 7.9 for $\mu > 0$ and $\mu - \frac{1}{2}\sigma^2 < 0$ where individual sample paths converge to 0 while their average grows exponentially.

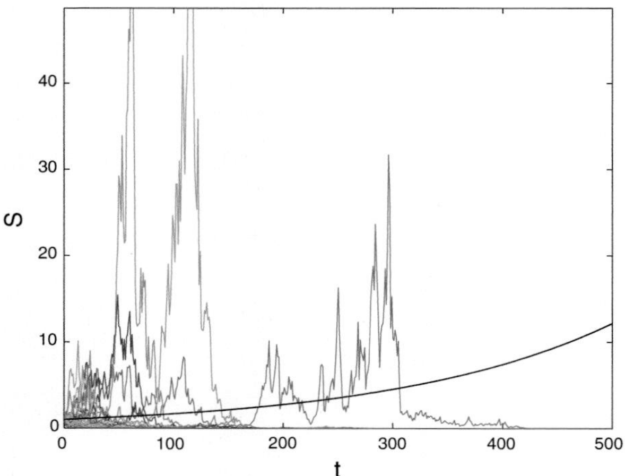

Fig. 7.9 The Law of Large Numbers for geometric Brownian motion for $\mu > 0$ and $\mu - \frac{1}{2}\sigma^2 < 0$ with exponentially increasing average

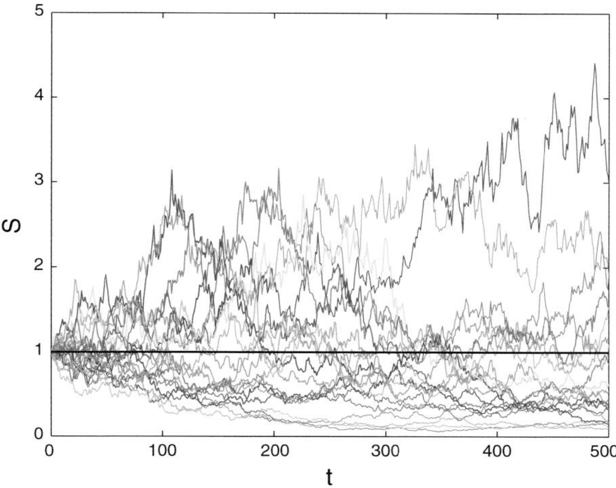

Fig. 7.10 The Law of Large Numbers for geometric Brownian motion and the martingale property for $\mu = 0$

Table 7.1 Limiting behaviors of geometric Brownian motion depending on the signs of μ and $\mu - \frac{1}{2}\sigma^2$

	$\mu > 0$	$\mu = 0$	$\mu < 0$
$\mu - \frac{1}{2}\sigma^2 > 0$	$S_t \to +\infty$ a.s. $\mathbb{E}[S_t] \to +\infty$	Not applicable	Not applicable
$\mu - \frac{1}{2}\sigma^2 = 0$	$S_t = e^{\sigma W_t}$ $\mathbb{E}[S_t] \to +\infty$	Not applicable	Not applicable
$\mu - \frac{1}{2}\sigma^2 < 0$	$S_t \to 0$ a.s. $\mathbb{E}[S_t] \to +\infty$	$S_t \to 0$ a.s. $\mathbb{E}[S_t] = S_0$	$S_t \to 0$ a.s. $\mathbb{E}[S_t] \to 0$

Finally, for $\mu = 0$ presented in Fig. 7.10, the martingale property makes the sample paths stay around the constant average S_0 for every t.

In Table 7.1 the limiting behaviors of geometric Brownian motion are classified by the signs of μ and $\mu - \frac{1}{2}\sigma^2$ under the assumption that $\sigma > 0$.

7.3 Brownian Motion and Martingales

In this section we present some of the most important examples of martingales defined by a Brownian motion $\{W_t\}_{t\geq 0}$.

Let $-\infty < \cdots < a_{k-1} < a_k < a_{k+1} < \cdots < \infty$, and let $J_k = (a_k, a_{k+1}]$. Fix $0 \leq s < t$. Note that the cylinder sets $C(s; J_k) = \{\omega : W_s(\omega) \in J_k\}$ partition the set of all Brownian paths Ω. In Fig. 7.11 the interval J_k is plotted along the line with the time coordinate equal to s. Let J be one such interval. If $a_{k+1} - a_k$ is close to

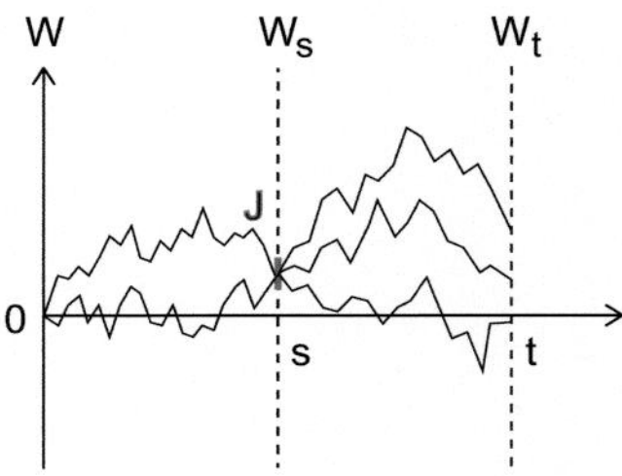

Fig. 7.11 Conditional expectation of W_t on a cylinder set $C(s; J)$

zero, then $\mathbb{E}[W_t|\mathcal{F}_s]$ is approximated by a number $\mathbb{E}[W_t|C(s;J)]$ on $C(s;J)$. Assume that J is of the form $J = (a, a + \delta x]$ for some a and $\delta x > 0$. We assume that δx is sufficiently small.

$$\begin{aligned}
\mathbb{E}[W_t|C(s;J)] &= \frac{\int_{C(s;J)} W_t \, \mathrm{d}\mathbb{P}}{\mathbb{P}((C(s;J))} \\
&= \frac{\int_{-\infty}^{\infty} \int_J x_2 p(s; 0, x_1) p(t-s; x_1, x_2) \, \mathrm{d}x_1 \mathrm{d}x_2}{\int_J p(s; 0, x_1) \, \mathrm{d}x_1} \\
&\approx \frac{\int_{-\infty}^{\infty} x_2 p(s; 0, a) p(t-s; a, x_2) \, \delta x \, \mathrm{d}x_2}{p(s; 0, a) \, \delta x} \\
&= \int_{-\infty}^{\infty} x_2 p(t-s; a, x_2) \, \mathrm{d}x_2 \\
&= a
\end{aligned}$$

where $p(u; x, y)$ denotes the transition probability density from x to y in time u. Since a is the representative value of W_s on $C(s;J)$, we observe that on each subset $C(s;J)$ two random variables $\mathbb{E}[W_t|\mathcal{F}_s]$ and W_s are sufficiently close to each other. In fact, $\mathbb{E}[W_t|\mathcal{F}_s] = W_s$, i.e., W_t is a martingale.

Lemma 7.3 *Given a real constant θ and $0 \le s < t$, we have*

$$\mathbb{E}[\mathrm{e}^{\theta W_t}|\mathcal{F}_s] = \mathrm{e}^{\frac{1}{2}\theta^2(t-s)} \mathrm{e}^{\theta W_s} \, .$$

In other words, if we let

$$L_t = \mathrm{e}^{-\frac{1}{2}\theta^2 t - \theta W_t} \, ,$$

then $\mathbb{E}[L_t|\mathcal{F}_s] = L_s$, that is, L_t is a martingale.

Proof Since $W_t - W_s$ is independent of $\mathcal{F}_s$ we have

$$\mathbb{E}[\mathrm{e}^{\theta(W_t - W_s)}|\mathcal{F}_s] = \mathbb{E}[\mathrm{e}^{\theta(W_t - W_s)}] = \mathbb{E}[\mathrm{e}^{\theta W_{t-s}}] = \mathrm{e}^{\frac{1}{2}\theta^2(t-s)} \, .$$

Hence $\mathbb{E}[\mathrm{e}^{\theta W_t - \frac{1}{2}\theta^2 t}|\mathcal{F}_s] = \mathrm{e}^{-\frac{1}{2}\theta^2 t}\mathbb{E}[\mathrm{e}^{\theta W_t}|\mathcal{F}_s] = \mathrm{e}^{-\frac{1}{2}\theta^2 s}\mathrm{e}^{\theta W_s}$. □

Here we present some examples of martingales.

Theorem 7.10 *The following stochastic processes are martingales:*

(i) *A Brownian motion* $\{W_t\}_{t\geq 0}$ *is a martingale.*
(ii) $\{W_t^2 - t\}_{t\geq 0}$ *is a martingale.*
(iii) $\{e^{W_t - \frac{1}{2}t}\}_{t\geq 0}$ *is a martingale.*
(iv) $\{e^{\frac{1}{2}t}\cos W_t\}_{t\geq 0}$ *is a martingale.*
(v) $\{W_t^3 - 3tW_t\}_{t\geq 0}$ *is a martingale.*

Proof Take $0 \leq s < t$.

(i) Since $W_t - W_s$ and $\mathcal{F}_s$ are independent, we have

$$\mathbb{E}[W_t - W_s|\mathcal{F}_s] = \mathbb{E}[W_t - W_s] = 0 ,$$

and hence

$$\mathbb{E}[W_t|\mathcal{F}_s] = \mathbb{E}[W_s|\mathcal{F}_s] = W_s .$$

(ii) To show that $\mathbb{E}[W_t^2 - t|\mathcal{F}_s] = W_s^2 - s$ we note that

$$\begin{aligned}
\mathbb{E}[W_t^2|\mathcal{F}_s] &= \mathbb{E}[(W_t - W_s)^2 + 2W_sW_t - W_s^2|\mathcal{F}_s] \\
&= \mathbb{E}[(W_t - W_s)^2|\mathcal{F}_s] + 2W_s\mathbb{E}[W_t|\mathcal{F}_s] - \mathbb{E}[W_s^2|\mathcal{F}_s] \\
&= \mathbb{E}[(W_t - W_s)^2] + 2W_sW_s - W_s^2 \\
&= \mathbb{E}[W_{t-s}^2] + W_s^2 \\
&= t - s + W_s^2 .
\end{aligned}$$

(iii) To show that

$$\mathbb{E}[e^{W_t}e^{-\frac{1}{2}t}|\mathcal{F}_s] = e^{W_s}e^{-\frac{1}{2}s}$$

for $s \leq t$, we choose $\theta = 1$ in Lemma 7.3.

(iv) Let $z = \alpha + i\beta \in \mathbb{C}$, $\alpha, \beta \in \mathbb{R}$ and $i^2 = -1$, be a complex number, and let $\Re(w)$ denote the real part of a complex number w. Then

$$\begin{aligned}
\mathbb{E}[\cos W_t|\mathcal{F}_s] &= \mathbb{E}\left[\Re\left(e^{iW_t}\right)|\mathcal{F}_s\right] \\
&= \Re\left(\mathbb{E}\left[e^{iW_t}|\mathcal{F}_s\right]\right) \\
&= \Re\left(\mathbb{E}\left[e^{iW_s}e^{i(W_t - W_s)}|\mathcal{F}_s\right]\right) \\
&= \Re\left(e^{iW_s}\mathbb{E}\left[e^{i(W_t - W_s)}|\mathcal{F}_s\right]\right) \\
&= \Re\left(e^{iW_s}\mathbb{E}\left[e^{i(W_t - W_s)}\right]\right) \\
&= \Re\left(e^{iW_s}\mathbb{E}\left[e^{iW_{t-s}}\right]\right) .
\end{aligned}$$

By Lemma 7.2 we have

$$\begin{aligned}\mathbb{E}[\cos W_t|\mathcal{F}_s] &= \Re\left(e^{iW_s}e^{-(t-s)/2}\right)\\ &= e^{-(t-s)/2}\cos W_s\end{aligned}$$

and $\mathbb{E}\left[e^{\frac{1}{2}t}\cos W_t|\mathcal{F}_s\right] = e^{\frac{1}{2}s}\cos W_s$.

(v) Since $W_t - W_s$ is independent of $\mathcal{F}_s$, $(W_t - W_s)^2$ and $(W_t - W_s)^3$ are also independent of $\mathcal{F}_s$. Hence we have

$$\begin{aligned}&\mathbb{E}[W_t^3|\mathcal{F}_s] - W_s^3\\ &= \mathbb{E}[(W_t - W_s + W_s)^3 - W_s^3|\mathcal{F}_s]\\ &= \mathbb{E}[(W_t - W_s)^3 + 3(W_t - W_s)^2\, W_s + 3(W_t - W_s)\, W_s^2|\mathcal{F}_s]\\ &= \mathbb{E}[(W_t - W_s)^3] + 3W_s\,\mathbb{E}[(W_t - W_s)^2] + 3W_s^2\,\mathbb{E}[W_t - W_s]\\ &= 0 + 3W_s(t-s) + 3W_s^2 \times 0\\ &= \mathbb{E}[3t\, W_t|\mathcal{F}_s] - 3sW_s\,.\end{aligned}$$

Hence $\mathbb{E}[W_t^3 - 3t\, W_t|\mathcal{F}_s] = W_s^3 - 3sW_s\,.$ □

Remark 7.9 As a by-product of the proof of Theorem 7.10 (iv), we obtain

$$\mathbb{E}[\cos W_t|\mathcal{F}_s] = e^{-(t-s)/2}\cos W_s\,.$$

For $s = 0$ the fact that $\mathbb{E}[\cos W_t] = e^{-t/2}$ has already been mentioned in Remark 7.4. (See Exercise 7.17 for a related result.)

Theorem 7.11 (Continuity of Brownian Motion) *For $\varepsilon > 0$ we have*

$$\lim_{\delta t\to 0}\mathbb{P}(\{\omega : |W_{t+\delta t}(\omega) - W_t(\omega)| > \varepsilon\}) = 0\,.$$

Proof Chebyshev's inequality implies that for any fixed $\varepsilon > 0$ we have

$$\begin{aligned}\mathbb{P}(|W_{t+\delta t} - W_t| > \varepsilon) &= \mathbb{P}(|W_{t+\delta t} - \mathbb{E}[W_{t+\delta t}|\mathcal{F}_t]\,| > \varepsilon)\\ &\le \frac{1}{\varepsilon^2}\mathrm{Var}[W_{t+\delta t}|\mathcal{F}_t]\\ &= \frac{1}{\varepsilon^2}\delta t \to 0\end{aligned}$$

as $\delta t \to 0$. □

The following facts due to Paul Lévy characterize Brownian motions using the concept of martingale.

Theorem 7.12 *Given a stochastic process $\{X_t\}_{t\geq 0}$ and the filtration $\mathcal{F}_t = \sigma(\{X_s : 0 \leq s \leq t\})$, the process X_t is a Brownian motion if and only if all of the following conditions hold:*

(i) $X_0 = 0$ *with probability 1.*
(ii) A sample path $t \mapsto X_t$ *is continuous with probability 1.*
(iii) $\{X_t\}_{t\geq 0}$ *is a martingale with respect to* $\{\mathcal{F}_t\}_{t\geq 0}$.
(iv) $\{X_t^2 - t\}_{t\geq 0}$ *is a martingale with respect to* $\{\mathcal{F}_t\}_{t\geq 0}$.

Theorem 7.13 *Let $\{M_t\}_{t\geq 0}$ be a martingale with respect to a filtration $\{\mathcal{F}_t\}_{t\geq 0}$ and a probability measure $\mathbb{Q}$. If M_t is continuous, $M_0 = 0$ and*

$$[M, M]_t = t ,$$

then it is a Brownian motion.

Proof Let

$$F(t, x) = \exp(\tfrac{1}{2}\lambda^2 t + \mathrm{i}\lambda x)$$

for some real constant λ where $\mathrm{i}^2 = -1$. Then

$$\begin{aligned} \mathrm{d}F(t, M_t) &= \tfrac{1}{2}\lambda^2 F(t, M_t)\,\mathrm{d}t + \mathrm{i}\lambda F(t, M_t)\,\mathrm{d}M_t - \tfrac{1}{2}\lambda^2 F(t, M_t)\,\mathrm{d}[M, M]_t \\ &= \mathrm{i}\lambda F(t, M_t)\,\mathrm{d}M_t . \end{aligned}$$

Hence $F(t, M_t) = \exp(\frac{1}{2}\lambda^2 t + \mathrm{i}\lambda M_t)$ is a martingale. Since, for $s \leq t$,

$$\mathbb{E}\left[\exp\left(\tfrac{1}{2}\lambda^2 t + \mathrm{i}\lambda M_t\right) \Big| \mathcal{F}_s\right] = \exp\left(\tfrac{1}{2}\lambda^2 s + \mathrm{i}\lambda M_s\right) ,$$

the characteristic function of $M_t - M_s$ is given by

$$\mathbb{E}\left[\exp\left(\mathrm{i}\lambda(M_t - M_s)\right) \Big| \mathcal{F}_s\right] = \exp\left(-\tfrac{1}{2}\lambda^2(t - s)\right) .$$

Hence $M_t - M_s$, $s \leq t$, is normally distributed with mean 0 and variance $t - s$ under $\mathbb{Q}$. As a by-product we obtain

$$\mathbb{E}\left[\mathrm{e}^{\mathrm{i}\lambda M_t} \Big| \mathcal{F}_s\right] = \mathrm{e}^{-\frac{1}{2}\lambda^2(t-s)}\mathrm{e}^{\mathrm{i}\lambda M_s} , \quad s < t .$$

Next, suppose that $0 \leq t_1 \leq t_2 \leq \cdots \leq t_n$. Then, for real constants $\lambda_1, \ldots, \lambda_n$,

$$\begin{aligned} &\mathbb{E}\left[\mathrm{e}^{\mathrm{i}\lambda_1 M_{t_1} + \mathrm{i}\lambda_2(M_{t_2} - M_{t_1}) + \cdots + \mathrm{i}\lambda_n(M_{t_n} - M_{t_{n-1}})}\right] \\ &= \mathbb{E}\left[\mathrm{e}^{\mathrm{i}(\lambda_1 - \lambda_2)M_{t_1} + \mathrm{i}(\lambda_2 - \lambda_3)M_{t_2} + \cdots + \mathrm{i}(\lambda_{n-1} - \lambda_n)M_{t_{n-1}} + \mathrm{i}\lambda_n M_{t_n}}\right] \\ &= \mathbb{E}\left[\mathbb{E}\left[\mathrm{e}^{\mathrm{i}(\lambda_1 - \lambda_2)M_{t_1} + \mathrm{i}(\lambda_2 - \lambda_3)M_{t_2} + \cdots + \mathrm{i}(\lambda_{n-1} - \lambda_n)M_{t_{n-1}} + \mathrm{i}\lambda_n M_{t_n}} | \mathcal{F}_{t_{n-1}}\right]\right] \end{aligned}$$

$$
\begin{aligned}
&= \mathbb{E}\left[\mathrm{e}^{\mathrm{i}(\lambda_1-\lambda_2)M_{t_1}+\mathrm{i}(\lambda_2-\lambda_3)M_{t_2}+\cdots+\mathrm{i}(\lambda_{n-1}-\lambda_n)M_{t_{n-1}}}\,\mathbb{E}\left[\mathrm{e}^{\mathrm{i}\lambda_n M_{t_n}}|\mathcal{F}_{t_{n-1}}\right]\right] \\
&= \mathbb{E}\left[\mathrm{e}^{\mathrm{i}(\lambda_1-\lambda_2)M_{t_1}+\mathrm{i}(\lambda_2-\lambda_3)M_{t_2}+\cdots+\mathrm{i}(\lambda_{n-1}-\lambda_n)M_{t_{n-1}}}\,\mathrm{e}^{-\frac{1}{2}\lambda_n^2(t_n-t_{n-1})}\mathrm{e}^{\mathrm{i}\lambda_n M_{t_{n-1}}}\right] \\
&= \mathrm{e}^{-\frac{1}{2}\lambda_n^2(t_n-t_{n-1})}\mathbb{E}\left[\mathrm{e}^{\mathrm{i}(\lambda_1-\lambda_2)M_{t_1}+\mathrm{i}(\lambda_2-\lambda_3)M_{t_2}+\cdots+\mathrm{i}\lambda_{n-1}M_{t_{n-1}}}\right] \\
&= \mathrm{e}^{-\frac{1}{2}\lambda_n^2(t_n-t_{n-1})}\mathbb{E}\left[\mathbb{E}\left[\mathrm{e}^{\mathrm{i}(\lambda_1-\lambda_2)M_{t_1}+\mathrm{i}(\lambda_2-\lambda_3)M_{t_2}+\cdots+\mathrm{i}\lambda_{n-1}M_{t_{n-1}}}|\mathcal{F}_{t_{n-2}}\right]\right] \\
&\qquad\vdots \\
&= \mathrm{e}^{-\frac{1}{2}\lambda_n^2(t_n-t_{n-1})}\mathrm{e}^{-\frac{1}{2}\lambda_{n-1}^2(t_{n-1}-t_{n-2})}\times\cdots\times\mathrm{e}^{-\frac{1}{2}\lambda_1^2 t_1}\,,
\end{aligned}
$$

thus the increments are independent. □

7.4 Computer Experiments

Simulation 7.1 (Sample Paths of Brownian Motion)
We plot 100 sample paths of Brownian motion. See Fig. 7.1.

```
N = 200;  % number of time steps
T = 20;
dt = T/N;
time = 0:dt:T;

M = 30; % number of sample paths
W = zeros(M,N+1); % dW and W are matrices.
dW = sqrt(dt)*randn(M,N);
for i=1:N
    W(:,i+1) = W(:,i) + dW(:,i); % Note W(j,1) = 0.0.
end

for j = 1:M
    plot(time,W(j,1:N+1),'b');
hold on;
end
```

Simulation 7.2 (Average of Geometric Brownian Motion)
We check the formula $\mathbb{E}[S_t] = S_0\mathrm{e}^{\mu t}$ given in Example 7.1. See Fig. 7.3.

```
mu = 0.25;
sigma = 0.3;
N = 100;  % number of time steps
T = 5;
dt = T/N;

time = 0:dt:T;
num = 200; % number of sample paths
W = zeros(num,N + 1);
dW = zeros(num,N);
```

```
S = zeros(num,N + 1);
S0 = 1;
S([1:num],1) = S0;

for i = 1:num
    dW(i,1:N) = sqrt(dt)*randn(1,N);
end

for i = 1:num
    for j=1:N
        S(i,j+1) = S(i,j) + mu*S(i,j)*dt + sigma*S(i,j)*dW(i,j);
    end
end

ave = zeros(1,N+1);
for j = 1: N+1
    ave(j) = mean(S(:,j));
end

plot(time,ave);
hold on
t = 0:0.01:T;
plot(t,S0*exp(mu*t),'r')
```

Simulation 7.3 (Brownian Motion with Boundary Condition)

Using the bisection method we generate a sample path of Brownian motion with a given condition at the final time T using conditional expectation. For an output see Fig. 7.5.

```
T = 1; % length of time interval
n = 8; % number of bisections
dt = T/2^n; % length of time step
time = 0:dt:T; % partition of the time interval
M = 2^n + 1;
W = zeros(1,M); % a sample path of Brownian motion
W(1) = 0; % initial condition
W(M) = 0.3; % condition on Brownian motion W at time T=dt*(M-1)

for i=1:n
Increment = 2^(n-i+1);
    for j=1:Increment:2^n
        index1 = j;
        index2 = j+Increment;
        t1 = time(index1);
        t2 = time(index2);
        W1 = W(index1);
        W2 = W(index2);
        ave = (W1 + W2)/2;
        var = (t2 - t1)/4;
        ind_mid = (index1 + index2)/2;
        W(ind_mid)=random('normal',ave,sqrt(var)); %conditional expectation
    end
end
```

```
for i=0:n
t_value = zeros(1,2^i+1);
W_value = zeros(1,2^i+1);
    for k=1:2^i+1
    t_value(k)=(k-1)*dt*2^(n-i);
    W_value(k) = W((k-1)*2^(n-i)+1);
    end
    plot(t_value,W_value,'-', 'color', hsv2rgb([1-i/n 1 1]));
    hold on;
    pause(0.3); % Pause between successive stages.
end
```

Simulation 7.4 (Law of Large Numbers for Brownian Motion)
We plot 10 sample paths of $\frac{1}{t}W_t$, $0 \leq t \leq 500$, to simulate the Law of Large Numbers for Brownian motion (Corollary 7.2). For an output see Fig. 7.8.

```
N = 200;  % number of time steps
T = 500;
dt = T/N;
time = 0:dt:T;
M = 10; % number of sample paths

W = zeros(M,N+1); % dW and W are matrices.
dW = sqrt(dt)*randn(M,N);

for i=1:N
    W(:,i+1) = W(:,i) + dW(:,i); % Note W(j,1) = 0.0.
end

X = zeros(M,N+1);
X(:,1) = 1;

for i = 2:N+1
    X(:,i) = W(:,i)/((i-1)*dt);
end

for j = 1:M
    plot(time,X(j,1:N+1));
hold on;
end
```

Exercises

7.1 Let Z be normally distributed with mean zero and variance under the measure $\mathbb{P}$. What is the distribution of $\sqrt{t}Z$? Is the process $X_t = \sqrt{t}Z$ a Brownian motion?

7.2 For a standard normal variable Z, show that $\mathbb{E}[\mathrm{e}^{W_t}] = \mathrm{e}^{\frac{1}{2}t}$.

7.3 Suppose that X is normally distributed with mean μ and variance σ^2. Calculate $\mathbb{E}[e^{\theta X}]$ and hence evaluate $\mathbb{E}[X^4]$. What is $\mathbb{E}[X^{2k}]$?

7.4 Show that $V_t = W_{t+T} - W_T$ is a Brownian motion for any $T > 0$.

7.5 Show that $\{W_t^3 - 3tW_t\}_{t\geq 0}$ is a martingale.

7.6 Prove that $\{W_t^3\}_{t\geq 0}$ is not a martingale even though $\mathbb{E}[W_t^3] = 0$ for every $t \geq 0$. (Hint: Use $\mathbb{E}[(W_t - W_s)^3|\mathcal{F}_s] = 0$.)

7.7 Compute (i) $\mathbb{E}[(W_t - W_s)^2]$, (ii) $\mathbb{E}[e^{-W_t}]$ and (iii) $\mathbb{E}[e^{2W_t}]$.

7.8 Given a Brownian motion W_t, $0 \leq t \leq 1$, define a stochastic process $X_t = W_t - tW_1$ which is called a Brownian bridge. Then clearly $X_0 = X_1 = 0$ and $\mathbb{E}[X_t] = 0$. Show that $\mathbb{E}[X_t^2] = t - t^2$.

7.9 Compute the following conditional expectations:

(i) $\mathbb{E}[W_t^2|W_s = x]$, $0 \leq s < t$.
(ii) $\mathbb{E}[W_t^2|W_r = x, W_s = y]$, $0 \leq r < s < t$.
(iii) $\mathbb{E}[e^{-W_t}|W_s = x]$, $0 \leq s < t$.
(iv) $\mathbb{E}[W_t^3|W_s = x]$, $0 \leq s < t$.
(v) $\mathbb{E}[W_s|W_t = 0]$, $0 \leq s < t$.
(vi) $\mathbb{E}[W_r|W_s = x, W_t = y]$, $0 \leq s < r < t$.
(vii) $\mathbb{E}[W_t^2|W_s = 0]$, $0 \leq s \leq t$.

7.10 For a real constant θ show that $\mathbb{E}\left[W_t e^{\theta W_t}\right] = \theta t\, e^{\frac{1}{2}\theta^2 t}$.

7.11 For a real constant θ show that $\mathbb{E}\left[W_t^2 e^{\theta W_t}\right] = (t + \theta^2 t^2)\, e^{\frac{1}{2}\theta^2 t}$.

7.12 Let $\{W_t\}_{t\geq 0}$ be a Brownian motion. Show that the processes $\{X_t\}_{t\geq 0}$ in the following are also Brownian motions.

(i) $X_t = \alpha W_{t/\alpha^2}$ for $\alpha \neq 0$.
(ii) $X_t = a^{-1/2} W_{at}$ for $a > 0$.
(iii) $X_t = W_{T+t} - W_T$ for $T > 0$.
(iv) $X_t = \begin{cases} W_t, & t \leq T, \\ 2W_T - W_t, & t > T, \end{cases}$ for $T > 0$.

7.13 Prove that for $0 \leq s < t$ we have $\mathbb{E}[W_s|W_t] = \frac{s}{t} W_t$. In other words, $\mathbb{E}[W_s|W_t = y] = \frac{s}{t} y$.

7.14 Show that the variation of the paths of W_t is infinite almost surely.

7.15 Show that $\mathbb{E}[W_s|W_t] = \frac{s}{t} W_t$ for $0 \leq s < t$ using the following idea: Define a process $\{X_u\}_{u\geq 0}$ by $X_0 = 0$ and $X_u = uW_{1/u}$ for $u > 0$. Then X_u is a Brownian motion.

7.16 Show that a process $X_t = e^{\sigma W_t - \frac{1}{2}\sigma^2 t}$, $0 \leq t < \infty$, is a martingale. Prove that $\lim_{t\to\infty} X_t = 0$ almost surely. Show that $\{X_t\}_{0\leq t<\infty}$ is not uniformly integrable directly from the definition of uniform integrability without using the law of large numbers for Brownian motion.

7.17 Find a constant a for which $X_t = e^{at} \cos W_t$, $t \geq 0$, is a martingale.

Chapter 8
Girsanov's Theorem

Let $\{W_t\}_{t\geq 0}$ be a Brownian motion with respect to a probability measure $\mathbb{P}$. Take a constant θ, and consider $X_t = W_t + \theta t$, $0 \leq t < \infty$, which is called a Brownian motion with drift. Our goal is to find a probability measure $\mathbb{Q}$ for which X_t, $0 \leq t \leq T$, is a Brownian motion for some fixed T. We require an additional condition that $\mathbb{Q}$ is equivalent to $\mathbb{P}$, i.e., $\mathbb{P}(A) = 0$ if and only if $\mathbb{Q}(A) = 0$. In other words, an event occurs with positive $\mathbb{P}$-probability if and only if it happens with positive $\mathbb{Q}$-probability. Such a condition is important in financial applications since we have to deal with the same set of asset price movements even when we switch to a new probability measure. Igor Girsanov proved the existence of such a measure $\mathbb{Q}$. We will find first a necessary condition for the existence of an equivalent probability measure $\mathbb{Q}$ for which a Brownian motion with drift is a Brownian motion. Such a necessary condition will turn out to be crucial in defining $\mathbb{Q}$.

8.1 Motivation

Let $\mathbb{E}$ or $\mathbb{E}^{\mathbb{P}}$ denote expectation with respect to a probability measure $\mathbb{P}$. Let $\{W_t\}_{t\geq 0}$ denote a $\mathbb{P}$-Brownian motion. For a given constant θ consider a new stochastic process

$$X_t = W_t + \theta t \ .$$

We want to find an equivalent probability measure $\mathbb{Q}$ such that $\{X_t\}_{0\leq t\leq T}$ is a $\mathbb{Q}$-Brownian motion for some fixed T.

First, we will see what a possible definition of $\mathbb{Q}$ would look like. Let Ω be the set of all continuous sample paths ω such that $\omega(0) = 0$, and let $\{\mathcal{F}_t\}$ be the filtration generated by the Brownian motion $\{W_t\}$. For $t \geq 0$ and an interval $I = [a, b] \subset \mathbb{R}$,

G.H. Choe, *Stochastic Analysis for Finance with Simulations*, Universitext,
DOI 10.1007/978-3-319-25589-7_8

consider a cylinder subset of Ω defined by

$$C(t;I) = \{\omega \in \Omega : W_t(\omega) \in I\} .$$

Note that $C(t;I)$ is $\mathcal{F}_t$-measurable since $C(t;I) = W_t^{-1}(I)$. Suppose that there exists an equivalent probability measure $\mathbb{Q}$ such that

$$\begin{aligned}\int_\Omega f(W_t(\omega))\,\mathrm{d}\mathbb{P}(\omega) &= \int_\Omega f(X_t(\omega))\,\mathrm{d}\mathbb{Q}(\omega) \\ &= \int_\Omega f(W_t(\omega) + \theta t)\,\mathrm{d}\mathbb{Q}(\omega)\end{aligned}$$

for every bounded measurable function $f : \mathbb{R} \to \mathbb{R}$. If we take an indicator function $f(x) = \mathbf{1}_I(x)$ for an interval $I = [a, b]$, then we have

$$\int_\Omega f(W_t)\,\mathrm{d}\mathbb{P} = \int_{C(t;I)} \mathrm{d}\mathbb{P} = \mathbb{P}(C(t;I))$$

$$\int_\Omega f(W_t + \theta t)\,\mathrm{d}\mathbb{Q} = \int_{C(t;I-\theta t)} \mathrm{d}\mathbb{Q} = \mathbb{Q}(C(t;I-\theta t)) .$$

Hence

$$\mathbb{P}(C(t;I)) = \mathbb{Q}(C(t;I-\theta t)) ,$$

or

$$\mathbb{P}(C(t;I+\theta t)) = \mathbb{Q}(C(t;I))$$

for every $t \geq 0$ and every I. By the definition of $\mathbb{P}$,

$$\begin{aligned}\mathbb{P}(C(t;I+\theta t)) &= \int_{a+\theta t}^{b+\theta t} \frac{1}{\sqrt{2\pi t}} \mathrm{e}^{-x^2/2t} \mathrm{d}x \\ &= \int_a^b \frac{1}{\sqrt{2\pi t}} \mathrm{e}^{-(y+\theta t)^2/2t} \mathrm{d}y .\end{aligned} \tag{8.1}$$

Let L be the Radon–Nikodym derivative of $\mathbb{Q}$ with respect to $\mathbb{P}$, i.e., $L = \frac{\mathrm{d}\mathbb{Q}}{\mathrm{d}\mathbb{P}}$, and denote its conditional expectation $\mathbb{E}[L|\mathcal{F}_t]$ by L_t. Later it will be shown that $L_t = \mathrm{e}^{-\frac{1}{2}\theta^2 t - \theta W_t}$, which is a $\mathbb{P}$-martingale. If we consider an option with finite expiry date $T < \infty$ and the set of all Brownian paths defined over $0 \leq t \leq T$, then $L = L_T$. Now we find a possible formula for L_t assuming that there exists a function $\rho_t : \mathbb{R} \to \mathbb{R}$

such that $L_t(\omega) = \rho_t(W_t(\omega))$. Then

$$\begin{aligned}
\mathbb{Q}(C(t;I)) &= \mathbb{E}^{\mathbb{Q}}[\mathbf{1}_{C(t;I)}] \\
&= \mathbb{E}^{\mathbb{P}}[\mathbf{1}_{C(t;I)}L] \\
&= \mathbb{E}^{\mathbb{P}}[\mathbb{E}^{\mathbb{P}}[\mathbf{1}_{C(t;I)}L|\mathcal{F}_t]] \qquad \text{(use the tower property)} \\
&= \mathbb{E}^{\mathbb{P}}[\mathbf{1}_{C(t;I)}\mathbb{E}^{\mathbb{P}}[L|\mathcal{F}_t]] \qquad \text{(take out what is known)} \\
&= \mathbb{E}^{\mathbb{P}}[\mathbf{1}_{C(t;I)}L_t] \\
&= \int_{C(t;I)} \rho_t(W_t)\,\mathrm{d}\mathbb{P} \\
&= \int_a^b \rho_t(x)\frac{1}{\sqrt{2\pi t}}\mathrm{e}^{-x^2/2t}\,\mathrm{d}x\,.
\end{aligned}$$

Using (8.1), we obtain

$$\rho_t(x)\,\mathrm{e}^{-x^2/2t} = \mathrm{e}^{-(x+\theta t)^2/2t}\,, \tag{8.2}$$

and hence $\rho_t(x) = \mathrm{e}^{-\theta x - \frac{1}{2}\theta^2 t}$ and

$$L_t = \rho_t(W_t) = \mathrm{e}^{-\theta W_t - \frac{1}{2}\theta^2 t}\,. \tag{8.3}$$

The same conclusion can be obtained using the Fourier transformation as follows: Since the probability density functions of X_t and W_t with respect to $\mathbb{Q}$ and $\mathbb{P}$, respectively, are identical, the corresponding Fourier transforms are equal. In fact, $\mathbb{E}^{\mathbb{Q}}[\mathrm{e}^{\mathrm{i}\xi X_t}] = \mathbb{E}^{\mathbb{P}}[\mathrm{e}^{\mathrm{i}\xi W_t}] = \mathrm{e}^{-\frac{1}{2}\xi^2 t}$ for real ξ. Since

$$\begin{aligned}
\mathbb{E}^{\mathbb{Q}}[\mathrm{e}^{\mathrm{i}\xi X_t}] &= \mathbb{E}^{\mathbb{P}}[\mathrm{e}^{\mathrm{i}\xi(W_t+\theta t)}\rho_t(W_t)] \\
&= \int_{-\infty}^{\infty} \mathrm{e}^{\mathrm{i}\xi(x+\theta t)}\rho_t(x)\frac{1}{\sqrt{2\pi t}}\mathrm{e}^{-\frac{x^2}{2t}}\,\mathrm{d}x \\
&= \int_{-\infty}^{\infty} \mathrm{e}^{\mathrm{i}\xi y}\rho_t(y-\theta t)\frac{1}{\sqrt{2\pi t}}\mathrm{e}^{-\frac{(y-\theta t)^2}{2t}}\,\mathrm{d}y
\end{aligned}$$

and since

$$\mathbb{E}^{\mathbb{P}}[\mathrm{e}^{\mathrm{i}\xi W_t}] = \int_{-\infty}^{\infty} \mathrm{e}^{\mathrm{i}\xi y}\frac{1}{\sqrt{2\pi t}}\mathrm{e}^{-\frac{y^2}{2t}}\,\mathrm{d}y\,,$$

by the uniqueness of the inverse, we have

$$\rho_t(y-\theta t)\frac{1}{\sqrt{2\pi t}}\mathrm{e}^{-\frac{(y-\theta t)^2}{2t}} = \frac{1}{\sqrt{2\pi t}}\mathrm{e}^{-\frac{y^2}{2t}}\,,$$

which is equivalent to (8.2).

8.2 Equivalent Probability Measure

Definition 8.1 Recall that $L_t = \mathrm{e}^{-\frac{1}{2}\theta^2 t - \theta W_t}$ is a martingale and $\mathbb{E}[L_t] = 1$ for $t \geq 0$. Define a probability measure $\mathbb{Q}$ on $(\Omega, \mathcal{F}_T)$ by $\mathrm{d}\mathbb{Q} = L_T\,\mathrm{d}\mathbb{P}$, i.e.,

$$\mathbb{Q}(A) = \mathbb{E}^{\mathbb{P}}[\mathbf{1}_A L_T] = \int_A L_T\,\mathrm{d}\mathbb{P}$$

for $A \in \mathcal{F}_T$. (Note that $\mathbb{Q}$ is equivalent to $\mathbb{P}$ since $L_T > 0$.)

Lemma 8.1 *For $0 \leq s \leq t$, and an $\mathcal{F}_t$-measurable random variable ϕ_t, we have*

$$\mathbb{E}^{\mathbb{Q}}[\phi_t|\mathcal{F}_s] = \mathbb{E}^{\mathbb{P}}\left[\phi_t \frac{L_t}{L_s}\,\middle|\,\mathcal{F}_s\right].$$

In particular, for $s = 0$, $\mathbb{E}^{\mathbb{Q}}[\phi_t] = \mathbb{E}^{\mathbb{P}}[\phi_t L_t]$.

Proof For $A \in \mathcal{F}_s$, we have

$$\begin{aligned}
\int_A \phi_t\,\mathrm{d}\mathbb{Q} &= \int_A \mathbb{E}^{\mathbb{Q}}[\phi_t|\mathcal{F}_s]\,\mathrm{d}\mathbb{Q} \quad \text{(definition of conditional expectation)}\\
&= \int_A \mathbb{E}^{\mathbb{Q}}[\phi_t|\mathcal{F}_s] L_T\,\mathrm{d}\mathbb{P}\\
&= \int_A \mathbb{E}^{\mathbb{P}}[\,\mathbb{E}^{\mathbb{Q}}[\phi_t|\mathcal{F}_s] L_T|\mathcal{F}_s]\,\mathrm{d}\mathbb{P} \quad \text{(the tower property)}\\
&= \int_A \mathbb{E}^{\mathbb{Q}}[\phi_t|\mathcal{F}_s]\,\mathbb{E}^{\mathbb{P}}[L_T|\mathcal{F}_s]\,\mathrm{d}\mathbb{P} \quad \text{(take out what is known)}\\
&= \int_A \mathbb{E}^{\mathbb{Q}}[\phi_t|\mathcal{F}_s]\,L_s\,\mathrm{d}\mathbb{P}\,.
\end{aligned}$$

On the other hand,

$$\begin{aligned}
\int_A \phi_t\,\mathrm{d}\mathbb{Q} &= \int_A \phi_t L_T\,\mathrm{d}\mathbb{P}\\
&= \int_A \mathbb{E}^{\mathbb{P}}[\phi_t L_T|\mathcal{F}_s]\,\mathrm{d}\mathbb{P} \quad \text{(definition of conditional expectation)}\\
&= \int_A \mathbb{E}^{\mathbb{P}}[\mathbb{E}^{\mathbb{P}}[\phi_t L_T|\mathcal{F}_t]|\mathcal{F}_s]\,\mathrm{d}\mathbb{P} \quad \text{(the tower property)}\\
&= \int_A \mathbb{E}^{\mathbb{P}}[\phi_t \mathbb{E}^{\mathbb{P}}[L_T|\mathcal{F}_t]|\mathcal{F}_s]\,\mathrm{d}\mathbb{P} \quad \text{(take out what is known)}\\
&= \int_A \mathbb{E}^{\mathbb{P}}[\phi_t L_t|\mathcal{F}_s]\,\mathrm{d}\mathbb{P}\,.
\end{aligned}$$

Thus $\int_A \mathbb{E}^{\mathbb{Q}}[\phi_t|\mathcal{F}_s]\,L_s\,\mathrm{d}\mathbb{P} = \int_A \mathbb{E}^{\mathbb{P}}[\phi_t L_t|\mathcal{F}_s]\,\mathrm{d}\mathbb{P}$ for every $A \in \mathcal{F}_s$, and hence $\mathbb{E}^{\mathbb{Q}}[\phi_t|\mathcal{F}_s]\,L_s = \mathbb{E}^{\mathbb{P}}[\phi_t L_t|\mathcal{F}_s]$. □

Lemma 8.2 *Let $\mathbb{Q}$ be the equivalent measure given in Definition 8.1, and let $X_t = W_t + \theta t$. Then $\mathbb{E}^{\mathbb{Q}}[X_t] = 0$ and $\mathbb{E}^{\mathbb{Q}}[X_t^2] = t$.*

Proof Recall the results in Exercises 7.10, 7.11. Then we have

$$
\begin{aligned}
\mathbb{E}^{\mathbb{Q}}[X_t] &= \mathbb{E}[(W_t + \theta t)\,\mathrm{e}^{-\frac{1}{2}\theta^2 t - \theta W_t}] \\
&= \mathrm{e}^{-\frac{1}{2}\theta^2 t}(\mathbb{E}[W_t \mathrm{e}^{-\theta W_t}] + \theta t\,\mathbb{E}[\mathrm{e}^{-\theta W_t}]) \\
&= \mathrm{e}^{-\frac{1}{2}\theta^2 t}(-\theta t\,\mathrm{e}^{\frac{1}{2}\theta^2 t} + \theta t\,\mathrm{e}^{\frac{1}{2}\theta^2 t}) \\
&= 0
\end{aligned}
$$

and

$$
\begin{aligned}
\mathbb{E}^{\mathbb{Q}}[X_t^2] &= \mathbb{E}[(W_t + \theta t)^2\,\mathrm{e}^{-\frac{1}{2}\theta^2 t - \theta W_t}] \\
&= \mathrm{e}^{-\frac{1}{2}\theta^2 t}\left(\mathbb{E}[W_t^2 \mathrm{e}^{-\theta W_t}] + 2\theta t\,\mathbb{E}[W_t \mathrm{e}^{-\theta W_t}] + \theta^2 t^2\,\mathbb{E}[\mathrm{e}^{-\theta W_t}]\right) \\
&= \mathrm{e}^{-\frac{1}{2}\theta^2 t}\{(t + \theta^2 t^2)\,\mathrm{e}^{\frac{1}{2}\theta^2 t} + 2\theta t(-\theta t)\,\mathrm{e}^{\frac{1}{2}\theta^2 t} + \theta^2 t^2\,\mathrm{e}^{\frac{1}{2}\theta^2 t}\} \\
&= t\,.
\end{aligned}
$$

□

8.3 Brownian Motion with Drift

Lemma 8.3 *Let $\mathbb{Q}$ be the equivalent measure given in Definition 8.1, and let $X_t = W_t + \theta t$. Then X_t is a $\mathbb{Q}$-martingale.*

Proof Since, by the result in Exercise 8.4,

$$
\mathbb{E}[W_t L_t|\mathcal{F}_s] = \mathrm{e}^{-\frac{1}{2}\theta^2 t}\mathbb{E}[W_t \mathrm{e}^{-\theta W_t}|\mathcal{F}_s] = (W_s - \theta(t-s))\,L_s\,,
$$

we have

$$
\begin{aligned}
\mathbb{E}^{\mathbb{Q}}[X_t|\mathcal{F}_s] &= \mathbb{E}[X_t L_t L_s^{-1}|\mathcal{F}_s] \quad \text{(by Lemma 8.1)} \\
&= L_s^{-1}\mathbb{E}[(W_t + \theta t)L_t|\mathcal{F}_s] \\
&= L_s^{-1}\mathbb{E}[W_t L_t|\mathcal{F}_s] + L_s^{-1}\theta t\,\mathbb{E}[L_t|\mathcal{F}_s] \\
&= (W_s - \theta(t-s)) + \theta t \\
&= W_s + \theta s = X_s\,.
\end{aligned}
$$

□

Theorem 8.1 (Girsanov) *Let W_t, $0 \le t \le T$, be a $\mathbb{P}$-Brownian motion with respect to a filtration $\mathcal{F}_t$, $0 \le t \le T$, on a probability space $(\Omega, \mathcal{F}, \mathbb{P})$ and let θ_t be an arbitrary real number. Let $X_t = W_t + \theta t$ and define an equivalent probability measure $\mathbb{Q}$ by $\mathrm{d}\mathbb{Q} = L_T\, \mathrm{d}\mathbb{P}$, $L_T = \mathrm{e}^{-\frac{1}{2}\theta^2 T - \theta W_T}$. Then X_t is a $\mathbb{Q}$-Brownian motion.*

Proof Two measures are equivalent since the Radon–Nikodym derivative is positive. Clearly, $X_0 = 0$ and $\mathrm{d}X\mathrm{d}X = \mathrm{d}t$. To prove that X_t is a $\mathbb{Q}$-Brownian motion, use Theorem 7.13 and Lemma 8.3. □

Note that our proof of Theorem 8.1 does not rely on Itô's lemma explicitly even though Itô's lemma is used via Theorem 7.13. A generalized version of Girsanov's theorem can be proved using Itô's lemma.

Theorem 8.2 (Generalized Girsanov's Theorem) *Fix $0 < T < \infty$. Let $\{W_t\}_{0 \le t \le T}$ be a $\mathbb{P}$-Brownian motion with respect to a filtration $\{\mathcal{F}_t\}_{0 \le t \le T}$ on a probability space $(\Omega, \mathcal{F}, \mathbb{P})$, and let $\{\theta_t\}_{0 \le t \le T}$ be an adapted process to $\{\mathcal{F}_t\}$. Let*

$$X_t = W_t + \int_0^t \theta_s \mathrm{d}s$$

and

$$L_t = \exp\left(-\frac{1}{2}\int_0^t \theta_s^2\, \mathrm{d}s - \int_0^t \theta_s\, \mathrm{d}W_s\right) .$$

Assume that the Novikov condition holds, i.e.,

$$\mathbb{E}^{\mathbb{P}}\left[\exp\left(\frac{1}{2}\int_0^T \theta_t^2\, \mathrm{d}t\right)\right] < \infty .$$

Then L_t is a martingale, and $\mathbb{E}[L_t] = \mathbb{E}[L_0] = 1$ for $t \ge 0$. Using L_T as the Radon–Nikodym derivative, define a probability measure $\mathbb{Q}$ by $\mathrm{d}\mathbb{Q} = L_T\, \mathrm{d}\mathbb{P}$. Then X_t, $0 \le t \le T$, is a Brownian motion with respect to $\mathbb{Q}$.

Proof Clearly, $\mathrm{d}X_t\mathrm{d}X_t = (\mathrm{d}W_t + \theta_t\mathrm{d}t)(\mathrm{d}W_t + \theta_t\mathrm{d}t) = \mathrm{d}t$. By the generalized Itô formula,

$$\mathrm{d}L_t = L_t\left(-\frac{1}{2}\theta_t^2\, \mathrm{d}t - \theta_t\, \mathrm{d}W_t\right) + \frac{1}{2}L_t\theta_t^2\mathrm{d}t = -L_t\theta_t\, \mathrm{d}W_t .$$

Hence L_t is a $\mathbb{P}$-martingale. Since

$$\begin{aligned}
&\mathrm{d}(X_tL_t)\\
&= L_t\mathrm{d}X_t + X_t\mathrm{d}L_t + (\mathrm{d}X_t)(\mathrm{d}L_t)\\
&= L_t(\mathrm{d}W_t + \theta_t\mathrm{d}t) + X_t(-L_t\theta_t\, \mathrm{d}W_t) + (\mathrm{d}W_t + \theta_t\mathrm{d}t)(-L_t\theta_t\, \mathrm{d}W_t)\\
&= (-X_t\theta_t + 1)L_t\mathrm{d}W_t ,
\end{aligned}$$

X_tL_t is also a $\mathbb{P}$-martingale. Hence, for $s \le t$,

$$\mathbb{E}^{\mathbb{Q}}[X_t|\mathcal{F}_s] = \mathbb{E}^{\mathbb{P}}[X_t \frac{L_t}{L_s}|\mathcal{F}_s] = \frac{1}{L_s}\mathbb{E}^{\mathbb{P}}[X_tL_t|\mathcal{F}_s] = \frac{1}{L_s}X_sL_s = X_s \,.$$

Now we apply Theorem 7.13. □

Theorem 8.3 (Multidimensional Girsanov's Theorem) *Fix $T < \infty$, and let $(W_t^1, \dots, W_t^d)$, $0 \le t \le T$, be a d-dimensional Brownian motion on a probability space $(\Omega, \mathcal{F}, \mathbb{P})$ with a filtration $\{\mathcal{F}_t\}_{0 \le t \le T}$, and let $(\theta_t^1, \dots, \theta_t^d)$ be an adapted process. Let*

$$X_t^i = W_t^i + \int_0^t \theta_s^i \, \mathrm{d}s$$

and

$$L_t = \exp\left(-\frac{1}{2}\sum_{i=1}^d \int_0^t (\theta_s^i)^2 \, \mathrm{d}s - \sum_{i=1}^d \int_0^t \theta_s^i \, \mathrm{d}W_s^i\right) \,.$$

Assume that the Novikov condition holds, i.e.,

$$\mathbb{E}\left[\exp\left(\frac{1}{2}\sum_{i=1}^d \int_0^T (\theta_t^i)^2 \, \mathrm{d}t\right)\right] < \infty \,.$$

Then L_t is a martingale, and hence $\mathbb{E}[L_t] = \mathbb{E}[L_0] = 1$ for $t \ge 0$. Define a probability measure $\mathbb{Q}$ by $\mathrm{d}\mathbb{Q} = L_T \, \mathrm{d}\mathbb{P}$. Then $(X_t^1, \dots, X_t^d)$, $0 \le t \le T$, is a d-dimensional Brownian motion with respect to $\mathbb{Q}$.

8.4 Computer Experiments

Simulation 8.1 (Brownian Motion with Drift)

We plot 30 sample paths of Brownian motion with drift. For the output see Fig. 8.1.

```
N = 300;  % number of time steps
T = 50.;
dt = T/N;
theta = 0.4;

time = 0:dt:T;
num_samples = 30; % number of sample paths
X = zeros(num_samples,N + 1); % dW and X are matrices.
dW = zeros(num_samples,N);
```

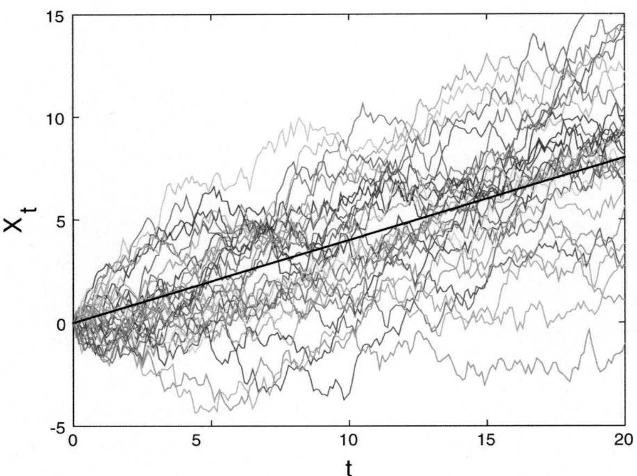

Fig. 8.1 Brownian motion with drift $X_t = W_t + \theta t$ with drift coefficient θ

```
for j = 1:num_samples
    dW(j,1:N) = sqrt(dt)*randn(1,N);
end

for j = 1:num_samples
    for i=1:N
        X(j,i+1) = X(j,i) + dW(j,i) + theta*dt; %  X(j,1) = 0.0
    end
end

for j = 1:num_samples
    plot(time,X(j,1:N + 1));
hold on;
end

x = 0:0.01:T;
plot(x,0,'k')
plot(x,theta*x,'r')
```

Exercises

8.1 Let Z denote a standard normal variable. (i) Show that $\mathbb{E}[e^{W_t}] = \mathbb{E}[e^{\sqrt{t}Z}] = e^{\frac{1}{2}t}$. (ii) For a real constant α, show that $e^{\alpha Z} \in L^p$ for $1 \leq p < \infty$.

8.2 Show that $\mathbb{E}^{\mathbb{Q}}[W_t] = -\theta t$ and $\mathbb{E}^{\mathbb{Q}}[W_t^2] = \theta^2 t^2 + t$ by direct computation.

8.3 Show directly that $\mathbb{E}^{\mathbb{P}}[e^{\alpha W_t}] = \mathbb{E}^{\mathbb{Q}}[e^{\alpha X_t}]$ for a real constant α using the Radon–Nikodym derivative.

8.4 For a real constant θ and for $s < t$, show that the following holds:

$$\mathbb{E}[W_t\, \mathrm{e}^{\theta W_t} | \mathcal{F}_s] = (W_s + \theta(t-s))\, \mathrm{e}^{\frac{1}{2}\theta^2(t-s)} \mathrm{e}^{\theta W_s} \,.$$

8.5 For a real constant θ and for $s < t$, show that the following holds:

$$\mathbb{E}[W_t^2\, \mathrm{e}^{\theta W_t} | \mathcal{F}_s] = \{(t-s) + (W_s + \theta(t-s))^2\}\, \mathrm{e}^{\frac{1}{2}\theta^2(t-s)} \mathrm{e}^{\theta W_s} \,.$$

8.6 Let $\mathbb{Q}$ be the equivalent measure given in Definition 8.1, and let $X_t = W_t + \theta t$. Show that $X_t^2 - t$ is a $\mathbb{Q}$-martingale without using the fact that X_t is a $\mathbb{Q}$-Brownian motion.

8.7 Let W_t be a Brownian motion with respect to a probability measure $\mathbb{P}$ for $0 \le t < \infty$. Take a constant $\theta \neq 0$. Let $\mathbb{Q}$ be a probability measure for which $X_t = W_t + \theta t$ is a Brownian motion for $0 \le t < \infty$. Show that $\mathbb{Q}$ is not equivalent to $\mathbb{P}$.

8.8 A possible discrete version of Girsanov's theorem can be regarded as a coin tossing problem using a biased coin. Assume that the probability of showing heads when we flip a given coin is equal to p, $\frac{1}{2} < p < 1$. If the outcome of a toss is heads, then the game under consideration pays a player the amount \$$A$, and if the outcome is tails it pays $-$\$$B$. What is a condition for A and B to have a fair game?

Chapter 9
The Reflection Principle of Brownian Motion

We investigate the reflection properties of Brownian motion. The results in this chapter will be used for the pricing of barrier options in Sect. 18.2. For the sake of simplicity of exposition we consider only one barrier problems.

9.1 The Reflection Property of Brownian Motion

Take $m > 0$ (the symbol m is chosen for 'maximum') and let $\{W_t\}_{t\geq 0}$ be a Brownian motion with respect to a probability measure $\mathbb{P}$. (The problem is symmetric with respect to m. For $m < 0$, we obtain corresponding equivalent results.) Let τ denote the *first hitting time* of a Brownian particle, i.e.,

$$\tau = \inf\{t \geq 0 : W_t = m\} .$$

See Fig. 9.1 where a sample Brownian path hits the level $m = 3$ and stops there. Consult Simulation 9.1.

In Fig. 9.2 we consider a Brownian path, Path A, and its reflection, Path B, after Path A hits a barrier of height m at $\tau < T$. Since a Brownian particle at the position m at time $\tau < T$ has the equal probability of being above or below m, we observe the following:

$$\mathbb{P}\{W_T < m, \tau < T\} = \mathbb{P}\{W_T > m, \tau < T\} \tag{9.1}$$

and

$$\mathbb{P}\{\tau < T\} = \mathbb{P}\{W_T < m, \tau < T\} + \mathbb{P}\{W_T > m, \tau < T\} . \tag{9.2}$$

G.H. Choe, *Stochastic Analysis for Finance with Simulations*, Universitext,
DOI 10.1007/978-3-319-25589-7_9

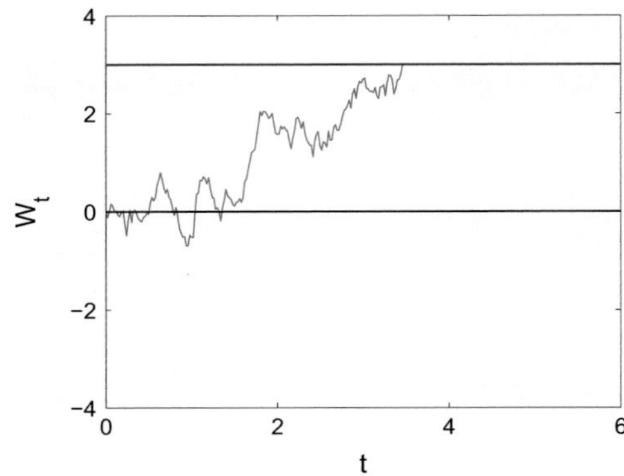

Fig. 9.1 A Brownian path is stopped when it hits a barrier $m = 3$

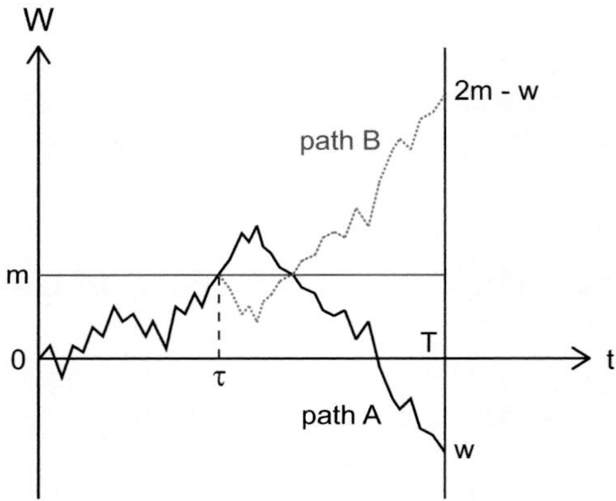

Fig. 9.2 A Brownian path is reflected after it hits a barrier

Substituting (9.1) into (9.2), we have

$$\mathbb{P}\{\tau < T\} = 2\,\mathbb{P}\{W_T > m, \tau < T\} .$$

Note that if $W_T > m$ then $\tau < T$, i.e.,

$$\mathbb{P}\{W_T > m, \tau < T\} = \mathbb{P}\{W_T > m\} .$$

Hence

$$\mathbb{P}\{\tau < T\} = 2\,\mathbb{P}\{W_T > m\} . \tag{9.3}$$

Thus

$$\mathbb{P}\{W_T > m | \tau < T\} = \frac{\mathbb{P}\{W_T > m, \tau < T\}}{\mathbb{P}\{\tau < T\}} = \frac{1}{2} .$$

In other words, once a Brownian particle is at the position m at time τ before T, then there is a fifty-fifty chance of being above or below m at T. Let

$$N(x) = \frac{1}{\sqrt{2\pi}} \int_{-\infty}^{x} \exp\left(-\frac{z^2}{2}\right) \mathrm{d}z$$

be the standard normal cumulative distribution function. Then we obtain the following result.

Lemma 9.1 *The first hitting time τ satisfies*

$$\mathbb{P}\{\tau < T\} = 2N\left(-\frac{m}{\sqrt{T}}\right) .$$

Hence the probability density function of τ is given by

$$f_\tau(T) = \frac{m}{T\sqrt{2\pi T}} \exp\left(-\frac{m^2}{2T}\right) .$$

Proof Note that (9.3) implies that

$$\begin{aligned}\mathbb{P}\{\tau < T\} &= \frac{2}{\sqrt{2\pi T}} \int_{m}^{\infty} \exp\left(-\frac{x^2}{2T}\right) \mathrm{d}x \\ &= \frac{2}{\sqrt{2\pi}} \int_{m/\sqrt{T}}^{\infty} \exp\left(-\frac{x^2}{2}\right) \mathrm{d}x .\end{aligned}$$

To obtain the pdf of τ, we take the derivative of $\mathbb{P}\{\tau < T\}$ with respect to T. □

9.2 The Maximum of Brownian Motion

Define

$$M_T = \max\{W_t : 0 \leq t \leq T\} .$$

Hence the values of the ordered pair (M_T, W_T) are distributed in the set

$$D = \{(m, w) : w \leq m, m \geq 0\} .$$

See Fig. 9.3. Note that the condition $\tau < T$ is equivalent to the condition $M_T > m$. Let $w < m$, where w will denote a value assumed by W_T. In Fig. 9.2 we observe that

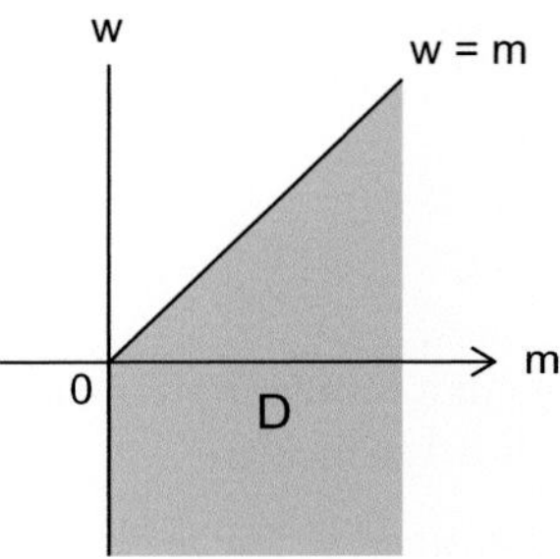

Fig. 9.3 Range of (M_T, W_T)

$W_T = w$ for Path A if and only if $W_T = 2m - w$ for Path B. Hence we make the following observations:

$$\mathbb{P}\{M_T > m, W_T < w\} = \mathbb{P}\{M_T > m, W_T > 2m - w\} \tag{9.4}$$

and

$$\mathbb{P}\{W_T < w\} = \mathbb{P}\{M_T > m, W_T < w\} + \mathbb{P}\{M_T < m, W_T < w\} . \tag{9.5}$$

Substituting (9.4) into (9.5), we have

$$\mathbb{P}\{M_T < m, W_T < w\} = \mathbb{P}\{W_T < w\} - \mathbb{P}\{M_T > m, W_T > 2m - w\} .$$

Since $w < m$, the condition $W_T > 2m - w$ implies that $M_T > m$. Hence

$$\mathbb{P}\{M_T < m, W_T < w\} = \mathbb{P}\{W_T < w\} - \mathbb{P}\{W_T > 2m - w\} . \tag{9.6}$$

Lemma 9.2

(i) *Let $w \le m$ and $m \ge 0$. The probability distribution of W_T below a barrier m is given by*

$$\mathbb{P}\{M_T < m, W_T < w\} = N\left(\frac{w}{\sqrt{T}}\right) - N\left(-\frac{2m - w}{\sqrt{T}}\right) .$$

(ii) *For $w \le m$ and $m \ge 0$, the joint probability density function of M_T and W_T is given by*

$$f_{M_T,W_T}(m, w) = \frac{2(2m - w)}{T\sqrt{2\pi T}} \exp\left(-\frac{(2m - w)^2}{2T}\right) .$$

Proof (i) Use (9.6). (ii) It suffices to take the second order partial derivative of the cumulative probability distribution $\mathbb{P}\{W_T < w, M_T < m\}$ with respect to w and m.

More precisely, we have

$$\frac{\partial}{\partial m} N\left(\frac{w}{\sqrt{T}}\right) = 0$$

and

$$\begin{aligned}
&\frac{\partial^2}{\partial w \partial m} N\left(-\frac{2m-w}{\sqrt{T}}\right) \\
&= \frac{\partial}{\partial w}\left[\frac{1}{\sqrt{2\pi}} \exp\left(-\frac{1}{2}\left(-\frac{2m-w}{\sqrt{T}}\right)^2\right)\left(-\frac{2}{\sqrt{T}}\right)\right] \\
&= \frac{1}{\sqrt{2\pi}} \exp\left(-\frac{1}{2}\left(-\frac{2m-w}{\sqrt{T}}\right)^2\right) \frac{\partial}{\partial w}\left[-\frac{1}{2}\left(-\frac{2m-w}{\sqrt{T}}\right)^2\right]\left(-\frac{2}{\sqrt{T}}\right).
\end{aligned}$$

Now note that

$$\frac{\partial}{\partial w}\left[-\frac{1}{2}\left(-\frac{2m-w}{\sqrt{T}}\right)^2\right] = \left(-\frac{1}{2}\right) 2\left(-\frac{2m-w}{\sqrt{T}}\right)\frac{1}{\sqrt{T}} = \frac{2m-w}{T}.$$

□

9.3 The Maximum of Brownian Motion with Drift

For $0 \leq t \leq T$ let $\widetilde{W}_t$ be a Brownian motion with respect to a probability measure $\widetilde{\mathbb{P}}$. For an arbitrary real number θ let $\widehat{W}_t$ be the Brownian motion with drift θ per unit time. More precisely, define

$$\widehat{W}_t = \widetilde{W}_t + \theta t, \quad 0 \leq t \leq T.$$

Define the maximum of $\widehat{W}_t$ by

$$\widehat{M}_T = \max_{0 \leq t \leq T} \widehat{W}_t.$$

Since $\widehat{M}_0 = 0$, we have $\widehat{M}_T \geq 0$ and $\widehat{M}_T \geq \widehat{W}_T$.

Theorem 9.1 *The joint probability density function $\widetilde{f}_{\widehat{M}_T, \widehat{W}_T}$ of $(\widehat{M}_T, \widehat{W}_T)$ with respect to $\widetilde{\mathbb{P}}$ is given by*

$$\widetilde{f}_{\widehat{M}_T, \widehat{W}_T}(m, w) = \frac{2(2m-w)}{T\sqrt{2\pi T}} \mathrm{e}^{\theta w - \frac{1}{2}\theta^2 T - \frac{1}{2T}(2m-w)^2}$$

in D, and 0 *elsewhere.*

Proof Define

$$\widehat{Z}_t = \mathrm{e}^{-\theta \widetilde{W}_t - \frac{1}{2}\theta^2 t} = \mathrm{e}^{-\theta \widehat{W}_t + \frac{1}{2}\theta^2 t}, \quad 0 \leq t \leq T ,$$

which is a martingale with respect to $\widetilde{\mathbb{P}}$. Note that there exists a probability measure $\widehat{\mathbb{P}}$ such that $\mathrm{d}\widehat{\mathbb{P}} = \widehat{Z}_T \, \mathrm{d}\widetilde{\mathbb{P}}$ for which $\widehat{W}_t$ is a Brownian motion. (See Theorem 8.1.) Then

$$\widehat{f}_{\widehat{M}_T, \widehat{W}_T}(m, w) = \frac{2(2m - w)}{T\sqrt{2\pi T}} \mathrm{e}^{-\frac{1}{2T}(2m-w)^2}$$

on D, and 0 elsewhere. Now the probability density function of $(\widehat{M}_T, \widehat{W}_T)$ with respect to $\widetilde{\mathbb{P}}$ is given by

$$\begin{aligned}
\widetilde{\mathbb{P}}\left\{\widehat{M}_T \leq m, \widehat{W}_T \leq w\right\} &= \mathbb{E}^{\widetilde{\mathbb{P}}}\left[\mathbf{1}_{\{\widehat{M}_T \leq m, \widehat{W}_T \leq w\}}\right] \\
&= \mathbb{E}^{\widehat{\mathbb{P}}}\left[\frac{1}{\widehat{Z}_T}\mathbf{1}_{\{\widehat{M}_T \leq m, \widehat{W}_T \leq w\}}\right] \\
&= \mathbb{E}^{\widehat{\mathbb{P}}}\left[\mathrm{e}^{\theta \widehat{W}_T - \frac{1}{2}\theta^2 T}\mathbf{1}_{\{\widehat{M}_T \leq m, \widehat{W}_T \leq w\}}\right] \\
&= \int_{-\infty}^{w}\int_{-\infty}^{m} \mathrm{e}^{\theta w - \frac{1}{2}\theta^2 T}\widehat{f}_{\widehat{M}_T, \widehat{W}_T}(m, w)\, \mathrm{d}m\, \mathrm{d}w .
\end{aligned}$$

By differentiating with respect to m and w, we have

$$\widetilde{f}_{\widehat{M}_T, \widehat{W}_T}(m, w) = \mathrm{e}^{\theta w - \frac{1}{2}\theta^2 T}\widehat{f}_{\widehat{M}_T, \widehat{W}_T}(m, w) .$$

□

Corollary 9.1 *We have*

$$\widetilde{f}_{\widehat{M}_T}(m) = \begin{cases} \dfrac{2}{\sqrt{2\pi T}}\mathrm{e}^{-\frac{1}{2T}(m-\theta T)^2} - 2\theta \mathrm{e}^{2\theta m} N\left(\dfrac{-m - \theta T}{\sqrt{T}}\right), & m \geq 0 , \\ 0 , & m < 0 . \end{cases}$$

Proof For $m < 0$, it is clear that $\widetilde{f}_{\widehat{M}_T}(m) = 0$. For $m \geq 0$ note that

$$\begin{aligned}
\widetilde{f}_{\widehat{M}_T}(m) &= \int_{-\infty}^{\infty} \widetilde{f}_{\widehat{M}_T, \widehat{W}_T}(m, w)\, \mathrm{d}w \\
&= \int_{-\infty}^{m} \frac{2(2m - w)}{T\sqrt{2\pi T}} \mathrm{e}^{\theta w - \frac{1}{2}\theta^2 T - \frac{1}{2T}(2m-w)^2} \mathrm{d}w \\
&= \mathrm{e}^{2\theta m} \frac{2}{T\sqrt{2\pi T}} \int_{-\infty}^{m} (2m - w)\mathrm{e}^{-\frac{1}{2T}(w-2m-\theta T)^2} \mathrm{d}w
\end{aligned}$$

$$
\begin{aligned}
&= e^{2\theta m} \frac{2}{T\sqrt{2\pi}} \int_{-\infty}^{(-m-\theta T)/\sqrt{T}} (-\sqrt{T}u - \theta T) e^{-\frac{1}{2}u^2} \, du \\
&= -\frac{2e^{2\theta m}}{\sqrt{T}\sqrt{2\pi}} \int_{-\infty}^{(-m-\theta T)/\sqrt{T}} u e^{-\frac{1}{2}u^2} \, du \\
&\quad - \frac{2\theta e^{2\theta m}}{\sqrt{2\pi}} \int_{-\infty}^{(-m-\theta T)/\sqrt{T}} e^{-\frac{1}{2}u^2} \, du \\
&= \frac{2e^{2\theta m}}{\sqrt{T}\sqrt{2\pi}} e^{-(m+\theta T)^2/2T} - 2\theta e^{2\theta m} N\left(\frac{-m-\theta T}{\sqrt{T}}\right)
\end{aligned}
$$

where we used the identity

$$
\theta w - \frac{1}{2}\theta^2 T - \frac{1}{2T}(2m-w)^2 = -\frac{1}{2T}(w-2m-\theta T)^2 + 2\theta m
$$

in the third equality, and we used the substitution

$$
u = \frac{w - 2m - \theta T}{\sqrt{T}}
$$

in the fourth equality. □

Corollary 9.2 *For $m \geq 0$, we have*

$$
\widetilde{\mathbb{P}}\{\widehat{M}_T \leq m\} = N\left(\frac{m-\theta T}{\sqrt{T}}\right) - e^{2\theta m} N\left(\frac{-m-\theta T}{\sqrt{T}}\right).
$$

Proof Note that

$$
\begin{aligned}
&\widetilde{\mathbb{P}}\{\widehat{M}_T \leq m\} \\
&= \int_{-\infty}^{m} \widetilde{f}_{\widehat{M}_T}(\mu) \, d\mu \\
&= \int_0^m \left[\frac{2}{\sqrt{2\pi T}} e^{-\frac{1}{2T}(\mu-\theta T)^2} - 2\theta e^{2\theta\mu} N\left(\frac{-\mu-\theta T}{\sqrt{T}}\right)\right] d\mu \\
&= \int_0^m \frac{2}{\sqrt{2\pi T}} e^{-\frac{1}{2T}(\mu-\theta T)^2} d\mu - \int_0^m 2\theta e^{2\theta\mu} N\left(\frac{-\mu-\theta T}{\sqrt{T}}\right) d\mu \\
&= I_1 - I_2
\end{aligned}
$$

where

$$\begin{aligned} I_1 &= \int_0^m \frac{2}{\sqrt{2\pi T}} \mathrm{e}^{-\frac{1}{2T}(\mu-\theta T)^2} \mathrm{d}\mu \\ &= 2\int_{-\theta\sqrt{T}}^{(m-\theta T)/\sqrt{T}} \frac{1}{\sqrt{2\pi}} \mathrm{e}^{-\frac{1}{2}z^2} \mathrm{d}z \\ &= 2\left\{N\left(\frac{m-\theta T}{\sqrt{T}}\right) - N\left(-\theta\sqrt{T}\right)\right\} \end{aligned}$$

and

$$\begin{aligned} I_2 &= \int_0^m 2\theta \mathrm{e}^{2\theta\mu} \int_{-\infty}^{(-\mu-\theta T)/\sqrt{T}} \frac{1}{\sqrt{2\pi}} \mathrm{e}^{-\frac{1}{2}z^2} \mathrm{d}z\, \mathrm{d}\mu \\ &= \int_{-\infty}^{(-m-\theta T)/\sqrt{T}} \int_0^m 2\theta \mathrm{e}^{2\theta\mu} \frac{1}{\sqrt{2\pi}} \mathrm{e}^{-\frac{1}{2}z^2} \mathrm{d}\mu\, \mathrm{d}z \\ &\quad + \int_{(-m-\theta T)/\sqrt{T}}^{-\theta\sqrt{T}} \int_0^{-\sqrt{T}z-\theta T} 2\theta \mathrm{e}^{2\theta\mu} \frac{1}{\sqrt{2\pi}} \mathrm{e}^{-\frac{1}{2}z^2} \mathrm{d}\mu\, \mathrm{d}z \\ &= (\mathrm{e}^{2\theta m} - 1) N\left(\frac{-m-\theta T}{\sqrt{T}}\right) \\ &\quad + \int_{(-m-\theta T)/\sqrt{T}}^{-\theta\sqrt{T}} (\mathrm{e}^{2\theta(-\sqrt{T}z-\theta T)} - 1) \frac{1}{\sqrt{2\pi}} \mathrm{e}^{-\frac{1}{2}z^2} \mathrm{d}z \quad \text{(See Fig. 9.4.)} \\ &= (\mathrm{e}^{2\theta m} - 1) N\left(\frac{-m-\theta T}{\sqrt{T}}\right) \\ &\quad + \int_{(-m-\theta T)/\sqrt{T}}^{-\theta\sqrt{T}} \frac{1}{\sqrt{2\pi}} \mathrm{e}^{2\theta(-\sqrt{T}z-\theta T)-\frac{1}{2}z^2} \mathrm{d}z \\ &\quad - \left\{N(-\theta\sqrt{T}) - N\left(\frac{-m-\theta T}{\sqrt{T}}\right)\right\} . \end{aligned}$$

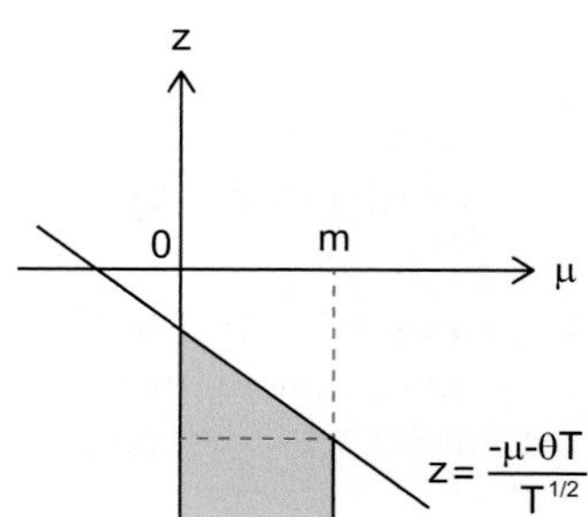

Fig. 9.4 Domain of integration for the double integral I_2

Since

$$2\theta(-\sqrt{T}z-\theta T)-\frac{1}{2}z^2=-\frac{1}{2}(z+2\theta\sqrt{T})^2\ ,$$

we have

$$\begin{aligned}I_2&=(\mathrm{e}^{2\theta m}-1)N\left(\frac{-m-\theta T}{\sqrt{T}}\right)+\left\{N(\theta\sqrt{T})-N\left(\frac{-m+\theta T}{\sqrt{T}}\right)\right\}\\&\quad-\left\{N(-\theta\sqrt{T})-N\left(\frac{-m-\theta T}{\sqrt{T}}\right)\right\}\ .\end{aligned}$$

Hence

$$\begin{aligned}&I_1-I_2\\&=2\left[N\left(\frac{m-\theta T}{\sqrt{T}}\right)-N\left(-\theta\sqrt{T}\right)\right]-(\mathrm{e}^{2\theta m}-1)N\left(\frac{-m-\theta T}{\sqrt{T}}\right)\\&\quad-\left[N(\theta\sqrt{T})-N\left(\frac{-m+\theta T}{\sqrt{T}}\right)\right]+\left[N(-\theta\sqrt{T})-N\left(\frac{-m-\theta T}{\sqrt{T}}\right)\right]\\&=2N\left(\frac{m-\theta T}{\sqrt{T}}\right)-N\left(-\theta\sqrt{T}\right)-\mathrm{e}^{2\theta m}N\left(\frac{-m-\theta T}{\sqrt{T}}\right)\\&\quad-N(\theta\sqrt{T})+N\left(\frac{-m+\theta T}{\sqrt{T}}\right)\\&=2N\left(\frac{m-\theta T}{\sqrt{T}}\right)-\mathrm{e}^{2\theta m}N\left(\frac{-m-\theta T}{\sqrt{T}}\right)-1+N\left(\frac{-m+\theta T}{\sqrt{T}}\right)\\&=2N\left(\frac{m-\theta T}{\sqrt{T}}\right)-\mathrm{e}^{2\theta m}N\left(\frac{-m-\theta T}{\sqrt{T}}\right)-N\left(\frac{m-\theta T}{\sqrt{T}}\right)\\&=N\left(\frac{m-\theta T}{\sqrt{T}}\right)-\mathrm{e}^{2\theta m}N\left(\frac{-m-\theta T}{\sqrt{T}}\right)\ ,\end{aligned}$$

where we used the identity $N(-x)+N(x)=1$ for the third and the fourth equalities. □

9.4 Computer Experiments

Simulation 9.1 (The First Hitting Time)

We plot a sample Brownian path hitting a barrier $m=3$. In the beginning of the program we set `hitting_time` $=N+1$ where N is the number of subintervals in the partition of the time interval $[0, T]$ just in case that the generated sample Brownian path does not hit the level $m = 3$ before or at T. See Fig. 9.1.

```
T = 6;
m = 3;
N = 200;
hitting_time = N+1;
dt = T/N;
t = 0:dt:T;
W = zeros(1,N+1);
W(1,1) = 0;
dW = sqrt(dt)*randn(1,N);

for j = 1:N;
    W(1,j+1) = W(1,j) + dW(1,j);
    if ( (W(1,j+1) >= m) )
        W(1,j+1) = m;
        hitting_time = j+1;
        break;
    end
end
plot(0:dt:(hitting_time-1)*dt,W(1,1:hitting_time));
```

Part IV
Itô Calculus

Chapter 10
The Itô Integral

We define the Itô integral of a stochastic process and investigate its properties. To define a Riemann–Stieltjes type integral $\int_0^T f(t)\,\mathrm{d}\alpha(t)$ using a function $\alpha : [0, T] \to \mathbb{R}$ as an integrator, we need the condition that the variation of α is bounded. (For the definition of variation, see Sect. A.3.) However, a sample path of a Brownian motion is of unbounded variation since the growth rate of δW is approximately equal to $\sqrt{\delta t}$, which is very large compared with δt as $\delta t \downarrow 0$. Therefore a Brownian sample path cannot be used as an integrator in a definition of a Riemann–Stieltjes type integral. K. Itô's idea is to take a suitable average over all possible Brownian paths. This idea will be explained gradually since it requires a considerable amount of preparation. For an elementary introduction to the Itô integral, see [82, 102].

10.1 Definition of the Itô Integral

If a stochastic process $\{f_t(\omega)\}_{t\geq 0}$ is measurable with respect to the filtration $\{\mathcal{F}_t\}$ for every $t \geq 0$, then the Itô integral of a stochastic process $f_t(\omega) = f(t, \omega)$ is integrated with respect to an integrator given by a Brownian motion W_t, and the resulting integral is a random variable. More precisely, if we let $t_i = \frac{T}{n} i$ we have

$$\int_0^T f_t(\omega)\,\mathrm{d}W_t(\omega) = \lim_{n\to\infty} \sum_{i=0}^{n-1} f_{t_i}(\omega) \left(W_{t_{i+1}}(\omega) - W_{t_i}(\omega)\right)$$

where the limit is defined in $L^2(\Omega, \mathbb{P})$ space. In other words,

$$\mathbb{E}\left[\left|\sum_{i=0}^{n-1} f_{t_i} \left(W_{t_{i+1}} - W_{t_i}\right) - \int_0^T f_t\,\mathrm{d}W_t\right|^2\right] \to 0$$

G.H. Choe, *Stochastic Analysis for Finance with Simulations*, Universitext,
DOI 10.1007/978-3-319-25589-7_10

as $n \to \infty$. Note that in the Itô integral, to approximate $f_t(\omega)$ on a subinterval $[t_i, t_{i+1}]$ we use the value at the left endpoint t_i. This means the Itô integral possesses many useful properties for financial mathematics.

Let $f_t(\omega) = f(t, \omega)$ be a stochastic process that is adapted to a filtration $\{\mathcal{F}_t\}_{0 \le t < \infty}$ such that

$$\mathbb{E}\left[\int_0^\infty |f_t(\omega)|^2 \, dt\right] < \infty .$$

Here $f_t(\omega)$ is regarded as a function $f : t \mapsto f(t, \omega)$, and is continuous for almost every ω, and

$$\int_0^\infty |f_t(\omega)|^2 \, dt$$

is a Riemann integral for almost every ω where the continuity holds.

Let V denote the collection of all such stochastic processes. It is a vector space with a norm defined by

$$||f||_V^2 = \mathbb{E}\left[\int_0^\infty |f_t(\omega)|^2 \, dt\right] .$$

On the right-hand side we write f_t instead of f to emphasize the fact that it is an integral with respect to t.

Definition 10.1 (Simple Process) Let $0 = t_0 < t_1 < \cdots < t_n = T$ be a partition of the time interval $[0, T]$. Suppose that for $0 \le i \le n-1$ there exist $\mathcal{F}_{t_i}$-measurable random variables $\zeta_i \in L^2(\Omega)$ such that a stochastic process $\{f_t\}$ is of the form

$$f_t(\omega) = \sum_{i=0}^{n-1} \zeta_i(\omega) \mathbf{1}_{[t_i, t_{i+1})}(t) .$$

Then f_t is called a *simple stochastic process.* (See Fig. 10.1.) The set of all simple stochastic processes is denoted by $\mathcal{H}_0^2$. Since $f \in \mathcal{H}_0^2$ satisfies

$$||f||_V^2 = \sum_{i=0}^{n-1} \mathbb{E}\left[\, |\zeta_i|^2\right] (t_{i+1} - t_i) < \infty ,$$

we have $f \in V$. In other words, $\mathcal{H}_0^2$ is a subspace of V.

Definition 10.2 (Itô Integral) For a simple process

$$f_t(\omega) = \sum_{i=0}^{n-1} \zeta_j(\omega) \mathbf{1}_{[t_i, t_{i+1})}(t) \in \mathcal{H}_0^2$$

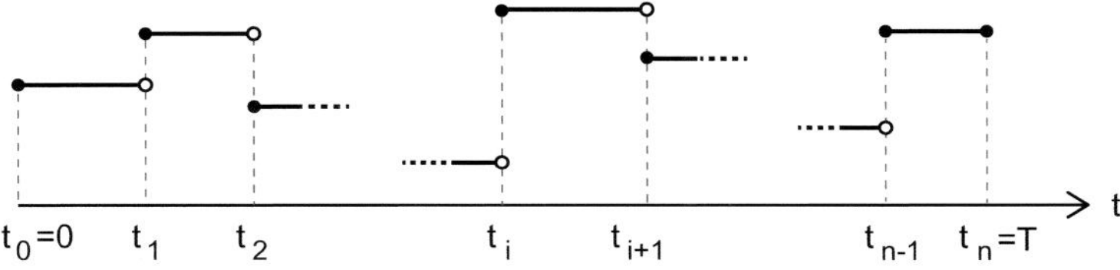

Fig. 10.1 A sample path of a simple process

we define its *Itô integral* $I(f)$ by

$$I(f)(\omega) = \sum_{i=0}^{n-1} \zeta_i(\omega) \left(W_{t_{i+1}}(\omega) - W_{t_i}(\omega) \right) .$$

Remark 10.1

(i) An Itô integral $I(f)$ is a random variable.
(ii) To avoid a subscript to a subscript as in W_{t_k}, which is rather difficult to read, we sometimes write $W(t_k)$. Similarly, sometimes $f(t)$ may denote a stochastic process f_t, not a deterministic function of t.
(iii) From time to time we write $\delta W_k = W_{t_{k+1}} - W_{t_k}$ and $\delta t_k = t_{k+1} - t_k$.
(iv) A simple process $f_t = \sum_{i=0}^{n-2} \zeta_j \mathbf{1}_{[t_i,t_{i+1})}(t) + \zeta_{n-1} \mathbf{1}_{[t_{n-1},t_n]}(t)$ is regarded as being defined over $[0, \infty)$ by taking $f_t = 0$ for $t \geq T$.

Theorem 10.1 (Itô Isometry) *If $f \in \mathcal{H}_0^2$, then $I(f) \in L^2(\Omega, \mathbb{P})$ and $I : \mathcal{H}_0^2 \to L^2(\Omega, \mathbb{P})$ is a norm-preserving linear transformation. In other words,*

$$\mathbb{E}\left[|I(f)|^2 \right] = \mathbb{E}\left[\int_0^\infty |f_t|^2 \mathrm{d}t \right] .$$

Proof The proof that I is a linear transformation is trivial, and omitted. If a step stochastic process f_t is of the form

$$f_t(\omega) = \sum_{i=0}^{n-1} \zeta_i(\omega) \, \mathbf{1}_{[t_i,t_{i+1})}(t) \in \mathcal{H}_0^2 ,$$

then

$$\begin{aligned} |I(f)|^2 &= \sum_{j=0}^{n-1} \sum_{k=0}^{n-1} \zeta_j \zeta_k \, \delta W_j \delta W_k \\ &= \sum_{k=0}^{n-1} |\zeta_k|^2 (\delta W_k)^2 + 2 \sum_{j<k} \zeta_j \zeta_k \, \delta W_j \delta W_k . \end{aligned}$$

Since ζ_k and δW_k are independent and $\mathbb{E}[(\delta W_k)^2] = \delta t_k$, we have

$$\mathbb{E}[\,|\zeta_k|^2(\delta W_k)^2] = \mathbb{E}[\,|\zeta_k|^2]\,\mathbb{E}[(\delta W_k)^2] = \mathbb{E}[\,|\zeta_k|^2]\,\delta t_k\ .$$

If $j < k$, then $\zeta_j\zeta_k\,\delta W_j$ and δW_k are independent, and hence

$$\mathbb{E}[\zeta_j\zeta_k\,\delta W_j\,\delta W_k] = \mathbb{E}[\zeta_j\zeta_k\,\delta W_j]\,\mathbb{E}[\delta W_k] = 0\ .$$

Thus

$$\mathbb{E}[\,|I(f)|^2] = \sum_{k=0}^{n-1}\mathbb{E}[\,|\zeta_k|^2]\delta t_k < \infty\ . \tag{10.1}$$

Therefore, $I(f) \in L^2(\Omega)$.

On the other hand, since

$$|f_t|^2 = \sum_{j=0}^{n-1}\sum_{k=0}^{n-1}\zeta_j\zeta_k\,\mathbf{1}_{[t_j,t_{j+1})}(t)\mathbf{1}_{[t_k,t_{k+1})}(t) = \sum_{k=0}^{n-1}|\zeta_k|^2\,\mathbf{1}_{[t_k,t_{k+1})}(t)\ ,$$

we have

$$\int_0^\infty |f_t|^2\mathrm{d}t = \sum_{k=0}^{n-1}|\zeta_k|^2\,\Delta t_k$$

and finally obtain

$$\mathbb{E}\left[\int_0^\infty |f_t|^2\mathrm{d}t\right] = \sum_{k=0}^{n-1}\mathbb{E}[\,|\zeta_k|^2]\,\Delta t_k\ . \tag{10.2}$$

Now we use (10.1) and (10.2) to complete the proof. □

Definition 10.3 (Extension of Itô Integral) Since the transformation $I : \mathcal{H}_0^2 \to L^2(\Omega, \mathbb{P})$ is linear and continuous, and since $L^2(\Omega, \mathbb{P})$ is a complete metric space, by Corollary A.1 the domain of I can be extended from $\mathcal{H}_0^2$ to a set, denoted by $\mathcal{H}^2$, which contains $\mathcal{H}_0^2$ as a dense subset. (When we emphasize the finite time interval $[0, T]$, we write $\mathcal{H}_T^2$ in place of $\mathcal{H}^2$.) Then the integral $I(f) \in L^2(\Omega, \mathbb{P})$ of a stochastic process $f(t, \omega)$ which is approximated by a sequence of simple processes $f_n(t, \omega)$ is determined by $I(f) = \lim_{n\to\infty} I(f_n)$ where the limit is taken in the L^2-sense. In other words,

$$\lim_{n\to\infty}\mathbb{E}\big[\,|I(f) - I(f_n)|^2\big] = 0\ .$$

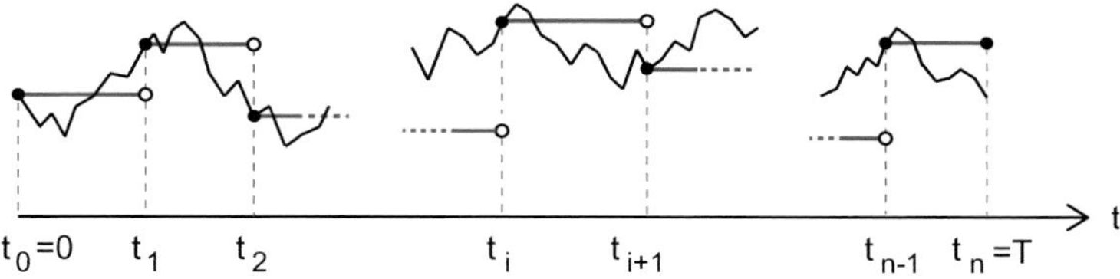

Fig. 10.2 Approximation of a general stochastic process by a simple process

Then $I(f)$ is called the *Itô integral* and is denoted by

$$\int_0^\infty f_t \, \mathrm{d}W_t \ .$$

Remark 10.2

(i) It is clear that Theorem 10.1 holds for $f \in \mathcal{H}^2$.
(ii) To approximate $f_t \in \mathcal{H}^2$ we use a simple process

$$f_n(t,\omega) = \sum_{i=0}^{n-1} f_{t_i}(\omega)\mathbf{1}_{[t_i,t_{i+1})}(t) \in \mathcal{H}_0^2 \ ,$$

and $I(f)$ is the L^2-limit of

$$I(f_n) = \sum_{i=0}^{n-1} f_{t_i}\left(W_{t_{i+1}} - W_{t_i}\right) \ .$$

(See Fig. 10.2.) Note that we choose the *left endpoint* t_i in each interval $[t_i, t_{i+1})$. This rule makes the Itô integral a martingale. See Theorem 10.4.

Theorem 10.2 (Itô Isometry) *For stochastic processes $f, g \in \mathcal{H}^2$ we have*

$$\mathbb{E}[I(f)I(g)] = \mathbb{E}\left[\int_0^\infty f_t \, g_t \, \mathrm{d}t\right] \ .$$

Using the inner products defined on $\mathcal{H}^2$ and $L^2(\Omega, \mathbb{P})$, we observe that the equation is equivalent to the relation $(I(f), I(g))_{L^2} = (f, g)_{\mathcal{H}^2}$.

Proof Using the inner products on the spaces $\mathcal{H}^2$ and $L^2(\Omega, \mathbb{P})$, we obtain

$$\begin{aligned}
& (f,f)_{\mathcal{H}^2} + (g,g)_{\mathcal{H}^2} + 2\,(f,g)_{\mathcal{H}^2} \\
&= (f+g, f+g)_{\mathcal{H}^2} \\
&= (I(f+g), I(f+g))_{L^2} \qquad \text{(by Theorem 10.1)}
\end{aligned}$$

$$\begin{aligned}&= (I(f) + I(g), I(f) + I(g))_{L^2}\\&= (I(f), I(f))_{L^2} + (I(g), I(g))_{L^2} + 2\,(I(f), I(g))_{L^2}\\&= (f,f)_{\mathcal{H}^2} + (g,g)_{\mathcal{H}^2} + 2\,(I(f), I(g))_{L^2}\,. \qquad \text{(by Theorem 10.1)}\end{aligned}$$

Hence $(f,g)_{\mathcal{H}^2} = (I(f), I(g))_{L^2}$. □

Definition 10.4 If $\{f_t\}_{t\geq 0}$ is adapted to $\{\mathcal{F}_t\}_{t\geq 0}$, then $\{\mathbf{1}_{[0,T]}(t)f_t\}_{t\geq 0}$ is also adapted to $\{\mathcal{F}_t\}_{t\geq 0}$. For arbitrary $T > 0$ we define the Itô integral on a finite interval $[0, T]$ by

$$\int_0^T f_t\,\mathrm{d}W_t = \int_0^\infty \mathbf{1}_{[0,T]}(t)f_t\,\mathrm{d}W_t\,.$$

Example 10.1 For $f_t = 1$, we have

$$\int_0^T \mathrm{d}W_t = W_T - W_0 = W_T\,.$$

Example 10.2 If $g(0) = 0$ then the Riemann integral satisfies

$$\int_0^T g(t)g'(t)\,\mathrm{d}t = \frac{1}{2}g(T)^2\,,$$

however, for the Itô integral we have

$$\int_0^T W_t\,\mathrm{d}W_t = \frac{1}{2}W_T^2 - \frac{1}{2}T\,.$$

Proof Put $f(t) = \mathbf{1}_{[0,T]}(t)W(t) \in \mathcal{H}^2$. Then

$$\int_0^T W_t\,\mathrm{d}W_t = \int_0^\infty f_t\,\mathrm{d}W_t\,.$$

Take $t_i = i \times \frac{T}{n}$, and partition the interval $[0, T]$ by $0 = t_0 < t_1 < \cdots < t_n = T$. Define

$$f_n(t) = \sum_{i=0}^{n-1} \mathbf{1}_{[t_i,t_{i+1})}(t)\,W(t_i) \in \mathcal{H}_0^2\,.$$

Since

$$\mathbb{E}\left[\int_0^\infty |f(t)-f_n(t)|^2 \mathrm{d}t\right] = \sum_{i=0}^{n-1}\int_{t_i}^{t_{i+1}} \mathbb{E}[\,|W(t)-W(t_i)|^2]\,\mathrm{d}t$$
$$= \sum_{i=0}^{n-1}\int_{t_i}^{t_{i+1}} (t-t_i)\,\mathrm{d}t$$
$$= \frac{1}{2}\sum_{i=0}^{n-1}(t_{i+1}-t_i)^2 = \frac{T^2}{2}\frac{1}{n} \to 0\,,$$

$f_n(t)$ converges to $f(t)$ as $n \to \infty$. Now we have

$$I(f_n) = \sum_{i=0}^{n-1} W(t_i)\{W(t_{i+1})-W(t_i)\}$$
$$= \frac{1}{2}\sum_{i=0}^{n-1}\{W(t_{i+1})^2 - W(t_i)^2\} - \frac{1}{2}\sum_{i=0}^{n-1}\{W(t_{i+1})-W(t_i)\}^2$$
$$= \frac{1}{2}W(T)^2 - \frac{1}{2}\sum_{i=0}^{n-1}\{W(t_{i+1})-W(t_i)\}^2\,,$$

which converges to $\frac{1}{2}W(T)^2 - \frac{1}{2}T$ in the L^2-sense by Theorem 7.4. □

As we have seen in Theorem 10.2, $I : \mathcal{H}^2 \to L^2(\Omega, \mathbb{P})$ is norm-preserving. Since $\mathcal{H}_0^2$ is dense in $\mathcal{H}^2$, most properties of I on $\mathcal{H}_0^2$ are extended to $\mathcal{H}^2$.

Remark 10.3 In summary, for stochastic processes f_t, g_t, the following holds:

(i) For a, b constant,

$$\int_0^t (af_u + bg_u)\,\mathrm{d}W_u = a\int_0^t f_u \mathrm{d}W_u + b\int_0^t g_u \mathrm{d}W_u\,.$$

(ii) For every $t \geq 0$,

$$\mathbb{E}\left[\,\left|\int_0^t f_u \mathrm{d}W_u\right|^2\right] = \mathbb{E}\left[\int_0^t |f_u|^2 du\right]\,.$$

Theorem 10.3 *For $f \in L^2[a,b]$, the Itô integral $\int_a^b f(t)\mathrm{d}W_t$ is a normally distributed random variable with mean 0 and variance $||f||^2 = \int_a^b |f(t)|^2 \mathrm{d}t$.*

Proof The statement holds true if f is a step function. For an arbitrary $f \in L^2[a,b]$, we take a sequence of step functions f_n converging to f in $L^2[a,b]$ and apply Theorem 4.2. □

10.2 The Martingale Property of the Itô Integral

Theorem 10.4 *Let $(\Omega, \mathbb{P}, \{\mathcal{F}_t\}_{t\geq 0})$ be a probability space with a filtration and let $\{W_t\}_{t\geq 0}$ be a $(\mathbb{P}, \{\mathcal{F}_t\}_{t\geq 0})$-Brownian motion. Let $g(t,\omega)$, $t \geq 0$, be an $\mathcal{F}_t$-predictable process such that $\mathbb{E}[g^2(t,\omega)] < \infty$, then a stochastic process*

$$M_t = \int_0^t g(s,\omega)\,\mathrm{d}W_s$$

is a martingale.

Proof First, we show that the given equation holds true for $\mathcal{H}_0^2$. Then we use the fact that the linear transformation defined by taking conditional expectation is continuous. □

Corollary 10.1 *The Itô integral has the following property:*

$$\mathbb{E}\left[\int_0^t f(s,\omega)\,\mathrm{d}W_s\right] = 0$$

where the expectation means the Lebesgue integral on Ω.

Proof Put

$$M_t = \int_0^t f(s,\omega)\,\mathrm{d}W_s\ .$$

Then

$$M_0 = \int_0^0 f(s,\omega)\,\mathrm{d}W_s = 0\ .$$

Since $\{M_t\}$ is a martingale, $\mathbb{E}[M_t]$ is constant, thus $\mathbb{E}[M_t] = \mathbb{E}[M_0] = 0$. □

Example 10.3 From the result in Example 10.2 we have

$$0 = \mathbb{E}\left[\int_0^T W_t\,\mathrm{d}W_t\right] = \frac{1}{2}\mathbb{E}\left[W_T^2\right] - \frac{1}{2}T\ ,$$

and hence $\mathbb{E}[W_T^2] = T$, which coincides with the fact that the variance of W_T is equal to T.

Example 10.4 The geometric Brownian motion

$$S_t = S_0 \mathrm{e}^{(\mu - \frac{1}{2}\sigma^2)t + \sigma W_t} ,$$

which is a model for stock price movement, satisfies

$$S_t - S_0 = \int_0^t \mu S_u \, \mathrm{d}u + \int_0^t \sigma S_u \, \mathrm{d}W_u .$$

Since

$$\mathbb{E}\left[\int_0^t S_u \, \mathrm{d}W_u\right] = 0 ,$$

we have

$$\mathbb{E}[S_t] - S_0 = \int_0^t \mu \, \mathbb{E}[S_u] \, \mathrm{d}u .$$

By differentiating both sides and putting $g(t) = \mathbb{E}[S_t]$ we obtain $g'(t) = \mu \, g(t)$, whose solution is $g(t) = S_0 \mathrm{e}^{\mu t}$. This result coincides with the result in Example 7.1. For $S_0 = 1$ and $\mu = \frac{1}{2}\sigma^2$ we have $\mathbb{E}\left[\mathrm{e}^{\sigma W_t}\right] = \mathrm{e}^{\frac{1}{2}\sigma^2 t}$.

10.3 Stochastic Integrals with Respect to a Martingale

Let $\{M_n\}_{n\geq 0}$ be a discrete time martingale and $\{H_n\}_{n\geq 0}$ a discrete time predictable process. If H_n is bounded, or if both H_n and M_n are L^2-integrable, then $X_0 = 0$ and

$$X_n = \sum_{k=1}^{n} H_k (M_k - M_{k-1})$$

for $n \geq 1$, defines a discrete time martingale, and we may informally write

$$X_n = \int H_k \mathrm{d}M_k .$$

Now we consider the continuous time case. Let M_t be a continuous martingale. As in the special case when $M_t = W_t$, using M_t as an integrator, we can define a stochastic integral

$$X_t = \int_0^t f(s, M_s) \, \mathrm{d}M_s$$

and derive the properties corresponding to the standard Itô integral. First, we define the stochastic integral for a simple process

$$f_t(\omega) = \sum_{i=0}^{n-1} \zeta_i(\omega)\, \mathbf{1}_{[t_i,t_{i+1})}(t)$$

by

$$\int_0^T f_s \,\mathrm{d}M_s = \sum_{i=0}^{n-1} \zeta_i \,(M_{t_{i+1}} - M_i) \ ,$$

then extend it to the general case by taking the limit and obtain X_t. It is known that $\{X_t\}_{t\geq 0}$ is also a martingale. More precisely, take a predictable process $f(t,\omega) \in L^2([0,T]\times\Omega)$ such that

$$\mathbb{E}\left[\int_0^T |f_t|^2 \,\mathrm{d}\langle M\rangle_t\right] < \infty \ .$$

Then

$$X_t = \int_0^t f_s \,\mathrm{d}M_s \ , \quad 0 \leq t \leq T \ ,$$

is a martingale and the following equality holds:

$$\mathbb{E}[X_t^2] = \mathbb{E}\left[\int_0^t |f_s|^2 \,\mathrm{d}\langle M\rangle_s\right] \ .$$

Furthermore, the compensator of X_t^2, defined in Definition 6.7, is given by

$$\langle X\rangle_t = \int_0^t |f_s|^2 \,\mathrm{d}\langle M\rangle_s \ .$$

Example 10.5 Let W_t be a Brownian motion, and take an adapted process $g(s,\omega) \in L^2([0,T],\Omega)$. Consider the martingale

$$M_t = \int_0^t g_s \,\mathrm{d}W_s \ , \quad 0 \leq t \leq T \ .$$

The compensator of M_t^2 is given by

$$\langle M\rangle_t = \int_0^t |g_s|^2 \,\mathrm{d}s \ .$$

Hence $f \in L^2_{\text{pred}}([0,T] \times \Omega)$ with the condition that

$$\mathbb{E}\left[\int_0^T |f_t|^2 \mathrm{d}\langle M\rangle_t\right] < \infty$$

if and only if

$$\mathbb{E}\left[\int_0^T |f_t|^2 |g_t|^2 \mathrm{d}t\right] < \infty .$$

Furthermore, we have

$$\int_0^T f_t \, \mathrm{d}M_t = \int_0^T f_t \, g_t \, \mathrm{d}W_t .$$

For more details see [58].

Theorem 10.5 *Let $\{X_t\}_{t\geq 0}$ be a continuous martingale. Then the process $\int_0^t f(u, X_u)\mathrm{d}X_u$ is a martingale.*

Proof Here is a sketch of the proof. For $0 \leq s < t$, we have

$$\begin{aligned}
&\mathbb{E}\left[\int_0^t f(u, X_u) \, \mathrm{d}X_u \,\middle|\, \mathcal{F}_s\right] \\
&= \mathbb{E}\left[\int_0^s f(u, X_u) \, \mathrm{d}X_u \,\middle|\, \mathcal{F}_s\right] + \mathbb{E}\left[\int_s^t f(u, X_u) \, \mathrm{d}X_u \,\middle|\, \mathcal{F}_s\right]
\end{aligned}$$

and the second term on the right is approximated by

$$\begin{aligned}
\mathbb{E}\left[\int_s^t f(u, X_u)\mathrm{d}X_u \,\middle|\, \mathcal{F}_s\right] &\approx \mathbb{E}\left[\sum_{i=0}^{n-1} f(u_i, X_{u_i})(X_{u_{i+1}} - X_{u_i}) \,\middle|\, \mathcal{F}_s\right] \\
&= \sum_{i=0}^{n-1} \mathbb{E}\left[f(u_i, X_{u_i})(X_{u_{i+1}} - X_{u_i}) \,\middle|\, \mathcal{F}_s\right] \\
&= \sum_{i=0}^{n-1} \mathbb{E}\left[\mathbb{E}\left[f(u_i, X_{u_i})(X_{u_{i+1}} - X_{u_i}) \,\middle|\, \mathcal{F}_{u_i}\right] \,\middle|\, \mathcal{F}_s\right] \\
&= \sum_{i=0}^{n-1} \mathbb{E}\left[f(u_i, X_{u_i})(X_{u_i} - X_{u_i}) \,\middle|\, \mathcal{F}_s\right] \\
&= 0
\end{aligned}$$

where $s = u_0 < \cdots < u_i < u_{i+1} < \cdots < u_n = t$. □

10.4 The Martingale Representation Theorem

Let $\{\mathcal{F}_t\}_{t\geq 0}$ be a filtration on a probability space $(\Omega, \mathcal{F}, \mathbb{P})$. Let $\{M_t\}_{t\geq 0}$ be a martingale adapted to $\{\mathcal{F}_t\}$. If $\mathbb{E}[M_t^2] < \infty$ for every $t \geq 0$, then $\{M_t\}_{t\geq 0}$ is called a *square-integrable* martingale.

The following is a special case of the Martingale Representation Theorem when the martingale under consideration is given by Brownian motion.

Theorem 10.6 (Martingale Representation Theorem) *Let $\{W_t\}_{t\geq 0}$ be a $\mathbb{P}$-Brownian motion, and let $\{\mathcal{F}_t\}_{t\geq 0}$ be the filtration generated by $\{W_t\}$. If a square-integrable process $\{M_t\}_{t\geq 0}$ is adapted to $\{\mathcal{F}_t\}_{t\geq 0}$, and it is a $\mathbb{P}$-martingale, then there exists an $\{\mathcal{F}_t\}$-predictable process $\{\alpha_t\}_{t\geq 0}$ satisfying*

$$M_t = M_0 + \int_0^t \alpha_s \, \mathrm{d}W_s$$

with probability 1.

Remark 10.4 Theorem 10.6 is the converse of Theorem 10.4. Its conclusion can be written as

$$\mathrm{d}M_t = \alpha_t \, \mathrm{d}W_t \; .$$

In other words, if M_t is a martingale, then $\mathrm{d}M_t$ has no $\mathrm{d}t$-term, and it looks as if M_t were differentiable with respect to W_t with its derivative equal to α_t.

Proof Since M_0 is $\mathcal{F}_0$-measurable, it is constant with probability 1. It suffices to prove the statement for $M_0 = 0$ since we can prove the above result for N_t using $N_t = M_t - M_0$ instead of M_t if necessary. By the properties of a martingale, we can easily see that $\mathbb{E}[M_t] = 0$ and $\mathbb{E}[M_T|\mathcal{F}_t] = M_t$.

Let V be the set of square-integrable stochastic processes adapted to the given filtration. For $X_t, Y_t \in V$ define

$$(X_t, Y_t)_V = \mathbb{E}\left[\int_0^T X_s Y_s \mathrm{d}s\right] ,$$

then $(\cdot\,,\,\cdot)_V$ is an inner product on a Hilbert space V. If we let

$$L_0^2(\Omega, \mathcal{F}_T, \mathbb{P}) = \{Z \in L^2(\Omega, \mathcal{F}_T, \mathbb{P}) : \mathbb{E}[Z] = 0\} \; ,$$

then the linear transformation $I : V \to L_0^2(\Omega, \mathcal{F}_T, \mathbb{P})$ defined by

$$I(\{X_t\}_{t\geq 0}) = \int_0^T X_s \mathrm{d}W_s$$

preserves distance. Now it remains to show that I is surjective. Then there would exist $\{\alpha_t\}_{0\le t\le T} \in V$ such that

$$M_T = \int_0^T \alpha_s \mathrm{d}W_s \ .$$

Hence

$$M_t = \mathbb{E}[M_T|\mathcal{F}_t] = \mathbb{E}\left[\int_0^T \alpha_s \mathrm{d}W_s \middle| \mathcal{F}_t\right] = \int_0^t \alpha_s \mathrm{d}W_s$$

and the proof is complete. For the details consult [54, 74]. □

Example 10.6 Let $X_t = \int_0^t a(s)\mathrm{d}W_s$ where $a(t)$ is a deterministic function. Then X_t is a martingale. For, if we apply the Itô formula to X_t^2, then

$$\mathrm{d}(X_t^2) = X_t\mathrm{d}X_t + X_t\mathrm{d}X_t + \mathrm{d}X_t\mathrm{d}X_t = 2X_t\mathrm{d}X_t + a(t)^2\mathrm{d}t$$

and $\mathrm{d}(X_t^2) - a(t)^2\mathrm{d}t = 2X_t\mathrm{d}X_t$. Hence $X_t^2 - \int_0^t a(s)^2\mathrm{d}s$ is a martingale.

10.5 Computer Experiments

Simulation 10.1 (Itô Integral)

We generate a single sample path of $\int_0^t W_s\mathrm{d}W_s$ and compare it with the theoretical formula $\frac{1}{2}W_t^2 - \frac{1}{2}t$. Recall that in the definition of the Itô integral we take the left endpoint from the subinterval $[t_i, t_{i+1}]$ to evaluate the integrand. In the following program the partition of the time interval $[0, T]$ is represented by an array `t = 0:dt:T` of length $N + 1$ in such a way that

$$\mathtt{t}(1) = t_0 = 0, \mathtt{t}(2) = t_1, \ldots, \mathtt{t}(i+1) = t_i, \ldots, \mathtt{t}(N+1) = t_N = T \ .$$

The array `Integral` represents the partial sum $\sum_{j=0}^{i} W_{t_j}\mathrm{d}W_{t_j}$, $0 \le i \le N-1$, as the time progresses while `Exact` is the exact formula $\frac{1}{2}W_t^2 - \frac{1}{2}t$. For the output see Fig. 10.3.

```
T= 3.0;
N = 300;
dt = T/N;
t = 0:dt:T;
dW = sqrt(dt)*randn(1,N);
W = zeros(1,N+1);
Integral = zeros(1,N+1);
Exact = zeros(1,N+1);

for i = 1:N
```

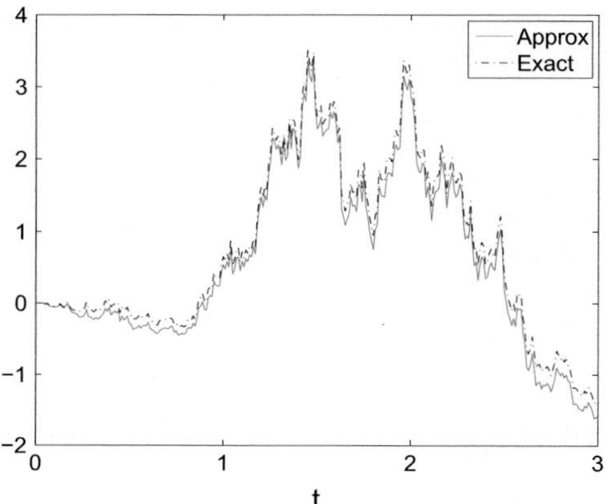

Fig. 10.3 Simulation of the Itô integral $\int_0^t W_s \mathrm{d}W_s$ and the exact answer $\frac{1}{2}W_t^2 - \frac{1}{2}t$

```
    W(i+1) = W(i) + dW(i);
    Integral(i+1) = Integral(i) + W(i)*dW(i); % Take the left endpoint.
    Exact(i+1)=W(i+1)^2/2 - i*dt/2;
end
plot(t,Integral,'r-',t,Exact,'k-.');
xlabel('t');
hlegend=legend('approx','exact');
```

Simulation 10.2 (Itô Integral)

We generate $M = 20$ sample paths of the stochastic process $\int_0^t W_s \mathrm{d}W_s$, $0 \le t \le T$. See Fig. 10.4 where the straight line $y = -\frac{1}{2}t$ is also plotted to check whether $\frac{1}{2}W_t^2 - \frac{1}{2}t \ge \frac{1}{2}t$. For small values of N the sample paths might go below the straight line.

```
T = 3.0;
M = 10;
N = 500;
dt = T/N;
t  = 0:dt:T;

dW = sqrt(dt)*randn(M,N);
W = [zeros(M,1), cumsum(dW,2)];
Integral = [zeros(M,1), cumsum(W(:,1:N).*dW,2)];
plot(t,Integral)
hold on
plot(t,-0.5*t)
xlabel('t');
```

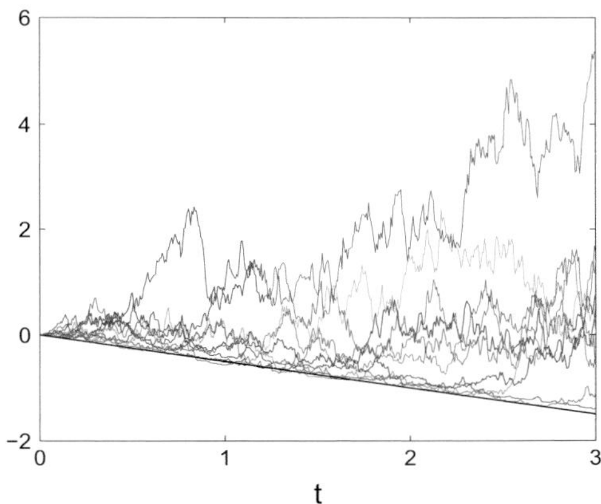

Fig. 10.4 Sample paths of the stochastic process $\int_0^t W_s \mathrm{d}W_s$

Simulation 10.3 (Convergence)

For the Itô integral $\int_0^T W_t \mathrm{d}W_t$ we check the L^2-convergence of the partial sum $\sum_{i=0}^{N-1} W_{t_i} \mathrm{d}W_{t_i}$ to the limit $\frac{1}{2}W_T^2 - \frac{1}{2}T$. In our simulation we obtained the `L2error` $=$ 0.0152 which is defined by

$$\mathbb{E}\left[\left|\sum_{i=0}^{N-1} W_{t_i} \mathrm{d}W_{t_i} - \left(\frac{1}{2}W_T^2 - \frac{1}{2}T\right)\right|^2\right].$$

The average was computed using $M = 10000$ samples.

```
T = 3.0;
M = 10000;
N = 300;
dt = T/N;
t  = 0:dt:T;

dW = sqrt(dt)*randn(M,N);
W = [zeros(M,1), cumsum(dW,2)];

Integral = [zeros(M,1), cumsum(W(:,1:N).*dW,2)];
Exact = 0.5*W(:,N+1).^2  - 0.5*T;

error = Integral(:,N+1) - Exact;
L2error = mean(error.^2)
```

Exercises

10.1 Compute $\int_0^T (W_t + 1)\, \mathrm{d}W_t$.

10.2 Without applying Corollary 10.1, show that $\mathbb{E}\left[\int_0^T (W_t + 1)^2\, \mathrm{d}W_t\right] = 0$.

10.3 Compute $\int_0^T \mathrm{e}^{W_t}\, \mathrm{d}W_t$.

10.4 Compute $\mathbb{E}[S_t]$ in Example 10.4 using the fact that $\{\mathrm{e}^{-\mu t} S_t\}_{t \geq 0}$ is a martingale.

10.5 Show that $\int_0^t s\, \mathrm{d}W_s = t\, W_t - \int_0^t W_s\, \mathrm{d}s$ from the definition of the Itô integral.

10.6 Show that $\int_0^t W_s^2\, \mathrm{d}W_s = \frac{1}{3} W_t^3 - \int_0^t W_s\, \mathrm{d}s$ from the definition of the Itô integral. (Hint: Use the identity $b^2(a-b) = \frac{1}{3}(b^3 - a^3) - \frac{1}{3}(b-a)^3 - b(b^2 - a^2)$.)

10.7 Show that

$$X_t = t^2 W_t - 2 \int_0^t s\, W_s\, \mathrm{d}s$$

is a martingale.

10.8 For a continuous deterministic function $f(t)$ define $X_t = \int_0^t f(s) \mathrm{d}W_s$. Show that $\mathrm{Cov}(X_t, X_{t+u}) = \int_0^t f(s)^2 \mathrm{d}s,\ u \geq 0$.

10.9 Let $X_t = \mathrm{e}^{W_t^2},\ t \geq 0$. Show that

$$\mathbb{E}\left[X_t^2\right] = \frac{1}{\sqrt{1-4t}}\,, \qquad 0 \leq t < \frac{1}{4}\,,$$

and hence $X_t \notin \mathcal{H}^2$. (Thus the identity in Theorem 10.1 does not hold.)

10.10 Recall that the Itô isometry in Theorem 10.1 implies that the following Itô integrals

$$X = \int_0^T t\, \mathrm{d}W_t \quad \text{and} \quad Y = \int_0^T (T-t) \mathrm{d}W_t$$

are normally distributed with mean 0 and variance equal to

$$\int_0^T t^2 \mathrm{d}t = \int_0^T (T-t)^2 \mathrm{d}t = \frac{T^3}{3}\,.$$

That is, $X \sim N(0, \frac{T^3}{3})$ and $Y \sim N(0, \frac{T^3}{3})$. Since

$$X + Y = \int_0^T t\,\mathrm{d}W_t + \int_0^T (T-t)\mathrm{d}W_t = \int_0^T T\,\mathrm{d}W_t = T\,W_T \sim N(0, T^3)\,,$$

we see that $\mathrm{Cov}(X, Y) = \frac{1}{6}T^3$. Prove this directly by applying Theorem 10.2.

10.11 Suppose that $X_t = f(t, W_t)$ is a martingale for some sufficiently smooth $f(t, x)$. Show that f satisfies the differential equation $f_t + \frac{1}{2}f_{xx} = 0$.

10.12 Does there exist $X_t = f(t, W_t)$ such that $\mathrm{d}X_t = X_t\,\mathrm{d}W_t$?

10.13 Let $\mathrm{d}X_t = a_t\mathrm{d}t + \mathrm{d}W_t$ where a_t is a bounded process. Show that X_t is not a martingale unless $a_t = 0$.

10.14 Fix $a > 0$. Define $f(t, x) = \mathbf{1}_{[0,a]}(t)$, $(t, x) \in [0, \infty) \times (-\infty, \infty)$.

(i) Compute $\int_0^t f(s, W_s)\mathrm{d}W_s$.
(ii) Prove that the stochastic process X_t defined by

$$X_t = \begin{cases} W_t, & 0 \le t \le a \\ W_a, & t > a \end{cases}$$

is a martingale.
(iii) Find α_t such that $\mathrm{d}X_t = \alpha_t\mathrm{d}W_t$.

10.15 Show that $W_t^3 - 3tW_t$ is a martingale by applying the Martingale Representation Theorem.

10.16

(i) Suppose that $\{M_t\}_{t\ge0}$ is a martingale with respect to $(\mathbb{P}, \{\mathcal{F}_t\}_{t\ge0})$. Show that $M_t^2 - [M]_t$ is also a martingale.
(ii) Suppose that M_t^1 and M_t^2 are two martingales for $t \ge 0$. Show that $M_t^1 M_t^2 - [M^1, M^2]_t$ is also a martingale.

10.17 Find a deterministic function $g(t)$ such that $X_t = \mathrm{e}^{W_t + g(t)}$, $t \ge 0$, is a martingale.

10.18 Show that $X_t = (W_t + t)\mathrm{e}^{-W_t - \frac{1}{2}t}$ is a martingale.

10.19 Let $\mathrm{d}X_t = a_t\mathrm{d}t + \mathrm{d}W_t$ where a_t is a bounded process. Define

$$L_t = \exp\Big(-\int_0^t a_s\mathrm{d}W_s - \frac{1}{2}\int_0^t a_s^2\mathrm{d}s\Big)\,.$$

Show that X_tL_t is a martingale.

10.20 Let $X_t = t^2 W_t - 2\int_0^t sW_s\mathrm{d}s$. Show that X_t is a martingale.

10.21 Find a constant a for which $X_t = \mathrm{e}^{at}\cos W_t$ is a martingale.

Chapter 11
The Itô Formula

The Itô formula, or the Itô lemma, is the most frequently used fundamental fact in stochastic calculus. It approximates a function of time and Brownian motion in a style similar to Taylor series expansion except that the closeness of approximation is measured in terms of probabilistic distribution of the increment in Brownian motion.

Let W_t denote a Brownian motion, and let S_t denote the price of a risky asset such as stock. The standard model for asset price movement is given by the geometric Brownian motion. More precisely, if δ represents small increment then S_t satisfies an approximate stochastic differential equation $\delta S_t = \mu S_t \delta t + \sigma S_t \delta W_t$ where δW represents uncertainty or risk in financial investment. Since $\mathbb{E}[\delta W] = 0$ and $\mathbb{E}[(\delta W)^2] = \delta t$, we may write $|\delta W| \approx \sqrt{\delta t}$. Thus, even when $\delta t \approx 0$ we cannot ignore $(\delta W)^2$ completely. Itô's lemma or Itô's formula regards $(\delta W)^2$ as δt in a mathematically rigorous way, and is the most fundamental tool in financial mathematics.

11.1 Motivation for the Itô Formula

The idea underlying the Itô formula is a probabilistic interpretation of the second order Taylor expansion. Given a sufficiently smooth function $f(t, x)$, consider the increment of the curve $t \mapsto f(t, W_t)$ using the second order Taylor expansion:

$$\begin{aligned}\delta f &= f(t + \delta t, W_{t+\delta t}) - f(t, W_t) \\ &= f_t \, \delta t + f_x \, \delta W_t + \frac{1}{2} \left\{ f_{tt}(\delta t)^2 + 2 f_{tx} \, \delta t \, \delta W_t + f_{xx}(\delta W_t)^2 \right\} .\end{aligned}$$

See Fig. 11.1. If $\delta t > 0$ is close to 0, then $(\delta t)^2$ is far smaller than δt. We note that δW_t is normally distributed with mean 0 and variance δt and it can be large, albeit with very small probability, which is troublesome. We adopt the practical rule of

G.H. Choe, *Stochastic Analysis for Finance with Simulations*, Universitext,
DOI 10.1007/978-3-319-25589-7_11

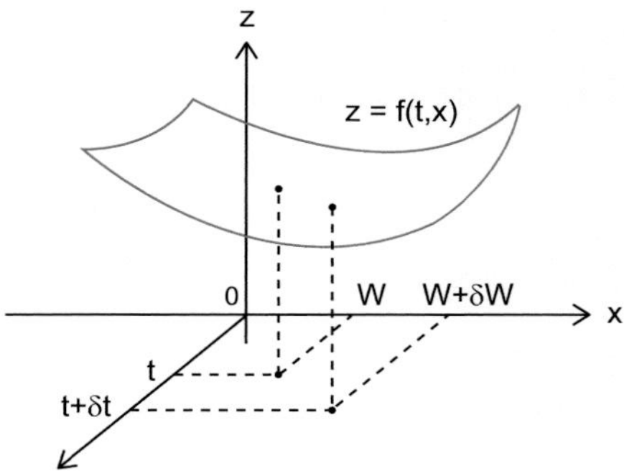

Fig. 11.1 Graph of a function and Itô's lemma

Table 11.1 Multiplication rule for Itô calculus

×	dt	dW
dt	0	0
dW	0	dt

thumb $(\delta t)^2 = 0$, $(\delta W)^2 = \delta t$, and take $\delta W \times \delta t = \pm(\delta t)^{3/2} = 0$, i.e., we ignore the δt-terms of order greater than 1. See Table 11.1 which is written in differential notation.

According to the new viewpoint we have

$$\delta f = \big(f_t + \frac{1}{2} f_{xx}\big)\delta t + f_x\, \delta W_t \; .$$

Before we prove the Itô formula in Sect. 11.2, we present a more detailed analysis of the rule $(\mathrm{d}W_t)^2 = \mathrm{d}t$ which is the essential idea in the proof of the formula. Take a partition $0 = t_0 < t_1 < \cdots < t_n = T$ of a given time interval $[0, T]$ and let $\delta t_i = t_{i+1} - t_i$ and $\delta W_i = W_{t_{i+1}} - W_{t_i}$, $0 \le i \le n-1$. Suppose we want to express $f(T, W_T) - f(0, W_0)$ as a sum of small increments

$$(\delta f)_i = f(t_{i+1}, W_{t_{i+1}}) - f(t_i, W_{t_i}) \; ,$$

i.e., we consider

$$f(T, W_T) - f(0, W_0) = \sum_{i=1}^{n-1} (\delta f)_i \; .$$

Since δW_i can be considerably large compared to δt_i, we ignore the terms corresponding to $(\delta t_i)^2$ and $(\delta t_i)(\delta W_i)$ in the second order Taylor expansion, then the approximation of $(\delta f)_i$ by the second order Taylor expansion is given by

$$(\delta f)_i^{\mathrm{Taylor}} = f_t\, \delta t_i + f_x\, \delta W_i + \frac{1}{2} f_{xx} (\delta W_i)^2$$

where the partial derivatives are evaluated at the point (t_i, W_{t_i}). It is known to be a good approximation for sufficiently small δt and δW. Compare it with the approximation by the Itô formula

$$(\delta f)_i^{\text{Itô}} = (f_t + \frac{1}{2} f_{xx})\,\delta t_i + f_x\,\delta W_i \ .$$

Observe that the essential difference between two approximations is given by

$$\varepsilon_i = \frac{1}{2} f_{xx} (\delta W_i)^2 - \frac{1}{2} f_{xx}\,\delta t_i = \frac{1}{2} f_{xx} \left((\delta W_i)^2 - \delta t_i \right) \ .$$

If this is small enough in some sense, then we can say that the approximation by the Itô formula is as good as Taylor approximation.

As a test we choose $f(t,x) = x^2$ and check the approximation errors for $\delta t \approx 0$ and $\delta W \approx 0$ resulting from two methods. We choose $i = 0$, and $t_i = 0$, $W_{t_i} = 0$, and plot the error of the second order Taylor approximation

$$f(t_i + \delta t_i, W_{t_i} + \delta W_i) - f(t_i, W_{t_i}) - (\delta f)_i^{\text{Taylor}}$$

and the error of the Itô approximation

$$f(t_i + \delta t_i, W_{t_i} + \delta W_i) - f(t_i, W_{t_i}) - (\delta f)_i^{\text{Itô}}$$

over the region $0 \le \delta t_i \le T$, $-1.5\sqrt{T} \le \delta W_i \le 1.5\sqrt{T}$, $T = 0.01$. See Fig. 11.2 where the approximation by the Itô formula is better and the corresponding error is close to zero. The second order Taylor expansion has negligible error so that the graph is almost flat with height 0 while the Itô formula does not approximate the given function well enough. Therefore we need to give a new interpretation for Itô

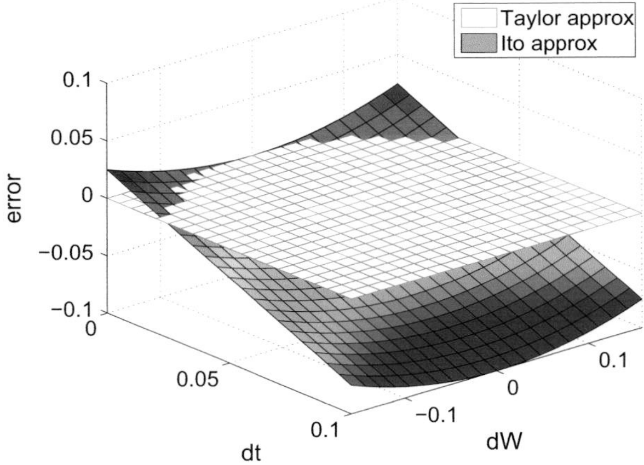

Fig. 11.2 Approximation errors of $f(t,x) = x^2$ by the second order Taylor expansion and the Itô formula

approximation, which is nothing but taking the average error in a suitable sense over all possible values of δW_i so that the average of ε_i would be close to 0 since δW_i is normally distributed with average 0 and variance δt_i.

In the following discussion we will compute the L^2-norm of ε_i to estimate its closeness to 0, and estimate the total error $\varepsilon_0 + \cdots + \varepsilon_{n-1}$.

For $i < j$, we have

$$\begin{aligned}\mathbb{E}[\varepsilon_i\varepsilon_j] &= \mathbb{E}\Big[\frac{1}{4}f_{xx}(t_i, W_{t_i})\{\delta t_i - (\delta W_i)^2\}f_{xx}(t_j, W_{t_j})\{\delta t_j - (\delta W_j)^2\}\Big] \\ &= \mathbb{E}\Big[\frac{1}{4}f_{xx}(t_i, W_{t_i})\{\delta t_i - (\delta W_i)^2\}f_{xx}(t_j, W_{t_j})\Big]\mathbb{E}\left[\delta t_j - (\delta W_j)^2\right] \\ &= 0\end{aligned}$$

by independence of

$$f_{xx}(t_i, W_{t_i})\{\delta t_i - (\delta W_i)^2\}f_{xx}(t_j, W_{t_j})$$

and $\delta t_j - (\delta W_j)^2$, and by the fact that $\mathbb{E}[\delta t_j - (\delta W_j)^2] = 0$.

For $i = j$ we have

$$\begin{aligned}\mathbb{E}[\varepsilon_i^2] &= \mathbb{E}\left[\frac{1}{4}(f_{xx}(t_i, W_{t_i}))^2\{\delta t_i - (\delta W_i)^2\}^2\right] \\ &\leq \frac{C^2}{4}\mathbb{E}\left[(\delta t_i)^2 - 2\,\delta t_i\,(\delta W_i)^2 + (\delta W_i)^4\right] \\ &= \frac{C^2}{4}\left\{(\delta t_i)^2 - 2(\delta t_i)^2 + 3(\delta t_i)^2\right\} = \frac{C^2}{2}(\delta t_i)^2\end{aligned}$$

where

$$C = \max\{|f_{xx}(t,x)| : t \geq 0, -\infty < x < \infty\} < \infty\,.$$

Hence

$$\mathbb{E}[(\varepsilon_0 + \cdots + \varepsilon_{n-1})^2] = \mathbb{E}[\varepsilon_0^2] + \cdots + \mathbb{E}[\varepsilon_{n-1}^2] \leq n \times \frac{C^2}{2}\left(\frac{T}{n}\right)^2 = \frac{C^2 T}{2}\delta t$$

if we take $\delta t = \frac{T}{n}$, $t_i = i \times \delta t$, $0 \leq i \leq n$. Thus the L^2-norm of the total error satisfies

$$\left(\mathbb{E}[(\varepsilon_0 + \cdots + \varepsilon_{n-1})^2]\right)^{\frac{1}{2}} \leq C\left(\frac{T}{2}\right)^{\frac{1}{2}}(\delta t)^{\frac{1}{2}}$$

which gives a bound for the speed of convergence to 0 of the total error in approximating $f(T, W_T) - f(0,0)$ as $\delta t \to 0$.

Example 11.1 Take $f(t,x) = x^2$ as a test function. Since $f_t = 0, f_x = 2x$ and $f_{tt} = 0$, $f_{xx} = 2, f_{tx} = 0$, we have $C = 2$ and

$$\varepsilon_i = (\delta W_i)^2 - \delta t_i \ .$$

Thus

$$\varepsilon_0 + \cdots + \varepsilon_{n-1} = \sum_{i=0}^{n-1}((\delta W_i)^2 - \delta t_i) = \sum_{i=0}^{n-1}(\delta W_i)^2 - T$$

and

$$\begin{aligned}
&\mathbb{E}\left[\left(\sum_{i=0}^{n-1}(\delta W_i)^2 - T\right)^2\right] \\
&= \mathbb{E}\left[\left(\sum_{i=0}^{n-1}(\delta W_i)^2\right)^2 - 2T\sum_{i=0}^{n-1}(\delta W_i)^2 + T^2\right] \\
&= \mathbb{E}\left[\sum_{i=0}^{n-1}(\delta W_i)^4 + \sum_{i\neq j}(\delta W_i)^2(\delta W_j)^2 - 2T\sum_{i=0}^{n-1}(\delta W_i)^2 + T^2\right] \\
&= \sum_{i=0}^{n-1} 3(\delta t_i)^2 + \sum_{i\neq j}(\delta t_i)(\delta t_j) - 2T\sum_{i=0}^{n-1}\delta t_i + T^2 \\
&= \sum_{i=0}^{n-1} 2(\delta t_i)^2 + \sum_{i=0}^{n-1}\sum_{j=0}^{n-1}(\delta t_i)(\delta t_j) - 2T\sum_{i=0}^{n-1}\delta t_i + T^2 \\
&= \sum_{i=0}^{n-1} 2(\delta t_i)^2 + \sum_{i=0}^{n-1}\delta t_i\sum_{j=0}^{n-1}\delta t_j - 2T\sum_{i=0}^{n-1}\delta t_i + T^2 \\
&= \sum_{i=0}^{n-1} 2(\delta t_i)^2
\end{aligned}$$

where we used $\sum_{i=0}^{n-1}\delta t_i = T$ for the last equality. In Fig. 11.3 we choose $T = 3$, $\delta t = 2^{-k}T$, $1 \leq k \leq 15$, and estimate the L^2-norm of $\sum_{i=0}^{n-1}(\delta W_i)^2 - T$ which has the theoretical value $(2n(\delta t)^2)^{1/2} = 2^{-(k-1)/2}T$ which is plotted as a function of k. Consult Simulation 11.1.

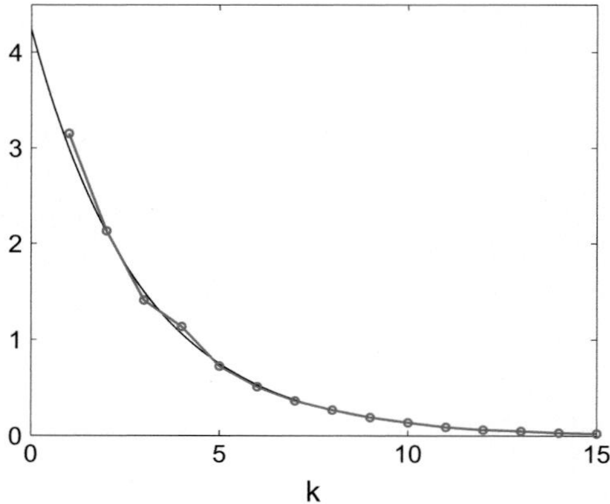

Fig. 11.3 Speed of L^2-convergence in Itô formula with $f(t,x) = x^2$

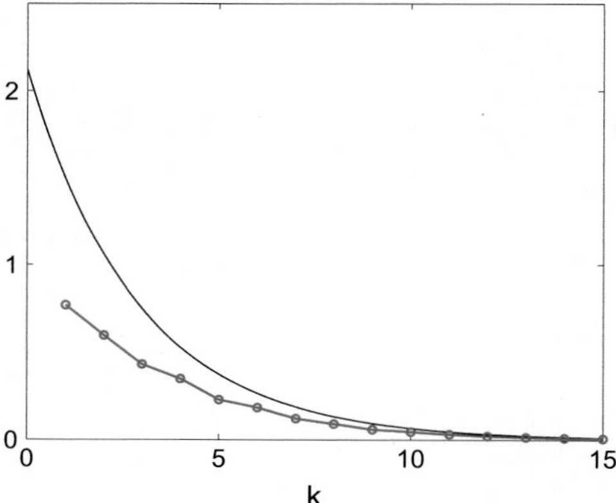

Fig. 11.4 Speed of L^2-convergence in Itô formula with $f(t,x) = \sin x$

Example 11.2 Take $f(t,x) = \sin x$ as a test function. Since $f_t = 0, f_x = \cos x$ and $f_{tt} = 0, f_{xx} = -\sin x, f_{tx} = 0$, we have $C = 1$ and

$$\varepsilon_i = \frac{1}{2}(\sin W_{t_i})\left(\delta t_i - (\delta W_i)^2\right) .$$

Thus

$$\mathbb{E}[(\varepsilon_1 + \cdots + \varepsilon_n)^2] \leq \frac{T}{2}\delta t .$$

In Fig. 11.4 we choose $T = 3$, $\delta t = 2^{-k}T$, $1 \leq k \leq 15$, and compute the L^2-norm of $\varepsilon_1 + \cdots + \varepsilon_n$ which is bounded by $(\frac{T}{2}\delta t)^{1/2} = 2^{-(k+1)/2}T$ which is plotted as a function of k, $0 \leq k \leq 15$. Consult Simulation 11.2.

11.2 The Itô Formula: Basic Form

The following fact is called the Itô formula or the Itô lemma.

Theorem 11.1 (Itô) *Let $f(t,x) : [0,\infty) \times \mathbb{R}^1 \to \mathbb{R}^1$ be a sufficiently smooth function such that its partial derivatives f_t, f_x, f_{xx} exist and are continuous. Fix a constant $0 < T < \infty$. Then almost surely the following equation holds:*

$$\begin{aligned} & f(T, W_T) - f(0, W_0) \\ &= \int_0^T f_t(s, W_s)\,\mathrm{d}s + \int_0^T f_x(s, W_s)\,\mathrm{d}W_s + \frac{1}{2}\int_0^T f_{xx}(s, W_s)\,\mathrm{d}s \end{aligned}$$

where the first and the third integrals are Riemann integrals for each fixed sample Brownian path.

Remark 11.1

(i) The formula in the theorem is written in integral notation, and it can also be written in differential notation as follows:

$$\mathrm{d}f(t, W_t) = f_t(t, W_t)\,\mathrm{d}t + f_x(t, W_t)\,\mathrm{d}W_t + \frac{1}{2}f_{xx}(t, W_t)\,\mathrm{d}t$$

or simply,

$$\mathrm{d}f = \left(f_t + \frac{1}{2}f_{xx}\right)\mathrm{d}t + f_x\,\mathrm{d}W_t\ .$$

(ii) For $f(t,x) = x^2$ in Example 11.1 a straightforward computation yields

$$\begin{aligned} f(t_{i+1}, W_{t_{i+1}}) - f(t_i, W_{t_i}) &= (W_{t_i} + \delta W_i)^2 - W_{t_i}^2 \\ &= 2W_{t_i}\delta W_i + (\delta W_i)^2 \end{aligned}$$

and

$$\begin{aligned} W_T^2 - W_0^2 &= f(T, W_T) - f(0, W_0) \\ &= \sum_{i=0}^{n-1} 2W_{t_i}\delta W_i + \sum_{i=0}^{n-1}(\delta W_i)^2 \end{aligned}$$

where the first sum converges to $\int_0^T 2W_t\mathrm{d}W_t$ as $n \to \infty$ in the L^2-sense and the second sum converges to T. On the other hand, since

$$(\delta f)_i^{\text{Itô}} = 2W_{t_i}\delta W_i + \delta t_i\ ,$$

we have

$$f(T,W_T)-f(0,W_0)\approx\sum_{i=0}^{n-1}(\delta f)_i^{\text{Itô}}=\sum_{i=0}^{n-1}2W_{t_i}\delta W_i+T$$

in agreement with the fact that $\sum_{i=0}^{n-1}(\delta W_i)^2$ converges to T in the L^2-sense.

Proof The essential idea of the proof is already presented in Sect. 11.1. Now we will proceed with mathematical rigour. First, consider the case when there exists a constant $C>0$ such that for every $(t,x)\in[0,\infty)\times\mathbb{R}^1$ we have

$$|f(t,x)|\le C\,,\quad |f_t(t,x)|\le C\,,\quad |f_x(t,x)|\le C\,,\quad |f_{xx}(t,x)|\le C\,.$$

The unbounded case will be considered at the end of the proof. Let $0=t_0<t_1<\cdots<t_n=T$ be a partition of $[0,T]$, and put $\delta W_i=W_{t_{i+1}}-W_{t_i}$, $\delta t_i=t_{i+1}-t_i$. We assume that $\max_{1\le i\le n}\delta t_i\to 0$ as $n\to\infty$.

On each subinterval $[t_i,t_{i+1}]$ by Taylor's theorem there exist $t_i^*\in[t_i,t_{i+1}]$ and $W_i^*\in[W_{t_i}(\omega),W_{t_{i+1}}(\omega)]$ such that the following holds:

$$\begin{aligned}
&f(T,W_T(\omega))-f(0,W_0(\omega))\\
&=\sum_{i=0}^{n-1}\big(f(t_{i+1},W_{t_{i+1}}(\omega))-f(t_i,W_{t_i}(\omega))\big)\\
&=\sum_{i=0}^{n-1}\big(f(t_{i+1},W_{t_{i+1}})-f(t_i,W_{t_{i+1}})\big)+\sum_{i=0}^{n-1}\big(f(t_i,W_{t_{i+1}})-f(t_i,W_{t_i})\big)\\
&=\sum_{i=0}^{n-1}f_t(t_i^*,W_{t_{i+1}})\,\delta t_i+\sum_{i=0}^{n-1}f_x(t_i,W_{t_i})\,\delta W_i+\frac{1}{2}\sum_{i=0}^{n-1}f_{xx}(t_i,W_i^*)\,(\delta W_i)^2\\
&=\sum_{i=0}^{n-1}f_t(t_i^*,W_{t_{i+1}})\,\delta t_i+\sum_{i=0}^{n-1}f_x(t_i,W_{t_i})\,\delta W_i+\frac{1}{2}\sum_{i=0}^{n-1}f_{xx}(t_i,W_{t_i})\,\delta t_i\\
&\quad+\frac{1}{2}\sum_{i=0}^{n-1}f_{xx}(t_i,W_{t_i})((\delta W_i)^2-\delta t_i)+\frac{1}{2}\sum_{i=0}^{n-1}[f_{xx}(t_i,W_i^*)-f_{xx}(t_i,W_{t_i})](\delta W_i)^2\\
&=\Sigma_1+\Sigma_2+\Sigma_3+\Sigma_4+\Sigma_5
\end{aligned}$$

where the symbol ω is omitted for the sake of notational simplicity if there is no danger of confusion. Note that if $W_{t_{i+1}}(\omega)<W_{t_i}(\omega)$, then $W_i^*\in[W_{t_{i+1}}(\omega),W_{t_i}(\omega)]$.

First, we consider Σ_1 and Σ_3, which are Riemann integrals of continuous functions for almost every ω. By the definition of the Riemann integral their limits for almost every ω are given by

$$\lim_{n\to\infty} \Sigma_1 = \int_0^T f_t(t, W_t(\omega))\,dt ,$$

$$\lim_{n\to\infty} \Sigma_3 = \int_0^T f_{xx}(t, W_t(\omega))\,dt .$$

For Σ_2, Σ_4, Σ_5 we first prove the L^2-convergence of the limits

$$\lim_{n\to\infty} \Sigma_2 = \int_0^T f_x(t, W_t)\,dW_t ,$$

$$\lim_{n\to\infty} \Sigma_4 = 0 ,$$

$$\lim_{n\to\infty} \Sigma_5 = 0 ,$$

and show that there exist subsequences which converge for almost every ω.

As for Σ_2 we use the boundedness condition $|f_x(t,x)| \le C$. Now, if we let

$$g(t,\omega) = f_x(t, W_t(\omega)) \in \mathcal{H}^2 ,$$

and define a simple stochastic process by

$$g_n(t,\omega) = \sum_{i=0}^{n-1} f_x(t_i, W_{t_i}(\omega))\mathbf{1}_{[t_i,t_{i+1})}(t) \in \mathcal{H}^2_{\text{simple}} ,$$

then $g_n(t,\omega)$ approximates g. Note that for every $0 \le t \le T$

$$\lim_{n\to\infty} |g_n(t,\omega) - g(t,\omega)|^2 = 0$$

for almost every ω by continuity.

Since

$$|g_n(t) - g(t)|^2 \le 4C^2 ,$$

by the Lebesgue Dominated Convergence Theorem we have

$$\lim_{n\to\infty} \int_0^T |g_n(t,\omega) - g(t,\omega)|^2 dt = 0$$

for almost every ω. Furthermore, since

$$\int_0^T |g_n(t,\omega) - g(t,\omega)|^2 dt \le 4TC^2 ,$$

we apply the Lebesgue Dominated Convergence Theorem once more to obtain

$$\lim_{n\to\infty} \mathbb{E}\left[\int_0^T |g_n(t) - g(t)|^2 \mathrm{d}t\right] = 0 ,$$

where $\int_0^T |g_n(t,\omega) - g(t,\omega)|^2 \mathrm{d}t$, $n \geq 1$, is regarded as a bounded sequence of functions of ω with an upper bound for all n given by $4TC^2$, and $\mathbb{E}$ denotes the Lebesgue integral on a probability space $(\Omega, \mathbb{P})$. Hence g_n converges to g in the L^2-sense in M_T^2, and hence

$$I(g_n) = \int_0^T g_n \mathrm{d}W = \sum_{i=0}^{n-1} f_x(t_i, W_{t_i}) \delta W_i$$

converges to

$$I(g) = \int_0^T f_x(t, W_t) \mathrm{d}W_t$$

where $I(\cdot)$ denotes the Itô integral of a function inside the parentheses.

Now we consider Σ_4. Recall that, for $i < j$, a function depending on W_{t_i}, δW_i, W_{t_j} and another function depending on δW_j are independent. In other words, for any measurable functions $\phi_1(\cdot,\cdot,\cdot)$ and $\phi_2(\cdot)$ we have

$$\mathbb{E}\left[\phi_1(W_{t_i}, \delta W_i, W_{t_j})\phi_2(\delta W_j)\right] = \mathbb{E}\left[\phi_1(W_{t_i}, \delta W_i, W_{t_j})\right]\mathbb{E}\left[\phi_2(\delta W_j)\right] .$$

Now since f_{xx} is bounded by C we have

$$\begin{aligned}
&\mathbb{E}\left[\left|\sum_{i=0}^{n-1} f_{xx}(t_i, W_{t_i})((\delta W_i)^2 - \delta t_i)\right|^2\right] \\
&= \mathbb{E}\left[\sum_{i=0}^{n-1}\sum_{j=0}^{n-1} f_{xx}(t_i, W_{t_i})((\delta W_i)^2 - \delta t_i) f_{xx}(t_j, W_{t_j})((\delta W_j)^2 - \delta t_j)\right] \\
&= \sum_{i=0}^{n-1}\sum_{j=0}^{n-1} \mathbb{E}\left[f_{xx}(t_i, W_{t_i})((\delta W_i)^2 - \delta t_i) f_{xx}(t_j, W_{t_j})((\delta W_j)^2 - \delta t_j)\right] \\
&= \sum_{i<j} \mathbb{E}\left[f_{xx}(t_i, W_{t_i})((\delta W_i)^2 - \delta t_i) f_{xx}(t_j, W_{t_j})\right]\mathbb{E}\left[((\delta W_j)^2 - \delta t_j)\right] \\
&\quad + \sum_{j<i} \mathbb{E}\left[((\delta W_i)^2 - \delta t_i)\right]\mathbb{E}\left[f_{xx}(t_i, W_{t_i}) f_{xx}(t_j, W_{t_j})((\delta W_j)^2 - \delta t_j)\right] \\
&\quad + \sum_{i=0}^{n-1} \mathbb{E}\left[f_{xx}(t_i, W_{t_i})^2 \left|(\delta W_i)^2 - \delta t_i\right|^2\right]
\end{aligned}$$

$$
\begin{aligned}
&= 0 + 0 + \sum_{i=0}^{n-1} \mathbb{E}\left[f_{xx}(t_i, W_{t_i})^2 \left|(\delta W_i)^2 - \delta t_i\right|^2\right] \\
&= \sum_{i=0}^{n-1} \mathbb{E}\left[f_{xx}(t_i, W_{t_i})^2\right] \mathbb{E}\left[|(\delta W_i)^2 - \delta t_i|^2\right] \\
&\leq C^2 \sum_{i=0}^{n-1} \mathbb{E}\left[|(\delta W_i)^2 - \delta t_i|^2\right] \\
&= C^2 \sum_{i=0}^{n-1} \left(\mathbb{E}\left[(\delta W_i)^4\right] - 2\,\mathbb{E}[(\delta W_i)^2]\,\delta t_i + (\delta t_i)^2\right) \\
&= C^2 \sum_{i=0}^{n-1} (3 - 2 + 1)(\delta t_i)^2 \\
&\leq 2C^2 \max_{1 \leq i \leq n} \delta t_i \sum_{i=0}^{n-1} \delta t_i \\
&= 2C^2 T \max_{1 \leq i \leq n} \delta t_i \to 0\,.
\end{aligned}
$$

The third equality holds since, for $i < j$, by independence we have

$$
\begin{aligned}
&\mathbb{E}\left[f_{xx}(t_i, W_{t_i})((\delta W_i)^2 - \delta t_i) f_{xx}(t_j, W_{t_j})((\delta W_j)^2 - \delta t_j)\right] \\
&= \mathbb{E}\left[f_{xx}(t_i, W_{t_i})((\delta W_i)^2 - \delta t_i) f_{xx}(t_j, W_{t_j})\right] \mathbb{E}\left[(\delta W_j)^2 - \delta t_j\right]
\end{aligned}
$$

and the fourth equality by $\mathbb{E}[(\delta W_j)^2 - \delta t_j] = 0$. The fifth equality holds since $|f_{xx}(t_i, W_{t_i})|^2$ and $|(\delta W_i)^2 - \delta t_i|^2$ are independent. Hence

$$
\sum_{i=0}^{n-1} f_{xx}(t_i, W_{t_i}) \left((\delta W_i)^2 - \delta t_i\right) \to 0
$$

in the L^2-sense.

As for Σ_5 we have

$$
\lim_{n \to \infty} \max_i \left| f_{xx}(t_i, W_i^*) - f_{xx}(t_i, W_{t_i}) \right| = 0
$$

for almost every ω by continuity. Since $\sum_{i=0}^{n-1} (\delta W_i)^2 \to T$ in the L^2-sense, by Fact 3.8 there exists a subsequence $\{n_k\}_{k=1}^{\infty}$ such that

$$
\lim_{n_k \to \infty} \sum_{i=0}^{n_k - 1} (\delta W_i(\omega))^2 = T
$$

for almost every ω. Hence for $\{n_k\}_{k=1}^\infty$ for almost every ω we have

$$\left|\sum_{i=0}^{n_k}(f_{xx}(t_i, W_i^*) - f_{xx}(t_i, W_{t_i}(\omega)))(\delta W_i(\omega))^2\right|$$

$$\leq \max_i \left|f_{xx}(t_i, W_i^*) - f_{xx}(t_i, W_{t_i}(\omega))\right| \sum_{i=0}^{n_k-1}(\delta W_i(\omega))^2 \to 0$$

as $k \to \infty$. Since Σ_2 converges in the L^2-sense along $\{n_k\}$, we can show the convergence for almost every ω by using a subsequence of $\{n_k\}$.

Now, we show that the same conclusion holds without the condition on boundedness of partial derivatives. Define $f_n(t,x) = \mathbf{1}_{[-n,n]}(x)f(t,x)$ where $\mathbf{1}_{[-n,n]}$ is the indicator of the interval $[-n,n]$. Since f_n and its derivatives are bounded, we may apply the previously proved result and obtain

$$\begin{aligned}
&f_n(T, W_T) - f_n(0, W_0)\\
&= \int_0^T (f_n)_t(s, W_s)\mathrm{d}s + \int_0^T (f_n)_x(s, W_s)\mathrm{d}W_s + \frac{1}{2}\int_0^T (f_n)_{xx}(s, W_s)\,\mathrm{d}s\,.
\end{aligned}$$

Now define a monotonically increasing sequence of subsets

$$B_n = \{\omega : \sup_{0\leq t\leq T} |W_t(\omega)| \leq n\}\,.$$

Since $f = f_n$ on B_n, the Itô formula holds on B_n for f. Since $\mathbb{P}(B_n) \to 1$, the Itô formula for f holds for almost every ω. □

In the following we obtain stochastic differential equations.

Example 11.3

(i) For $X_t = W_t$, take $f(t,x) = x$, then $X_t = f(t, W_t)$ and $X_0 = 0$. Since $f_x = 1$ and $f_{xx} = 0$, we have $\mathrm{d}X_t = \mathrm{d}W_t$, and $\int_0^T \mathrm{d}W_t = W_T$ in integral form.

(ii) For $X_t = (W_t)^2$, take $f(t,x) = x^2$, then $X_t = f(t, W_t)$ and $X_0 = 0$. Since $f_x = 2x$ and $f_{xx} = 2$, we have $\mathrm{d}X_t = 2W_t\mathrm{d}W_t + \frac{1}{2}\times 2\mathrm{d}t = 2W_t\mathrm{d}W_t + \mathrm{d}t$, and

$$\int_0^T W_t\mathrm{d}W_t = \frac{1}{2}(W_T)^2 - \frac{1}{2}T$$

in integral form. (See also Remark 11.1.)

(iii) For $X_t = \frac{1}{3}(W_t)^3$, take $f(t,x) = \frac{1}{3}x^3$, then $X_t = f(t, W_t)$ and $X_0 = 0$. Since $f_x = x^2$ and $f_{xx} = 2x$, we have $\mathrm{d}X_t = (W_t)^2\mathrm{d}W_t + \frac{1}{2}\times 2W_t\mathrm{d}t$ and

$$\int_0^T (W_t)^2\mathrm{d}W_t = \frac{1}{3}(W_T)^3 - \int_0^T W_t\mathrm{d}t\,.$$

(iv) For $X_t = tW_t$, take $f(t,x) = tx$, then $X_t = f(t, W_t)$ and $X_0 = 0$. Since $f_t = x$, $f_x = t$ and $f_{xx} = 0$, we have $\mathrm{d}X_t = W_t\mathrm{d}t + t\,\mathrm{d}W_t$ and

$$\int_0^T t\,\mathrm{d}W_t = TW_T - \int_0^T W_t\mathrm{d}t\,.$$

So far, we have considered examples when stochastic processes X_t are given and we derive the stochastic differential equations satisfied by X_t. We can evaluate a given Itô integral directly as in Exercises 11.6, 11.7, 11.8, 11.10.

11.3 The Itô Formula: General Form

Now we define an Itô integral with respect to a general Itô process other than a Brownian motion $\{W_t\}_{t\geq 0}$.

Definition 11.1 (Itô Process) A stochastic process $\{X_t\}_{t\geq 0}$ of the form

$$X_t = X_0 + \int_0^t a_u\,\mathrm{d}u + \int_0^t b_u\,\mathrm{d}W_u$$

is called an *Itô process* if it is adapted to the filtration $\{\mathcal{F}_t\}_{t\geq 0}$ associated with a Brownian motion, i.e.,

$$\mathrm{d}X_t = a_t\,\mathrm{d}t + b_t\,\mathrm{d}W_t$$

where X_0 is a constant, a_t and b_t are stochastic processes adapted to the filtration $\{\mathcal{F}_t\}_{t\geq 0}$ such that

$$\int_0^T |a_t(\omega)|\mathrm{d}t < \infty \quad \text{for almost every } \omega$$

and $\{b_t\}_{t\geq 0} \in \mathcal{H}_T^2$ for every T.

Example 11.4

(i) The simplest Itô process is given by

$$X_t = X_0 + \int_0^t a_s\mathrm{d}s$$

where $b_t = 0$.

(ii) For $X_t = W_t$ we have $\mathrm{d}X_t = \mathrm{d}W_t$, and hence $a_t = 0$, $b_t = 1$.

(iii) For $X_t = (W_t)^2$ we have $\mathrm{d}X_t = \mathrm{d}t + 2W_t\mathrm{d}W_t$, and hence $a_t = 1$, $b_t = 2W_t$.

(iv) For $X_t = \frac{1}{3}(W_t)^3$ we have $\mathrm{d}X_t = W_t\mathrm{d}t + (W_t)^2\mathrm{d}W_t$, and hence $a_t = W_t$, $b_t = (W_t)^2$.

Example 11.5 For $X_t = t\,W_t$ we have $\mathrm{d}X_t = W_t\mathrm{d}t + t\,\mathrm{d}W_t$, and hence $a_t = W_t$, $b_t = t$. This formula looks as if it were the usual product rule for differentiation in calculus.

Definition 11.2 (Itô Integral with Respect to Itô Processes) Let $\{X_t\}_{t\geq 0}$ be an Itô process given by

$$\mathrm{d}X_t = a_t\,\mathrm{d}t + b_t\,\mathrm{d}W_t \; .$$

For a process $\{Y_t\}_{t\geq 0}$ adapted to $\{\mathcal{F}_t\}_{t\geq 0}$ define the stochastic integral of Y_t with respect to X_t by

$$\int_0^t Y_u\,\mathrm{d}X_u = \int_0^t Y_u\,a_u\,\mathrm{d}u + \int_0^t Y_u\,b_u\,\mathrm{d}W_u \; ,$$

which can be rewritten as

$$Y_t\,\mathrm{d}X_t = Y_t\,a_t\,\mathrm{d}t + Y_t\,b_t\,\mathrm{d}W_t \; .$$

Remark 11.2 (Itô Integral with Respect to a Martingale) If X_t is a martingale then it is represented as an Itô process by Theorem 10.6, and we can define a stochastic integral $\int_0^t Y_u\,\mathrm{d}X_u$ using X_t as an integrator.

Definition 11.3 (Quadratic Variation of a Martingale) Let M_t be a continuous square-integrable martingale. Define the mesh size of the partition $0 = t_0 < t_1 < \cdots < t_n = t$ by

$$\delta_n = \max_{0\leq j\leq n-1} |t_{j+1} - t_j| \; .$$

The *quadratic variation* of M_t, denoted by $[M,M]_t$ or simply $[M]_t$, is defined by

$$[M,M]_t = \lim_{n\to\infty} \sum_{j=0}^{n-1} |M_{t_{j+1}} - M_{t_j}|^2$$

where the limit is in probability as δ_n decreases to 0 as $n \to \infty$. If N_t is another martingale, then the *covariation* of M_t and N_t is given by

$$[M,N]_t = \lim_{n\to\infty} \sum_{j=0}^{n-1} (M_{t_{j+1}} - M_{t_j})(N_{t_{j+1}} - N_{t_j})$$

and satisfies

$$[M,N]_t = \frac{1}{2}([M+N,M+N]_t - [M,M]_t - [N,N]_t) \; .$$

Theorem 11.2 (Product Rule) *Let X_t and Y_t be Itô processes. Then*

$$[X,Y]_t = X_tY_t - X_0Y_0 - \int_0^t X_s \mathrm{d}Y_s - \int_0^t Y_s \mathrm{d}X_s \,,$$

or in differential notation

$$\mathrm{d}(X_tY_t) = X_t\mathrm{d}Y_t + Y_t\mathrm{d}Y_t + \mathrm{d}[X,Y]_t \,.$$

Proof Since

$$\begin{aligned}
&\sum_{j=0}^{n-1}(X_{t_{j+1}} - X_{t_j})(Y_{t_{j+1}} - Y_{t_j}) \\
&= \sum_{j=0}^{n-1}(X_{t_{j+1}}Y_{t_{j+1}} - X_{t_j}Y_{t_j}) - \sum_{j=0}^{n-1}X_{t_j}(Y_{t_{j+1}} - Y_{t_j}) - \sum_{j=0}^{n-1}Y_{t_j}(X_{t_{j+1}} - X_{t_j}) \\
&= X_tY_t - X_0Y_0 - \sum_{j=0}^{n-1}X_{t_j}(Y_{t_{j+1}} - Y_{t_j}) - \sum_{j=0}^{n-1}Y_{t_j}(X_{t_{j+1}} - X_{t_j}) \,.
\end{aligned}$$

The last two terms in the above converge in probability to Itô integrals $\int_0^t X_s\mathrm{d}Y_s$ and $\int_0^t Y_s\mathrm{d}X_s$. □

Example 11.6 For a Brownian motion W_t we have $[W]_t = t$.

Remark 11.3 For an Itô process X_t given by $\mathrm{d}X_t = a(t,X_t)\,\mathrm{d}t + b(t,X_t)\,\mathrm{d}W_t$, consider the martingale

$$M_t = \int_0^t b(s,X_s)\,\mathrm{d}W_s \,.$$

Then

$$[M,M]_t = \int_0^t b(s,X_s)^2\,\mathrm{d}s \,.$$

Now we present the Itô formula for an Itô process $\{X_t\}_{t\geq 0}$. Its proof is similar to the one for Theorem 11.1, and is omitted.

Theorem 11.3 (General Itô Formula) *Consider an Itô process $\{X_t\}_{t\geq 0}$ given by*

$$\mathrm{d}X_t = a_t\,\mathrm{d}t + b_t\,\mathrm{d}W_t \,.$$

Let $f(t,x) : [0,\infty) \times \mathbb{R} \to \mathbb{R}^1$ be a function whose partial derivatives f_t, f_x and f_{xx} exist and are continuous. Suppose that

$$f_x(t, X_t)\, b_t \in \mathcal{H}_T^2$$

for every T. Then $f(t, X_t)$ is an Itô process, and with probability 1 we have

$$\begin{aligned} & f(t, X_t) - f(0, X_0) \\ &= \int_0^t f_t(u, X_u)\,\mathrm{d}u + \int_0^t f_x(u, X_u)\,\mathrm{d}X_u + \frac{1}{2}\int_0^t f_{xx}(u, X_u)\,\mathrm{d}[X, X]_u \\ &= \int_0^t \left(f_t(u, X_u) + f_x(u, X_u)\, a_u + \frac{1}{2} f_{xx}(u, X_u)\, b_u^2 \right) \mathrm{d}u \\ &\quad + \int_0^t f_x(u, X_u)\, b_u\,\mathrm{d}W_u \end{aligned}$$

for $t \geq 0$.

The above formula can be rewritten in a differential form as

$$\mathrm{d}f(t, X_t) = \left(f_t + f_x\, a_t + \frac{1}{2} f_{xx}\, b_t^2 \right) \mathrm{d}t + f_x\, b_t\,\mathrm{d}W_t\ .$$

See Table 11.2.

Example 11.7 Consider the geometric Brownian motion

$$\mathrm{d}S_t = \mu S_t \mathrm{d}t + \sigma S_t \mathrm{d}W_t\ .$$

First, we look for a solution of the form $S_t = f(t, W_t)$. By the Itô formula we have

$$\mathrm{d}S_t = \left(f_t(t, W_t) + \frac{1}{2} f_{xx}(t, W_t) \right) \mathrm{d}t + f_x(t, W_t)\,\mathrm{d}W_t$$

and

$$\begin{cases} f_t(t, W_t) + \frac{1}{2} f_{xx}(t, W_t) = \mu f(t, W_t) \\ f_x(t, W_t) = \sigma f(t, W_t)\ . \end{cases}$$

Table 11.2 Multiplication rule for general Itô calculus

$\times$	$\mathrm{d}t$	$a\,\mathrm{d}t + b\,\mathrm{d}W$
$\mathrm{d}t$	0	0
$a\,\mathrm{d}t + b\,\mathrm{d}W$	0	$b^2\,\mathrm{d}t$

Since these equations must hold for an arbitrary real number $W_t(\omega)$ we have

$$\begin{cases} f_t(t,x) + \frac{1}{2} f_{xx}(t,x) = \mu f(t,x) , \\ f_x(t,x) = \sigma f(t,x) . \end{cases}$$

From the second equation we see that $f(t,x) = g(t)\mathrm{e}^{\sigma x}$ for some function $g(t)$. A necessary condition for g is

$$g'(t) + \frac{1}{2}\sigma^2 g(t) = \mu\, g(t) .$$

Hence for some constant C we have $g(t) = C\mathrm{e}^{(\mu-\frac{1}{2}\sigma^2)t}$. Thus $f(t,x) = C\mathrm{e}^{(\mu-\frac{1}{2}\sigma^2)t+\sigma x}$ and

$$S_t = S_0\, \mathrm{e}^{(\mu-\frac{1}{2}\sigma^2)t+\sigma W_t} .$$

In Fig. 11.5 are given some sample paths of geometric Brownian motion with $\mu = 0.15$, $\sigma = 0.2$, $S_0 = 1$ together with the curve $S = S_0\mathrm{e}^{\mu t}$ (left) and the probability density functions for S_t, $t = 1, \ldots, 10$ (right).

Remark 11.4 Let Ω be a set of Brownian sample paths ω, and let $\mathbb{P}$ be a Wiener measure on Ω. For each t we may regard S_t as a function in L^p. For $p = 1$ we have

$$||S_t - 0||_1 = \int_\Omega |S_t(\omega) - 0| \,\mathrm{d}\mathbb{P} = \mathbb{E}[S_t] = S_0\mathrm{e}^{\mu t} \to \infty ,$$

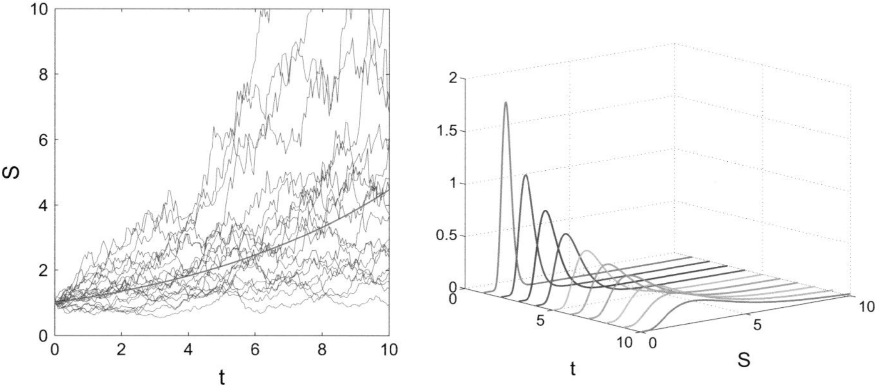

Fig. 11.5 Sample paths of geometric Brownian motion together with the average given by the curve $S = S_0\mathrm{e}^{\mu t}$ (*left*) and the probability density functions for S_t (*right*)

and for $p = 2$

$$||S_t - 0||_2^2 = \int_{\Omega} |S_t(\omega) - 0|^2 \, \mathrm{d}\mathbb{P} = \mathbb{E}[S_t^2] = S_0^2 \mathrm{e}^{(2\mu+\sigma^2)t} \to \infty$$

as $t \to \infty$. However, with probability 1,

$$\lim_{t\to\infty} S_t(\omega) = \begin{cases} \infty, & \mu - \frac{1}{2}\sigma^2 > 0 \,, \\ 0, & \mu - \frac{1}{2}\sigma^2 < 0 \,. \end{cases}$$

Example 11.8 Solve $\mathrm{d}X_t = X_t^3\mathrm{d}t - X_t^2\mathrm{d}W_t$, $X_0 = 1$. Assuming $X_t = f(t, W_t)$, we have

$$\mathrm{d}X_t = \big(f_t + \frac{1}{2}f_{xx}\big)\mathrm{d}t + f_x\mathrm{d}W_t \,,$$

and hence

$$\begin{cases} f_t + \dfrac{1}{2}f_{xx} = f^3 \,, \\ f_x = -f^2 \,. \end{cases}$$

From the second equation we have

$$-\frac{1}{f^2}f_x = 1 \,,$$

and hence

$$\frac{\partial}{\partial x}\left(\frac{1}{f}\right) = 1 \,,$$

and

$$\frac{1}{f(t,x)} = x + C(t) \,,$$

and finally

$$f(t,x) = \frac{1}{x + C(t)} \,.$$

From the first equation in the above, we have

$$-\frac{C'(t)}{(x + C(t))^2} + \frac{1}{2}\frac{(-1)(-2)}{(x + C(t))^3} = \frac{1}{(x + C(t))^3} \,.$$

Since

$$-C'(t)(x + C(t)) + 1 = 1$$

holds for every (t, x), we conclude that C is constant. Thus

$$f = \frac{1}{x + C} .$$

Since $f(0, 0) = \frac{1}{C} = 1$, we have $C = 1$. Therefore $f(t, x) = \frac{1}{x+1}$, and finally we obtain

$$f(t, W_t) = \frac{1}{W_t + 1} .$$

Example 11.9 (Brownian Bridge) A stochastic process X_t satisfying the SDE

$$\mathrm{d}X_t = \frac{b - X_t}{T - t}\mathrm{d}t + \mathrm{d}W_t , \quad 0 \le t \le T , \quad X(0) = a ,$$

is called a *Brownian bridge*. It can be written as

$$X_t = a\big(1 - \frac{t}{T}\big) + b\frac{t}{T} + (T - t)\int_0^t \frac{1}{T - s}\mathrm{d}W_s .$$

Since $\int_0^t 1/(T - s)^2 \,\mathrm{d}s < \infty$, the integral $\int_0^t 1/(T - s)\,\mathrm{d}W_s$ is a martingale.

Example 11.10 (Tanaka) Let $f(t, x) = |x|$. Then f is continuous, and

$$f_x(t, x) = \begin{cases} +1 , & x > 0 , \\ -1 , & x < 0 . \end{cases}$$

The Itô formula cannot be directly applied in this case since f_x does not exist at $x = 0$. However, we may regard $2 \times \delta_0(x)$ as the second order derivative of f at $x = 0$ where $\delta_0(x)$ is the Dirac delta measure at 0. If the Itô formula were to hold for

$$f(t, x) = |x| ,$$

then we would have

$$\mathrm{d}f = (f_t + \frac{1}{2}f_{xx})\,\mathrm{d}t + f_x\,\mathrm{d}W_t = \delta_0(x)\,\mathrm{d}t + \mathrm{sign}(x)\,\mathrm{d}W_t$$

and

$$|W_t| = \int_0^t \delta_0(W_s)\,\mathrm{d}s + \int_0^t \mathrm{sign}(W_s)\,\mathrm{d}W_s$$

in a suitable sense where

$$\mathrm{sign}(x) = \begin{cases} 1, & x \geq 0 \\ -1, & x < 0 \,. \end{cases}$$

To proceed rigorously, we define the *local time* for Brownian motion at 0 by

$$L_t(\omega) = \lim_{\varepsilon \to 0} \frac{1}{2\varepsilon} \lambda(\{s \in [0,t] : W_s(\omega) \in (-\epsilon, \epsilon)\})$$

where the limit is taken in the L^2-sense, and obtain *Tanaka's formula*

$$|W_t| = L_t + \int_0^t \mathrm{sign}(W_s)\,\mathrm{d}W_s \,.$$

For a sketch of the proof, we modify f near $x = 0$ as follows:

$$f_\varepsilon(t,x) = \begin{cases} |x| \,, & |x| \geq \varepsilon \\ \frac{1}{2}\left(\varepsilon + \dfrac{x^2}{\varepsilon}\right), & |x| < \varepsilon \,. \end{cases}$$

Then f_ε is continuous, and

$$\frac{\partial}{\partial x} f_\varepsilon(t,x) = \begin{cases} +1 \,, & x \geq \varepsilon \\ \dfrac{x}{\varepsilon} \,, & |x| < \varepsilon \\ -1 \,, & x \leq -\varepsilon \,. \end{cases}$$

It can be shown that

$$f_\varepsilon(W_t) = f_\varepsilon(W_0) + \int_0^t f_\varepsilon'(W_s)\mathrm{d}W_s + \frac{1}{2\varepsilon}\lambda(\{s \in [0,t] : W_s \in (-\varepsilon, \varepsilon)\})$$

where λ denotes the Lebesgue measure on the real line. Applying the Itô isometry to

$$\mathbb{E}\left[\left(\int_0^t \frac{W_s}{\varepsilon} \mathbf{1}_{\{W_s \in (-\varepsilon,\varepsilon)\}} \mathrm{d}W_s\right)^2\right],$$

we obtain

$$\int_0^t f_\varepsilon'(W_s) \mathbf{1}_{\{W_s \in (-\varepsilon,\varepsilon)\}} \mathrm{d}W_s = \int_0^t \frac{W_s}{\varepsilon} \mathbf{1}_{\{W_s \in (-\varepsilon,\varepsilon)\}} \mathrm{d}W_s \to 0$$

in L^2 as $\varepsilon \to 0$. Finally, we let $\varepsilon \to 0$. For more details, consult [85]. See also Example 12.3.

11.4 Multidimensional Brownian Motion and the Itô Formula

In this section we investigate multidimensional stochastic calculus used in multi-asset continuous financial models.

Theorem 11.4 *Suppose that* $\{W_t^1\}_{t\geq 0}$ *and* $\{W_t^2\}_{t\geq 0}$ *are two independent Brownian motions and let* ρ, $-1 \leq \rho \leq 1$, *be a constant. Then the process*

$$X_t = \rho\, W_t^1 + \sqrt{1-\rho^2}\, W_t^2$$

is a Brownian motion.

Proof

(i) $X_0 = 0$, X_t is continuous.
(ii) For $s < t$, we have

$$\begin{aligned} X_t - X_s &= (\rho\, W_t^1 + \sqrt{1-\rho^2}\, W_t^2) - (\rho\, W_s^1 + \sqrt{1-\rho^2}\, W_s^2) \\ &= \rho\,(W_t^1 - W_s^1) + \sqrt{1-\rho^2}\,(W_t^2 - W_s^2)\,, \end{aligned}$$

which is a sum of two normal random variables of mean 0 and variance $\rho^2(t-s)$ and $(1-\rho^2)(t-s)$, respectively. Hence $X_t - X_s$ is also normally distributed with mean $0+0=0$ and variance $\rho^2(t-s) + (1-\rho^2)(t-s) = t-s$.
(iii) For $0 \leq t_1 < t_2 \leq t_3 < t_4 \leq \cdots \leq t_{2n-1} < t_{2n}$ the random variables

$$X(t_{2k}) - X(t_{2k-1}) = \rho(W^1(t_{2k}) - W^1(t_{2k-1})) + \sqrt{1-\rho^2}(W^2(t_{2k}) - W^2(t_{2k-1}))$$

are independent for $1 \leq k \leq n$. □

Here is the uncorrelated multidimensional Itô formula.

Fact 11.5 *Given two independent Brownian motions* W^1, W^2, *we consider two stochastic processes*

$$\mathrm{d}X_t = \mu_X \mathrm{d}t + \sigma_{X,1}\,\mathrm{d}W_t^1 + \sigma_{X,2}\,\mathrm{d}W_t^2$$

and

$$\mathrm{d}Y_t = \mu_Y \mathrm{d}t + \sigma_{Y,1}\,\mathrm{d}W_t^1 + \sigma_{Y,2}\,\mathrm{d}W_t^2\,.$$

If $f(x,y)$ is sufficiently smooth, then

$$\mathrm{d}f(X,Y) = f_x\,\mathrm{d}X + f_y\,\mathrm{d}Y + \frac{1}{2}\{f_{xx}(\sigma_{X,1}^2 + \sigma_{X,2}^2) + f_{yy}(\sigma_{Y,1}^2 + \sigma_{Y,2}^2) + 2f_{xy}(\sigma_{X,1}\sigma_{Y,1} + \sigma_{X,2}\sigma_{Y,2})\}\,\mathrm{d}t\,.$$

Remark 11.5 Using $\mathrm{d}W_1\,\mathrm{d}W_2 = 0$ we may rewrite the above formula as

$$\mathrm{d}f(X,Y) = f_x\,\mathrm{d}X + f_y\,\mathrm{d}Y + \frac{1}{2}\{f_{xx}\mathrm{d}X\,\mathrm{d}X + f_{yy}\mathrm{d}Y\,\mathrm{d}Y + 2f_{xy}\mathrm{d}X\,\mathrm{d}Y\}\,.$$

In general, two Brownian motions W_1 and W_2 are not independent but correlated, i.e., there exists a constant ρ such that for every t we have $\mathbb{E}[W_t^1 W_t^2] = \rho t$, which may be rewritten simply as $\mathrm{d}W^1\mathrm{d}W^2 = \rho\,\mathrm{d}t$. Then we obtain the following correlated multidimensional Itô formula.

Fact 11.6 (Product Rule) *Suppose that there exists a constant ρ such that $\mathbb{E}[W_t^1 W_t^2] = \rho t$ for $t \geq 0$. If $f(x,y)$ is differentiable sufficiently many times as needed, then we have*

$$\mathrm{d}f(X,Y) = f_x\,\mathrm{d}X + f_y\,\mathrm{d}Y + \frac{1}{2}\Big\{f_{xx}(\sigma_{X,1}^2 + \sigma_{X,2}^2) + f_{yy}(\sigma_{Y,1}^2 + \sigma_{Y,2}^2) + 2f_{xy}(\sigma_{X,1}\sigma_{Y,1} + \sigma_{X,2}\sigma_{Y,2} + \rho\,\sigma_{X,1}\sigma_{Y,2} + \rho\,\sigma_{X,2}\sigma_{Y,1})\Big\}\,\mathrm{d}t\,.$$

Remark 11.6 Using $\mathrm{d}W^1\,\mathrm{d}W^2 = \rho\,\mathrm{d}t$, we rewrite the above formula as

$$\mathrm{d}f(X,Y) = f_x\,\mathrm{d}X + f_y\,\mathrm{d}Y + \frac{1}{2}\Big(f_{xx}\mathrm{d}X\,\mathrm{d}X + f_{yy}\mathrm{d}Y\,\mathrm{d}Y + 2f_{xy}\mathrm{d}X\,\mathrm{d}Y\Big)\,.$$

Example 11.11 For $f(x,y) = xy$ we have

$$\mathrm{d}(XY) = X\mathrm{d}Y + Y\mathrm{d}X + (\sigma_{X,1}\sigma_{Y,1} + \sigma_{X,2}\sigma_{Y,2} + \rho\,\sigma_{X,1}\sigma_{Y,2} + \rho\,\sigma_{X,2}\sigma_{Y,1})\,\mathrm{d}t\,.$$

Example 11.12 As a special case of Example 11.11 we consider $X_t = Y_t = W_t$. In this case, we have $\mu_X = 0$, $\sigma_{X,1} = 1$, $\sigma_{X,2} = 0$, $\mu_Y = 0$, $\sigma_{Y,1} = 0$, $\sigma_{Y,2} = 1$. Since $\mathrm{d}X\mathrm{d}Y = \mathrm{d}t$ and $\rho = 1$, we obtain the formula

$$\mathrm{d}((W_t)^2) = 2W_t\,\mathrm{d}W_t + \mathrm{d}t\,,$$

which coincides with Example 11.3(ii).

Remark 11.7 Given two stochastic differential equations

$$\mathrm{d}X_t^i = \mu_i(t, X_t^i)\,\mathrm{d}t + \sigma_i(t, X_t^i)\,\mathrm{d}W_t\,, \quad i = 1, 2\,,$$

we consider two martingales M^1 and M^2 defined by

$$M_t^i = \int_0^t \sigma_i(s, X_s^i)\,\mathrm{d}W_s\ , \quad i = 1, 2\ .$$

Then

$$\left[M^1, M^2\right]_t = \int_0^t \sigma_1(s, X_s^1)\,\sigma_2(s, X_s^2)\,\mathrm{d}s\ .$$

Remark 11.8 The stochastic differential equation in the multidimensional Black–Scholes–Merton model consisting of N risky assets and d Brownian motions is of the form

$$\mathrm{d}S_t^i = \mu^i S_t^i\,\mathrm{d}t + \sum_{j=1}^{d} \sigma^{ij} S_t^i\,\mathrm{d}W_t^j, \quad i = 1, \ldots, N$$

where the matrix $\left[\sigma^{ij}\right]_{1\leq i\leq N, 1\leq j\leq d}$ is called a volatility matrix.

11.5 Computer Experiments

Simulation 11.1 (Itô Formula)

We test the rule $(\mathrm{d}W)^2 = \mathrm{d}t$. See Fig. 11.3.

```
T = 3;
K = 15;
M = 300; % number of Brownian paths
L2 = zeros(1,K);

for k = 1:K
    dt = T*2^(-k);
    dW = sqrt(dt)*randn(M,2^k);
    for m = 1:M
    error(m) = sum(dW(m,1:2^k).^2) - T;
    end
    L2(k) = sqrt(mean(error(1:M).^2)); % L2-norm
end

x = 0:0.01:K;
plot(x, T*2.^((1-x)/2),'b-');
hold on;
plot(L2,'or-')
```

Simulation 11.2 (Itô Formula)
We test the rule $(\mathrm{d}W)^2 = \mathrm{d}t$. See Fig. 11.4.

```
T = 3;
K = 15;
M = 300; % number of Brownian paths
L2 = zeros(1,K);

for k = 1:K
    dt = T*2^(-k);
    dW = sqrt(dt)*randn(M,2^k);
    W = zeros(M,2^k+1);
    error = zeros(M,1);
    epsilon = zeros(M,2^k);
    for m = 1:M
        for i = 1:2^k
        W(m,i+1) = W(m,i) + dW(m,i);
        epsilon(m,i)=0.5*sin(W(m,i))*(dt-dW(m,i)^2);
        end
    error(m,1) = sum( epsilon(m,1:2^k) );
    end
    L2(k) = sqrt(mean(error(1:M,1).^2)); % L2-norm
end

x = 0:0.01:K;
plot(x,T*2.^(-(x+1)/2),'b');
hold on;
plot(L2,'or-')
```

Simulation 11.3 (Geometric Brownian Motion)
We plot sample paths of geometric Brownian motion. For an output see Fig. 11.5.

```
N = 200;  % number of time steps
T = 10;
dt = T/N;
mu = 0.15;
sigma = 0.25;
time = 0:dt:T;
M = 30; % number of sample paths
S = zeros(M,N + 1); % dW and S are matrices.
S0 = 1;
S(:,1)=S0;

dW = sqrt(dt)*randn(M,N);

for i=1:N
    S(:,i+1) = S(:,i) + mu*S(:,i)*dt + sigma*S(:,i).*dW(:,i);
end

for j = 1:M
    plot(time,S(j,1:N + 1),'b');
hold on;
end
```

Exercises

11.1 Solve the stochastic differential equation $\mathrm{d}X_t = 5X_t\,\mathrm{d}t + 3X_t\,\mathrm{d}W_t$ where $X_0 = 1$.

11.2 If a stock S_t pays dividend continuously, what would be a modification of the geometric Brownian motion model for S_t?

11.3

(i) Let $X_t > 0$ for every t. Using the relationship $0 = \mathrm{d}\left(X_t \frac{1}{X_t}\right)$, express $\mathrm{d}\left(\frac{1}{X_t}\right)$ in terms of X_t and $\mathrm{d}X_t$.
(ii) Compute $\mathrm{d}\left(\mathrm{e}^{W_t}\right)$.
(iii) Compute $\mathrm{d}\left(\mathrm{e}^{-W_t}\right)$.

11.4

(i) Let X_t be a stochastic process given by $\mathrm{d}X_t = a_t\,\mathrm{d}t + b_t\,\mathrm{d}W_t$ for some processes a_t and b_t. Compute $\mathrm{d}(\mathrm{e}^{X_t})$.
(ii) Let r_t, $t \geq 0$, be a stochastic process, and define $Z_t = \exp\left(-\int_0^t r_s \mathrm{d}s\right)$. Compute $\mathrm{d}Z_t$.
(iii) Suppose that a process $B_t > 0$ satisfies $\mathrm{d}B_t = r_t B_t \mathrm{d}t$ for some process r_t. Find B_t.

11.5 Solve the stochastic differential equation $\mathrm{d}X_t = \mathrm{d}t + 2\sqrt{X_t}\,\mathrm{d}W_t$ where $X_0 = 1$.

11.6 Compute $\int_0^t W_s\,\mathrm{d}W_s$ by assuming that $\int_0^t W_s\,\mathrm{d}W_s = f(t, W_t)$ for some f. (If this assumption is wrong, then we may try other candidates.)

11.7

(i) Show that there exists no sufficiently smooth function $f(t,x)$ such that $\int_0^t W_s^2\,\mathrm{d}W_s = f(t, W_t)$.
(ii) Assume that $\int_0^t W_s^2\,\mathrm{d}W_s = f(t, W_t) + \int_0^t g(s, W_s)\,\mathrm{d}s$ for some f and g. Compute $\int_0^t W_s^2\,\mathrm{d}W_s$.

11.8 Under the assumption that $\int_0^t s\,\mathrm{d}W_s = f(t, W_t) + \int_0^t g(s, W_s)\,\mathrm{d}s$ for some f and g, find $\int_0^t s\,\mathrm{d}W_s$.

11.9 For a deterministic function $f(t)$, show that

$$\int_0^t f(s)\,\mathrm{d}W_s = f(t)W_t - \int_0^t W_s f'(s)\,\mathrm{d}s\ .$$

11.10 Compute $\int_0^t \mathrm{e}^{W_s}\,\mathrm{d}W_s$ by assuming that $\int_0^t \mathrm{e}^{W_s}\,\mathrm{d}W_s = f(t, W_t) + \int_0^t g(s, W_s)\,\mathrm{d}s$ for some f and g.

11.11 Let $a_k(t) = \mathbb{E}[W_t^k]$ for $k \geq 0$ and $t \geq 0$. (i) Apply the Itô formula to show that

$$a_k(t) = \frac{1}{2}k(k-1)\int_0^t a_{k-2}(s)\,\mathrm{d}s\,, \quad k \geq 2\,.$$

(ii) Prove that $\mathbb{E}[W_t^{2k+1}] = 0$ and

$$\mathbb{E}[W_t^{2k}] = \frac{(2k)!t^k}{2^k k!}\,.$$

11.12 Let $X_t = \int_0^t a_s\,\mathrm{d}W_s$ for some process a_t. Prove that if a_t is bounded, then $X_t^2 - \int_0^t a_s^2\,\mathrm{d}s$ is a martingale. For example, $W_t^2 - t$ is a martingale.

11.13 Let $\mathrm{d}S_t = \mu S_t\,\mathrm{d}t + \sigma S_t\,\mathrm{d}W_t$. Compute the following without solving for S_t explicitly: (i) $\mathrm{d}(\mathrm{e}^{-rt}S_t)$, (ii) $\mathrm{d}(S_t^2)$, (iii) $\mathrm{d}(\log S_t)$, (iv) $\mathrm{d}\,(1/S_t)$, (v) $\mathrm{d}(\sqrt{S_t})$

11.14 Let $a(t)$, $b(t)$, $c(t)$, $d(t)$ be continuous deterministic functions, and consider a stochastic differential equation

$$X_t = \int_0^t (a(s)X_s + b(s))\,\mathrm{d}s + \int_0^t (c(s)X_s + d(s))\,\mathrm{d}W_s\,.$$

Let $\alpha(t) = \mathbb{E}[X_t]$ and $\beta(t) = \mathbb{E}[X_t^2]$. Show that $\alpha' = a\alpha + b$ and $\beta' = (2a + c^2)\beta + 2(b + cd)\alpha + d^2$.

Chapter 12
Stochastic Differential Equations

Let $\mathbf{W} = (W^1, \ldots, W^m)$ be an m-dimensional Brownian motion, and let

$$\boldsymbol{\sigma} = (\sigma_{ij})_{1 \le i \le d, 1 \le j \le m} : [0, \infty) \times \mathbb{R}^d \to \mathbb{R}^d \times \mathbb{R}^m$$

and

$$\boldsymbol{\mu} = (\mu^1, \ldots, \mu^d) : [0, \infty) \times \mathbb{R}^d \to \mathbb{R}^d$$

be continuous functions where $\boldsymbol{\sigma}$ is regarded as a $d \times m$ matrix.

Consider a stochastic differential equation (SDE) given by

$$X_t^i = x_0^i + \int_0^t \mu^i(s, \mathbf{X}_s) \, \mathrm{d}s + \sum_{j=1}^m \int_0^t \sigma_{ij}(s, \mathbf{X}_s) \, \mathrm{d}W_s^j ,$$

or equivalently,

$$\mathbf{X}_t = \mathbf{x}_0 + \int_0^t \boldsymbol{\mu}(s, \mathbf{X}_s) \, \mathrm{d}s + \int_0^t \boldsymbol{\sigma}(s, \mathbf{X}_s) \, \mathrm{d}\mathbf{W}_s$$

where $\boldsymbol{\mu}$ and σ are called coefficients. By a solution of the preceding SDE we mean a stochastic process $\mathbf{X}_t$ satisfying the SDE. We will show that there exists a unique solution. In the differential notation the given SDE is written as

$$\mathrm{d}X_t^i = \mu^i(t, \mathbf{X}_t) \, \mathrm{d}t + \sum_{j=1}^m \sigma_{ij}(t, \mathbf{X}_t) \, \mathrm{d}W_t^j ,$$

or

$$\mathrm{d}\mathbf{X}_t = \boldsymbol{\mu}(t, \mathbf{X}_t) \, \mathrm{d}t + \boldsymbol{\sigma}(t, \mathbf{X}_t) \, \mathrm{d}\mathbf{W}_t . \quad (12.1)$$

G.H. Choe, *Stochastic Analysis for Finance with Simulations*, Universitext,
DOI 10.1007/978-3-319-25589-7_12

Throughout the chapter we consider only the case that $d = 1$ and $m = 1$ for the sake of notational simplicity. The statements of facts and the corresponding proofs for multidimensional case are almost identical to those for the one-dimensional case.

12.1 Strong Solutions

Definition 12.1 A function $g : \mathbb{R}^k \to \mathbb{R}^\ell$ is *Lipschitz continuous* if there exists a constant $0 < L < \infty$ such that

$$||g(\mathbf{x}) - g(\mathbf{y})|| \leq L||\mathbf{x} - \mathbf{y}|| .$$

Definition 12.2 We are given a Brownian motion $\{W_t\}$ and a stochastic process $\{X_t\}$. If the integrals $\int_0^t \mu(s, X_s)\mathrm{d}s$ and $\int_0^t \sigma(s, X_s)\mathrm{d}W_s$ exist for every $t \geq 0$, and if

$$X_t = X_0 + \int_0^t \mu(s, X_s)\,\mathrm{d}s + \int_0^t \sigma(s, X_s)\,\mathrm{d}W_s ,$$

then X_t is called a *strong solution* of the SDE given by (12.1).

Theorem 12.1 (Existence and Uniqueness) *We are given a stochastic differential equation*

$$\mathrm{d}X_t = \mu(t, X_t)\,\mathrm{d}t + \sigma(t, X_t)\,\mathrm{d}W_t , \quad X_0 = x_0 .$$

Suppose that μ and σ are Lipschitz continuous. Then there exists a unique solution X_t that is continuous and adapted.

Proof Recall that for a locally bounded predictable process Y_t, we have by Doob's L^2-inequality

$$\mathbb{E}\left[\sup_{0\leq s\leq t}\left|\int_0^s Y_u \mathrm{d}W_u\right|^2\right] \leq 4\,\mathbb{E}\left[\int_0^t Y_u^2 \mathrm{d}u\right]$$

and by the Cauchy–Schwarz inequality

$$\sup_{0\leq s\leq t}\left|\int_0^s Y_u \mathrm{d}u\right|^2 \leq t\int_0^t Y_u^2 \mathrm{d}u .$$

Start with $X_t^0 = x_0$, $t \geq 0$, and define X^n inductively for $n \geq 0$ by

$$X_t^{n+1} = x_0 + \int_0^t \mu(X_s^n)\,\mathrm{d}s + \int_0^t \sigma(X_s^n)\,\mathrm{d}W_s . \tag{12.2}$$

For $n \geq 1$ let

$$\mu^n(t) = \mathbb{E}\left[\sup_{0\leq s\leq t} |X_s^n - X_s^{n-1}|^2\right].$$

Fix $T \geq 1$. Then for $0 \leq t \leq T$ we have

$$\begin{aligned}
\mu^1(t) &= \mathbb{E}\left[\sup_{0\leq s\leq t}\left|\int_0^s \mu(x_0)\,\mathrm{d}r + \int_0^s \sigma(x_0)\,\mathrm{d}W_r\right|^2\right] \\
&\leq 2\,\mathbb{E}\left[\sup_{0\leq s\leq t}\left|\int_0^s \mu(x_0)\,\mathrm{d}r\right|^2\right] + 2\,\mathbb{E}\left[\sup_{0\leq s\leq t}\left|\int_0^s \sigma(x_0)\,\mathrm{d}W_r\right|^2\right] \\
&\leq 2t\,\mathbb{E}\left[\int_0^t |\mu(x_0)|^2\mathrm{d}r\right] + 8\,\mathbb{E}\left[\int_0^t |\sigma(x_0)|^2\mathrm{d}s\right] \\
&\leq 2t^2|\mu(x_0)|^2 + 8t|\sigma(x_0)|^2 \\
&\leq 10Tt\,(|\mu(x_0)|^2 + |\sigma(x_0)|^2)
\end{aligned}$$

where the first inequality holds since $(a+b)^2 \leq 2a^2+2b^2$ and the second inequality is due to the Cauchy–Schwarz inequality and Doob's L^2-inequality. For $n \geq 1$ we have

$$\begin{aligned}
\mu^{n+1}(t) &\leq 2\,\mathbb{E}\left[\sup_{0\leq s\leq t}\left|\int_0^s (\mu(X_r^n) - \mu(X_r^{n-1}))\,\mathrm{d}r\right|^2\right] \\
&\quad + 2\,\mathbb{E}\left[\sup_{0\leq s\leq t}\left|\int_0^s (\sigma(X_r^n) - \sigma(X_r^{n-1}))\,\mathrm{d}r\right|^2\right] \\
&\leq 10K^2T\int_0^t \mu^n(r)\,\mathrm{d}r\,.
\end{aligned}$$

By mathematical induction, we have

$$\mu^n(t) \leq C\frac{(10TKt)^n}{n!}$$

for $0 \leq t \leq T$ where

$$C = \frac{\mu(x_0)^2 + \sigma(x_0)^2}{K^2}\,.$$

Hence

$$\left\|\sum_{n=1}^{\infty}\sup_{0\leq s\leq T}\left|X_s^n - X_s^{n-1}\right|\right\|_2 \leq \sum_{n=1}^{\infty}\sqrt{\mu^n(t)} < \infty$$

where $||\cdot||_2$ denotes the L^2-norm. Therefore there exists a continuous adapted process X_t such that

$$\lim_{n\to\infty} \sup_{0\le s\le T} \left|X_s^n - X_s\right| = 0$$

with probability 1 and in L^2. Thus we have

$$\lim_{n\to\infty} \mathbb{E}\left[\sup_{0\le s\le T}\left|\int_0^s \mu(X_r^n)\,\mathrm{d}r - \int_0^s \mu(X_r)\,\mathrm{d}r\right|^2\right] = 0\,,$$

$$\lim_{n\to\infty} \mathbb{E}\left[\sup_{0\le s\le t}\left|\int_0^s \sigma(X_r^n)\,\mathrm{d}W_r - \int_0^s \sigma(X_r)\,\mathrm{d}W_r\right|^2\right] = 0\,.$$

Letting $n\to\infty$ in (12.2), we observe that X must satisfy the given SDE.

As for the uniqueness, suppose that X and Y are solutions of the given SDE. Define

$$g(t) = \mathbb{E}\left[\sup_{0\le s\le t} |X_s - Y_s|^2\right]\,.$$

Proceeding as before we deduce

$$g(t) \le 10K^2T\int_0^t g(s)\,\mathrm{d}s\,, \quad 0\le t\le T$$

and conclude that $g = 0$ by Gronwall's inequality. (See Lemma C.1.) □

Example 12.1 (Langevin Equation) Let α and σ be constants. The solution of the Langevin equation

$$\mathrm{d}X_t = -\alpha X_t\,\mathrm{d}t + \sigma\,\mathrm{d}W_t\,, \quad X_0 = x_0\,,$$

is called the Ornstein–Uhlenbeck process. If $\sigma = 0$, then the SDE becomes an ordinary differential equation, and the solution is deterministic and given by $X_t = x_0\mathrm{e}^{-\alpha t}$. Hence we see that $X_t\mathrm{e}^{\alpha t} = x_0$ is constant. For $\sigma \neq 0$ the process $Y_t = X_t\mathrm{e}^{\alpha t}$ will move up and down about the constant x_0. Since $\mathrm{d}Y_t = \sigma\mathrm{e}^{\alpha t}\mathrm{d}W_t$, we have a solution $Y_t = Y_0 + \sigma\int_0^t \mathrm{e}^{\alpha s}\mathrm{d}W_s$, and hence

$$X_t = e^{-\alpha t}x_0 + e^{-\alpha t}\sigma\int_0^t e^{\alpha s}\mathrm{d}W_s\,. \tag{12.3}$$

For $\alpha > 0$, the mean and the variance are given by

$$\mathbb{E}[X_t] = \mathrm{e}^{-\alpha t}x_0$$

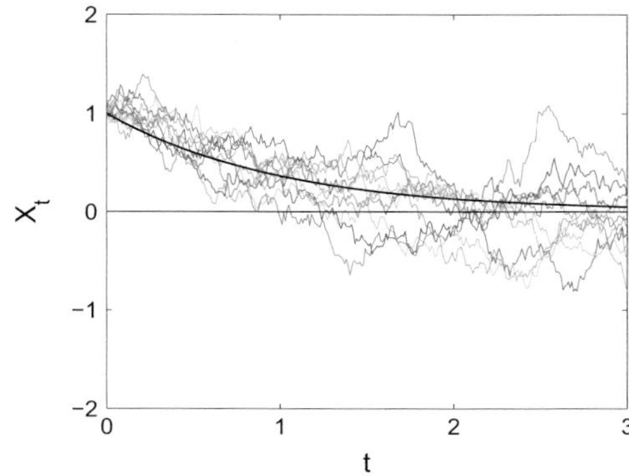

Fig. 12.1 Sample paths of the Ornstein–Uhlenbeck process

and

$$\begin{aligned}\mathrm{Var}(X_t) &= \mathbb{E}[(X_t - \mathrm{e}^{-\alpha t}x_0)^2] \\ &= \mathbb{E}\left[\left(\mathrm{e}^{-\alpha t}\sigma \int_0^t \mathrm{e}^{\alpha s}\mathrm{d}W_s\right)^2\right] \\ &= \mathrm{e}^{-2\alpha t}\sigma^2\,\mathbb{E}\left[\int_0^t \mathrm{e}^{2\alpha s}\mathrm{d}s\right] \qquad \text{(Itô isometry)} \\ &= \sigma^2 \frac{1-\mathrm{e}^{-2\alpha t}}{2\alpha}\ .\end{aligned}$$

In Fig. 12.1 are plotted sample paths of the Ornstein–Uhlenbeck process with $\alpha = 1$, $\sigma = 0.5$ and $x_0 = 1$ together with its mean $\mathrm{e}^{-\alpha t}x_0$ for $0 \le t \le 3$. See Simulation 12.2.

Theorem 12.2 *For a continuous stochastic process X_t of finite variation, suppose that $\mathcal{E}_t$ satisfies*

$$\mathrm{d}\mathcal{E}_t = \mathcal{E}_t\,\mathrm{d}X_t\ , \quad \mathcal{E}_0 = 1\ .$$

Then the solution, called a stochastic exponential, is given by

$$\mathcal{E}_t = \exp\left(X_t - X_0 - \frac{1}{2}[X,X]_t\right)\ .$$

Proof Let $Y_t = X_t - X_0 - \frac{1}{2}[X,X]_t$, then

$$\mathrm{d}\left(\mathrm{e}^{Y_t}\right) = \mathrm{e}^{Y_t}\mathrm{d}Y_t + \frac{1}{2}\mathrm{e}^{Y_t}\mathrm{d}[Y,Y]_t\ .$$

Since $[X,[X,X]]_t = 0$, we have $[Y,Y]_t = [X,X]_t$, and hence

$$d\mathcal{E}_t = e^{Y_t} dX_t - \frac{1}{2} e^{Y_t} d[X,X]_t + \frac{1}{2} e^{Y_t} d[X,X]_t = e^{Y_t} dX_t \,.$$

To prove the uniqueness, suppose that V_t is another solution, and show that $d(V_t/\mathcal{E}_t) = 0$. □

Example 12.2

(i) If $X_t = W_t$ is a Brownian motion in Theorem 12.2, then

$$\mathcal{E}_t = e^{-\frac{1}{2}t + W_t} \,.$$

(ii) If $\mathcal{E}_t = S_t$ is the price of a stock and $X_t = R_t$ is the return on the investment in the stock defined by $dR_t = \frac{dS_t}{S_t}$, then $dS_t = S_t dR_t$, and hence

$$S_t = S_0 e^{R_t - R_0 - \frac{1}{2}[R,R]_t} \,.$$

If S_t is given by a geometric Brownian motion $dS_t = \mu S_t dt + \sigma S_t dW_t$, then $R_t = \mu t + \sigma W_t$ and

$$S_t = S_0^{(\mu - \frac{1}{2}\sigma^2)t + \sigma W_t} \,.$$

(iii) If $\mathcal{E}_t = B_t$ is the price of a risk-free bond and $X_t = R_t$ the return on the investment in the bond defined by $dR_t = \frac{dB_t}{B_t}$, then $dB_t = B_t \, dR_t$ and R_t satisfies $dR_t = g(t)\, dt$ for some deterministic function $g(t)$. Hence $R(t) - R(0) = \int_0^t g(t)\, dt$ and $[R,R]_t = 0$. Thus

$$B_t = B_0 e^{R_t - R_0 - \frac{1}{2}[R,R]_t} = B_0 e^{\int_0^t g(t)\, dt} \,.$$

Remark 12.1 The Markov property means that given a present state of a process, the future is independent of the past. A solution X_t of the SDE in Theorem 12.1 has the Markov property. Consult [74].

12.2 Weak Solutions

Definition 12.3 Suppose that there exists a probability space with a filtration $\widetilde{\mathcal{F}}_t$, a Brownian motion $\widetilde{W}_t$ and a process $\widetilde{X}_t$ adapted to $\widetilde{\mathcal{F}}_t$ such that $\widetilde{X}_0$ has the given distribution, and $\widetilde{X}_t$ satisfies

$$\widetilde{X}_t = \widetilde{X}_0 + \int_0^t \mu(s, \widetilde{X}_s)\, ds + \int_0^t \sigma(s, \widetilde{X}_s)\, d\widetilde{W}_s \,.$$

Then $\widetilde{X}_t$ is called a *weak solution* of the SDE

$$\mathrm{d}X_t = \mu(t, X_t)\,\mathrm{d}t + \sigma(t, X_t)\,\mathrm{d}W_t \;.$$

By definition, a strong solution is a weak solution. In the following we consider an SDE with a discontinuous coefficient for which Theorem 12.1 is not applicable since the coefficient is not Lipschitz.

Example 12.3 Consider the stochastic differential equation

$$\mathrm{d}X_t = \operatorname{sign}(X_t)\,\mathrm{d}W_t$$

where

$$\operatorname{sign}(x) = \begin{cases} 1, & x \geq 0 \\ -1, & x < 0 \;. \end{cases}$$

The function $\operatorname{sign}(x)$ does not satisfy the Lipschitz condition, and it is known that the equation does not have a strong solution. However, the Brownian motion is the unique weak solution. For, if we take any Brownian motion $\widetilde{W}_t$ for X_t, and define

$$Y_t = \int_0^t \operatorname{sign}(\widetilde{W}_s)\,\mathrm{d}\widetilde{W}_s = \int_0^t \operatorname{sign}(X_s)\,\mathrm{d}X_s \;,$$

then $\operatorname{sign}(\widetilde{W}_s)$ is adapted,

$$\int_0^T \operatorname{sign}(\widetilde{W}_s)^2 \mathrm{d}t = \int_0^T 1\,\mathrm{d}t = T < \infty \;,$$

and Y_t is a continuous martingale such that

$$[Y, Y]_t = \int_0^t \operatorname{sign}(\widetilde{W}_s)^2 \mathrm{d}[\widetilde{W}, \widetilde{W}]_s = \int_0^t 1\,\mathrm{d}s = t \;.$$

Then Y_t is a Brownian motion by Lévy's theorem. Finally, since

$$\mathrm{d}Y_t = \operatorname{sign}(X_t)\,\mathrm{d}X_t \;,$$

multiplying both sides by $\operatorname{sign}(X_t)$, we have

$$\mathrm{d}X_t = \operatorname{sign}(X_t)\,\mathrm{d}Y_t \;.$$

See also Example 11.10. For the proof of the uniqueness of the weak solution, consult [74].

12.3 Brownian Bridges

In Theorem 7.6 we showed how to construct sample paths of a Brownian motion constrained at $t = 1$ using successive approximations. In this section we present a constrained Brownian motion as a stochastic process.

Definition 12.4 A stochastic process $\{X_t\}_{t\geq 0}$ is called a *Gaussian process* if for $0 < t_1 < \cdots < t_n$ any linear combination of $X_{t_1}, \ldots, X_{t_n}$ is jointly normally distributed.

Example 12.4 A Brownian motion W_t is a Gaussian process. To see why, note that for $0 < t_1 < t_2 < \cdots < t_n$, the increments

$$\delta W_1 = W_{t_1} - W_0 \,, \ \ldots \,, \delta W_n = W_{t_n} - W_{t_{n-1}}$$

are independent and normal. Since

$$W_{t_1} = \delta W_1 \,, \ W_{t_2} = \delta W_1 + \delta W_2 \,, \ \ldots \,, \ W_{t_n} = \delta W_1 + \cdots + \delta W_n \,,$$

the random variables $W_{t_1}, W_{t_2}, \ldots, W_{t_n}$ are jointly normally distributed. (See Remark 4.10.) Recall that the covariance function is given by $c(s,t) = \min\{s,t\}$, $0 \leq s \leq t$.

Example 12.5 Let $f(t)$ be a deterministic function of time t, and define $X_t = \int_0^t f(s)\mathrm{d}W_s$ where W_t is a Brownian motion. Then X_t is a Gaussian process. For the proof, recall that for a real constant θ the process

$$M_t = \exp\left(\theta X_t - \frac{1}{2}\theta^2 \int_0^t f(s)^2 \mathrm{d}s\right)$$

is a martingale. Hence

$$1 = M_0 = \mathbb{E}[M_t] = \exp\left(-\frac{1}{2}\theta^2 \int_0^t f(s)^2 \mathrm{d}s\right) \mathbb{E}[\mathrm{e}^{\theta X_t}] \,,$$

and hence the moment generating function is given by

$$\mathbb{E}[\mathrm{e}^{\theta X_t}] = \exp\left(\frac{1}{2}\theta^2 \int_0^t f(s)^2 \mathrm{d}s\right) \,,$$

which is a moment generating function for a normal distribution with mean 0 and variance $\int_0^t f(s)^2 \mathrm{d}s$. Thus X_t is normally distributed with the same mean and variance. Now it remains to prove that for $0 < t_1 < \cdots < t_n$ the random variables $X_{t_1}, \ldots, X_{t_n}$ are jointly normal. Note that the increments

$$\delta X_1 = X_{t_1} - X_0 = X_{t_1} \,, \ \delta X_2 = X_{t_2} - X_{t_1} \,, \ \ldots \,, \ \delta X_n = X_{t_n} - X_{t_{n-1}}$$

are normally distributed and independent. Since

$$X_{t_1} = \delta X_1 \,,\ X_{t_2} = \delta X_1 + \delta X_2 \,,\ \ldots\,,\ X_{t_n} = \delta X_1 + \cdots + \delta X_n \,,$$

the random variables $X_{t_1}, X_{t_2}, \ldots, X_{t_n}$ are jointly normally distributed. (See Remark 4.10.) For the details, consult [94].

Definition 12.5 (Brownian Bridge) Let W_t be a Brownian motion. For $T > 0$, the *Brownian bridge from* 0 *to* 0 is defined by

$$X_t = W_t - \frac{t}{T} W_T \,, \quad 0 \le t \le T \,.$$

More generally, the *Brownian bridge from a to b* defined by

$$X_t^{a\to b} = a + \frac{t}{T}(b-a) + W_t - \frac{t}{T} W_T \,, \quad 0 \le t \le T \,.$$

See Simulation 12.3 and Fig. 12.2 for ten sample paths of the Brownian bridge from $a = 0$ to $b = 2$ for $T = 3$.

Remark 12.2

(i) Note that $X_t = X_t^{0\to 0}$ and that $X_0^{a\to b} = a$, $X_T^{a\to b} = b$.
(ii) Since we have to know the values of W_T to define X_t, $0 \le t \le T$, the Brownian bridge X_t is not adapted to the filtration generated by W_t.
(iii) For $0 < t_1 < \cdots < t_n < T$, the random variables

$$X_{t_1} = W_{t_1} - \frac{t_1}{T} W_T \,,\ \ldots\,,\ X_{t_n} = W_{t_n} - \frac{t_n}{T} W_T$$

are jointly normal since $W_{t_1}, \ldots, W_{t_n}, W_T$ are jointly normal. Hence the Brownian bridge from 0 to 0 is a Gaussian process, and so is $X_t^{a\to b}$ since

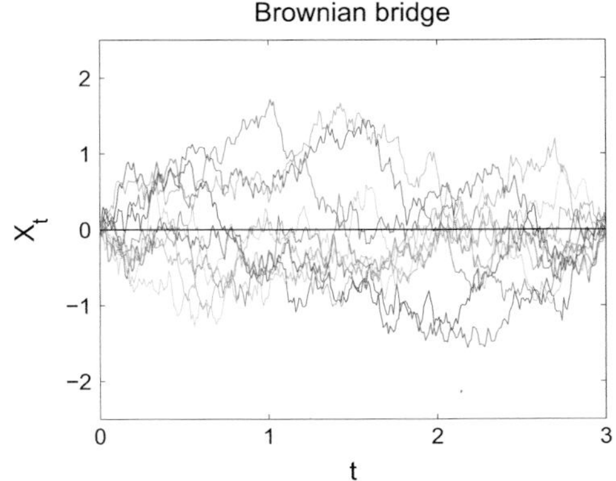

Fig. 12.2 Sample paths of the Brownian bridge $X_t = X_t^{0\to 0}$ from $a = 0$ to $b = 0$ for $T = 3$

addition of a deterministic function to a Gaussian process X_t gives a new Gaussian process.

(iv) $\mathbb{E}[X_t] = \mathbb{E}[W_t - \frac{t}{T}W_T] = 0$ and $\mathbb{E}[X_t^{a\to b}] = a + \frac{t}{T}(b-a)$.

(v) For $0 < s \le t < T$, the covariance function $c(s,t) = c(t,s)$ is given by

$$\begin{aligned} c(s,t) &= \mathbb{E}\left[(W_s - \frac{s}{T}W_T)\,(W_t - \frac{t}{T}W_T)\right] \\ &= \mathbb{E}\left[W_sW_t - \frac{s}{T}W_tW_T - \frac{t}{T}W_sW_T + \frac{st}{T^2}W_T^2\right] \\ &= s - \frac{s}{T}t - \frac{t}{T}s + \frac{st}{T^2}T \\ &= \frac{s\,(T-t)}{T}\,. \end{aligned}$$

The covariance function of $X_t^{a\to b}$ is the same.

(vi) The Brownian bridge cannot be expressed as an Itô integral $\int_0^t f(s)\,\mathrm{d}W_s$ of a deterministic function $f(t)$ since

$$\mathrm{Var}(X_t) = \mathbb{E}[X_t^2] = \frac{t\,(T-t)}{T}$$

which increases as t increases for $0 \le t \le \frac{T}{2}$ and then decreases to zero for $\frac{T}{2} \le t \le T$. However, the Itô isometry implies that the variance of the Itô integral is given by

$$\mathrm{Var}\left(\int_0^t f(s)\,\mathrm{d}W_s\right) = \mathbb{E}\left[\left(\int_0^t f(s)\,\mathrm{d}W_s\right)^2\right] = \int_0^t f(s)^2\mathrm{d}s\,,$$

which increases monotonically for $0 \le t \le T$.

(vii) For an application of the Brownian bridge in simulating an at-the-money digital option, see Simulation 17.2.

Theorem 12.3 (Brownian Bridge) *Define a process Y_t, $0 \le t \le T$, by*

$$Y_t = (T-t)\int_0^t \frac{1}{T-s}\mathrm{d}W_s\,, \quad 0 \le t < T\,,$$

and $Y_T = 0$. Then the following facts hold:

(i) Y_t is adapted to the filtration generated by the Brownian motion W_t.

(ii) Y_t satisfies the stochastic differential equation

$$\mathrm{d}Y_t = -\frac{1}{T-t}\,Y_t\,\mathrm{d}t + \mathrm{d}W_t\,.$$

(iii) *Y_t is a continuous Gaussian process on $[0, T]$ and has mean $m(t) = 0$ for every t and covariance*

$$c(s,t) = \frac{s(T-t)}{T}, \quad 0 \le s \le t \le T .$$

(iv) *Y_t has the same distribution as the Brownian bridge X_t from 0 to 0.*

Proof

(i) Since Y_t is defined by W_s, $0 \le s \le t$, it is adapted to the filtration generated by W_s, $0 \le s \le t$.

(ii) Use the fact

$$\mathrm{d}Y_t = \left(\int_0^t \frac{1}{T-s} \mathrm{d}W_s \right) (-\mathrm{d}t) + (T-t) \frac{1}{T-t} \mathrm{d}W_t .$$

(iii) For $0 \le t < T$, the process

$$I_t = \int_0^t \frac{1}{T-s} \mathrm{d}W_s$$

is a Gaussian process by the argument in Example 12.5. Since $I_{t_1}, \ldots, I_{t_n}$ are jointly normal, the random variables

$$Y_{t_1} = (T-t_1) I_{t_1} , \ \ldots \ , \ Y_{t_n} = (T-t_n) I_{t_n}$$

are jointly normal. Hence Y_t is a Gaussian process for $0 \le t < T$. The mean of Y_t is equal to 0 since the expectation of an Itô integral is 0. For $0 \le s \le t < T$, the Itô isometry implies that the covariance of Y is given by

$$\begin{aligned}
c(s,t) &= \mathbb{E}\left[(T-s) \int_0^s \frac{1}{T-u} \mathrm{d}W_u \times (T-t) \int_0^t \frac{1}{T-u} \mathrm{d}W_u \right] \\
&= (T-s)(T-t)\, \mathbb{E}\left[\int_0^t \mathbf{1}_{[0,s]}(u) \frac{1}{T-u} \mathrm{d}W_u \times \int_0^t \frac{1}{T-u} \mathrm{d}W_u \right] \\
&= (T-s)(T-t)\, \mathbb{E}\left[\int_0^t \mathbf{1}_{[0,s]}(u) \frac{1}{(T-u)^2} \mathrm{d}u \right] \\
&= (T-s)(T-t) \int_0^t \mathbf{1}_{[0,s]}(u) \frac{1}{(T-u)^2} \mathrm{d}u \\
&= (T-s)(T-t) \int_0^s \frac{1}{(T-u)^2} \mathrm{d}u \\
&= (T-s)(T-t) \left(\frac{1}{T-s} - \frac{1}{T} \right) \\
&= \frac{s(T-t)}{T} .
\end{aligned}$$

The variance of Y_t, which is equal to $\frac{t(T-t)}{T}$, $0 \le t < T$, converges to 0 as $t \uparrow T$. In combination with the fact that the mean of Y_t is always equal to 0, the definition $Y_T = 0$ enables Y_t to be continuous at $t = T$.

(iv) Since Y_t has the same mean and covariance functions $m(t)$ and $c(s,t)$ as the Brownian bridge X_t from 0 to 0 and since a Gaussian process is completely determined by $m(t)$ and $c(s,t)$, the process Y_t has the same distribution as X_t.

□

See Simulation 12.4 and Fig. 12.4 for ten sample paths of the process

$$Y_t = (T-t)\int_0^t \frac{1}{T-s}\,\mathrm{d}W_s$$

with $a = 0$, $b = 2$ and $T = 3$, and see Simulation 12.5 and Fig. 12.5 for the corresponding SDE

$$\mathrm{d}Y_t = -\frac{1}{T-t}\,Y_t\,\mathrm{d}t + \mathrm{d}W_t\ .$$

Theorem 12.4 (Probability Density of Brownian Bridge) *Let X_t be the Brownian bridge from 0 to 0 for $0 \le t \le T$ in Definition 12.5. Take $0 = t_0 < t_1 < \cdots < t_n < T$ and put $\delta t_i = t_i - t_{i-1}$. The joint density $f_{X_{t_1},\dots,X_{t_n}}$ for $X_{t_1},\dots,X_{t_n}$ is given by*

$$f_{X_{t_1},\dots,X_{t_n}}(x_1,\dots,x_n) = \frac{p(T-t_n, x_n, 0)}{p(T,0,0)}\prod_{i=1}^{n} p(\delta t_i, x_{i-1}, x_i)$$

where $x_0 = 0$ and

$$p(\delta t, x, x') = \frac{1}{\sqrt{2\pi\,\delta t}}\exp\left(-\frac{(x'-x)^2}{2\,\delta t}\right)$$

is the transition density for Brownian motion. Note that $f_{X_{t_1},\dots,X_{t_n}}$ is equal to the joint density of a Brownian motion W_t at $t_1,\dots,t_n$ conditional on $W_T = 0$, and hence X_t has the same probability distribution as W_t conditional on $W_T = 0$.

Proof Put $\tau_0 = T$ and $\tau_i = T - t_i$. Define

$$Y_i = \frac{1}{\tau_i}X_{t_i} - \frac{1}{\tau_{i-1}}X_{t_{i-1}}\ .$$

Since each of $Y_{t_1},\dots,Y_{t_n}$ is a linear combination of $X_{t_1},\dots,X_{t_n}$, which are jointly normal, we see that $Y_{t_1},\dots,Y_{t_n}$ are also jointly normal. Then

$$\mathbb{E}[Y_i] = \frac{1}{\tau_i}\mathbb{E}[X_{t_i}] - \frac{1}{\tau_{i-1}}\mathbb{E}[X_{t_{i-1}}] = 0\ ,$$

$$
\begin{aligned}
\mathrm{Var}(Y_i) &= \mathbb{E}[Y_i^2] \\
&= \frac{1}{\tau_i^2}\mathrm{Var}(X_{t_i}) - \frac{2}{\tau_i\tau_{i-1}}\mathrm{Cov}(X_{t_i}, X_{t_{i-1}}) + \frac{1}{\tau_{i-1}^2}\mathrm{Var}(X_{t_{i-1}}) \\
&= \frac{1}{\tau_i^2}\frac{t_i(T-t_i)}{T} - \frac{2}{\tau_i\tau_{i-1}}\frac{t_{i-1}(T-t_i)}{T} + \frac{1}{\tau_{i-1}^2}\frac{t_{i-1}(T-t_{i-1})}{T} \\
&= \frac{1}{\tau_i}\frac{t_i}{T} - \frac{2}{\tau_{i-1}}\frac{t_{i-1}}{T} + \frac{1}{\tau_{i-1}}\frac{t_{i-1}}{T} \\
&= \frac{\tau_{i-1}t_i - 2t_{i-1}\tau_i + t_{i-1}\tau_i}{\tau_{i-1}\tau_i T} \\
&= \frac{\tau_{i-1}t_i - t_{i-1}\tau_i}{\tau_{i-1}\tau_i T} \\
&= \frac{(T-t_{i-1})t_i - t_{i-1}(T-t_i)}{\tau_{i-1}\tau_i T} \qquad \text{(since } \tau_i = T - t_i) \\
&= \frac{\delta t_i}{\tau_{i-1}\tau_i} \,.
\end{aligned}
$$

For $i < j$, we have $t_{i-1} < t_i \le t_{j-1} < t_j$ and

$$
\begin{aligned}
&\mathrm{Cov}(Y_i, Y_j) \\
&= \mathrm{Cov}\left(\frac{1}{\tau_i}X_{t_i} - \frac{1}{\tau_{i-1}}X_{t_{i-1}}, \frac{1}{\tau_j}X_{t_j} - \frac{1}{\tau_{j-1}}X_{t_{j-1}}\right) \\
&= \frac{1}{\tau_i\tau_j}c(t_i, t_j) - \frac{1}{\tau_i\tau_{j-1}}c(t_i, t_{j-1}) - \frac{1}{\tau_{i-1}\tau_j}c(t_{i-1}, t_j) + \frac{1}{\tau_{i-1}\tau_{j-1}}c(t_{i-1}, t_{j-1}) \\
&= \frac{t_i}{\tau_i T} - \frac{t_i}{\tau_i T} - \frac{t_{i-1}}{\tau_{i-1}T} + \frac{t_{i-1}}{\tau_{i-1}T} = 0 \,.
\end{aligned}
$$

Hence the jointly normal variables $Y_{t_1}, \ldots, Y_{t_n}$ are independent and their joint density is given by

$$
f_{Y_{t_1},\ldots,Y_{t_n}}(y_1, \ldots, y_n) = \prod_{i=1}^{n} \frac{1}{\sqrt{2\pi \frac{\delta t_i}{\tau_{i-1}\tau_i}}} \exp\left(-\frac{1}{2\frac{\delta t_i}{\tau_{i-1}\tau_i}} y_i^2\right) \,.
$$

Now we change the variables from $(y_1, \ldots, y_n)$ to $(x_1, \ldots, x_n)$ by the rule

$$
y_i = \frac{x_i}{\tau_i} - \frac{x_{i-1}}{\tau_{i-1}} \,, \; 1 \le i \le n
$$

where x_i represents the value taken by X_{t_i}, the Brownian bridge from 0 to 0 at time t_i. Since

$$\begin{aligned}
\frac{1}{\frac{\delta t_i}{\tau_{i-1}\tau_i}} y_i^2 &= \frac{\tau_{i-1}\tau_i}{\delta t_i}\left(\frac{x_i^2}{\tau_i^2} - 2\frac{x_i x_{i-1}}{\tau_{i-1}\tau_i} + \frac{x_{i-1}^2}{\tau_{i-1}^2}\right) \\
&= \frac{1}{\delta t_i}\left(\frac{\tau_{i-1}}{\tau_i}x_i^2 - 2x_i x_{i-1} + \frac{\tau_i}{\tau_{i-1}}x_{i-1}^2\right) \\
&= \frac{1}{\delta t_i}\left((1+\frac{\delta t_i}{\tau_i})x_i^2 - 2x_i x_{i-1} + (1-\frac{\delta t_i}{\tau_{i-1}})x_{i-1}^2\right) \\
&= \frac{1}{\delta t_i}\left((x_i - x_{i-1})^2 + \frac{\delta t_i}{\tau_i}x_i^2 - \frac{\delta t_i}{\tau_{i-1}}x_{i-1}^2\right) \\
&= \frac{1}{\delta t_i}(x_i - x_{i-1})^2 + \frac{1}{\tau_i}x_i^2 - \frac{1}{\tau_{i-1}}x_{i-1}^2 \,,
\end{aligned}$$

we have

$$\begin{aligned}
\prod_{i=1}^n \exp\left(-\frac{1}{2\frac{\delta t_i}{\tau_{i-1}\tau_i}} y_i^2\right) &= \exp\left(-\frac{1}{2}\sum_{i=1}^n \frac{1}{\frac{\delta t_i}{\tau_{i-1}\tau_i}} y_i^2\right) \\
&= \exp\left(-\frac{1}{2}\sum_{i=1}^n \frac{1}{\delta t_i}(x_i - x_{i-1})^2 - \frac{1}{2}\frac{1}{T-t_n}x_n^2\right) .
\end{aligned}$$

Also note that

$$\prod_{i=1}^n \frac{1}{\sqrt{2\pi\frac{\delta t_i}{\tau_{i-1}\tau_i}}} = \frac{\sqrt{T}}{\sqrt{T-t_n}}\prod_{i=1}^n \frac{1}{\sqrt{2\pi\delta t_i}}\prod_{i=1}^n \tau_i \,.$$

Since

$$\det\left[\frac{\partial y_i}{\partial x_j}\right] = \prod_{i=1}^n \frac{1}{\tau_i}$$

and $p(T,0,0) = \sqrt{2\pi T}$, we have

$$\begin{aligned}
&f_{X_{t_1},\ldots,X_{t_n}}(x_1,\ldots,x_n) \\
&= f_{Y_{t_1},\ldots,Y_{t_n}}(y_1,\ldots,y_n)\prod_{i=1}^n \frac{1}{\tau_i} \\
&= \frac{\sqrt{T}}{\sqrt{T-t_n}}\exp\left(-\frac{1}{2}\frac{x_n^2}{T-t_n}\right)\prod_{i=1}^n \frac{1}{\sqrt{2\pi\delta t_i}}\exp\left(-\frac{1}{2}\sum_{i=1}^n \frac{(x_i-x_{i-1})^2}{\delta t_i}\right)
\end{aligned}$$

$$= \frac{\sqrt{2\pi T}}{\sqrt{2\pi(T-t_n)}} \exp\left(-\frac{1}{2}\frac{x_n^2}{T-t_n}\right) \prod_{i=1}^{n} \frac{1}{\sqrt{2\pi\delta t_i}} \exp\left(-\frac{1}{2}\sum_{i=1}^{n}\frac{(x_i-x_{i-1})^2}{\delta t_i}\right)$$

$$= \frac{p(T-t_n, x_n, 0)}{p(T,0,0)} \prod_{i=1}^{n} p(\delta t_i, x_{i-1}, x_i) \ .$$

□

Corollary 12.1 *The Brownian bridge $X_t^{a\to b}$ from a to b in Definition 12.5 has the same probability distribution as the Brownian motion conditional on $W_0 = a$ and $W_T = b$.*

Proof Let

$$X_t^{a\to b} = a + \frac{t}{T}(b-a) + W_t - \frac{t}{T}W_T$$

be the Brownian bridge from a to b in Definition 12.5 for some probability measure $\mathbb{P}$ and a $\mathbb{P}$-Brownian motion W_t. Then

$$X_t^{0\to b-a} = \frac{t}{T}(b-a) + W_t - \frac{t}{T}W_T$$

is the Brownian bridge from 0 to $b-a$. Recall that

$$X_t = X_t^{0\to 0} = W_t - \frac{t}{T}W_T$$

is the Brownian bridge from 0 to 0, and has the same probability distribution as W_t for $0 \le t \le T$ conditional on $W_T = 0$. Let $\theta = \frac{b-a}{T}$. Define a new probability measure $\mathbb{Q}$ by

$$\frac{d\mathbb{Q}}{d\mathbb{P}} = e^{-\frac{1}{2}\theta^2 T - \theta W_T} \ .$$

By Girsanov's theorem, $\widetilde{W}_t = W_t + \theta t$, $0 \le t \le T$, is a $\mathbb{Q}$-Brownian motion. Note that

$$X_t^{0\to b-a} = X_t + \theta t$$

has the same probability distribution as $W_t + \theta t$ for $0 \le t \le T$ conditional on $W_T = 0$. Since $W_T = 0$ if and only if $\widetilde{W}_T = b-a$, we note that $X_t^{0\to b-a}$ has the same probability distribution as a Brownian motion $\widetilde{W}_t$ conditional on $\widetilde{W}_T = b-a$. Therefore

$$X_t^{a\to b} = a + X_t^{0\to b-a}$$

has the same probability distribution as a generalized Brownian motion $\widehat{W}_t = a + \widetilde{W}_t$ conditional on $\widehat{W}_T = b$. □

Remark 12.3

(i) In the proof of Corollary 12.1 we may directly derive the probability distribution of the Brownian bridge from a to b and show that it is equal to the probability distribution of the Brownian motion conditional on $W_0 = a$ and $W_T = b$ as done in [94]. Our approach is to simplify the notational burden in doing so and to prove Theorem 12.4 first, because its proof is relatively simple.
(ii) In Monte Carlo simulations for option pricing, Corollary 12.1 enables us to use Definition 12.5 to simulate the Brownian motion conditional on $W_T = b$ instead of constructing a Brownian motion using the successive approximations given in Theorem 7.6.

12.4 Computer Experiments

Simulation 12.1 (Numerical Solution of an SDE)

We compare a numerical solution of the SDE given by

$$dX_t = 2W_t dW_t + dt$$

with the exact solution $X_t = W_t^2 + X_0$. See Fig. 12.3.

```
T= 3.0;
N = 500;
dt = T/N;
t = 0:dt:T;
dW = sqrt(dt)*randn(1,N);
W = zeros(1,N+1);
```

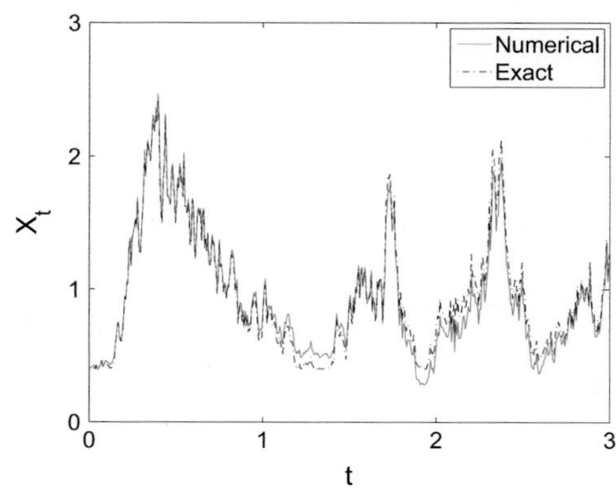

Fig. 12.3 A numerical solution of $dX_t = 2W_t dW_t + dt$ and the exact solution $X_t = W_t^2 + X_0$

```
X = zeros(1,N+1);
X0 = 0.4;
X(1) = X0;

Exact = zeros(1,N+1);
Exact(1) = X0;

for i = 1:N
    W(i+1) = W(i) + dW(i);
    X(i+1) = X(i) + 2*W(i)*dW(i) + dt;
    Exact(i+1)=W(i+1)^2 + X0;
end
plot(t,X,'r-',t,Exact,'k-.');
hlegend=legend('approx','exact');
```

Simulation 12.2 (Ornstein–Uhlenbeck Process)
We generate M sample paths of the Ornstein–Uhlenbeck process

$$dX_t = -\alpha X_t dt + \sigma dW_t$$

together with the curve $\mathbb{E}[X_t] = e^{-\alpha t}x_0$, $0 \le t \le T$. See Fig. 12.1.

```
T= 3;
N = 200;
dt = T/N;
t = 0:dt:T;
M = 10;
X = zeros(M,N);
alpha = 1;
sigma = 0.5;
x0 = 1;
X(:,1) = x0 ;
dW = sqrt(dt)*randn(M,N);
for i = 1:N
    X(:,i+1) = X(:,i) - alpha*X(:,i)*dt + sigma*dW(:,i);
end
for j = 1:M
    plot(t,X(j,:));
    hold on
end
t = 0:0.01:T;
plot(t,x0*exp(-alpha*t))
```

Simulation 12.3 (Brownian Bridge)
We generate numerical realizations of the Brownian bridge

$$X_t = X_t^{0\to 0} = W_t - \frac{t}{T}W_T$$

from $a = 0$ to $b = 0$ for $T = 3$ in Definition 12.5. We take $N = 100$ for the number of subintervals, $\delta t = \frac{T}{N}$, and $M = 10$ for the number of sample paths. See Fig. 12.2.

```
T= 3.0;
N = 100;
dt = T/N;
t = 0:dt:T;
M = 10;
W = zeros(M,N);
X = zeros(M,N);
a = 0;
b = 0;
X(:,1) = a;
dW = sqrt(dt)*randn(M,N);

for i = 1:N
    W(:,i+1) =  W(:,i) + dW(:,i);
end
for i = 1:N+1
    X(:,i) = a*(1-(i-1)*dt)/T + b*(i-1)*dt/T + W(:,i) -(i-1)*dt/T*W(:,N+1);
end

for j = 1:M
    plot(t,X(j,:));
    hold on
end
```

Simulation 12.4 (Brownian Bridge Given by an Integral)

We generate the sample paths of the stochastic process Y_t defined in Theorem 12.3. As in Simulation 12.3 we take $a = 0$, $b = 2$, $T = 3$, $N = 100$, $\delta t = \frac{T}{N}$ and $M = 10$. See Fig. 12.4.

```
Y = zeros(M,N+1);
Y(:,1) = a;
Integral = zeros(M,N+1);
Integral(:,1) = 0;
dW = sqrt(dt)*randn(M,N);

for i = 1:N
    Integral(:,i+1) =  Integral(:,i) + (1/(T-(i-1)*dt))*dW(:,i);
end
for i = 1:N+1
    Y(:,i) = a*(1-(i-1)*dt/T) + b*(i-1)*dt/T + (T-(i-1)*dt)*Integral(:,i);
end

for j = 1:M
    plot(t,Y(j,:));
    hold on
end
```

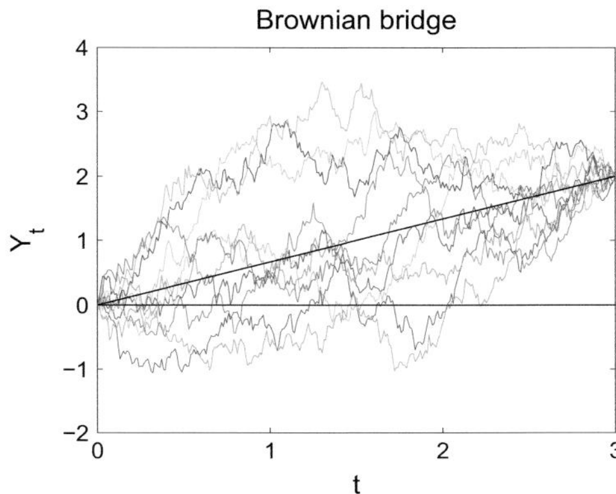

Fig. 12.4 Sample paths of the process $Y_t = (T-t)\int_0^t \frac{1}{T-s} \mathrm{d}W_s$ for $0 \le t \le 3$

Simulation 12.5 (Brownian Bridge Given by an SDE)

We generate the sample paths of the SDE given in Theorem 12.3. As in Simulations 12.3, 12.4 we take $a = 0$, $b = 2$, $T = 3$, $N = 100$, $\delta t = \frac{T}{N}$ and $M = 10$.

```
Y = zeros(M,N+1);
Y(:,1) = a;
dW = sqrt(dt)*randn(M,N);

for i = 1:N
    Y(:,i+1) =  Y(:,i) + (b-Y(:,i))/(T-(i-1)*dt)*dt + dW(:,i);
    hold on
end

for j = 1:M
    plot(t,Y(j,:));
    hold on
end
```

See Fig. 12.5 for ten sample paths of the SDE given by

$$\mathrm{d}Y_t = \frac{1}{T-t}(b - Y_t)\,\mathrm{d}t + \mathrm{d}W_t\,, \quad Y_0 = a$$

with $a = 0$, $b = 2$ and $T = 3$. In the numerical realizations of the SDE we take $\delta t = \frac{T}{N} = \frac{3}{100}$, and due to the discretization error in the final time step, Y_T cannot be uniquely determined, but is distributed around $b = 2$. As N becomes larger, Y_T becomes more concentrated around b.

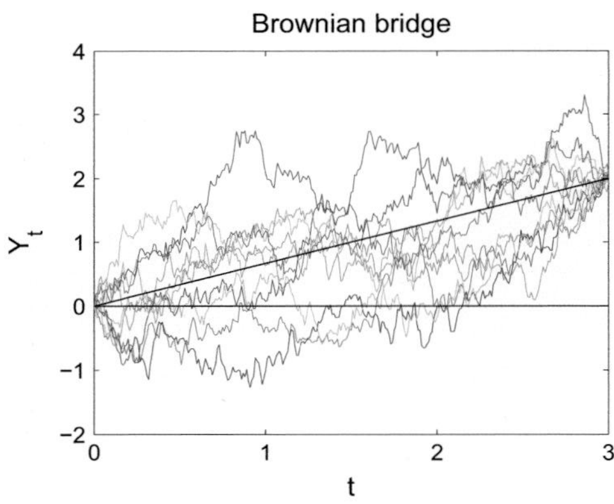

Fig. 12.5 Brownian bridge: sample paths of the SDE $\mathrm{d}Y_t = \frac{1}{T-t}(b - Y_t)\,\mathrm{d}t + \mathrm{d}W_t$, $Y_0 = a$ and $Y_T = b$, for $0 \le t \le T = 3$

Exercises

12.1 What is a stochastic differential equation satisfied by $Y_t = {W_t}^3$?

12.2 Show that $\mathrm{d}X_t = 3X_t^{1/3}\mathrm{d}t + 3X_t^{2/3}\mathrm{d}W_t$, $X_0 = 1$, has infinitely many solutions.

12.3 Show that the solution of

$$\mathrm{d}X_t = \mathrm{d}t + 2\sqrt{X_t}\,\mathrm{d}W_t$$

is given by

$$X_t = (W_t + \sqrt{X_0}\,)^2 \ .$$

12.4 Show that the solution of

$$\mathrm{d}X_t = -X_t(2\log X_t + 1)\,\mathrm{d}t - 2X_t\sqrt{-\log X_t}\,\mathrm{d}W_t$$

is given by

$$X_t = \exp\left(-\left(W_t + \sqrt{-\log X_0}\right)^2\right) \ .$$

12.5 Show that the solution of the generalized Langevin equation or the mean-reverting Ornstein–Uhlenbeck process

$$\mathrm{d}X_t = \alpha(m - X_t)\mathrm{d}t + \sigma\mathrm{d}W_t$$

is given by

$$X_t = m + e^{-\alpha t}(X_0 - m) + \sigma \int_0^t e^{-\alpha(t-s)} \mathrm{d}W_s \,.$$

(Hint: Take $Y_t = X_t - m$ and apply the result in Example 12.1.)

12.6 Find the solution of the SDE

$$\mathrm{d}X_t = (\alpha_t + \beta_t X_t)\mathrm{d}t + (\gamma_t + \delta_t X_t)\mathrm{d}W_t$$

where α_t, β_t, γ_t, δ_t are adapted continuous processes. (Hint: Consider first the case that $\alpha_t = \gamma_t = 0$.)

12.7 Solve the SDE

$$\mathrm{d}X_t = X_t^3 \mathrm{d}t - X_t^2 \mathrm{d}W_t$$

where $X_0 = 1$.

12.8 Suppose that X_t is the Ornstein–Uhlenbeck process $\mathrm{d}X_t = -\alpha X_t \mathrm{d}t + \sigma \mathrm{d}W_t$ given in (12.3). Does $Y_t = X_t^2$ satisfy

$$\mathrm{d}Y_t = (\sigma^2 - 2\alpha Y_t)\,\mathrm{d}t + 2\sigma\sqrt{Y_t}\,\mathrm{d}W_t \;?$$

12.9 Solve the SDE

$$\mathrm{d}X_t = \frac{3}{4} X_t^2 \,\mathrm{d}t - X_t^{3/2}\,\mathrm{d}W_t \,, \quad X_0 > 0 \,.$$

Chapter 13
The Feynman–Kac Theorem

As an alternative method for the derivation of the Schrödinger differential equation in quantum mechanics, the path integral approach was introduced in the 1960s by the physicist Richard Feynman. Along a similar line a certain type of partial differential equation can be solved using the expectation over the sample paths of a stochastic process. Its mathematical formulation, the Feynman–Kac Theorem, provides a link between two option pricing methods, one based on the Black–Scholes–Merton partial differential equation and the other the martingale method.

13.1 The Feynman–Kac Theorem

Let $F(t, x)$ be a function of t and x representing time and space, respectively. Given a stochastic process X_t, when is $F(t, X_t)$ a martingale? The Martingale Representation Theorem implies that $\mathrm{d}F(t, X_t)$ should have no drift term. If that is the case, we have $F(t, X_t) = \mathbb{E}[F(T, X_T)|\mathcal{F}_t]$ for $0 \leq t \leq T$ by the definition of a martingale, and hence $F(t, x) = \mathbb{E}[F(T, X_T)|X_t = x]$. The following theorem shows how to find a solution of a differential equation using a probabilistic idea.

Theorem 13.1 (Feynman–Kac) *Let $F, \mu, \sigma : [0, T] \times \mathbb{R} \to \mathbb{R}$ be functions of t and x, and let $h : \mathbb{R} \to \mathbb{R}$ be a function of x. Consider a partial differential equation*

$$\begin{cases} \dfrac{\partial F}{\partial t}(t, x) + \mu(t, x)\dfrac{\partial F}{\partial x}(t, x) + \dfrac{1}{2}\sigma^2(t, x)\dfrac{\partial^2 F}{\partial x^2}(t, x) = 0 \ , & 0 < t < T \ , \\ F(T, x) = h(x) \ . & \end{cases}$$

If a stochastic process X_t satisfies a stochastic differential equation

$$\mathrm{d}X_t = \mu(t, X_t)\,\mathrm{d}t + \sigma(t, X_t)\,\mathrm{d}W_t \ ,$$

G.H. Choe, *Stochastic Analysis for Finance with Simulations*, Universitext,
DOI 10.1007/978-3-319-25589-7_13

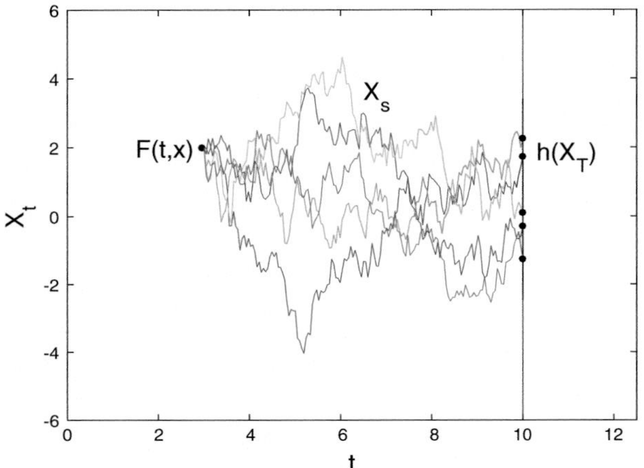

Fig. 13.1 The Feynman–Kac Theorem: Expectation of the final value $h(X_T)$ is the present value $F(t,x)$

then we have

$$F(t,x) = \mathbb{E}[h(X_T)|\mathcal{F}_t]\big|_{X_t=x} = \mathbb{E}[h(X_T)|X_t = x]\ .$$

See Fig. 13.1 where we take $(t,x) = (3,2)$ *and* $T = 10$.

Proof By the Itô formula we have

$$\begin{aligned}
&F(T,X_T) - F(t,X_t)\\
&= \int_t^T \left\{ \frac{\partial F}{\partial s}(s,X_s) + \mu(s,X_s)\frac{\partial F}{\partial x}(s,X_s) + \frac{1}{2}\sigma^2(s,X_s)\frac{\partial^2 F}{\partial x^2}(s,X_s) \right\} \mathrm{d}s\\
&\quad + \int_t^T \sigma(s,X_s)\frac{\partial F}{\partial x}(s,X_s)\,\mathrm{d}W_s\\
&= \int_t^T \sigma(s,X_s)\frac{\partial F}{\partial x}(s,X_s)\,\mathrm{d}W_s\ .
\end{aligned}$$

Taking the conditional expectation $\mathbb{E}[\,\cdot\,|\mathcal{F}_t]$ on both sides, we obtain

$$\mathbb{E}[F(T,X_T)|\mathcal{F}_t] - F(t,X_t) = 0\ .$$

Thus $F(t,X_t) = \mathbb{E}[F(T,X_T)|\mathcal{F}_t] = \mathbb{E}[h(X_T)|\mathcal{F}_t]$. □

Example 13.1 If $\mu = \sigma = 0$, then X_t is constant. In this case the conclusion of the theorem implies that F depends only on x and $F(t,x) = h(x)$.

Example 13.2 If $\sigma = 0$, then X_t is a solution of an ordinary differential equation $\frac{\mathrm{d}X}{\mathrm{d}t} = \mu(t, X)$, and may be regarded as a deterministic function $\phi(t)$. Then

$$\begin{aligned}
\frac{\mathrm{d}}{\mathrm{d}t} F(t, \phi(t)) &= \frac{\partial F}{\partial t}(t, \phi(t)) + \frac{\partial F}{\partial x}(t, \phi(t))\, \phi'(t) \\
&= \frac{\partial F}{\partial t}(t, \phi(t)) + \frac{\partial F}{\partial x}(t, \phi(t))\, \mu(t, \phi(t)) = 0
\end{aligned}$$

by the chain rule. Hence F is constant along the curve $t \mapsto (t, \phi(t))$, and

$$F(t, X_t) = F(t, \phi(t)) = F(T, \phi(T)) = F(T, X_T) = h(X_T)$$

for all t, i.e., $F(t, \phi(t)) = F(T, \phi(T))$.

Example 13.3 (Brownian Motion) For the case that $\mu = 0$ and $\sigma = 1$, we have the simplest stochastic differential equation $\mathrm{d}X_t = \mathrm{d}W_t$. Then

$$X_t = X_0 + \int_0^t \mathrm{d}W_u = X_0 + W_t \ ,$$

and

$$\begin{aligned}
&\mathbb{E}[h(X_T)|\mathcal{F}_t]\big|_{X_t=x} \\
&= \mathbb{E}[h(W_T + X_0)|\mathcal{F}_t]\big|_{W_t+X_0=x} \\
&= \mathbb{E}[h(W_T - W_t + x)|\mathcal{F}_t]\big|_{W_t=x-X_0} \\
&= \mathbb{E}[h(W_T - W_t + x)] \qquad \text{(by the independence of } W_T - W_t \text{ and } \mathcal{F}_t) \\
&= \mathbb{E}[h(W_{T-t} + x)] \qquad (W_T - W_t \text{ and } W_{T-t} \text{ have the same distribution}) \\
&= \int_{-\infty}^{\infty} h(z + x) \frac{1}{\sqrt{2\pi(T-t)}} \exp\left(-\frac{z^2}{2(T-t)}\right) \mathrm{d}z \\
&= \int_{-\infty}^{\infty} h(y) \frac{1}{\sqrt{2\pi(T-t)}} \exp\left(-\frac{(x-y)^2}{2(T-t)}\right) \mathrm{d}y \ .
\end{aligned}$$

The last integral, denoted by $F(t, x)$, is a convolution of h and the heat kernel

$$\frac{1}{\sqrt{2\pi(T-t)}} \exp\left(-\frac{x^2}{2(T-t)}\right)$$

and hence it satisfies a partial differential equation

$$\frac{\partial F}{\partial t} + \frac{1}{2}\frac{\partial^2 F}{\partial x^2} = 0 \ , \qquad F(T, x) = h(x) \ .$$

Example 13.4 (Geometric Brownian Motion) Consider

$$\mathrm{d}X_t = \mu_0 X_t \,\mathrm{d}t + \sigma_0 X_t \,\mathrm{d}W_t$$

for which $\mu(t,x) = \mu_0 x$ and $\sigma(t,x) = \sigma_0 x$. Then

$$X_T = X_t \exp\left((\mu_0 - \frac{1}{2}\sigma_0^2)(T-t) + \sigma_0 W_{T-t}\right) .$$

Consider

$$F_t + \mu_0 x F_x + \frac{1}{2}\sigma_0^2 x^2 F_{xx} = 0 .$$

Then the Feynman–Kac Theorem implies

$$F(t,x) = \mathbb{E}[F(T, X_T)|X_t = x] .$$

Hence

$$F(t,x) = \mathbb{E}\Big[F(T, x\exp\left((\mu_0 - \frac{1}{2}\sigma_0^2)(T-t) + \sigma_0 W_{T-t})\right)\Big] .$$

Definition 13.1 (Infinitesimal Generator) For a stochastic differential equation

$$\mathrm{d}X_t = \mu(t, X_t)\,\mathrm{d}t + \sigma(t, X_t)\,\mathrm{d}W_t \tag{13.1}$$

we define an *infinitesimal generator* by

$$\mathcal{A} = \mu(s,x)\frac{\partial}{\partial x} + \frac{1}{2}\sigma^2(s,x)\frac{\partial^2}{\partial x^2} .$$

13.2 Application to the Black–Scholes–Merton Equation

For a function $F(t,x)$ satisfying the partial differential equation given in Theorem 13.1 we define a new function

$$V(t,x) = \mathrm{e}^{-r(T-t)}F(t,x) .$$

Since

$$\frac{\partial}{\partial t}V(t,x) = r\mathrm{e}^{-r(T-t)}F(t,x) + \mathrm{e}^{-r(T-t)}\frac{\partial}{\partial t}F(t,x) ,$$

V satisfies a partial differential equation

$$\frac{\partial V}{\partial t}(t,x) + \mu(t,x)\frac{\partial V}{\partial x}(t,x) + \frac{1}{2}\sigma^2(t,x)\frac{\partial^2 V}{\partial x^2}(t,x) = rV(t,x)$$

with a final condition

$$V(T,x) = \mathrm{e}^{-r(T-T)}F(T,x) = h(x) .$$

Suppose that a process S_t follows the geometric Brownian motion

$$\mathrm{d}S_t = \mu_0 S_t\,\mathrm{d}t + \sigma_0 S_t\,\mathrm{d}W_t$$

where μ_0 and σ_0 are constant, i.e., $\mu(t,x) = \mu_0 x$ and $\sigma(t,x) = \sigma_0 x$. By Theorem 13.1 we have

$$V(t,x) = \mathrm{e}^{-r(T-t)}F(t,x) = \mathrm{e}^{-r(T-t)}\mathbb{E}[h(S_T)|S_t = x] .$$

In conclusion, the partial differential equation

$$\frac{\partial V}{\partial t}(t,x) + \mu(t,x)\frac{\partial V}{\partial x}(t,x) + \frac{1}{2}\sigma^2(t,x)\frac{\partial^2 V}{\partial x^2}(t,x) = rV(t,x)$$

with a final condition $V(T,x) = h(x)$ has a solution of the form

$$V(t,x) = \mathrm{e}^{-r(T-t)}\mathbb{E}[h(S_T)|S_t = x] . \tag{13.2}$$

Remark 13.1 When V denotes the price of a European option, if we take $\mu_0 = r$ by risk-neutrality, we obtain the Black–Scholes–Merton equation

$$\frac{\partial V}{\partial t}(t,x) + rx\frac{\partial V}{\partial x}(t,x) + \frac{1}{2}\sigma_0^2 x^2\frac{\partial^2 V}{\partial x^2}(t,x) = rV(t,x) ,$$

whose solution is given by (13.2). For more details see Sect. 15.1 and Sect. 16.4.

13.3 The Kolmogorov Equations

In this section we compute the density function of the transition probability of Brownian motion.

I. The Kolmogorov Backward Equation

Theorem 13.2 (Kolmogorov Backward Equation) *If X_t satisfies (13.1), then for a Borel subset $B \subset \mathbb{R}^1$ the transition probability*

$$P(s,x;t,B) = \mathbb{P}(X_t \in B|X_s = x)$$

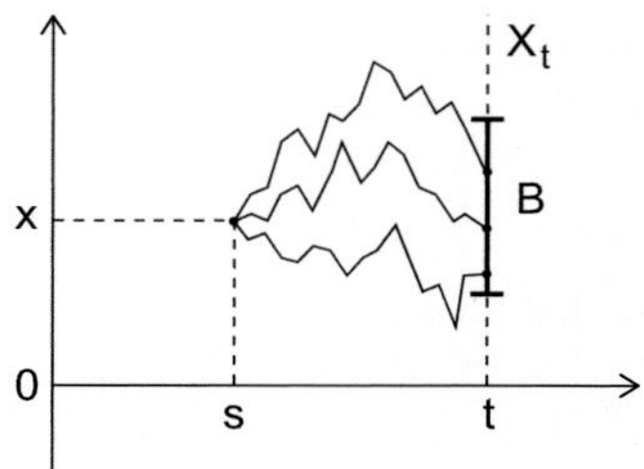

Fig. 13.2 Derivation of the Kolmogorov backward equation

satisfies the Kolmogorov backward equation given by

$$\left(\frac{\partial}{\partial s} + \mathcal{A}\right) P(s, x; t, B) = 0 \,, \quad 0 < s < t \,,$$

$$P(t, x; t, B) = \mathbf{1}_B(x) \,.$$

In other words,

$$\frac{\partial}{\partial s} P = -\mathcal{A} P$$

and

$$P = \mathrm{e}^{-t\mathcal{A}} P_0 \,.$$

Proof In Theorem 13.1 use

$$\mathbb{P}(X_t \in B | X_s = x) = \mathbb{E}[\mathbf{1}_B(X_t) | X_s = x] \,.$$

Consult Fig. 13.2 where B is represented by a vertical segment. □

Corollary 13.1 *For an infinitesimal case, we choose an infinitesimally short interval $B = [y, y + \mathrm{d}y]$. Suppose that the probability density is given by*

$$P(s, x; t, [y, y + \mathrm{d}y]) = p(s, x; t, y) \mathrm{d}y \,.$$

Then p is a solution of the differential equation

$$\left(\frac{\partial}{\partial s} + \mathcal{A}\right) p(s, x; t, y) = 0 \,, \quad 0 < s < t \,,$$

and $p(s, x; t, y)$ converges to the Dirac delta measure $\delta_x(y)$ as $s \to t$ in a suitable sense.

Proof Since

$$p(s,x;t,y)\,\mathrm{d}y = P(s,x;t,[y,y+\mathrm{d}y]) \approx \mathbf{1}_{[y,y+\mathrm{d}y]}(x)$$

as $s \to t$, we use

$$p(s,x;t,y) \approx \frac{1}{\mathrm{d}y}\mathbf{1}_{[y,y+\mathrm{d}y]}(x) \approx \delta_y(x)\ . \qquad \square$$

Example 13.5 Consider the Ornstein–Uhlenbeck process

$$\mathrm{d}X_t = -\alpha\, X_t\,\mathrm{d}t + \sigma\,\mathrm{d}W_t\ .$$

Take $B = (-\infty, y]$ and let

$$F(s,x;t,y) = P(s,x;t,B) = \mathbb{P}(X_t \in B | X_s = x)\ .$$

By the Kolmogorov backward equation, we have

$$\frac{\partial F}{\partial s} - \alpha x \frac{\partial F}{\partial x} + \frac{1}{2}\sigma^2 \frac{\partial^2 F}{\partial x^2} = 0\ , \quad 0 < s < t\ , \tag{13.3}$$

with the final condition

$$F(t,x;t,y) = \mathbf{1}_{(-\infty,y]}(x)\ .$$

Define

$$p(s,x;t,y) = \frac{\partial}{\partial y}F(s,x;t,y)\ .$$

By differentiating the left-hand side of (13.3) with respect to y, we obtain

$$\frac{\partial p}{\partial s} - \alpha x \frac{\partial p}{\partial x} + \frac{1}{2}\sigma^2 \frac{\partial^2 p}{\partial x^2} = 0\ , \quad 0 < s < t\ , \tag{13.4}$$

with the final condition

$$p(t,x;t,y) = \delta_y(x)$$

where δ_y denotes the Dirac delta measure at y. From (13.4), we obtain

$$p(s,y;t,x) = \frac{1}{\sqrt{\pi(1-\mathrm{e}^{-2\alpha(t-s)})\frac{\sigma^2}{\alpha}}} \exp\left(-\frac{(x-\mathrm{e}^{-\alpha(t-s)}y)^2}{(1-\mathrm{e}^{-2\alpha(t-s)})\frac{\sigma^2}{\alpha}}\right)\ .$$

II. The Kolmogorov Forward Equation

Now we derive the Kolmogorov forward equation, which is also called the Fokker–Planck equation.

Definition 13.2 (Adjoint Operator) For an infinitesimal generator

$$\mathcal{A} = \mu(s,x)\frac{\partial}{\partial x} + \frac{1}{2}\sigma^2(s,x)\frac{\partial^2}{\partial x^2}$$

its *adjoint operator* $\mathcal{A}^*$ is defined by

$$(\mathcal{A}^* f)(t,x) = -\frac{\partial}{\partial x}\left(\mu(t,x)f(t,x)\right) + \frac{1}{2}\frac{\partial^2}{\partial x^2}\left(\sigma^2(t,x)f(t,x)\right) .$$

Theorem 13.3 (Kolmogorov Forward Equation) *Let $p(s,x;t,y)$ be the transition probability density of the Itô process X_t introduced previously. Then p satisfies the Kolmogorov forward equation given by*

$$\frac{\partial}{\partial t}p(s,x;t,y) = \mathcal{A}^* p(s,x;t,y) , \quad 0 < t < T$$

$$p(s,x;t,y) \to \delta_x(y) , \quad s \to t .$$

Proof For any function $\phi(t,x)$ that is sufficiently smooth and compactly supported in the domain $(s,T) \times \mathbb{R}^1$ we have

$$\phi(T,X_T) = \phi(s,X_s) + \int_s^T \left(\frac{\partial \phi}{\partial t} + \mathcal{A}\phi\right)(t,X_t)\,\mathrm{d}t + \sigma \int_s^T \frac{\partial \phi}{\partial x}(t,X_t)\,\mathrm{d}W_t$$

by the Itô formula. Here, since $t = T$ and $t = s$ are a part of the boundary of the domain $(s,T) \times \mathbb{R}^1$, we have $\phi(T,X_T) = \phi(s,X_s) = 0$. Furthermore, if we take the conditional expectation $\mathbb{E}[\,\cdot\,|X_s = x]$, then the last integral on the right-hand side is zero. Hence

$$\mathbb{E}\left[\int_s^T \left(\frac{\partial \phi}{\partial t} + \mathcal{A}\phi\right)(t,X_t)\,\mathrm{d}t \,\middle|\, X_s = x\right] = 0 ,$$

and by the definition of a transition probability density $p(s,x;t,y)$ we obtain

$$\int_{-\infty}^{\infty} p(s,x;t,y) \int_s^T \left(\frac{\partial \phi}{\partial t} + \mathcal{A}\phi\right)(t,y)\,\mathrm{d}t\,\mathrm{d}y = 0 .$$

Now by integration by parts with respect to t and y we obtain

$$\int_{-\infty}^{\infty}\int_s^T \phi(t,y)\left(-\frac{\partial}{\partial t} + \mathcal{A}^*\right)p(s,x;t,y)\,\mathrm{d}t\,\mathrm{d}y = 0 .$$

Since the equation holds for arbitrary ϕ we conclude that

$$\left(-\frac{\partial}{\partial t} + \mathcal{A}^*\right) p(s, x; t, y) = 0 \ .$$

Since $s > 0$ is arbitrary, the equation holds for $0 < t < T$. □

Example 13.6 Consider the Ornstein–Uhlenbeck process

$$\mathrm{d}X_t = -\alpha X_t \mathrm{d}t + \sigma \mathrm{d}W_t \ .$$

From the Kolmogorov forward equation

$$\frac{\partial p}{\partial t} = -\frac{\partial}{\partial x}(-\alpha x p) + \frac{\partial^2}{\partial x^2}\left(\frac{1}{2}\sigma^2 p\right) ,$$

we obtain

$$p(s, y; t, x) = \frac{1}{\sqrt{\pi(1 - \mathrm{e}^{-2\alpha(t-s)})\frac{\sigma^2}{\alpha}}} \exp\left(-\frac{(x - \mathrm{e}^{-\alpha(t-s)}y)^2}{(1 - \mathrm{e}^{-2\alpha(t-s)})\frac{\sigma^2}{\alpha}}\right) .$$

This can be obtained directly from the closed form solution of the given SDE. For $0 \le s \le t$, we have

$$\begin{aligned} X_t &= \mathrm{e}^{-\alpha t} x_0 + \mathrm{e}^{-\alpha t}\sigma \int_0^t \mathrm{e}^{\alpha u} \mathrm{d}W_u \\ &= \mathrm{e}^{-\alpha t} x_0 + \mathrm{e}^{-\alpha t}\sigma \left(\int_0^s \mathrm{e}^{\alpha u} \mathrm{d}W_u + \int_s^t \mathrm{e}^{\alpha u} \mathrm{d}W_u\right) \\ &= \mathrm{e}^{-\alpha(t-s)}\left(\mathrm{e}^{-\alpha s} x_0 + \mathrm{e}^{-\alpha s}\sigma \int_0^s \mathrm{e}^{\alpha u} \mathrm{d}W_u\right) + \mathrm{e}^{-\alpha t}\sigma \int_s^t \mathrm{e}^{\alpha u} \mathrm{d}W_u \\ &= \mathrm{e}^{-\alpha(t-s)} X_s + \mathrm{e}^{-\alpha t}\sigma \int_s^t \mathrm{e}^{\alpha u} \mathrm{d}W_u \ . \end{aligned}$$

Let $y = X_s$. Then

$$\mathbb{E}[X_t | X_s = y] = \mathrm{e}^{-\alpha(t-s)} y$$

and

$$\mathrm{Var}(X_t | X_s = y) = \mathrm{e}^{-2\alpha t}\sigma^2 \int_s^t \mathrm{e}^{2\alpha u} \mathrm{d}u = \sigma^2 \frac{1 - \mathrm{e}^{-2\alpha(t-s)}}{2\alpha}$$

by the Itô isometry. Since X_t is Gaussian, we have

$$p(s, y; t, x) = \frac{1}{\sqrt{2\pi(1 - e^{-2\alpha(t-s)})\frac{\sigma^2}{2\alpha}}} \exp\left(-\frac{(x - e^{-\alpha(t-s)}y)^2}{2(1 - e^{-2\alpha(t-s)})\frac{\sigma^2}{2\alpha}}\right) .$$

By letting $t \to \infty$, we can obtain the limiting probability density

$$f(x) = \frac{1}{\sqrt{2\pi\frac{\sigma^2}{2\alpha}}} \exp\left(-\frac{x^2}{2\frac{\sigma^2}{2\alpha}}\right) . \tag{13.5}$$

If we want to find only the limiting probability density directly, then we take $\frac{\partial p}{\partial t} = 0$ in the Kolmogorov forward equation, and solve

$$0 = -\frac{\partial}{\partial x}(-\alpha x f(x)) + \frac{\partial^2}{\partial x^2}\left(\frac{1}{2}\sigma^2 f(x)\right) .$$

Hence

$$C = \alpha x f(x) + \frac{\partial}{\partial x}\left(\frac{1}{2}\sigma^2 f(x)\right)$$

for some constant C. If we look for a probability density $f(x)$ satisfying $\lim_{x\to\pm\infty} x f(x) = 0$, then $C = 0$. Thus we obtain (13.5).

13.4 Computer Experiments

Simulation 13.1 (Feynman–Kac Theorem)

We produce sample paths of the Ornstein–Uhlenbeck process $dX_t = -\alpha X_t dt + \sigma dW_t$ in Fig. 13.1.

```
t0 = 3;        % initial time
X0 = 2 ;       % initial space point
T = 10;        % terminal time
N = 150;       % number of time steps from t0 to T
dt = (T-t0)/N; % length of time interval
t = t0:dt:T;
M = 5;         % number of sample paths
alpha = 0.5;
sigma = 1.5;

X = zeros(M,N);
X(:,1) = X0;
dW = sqrt(dt)*randn(M,N);
for i = 1:N
```

```
    X(:,i+1) = X(:,i) - alpha*X(:,i)*dt + sigma*dW(:,i);
end

for j = 1:M
    plot(t(:),X(j,:),'color',hsv2rgb([1-j/M 1 1]));
    plot(T,X(j,N+1),'.') ;
    hold on
end
```

Now we choose $h(x) = \mathbf{1}_B$ where $B = [a, b]$, and evaluate $\mathbb{E}[h(X_T)]$.

```
a = 1;
b = 3;
X = zeros(M,N);
X(:,1) = X0;
dW = sqrt(dt)*randn(M,N);
for i = 1:N
    X(:,i+1) = X(:,i) - alpha*X(:,i)*dt + sigma*dW(:,i);
end
mean(heaviside(X(:,N)-a).*heaviside(b-X(:,N)))
```

The command `heaviside` is the Heaviside function $H(x) = \mathbf{1}_{[0,\infty)}(x)$, and $H(x-a)H(b-x) = \mathbf{1}_{[a,b]}(x)$. Now we obtain an estimate for $F(t, x)$.

```
ans = 0.2409
```

Part V
Option Pricing Methods

Chapter 14
The Binomial Tree Method for Option Pricing

Not long after the partial differential equation approach was developed for option pricing by Black, Scholes [6] and Merton [65] in 1973, the binomial tree method was introduced by Cox, Ross and Rubinstein [23] in 1979, which is a discrete time model and much easier to understand and implement in practice. The method is quite flexible and allows us to use parameters which may depend on time and asset prices. In this chapter only the binomial tree method is introduced, and the Black–Scholes–Merton differential equation will be presented in Chap. 15.

14.1 Motivation for the Binomial Tree Method

Consider a European call option on an underlying stock with its present price $S_0 =$ \$50 per share. Suppose that at the expiry date T the stock has only two values $S^u =$ \$80 and $S^d =$ \$40 with probabilities $p_u = \frac{1}{2}$, $p_d = \frac{1}{2}$. See Fig. 14.1.

If the strike price is given by $K =$ \$60 and the risk-free interest rate is $r = 0$, then we might conclude that the present value V_0 of the option is given by taking the average of payoffs $\max(S^u - K, 0) =$ \$20 and $\max(S^d - K, 0) =$ \$0 when the stock price rises and falls, respectively, and hence $V_0 = p_u \times \$20 + p_d \times \$0 = \$10$, which will turn out to be wrong.

For, if we construct a portfolio consisting of a borrowed cash \$20 at time $t = 0$ and $\frac{1}{2}$ shares of the stock, then its value is equal to

$$-\$20 + \frac{1}{2} \times \$S_0 = -\$20 + \frac{1}{2} \times \$50 = \$5 \,.$$

At the expiry date the value of the portfolio will be equal to either

$$-\$20 + \frac{1}{2} \times \$S^u = -\$20 + \frac{1}{2} \times \$80 = \$20$$

G.H. Choe, *Stochastic Analysis for Finance with Simulations*, Universitext,
DOI 10.1007/978-3-319-25589-7_14

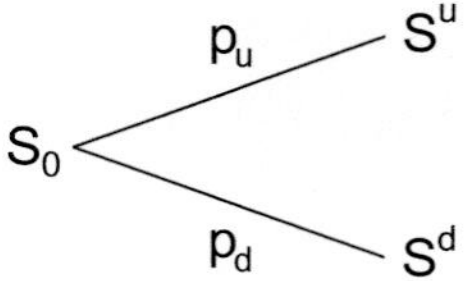

Fig. 14.1 A one period binomial tree and real probabilities

if the stock price goes up, or

$$-\$20 + \frac{1}{2} \times \$S^d = -\$20 + \frac{1}{2} \times \$40 = \$0$$

if the stock price goes down. In either case, the value of the portfolio will be equal to the value of the option. Hence the price of the option at $t = 0$ is equal to the value of the portfolio, which is \$5. We say that such a portfolio replicates the given option.

14.2 The One Period Binomial Tree Method

Now we give a more systematic analysis. Let r denote the risk-free interest rate, and let T be the expiry date of the option. Assume that the underlying asset S_t can have two values S^u and S^d at time T depending on the up and the down states, respectively. See Fig. 14.2 where the probability space is given by $\Omega = \{\text{up}, \text{down}\}$ and the bond price increases from 1 at time $t = 0$ to e^{rT} at time T regardless of the state. Similarly, the option price V_t has two values V^u and V^d at time T depending on the up and the down states, respectively. Under the no arbitrage principle, we have $S^d < \mathrm{e}^{rT} < S^u$. For, if $\mathrm{e}^{rT} \le S^d$ then no investor would deposit his/her money in the bank, and if $S^u \le \mathrm{e}^{rT}$ then no investor would invest in the stock market.

14.2.1 *Pricing by Hedging*

We construct a risk-free, or hedged, portfolio to compute the option price. Consider a portfolio Π consisting of three assets: an option sold, a bond (or a bank deposit), and Δ shares of the underlying risky asset or stock. More precisely, Π is given by

$$\Pi_t = -V_t + B_t + \Delta S_t$$

for $t = 0, T$, which is held by an option seller to hedge the risk. Note that Δ is decided at time $t = 0$, i.e., predictable, and is constant in the time interval $[0, T]$.

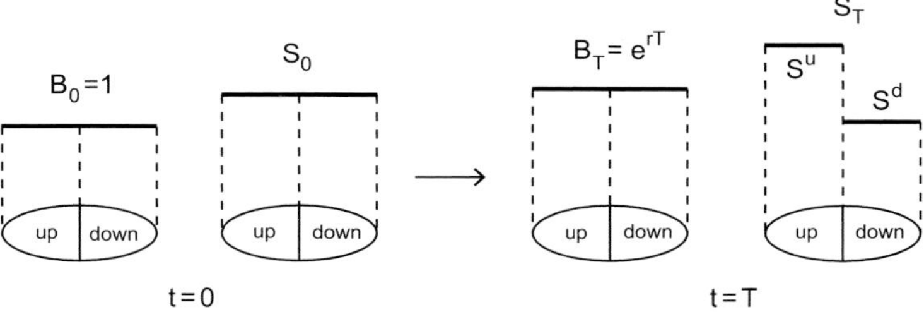

Fig. 14.2 Prices of risk-free and risky assets in the one period binomial tree model

Since short-selling is allowed, Δ can be negative. We find Δ for which Π becomes a risk-free portfolio by solving

$$\begin{cases} -V^u + \mathrm{e}^{rT}B_0 + \Delta S^u = \mathrm{e}^{rT}(-V_0 + B_0 + \Delta S_0) \\ \\ -V^d + \mathrm{e}^{rT}B_0 + \Delta S^d = \mathrm{e}^{rT}(-V_0 + B_0 + \Delta S_0) \end{cases}$$

where V^u and V^d are values taken by the payoff V_T depending on the up and down states of S_T, respectively. Hence we have

$$\begin{bmatrix} \mathrm{e}^{rT} & S^u - \mathrm{e}^{rT}S_0 \\ \mathrm{e}^{rT} & S^d - \mathrm{e}^{rT}S_0 \end{bmatrix} \begin{bmatrix} V_0 \\ \Delta \end{bmatrix} = \begin{bmatrix} V^u \\ V^d \end{bmatrix} ,$$

and

$$\begin{bmatrix} V_0 \\ \Delta \end{bmatrix} = \frac{1}{\mathrm{e}^{rT}(S^d - S^u)} \begin{bmatrix} S^d - \mathrm{e}^{rT}S_0 & -(S^u - \mathrm{e}^{rT}S_0) \\ -\mathrm{e}^{rT} & \mathrm{e}^{rT} \end{bmatrix} \begin{bmatrix} V^u \\ V^d \end{bmatrix}$$
$$= \begin{bmatrix} \dfrac{1}{\mathrm{e}^{rT}} \left(\dfrac{\mathrm{e}^{rT}S_0 - S^d}{S^u - S^d} V^u + \dfrac{S^u - \mathrm{e}^{rT}S_0}{S^u - S^d} V^d \right) \\ \dfrac{V^u - V^d}{S^u - S^d} \end{bmatrix} .$$

Put

$$q_u = \frac{\mathrm{e}^{rT}S_0 - S^d}{S^u - S^d} , \quad q_d = \frac{S^u - \mathrm{e}^{rT}S_0}{S^u - S^d} .$$

Then

$$V_0 = \mathrm{e}^{-rT}(q_u V^u + q_d V^d)$$

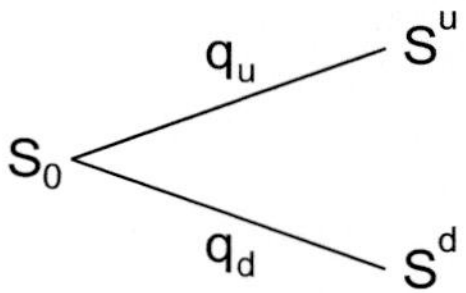

Fig. 14.3 A single-period binomial tree and the risk-neutral probability

and

$$\Delta = \frac{V^u - V^d}{S^u - S^d}\,.$$

Note that $0 < q_u < 1, 0 < q_d < 1$ under the no arbitrage assumption, and $q_u + q_d = 1$. Then the pair (q_u, q_d) represents a probability distribution $\mathbb{Q}$, which is called a *risk-neutral probability*. See Fig. 14.3.

The expectation with respect to $\mathbb{Q}$ is denoted by $\mathbb{E}^{\mathbb{Q}}$. The option price at $t = 0$ is given by

$$V_0 = \mathrm{e}^{-rT}(q_u V^u + q_d V^d) = \mathrm{e}^{-rT}\mathbb{E}^{\mathbb{Q}}[V_T]$$

where V_T is a random variable defined by the payoff at the expiry date. This is called the risk-neutral pricing.

14.2.2 Pricing by Replication

We replicate a given payoff at T and apply the no arbitrage principle to compute the European option price. Consider a portfolio Π consisting of a risk-free asset and an underlying stock. More precisely,

$$\Pi_t = B_t + \Delta S_t$$

where Δ is the number of shares of the underlying stock. We find Δ for which $\Pi_T = V_T$ by solving

$$\begin{cases} \mathrm{e}^{rT} B_0 + \Delta S^u = V^u \\ \mathrm{e}^{rT} B_0 + \Delta S^d = V^d \end{cases}$$

or equivalently,

$$\begin{bmatrix} \mathrm{e}^{rT} & S^u \\ \mathrm{e}^{rT} & S^d \end{bmatrix} \begin{bmatrix} B_0 \\ \Delta \end{bmatrix} = \begin{bmatrix} V^u \\ V^d \end{bmatrix}.$$

Hence

$$\begin{bmatrix} B_0 \\ \Delta \end{bmatrix} = \frac{1}{e^{rT}(S^d - S^u)} \begin{bmatrix} S^d & -S^u \\ -e^{rT} & e^{rT} \end{bmatrix} \begin{bmatrix} V^u \\ V^d \end{bmatrix} .$$

Thus

$$B_0 = \frac{1}{e^{rT}} \left(\frac{-S^d}{S^u - S^d} V^u + \frac{S^u}{S^u - S^d} V^d \right) ,$$

$$\Delta = \frac{V^u - V^d}{S^u - S^d} .$$

Note that the option and the portfolio that replicates the option at $t = T$ have the same value at $t = 0$ by the no arbitrage principle. Hence

$$\begin{aligned} V_0 &= \Pi_0 \\ &= B_0 + \Delta S_0 \\ &= \frac{1}{e^{rT}} \left(\frac{-S^d}{S^u - S^d} V^u + \frac{S^u}{S^u - S^d} V^d \right) + \frac{V^u - V^d}{S^u - S^d} S_0 \\ &= \frac{1}{e^{rT}} \left(\frac{e^{rT} S_0 - S^d}{S^u - S^d} V^u + \frac{S^u - e^{rT} S_0}{S^u - S^d} V^d \right) , \end{aligned}$$

which is identical to the result obtained by the hedging method.

14.3 The Multiperiod Binomial Tree Method

Given an option with expiry date T, we consider a multiperiod binomial tree of length N, obtained by stringing together single-period binomial trees, where the length of the time interval for each single-period binomial tree is $\delta t = \frac{T}{N}$. At time $T = N \times \delta t$, the asset price can take one of $N + 1$ possible values. In Sect. 14.4 it will be shown that the European option price obtained by the binomial tree method converges to a formula from the continuous time model. In Fig. 14.4 a binomial tree of length $N = 4$ is plotted where $u = \frac{3}{2}$, $d = \frac{1}{2}$, and in Table 14.1 the corresponding spreadsheet is represented by an upper triangular matrix.

Example 14.1 We now compute the price of a European call option with exercise price $K = 10$ using a three-step binomial tree. For the sake of computational convenience we assume $r = 0$. Other parameters are given by $S^u = uS_0$, $S^d = dS_0$ where $u = \frac{3}{2}$, $d = \frac{1}{2}$. Note that $q_u = q_d = \frac{1}{2}$.

In Figs. 14.5, 14.6, 14.7 the numbers in the circles represent stock prices while the numbers in the rectangles are call option prices for a given time and stock price. In

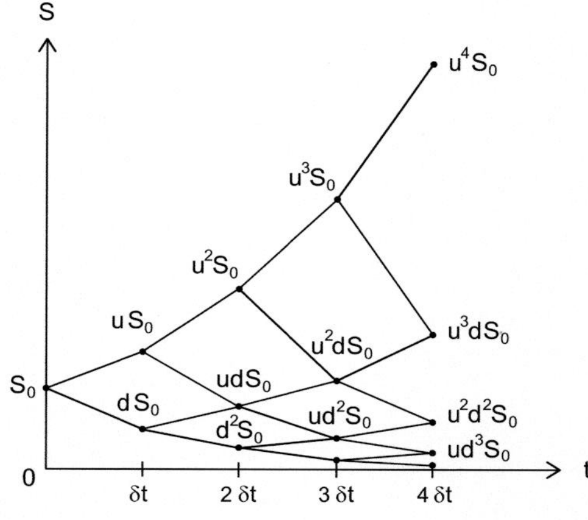

Fig. 14.4 A multiperiod binomial tree

Table 14.1 A spreadsheet for a multiperiod binomial tree

32	48	72	108	162
	16	24	36	54
		8	12	18
			4	6
				2

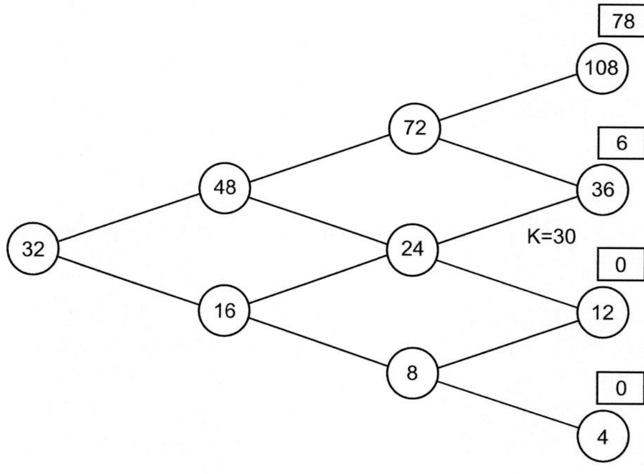

Fig. 14.5 Payoff of a European call option in a binomial tree

Fig. 14.5 the numbers in the rectangles are payoffs at maturity date. As in Fig. 14.6, from the payoff specified at time $t = 3$ we obtain the option price at time $t = 2$ by the single-period binomial tree method, and from the option price at $t = 2$ we obtain the option price at $t = 1$ by the single-period binomial tree method, and finally from the price at $t = 1$ we obtain the price at $t = 0$. For example, at $t = 2$ and with $S_2 = 36$ the option price is equal to 26. The option price at $t = 0$ is 8.5.

The binomial model, which is based on the risk-neutral valuation, is a popular approach to option pricing. It is based on the simplified assumption that over a

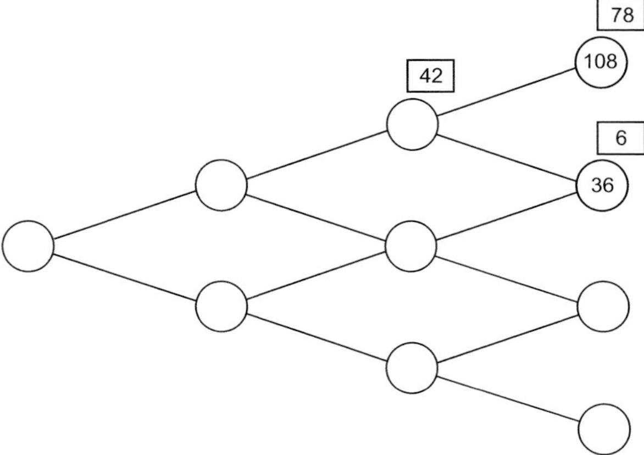

Fig. 14.6 The first step in pricing a European call option by a binomial tree

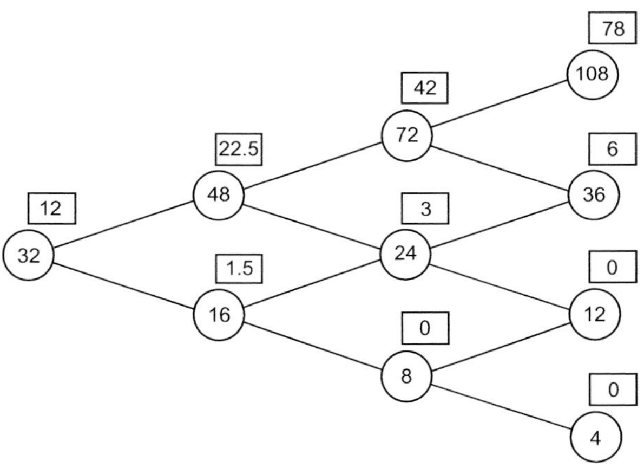

Fig. 14.7 Pricing of a European call option by a binomial tree

single period (of very short duration δt), the underlying asset can only move from its current price to two possible levels. As will be shown in Sect. 14.4, the Black–Scholes–Merton formula is the limiting value resulting from taking $N \to \infty$ in the binomial period where $\delta t = \frac{T}{N}$. See Table 14.2 for the formal comparison of continuous time and discrete time cases.

Recall that to compute the values $\binom{N}{i}$, $1 \leq i \leq N$, we consider the binomial coefficients in the expansion of $(x + y)^N$ and *Pascal's triangle*, which is a recombining tree. That is why the tree employed in option pricing is called a *binomial tree*.

It is impossible to use a *binary tree* as a model for an asset price movement. The fact that a binomial tree is recombinant reflects the usual assumption that future asset price movement is not dependent on the past history, while a binary tree tracks all the price history and the price movement depends not only on the present but also on the past information. Furthermore, the number of nodes of a binomial tree is equal to $N + 1$ where N is the length of the tree, while the number of nodes for a binary

Table 14.2 Option pricing by risk-neutral method

	Continuous time	Discrete time
Observation time	$0 \le t \le T$	$n = 0, 1, \ldots, N$
Expiry date	T	$N \times \delta t$
Random motion	Brownian motion W_t	Random walk
Range of random motion	$\mathbb{R}^1$	Integral multiples of δx
Probability distribution	Normal distribution	Binomial distribution
Asset price movement	Geometric Brownian motion	Binomial tree
Size of randomness	Volatility	Standard deviation
Physical probability	$\mathbb{P}$	(p_u, p_d)
Risk-neutral probability	$\mathbb{Q}$	(q_u, q_d)
Payoff	C_T	C_N
Option price	$\mathrm{e}^{-rT}\mathbb{E}[C_T]$	$\mathrm{e}^{-rT}\mathbb{E}[C_N]$

tree grows exponentially, i.e., the number of nodes is equal to 2^N. For example, for an option with expiry date $T = 1$ year and $\delta t = 1$ day, we have $N = 252$, in which case $2^{252} \approx 10^{75}$ is too large to deal with computationally.

Remark 14.1 A rigorous presentation of the binomial tree method may be given as follows: Let Ω be the set of all paths ω. Consider a subset, called a cylinder set, $[a_1, \ldots, a_M]$, $1 \le M \le N$, defined by

$$[a_1, \ldots, a_M] = \{w = a_1 \cdots a_M \omega_{M+1} \cdots \omega_N | \, \omega_{M+1}, \ldots, \omega_N \in \{\mathrm{u}, \mathrm{d}\}\}$$

where we write 'u' and 'd' to denote up and down movements, respectively. Note that the number of paths in $[a_1, \ldots, a_M]$ is equal to 2^{N-M} and that

$$\begin{aligned}\Omega &= [\mathrm{u}] \cup [\mathrm{d}] \\ &= [\mathrm{u,u}] \cup [\mathrm{u,d}] \cup [\mathrm{d,u}] \cup [\mathrm{d,d}] \\ &= [\mathrm{u,u,u}] \cup [\mathrm{u,u,d}] \cup [\mathrm{u,d,u}] \cup [\mathrm{u,d,d}][\mathrm{d,u,u}] \cup [\mathrm{d,u,d}] \cup [\mathrm{d,d,u}] \cup [\mathrm{d,d,d}] \\ &= \cdots\end{aligned}$$

Let $\mathcal{F}_0 = \{\emptyset, \Omega\}$, and let $\mathcal{F}_n$ be the σ-algebra on Ω consisting of cylinder sets $[a_1, \ldots, a_M]$. Then we have a filtration $\mathcal{F}_0 \subset \mathcal{F}_1 \subset \mathcal{F}_2 \subset \cdots \subset \mathcal{F}_N$. The payoff function of a European option is measurable with respect to $\mathcal{F}_N$, and hence is denoted by C_N to emphasize the fact. Define

$$C_{N-1} = \mathrm{e}^{-r\delta t}\mathbb{E}^{\mathbb{Q}}[C_N | \mathcal{F}_{N-1}] \,,$$

which is the option price at time $(N-1)\delta t$ and depends on the asset price at time $(N-1)\delta t$. This is measurable with respect to $\mathcal{F}_{N-1}$. We need to define $\mathbb{Q}$ on the whole space Ω by

$$\mathbb{Q}([a_1, \cdots, a_N]) = q_u^k q_d^{N-k}$$

where k is equal to the number of up movements.

14.4 Convergence to the Black–Scholes–Merton Formula

In this section we show that the European call option price obtained from the N-period binomial tree model converges to the solution from the Black–Scholes–Merton partial differential equation in Chap. 15 as $N \to \infty$.

Let $\delta t = \frac{T}{N}$, and use $r\delta t$ as the interest rate in each single-period. If we take

$$S^u = uS_0 \ , \quad S^d = dS_0$$

in the single period binomial model, then

$$\frac{S^u - S_0}{S_0} = u - 1 \ , \quad \frac{S^d - S_0}{S_0} = d - 1 \ .$$

Then, by the binomial tree method, the price of the European call option is given by

$$V_0 = \mathrm{e}^{-rT} \sum_{i=0}^{N} \binom{N}{i} q^i (1-q)^{N-i} (S_0 u^i d^{N-i} - K)^+$$

where N is the number of time steps, and

$$q = \frac{\mathrm{e}^{r\delta t} - d}{u - d} \ .$$

Note that

$$\mathrm{e}^{-r\delta t} qu + \mathrm{e}^{-r\delta t}(1-q)d = 1 \ .$$

Let m be the smallest integer such that

$$S_0 u^m d^{N-m} > K \ ,$$

or equivalently, m is the unique integer such that

$$S_0 u^{m-1} d^{N-m+1} \leq K < S_0 u^m d^{N-m} \ .$$

Then

$$\frac{\log \frac{K}{S_0} - N \log d}{\log \frac{u}{d}} < m \leq \frac{\log \frac{K}{S_0} - N \log d}{\log \frac{u}{d}} + 1$$

and

$$V_0 = \mathrm{e}^{-rT} \sum_{i=m}^{N} \binom{N}{i} q^i (1-q)^{N-i} (S_0 u^i d^{N-i} - K) .$$

Put

$$A = \mathrm{e}^{-rT} \sum_{i=m}^{N} \binom{N}{i} q^i (1-q)^{N-i} u^i d^{N-i}$$

and

$$B = \sum_{i=m}^{N} \binom{N}{i} q^i (1-q)^{N-i} .$$

Then

$$V_0 = A S_0 - K \mathrm{e}^{-rT} B .$$

To find the limit of B as $N \to \infty$, recall that the binomial distribution $B(N, q)$ is approximated by the normal distribution $N(Nq, Nq(1-q))$ with mean Nq and variance $Nq(1-q)$ for sufficiently large N, and note that

$$\begin{aligned}
\sum_{i=m}^{N} \binom{N}{i} q^i (1-q)^{N-i} &= \Pr(B(N,q) \geq m) \\
&\approx \Pr(N(Nq, Nq(1-q)) \geq m) \\
&= \int_m^{\infty} \frac{1}{\sqrt{2\pi Nq(1-q)}} \exp\left(-\frac{1}{2} \frac{(x-Nq)^2}{Nq(1-q)}\right) \mathrm{d}x \\
&= \int_{-d_2}^{\infty} \frac{1}{\sqrt{2\pi}} \exp\left(-\frac{y^2}{2}\right) \mathrm{d}y \\
&= N(d_2)
\end{aligned}$$

where N denotes the standard normal distribution function,

$$d_2 = \lim_{N\to\infty} \frac{-m + Nq}{\sqrt{Nq(1-q)}} = \lim_{N\to\infty} \frac{\dfrac{\log \frac{S_0}{K} + N\log d}{\log \frac{u}{d}} + Nq}{\sqrt{Nq(1-q)}}$$

and we used the substitution $y = \frac{x-Nq}{\sqrt{Nq(1-q)}}$. Similarly,

$$\begin{aligned}
\lim_{N\to\infty} A &= \lim_{N\to\infty} \mathrm{e}^{-rT} \sum_{i=m}^{N} \binom{N}{i} q^i (1-q)^{N-i} u^i d^{N-i} \\
&= \lim_{N\to\infty} \sum_{i=m}^{N} \binom{N}{i} \left(\mathrm{e}^{-r\delta t} qu\right)^i \left(\mathrm{e}^{-r\delta t}(1-q)d\right)^{N-i} \\
&= \int_{d_1}^{\infty} \frac{1}{\sqrt{2\pi}} \exp\left(-\frac{y^2}{2}\right) \mathrm{d}y \\
&= 1 - N(d_1)
\end{aligned}$$

where

$$d_1 = \lim_{N\to\infty} \frac{\dfrac{\log \frac{S_0}{K} + N\log d}{\log \frac{u}{d}} + Nq^*}{\sqrt{Nq^*(1-q^*)}}$$

and $q^* = \mathrm{e}^{-r\delta t} qu$. Hence we conclude that

$$\lim_{N\to\infty} V_0 = N(d_1) S_0 - K\mathrm{e}^{-rT} N(d_2) \; .$$

In the following we show that by choosing suitable values for u and d the option price formula obtained from the binomial tree method converges to the Black–Scholes–Merton formula.

Theorem 14.1 *Take*

$$u = \mathrm{e}^{\sigma\sqrt{\delta t}} \; , \quad d = \mathrm{e}^{-\sigma\sqrt{\delta t}} \; .$$

Then the aforementioned d_1, d_2 *converge to*

$$\frac{\log \frac{S_0}{K} \pm \left(r + \frac{1}{2}\sigma^2\right) T}{\sigma\sqrt{T}}$$

defined in the Black–Scholes–Merton formula.

Proof We will show that d_1 and d_2 given in the above converge to the usual d_1 and d_2 in the closed form solution of the Black–Scholes–Merton equation. Note that

$$\begin{aligned}
q &= \frac{e^{r\delta t} - e^{-\sigma\sqrt{\delta t}}}{e^{\sigma\sqrt{\delta t}} - e^{-\sigma\sqrt{\delta t}}} \\
&= \frac{(1 + r\delta t) - (1 - \sigma\sqrt{\delta t} + \frac{1}{2}\sigma^2\delta t + \cdots)}{(1 + \sigma\sqrt{\delta t} + \frac{1}{2}\sigma^2\delta t + \cdots) - (1 - \sigma\sqrt{\delta t} + \frac{1}{2}\sigma^2\delta t + \cdots)} \\
&\approx \frac{\sigma\sqrt{\delta t} + (r - \frac{1}{2}\sigma^2)\delta t}{2\sigma\sqrt{\delta t}} \\
&= \frac{1}{2} + \frac{r - \frac{1}{2}\sigma^2}{2\sigma}\sqrt{\delta t}\,.
\end{aligned}$$

Note that

$$\begin{aligned}
d_2 &\approx \frac{\log\frac{S_0}{K} + N\log d + Nq\log\frac{u}{d}}{\sqrt{Nq(1-q)}\log\frac{u}{d}} \\
&= \frac{\log\frac{S_0}{K} - N\sigma\sqrt{\delta t} + N(\frac{1}{2} + \frac{r-\frac{1}{2}\sigma^2}{2\sigma}\sqrt{\delta t})2\sigma\sqrt{\delta t}}{\sqrt{N\left(\frac{1}{4} - \left(\frac{r-\frac{1}{2}\sigma^2}{2\sigma}\right)^2\delta t\right)}2\sigma\sqrt{\delta t}} \\
&\approx \frac{\log\frac{S_0}{K} + (r - \frac{1}{2}\sigma^2)\,T}{\sigma\sqrt{T}}
\end{aligned}$$

and that

$$B \approx -K\left(e^{-r\delta t}\right)^N \int_{d_2}^{\infty} \frac{1}{2\pi}\exp\left(-\frac{y^2}{2}\right)dy = -Ke^{-rT}N(d_2)\,.$$

As for A, note that

$$q^* = e^{-r\delta t}\left(\frac{1}{2} + \frac{r + \frac{1}{2}\sigma^2}{2\sigma}\sqrt{\delta t}\right) \approx \frac{1}{2} + \frac{r + \frac{1}{2}\sigma^2}{2\sigma}\sqrt{\delta t}\,.$$

Hence

$$d_1 \approx \frac{\log \frac{S_0}{K} + N \log d + Nq^* \log \frac{u}{d}}{\sqrt{Nq^*(1-q^*)} \log \frac{u}{d}}$$

$$\approx \frac{\log \frac{S_0}{K} - N\sigma\sqrt{\delta t} + N(\frac{1}{2} + \frac{r+\frac{1}{2}\sigma^2}{2\sigma}\sqrt{\delta t})2\sigma\sqrt{\delta t}}{\sqrt{N\left(\frac{1}{4} - \left(\frac{r+\frac{1}{2}\sigma^2}{2\sigma}\right)^2 \delta t\right)}2\sigma\sqrt{\delta t}}$$

$$\approx \frac{\log \frac{S_0}{K} + \left(r + \frac{1}{2}\sigma^2\right) T}{\sigma\sqrt{T}}$$

as $N \to \infty$. □

Remark 14.2 Note that for $u, d = e^{\pm\sigma\sqrt{\delta t}} \approx 1 \pm \sigma\sqrt{\delta t} + \frac{1}{2}\sigma^2\delta t$, we have

$$\frac{\delta S_t}{S_t} = \frac{S_{t+\delta t} - S_t}{S_t} \approx \pm\sigma\sqrt{\delta t} + \frac{1}{2}\sigma^2\delta t\ .$$

This result is consistent with the geometric Brownian motion for continuous time model for the derivation of the Black–Scholes–Merton differential equation in Chap. 15 if we regard the first term $\pm\sigma\sqrt{\delta t}$ as $\sigma\delta W_t$ and the second term $\frac{1}{2}\sigma^2\delta t$ as $\mu\delta t$. Since the drift coefficient μ does not appear in the solution of the Black–Scholes–Merton equation, it does not matter that the above choice of u and d produces μ as a function of σ instead of an unrelated independent parameter. Note that the continuous limit of $\Delta = \frac{V^u - V^d}{S^u - S^d}$ is $\frac{\partial V}{\partial S}$.

14.5 Computer Experiments

Simulation 14.1 (Binomial Tree Method)

For a European call option we apply the binomial tree method and obtain the value 10.4043 which is a rough approximation to the value 10.0201 that is obtained by the Black–Scholes–Merton formula (Tables 14.3 and 14.4). For a better estimation we need to increase the number of time steps. The following MATLAB

Table 14.3 The binomial tree for the underlying asset

100	119.1093	141.8704	168.9809
	84.2368	100.3339	119.5070
		70.9584	84.5180
			59.7731

Table 14.4 The binomial tree for the call option price

10.4043	18.8595	33.6779	58.9809
	2.2988	4.6750	9.5070
		0	0
			0

code produces two binomial trees for the asset price and a European call option price. (For the simulations for American options, see Chap. 19.)

```
S0 = 100;
K = 110;
T = 1;
r = 0.05;
sigma = 0.3;
M = 3; % number of time steps
dt = T/M;

u = exp(sigma*sqrt(dt) + (r-0.5*sigma^2)*dt);
d = exp(-sigma*sqrt(dt) + (r-0.5*sigma^2)*dt);
q = 1/2;

% We compute asset prices.
for i = 0:1:M
    fprintf('time = %i\n', i)
    S = S0*u.^([i:-1:0]').*d.^([0:1:i]')
end

fprintf('Payoff at expiry\n')
Call = max(S0*u.^([M:-1:0]').*d.^([0:1:M]')- K,0)

% We proceed backward to compute option value at time 0.
for i = M:-1:1
    fprintf('time = %i\n', i-1)
    Call = exp(-r*dt)*(q*Call(1:i) + (1-q)*Call(2:i+1))
end
```

Simulation 14.2 (Convergence of the Binomial Tree Method)

We take the same set of parameter values for S_0, K, T, r, σ and p as in Simulation 14.1. The option price obtained by the binomial tree method converges to the price from the Black–Scholes–Merton formula, represented by the horizontal line in Fig. 14.8, as $M \to \infty$. The oscillating behavior of the graph is inherent in the binomial tree method due to the imperfect stability of the numerical algorithm as explained in Chap. 29.

```
M_values = [50:1:1000];
Call_prices = zeros(length(M_values),1);

for j = 1:length(M_values)
    M = M_values(j);
    dt = T/M;
    u = exp(sigma*sqrt(dt) + (r-0.5*sigma^2)*dt);
    d = exp(-sigma*sqrt(dt) + (r-0.5*sigma^2)*dt);
```

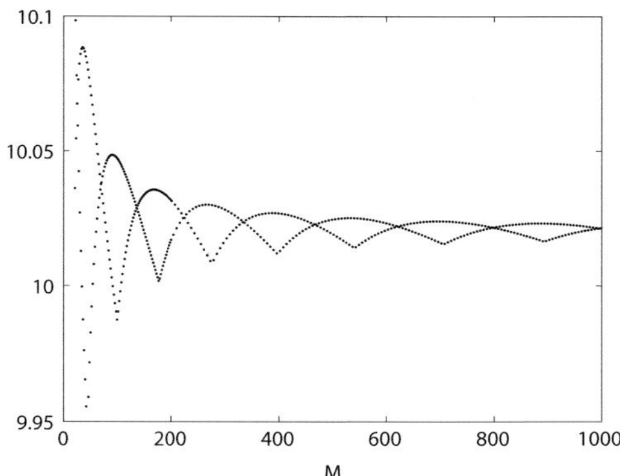

Fig. 14.8 Convergence to the Black–Scholes–Merton price

```
        Call = max(S0 *u.^([M:-1:0]') .* d.^([0:1:M]') - K,0);
        for i = M:-1:1
            Call = exp(-r*dt)*(p*Call(1:i)+(1-p)*Call(2:i+1));
        end
        Call_prices(j) = Call;
    end
    plot(M_values,Call_prices,'.');
    hold on;

    % the Black-Scholes-Merton formula.
    d1 = (log(S0/K)+(r+0.5*sigma^2)*T)/(sigma*sqrt(T));
    d2 = d1-sigma*sqrt(T);
    Call_BSM = S0*normcdf(d1) - K*exp(-r*T)*normcdf(d2)
    x=0:1000;
    plot(x,Call_BSM,'r');
```

Exercises

14.1 Suppose that we want to use the simple interest rate r in Sect. 14.2, i.e., $B_T = (1+r)^T B_0$. Show that under the assumption that $S^u = uS_0$ and $S^d = dS_0$ we have

$$V_0 = \frac{1}{1+r}\left(\frac{1+r-d}{u-d}V^u + \frac{u-(1+r)}{u-d}V^d\right).$$

In this case, the no arbitrage principle implies that $0 < d < 1+r < u$.

Chapter 15
The Black–Scholes–Merton Differential Equation

The simultaneous publications of Black and Scholes [6] and Merton [65] in 1973 mark the beginning of the theory of option pricing. Using the theory of stochastic calculus, they derived the so-called Black–Scholes–Merton differential equation. It is essentially a heat equation with the direction of time reversed.

15.1 Derivation of the Black–Scholes–Merton Differential Equation

Here is a list of the assumptions for the Black–Scholes–Merton model.

Assumptions on the Underlying Asset

(i) The asset follows a geometric Brownian motion with constant volatility.
(ii) There are no dividends or stock splits.

Assumptions on the Financial Market

(iii) It is possible to buy and sell any amount of the asset at any time.
(iv) The bid and the ask prices are equal, i.e., the bid-ask spread is zero.
(v) There are no transaction costs or taxes.
(vi) Short selling is allowed without any cost. Borrowing money is possible at any time.
(vii) The risk-free interest rate is known and constant.

G.H. Choe, *Stochastic Analysis for Finance with Simulations*, Universitext,
DOI 10.1007/978-3-319-25589-7_15

Recall the general version of Itô's lemma. Its idea can be summarized in the following heuristic argument. Since the option price V depends on t and S we will write $V(S,t)$. As time changes from t to $t+\delta t$, V also changes. Using the second order Taylor series expansion we obtain

$$\delta V = \frac{\partial V}{\partial t}\delta t + \frac{\partial V}{\partial S}\delta S + \frac{1}{2}\frac{\partial^2 V}{\partial S^2}(\delta S)^2 + \frac{\partial^2 V}{\partial S \partial t}\delta S\,\delta t + \frac{1}{2}\frac{\partial^2 V}{\partial t^2}(\delta t)^2 .$$

The increment δt is close to 0, and we ignore any term whose order is greater than 1 by Itô's lemma. Consult Table 11.1. Since

$$(\delta S)^2 = \sigma^2 S^2 (\delta W)^2 + \text{higher order terms} \approx \sigma^2 S^2 \delta t$$

we can classify the increments as follows:

$$\delta V = \underbrace{\left(\frac{\partial V}{\partial t} + \frac{1}{2}\sigma^2 S^2 \frac{\partial^2 V}{\partial S^2}\right)\delta t}_{\text{risk-free}} + \underbrace{\frac{\partial V}{\partial S}\delta S}_{\text{risky}} .$$

Now we are ready to derive a partial differential equation called the Black–Scholes–Merton equation for a European option price V.

Theorem 15.1 (Black–Scholes–Merton Equation) *The price $V(S,t)$ of a European option at time t with maturity T and strike price K satisfies*

$$\frac{\partial V}{\partial t} + \frac{1}{2}\sigma^2 S^2 \frac{\partial^2 V}{\partial S^2} + rS\frac{\partial V}{\partial S} = rV .$$

The final condition $V(S,T)$ is given by the payoff function of the option.

Proof To derive a partial differential equation for V we take the viewpoint of a fund manager of the portfolio Π based on the idea of hedging, and construct a portfolio Π that is self-financing and risk-free as follows:

$$\Pi(S,t) = -V(S,t) + D(S,t) + \Delta(S,t)S .$$

In other words, Π consists of an option that has been sold, a bank deposit or risk-free asset D, and Δ shares of risky asset S. Here Δ is a function of t and S, and is called the hedge ratio. If $\Delta < 0$, it represents short selling.

As a fund manager we maintain the same number of shares of a stock from time t to $t+\delta t$, which fits common sense. More precisely, since we do not know how much S would change, we wait until we obtain the information on the stock value at time $t+\delta t$, and make an investment decision upon that information. Suppose that while the hedge ratio Δ_t is fixed, S_t changes to $S_{t+\delta t}$, and Π_t changes to $\Pi_t + \delta\Pi$ (Fig. 15.1).

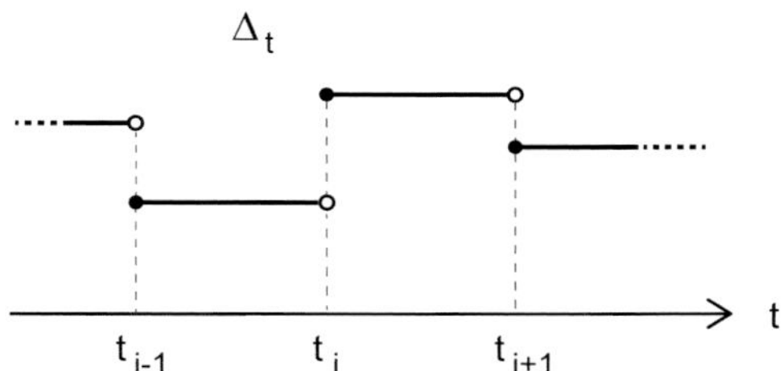

Fig. 15.1 Adjustment of Δ_t in discrete time hedging

The risk-free asset D_t gains interest

$$\delta D_t = rD_t\delta t \; .$$

Hence

$$\begin{aligned}\delta\Pi &= -\delta V + \delta D + \Delta\,\delta S \\ &= -\delta V + rD\delta t + \Delta\,\delta S \\ &= \left(-\frac{\partial V}{\partial t} - \frac{1}{2}\sigma^2 S^2 \frac{\partial^2 V}{\partial S^2} + rD\right)\delta t + \left(\Delta - \frac{\partial V}{\partial S}\right)\delta S \; .\end{aligned}$$

If we take

$$\Delta_t = \frac{\partial V}{\partial S}(t, S_t)$$

for every t, then we have

$$\delta\Pi = \left(-\frac{\partial V}{\partial t} - \frac{1}{2}\sigma^2 S^2 \frac{\partial^2 V}{\partial S^2} + rD\right)\delta t \; . \tag{15.1}$$

Now the δS term has disappeared and $\delta\Pi$ is risk-free, and hence it is equivalent to a bank deposit for a time duration δt. Thus we obtain

$$\delta\Pi = r\,\Pi\,\delta t \; . \tag{15.2}$$

By the no arbitrage principle, the right-hand sides of (15.1) and (15.2) are equal, and we have

$$\left(-\frac{\partial V}{\partial t} - \frac{1}{2}\sigma^2 S^2 \frac{\partial^2 V}{\partial S^2} + rD\right)\delta t = r\,\Pi\,\delta t \; .$$

Therefore

$$\left(-\frac{\partial V}{\partial t} - \frac{1}{2}\sigma^2 S^2 \frac{\partial^2 V}{\partial S^2} + rD\right) = r\left(-V + D + \frac{\partial V}{\partial S} S\right)$$

and we obtain the Black–Scholes–Merton equation after canceling the rD terms.

□

For a different derivation of the Black–Scholes-Merton equation based on the martingale method see Sect. 16.4. For the logical implications among the facts related to option pricing, see the diagram in Fig. 2.7.

15.2 Price of a European Call Option

Now we solve the Black–Scholes–Merton equation and find the price $V(S, t)$ of a European call option with expiry T. The domain of V is given by $0 \leq t \leq T$, $0 \leq S < \infty$. The boundary conditions are given by

$$V(0, t) = 0 ,$$
$$V(S, t) \approx S - K$$

for sufficiently large S, and the final condition is given by

$$V(S, T) = \max\{S - K, 0\} .$$

(See Fig. 15.2.) The Black–Scholes–Merton equation resembles the heat equation

$$\frac{\partial V}{\partial t} = \frac{1}{2}\frac{\partial^2 V}{\partial S^2}$$

except for the sign of the time derivative, and instead of an initial condition at time $t = 0$ a final condition at $t = T$ is given. The reason is that the payoff at the expiry is fixed from the beginning of the life of an option.

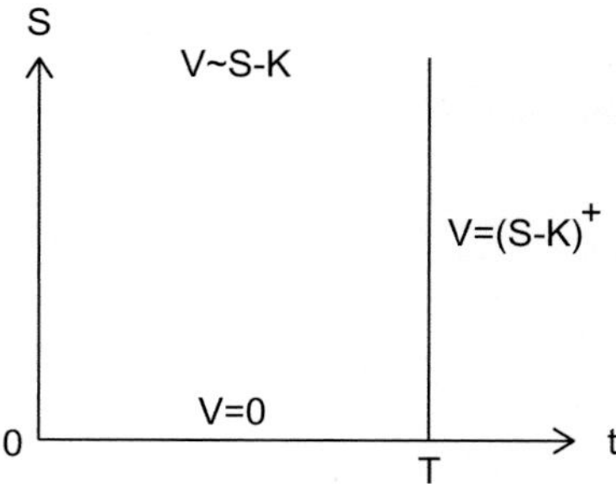

Fig. 15.2 Domain and a boundary condition for a payoff of a European call option

Theorem 15.2 (Black–Scholes–Merton Formula) *The price V_t at time $0 \le t \le T$ of a European call option with expiry T and strike price K is given by*

$$V_t = S_t N(d_1) - K\mathrm{e}^{-r(T-t)} N(d_2)$$

where

$$d_{1,2} = \frac{\log \frac{S_t}{K} + (r \pm \frac{1}{2}\sigma^2)(T-t)}{\sigma\sqrt{T-t}}$$

and $N(\cdot)$ denotes the standard normal cumulative distribution function.

Proof We convert the Black–Scholes–Merton equation into the heat equation with an initial condition going through several steps of change of variables. Put

$$\begin{cases} S &= K\mathrm{e}^x \\ t &= T - \dfrac{1}{\sigma^2}\tau \\ V(S,t) &= Kv(x,\tau) \ . \end{cases}$$

In other words,

$$v(x,\tau) = \frac{1}{K}V(S,t) = \frac{1}{K}V(K\mathrm{e}^x, T - \frac{1}{\sigma^2}\tau) \ .$$

Then $-\infty < x < \infty$, $0 \le \tau \le \sigma^2 T$ and

$$\sigma^2 \frac{\partial v}{\partial \tau} = \frac{1}{2}\sigma^2 \frac{\partial^2 v}{\partial x^2} + \left(r - \frac{1}{2}\sigma^2\right)\frac{\partial v}{\partial x} - rv \ .$$

If we let $C = \dfrac{r}{\sigma^2}$, then

$$\frac{\partial v}{\partial \tau} = \frac{1}{2}\frac{\partial^2 v}{\partial x^2} + \left(C - \frac{1}{2}\right)\frac{\partial v}{\partial x} - Cv \ .$$

Since $V(S,T) = \max\{S-K, 0\}$, the initial condition for v is given by

$$v(x,0) = \frac{1}{K}V(K\mathrm{e}^x, T) = \frac{1}{K}\max\{K\mathrm{e}^x - K, 0\} = \max\{\mathrm{e}^x - 1, 0\} \ .$$

Observe that the parameters K, T, σ^2, r have disappeared, and only one parameter C remains in the equation. When several parameters appear in one equation simultaneously and make the equation look complicated, we can sometimes combine some of parameters into one new parameter so that the new equation becomes simpler and easier to solve. If such a new parameter is dimensionless, then we may say that it is an essential parameter in describing the equation. For example, the dimensions of r and σ^2 are both time^{-1}, and hence $C = r/\sigma^2$ is dimensionless.

Now we look for v of the form

$$v(x,\tau) = \mathrm{e}^{\alpha x+\beta\tau} u(x,\tau)$$

for some α, β, $u(x,\tau)$. If such a solution exists, then we have

$$\beta u + \frac{\partial u}{\partial \tau} = \frac{1}{2}\frac{\partial^2 u}{\partial x^2} + \left(C - \frac{1}{2} + \alpha\right)\frac{\partial u}{\partial x} + \left(\frac{1}{2}\alpha^2 + (C - \frac{1}{2})\alpha - C\right)u \,.$$

If we take

$$\alpha = -C + \frac{1}{2}$$

and

$$\beta = \frac{1}{2}\alpha^2 + (C - \frac{1}{2})\alpha - C = -\frac{1}{2}(C + \frac{1}{2})^2 \,,$$

then a necessary condition for u becomes the heat equation

$$\frac{\partial u}{\partial \tau} = \frac{1}{2}\frac{\partial^2 u}{\partial x^2} \,.$$

Note that the initial condition for u is given by

$$u(x,0) = u_0(x) = \mathrm{e}^{-\alpha x} v(x,0) = \mathrm{e}^{(C-\frac{1}{2})x} \max\{\mathrm{e}^x - 1, 0\} \,.$$

Hence

$$\begin{aligned} u(x,\tau) &= \frac{1}{\sqrt{2\pi\tau}} \int_{-\infty}^{\infty} u_0(\xi)\mathrm{e}^{-(x-\xi)^2/(2\tau)} \mathrm{d}\xi \\ &= \frac{1}{\sqrt{2\pi\tau}} \int_0^{\infty} \mathrm{e}^{(C-\frac{1}{2})\xi}(\mathrm{e}^{\xi} - 1)\mathrm{e}^{-(x-\xi)^2/(2\tau)} \mathrm{d}\xi \,. \end{aligned}$$

Put

$$I_1, I_2 = \frac{1}{\sqrt{2\pi\tau}} \int_0^{\infty} \mathrm{e}^{(C\pm\frac{1}{2})\xi} \mathrm{e}^{-(x-\xi)^2/(2\tau)} \mathrm{d}\xi \,,$$

then $u(x,\tau) = I_1 - I_2$. Note that

$$
\begin{aligned}
&e^{\alpha x+\beta\tau} I_1 \\
&= e^{(-C+\frac{1}{2})x-\frac{1}{2}(C+\frac{1}{2})^2\tau} \frac{1}{\sqrt{2\pi\tau}} \int_0^\infty e^{-(x-\xi)^2/(2\tau)+(C+\frac{1}{2})\xi} d\xi \qquad (\text{take } \eta = -\xi) \\
&= e^{(-C+\frac{1}{2})x-\frac{1}{2}(C+\frac{1}{2})^2\tau} \frac{1}{\sqrt{2\pi\tau}} \int_0^{-\infty} e^{-(x+\eta)^2/(2\tau)-(C+\frac{1}{2})\eta} (-1)\, d\eta \\
&= e^{(-C+\frac{1}{2})x-\frac{1}{2}(C+\frac{1}{2})^2\tau} \frac{1}{\sqrt{2\pi\tau}} \int_{-\infty}^0 e^{-(x+\eta)^2/(2\tau)-(C+\frac{1}{2})\eta} d\eta \\
&= e^{(-C+\frac{1}{2})x-\frac{1}{2}(C+\frac{1}{2})^2\tau} \frac{1}{\sqrt{2\pi\tau}} \int_{-\infty}^0 e^{-(\eta+x+(C+\frac{1}{2})\tau)^2/(2\tau)+(C+\frac{1}{2})x+\frac{1}{2}(C+\frac{1}{2})^2\tau} d\eta \\
&= e^x \frac{1}{\sqrt{2\pi\tau}} \int_{-\infty}^0 e^{-(\eta+x+(C+\frac{1}{2})\tau)^2/(2\tau)} d\eta \\
&= e^x \frac{1}{\sqrt{2\pi\tau}} \int_{-\infty}^{x+(C+\frac{1}{2})\tau} e^{-\xi^2/(2\tau)} d\xi \qquad (\text{take } \zeta = \frac{\xi}{\sqrt{\tau}}) \\
&= e^x \frac{1}{\sqrt{2\pi}} \int_{-\infty}^{(x+(C+\frac{1}{2})\tau)/\sqrt{\tau}} e^{-\zeta^2/2} d\zeta = e^x N(d_1)
\end{aligned}
$$

where

$$
d_1 = \frac{x+(C+\frac{1}{2})\tau}{\sqrt{\tau}} = \frac{\log\frac{S}{K} + (\frac{r}{\sigma^2}+\frac{1}{2})\sigma^2(T-t)}{\sigma\sqrt{T-t}} .
$$

Similarly, for the second term we obtain

$$
\begin{aligned}
&e^{\alpha x+\beta\tau} I_2 \\
&= e^{(-C+\frac{1}{2})x-\frac{1}{2}(C+\frac{1}{2})^2\tau} \frac{1}{\sqrt{2\pi\tau}} \int_{-\infty}^0 e^{-(\eta+x+(C-\frac{1}{2})\tau)^2/2\tau+(C-\frac{1}{2})x+\frac{1}{2}(C-\frac{1}{2})^2\tau} d\eta \\
&= e^{-C\tau} \frac{1}{\sqrt{2\pi}} \int_{-\infty}^{(x+(C-\frac{1}{2})\tau)/\sqrt{\tau}} e^{-\eta^2/2} d\zeta \\
&= e^{-C\tau} N(d_2)
\end{aligned}
$$

where

$$
d_2 = \frac{x+(C-\frac{1}{2})\tau}{\sqrt{\tau}} = \frac{\log\frac{S}{K} + (\frac{r}{\sigma^2}-\frac{1}{2})\sigma^2(T-t)}{\sigma\sqrt{T-t}} .
$$

Since $\tau = \sigma^2(T-t)$, we have $C\tau = \frac{r}{\sigma^2}\sigma^2(T-t) = r(T-t)$. Hence

$$\begin{aligned} V(S,t) &= Kv(x,\tau) \\ &= Ke^{x}N(d_1) - Ke^{-C\tau}N(d_2) \\ &= SN(d_1) - Ke^{-r(T-t)}N(d_2)\ . \end{aligned}$$

□

Remark 15.1 The formula is still meaningful in some extreme cases.

(i) If $S_0 = 0$ then $S_t = 0$ for every $0 \le t \le T$, and hence the payoff is 0 for any K. Therefore $V_t = 0$, which can be seen from the formula since

$$\lim_{S_0 \to 0+} d_1 = \lim_{S_0 \to 0+} d_2 = -\infty$$

and

$$\lim_{S_0 \to 0+} N(d_1) = \lim_{S_0 \to 0+} N(d_2) = 0\ .$$

(ii) If $K = 0$, then the payoff is equal to the stock price at expiry, and hence $V_T = S_T$, thus $V_t = S_t$ for every $0 \le t \le T$. This is obvious from the formula since $\lim_{K \to 0+} d_1 = +\infty$ and $\lim_{K \to 0+} N(d_1) = 1$.

(iii) If $\sigma = 0$, then there is no risk and hence $\mu = r$. Note that $S_t = S_0 e^{rt}$ is deterministic. Hence $V_T = (S_T - K)^+ = (S_0 e^{rT} - K)^+$ is also deterministic, and

$$V_t = e^{-r(T-t)}(S_0 e^{rT} - K)^+\ .$$

This is obvious from the formula since if $S_t > Ke^{-r(T-t)}$ then

$$\lim_{\sigma \to 0+} d_1 = \lim_{\sigma \to 0+} d_2 = +\infty\ ,$$

$$\lim_{\sigma \to 0+} N(d_1) = \lim_{\sigma \to 0+} N(d_2) = 1\ ,$$

thus $V_t = S_t - Ke^{-r(T-t)}$ for every $0 \le t \le T$. If $S_t < Ke^{-r(T-t)}$ then

$$\lim_{\sigma \to 0+} d_1 = \lim_{\sigma \to 0+} d_2 = -\infty$$

$$\lim_{\sigma \to 0+} N(d_1) = \lim_{\sigma \to 0+} N(d_2) = 0\ ,$$

and $V_t = 0$.

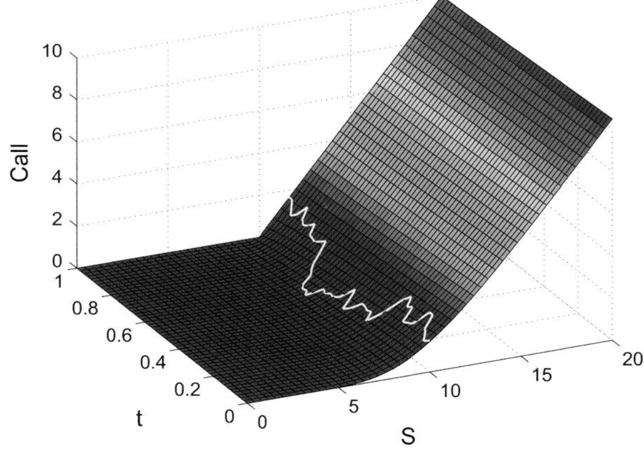

Fig. 15.3 Price of a European call option

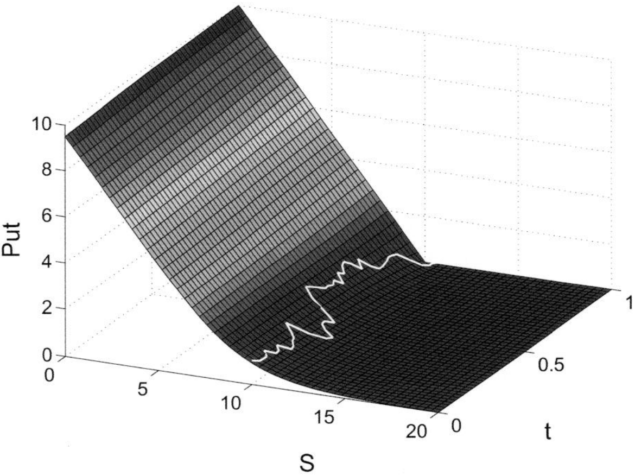

Fig. 15.4 Price of a European put option

In Figs. 15.3 and 15.4 the call price and put price surfaces are plotted for the values of $0 \leq t \leq 2$, $0 \leq S_0 \leq 20$.

In all experiments we take strike price $K = 100$, interest rate $r = 0.05$ and volatility $\sigma = 0.3$. In Fig. 15.5 we take present date $t_0 = 0$, expiry date $T = \frac{1}{12}, 1, 2$ and plot the graphs on the interval $40 \leq S \leq 160$.

In Fig. 15.6 we plot the option prices as functions of time to expiry $T - t$ for various strike prices.

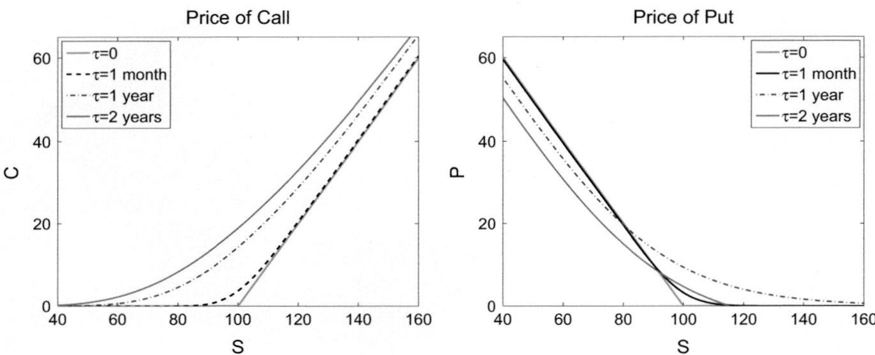

Fig. 15.5 Prices of European call and put options as functions of asset price with different times to expiration

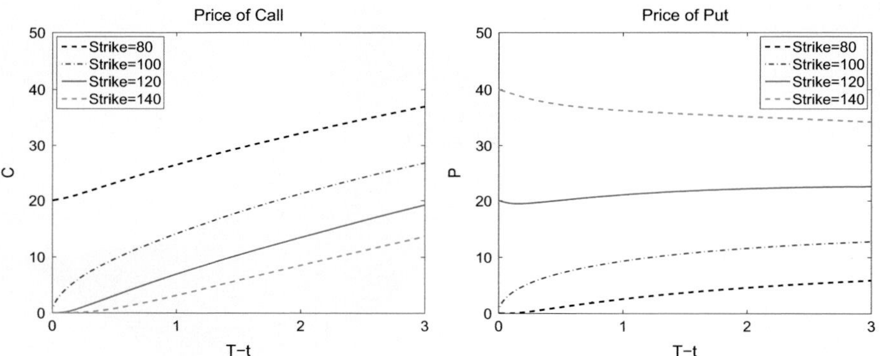

Fig. 15.6 Prices of European call and put options as functions of time to expiry with different strike prices

15.3 Greeks

The formula for the option price contains several parameters, and we can take partial derivatives with respect to these parameters. Those partial derivatives of the option price are denoted by various Greek letters, called *Greeks*, and we list some of them in the following:

$$\Delta = \frac{\partial V}{\partial S}$$

$$\Gamma = \frac{\partial^2 V}{\partial S^2}$$

$$\rho = \frac{\partial V}{\partial r}$$

$$\Theta = \frac{\partial V}{\partial t}$$

$$\text{Vega} = \frac{\partial V}{\partial \sigma}$$

Here is a convenient way to memorize which Greek letter represents which partial derivative: The Greek letter ρ corresponds to the Roman letter r, and θ to t. However, Vega is an invented letter and not in the Greek alphabet. To represent the English word 'volatility', we need a letter corresponding to the Roman symbol 'v' but the Greek alphabet does not have such a symbol. Sometimes we write $\mathcal{V}$ to denote Vega.

Uppercase delta (Δ) represents the number of shares needed in hedging. To avoid the possibility of confusion, we use a lowercase delta (δ) to denote the increment. For the list of Greek letters consult Table 15.1, in which Greek letters are given with their sounds written in Roman letters.

Table 15.1 The Greek alphabet

Uppercase	Lowercase	Name	Pronunciation
A	α	alpha	a
B	β	beta	b (v in modern Greek)
Γ	γ	gamma	g
Δ	δ	delta	d
E	ϵ	epsilon	e
Z	ζ	zeta	z
H	η	eta	e (i in modern Greek)
Θ	θ	theta	th
I	ι	iota	i
K	κ	kappa	k
Λ	λ	lambda	l
M	μ	mu	m
N	ν	nu	n
Ξ	ξ	xi	ks
O	o	omicron	o
Π	π	pi	p
P	ρ	rho	r
Σ	σ	sigma	s
T	τ	tau	t
Υ	υ	upsilon	y (i in modern Greek)
Φ	ϕ	phi	ph
X	χ	chi	kh
Ψ	ψ	psi	ps
Ω	ω	omega	o

Theorem 15.3 *Consider a European call option, and let*

$$d_1, d_2 = \frac{\log \frac{S}{K} + (r \pm \frac{1}{2}\sigma^2)(T-t)}{\sigma\sqrt{T-t}} .$$

Then the Greeks are given by

$$\begin{aligned}
\Delta &= N(d_1) \\
\Gamma &= \frac{N'(d_1)}{S\sigma\sqrt{T-t}} \\
\rho &= (T-t)K\mathrm{e}^{-r(T-t)}N(d_2) \\
\Theta &= \frac{-S\sigma}{2\sqrt{T-t}}N'(d_1) - rK\mathrm{e}^{-r(T-t)}N(d_2) \\
\text{Vega} &= S\sqrt{T-t}\,N'(d_1)
\end{aligned}$$

Proof By direct computation, we obtain

$$\frac{\partial V}{\partial S} = N(d_1) + SN'(d_1)\frac{1}{S}\frac{1}{\sigma\sqrt{T-t}} - K\mathrm{e}^{-r(T-t)}N'(d_2)\frac{1}{S}\frac{1}{\sigma\sqrt{T-t}} .$$

Now it suffices to show

$$SN'(d_1) = K\mathrm{e}^{-r(T-t)}N'(d_2) .$$

Since $N'(x) = \frac{1}{\sqrt{2\pi}}\mathrm{e}^{-x^2/2}$, it suffices to show

$$\log\frac{S}{K} + r(T-t) = \frac{1}{2}(d_1^2 - d_2^2) .$$

Now we use

$$\frac{1}{2}(d_1^2 - d_2^2) = \frac{1}{2}(d_1 + d_2)(d_1 - d_2) = \frac{1}{2} \times 2\frac{\log\frac{S}{K} + r(T-t)}{\sigma\sqrt{T-t}}\sigma\sqrt{T-t} .$$

For the others, the proofs are omitted. □

Remark 15.2 For a European call option the following statements are true:

(i) When the underlying stock price rises, the price of the call option also rises, and hence $\Delta > 0$, which is observed in the formula for Δ.
(ii) As the interest rate increases, there is more discount, and the discounted strike price decreases, and hence $\rho > 0$ as seen in the formula.
(iii) As the time to expiry decreases, $\Theta < 0$ as seen in the formula, thus the option price decreases. For European put options, this need not be true. See

Figs. 15.5, 15.6 where for put options the price is not necessarily monotone with respect to time. See also Fig. 15.11 for the plot of Theta.

(iv) Note that Vega is always positive since as volatility increases the possibility that the stock price goes up above the strike price, and therefore the chance that the option can be exercised, increases.

(v) Since $\Gamma > 0$, V as a function of S, the graph of $V(t, S)$ is convex.

(vi) Note that $\frac{\partial V}{\partial K} < 0$ since the call option price decreases as K increases. In fact, we have

$$\frac{\partial V}{\partial K} = -\mathrm{e}^{-r(T-t)} N(d_2) \; .$$

Consult Exercise 15.7 for a proof. For an application see Theorem 17.2.

We plot the Greeks using the Black–Scholes–Merton formula. In all experiments we take strike price $K = 100$, interest rate $r = 0.05$ and volatility $\sigma = 0.3$. In Fig. 15.7 we take $T - t = \frac{1}{12}, 1, 2$ and plot the graphs on the interval $40 \le S \le 160$. Let Δ_C and Δ_P denote the Deltas of call and put options with the same K and T. Due to the put-call parity $C - P = S - K\mathrm{e}^{-rT}$, they satisfy

$$\Delta_C - \Delta_P = 1 \; .$$

The slopes of the curves in Fig. 15.5 are plotted in Fig. 15.7.

For Gamma we take $T - t = \frac{1}{2}, 1, 2$. Since $\Delta_C - \Delta_P = 1$, we know that Γ of a call and a put is identical since $\Gamma = \partial\Delta/\partial S$. Recall that Γ is positive. See Fig. 15.8.

Rho is positive for a call option while it is negative for a put option. For, if the interest rate rises, the present value of the strike price decreases, and the call option value increases while the put option value decreases. Rho has large absolute value, as seen in Fig. 15.9, which is compensated by the fact that δr is usually very small in the relation $V(r + \delta r) - V(r) \approx \rho \times \delta r$. Note that if $r = 5\%$ moves to 5.1% then $\delta r = 0.1\% = 0.001$.

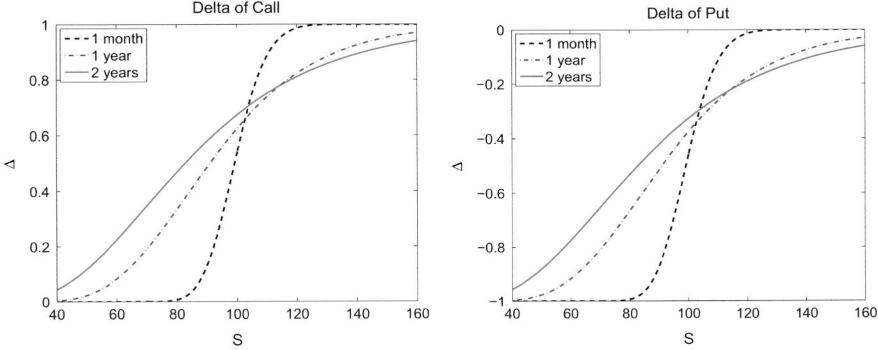

Fig. 15.7 Delta of European call and put options

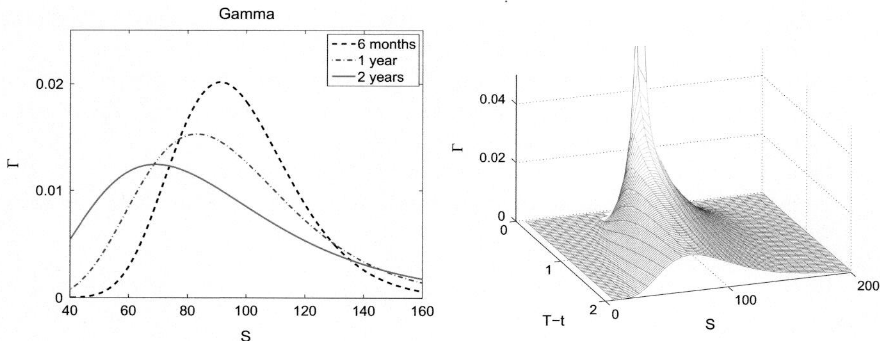

Fig. 15.8 Gamma of European call and put options

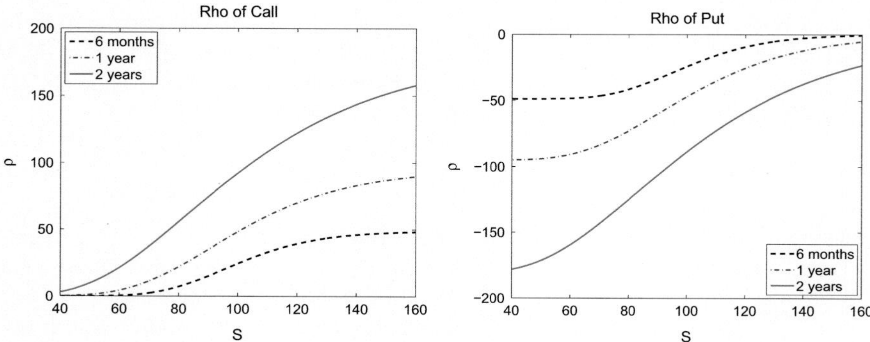

Fig. 15.9 Rho of European call and put options

Vega is large for asset values close to the strike price, and at-the-money options have large values of Vega. Out-of-the-money options have low values of Vega since their prices are low. Deep in-the-money options also have low values of Vega since their values are mostly intrinsic values. See Fig. 15.10.

For the rate of change of the option value as time changes, we consider Theta, which is negative for European call options. Let Θ_C and Θ_P denote the Thetas of call and put options. From the put-call parity

$$C(t) - P(t) = S - K\mathrm{e}^{-r(T-t)} ,$$

we have

$$\Theta_C - \Theta_P = -rK\mathrm{e}^{-r(T-t)} .$$

In our example, $K = 100$, $r = 0.05$ and $T - t = 0.5, 1, 2$. Hence $\Theta_P = \Theta_C + rK\mathrm{e}^{-r(T-t)}$ and $rK\mathrm{e}^{-r(T-t)} = 4.8765, 4.7561, 4.5242$. See Fig. 15.11.

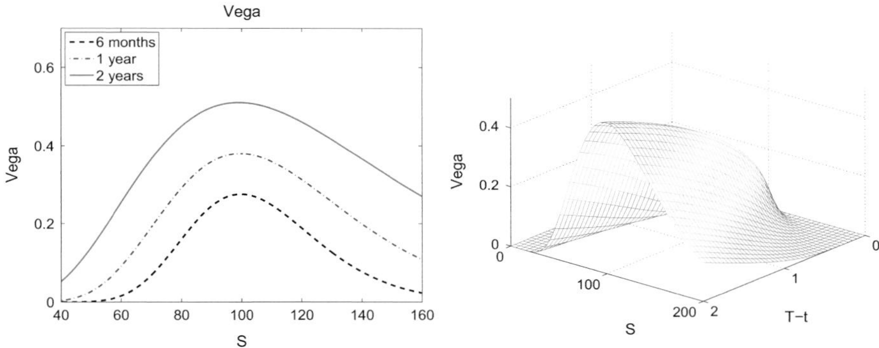

Fig. 15.10 Vega of European call and put options

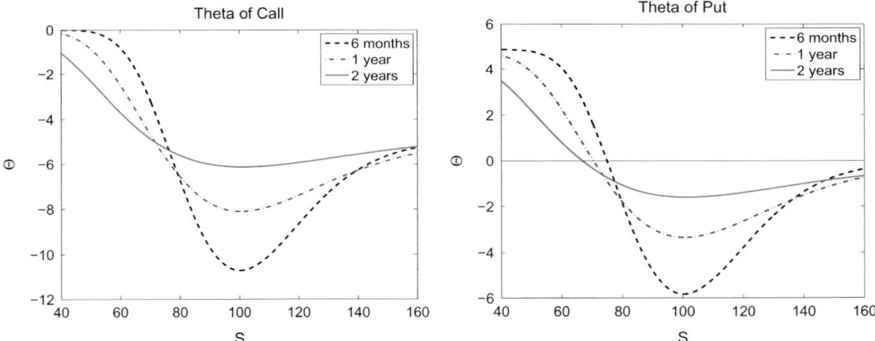

Fig. 15.11 Theta of European call and put options

15.4 Solution by the Laplace Transform

In Sect. 15.2 we employed a more or less direct method to solve the Black–Scholes–Merton partial differential equation. In this section we use another method based on the Laplace transformation. The Black–Scholes–Merton equation is essentially a heat equation (or a diffusion equation), and we modify the Laplace transformation method for the heat equation, which is given in Appendix D.3.

To convert the given problem into an initial value problem, we let

$$\tau = T - t$$

and put

$$v(S, \tau) = V(S, T - \tau) \ .$$

By Theorem 15.1 we have

$$\frac{1}{2}\sigma^2 S^2 \frac{\partial^2 v}{\partial S^2} + rS\frac{\partial v}{\partial S} - \frac{\partial v}{\partial \tau} = rv \ . \tag{15.3}$$

The initial data is given by $v(S,0) = V(S,T)$. Let $f(S,s)$ be the Laplace transformation $\mathscr{L}[v]$ of $v(S,\tau)$ with respect to τ. Since

$$\mathscr{L}\left[\frac{\partial v}{\partial \tau}\right](S,s) = s\mathscr{L}[v](S,s) - v(S,0) \ ,$$

we take the Laplace transformations of both sides of (15.3) and obtain

$$\frac{1}{2}\sigma^2 S^2 \frac{\partial^2 f}{\partial S^2} + rS\frac{\partial f}{\partial S} - (s+r)f = -v(S,0) \ . \tag{15.4}$$

To solve it, find first a homogeneous solution of the homogeneous equation

$$\frac{1}{2}\sigma^2 S^2 \frac{\partial^2 f}{\partial S^2} + rS\frac{\partial f}{\partial S} - (s+r)f = 0 \ . \tag{15.5}$$

In this case there exist two linearly independent solutions of the form

$$f(S,s) = S^\lambda \ . \tag{15.6}$$

(Note that since the coefficients of the given differential equation are not constant, it is hard to guess a general form. To make the differential equation a little simpler, we put $y = \log S$, and obtain a constant coefficient differential equation

$$\frac{1}{2}\sigma^2 \frac{\partial^2 f}{\partial y^2} + (-\frac{1}{2}\sigma^2 + r)\frac{\partial f}{\partial y} - (s+r)f = 0$$

and find a solution of the form $c_1 e^{\lambda_1 y} + c_2 e^{\lambda_2 y}$ using (15.7).)

Substituting (15.6) into (15.5), we obtain

$$\frac{1}{2}\sigma^2 S^2 \lambda(\lambda-1)S^{\lambda-2} + rS\lambda S^{\lambda-1} - (s+r)S^\lambda = 0 \ .$$

Eliminating S^λ from the both sides, we obtain the quadratic equation

$$\frac{1}{2}\sigma^2\lambda^2 + (-\frac{1}{2}\sigma^2 + r)\lambda - (s+r) = 0 \ . \tag{15.7}$$

Hence there exist two zeros

$$\lambda_{1,2} = \frac{-(r - \frac{1}{2}\sigma^2) \pm \sqrt{(r - \frac{1}{2}\sigma^2)^2 + 2(s+r)\sigma^2}}{\sigma^2} .$$

For $s > 0$ we have $\lambda_1 > 1$, $\lambda_2 < -\dfrac{2r}{\sigma^2} < 0$. In summary a homogeneous solution f_H of (15.5) is of the form

$$f_H = c_1 S^{\lambda_1} + c_2 S^{\lambda_2}$$

where $c_1(s)$ and $c_2(s)$ are constants with respect to S.

Now we find the prices of financial derivatives.

Example 15.1 (Forward Contract) Consider a forward contract to buy a share of a stock at price K on a future date T. Let us find its price v. Note that the initial condition is given by

$$v(S, 0) = V(S, T) = S - K .$$

Hence (15.4) becomes

$$\frac{1}{2}\sigma^2 S^2 \frac{\partial^2 f}{\partial S^2} + rS\frac{\partial f}{\partial S} - (s+r)f = K - S . \tag{15.8}$$

Let us find its particular solution f_P of the form

$$f_P(S, s) = c_1(s)S^{\lambda_1} + c_2(s)S^{\lambda_2} + \frac{S}{s} - \frac{K}{s+r} .$$

To determine the constants we check the behavior of v on the boundary of the domain. As $S \approx +\infty$, we have $v(S, \tau) \sim S$, and hence

$$f_P(S, s) \sim S\frac{1}{s}$$

and $c_1 = 0$. On the other hand, for $S \approx 0$, we have $v(S, \tau) \sim -K\mathrm{e}^{-r\tau}$, and hence

$$f_P(S, s) \sim -\frac{K}{s+r} ,$$

thus $c_2 = 0$. Hence

$$f_P(S, s) = \frac{S}{s} - \frac{K}{s+r} .$$

Now we take the inverse Laplace transformation and obtain

$$v(S,\tau) = S - K\mathrm{e}^{-r\tau}$$

and

$$V(S,t) = S - K\mathrm{e}^{-r(T-t)} .$$

Example 15.2 (European Call Option) The initial condition is given by

$$v(S,0) = V(S,T) = \max\{S-K,0\} .$$

Let $f(S,s)$ be the Laplace transformation $\mathscr{L}[v]$ of $v(S,\tau)$ with respect to τ. Then

$$\frac{1}{2}\sigma^2 S^2 \frac{\partial^2 f}{\partial S^2} + rS\frac{\partial f}{\partial S} - (s+r)f = \min\{K-S,0\} . \tag{15.9}$$

The homogeneous solution f_H of the above equation is of the form

$$f_H = c_1 S^{\lambda_1} + c_2 S^{\lambda_2} .$$

Now we try to find a particular solution f_P of (15.9) of the following form

$$f_P(S,s) = \begin{cases} aS^{\lambda_1} + bS^{\lambda_2} + \dfrac{S}{s} - \dfrac{K}{s+r}, & K < S < \infty , \\ cS^{\lambda_1} + dS^{\lambda_2}, & 0 \le S < K . \end{cases}$$

To determine the constants let us check the property of v on the boundary of the domain. As $S \approx +\infty$, we have $v(S,\tau) \sim S$, and hence $f_P(S,s) \sim S\frac{1}{s}$ and $a = 0$. Similarly, as $S \approx 0$, we have $v(S,\tau) \sim 0$, and hence $f_P(S,s) \approx 0$ and $d = 0$. Thus

$$f_P(S,s) = \begin{cases} bS^{\lambda_2} + \dfrac{S}{s} - \dfrac{K}{s+r}, & K < S < \infty , \\ cS^{\lambda_1}, & 0 \le S < K . \end{cases}$$

See Fig. 15.12.

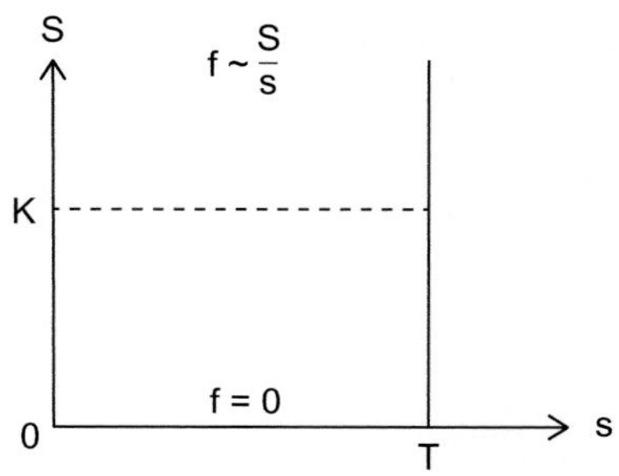

Fig. 15.12 Boundary condition for the Laplace transformation of a European call price

Since $f_P(S, s)$ is continuous and differentiable along the boundary $S = K$ of two domains, it satisfies the following two conditions:

$$bK^{\lambda_2} + \frac{K}{s} - \frac{K}{s+r} = cK^{\lambda_1} ,$$

$$\lambda_2 b K^{\lambda_2 - 1} + \frac{1}{s} = \lambda_1 c K^{\lambda_1 - 1} .$$

Now we find b and c, and then take the inverse Laplace transformation of $f_P(S, s)$. For more algebraic details consult [57, 95].

15.5 Computer Experiments

Simulation 15.1 (Delta Hedging)

Let us review **discrete time** hedging in the derivation of the Black–Scholes–Merton equation. What is the meaning of $\mathrm{d}D_t = rD_t\mathrm{d}t$? In contrast to $\mathrm{d}\Pi_t = r\Pi_t\mathrm{d}t$ over the time interval $[0, T]$, which has a continuous global solution $\Pi_t = \Pi_0 \mathrm{e}^{rt}$, the bank deposit D_t does not satisfy $D_t = D_0\mathrm{e}^{rt}$ for $0 \le t \le T$. The reason is that we readjust the investment into the risky asset upon receiving new stock market information by using a new value of Δ at t_i, and the investment into risk-free asset is also readjusted. More precisely, $D_t = D_{t_i} \times \mathrm{e}^{r(t-t_i)}$ for $t_i \le t < t_{i+1}$. Consult Fig. 15.13.

Since Π_t is self-financing, it is continuous at time t_i, and we have

$$\Pi_{t_i} = -V_{t_i} + D_{t_{i-1}}\mathrm{e}^{r\delta t} + \Delta_{t_{i-1}}S_{t_i} = -V_{t_i} + D_{t_i} + \Delta_{t_i}S_{t_i} ,$$

which implies that

$$D_{t_i} = D_{t_{i-1}}\mathrm{e}^{r\delta t} + (\Delta_{t_{i-1}} - \Delta_{t_i})S_{t_i} .$$

We present in-the-money and out-of-the-money cases for a European call option in Figs. 15.14 and 15.15, respectively. The values of Δ approaches 1 or 0 depending on whether the option is in-the-money or out-of-the-money. In the first case, the option will be exercised with certainty, so the seller must hedge it fully, while in

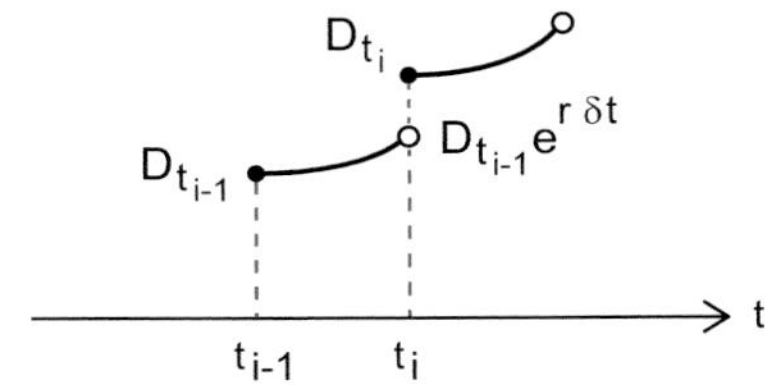

Fig. 15.13 Readjustment of the bank deposit in discrete time hedging

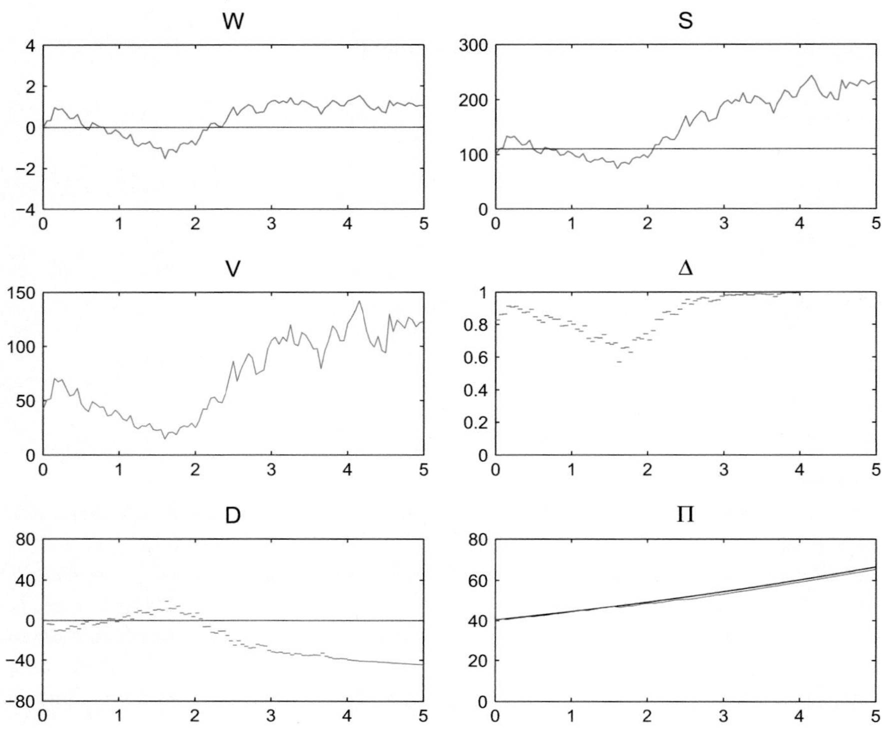

Fig. 15.14 In-the money option: delta hedging using a portfolio $\Pi = -V + D + \Delta S$

the second case there is not much chance to be exercised and the need to hedge is almost zero. In both cases the plots for Π_t are close to the exponential curve $\Pi_0 e^{rt}$ as expected. The graphs for Δ and D are step functions which will converge to continuous functions as the number of time steps increases to ∞.

```
T = 5;
r = 0.10; % interest rate
mu = 0.15; % drift coefficient
sigma = 0.3; % volatility
S0 = 100; % asset price at time t=0
K = 110; % strike price

N = 100 ; % number of time steps
dt = T/N;
t_value = [0:dt:T];

W = zeros(1,N+1); % Brownian motion
S = zeros(1,N+1); % asset price
V = zeros(1,N+1); % option price
Delta = zeros(1,N+1); % Delta
D = zeros(1,N+1); % bank deposit
```

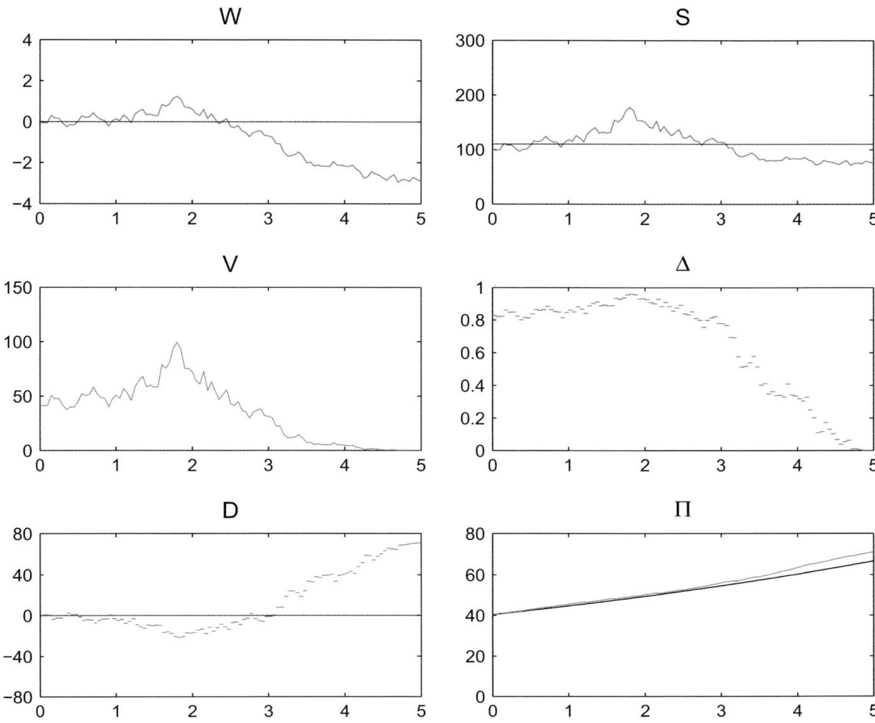

Fig. 15.15 Out-of-the-money option: delta hedging using a portfolio $\Pi = -V + D + \Delta S$

```
Pi = zeros(1,N+1); % portfolio

D(1) = 0.0; % Choose any number for the initial cash amount.
S(1)= S0;

for i=2:N+1
    dW = sqrt(dt)*randn;
    W(i) = W(i-1) + dW;
    S(i) = S(i-1) + mu*S(i-1)*dt + sigma*S(i-1)*dW;
end

for i=1:N+1
    tau = T-(i-1)*dt;
    d1 = (log(S(i)/K) + (r+0.5*sigma^2)*tau)/sigma/sqrt(tau);
    d2 = d1 - sigma*sqrt(tau);
    V(i)= S(i)*normcdf(d1) - K*exp(-r*tau)*normcdf(d2);
    Delta(i)=normcdf(d1);
end

for i = 1:N
    D(i+1) = exp(r*dt)*D(i) + (Delta(i)-Delta(i+1))*S(i+1);
```

```
            %self-financing
end

for i = 1:N+1
Pi(i) = -V(i) + D(i) + Delta(i)*S(i);
end

subplot(3,2,1);
plot([0:dt:T],W)
title('W')

subplot(3,2,2);
plot([0:dt:T],S)
title('S')
hold on
plot([0:dt:T],K,'--')

subplot(3,2,3);
plot([0:dt:T],V)
title('V')

subplot(3,2,4);
for i=1:N
    x = (i-1)*dt:dt/(500/N):i*dt;
    y = Delta(i)*exp(0*x);
plot(x,y)  % Plot the graph on each subinterval.
hold on
end
title('\Delta')

subplot(3,2,5);
for i=1:N
    x = (i-1)*dt:dt/(500/N):i*dt;
    y = D(i);
plot(x,y*exp(r*(x-(i-1)*dt))) % Plot the graph on each subinterval.
hold on
end
title('D')

subplot(3,2,6);
plot([0:dt:T],Pi(1)*exp(r*[0:dt:T]))
title('\Pi')
hold on
plot([0:dt:T],Pi)
```

Simulation 15.2 (Option Price Surface)

We regard the European call and put option prices as functions of t and S, and plot their graphs. See Figs. 15.3, 15.4.

```
K = 10;
S0 = 10;
r = 0.05;
sigma = 0.3;
mu = 0.1;
```

```
T = 1;
N = 51;
dt = T/(N-1);

S_path = zeros(N,1);
S_path(1)= S0; % asset price at time 0

for i=1:N-1
    dW = sqrt(dt)*randn;
    S_path(i+1) = S_path(i) + mu*S_path(i)*dt + sigma*S_path(i)*dW;
end

C_path = zeros(N,1);
P_path = zeros(N,1);

for i = 1:N
    S = S_path(i);
    tau = T-(i-1)*dt;
    d1 = (log(S/K) + (r+0.5*sigma^2)*tau)/(sigma*sqrt(tau));
    d2 = d1 - sigma*sqrt(tau);
    C_path(i) = S*normcdf(d1) - K*exp(-r*tau)*normcdf(d2);
    P_path(i) = C_path(i) - S + K*exp(-r*tau);
end

t_value = linspace(0,T,N);
S_value = linspace(0,20,N);
C = zeros(N,N);
P = zeros(N,N);

for j=1:N
    S = S_value(j);
    for i = 1:N-1
        tau = T-t_value(i);
        d1 = (log(S/K)+(r+0.5*sigma^2)*tau)/(sigma*sqrt(tau));
        d2 = d1-sigma*sqrt(tau);
        N1 = normcdf(d1);
        N2 = normcdf(d2);
        C(i,j) = S*N1 - K*exp(-r*tau)*N2;
        P(i,j) = C(i,j) + K*exp(-r*tau) - S;
    end
    C(N,j) = max(S-K,0); % payoff at T
    P(N,j) = max(K-S,0); % payoff at T
end

% Plot the superimposed image of asset price movement.
[S_grid,t_grid] = meshgrid(S_value,t_value);

figure(1);
surf(S_grid,t_grid,C)
hold on
plot3(S_path,t_grid,C_path);
xlabel('S'), ylabel('t'), zlabel('Call')
hold off
```

```
figure(2);
surf(S_grid,t_grid,P)
hold on
plot3(S_path,t_grid,P_path);
xlabel('S'), ylabel('t'), zlabel('Put')
hold off
```

Exercises

15.1 What is the boundary condition of the Black–Scholes–Merton equation for a European put option? Consult Fig. 15.16.

15.2 Define d_1 and d_2 as usual. Show that $d_1 - d_2 = \sigma\sqrt{T-t}$ and hence $d_2 = d_1 - \sigma\sqrt{T-t}$.

15.3 Recall that the value of a European call option C is given by $C(S) = SN(d_1) - Ke^{-rT}N(d_2)$ at time $t = 0$, where d_1 and d_2 are defined as usual. Using the idea in Theorem 2.1, find the value of a European put option at $t = 0$.

15.4 Let $C(S,t)$ denote the value of a European call option where S is the underlying asset price.

(i) Find $\lim_{K\to 0+} C(S)$.
(ii) Find $\lim_{\sigma\to 0+} C(S,t)$.
(iii) How about the case when the volatility σ is very large? Does it agree with common sense?

15.5 Recall that the value of a European put option is given by

$$\begin{aligned} P(S_t,t) &= S_tN(d_1) - Ke^{-r(T-t)}N(d_2) - S_t + Ke^{-r(T-t)} \\ &= Ke^{-r(T-t)}N(-d_2) - S_tN(-d_1)\ . \end{aligned}$$

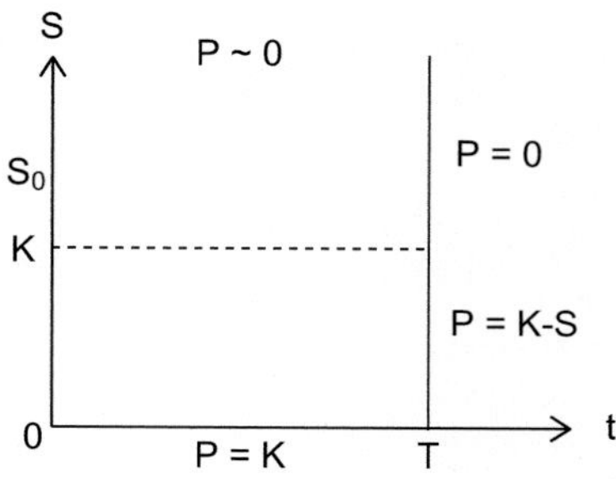

Fig. 15.16 Boundary condition for the price of a European put option

(i) Find $\lim_{K\to 0+} P(S)$.
(ii) Find $\lim_{\sigma\to 0+} P(S,t)$.
(iii) How about the case when the volatility σ is very large? Does it agree with common sense?

15.6 Let $C(S,t)$ denote the value of a European call option with strike price K. When $S_T > K$, compute

$$\lim_{t\to T-} \frac{\partial C(S,t)}{\partial S} .$$

15.7 Prove that for a European call option price C we have

$$\frac{\partial C}{\partial K} = -\mathrm{e}^{-r(T-t)} N(d_2) .$$

15.8 Explain the behavior of the solution of the Black–Scholes–Merton equation for a European put option in the case when the volatility σ is zero. How about the case when σ is very large? Does it agree with common sense?

15.9 The Black–Scholes–Merton equation can be derived using the first form of Itô's lemma. First, consider a portfolio

$$\Pi(t, S(t, W_t)) = -V(t, S(t, W_t)) + D(t, S(t, W_t)) + \Delta(t, S(t, W_t))S(t, W_t) .$$

Define $\widehat{\Pi}$, $\widehat{V}$, $\widehat{D}$, $\widehat{\Delta}$ by

$$\begin{aligned}
\widehat{\Pi}(t, W_t) &= \Pi(t, S(t, W_t)) , \\
\widehat{V}(t, W_t) &= V(t, S(t, W_t)) , \\
\widehat{D}(t, W_t) &= D(t, S(t, W_t)) , \\
\widehat{\Delta}(t, W_t) &= \Delta(t, S(t, W_t)) .
\end{aligned}$$

Then

$$\widehat{\Pi}(t, W_t) = -\widehat{V}(t, W_t) + \widehat{D}(t, W_t) + \widehat{\Delta}(t, W_t)S(t, W_t) .$$

15.10 The *elasticity* Ω of an option price is the ratio of the percentage change in the option price V with respect to the percentage change in the underlying asset price S. More precisely,

$$\Omega = \lim_{\delta S\to 0} \frac{\frac{\delta V}{V}}{\frac{\delta S}{S}} .$$

(It measures the leverage in the option investment.)

(i) Show that $\Omega = \Delta \frac{S}{V}$. (This means that for a deep out-of-the-money option, i.e., an option with $V \approx 0$, the elasticity can be large.)
(ii) Show that for a European call option, $\Omega \geq 1$. (This means that the investment in an option is riskier than that in the underlying asset.)
(iii) Show that for a European put option, $\Omega \leq 0$.

Chapter 16
The Martingale Method

In this chapter we introduce two proofs of the option pricing formula given by (16.2) by applying martingale theory. The first method is based on hedging of a portfolio process, and the second on replication of the payoff at expiry date T.

16.1 Option Pricing by the Martingale Method

Let $\{W_t\}_{t\geq 0}$ be a Brownian motion with respect to a probability measure $\mathbb{P}$. Assume that the asset price movement $\{S_t\}_{t\geq 0}$ follows geometric Brownian motion

$$\mathrm{d}S_t = \mu S_t \,\mathrm{d}t + \sigma S_t \,\mathrm{d}W_t$$

where μ and σ are constant, which has a solution

$$S_t = S_0 \,\mathrm{e}^{\sigma W_t + (\mu - \frac{1}{2}\sigma^2)t} \ .$$

Let $r > 0$ be risk-free interest rate, and put $\theta = \frac{\mu - r}{\sigma}$ and define X_t by

$$X_t = W_t + \theta t \ .$$

Then $\mathrm{d}X_t = \mathrm{d}W_t + \theta \mathrm{d}t$ and

$$\begin{aligned}\mathrm{d}S_t &= \mu S_t \,\mathrm{d}t + \sigma S_t(\mathrm{d}X_t - \theta \mathrm{d}t) \\ &= (\mu - \sigma\theta) S_t \,\mathrm{d}t + \sigma S_t \,\mathrm{d}X_t \\ &= r S_t \,\mathrm{d}t + \sigma S_t \,\mathrm{d}X_t \ .\end{aligned}$$

G.H. Choe, *Stochastic Analysis for Finance with Simulations*, Universitext,
DOI 10.1007/978-3-319-25589-7_16

Let $\mathbb{Q}$ be a probability measure such that $\mathbb{E}\left[\frac{d\mathbb{Q}}{d\mathbb{P}}\middle|\mathcal{F}_t\right] = e^{-\frac{1}{2}\theta^2 t-\theta W_t}$ for $0 \le t \le T$. Girsanov's theorem states that $\mathbb{P}$ and $\mathbb{Q}$ are equivalent, and $\{X_t\}$ is a $\mathbb{Q}$-Brownian motion. Let $\widetilde{S}_t = e^{-rt}S_t$ be a discounted asset price. Since

$$\widetilde{S}_t = S_0 e^{(\mu-r-\frac{1}{2}\sigma^2)t+\sigma W_t} = \widetilde{S}_0\, e^{-\frac{1}{2}\sigma^2 t+\sigma X_t} ,$$

we have

$$d\widetilde{S}_t = \sigma \widetilde{S}_t\, dX_t$$

and $\{\widetilde{S}_t\}$ is a $\mathbb{Q}$-martingale.

Recall that if a payoff function at T depends on a process $\{S_t\}_{0\le t\le T}$, then it is $\mathcal{F}_T$-measurable.

Method I: Option Pricing by Hedging

Consider a portfolio process

$$\Pi_t = -V_t + D_t + \Delta_t S_t$$

as given in Sect. 15.1. Define the discounted processes

$$\widetilde{\Pi}_t = e^{-rt}\Pi_t , \quad \widetilde{V}_t = e^{-rt}V_t \quad \text{and} \quad \widetilde{D}_t = e^{-rt}D_t .$$

Then

$$\widetilde{\Pi}_t = -\widetilde{V}_t + \widetilde{D}_t + \Delta_t \widetilde{S}_t .$$

Since, with the choice of

$$\Delta_t = \frac{\partial V}{\partial S} ,$$

the portfolio Π_t becomes risk-free, we have $\Pi_t = \Pi_0 e^{rt}$, $0 \le t \le T$. Hence

$$\widetilde{\Pi}_t = \Pi_0$$

and

$$d\widetilde{\Pi}_t = 0 .$$

Let Y_{t-} denote the limit from the left of the process Y at time $t > 0$. Take a partition $0 = t_0 < t_1 < \cdots < t_{N-1} < t_N = T$. Since the portfolio is self-financing, we have

$$\widetilde{\Pi}_{t_{i+1}-} - \widetilde{\Pi}_{t_i} = -(\widetilde{V}_{t_{i+1}-} - \widetilde{V}_{t_i}) + (\widetilde{D}_{t_{i+1}-} - \widetilde{D}_{t_i}) + \Delta_{t_i}(\widetilde{S}_{t_{i+1}-} - \widetilde{S}_{t_i}) .$$

Since $\widetilde{\Pi}_t$, $\widetilde{V}_t$ and $\widetilde{S}_t$ are continuous (with probability 1), we have

$$\widetilde{\Pi}_{t_{i+1}} - \widetilde{\Pi}_{t_i} = -(\widetilde{V}_{t_{i+1}} - \widetilde{V}_{t_i}) + (\widetilde{D}_{t_{i+1}-} - \widetilde{D}_{t_i}) + \Delta_{t_i}(\widetilde{S}_{t_{i+1}} - \widetilde{S}_{t_i}) .$$

Since $\widetilde{\Pi}_t = \Pi_0$ for $0 \le t \le T$ and since $\widetilde{D}_t = D_{t_i}\mathrm{e}^{-rt_i}$ for $t_i \le t < t_{i+1}$, we have

$$0 = -(\widetilde{V}_{t_{i+1}} - \widetilde{V}_{t_i}) + 0 + \Delta_{t_i}(\widetilde{S}_{t_{i+1}} - \widetilde{S}_{t_i}) .$$

(See Fig. 16.1.) Hence

$$\widetilde{V}_T - \widetilde{V}_0 = \sum_{i=0}^{N-1}(\widetilde{V}_{t_{i+1}} - \widetilde{V}_{t_i}) = \sum_{i=0}^{N-1} \Delta_{t_i}(\widetilde{S}_{t_{i+1}} - \widetilde{S}_{t_i}) .$$

By taking the limit, we have

$$\widetilde{V}_T - \widetilde{V}_0 = \int_0^T \Delta_t \, \mathrm{d}\widetilde{S}_t . \tag{16.1}$$

The stochastic integral on the right-hand side of (16.1) is meaningfully defined since the integrator $\widetilde{S}_t$ is a continuous martingale with respect to $\mathbb{Q}$. By the Martingale Representation Theorem there exists an α_t such that

$$\mathrm{d}\widetilde{S}_t = \alpha_t \, \mathrm{d}X_t .$$

In fact,

$$\alpha_t = \sigma \widetilde{S}_t .$$

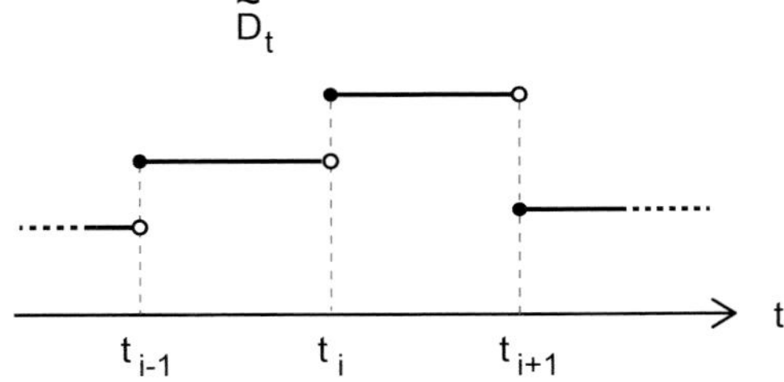

Fig. 16.1 Time discretization for discounted bond price $\widetilde{D}_t$

Hence

$$\widetilde{V}_T - \widetilde{V}_0 = \int_0^T \Delta_t \sigma \widetilde{S}_t \, \mathrm{d}X_t \, .$$

By taking the expectations of both sides, we have

$$\mathbb{E}^{\mathbb{Q}}[\widetilde{V}_T] - \widetilde{V}_0 = 0$$

since the expectation of the Itô integral is zero. Therefore,

$$V_0 = \mathrm{e}^{-rT} \mathbb{E}^{\mathbb{Q}}[V_T] \, .$$

If we consider the time interval $[t, T]$ in the beginning of the derivation of the formula, we obtain

$$\widetilde{V}_T - \widetilde{V}_t = \int_t^T \Delta_t \sigma \widetilde{S}_u \, \mathrm{d}X_u$$

and

$$V_t = \mathrm{e}^{-r(T-t)} \mathbb{E}^{\mathbb{Q}}[V_T | \mathcal{F}_t] \, . \tag{16.2}$$

Method II: Option Pricing by Replication

Let C_T denote the payoff of a given European option, and

$$B_t = \mathrm{e}^{rt}$$

be the value of risk-free deposit of unit amount. Define V_t and $\widetilde{V}_t$ by

$$V_t = \mathbb{E}^{\mathbb{Q}}[\, \mathrm{e}^{-r(T-t)} C_T | \mathcal{F}_t \,]$$

and

$$\widetilde{V}_t = \mathrm{e}^{-rt} V_t = \mathbb{E}^{\mathbb{Q}}[\, \mathrm{e}^{-rT} C_T | \mathcal{F}_t \,] \, .$$

Then we have a replication of the payoff since

$$V_T = \mathrm{e}^{rT} \widetilde{V}_T = \mathbb{E}^{\mathbb{Q}}[\, C_T | \mathcal{F}_T \,] = C_T \, .$$

It remains to show that V_t is self-financing. Since $\widetilde{V}_t$ is a $\mathbb{Q}$-martingale, by the Martingale Representation Theorem there exists a predictable process $\{\alpha_t\}_{t\geq 0}$ such that

$$\widetilde{V}_t = \widetilde{V}_0 + \int_0^t \alpha_u \, \mathrm{d}X_u = \widetilde{V}_0 + \int_0^t \alpha_u \frac{1}{\sigma \widetilde{S}_u} \, \mathrm{d}\widetilde{S}_u$$

where the continuous martingale $\widetilde{S}_t$ is the integrator for the given stochastic integral. Now define ϕ_t by

$$\phi_t = \alpha_t \frac{1}{\sigma \widetilde{S}_t} \ .$$

Then we have

$$\widetilde{V}_t = \widetilde{V}_0 + \int_0^t \phi_u \mathrm{d}\widetilde{S}_u \ .$$

Since

$$\mathrm{d}\widetilde{V}_t = \phi_t \, \mathrm{d}\widetilde{S}_t \ , \tag{16.3}$$

we have

$$\mathrm{d}(\mathrm{e}^{-rt} V_t) = \phi_t \, \mathrm{d}(\mathrm{e}^{-rt} S_t) \ ,$$

and hence

$$-r\mathrm{e}^{-rt} V_t \, \mathrm{d}t + \mathrm{e}^{-rt} \mathrm{d}V_t = \phi_t(-r\mathrm{e}^{-rt} S_t \, \mathrm{d}t + \mathrm{e}^{-rt} \mathrm{d}S_t) \ ,$$

and after multiplying by e^{rt} we obtain

$$-rV_t \, \mathrm{d}t + \mathrm{d}V_t = \phi_t(-rS_t \, \mathrm{d}t + \mathrm{d}S_t) \ .$$

Note that since $\mathrm{e}^{-rt} \, \mathrm{d}B_t = r \, \mathrm{d}t$,

$$\mathrm{d}V_t = (V_t - \phi_t S_t) \, \mathrm{e}^{-rt} \mathrm{d}B_t + \phi_t \, \mathrm{d}S_t \ .$$

Now define ψ_t by

$$\psi_t = (V_t - \phi_t S_t) \, \mathrm{e}^{-rt} \ ,$$

or equivalently,

$$V_t = \psi_t B_t + \phi_t S_t \ . \tag{16.4}$$

Since

$$\mathrm{d}V_t = \psi_t \,\mathrm{d}B_t + \phi_t \,\mathrm{d}S_t \ ,$$

ψ_t and ϕ_t produce a self-financing portfolio V_t. Thus V_t is the option price at time t.

The following results shows that the option price does not depend on the drift coefficient in the geometric Brownian motion, which is also observed in the Black–Scholes–Merton formula.

Theorem 16.1 (Risk Neutrality) *For a European option with its expiry date T and payoff C_T, its price V_0 at $t = 0$ is given by*

$$V_0 = \mathbb{E}^{\mathbb{P}}[\,\mathrm{e}^{-rT} C_T(S_0\,\mathrm{e}^{\sigma W_T + (r-\frac{1}{2}\sigma^2)T})\,] \ .$$

In other words, the option price is the expected discounted payoff taken over the sample paths of a geometric Brownian motion with the drift coefficient μ replaced by the risk-free interest rate r.

Proof Note that (16.2) implies

$$\begin{aligned} V_0 &= \mathbb{E}^{\mathbb{Q}}[\,\mathrm{e}^{-rT} C_T(S_T)\,] \\ &= \mathbb{E}^{\mathbb{Q}}[\,\mathrm{e}^{-rT} C_T(\mathrm{e}^{rT}\widetilde{S}_T)\,] \\ &= \mathbb{E}^{\mathbb{Q}}[\,\mathrm{e}^{-rT} C_T(\mathrm{e}^{rT} S_0\,\mathrm{e}^{\sigma X_T - \frac{1}{2}\sigma^2 T})\,] \\ &= \mathbb{E}^{\mathbb{Q}}[\,\mathrm{e}^{-rT} C_T(S_0\,\mathrm{e}^{\sigma X_T + (r-\frac{1}{2}\sigma^2)T})\,] \\ &= \mathbb{E}^{\mathbb{P}}[\,\mathrm{e}^{-rT} C_T(S_0\,\mathrm{e}^{\sigma W_T + (r-\frac{1}{2}\sigma^2)T})\,] \end{aligned}$$

where the last equality is merely a change of symbols X_T and W_T which represent Brownian motions with respect to $\mathbb{Q}$ and $\mathbb{P}$, respectively. □

16.2 The Probability Distribution of Asset Price

Now we find the price of a European option with expiry date T and payoff $C_T(S_T)$ where S_T is the asset price at T by using the probability density function of S_T.

If the asset price S_t follows the geometric Brownian motion, then at time $t = T$ we have

$$S_T = S_0\,\mathrm{e}^{\sigma W_T + (\mu - \frac{1}{2}\sigma^2)T} \ .$$

If x denotes a value taken by S_T, then we have

$$\frac{\log \frac{x}{S_0} - (\mu - \frac{1}{2}\sigma^2)T}{\sigma} = W_T \ .$$

If y denotes a value taken by W_T, the pdf of y is given by

$$f(y) = \frac{1}{\sqrt{2\pi T}} \exp\left(- \frac{1}{2T} \left(\frac{\log \frac{x}{S_0} - (\mu - \frac{1}{2}\sigma^2)T}{\sigma} \right)^2 \right) .$$

Since $\frac{\mathrm{d}y}{\mathrm{d}x} = \frac{1}{\sigma x}$, the price of the option is given by

$$\begin{aligned} \mathrm{e}^{-rT}\mathbb{E}[C_T(S_T)] &= \mathrm{e}^{-rT} \int_0^\infty C_T(x) f(y(x)) \frac{\mathrm{d}y}{\mathrm{d}x} \mathrm{d}x \\ &= \mathrm{e}^{-rT} \int_0^\infty \frac{C_T(x)}{x\sigma\sqrt{2\pi T}} \exp\left(- \frac{(\log \frac{x}{S_0} - (\mu - \frac{1}{2}\sigma^2)T)^2}{2T\sigma^2} \right) \mathrm{d}x . \end{aligned}$$

To apply the risk-neutral method, we put $\mu = r$. Then the option price equals

$$\mathrm{e}^{-rT} \int_0^\infty \frac{C_T(x)}{x\sigma\sqrt{2\pi T}} \exp\left(- \frac{(\log \frac{x}{S_0} - (r - \frac{1}{2}\sigma^2)T)^2}{2T\sigma^2} \right) \mathrm{d}x .$$

16.3 The Black–Scholes–Merton Formula

In this section we compute the price of a European call option using the martingale method. Since the payoff is given by $C(S_T) = (S_T - K)^+$ at expiry date T, we have

$$\begin{aligned} V_0 &= \mathrm{e}^{-rT}\mathbb{E}^{\mathbb{Q}}[\,(S_T - K)^+\,] \\ &= \mathrm{e}^{-rT}\mathbb{E}^{\mathbb{Q}}[\,(\mathrm{e}^{rT}\widetilde{S}_T - K)^+\,] \\ &= \mathrm{e}^{-rT}\mathbb{E}^{\mathbb{Q}}[\,(\mathrm{e}^{rT} S_0\, \mathrm{e}^{\sigma X_T - \frac{1}{2}\sigma^2 T} - K)^+\,] \\ &= \mathrm{e}^{-rT}\mathbb{E}^{\mathbb{Q}}[\,(S_0\, \mathrm{e}^{\sigma\sqrt{T}Z + (r - \frac{1}{2}\sigma^2)T} - K)^+\,] \\ &= \mathrm{e}^{-rT} \int_{-\infty}^\infty (S_0\, \mathrm{e}^{\sigma\sqrt{T}x + (r - \frac{1}{2}\sigma^2)T} - K)^+ \frac{1}{\sqrt{2\pi}} \mathrm{e}^{-x^2/2}\, \mathrm{d}x \end{aligned}$$

where the symbol Z denotes the standard normal variable. If we consider a European call option with exercise price K, then the domain of integration is given by

$$\{x : S_0 \mathrm{e}^{\sigma\sqrt{T}x + (r - \frac{1}{2}\sigma^2)T} \geq K\} .$$

Hence

$$x \geq x_0 = \frac{\log \frac{K}{S_0} - (r - \frac{1}{2}\sigma^2)T}{\sigma\sqrt{T}} . \tag{16.5}$$

Therefore the given integral is equal to

$$e^{-rT} \int_{x_0}^{\infty} (S_0 \, e^{\sigma\sqrt{T}x+(r-\frac{1}{2}\sigma^2)T} - K) \frac{1}{\sqrt{2\pi}} e^{-x^2/2} \, dx \, . \tag{16.6}$$

Now we put

$$d_1, d_2 = \frac{\log \frac{S_0}{K} + (r \pm \frac{1}{2}\sigma^2)T}{\sigma\sqrt{T}} \, .$$

If we let $y = -x$, then $dy = -dx$, and (16.6) is equal to

$$\begin{aligned} & e^{-rT} \int_{-\infty}^{d_2} (S_0 \, e^{-\sigma\sqrt{T}y+(r-\frac{1}{2}\sigma^2)T} - K) \frac{1}{\sqrt{2\pi}} e^{-y^2/2} \, dy \\ &= S_0 \int_{-\infty}^{d_2} e^{-\sigma\sqrt{T}y-\frac{1}{2}\sigma^2 T} \frac{1}{\sqrt{2\pi}} e^{-y^2/2} \, dy - Ke^{-rT} \int_{-\infty}^{d_2} \frac{1}{\sqrt{2\pi}} e^{-y^2/2} \, dy \, . \end{aligned}$$

Then we have

$$\begin{aligned} \int_{-\infty}^{d_2} \frac{1}{\sqrt{2\pi}} e^{-\sigma\sqrt{T}y-\frac{1}{2}\sigma^2 T-y^2/2} \, dy &= \int_{-\infty}^{d_2} \frac{1}{\sqrt{2\pi}} e^{-(y+\sigma\sqrt{T})^2/2} \, dy \\ &= N(d_2 + \sigma\sqrt{T}) \\ &= N(d_1) \, , \end{aligned}$$

and the original integral representing the option price is given by

$$S_0 N(d_1) - Ke^{-rT} N(d_2) \, ,$$

which is identical with the solution given in Sect. 15.2.

The martingale method can derive a closed form solution in some cases, but in general the Monte Carlo method is used to find an approximate solution, and it is more widely used than the partial differential equation approach. For an application of the restricted normal distribution in the Monte Carlo method for the estimation of (16.6), see Exercise 27.2.

16.4 Derivation of the Black–Scholes–Merton Equation

In this section, using the martingale method combined with the Feynman–Kac Theorem, we derive the Black–Scholes–Merton equation for European options. Let V_t denote the option price at time t, and assume that the payoff is of the form $C_T(S_T)$.

If a function $F : [0,T] \times [0,\infty) \to \mathbb{R}$ satisfies

$$V_t(\omega) = F(t, S_t(\omega)) ,$$

then

$$F(t,x) = \mathbb{E}^{\mathbb{Q}}[\mathrm{e}^{-r(T-t)} C_T(S_T) | S_t = x] .$$

Now if we put

$$G(t,x) = \mathbb{E}^{\mathbb{Q}}[C_T(S_T) | S_t = x] ,$$

then

$$G(t,x) = \mathrm{e}^{r(T-t)} F(t,x) .$$

Now we apply the Feynman–Kac Theorem to $G(t,x)$ and obtain

$$\frac{\partial}{\partial t}(\mathrm{e}^{r(T-t)} F(t,x))\bigg|_{x=S_t} + rS_t \frac{\partial}{\partial x}(\mathrm{e}^{r(T-t)} F(t,x))\bigg|_{x=S_t} + \frac{1}{2}\sigma^2 S_t^2 \frac{\partial^2}{\partial x^2}(\mathrm{e}^{r(T-t)} F(t,x))\bigg|_{x=S_t} = 0 .$$

After dividing by e^{rT}, then taking derivatives, and finally multiplying by e^{rt}, we obtain

$$-rF(t,S_t) + \frac{\partial F(t,x)}{\partial t}\bigg|_{x=S_t} + rS_t \frac{\partial F(t,x)}{\partial x}\bigg|_{x=S_t} + \frac{1}{2}\sigma^2 S_t^2 \frac{\partial^2 F(t,x)}{\partial x^2}\bigg|_{x=S_t} = 0 .$$

Since S_t can take any arbitrary value, the following equation should hold:

$$-rF(t,x) + \frac{\partial F(t,x)}{\partial t} + rx\frac{\partial F(t,x)}{\partial x} + \frac{1}{2}\sigma^2 x^2 \frac{\partial^2 F(t,x)}{\partial x^2} = 0 , \qquad (16.7)$$

which is nothing but the Black–Scholes–Merton equation.

16.5 Delta Hedging

For hedging, we will find the Delta (Δ) of a European call option. From (16.4) it is equal to ϕ_t, which is given by (16.3).

First, put

$$\widetilde{F}(t,y) = \mathrm{e}^{-rt} F(t, y\mathrm{e}^{rt}) \ ,$$

then we have

$$\begin{aligned}
&\frac{\partial}{\partial t}\widetilde{F}(t,y) \\
&= -r\mathrm{e}^{-rt}F(t,y\mathrm{e}^{rt}) + \mathrm{e}^{-rt}\left\{\frac{\partial F}{\partial t}(t,y\mathrm{e}^{rt}) + \frac{\partial F}{\partial x}(t,y\mathrm{e}^{rt})ry\mathrm{e}^{rt}\right\} \\
&= \mathrm{e}^{-rt}\left\{-rF(t,y\mathrm{e}^{rt}) + \frac{\partial F}{\partial t}(t,y\mathrm{e}^{rt}) + ry\mathrm{e}^{rt}\frac{\partial F}{\partial x}(t,y\mathrm{e}^{rt})\right\} \\
&= \mathrm{e}^{-rt}\left\{-\frac{1}{2}\sigma^2 y^2 \mathrm{e}^{2rt}\frac{\partial^2 F}{\partial x^2}(t,y\mathrm{e}^{rt})\right\} \qquad \text{(by (16.7))} \\
&= -\frac{1}{2}\sigma^2 y^2 \mathrm{e}^{rt}\frac{\partial^2 F}{\partial x^2}(t,y\mathrm{e}^{rt}) \ ,
\end{aligned}$$

$$\frac{\partial \widetilde{F}(t,x)}{\partial x} = \frac{\partial}{\partial x}\left(\mathrm{e}^{-rt}F(t,x\mathrm{e}^{rt})\right) = \left.\frac{\partial F}{\partial x}\right|_{(t,x\mathrm{e}^{rt})} ,$$

and

$$\frac{\partial^2 \widetilde{F}(t,x)}{\partial x^2} = \frac{\partial^2}{\partial x^2}\left(\mathrm{e}^{-rt}F(t,x\mathrm{e}^{rt})\right) = \mathrm{e}^{rt}\left.\frac{\partial^2 F}{\partial x^2}\right|_{(t,x\mathrm{e}^{rt})} .$$

Hence

$$\frac{\partial \widetilde{F}}{\partial t}(t,x) = -\frac{1}{2}\sigma^2 x^2 \frac{\partial^2 \widetilde{F}}{\partial x^2}(t,x) \ . \tag{16.8}$$

On the other hand, from

$$\mathrm{d}\widetilde{S}_t = \sigma\, \widetilde{S}_t\, \mathrm{d}X_t$$

we obtain

$$(\mathrm{d}\widetilde{S}_t)^2 = \sigma^2 \widetilde{S}_t^2\, \mathrm{d}t \ ,$$

and by applying Itô's lemma to $\widetilde{F}(t, \widetilde{S}_t)$, and using (16.8), we obtain

$$\begin{aligned} \mathrm{d}\widetilde{F}(t, \widetilde{S}_t) &= \frac{\partial \widetilde{F}}{\partial t}(t, \widetilde{S}_t)\mathrm{d}t + \frac{\partial \widetilde{F}}{\partial x}(t, \widetilde{S}_t)\mathrm{d}\widetilde{S}_t + \frac{1}{2}\frac{\partial^2 \widetilde{F}}{\partial x^2}(t, \widetilde{S}_t)\sigma^2 \widetilde{S}_t^2 \mathrm{d}t \\ &= \frac{\partial \widetilde{F}}{\partial x}(t, \widetilde{S}_t)\mathrm{d}\widetilde{S}_t \ . \end{aligned}$$

Since

$$\widetilde{V}_t = \mathrm{e}^{-rt} V_t = \mathrm{e}^{-rt} F(t, S_t) = \widetilde{F}(t, \mathrm{e}^{-rt} S_t) = \widetilde{F}(t, \widetilde{S}_t) \ ,$$

we have

$$\mathrm{d}\widetilde{V}_t = \mathrm{d}\widetilde{F}(t, \widetilde{S}_t) = \frac{\partial \widetilde{F}}{\partial x}(t, \widetilde{S}_t)\mathrm{d}\widetilde{S}_t \ .$$

Hence (16.3) implies that

$$\Delta = \phi_t = \frac{\partial \widetilde{F}}{\partial x}(t, \widetilde{S}_t) = \frac{\partial F}{\partial x}(t, S_t) \ .$$

Remark 16.1

(i) In the case of a European call option we can show easily

$$\Delta = N(d_1)$$

using the martingale method as done in Theorem 15.3. First, put

$$d_1, d_2 = \frac{\log \frac{S_t}{K} + (r \pm \frac{1}{2}\sigma^2)(T - t)}{\sigma \sqrt{T - t}} \ .$$

Now note that

$$V_t = F(t, S_t) = S_t N(d_1) - K\mathrm{e}^{-rT} N(d_2)\mathrm{e}^{rt}$$

and $B_t = \mathrm{e}^{rt}$. Then by checking the coefficients of S_t and B_t, we find

$$\begin{cases} \phi_t = N(d_1) \ , \\ \psi_t = -K\mathrm{e}^{-rT} N(d_2) \ . \end{cases}$$

(ii) In the standard continuous hedging strategy (as in the derivation of the Black–Scholes–Merton equation) we consider a portfolio

$$\Pi_t = -V_t + D_t + \Delta S_t$$

from the viewpoint of the option seller where D_t is the bank deposit. Then

$$Ke^{-r(T-t)}N(d_2) = \Delta_t S_t - V_t = \Pi_t - D_t ,$$

which is risk-free. Suppose that we, as a seller of the option, start with no money, i.e., $\Pi_0 = 0$. This means that we sell an option and using the option premium received, we buy some shares of the underlying asset and invest the left-over cash in risk-free bond. Note that

$$0 = -V_0 + D_0 + \Delta_0 S_0 ,$$

and hence

$$D_0 = V_0 - \Delta_0 S_0 = -Ke^{-rT}N(d_2)$$

which is the borrowed amount at the start.

16.6 Computer Experiments

Simulation 16.1 (Martingale Method)
We compare the martingale method with the Black–Scholes–Merton formula. See Figs. 16.2, 16.3 where the call price and Delta are approximately equal to 5.5871 and 0.4108, respectively.

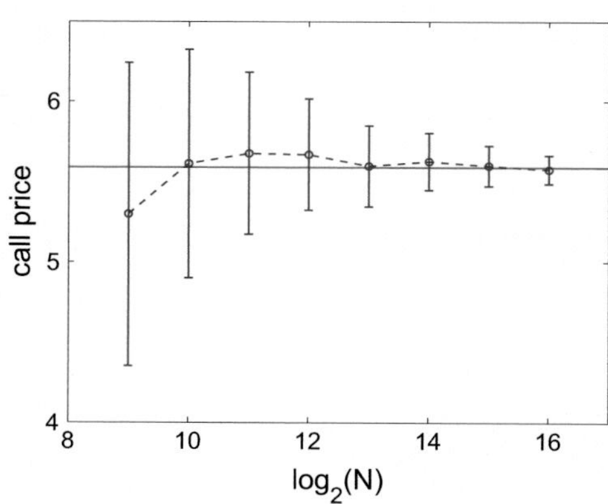

Fig. 16.2 Convergence to the Black–Scholes–Merton price

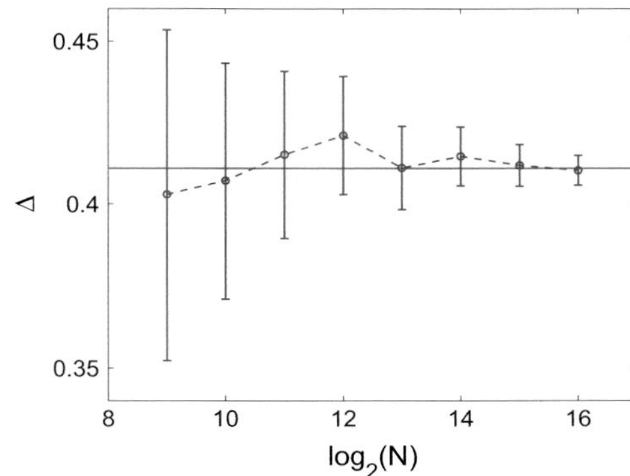

Fig. 16.3 Convergence to the Black–Scholes–Merton formula for Delta

```
S0 = 100;
K = 110;
sigma = 0.3;
r = 0.05;
T = 0.5;

% Black-Scholes-Merton formula
d1 = (log(S0/K)+(r+0.5*sigma^2)*T)/(sigma*sqrt(T));
d2 = d1 - sigma*sqrt(T);
N1 = normcdf(d1);
N2 = normcdf(d2);

Call_formula = S0*N1 - K*exp(-r*T)*N2;
Delta_formula = N1;

% option price
J = 16;
L = 2^J;
W = sqrt(T)*randn(L,1);
S = S0*exp((r-0.5*sigma^2)*T + sigma*W); % S_T
V = exp(-r*T)*max(S - K,0);

figure(1)
for j = 9:J
M(j) = 2^j; % number of samples
a(j) = mean(V(1:M(j)));
b(j) = 1.96*std(V(1:M(j)))/sqrt(M(j));
end
x = 8:0.01:J+1;
plot(x,Call_formula,'r')
hold on
errorbar(9:J, a(9:J),b(9:J));
xlabel('log_2(N)')
ylabel('call price');
hold off

% Delta
```

```
dS = 0.001;
S1 = (S0+dS)*exp((r-0.5*sigma^2)*T + sigma*W);
dVdS = exp(-r*T)*(max(S1-K,0)-max(S-K,0))/dS;
hold off

figure(2)
for j = 9:J
M(j) = 2^j; % number of samples
a2(j) = mean(dVdS(1:M(j)));
b2(j) = 1.96*std(dVdS(1:M(j)))/sqrt(M(j));
end
x = 8:0.01:J+1;
plot(x,Delta_formula,'-r')
hold on
errorbar(9:J, a2(9:J),b2(9:J));
xlabel('log_2(N)')
ylabel('\Delta');
```

Exercises

16.1 (A trivial application of the martingale method) Consider a European option that pays \$1 at time T. Find its price V_0 at $t = 0$. Its price V_0 at $t = 0$ is given by the expectation with respect to a martingale measure $\mathbb{Q}$, and $V_0 = e^{-rT}\mathbb{E}^{\mathbb{Q}}[1] = e^{-rT}$, which is nothing but the risk-free bond price. If we follow the rule that μ should be replaced by r in using the expectation with respect to a physical measure $\mathbb{P}$, we would have the same answer since $e^{-rT}\mathbb{E}^{\mathbb{P}}[1] = e^{-rT}$.

Part VI
Examples of Option Pricing

Chapter 17
Pricing of Vanilla Options

The Black–Scholes–Merton formula for a European call option is derived in Sect. 16.3 using the martingale method. In this chapter we present more examples of pricing of vanilla options. For more information consult [9] and [29].

17.1 Stocks with a Dividend

A dividend is a distribution of a company's profits to its shareholders. Dividends are usually issued as cash payments or shares of stock. The dividend rate may mean the cash amount each share receives, or a percent of the current market price.

When a stock pays a dividend, it is periodically paid. Although dividends are not paid continuously in the real financial world, for theoretical analysis we also consider the case when the dividend is paid continuously.

1. Continuous Dividends

Suppose that the asset price follows the geometric Brownian motion $S_t = S_0 \exp(\nu t + \sigma W_t)$. Over the time interval $[t, t + \mathrm{d}t)$ for a sufficiently short time length $\mathrm{d}t > 0$, we receive a dividend $\delta\, S_t\, \mathrm{d}t$ per share for some constant $\delta > 0$, and buy $\delta\, \mathrm{d}t$ shares of stock and add them to the existing stocks. If we let $g(t)$ be the number of stocks at t, then $g(t + \mathrm{d}t) - g(t) = g(t)\delta \mathrm{d}t$ since we can use the dividend $g(t)\delta\, S_t\, \mathrm{d}t$ to buy $g(t)\delta\, \mathrm{d}t$ shares of stock. In other words, $g(t)$ satisfies an ordinary differential equation $g'(t) = g(t)\delta$. Hence $g(t) = g(0)\mathrm{e}^{\delta t}$. If we begin with one share at $t = 0$, i.e., $g(0) = 1$, then $g(t) = \mathrm{e}^{\delta t}$. If we reinvest the dividend into stock continuously, then the total value of stock is equal to

$$Z_t = g(t)S_0 \exp(\nu t + \sigma W_t) = S_0 \exp((\nu + \delta)t + \sigma W_t) \; .$$

G.H. Choe, *Stochastic Analysis for Finance with Simulations*, Universitext,
DOI 10.1007/978-3-319-25589-7_17

Regarding Z_t as a new underlying asset, we apply the martingale method to compute the option price.

First, let

$$\widetilde{Z}_t = \mathrm{e}^{-rt} Z_t = S_0 \exp((\nu + \delta - r)t + \sigma W_t) .$$

Then

$$\mathrm{d}\widetilde{Z}_t = (\nu + \delta + \frac{1}{2}\sigma^2 - r)\widetilde{Z}_t \mathrm{d}t + \sigma \widetilde{Z}_t \mathrm{d}W_t .$$

If we let

$$\theta = \frac{\nu + \delta + \frac{1}{2}\sigma^2 - r}{\sigma}$$

and define a probability measure $\mathbb{Q}$ by

$$\left.\frac{\mathrm{d}\mathbb{Q}}{\mathrm{d}\mathbb{P}}\right|_{\mathcal{F}_t} = \exp\left(-\theta W_t - \frac{1}{2}\theta^2 t\right) ,$$

then $X_t = W_t + \theta t$ is a $\mathbb{Q}$-Brownian motion, and $\{\widetilde{Z}_t\}_{t \geq 0}$ is a $\mathbb{Q}$-martingale.

Thus the price of a call option is given by

$$\begin{aligned} V_t &= \mathrm{e}^{-r(T-t)} \mathbb{E}^{\mathbb{Q}}[(S_T - K)^+ | \mathcal{F}_t] \\ &= \mathrm{e}^{-r(T-t)} \mathbb{E}^{\mathbb{Q}}[(\mathrm{e}^{-\delta T} Z_T - K)^+ | \mathcal{F}_t] \\ &= \mathrm{e}^{-\delta T} \mathrm{e}^{-r(T-t)} \mathbb{E}^{\mathbb{Q}}[(Z_T - \mathrm{e}^{\delta T} K)^+ | \mathcal{F}_t] . \end{aligned}$$

This value is equal to the price of $\mathrm{e}^{-\delta T}$ contracts of call options on the underlying asset $\{Z_t\}_{\geq 0}$ with maturity date T and exercise price $\mathrm{e}^{\delta T} K$. If we substitute the result in the previous formula, then

$$\begin{aligned} V_t &= \mathrm{e}^{-\delta T} (Z_t N(c_1) - \mathrm{e}^{\delta T} K \mathrm{e}^{-r(T-t)} N(c_2)) \\ &= S_T N(c_1) - K \mathrm{e}^{-r(T-t)} N(c_2) \end{aligned}$$

where

$$c_1, c_2 = \frac{\log \frac{Z_t}{\mathrm{e}^{\delta T} K} + (r \pm \frac{1}{2}\sigma^2)(T-t)}{\sigma\sqrt{T-t}} = \frac{\log \frac{S_t}{K} + (r \pm \frac{1}{2}\sigma^2 - \delta)(T-t)}{\sigma\sqrt{T-t}} .$$

Now, if we let

$$d_1, d_2 = \frac{\log \frac{F_t}{K} \pm \frac{1}{2}\sigma^2 (T-t)}{\sigma\sqrt{T-t}} = \frac{\log \frac{S_t}{K} + (r - \delta \pm \frac{1}{2}\sigma^2)(T-t)}{\sigma\sqrt{T-t}} = c_1, c_2$$

and

$$F_t = e^{(r-\delta)(T-t)} S_t ,$$

i.e., if we let F_t be the forward price of the stock, then we can write

$$V_t = e^{-r(T-t)}(F_t N(d_1) - KN(d_2)) .$$

Remark 17.1 Let us find the replicating portfolio for $V_t = D_t + \Delta_t S_t$. If

$$\phi_t = N(c_1) = N(d_1)$$

then we have to keep $e^{-\delta T}\phi_t$ units of Z_t. In other words, we have to keep $e^{-\delta(T-t)}\phi_t$ units of S_t. On the other hand, since

$$V_t - Z_t e^{-\delta T}\phi_t = -Ke^{-rT}N(c_2)e^{rt} ,$$

we have to keep $-Ke^{-rT}N(c_2) = -Ke^{-rT}N(d_2)$ units of bond.

2. Periodic Dividends

In the real world dividends are paid periodically, not continuously. If a dividend δS_{T_i} is paid at predetermined dates T_i, $i \geq 1$, then under the assumption that the price of the stock S_t is continuous from the left,

$$S_{T_i+} - S_{T_i-} = -\delta S_{T_i-}$$

since S_t decreases instantaneously by δS_{T_i-} as time passes by across T_i. (As for the notation, we write $\lim_{t\to T_i\pm} S_t(\omega) = S_{T_i\pm}(\omega)$ for every Brownian sample path ω.) Hence

$$S_{T_i+} = (1-\delta)S_{T_i-} ,$$

and every time when dividend is paid the stock price is multiplied by $1-\delta$. If we let $n[t]$ be the number of times when dividend is paid until time t, i.e.,

$$n[t] = \max\{i : T_i \leq t\} ,$$

then the stock price is given by

$$S_t = S_0(1-\delta)^{n[t]} \exp(\nu t + \sigma W_t) .$$

If at every time the dividend is paid we increase the number of shares of the stock by buying the stock using the dividend, the instantaneous increment in the number of shares across the time T_i is equal to

$$\frac{\delta S_{T_i-}}{S_{T_i+}} = \frac{\delta}{1-\delta} .$$

Hence immediately after the time T_i the number of shares of the stock increases by the ratio of $1 + \frac{\delta}{1-\delta} = \frac{1}{1-\delta}$. If we denote such a portfolio by Z_t, then it satisfies

$$Z_t = \frac{1}{(1-\delta)^{n[t]}} S_t = S_0 \exp(\nu t + \sigma W_t) .$$

Thus defined Z_t may be regarded as a stock without dividend. Put

$$\theta = \frac{\nu + \frac{1}{2}\sigma^2 - r}{\sigma}$$

and define a probability measure $\mathbb{Q}$ by

$$\left.\frac{d\mathbb{Q}}{d\mathbb{P}}\right|_{\mathcal{F}_t} = \exp\left(-\theta W_t - \frac{1}{2}\theta^2 t\right) .$$

Then $X_t = W_t + \theta t$ is a $\mathbb{Q}$-Brownian motion, and the discounted process of Z_t, $\{\widetilde{Z}_t\}_{t\geq 0}$, is a $\mathbb{Q}$-martingale.

Theorem 17.1 *The fair price for a forward contract on a stock that periodically pays dividend is given by*

$$K = e^{rT}(1-\delta)^{n[T]} S_0 .$$

Proof Since the value of the forward contract at expiry date T is equal to $C_T = S_T - K$, its value at time t is equal to

$$\begin{aligned} V_t &= \mathbb{E}^{\mathbb{Q}}[e^{-r(T-t)}(S_T - K)|\mathcal{F}_t] \\ &= \mathbb{E}^{\mathbb{Q}}[e^{-r(T-t)}((1-\delta)^{n[T]} Z_T - K)|\mathcal{F}_t] \\ &= (1-\delta)^{n[T]} Z_t - e^{-r(T-t)} K \\ &= (1-\delta)^{n[T]-n[t]} S_t - e^{-r(T-t)} K . \end{aligned}$$

Since $V_0 = 0$ for a fair contract, we have $(1-\delta)^{n[T]} S_0 - e^{-rT} K = 0$. □

Remark 17.2 As seen in the proof, we have

$$V_t = (1-\delta)^{n[T]} Z_t - e^{-r(T-t)} K ,$$

and hence we keep $(1-\delta)^{n[T]}$ units of Z_t, i.e., $(1-\delta)^{n[T]-n[t]}$ shares of the stock for hedging.

17.2 Bonds with Coupons

Consider a bond whose price changes stochastically in the financial market. Suppose that the bond pays coupons at predetermined dates $0 < T_1 < \cdots < T_n < T$. Let C denote the coupon amount, and $r > 0$ the risk-free interest rate. If a bond investor saves the coupons in the form of a risk-free deposit at a bank, the sum of present values of the coupons until maturity is equal to $\sum_{i=1}^{n} C\mathrm{e}^{-rT_i}$. Let

$$J(t) = \min\{i : t < T_i\} , \quad 0 \le t \le T .$$

That is, $J(t)$ is the first time a coupon payment is received after time t. For example, $J(T_k) = k + 1$ for $k = 1, \ldots, n - 1$. Hence

$$\sum_{i=J(t)}^{n} C\mathrm{e}^{-r(T_i-t)}$$

is the sum of values at the time t of coupons which will be received after t. Assume that the bond price S_t satisfies

$$S_t = \sum_{i=J(t)}^{n} C\mathrm{e}^{-r(T_i-t)} + A\mathrm{e}^{\nu t+\sigma W_t} \tag{17.1}$$

for some constants A, ν and σ. Suppose that the bond investor deposits a coupon payment into a bank account immediately after it is paid. Then the value of the coupon payments until time t is equal to

$$\sum_{i=1}^{J(t)-1} C\mathrm{e}^{-r(T_i-t)} .$$

Since the bond itself has value given by (17.1), the portfolio Z_t consisting of the bond itself and the coupons received until time t is given by

$$Z_t = \sum_{i=1}^{n} C\mathrm{e}^{-r(T_i-t)} + A\mathrm{e}^{\nu t+\sigma W_t} .$$

Now we look for a measure $\mathbb{Q}$ for which the discounted price $\widetilde{Z}_t = \mathrm{e}^{-rt} Z_t$ is a martingale. Since

$$\widetilde{Z}_t = \sum_{i=1}^{n} C\mathrm{e}^{-rT_i} + A\mathrm{e}^{(\nu-r)t+\sigma W_t} ,$$

it suffices to find $\mathbb{Q}$ for which $Ae^{(\nu-r)t+\sigma W_t}$ is a martingale. To find it, put

$$\theta = \frac{\nu + \frac{1}{2}\sigma^2 - r}{\sigma}$$

and let $\mathbb{Q}$ be the measure such that

$$\left.\frac{d\mathbb{Q}}{d\mathbb{P}}\right|_{\mathcal{F}_t} = e^{-\theta W_t - \frac{1}{2}\theta^2 t} ,$$

then $X_t = W_t + \theta t$ is a $\mathbb{Q}$-Brownian motion, and $\{\widetilde{Z}_t\}$ is a $\mathbb{Q}$-martingale.

The value at time t of the option whose payoff is given by $C_T \in \mathcal{F}_T$ is given by $\mathbb{E}^{\mathbb{Q}}[e^{-r(T-t)} C_T | \mathcal{F}_t]$. At maturity T there remains no coupon to receive, and the bond price is equal to

$$S_T = Ae^{\nu T + \sigma W_T} = Ae^{(r - \frac{1}{2}\sigma^2)T + \sigma X_T} .$$

17.3 Binary Options

A European cash-or-nothing binary call option with strike price K pays an agreed amount A at expiry date T if the asset price S_T satisfies $S_T > K$, pays nothing if $S_T < K$, and $\frac{1}{2}A$ if $S_T = K$. If we ignore the probability zero case that $S_T = K$ in a continuous asset price model, then the payoff may be written as $A \times \mathbf{1}_{\{S_T \geq K\}}$. A European asset-or-nothing binary call option pays S_T if $S_T \geq K$, and pays nothing if $S_T < K$. Binary put options are defined similarly. A binary option is also called a *digital* option. The put-call parity relation for cash-or-nothing options is given by

$$C(S_t, t) + P(S_t, t) = Ae^{-r(T-t)}$$

since

$$C(S_T, T) + P(S_T, T) = A$$

where $C(S, t)$ and $P(S, t)$ denote the prices of binary call and binary put with the same strike price (Fig. 17.1).

If we combine a cash-or-nothing binary option and a European call option, we can construct an asset-or-nothing binary option since

$$K \times \mathbf{1}_{\{S_T \geq K\}} + (S_T - K)^+ = S_T \times \mathbf{1}_{\{S_T \geq K\}} .$$

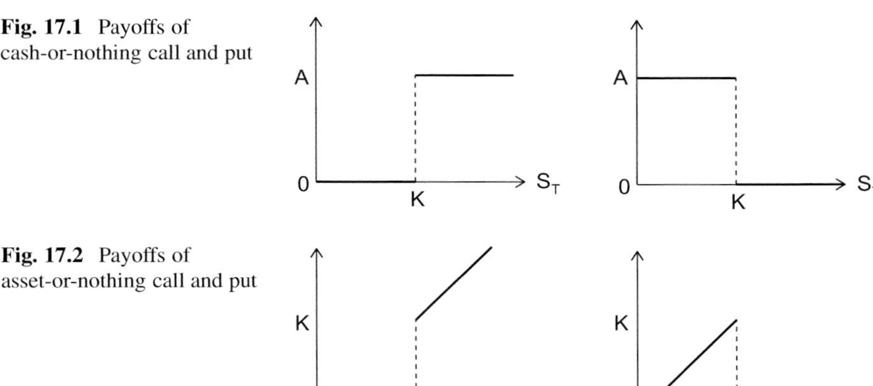

Fig. 17.1 Payoffs of cash-or-nothing call and put

Fig. 17.2 Payoffs of asset-or-nothing call and put

The put-call parity for asset-or-nothing options is given by

$$C(S_t, t) + P(S_t, t) = S_t \ .$$

(See Fig. 17.2.) In continuous asset price models such as the geometric Brownian motion model, it does not make any difference what value we assign to the payoff when $S_T = K$ since the probability of such an event is zero. However, in the real financial market stock price can have integral multiples of tick size which is the minimum amount that the price of the stock can change. Thus we have to define the payoff for the event $S_T = K$. For example, we may take $\frac{1}{2}A$ as the payoff for $S_T = K$ in the case of a cash-or-nothing binary option.

A cash-or-nothing binary option can be dynamically hedged. However, if near the expiry date the asset price is close to the exercise price then hedging is difficult in practice.

Theorem 17.2 *The price C of a cash-or-nothing call option is given by*

$$C(S, t) = \mathrm{e}^{-r(T-t)} N(d_2)$$

where d_2 is identical as in the formula for the European call. (See Fig. 17.3.)

We present five different proofs. For the sake of notational simplicity we find $C(S, t)$ for $t = 0$.

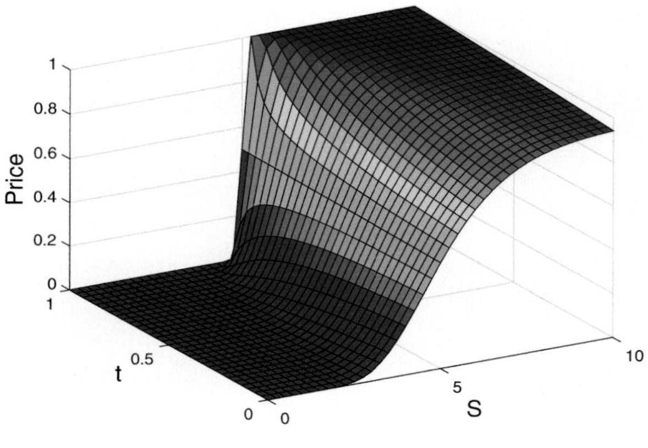

Fig. 17.3 Price of a cash-or-nothing call option

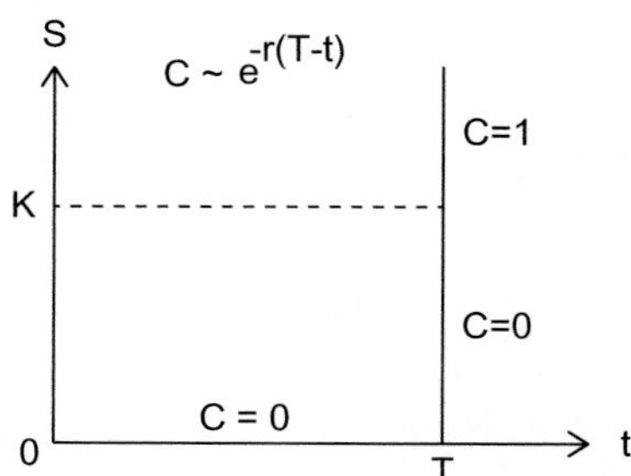

Fig. 17.4 Domain and boundary conditions of a cash-or-nothing call option

Proof (Method 1: Partial Differential Equation) To derive the option price formula, we solve the Black–Scholes–Merton equation with the boundary conditions

$$\begin{cases} C(0,t) = 0 \,, & 0 \le t \le T \,, \\ C(S,T) = 0 \,, & S < K \,, \\ C(S,T) = 1 \,, & K < S \,, \\ C(S,t) \approx \mathrm{e}^{-r(T-t)} \,, & S \approx \infty \,. \end{cases}$$

See Fig. 17.4. The last condition means that if the stock price is very high then the probability of exercising the option at expiry date is close to 1 and its discounted value is the present option price.

As in the case for a European call option, put

$$\begin{cases} S &= K\mathrm{e}^x \,, \\ t &= T - \frac{1}{\sigma^2}\tau \,, \\ V(S,t) &= Kv(x,\tau) \,. \end{cases}$$

Then

$$\sigma^2 \frac{\partial v}{\partial \tau} = \frac{1}{2}\sigma^2 \frac{\partial^2 v}{\partial x^2} + \left(r - \frac{1}{2}\sigma^2\right) \frac{\partial v}{\partial x} - rv \ .$$

If we let $C = \dfrac{r}{\sigma^2}$, then

$$\frac{\partial v}{\partial \tau} = \frac{1}{2} \frac{\partial^2 v}{\partial x^2} + \left(C - \frac{1}{2}\right) \frac{\partial v}{\partial x} - Cv \ .$$

Since $V(S,T) = \mathbf{1}_{\{S \geq K\}}$, the initial condition for v is given by

$$v(x,0) = \frac{1}{K} V(K\mathrm{e}^x, T) = \frac{1}{K} \mathbf{1}_{\{x \geq 0\}} \ .$$

Now we look for v of the form

$$v(x,\tau) = \mathrm{e}^{\alpha x + \beta \tau} u(x,\tau)$$

for some α, β, u. If such a solution exists, then we have

$$\beta u + \frac{\partial u}{\partial \tau} = \frac{1}{2} \frac{\partial^2 u}{\partial x^2} + \left(C - \frac{1}{2} + \alpha\right) \frac{\partial u}{\partial x} + \left(\frac{1}{2}\alpha^2 + (C - \frac{1}{2})\alpha - C\right) u \ .$$

If we take $\alpha = -C + \frac{1}{2}$ and

$$\beta = \frac{1}{2}\alpha^2 + \left(C - \frac{1}{2}\right)\alpha - C = -\frac{1}{2}\left(C + \frac{1}{2}\right)^2 \ ,$$

then

$$\frac{\partial u}{\partial \tau} = \frac{1}{2} \frac{\partial^2 u}{\partial x^2}$$

and

$$u(x,0) = u_0(x) = \mathrm{e}^{-\alpha x} v(x,0) = \frac{1}{K} \mathrm{e}^{(C - \frac{1}{2})x} \mathbf{1}_{\{x \geq 0\}} \ .$$

Hence

$$\begin{aligned} &\mathrm{e}^{\alpha x + \beta \tau} u(x,\tau) \\ &= \mathrm{e}^{\alpha x + \beta \tau} \frac{1}{\sqrt{2\pi\tau}} \int_{-\infty}^{\infty} u_0(\xi) \mathrm{e}^{-(x-\xi)^2/(2\tau)} \mathrm{d}\xi \end{aligned}$$

$$= e^{\alpha x+\beta\tau}\frac{1}{\sqrt{2\pi\tau}}\int_0^\infty \frac{1}{K}e^{(C-\frac{1}{2})\xi}e^{-(x-\xi)^2/(2\tau)}d\xi$$

$$= \frac{1}{K}e^{(-C+\frac{1}{2})x-\frac{1}{2}(C+\frac{1}{2})^2\tau}\frac{1}{\sqrt{2\pi\tau}}\int_{-\infty}^0 e^{-(\eta+x+(C-\frac{1}{2})\tau)^2/2\tau+(C-\frac{1}{2})x+\frac{1}{2}(C-\frac{1}{2})^2\tau}d\eta$$

$$= \frac{1}{K}e^{-C\tau}\frac{1}{\sqrt{2\pi}}\int_{-\infty}^{(x+(C-\frac{1}{2})\tau)/\sqrt{\tau}} e^{-\zeta^2/2}d\zeta$$

$$= \frac{1}{K}e^{-C\tau}N(d_2)$$

where

$$d_2 = \frac{x+(C-\frac{1}{2})\tau}{\sqrt{\tau}} = \frac{\log\frac{S}{K}+(r-\frac{1}{2}\sigma^2)(T-t)}{\sigma\sqrt{T-t}} .$$

Since $C\tau = r(T-t)$, we conclude

$$V(S,t) = Kv(x,\tau) = e^{-r(T-t)}N(d_2) .$$

□

Proof (Method 2: Risk-Neutrality) Suppose that the underlying asset price S_t follows a geometric Brownian motion

$$dS_t = rS_t dt + \sigma S_t dW_t$$

under a risk-neutral measure $\mathbb{Q}$. The option price is given by

$$C(S,t) = \mathbb{E}[e^{-r(T-t)}\mathbf{1}_{\{S_T\geq K\}}|S_t]$$

where the conditional expectation is taken with respect to $\mathbb{Q}$. Since $S_T = S_t e^{(r-1/2\sigma^2)(T-t)+\sigma W_{T-t}}$ conditional on S_t at time t, we have

$$\begin{aligned}
&\mathbb{E}[\mathbf{1}_{\{S_T\geq K\}}|S_t] \\
&= \mathbb{E}[\mathbf{1}_{\{S_t e^{(r-1/2\sigma^2)(T-t)+\sigma W_{T-t}}\geq K\}}] \\
&= \mathbb{Q}\{S_t e^{(r-\frac{1}{2}\sigma^2)(T-t)+\sigma W_{T-t}} \geq K\} \\
&= \mathbb{Q}\{(r-\frac{1}{2}\sigma^2)(T-t)+\sigma W_{T-t} \geq \log\frac{K}{S_t}\} \\
&= \mathbb{Q}\{W_{T-t} > \frac{\log\frac{K}{S_t}-(r-\frac{1}{2}\sigma^2)(T-t)}{\sigma}\} \\
&= \mathbb{Q}\{Z > \frac{\log\frac{K}{S_t}-(r-\frac{1}{2}\sigma^2)(T-t)}{\sigma\sqrt{T-t}}\} \quad (\text{where } Z\sim N(0,1))
\end{aligned}$$

$$= \mathbb{Q}\{Z < \frac{\log \frac{S_t}{K} + (r - \frac{1}{2}\sigma^2)(T-t)}{\sigma\sqrt{T-t}}\} \quad (\text{since } -Z \sim N(0,1))$$

$$= N(d_2)\ .$$

Hence $C(S_t, t) = e^{-r(T-t)}N(d_2)$. □

Proof (Method 3: Differentiation of a European Call) There is a simpler method of finding the price of a digital option if we already have the classical Black–Scholes–Merton formula for a European call option. Let $C(K)$ denote the price of a European call option with exercise price K. Assume $K < K_1$. Then the price of a European call option with exercise price K_1 is equal to $C(K_1)$. If we consider a new option defined by the difference of two European call options with exercise prices K and K_1, respectively, then its payoff at expiry T is given by

$$\begin{cases} 0 & \text{if} \quad 0 \le S_T < K\ , \\ S_T - K & \text{if} \quad K \le S_T < K_1\ , \\ K_1 - K & \text{if} \quad K_1 \le S_T\ , \end{cases}$$

and its price is given by $C(K) - C(K_1)$. See Fig. 17.5. Now note that the cash-or-nothing call option with payoff $\mathbf{1}_{\{S_T \ge K\}}$ can be approximated by a European option whose payoff at expiry date T is given by the second graph in Fig. 17.6. Hence the price of the cash-or-nothing call option with exercise price K can be approximated by $\frac{1}{K_1-K}(C(K) - C(K_1))$ as shown in Fig. 17.6. Therefore the price of the digital option is equal to the limit

$$\lim_{K_1 \to K} \frac{C(K) - C(K_1)}{K_1 - K} = -\frac{\partial C}{\partial K} = e^{-r(T-t)}N(d_2)\ .$$

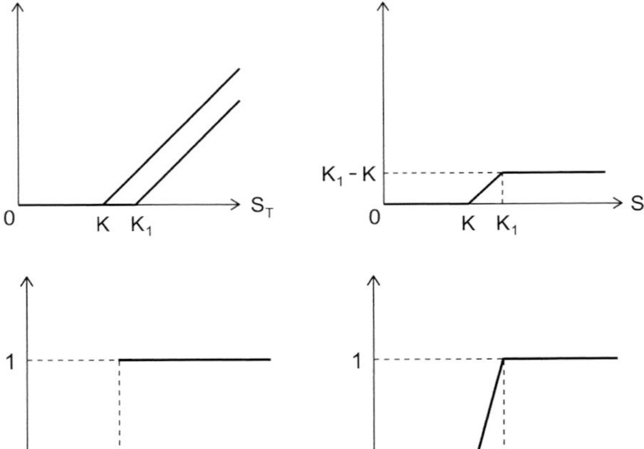

Fig. 17.5 Difference of two European call options

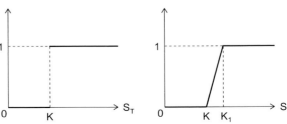

Fig. 17.6 Approximation of a cash-or-nothing call option

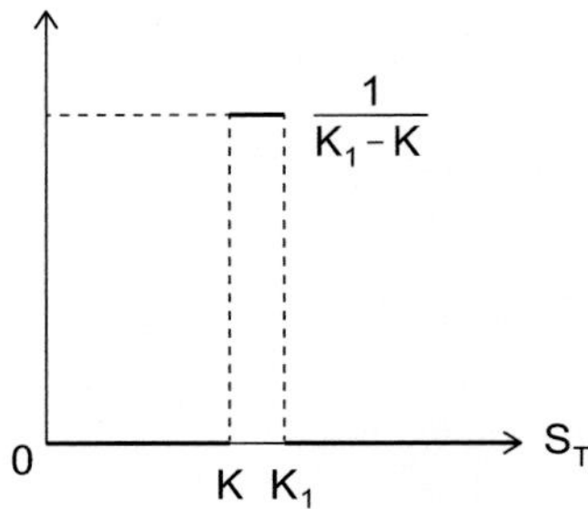

Fig. 17.7 Difference of two binary call options

See Remark 15.2 for the second equality.

Proof (Method 4: Differentiation of a Binary Call)
Let B denote the price of a cash-or-nothing binary call option. It will be regarded as a function of strike price K while the other parameters are fixed. Then the price of the European option with the payoff given in Fig. 17.7 is equal to

$$\frac{1}{K_1 - K}(B(K) - B(K_1))$$

which converges to $-\dfrac{\partial B}{\partial K}$ as $K_1 \to K$.

On the other hand, according to the risk-neutral method the option price is equal to

$$e^{-r(T-t)}\mathbb{E}^{\mathbb{Q}}\left[\frac{1}{K_1 - K}\mathbf{1}_{\{K \le S_T \le K_1\}}(S_T)\middle| S_t\right] . \tag{17.2}$$

Since Fig. 17.7 is an approximation of a point mass $\delta_{\{S_T=K\}}$ called the Dirac delta functional at K, the conditional expectation in (17.2) converges to

$$e^{-r(T-t)}\mathbb{E}^{\mathbb{Q}}[\delta_{\{S_T=K\}}(S_T)|S_t] \tag{17.3}$$

as $K_1 \to K$.

By the formula in Example 4.16 the conditional pdf $f(y)$, $0 < y < \infty$, of S_T, under the condition defined by S_t, is given by

$$f(y) = \frac{1}{\sigma y\sqrt{2\pi(T-t)}}\exp\left(-\frac{(\log y - \log S_t - (r - \frac{1}{2}\sigma^2)(T-t))^2}{2\sigma^2(T-t)}\right) .$$

Then the conditional expectation in (17.3) is equal to $e^{-r(T-t)}f(K)$. Since

$$\begin{aligned} f(K) &= \frac{1}{\sigma K\sqrt{2\pi(T-t)}} \exp\left(-\frac{(\log K - \log S_t - (r-\frac{1}{2}\sigma^2)(T-t))^2}{2\sigma^2(T-t)}\right) \\ &= \frac{1}{\sigma K\sqrt{2\pi(T-t)}} \exp\left(-\frac{1}{2}d_2^2\right) \end{aligned}$$

and since two limits should be equal, we have

$$-\frac{\partial B}{\partial K} = e^{-r(T-t)} \frac{1}{\sigma K\sqrt{2\pi(T-t)}} \exp\left(-\frac{1}{2}d_2^2\right) .$$

Now, since

$$\frac{\partial N(d_2)}{\partial K} = \frac{1}{\sqrt{2\pi}} \exp\left(-\frac{1}{2}d_2^2\right)\left(-\frac{1}{K\sigma\sqrt{T-t}}\right) ,$$

we conclude that

$$\frac{\partial B}{\partial K} = e^{-r(T-t)} \frac{\partial N(d_2)}{\partial K}$$

and hence $B(K) = e^{-r(T-t)}N(d_2) + c$ for some constant c. Since $B(0) = e^{-r(T-t)}$ and $N(d_2(0)) = 1$, we have $c = 0$. □

Proof (Method 5: Laplace Transformation Approach) As in Sect. 15.4, we let $\tau = T - t$ and put $v(S,\tau) = V(S, T-\tau)$, and obtain

$$\frac{1}{2}\sigma^2 S^2 \frac{\partial^2 v}{\partial S^2} + rS\frac{\partial v}{\partial S} - \frac{\partial v}{\partial \tau} = rv \tag{17.4}$$

with the initial data

$$v(S,0) = V(S,T) .$$

Let $f(S,s)$ be the Laplace transformation $\mathscr{L}[v]$ of $v(S,\tau)$. Since

$$\mathscr{L}\left[\frac{\partial v}{\partial \tau}\right](S,s) = s\mathscr{L}[v](S,s) - v(S,0) ,$$

after taking the Laplace transformations of both sides of (17.4), we obtain

$$\frac{1}{2}\sigma^2 S^2 \frac{\partial^2 f}{\partial S^2} + rS\frac{\partial f}{\partial S} - (s+r)f = -v(S,0) . \tag{17.5}$$

To solve it, first find a solution of the homogeneous equation

$$\frac{1}{2}\sigma^2 S^2 \frac{\partial^2 f}{\partial S^2} + rS\frac{\partial f}{\partial S} - (s+r)f = 0 \,. \tag{17.6}$$

In this case there exist two linearly independent solutions, and we look for solutions of the form

$$f(S,s) = S^{\lambda} \,. \tag{17.7}$$

Note that $\lambda \neq 1, 2$. Substituting (17.7) into (17.6), we obtain

$$\frac{1}{2}\sigma^2 S^2 \lambda(\lambda-1)S^{\lambda-2} + rS\lambda S^{\lambda-1} - (s+r)S^{\lambda} = 0 \,.$$

Eliminating S^{λ} from the both sides, we obtain the quadratic equation

$$\frac{1}{2}\sigma^2\lambda^2 + (-\frac{1}{2}\sigma^2 + r)\lambda - (s+r) = 0 \,. \tag{17.8}$$

Hence there exist two zeros

$$\lambda_1, \lambda_2 = \frac{-(r-\frac{1}{2}\sigma^2) \pm \sqrt{(r-\frac{1}{2}\sigma^2)^2 + 2(s+r)\sigma^2}}{\sigma^2} \,.$$

For $s > 0$ we have

$$\lambda_1 > \frac{-(r-\frac{1}{2}\sigma^2) + \sqrt{(r-\frac{1}{2}\sigma^2)^2 + 2r\sigma^2}}{\sigma^2} = \frac{\sigma^2}{\sigma^2} = 1 \,,$$

$$\lambda_2 < \frac{-(r-\frac{1}{2}\sigma^2) - \sqrt{(r-\frac{1}{2}\sigma^2)^2 + 2r\sigma^2}}{\sigma^2} = -\frac{2r}{\sigma^2} < 0 \,.$$

In summary, a homogeneous solution f_H of (17.6) is of the form

$$f_H = c_1(s)S^{\lambda_1} + c_2(s)S^{\lambda_2}$$

where $c_1(s)$ and $c_2(s)$ depend only on s but not on S.

Now we recall that the initial condition is given by

$$v(S,0) = V(S,T) = \mathbf{1}_{\{S \geq K\}} \,.$$

Hence (17.5) becomes

$$\frac{1}{2}\sigma^2 S^2 \frac{\partial^2 f}{\partial S^2} + rS\frac{\partial f}{\partial S} - (s+r)f = -\mathbf{1}_{\{S\geq K\}} \,. \tag{17.9}$$

In the region $S \geq K$, we have

$$\frac{1}{2}\sigma^2 S^2 \frac{\partial^2 f}{\partial S^2} + rS\frac{\partial f}{\partial S} - (s+r)f = -1 \,,$$

and we have a particular solution

$$f_P(S,s) = \frac{1}{s+r} \,.$$

Therefore we try to find a general solution f of (17.9) of the form

$$f(S,s) = \begin{cases} a_1 S^{\lambda_1} + a_2 S^{\lambda_2} + \dfrac{1}{s+r} \,, & S \geq K \,, \\ a_3 S^{\lambda_1} + a_4 S^{\lambda_2} \,, & S < K \,, \end{cases}$$

where a_i may depend on s. To determine a_i in the region $S \geq K$ we note that $v(S,\tau) \to 1$ as $S \to +\infty$ and hence $f(S,s)$ must be bounded as $S \to +\infty$, which is possible only when $a_1 = 0$. On the other hand, if $S < K$, we have $v(S,\tau) \to 0$ as $S \downarrow 0$, and hence $f(S,s) \to 0$ as $S \downarrow 0$, thus we conclude that $a_4 = 0$. If we summarize what has been obtained so far, we have

$$f(S,s) = \begin{cases} c_2(s) S^{\lambda_2} + \dfrac{1}{s+r} \,, & S \geq K \,, \\ c_1(s) S^{\lambda_1} \,, & S < K \,, \end{cases}$$

for some $c_1(s)$ and $c_2(s)$. Since $f(S,s)$ is continuous and differentiable at (K,s), we have

$$\begin{cases} c_2(s) K^{\lambda_2} + \dfrac{1}{s+r} = c_1(s) K^{\lambda_1} \,, \\ c_2(s) \lambda_2 K^{\lambda_2 - 1} = c_1(s) \lambda_1 K^{\lambda_1 - 1} \,. \end{cases}$$

Hence

$$\begin{cases} c_1(s) = \dfrac{\lambda_2}{\lambda_1 - \lambda_2} K^{-\lambda_1} \dfrac{1}{s+r} \,, \\ c_2(s) = \dfrac{\lambda_1}{\lambda_1 - \lambda_2} K^{-\lambda_2} \dfrac{1}{s+r} \,. \end{cases}$$

Note that

$$\begin{aligned}
&\frac{\lambda_1}{\lambda_1 - \lambda_2} \\
&= \frac{-(r - \frac{1}{2}\sigma^2) + \sqrt{(r - \frac{1}{2}\sigma^2)^2 + 2(s + r)\sigma^2}}{2\sqrt{(r - \frac{1}{2}\sigma^2)^2 + 2(s + r)\sigma^2}} \\
&= \frac{(s + r)\sigma^2}{\sqrt{(r - \frac{1}{2}\sigma^2)^2 + 2(s + r)\sigma^2}} \frac{1}{r - \frac{1}{2}\sigma^2 + \sqrt{(r - \frac{1}{2}\sigma^2)^2 + 2(s + r)\sigma^2}}
\end{aligned}$$

and

$$\begin{aligned}
&\frac{\lambda_2}{\lambda_1 - \lambda_2} \\
&= \frac{-(r - \frac{1}{2}\sigma^2) - \sqrt{(r - \frac{1}{2}\sigma^2)^2 + 2(s + r)\sigma^2}}{2\sqrt{(r - \frac{1}{2}\sigma^2)^2 + 2(s + r)\sigma^2}} \\
&= \frac{(s + r)\sigma^2}{\sqrt{(r - \frac{1}{2}\sigma^2)^2 + 2(s + r)\sigma^2}} \frac{1}{r - \frac{1}{2}\sigma^2 - \sqrt{(r - \frac{1}{2}\sigma^2)^2 + 2(s + r)\sigma^2}} .
\end{aligned}$$

Since, for $S < K$, we have

$$\begin{aligned}
&f(S, s) \\
&= \frac{\lambda_2}{\lambda_1 - \lambda_2} \left(\frac{S}{K}\right)^{\lambda_1} \frac{1}{s + r} \\
&= \frac{\sigma^2}{\sqrt{(r - \frac{1}{2}\sigma^2)^2 + 2(s + r)\sigma^2}} \frac{\exp\left(\lambda_1 \log \frac{S}{K}\right)}{r - \frac{1}{2}\sigma^2 - \sqrt{(r - \frac{1}{2}\sigma^2)^2 + 2(s + r)\sigma^2}} .
\end{aligned}$$

Now we take the inverse Laplace transformation and obtain $v(S, \tau) = \mathrm{e}^{-r\tau} N(d_2)$ and finally we have $V(S, t) = \mathrm{e}^{-r(T-t)} N(d_2)$. □

Theorem 17.3 *Consider a cash-or-nothing call with payoff* $\mathbf{1}_{\{S_T \geq K\}}$. *The hedge ratio* Δ *is given by*

$$\Delta = \frac{\partial C}{\partial S} = \frac{\mathrm{e}^{-r(T-t)} N'(d_2)}{\sigma S \sqrt{T - t}}$$

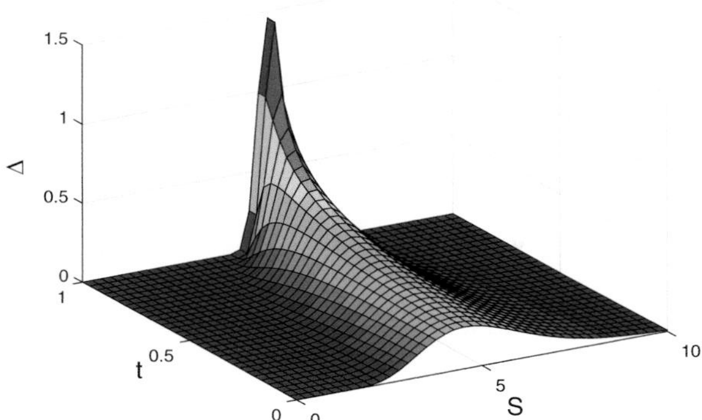

Fig. 17.8 Delta of a cash-or-nothing call option

and satisfies

$$\lim_{t \to T-} \Delta = \begin{cases} 0, & S_T > K \\ \infty, & S_T = K \\ 0, & S_T < K \end{cases}$$

with probability 1 where the limit is taken as t converges to T from the left. (See Fig. 17.8.)

Proof In the following argument, we consider continuous sample paths for S_t since it is continuous with probability 1.

(i) Suppose $S_T > K$. Since

$$d_2(t, S_t) \approx \frac{\log \frac{S_T}{K} + (r - \frac{1}{2}\sigma^2)(T-t)}{\sigma\sqrt{T-t}} \approx C_1 \frac{1}{\sqrt{T-t}}$$

for $T - t \approx 0$, we have

$$N'(d_2) = \frac{1}{\sqrt{2\pi}} \mathrm{e}^{-\frac{1}{2}d_2^2} \approx C_2 \mathrm{e}^{-C_3/(T-t)}$$

and

$$\Delta \approx C_4 \frac{\mathrm{e}^{-C_3/(T-t)}}{\sqrt{T-t}} \approx 0$$

for $T - t \approx 0$ where C_i, $1 \le i \le 4$, are positive constants.

(ii) Suppose $S_T = K$. Then $S_t \approx K$ for $t \approx T$. Since

$$d_2(t, S_t) \approx \frac{\log 1 + (r - \frac{1}{2}\sigma^2)(T-t)}{\sigma\sqrt{T-t}} \approx C_1\sqrt{T-t}\,,$$

we have

$$N'(d_2) = \frac{1}{\sqrt{2\pi}} e^{-\frac{1}{2}d_2^2} \approx C_2 e^{-C_3(T-t)} \approx C_4$$

and

$$\Delta \approx C_5 \frac{1}{\sqrt{T-t}} \approx \infty$$

for $T - t \approx 0$ where C_i, $1 \le i \le 5$, are positive constants.

(iii) Suppose $S_T < K$. Since there exists a constant $C > 0$ such that

$$d_2(t, S_t) \approx \frac{\log \frac{S_T}{K} + (r - \frac{1}{2}\sigma^2)(T-t)}{\sigma\sqrt{T-t}} \approx -C\frac{1}{\sqrt{T-t}}$$

for $T - t \approx 0$, we have $\Delta \approx 0$ for $T - t \approx 0$ as in the case (i). □

Remark 17.3

(i) The result implies that if $S > K$ or $S < K$ almost at expiry date with almost certainty the seller of the option has to pay to the buyer of the option either \$1 or nothing, and hence, close to expiry date, either there is no need for the option seller to hedge by keeping stocks or there is no way to hedge.
(ii) Note that delta is inversely proportional to the asset price as expected since delta represents the number of shares of the underlying asset in hedging the option sold.

Remark 17.4 (European Call Option Price) We can derive the formula of the price of a European call option from that of the price of a binary option. Let K be the strike price of a given European call option. Its payoff is approximated by a sum of payoffs of infinitely many binary options. More precisely, the payoff $(S_T - K)^+$ is approximated by

$$\sum_{i=0}^{\infty} \delta K \times \mathbf{1}_{\{S_T \ge K + i \times \delta K\}}$$

if $\delta K > 0$ is sufficiently small. See Fig. 17.9. Let $B(x)$ denote the time $t = 0$ price of a binary option with strike price x and payoff $\mathbf{1}_{\{S_T \ge x\}}$. Then the price of a European

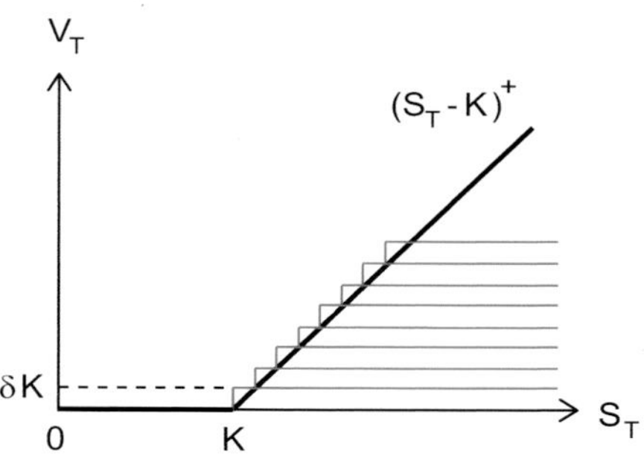

Fig. 17.9 Difference of two binary options

call option with strike price K, denoted by $C(K)$, is approximated by

$$\sum_{i=0}^{\infty} B(K + i \times \delta K)\, \delta K\ .$$

Hence

$$\begin{aligned} C(K) &= \lim_{\delta K \to 0} \sum_{i=0}^{\infty} B(K + i \times \delta K)\, \delta K \\ &= \int_K^{\infty} B(x)\, \mathrm{d}x \\ &= \int_K^{\infty} \mathrm{e}^{-rT} N(d_2)\, \mathrm{d}x \end{aligned}$$

where

$$d_{1,2}(x) = \frac{-\log x + \log S_0 + (r \pm \frac{1}{2}\sigma^2)T}{\sigma\sqrt{T}}\ .$$

Note that

$$\begin{aligned} \int_K^{\infty} N(d_2)\, \mathrm{d}x &= \int_K^{\infty} \int_{-\infty}^{d_2(x)} \frac{1}{\sqrt{2\pi}} \mathrm{e}^{-z^2/2}\, \mathrm{d}z\, \mathrm{d}x \\ &= \int_{-\infty}^{d_2(K)} \int_K^{x^*} \frac{1}{\sqrt{2\pi}} \mathrm{e}^{-z^2/2}\, \mathrm{d}x\, \mathrm{d}z \quad (x^* = S_0 \mathrm{e}^{(r-\frac{1}{2}\sigma^2)T - \sigma\sqrt{T}z}) \\ &= \int_{-\infty}^{d_2(K)} \frac{1}{\sqrt{2\pi}} \mathrm{e}^{-z^2/2} \left(x^* - K\right) \mathrm{d}z \\ &= \int_{-\infty}^{d_2(K)} \frac{1}{\sqrt{2\pi}} \mathrm{e}^{-z^2/2} \left(S_0 \mathrm{e}^{(r-\frac{1}{2}\sigma^2)T - \sigma\sqrt{T}z} - K\right) \mathrm{d}z \end{aligned}$$

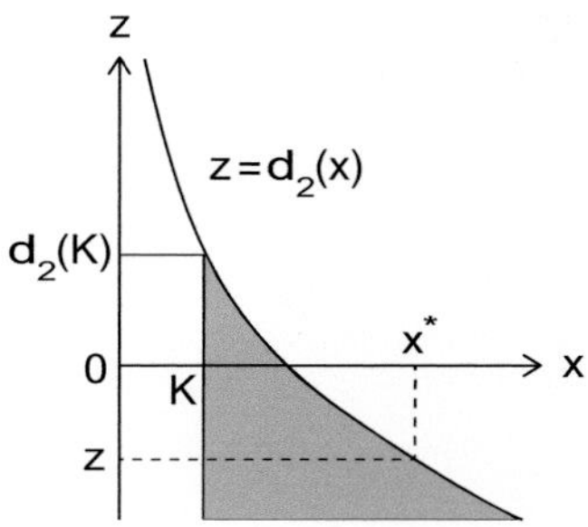

Fig. 17.10 d_2 as a function of strike price

$$= S_0 e^{rT} \int_{-\infty}^{d_2(K)} \frac{1}{\sqrt{2\pi}} e^{-(z+\sigma\sqrt{T})^2/2} dz - KN(d_2(K))$$

$$= S_0 e^{rT} N(d_2(K) + \sigma\sqrt{T}) - KN(d_2(K)) \; .$$

For the domain of integration for the double integral see Fig. 17.10. Now we use the fact that $d_2(K) + \sigma\sqrt{T} = d_1(K)$.

Exercises

17.1 Show that the fair price for a forward contract on S_T at $t = 0$ is equal to $F = Ae^{rT}$.

17.2 What is the price of an asset-or-nothing binary call option?

17.4 Computer Experiments

Simulation 17.1 (Cash-or-Nothing Call)

The price surface defined by the Black–Scholes–Merton formula for a cash-or-nothing call option is given in Fig. 17.3. Also presented in Fig. 17.8 is the delta for the option where the graph is unbounded at $t = T$ as proved in Theorem 17.3.

```
K = 5;
A = 1;
r = 0.05;
sigma = 0.3;

M= 70; % number of points on the asset axis including endpoints
N= 50; % number of points on the time axis including endpoints

mu = 0.1;

T = 1;
```

```
dt=T/N;

S0 = 4;
S_max = 10;
dS = S_max/M;

S_range = 0:dS:S_max;
t_range = 0:dt:T;

call = zeros(M+1,N+1);
Delta= zeros(M+1,N+1);

for i = 1:M+1
    S = S_range(i);
    for j = 1:N
        tau = T-t_range(j);
        d2 = (log(S/K)+(r-0.5*sigma^2)*(tau))/(sigma*sqrt(tau));
        call(i,j) = A*exp(-r*tau)*normcdf(d2);
        Delta(i,j) = (A*exp(-r*tau)*normpdf(d2))/(sigma*S*sqrt(tau));
    end
    Delta(i,N+1) = (A*normpdf(d2))/(sigma*S*sqrt(tau));
    call(i,N+1) = A*heaviside(S-K); % payoff at T
end

Delta(1,:) = 0;
[t_grid,S_grid] = meshgrid(t_range,S_range);

figure(1)
surf(S_grid,t_grid,call);
hold off;

figure(2);
surf(S_grid,t_grid,delta);
```

Simulation 17.2 (Brownian Bridge and At-the-Money Option)

The delta for an at-the-money cash-or-nothing call is computed, and the output is given in Fig. 17.11 where $\Delta \to \infty$ as $t \to T$. To generate a sample asset path S_t such that

$$S_T = S_0 \exp\left((\mu - \frac{1}{2}\sigma^2)T + \sigma W_T\right) = K ,$$

we use a Brownian bridge path $X_t^{0\to b}$ from 0 to b where b satisfies

$$S_0 \exp\left((\mu - \frac{1}{2}\sigma^2)T + \sigma b\right) = K ,$$

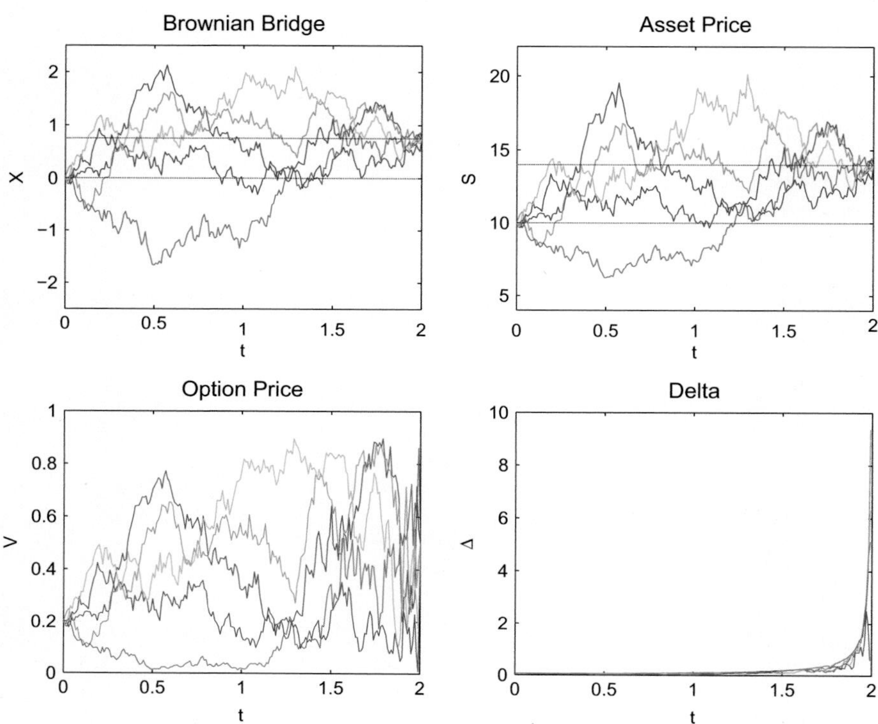

Fig. 17.11 Delta of an at-the-money digital option

i.e.,

$$b = \frac{1}{\sigma}\left\{\log\frac{K}{S0} - (\mu - \frac{1}{2}\sigma^2)T\right\} \ .$$

We take $\frac{1}{2}A$ as the payoff for $S_T = K$

```
mu = 0.1; % drift coefficient
sigma = 0.3; % volatility
T = 2; % expiry date
r = 0.05; % risk-free interest rate

N = 200; % number of time subintervals
dt = T/N;
M = 5; % number of sample paths
K = 14; % strike price
S0 = 10;
A = 1; % payoff
t = 0:dt:T;

% Brownian Motion
```

```
W = zeros(M,N + 1);
dW = sqrt(dt)*randn(M,N);
for i=1:N
    W(:,i+1) = W(:,i) + dW(:,i);
end

% Brownian Bridge
b = ( log(K/S0) - (mu-(sigma^2)/2)*T )/sigma
X = zeros(M,N+1);
for i=1:N+1
    X(:,i) = b*(i-1)*dt/T + W(:,i) - W(:,N+1)*(i-1)*dt/T;
end

subplot(2,2,1)
for i = 1:M
    plot(t,X(i,1:N+1));
hold on;
end

title('Brownian Bridge');

% Asset Price
S = zeros(M,N+1);
S(:,1) = S0;
for i=1:N+1
    S(:,i) = S0*exp((mu-(sigma^2)/2)*(i-1)*dt + sigma*X(:,i));
end

subplot(2,2,2);
for i=1:M
    plot(t,S(i,:))
    hold on;
end

title('Asset Price');

% Option Price
V=zeros(M,N+1);
d2=zeros(M,N+1);
for i=1:N+1
    tau = T-(i-1)*dt;
    d2(:,i) = (log(S(:,i)/K) + (r-0.5*sigma^2)*tau)/sigma/sqrt(tau);
    V(:,i)= A*exp(-r*tau)*normcdf(d2(:,i));
end
V(:,N+1) = A/2;

subplot(2,2,3)
for i=1:M
    plot(t,V(i,:));
    hold on;
end
title('Option Price');

% Delta
```

```
Delta = zeros(M,N+1);
for i=1:N+1
    tau = T-(i-1)*dt;
    d2 = (log(S(:,i)/K) + (r-0.5*sigma^2)*tau)/sigma/sqrt(tau);
    Delta(:,i) = A*exp(-r*tau)*normpdf(d2(:))./(S(:,i)*sigma*tau);
end

subplot(2,2,4)
for i=1:M
    plot(t, Delta(i,:));
    hold on;
end
title('Delta');
```

Chapter 18
Pricing of Exotic Options

Options with nonstandard features are called exotic options. In this chapter we introduce exotic options such as Asian options and barrier options. For an encyclopedic collection of option pricing formulas consult [36].

18.1 Asian Options

The payoff of an Asian option with expiry date T is determined by a suitably defined average of underlying asset prices $S_1, \ldots, S_n$ measured on predetermined dates $0 < t_1 < t_2 < \cdots < t_n = T$. For example, we can choose any of the following definitions of average to define a payoff of an Asian option.

$$A_1 = \text{arithmetic average} = \frac{S_1 + \cdots + S_n}{n}$$

$$A_2 = \text{geometric average} = (S_1 \times \cdots \times S_n)^{1/n}$$

Or using weights w_i we may define weighted averages by

$$\text{weighted arithmetic average} = \frac{w_1 S_1 + \cdots + w_n S_n}{w_1 + \cdots + w_n} ,$$

$$\text{weighted geometric average} = (S_1^{w_1} \times \cdots \times S_n^{w_n})^{1/(w_1+\cdots+w_n)} .$$

We can also define the corresponding continuous versions of averages. If we consider only equal weights in computing the average, there are four types of averages as given in Table 18.1.

In Fig. 18.1 a sample asset price path is presented for Asian option with $n = 6$ monitoring times for the period $0 \leq t \leq T = 3$.

G.H. Choe, *Stochastic Analysis for Finance with Simulations*, Universitext,
DOI 10.1007/978-3-319-25589-7_18

Table 18.1 Definitions of average for Asian options

Average	Discrete time	Continuous time
Arithmetic	$\frac{1}{n}\sum_{i=1}^{n} S_{t_i}$	$\frac{1}{T}\int_0^T S_t\,\mathrm{d}t$
Geometric	$(S_{t_1}\cdots S_{t_n})^{1/n}$	$\exp\left(\frac{1}{T}\int_0^T \log S_t\,\mathrm{d}t\right)$

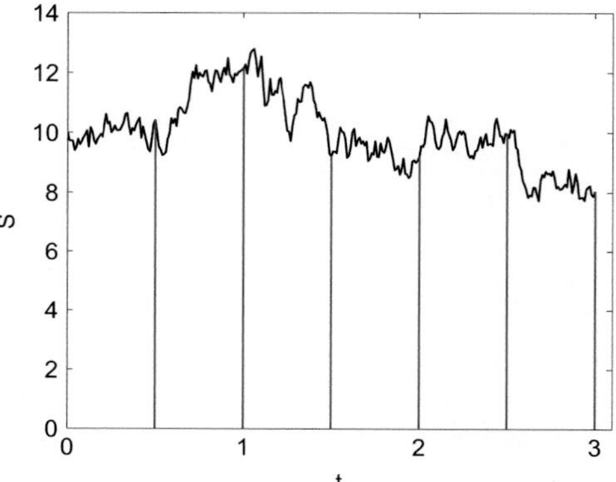

Fig. 18.1 An asset price path for Asian option with six monitoring times

For a call option with exercise price K the payoff is given by $(A - K)^+$ for some average A. Asian options are used to hedge the risk from fluctuations of metal prices, crude oil price or foreign currency rates. Due to lower volatility of an average compared to the original underlying asset price, Asian option prices are lower than those of vanilla options, in general.

Theorem 18.1 *Suppose that the price of the underlying asset S_t follows a geometric Brownian motion, i.e., $S_t = S_0 \exp((\mu - \frac{1}{2}\sigma^2)t + \sigma W_t)$ for some μ and σ. Then the price of an Asian call option with geometric average, with strike price K and expiry date T, is given by*

$$S_0 \mathrm{e}^{(\bar{\mu}-r)T} N(d_1) - K\mathrm{e}^{-rT} N(d_2)$$

where

$$d_1, d_2 = \frac{\log \frac{S_0}{K} + (\bar{\mu} \pm \frac{1}{2}\bar{\sigma}^2)T}{\bar{\sigma}\sqrt{T}} ,$$

$$\bar{\sigma}^2 = \frac{1}{6}\sigma^2(1 + \frac{1}{n})(2 + \frac{1}{n}) ,$$

$$\bar{\mu} = \frac{1}{2}\bar{\sigma}^2 + \frac{1}{2}(r - \frac{1}{2}\sigma^2)(1 + \frac{1}{n}) .$$

Proof We apply the risk-neutral method in Sect. 16.3 and take $\mu = r$. Put $\delta t = \frac{T}{n}$, then

$$A_2 = S_0 \exp\left(\frac{1}{n}\left\{(r - \frac{1}{2}\sigma^2)\sum_{i=1}^{n} i\delta t + \sigma \sum_{i=1}^{n} W_{i\delta t}\right\}\right).$$

Since

$$\begin{aligned}\sum_{i=1}^{n} W_{i\delta t} = n(W_{\delta t} - W_0) & \\ + (n-1)(W_{2\delta t} - W_{\delta t}) & \\ + \qquad \vdots \qquad & \\ + 3(W_{(n-2)\delta t} - W_{(n-3)\delta t}) & \\ + 2(W_{(n-1)\delta t} - W_{(n-2)\delta t}) & \\ + (W_{n\delta t} - W_{(n-1)\delta t}) &\end{aligned}$$

and since the increments $W_{i\delta t} - W_{(i-1)\delta t}$ are independent on the right-hand side, $\sum_{i=1}^{n} W_{i\delta t}$ is normally distributed with mean 0 and variance

$$\sum_{i=1}^{n} i^2\delta t = \frac{1}{6}n(n+1)(2n+1)\delta t = \frac{1}{6}(n+1)(2n+1)T .$$

If we put

$$X = \frac{1}{n}\left\{(r - \frac{1}{2}\sigma^2)\sum_{i=1}^{n} i\delta t + \sigma \sum_{i=1}^{n} W_{i\delta t}\right\} ,$$

then

$$A_2 = S_0\, \mathrm{e}^X .$$

Note that X has normal distribution with mean $(\bar{\mu} - \frac{1}{2}\bar{\sigma}^2)T$ and variance $\bar{\sigma}^2 T$. Hence

$$Y = X - (\bar{\mu} - r)T$$

has normal distribution with mean $(r - \frac{1}{2}\bar{\sigma}^2)T$ and variance $\bar{\sigma}^2 T$. Furthermore, we have

$$A_2 = S_0 \mathrm{e}^{(\bar{\mu}-r)T} \mathrm{e}^Y .$$

Now we find the price of an Asian call option where A_2 is used as the definition of average and the payoff is given by $\max\{0, A_2 - K\}$. Take $S_0 e^{(\bar{\mu}-r)T}$ as the initial price in the Black–Scholes–Merton formula, and we obtain the Asian call option price. □

Remark 18.1

(i) Note that for sufficiently large n, we have

$$\bar{\sigma}^2 \approx \frac{1}{3}\sigma^2$$

and

$$\bar{\mu} - r \approx \frac{1}{2} \times \frac{1}{3}\sigma^2 + \frac{1}{2}(r - \frac{1}{2}\sigma^2) - r = -\frac{1}{12}\sigma^2 - \frac{1}{2}r < 0 ,$$

which implies that the Asian call option price is cheaper since $\bar{\sigma} < \sigma$ and

$$S_0 e^{(\bar{\mu}-r)T} < S_0 .$$

(ii) In the above proof, we used the identity

$$\sum_{i=1}^{n} W_{i\delta t} = \sum_{k=0}^{n-1} (n-k)\,(W_{(k+1)\delta t} - W_{k\delta t}) ,$$

which is nothing but a discretized version of the Itô integral: After multiplying both sides by $\delta t = \frac{T}{n}$, we obtain

$$\sum_{i=1}^{n} W_{i\delta t} \times \frac{T}{n} = \sum_{k=0}^{n-1} (T - k\frac{T}{n})\,(W_{(k+1)\delta t} - W_{k\delta t}) ,$$

whose continuous limit is given by

$$\int_0^T W_t \,\mathrm{d}t = \int_0^T (T-t)\,\mathrm{d}W_t .$$

The right-hand side is equal to

$$TW_T - \int_0^T t\,\mathrm{d}W_t .$$

Hence we obtain the formula in Example 11.3(iv).

18.2 Barrier Options

A barrier option has a barrier or barriers, and becomes effective only when the asset price stays within the barriers or goes outside the boundary depending on the type of the option during the lifetime of the option. Figure 18.2 illustrates a barrier option with an upper barrier. For a knock-out option, if the asset price crosses the barrier or barriers, the option loses its value immediately. For a knock-in option the barrier option has value only when the asset price reaches a certain level before or at expiry date. Barriers may be defined only on a part of the time to expiry date, and barriers of different heights can be set up.

For example, consider a call option with its exercise price equal to \$10 and knock-in barrier equal to \$11. The asset price at $t = 0$ was \$9, and the price at expiry date is \$10.5. If the asset price has never been equal to \$11, then the option at expiry date will be useless. However, the value of the option at expiry date is worth \$0.5 if the asset price has been at \$11 at least once. Or, we may consider a knock-out barrier option which becomes worthless if the asset price hits a given barrier at any time t, $0 \leq t \leq T$. See Fig. 18.2 for several scenarios of asset price movement. Consult Simulation 18.1.

A barrier option can have two barriers $L < U$. In Fig. 18.3 a double barrier option is shown with its lower and upper barriers L and U. In this case, knock-out barrier

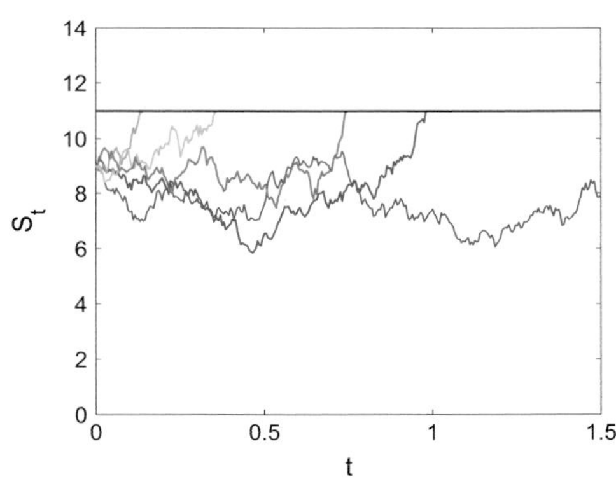

Fig. 18.2 A knock-out barrier option with five sample asset paths

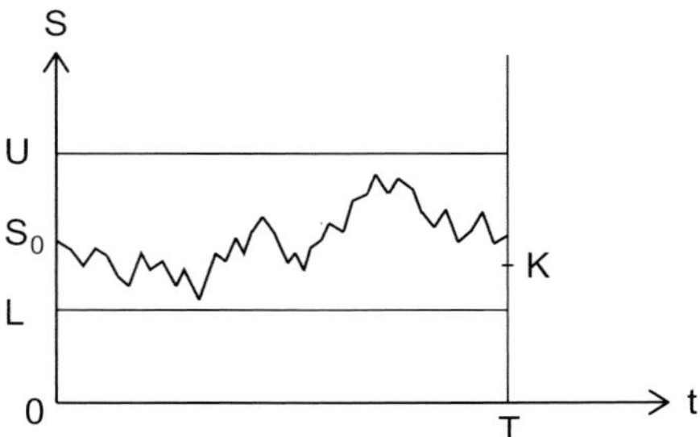

Fig. 18.3 A double barrier option

options can have many different kinds of knock-outs and knock-ins: up-and-out, down-and-out, up-and-in, down-and-in, and their various combinations.

Let $\widetilde{W}_t$ is a $\widetilde{\mathbb{P}}$-Brownian motion. Let $\widehat{W}_t = \widetilde{W}_t + \theta t$ where

$$\theta = \frac{r - \frac{1}{2}\sigma^2}{\sigma} = \frac{r}{\sigma} - \frac{1}{2}\sigma \; .$$

Then

$$S_t = S_0 \mathrm{e}^{\sigma \widetilde{W}_t + (r - \frac{1}{2}\sigma^2)t} = S_0 \mathrm{e}^{\sigma \widehat{W}_t} \; .$$

Let us compute the price of an up-and-out call option where there is one barrier $B = B_1 > S_0$. Define

$$\widehat{M}_T = \max_{0 \le t \le T} \widehat{W}_t \; .$$

Then

$$\max_{0 \le t \le T} S_t = S_0 \mathrm{e}^{\sigma \widehat{M}_T} \; .$$

The option knocks out if and only if

$$S_0 \mathrm{e}^{\sigma \widehat{M}_T} > B \; .$$

Otherwise, the option pays off

$$(S_T - K)^+ = (S_0 \mathrm{e}^{\sigma \widehat{W}_T} - K)^+ \; .$$

Hence the payoff at expiry date T is given by

$$\begin{aligned}
V_T &= (S_0 \mathrm{e}^{\sigma \widehat{W}_T} - K)^+ \mathbf{1}_{\{S_0 \mathrm{e}^{\sigma \widehat{M}_T} \le B\}} \\
&= (S_0 \mathrm{e}^{\sigma \widehat{W}_T} - K) \mathbf{1}_{\{S_0 \mathrm{e}^{\sigma \widehat{W}_T} \ge K,\; S_0 \mathrm{e}^{\sigma \widehat{M}_T} \le B\}} \\
&= (S_0 \mathrm{e}^{\sigma \widehat{W}_T} - K) \mathbf{1}_{\{\widehat{W}_T \ge k,\; \widehat{M}_T \le b\}}
\end{aligned}$$

where

$$k = \frac{1}{\sigma} \log \frac{K}{S_0}$$

and

$$b = \frac{1}{\sigma} \log \frac{B}{S_0} .$$

The option value at $t \le T$ is given by

$$V_t = \mathbb{E}^{\widetilde{\mathbb{P}}}[\mathrm{e}^{-r(T-t)} V_T | \mathcal{F}_t] ,$$

and hence

$$\mathrm{e}^{-rt} V_t = \mathbb{E}^{\widetilde{\mathbb{P}}}[\mathrm{e}^{-rT} V_T | \mathcal{F}_t]$$

is a martingale.

Let τ denote the first time at which the asset price reaches the barrier B, i.e.,

$$\tau = \begin{cases} \inf\{t \ge 0 : S_t = B\} & \text{if } S_t = B \text{ for some } 0 \le t < \infty , \\ \infty , & \text{otherwise} . \end{cases}$$

(Recall that a sample path of S_t is continuous with probability 1.) Then τ is a stopping time. Define a stochastic process by

$$\mathrm{e}^{-r(t \wedge \tau)} V_{t \wedge \tau} = \begin{cases} \mathrm{e}^{-rt} V_t , & 0 \le t \le \tau , \\ \mathrm{e}^{-r\tau} V_\tau , & \tau < t \le T . \end{cases}$$

Since a stopped martingale is also a martingale, we see that $\mathrm{e}^{-r(t \wedge \tau)} V_{t \wedge \tau}$ is a $\widetilde{\mathbb{P}}$-martingale. (Consult Theorem 6.8.) More precisely, let $V_t = v(t, S_t)$, $0 \le t \le \tau$, then $\mathrm{e}^{-r(t \wedge \tau)} v(t \wedge \tau, S_{t \wedge \tau})$ is a $\widetilde{\mathbb{P}}$-martingale.

Now we derive the Black–Scholes–Merton equation for an up-and-out call barrier option.

Theorem 18.2 (Up-and-Out Call) *Let $v(t, x)$ denote the option price at time t, $0 \le t \le T$, of an up-and-out call under the assumption that the call has not knocked out prior to time t and $S_t = x$. Then we have the Black–Scholes–Merton equation*

$$v_t + rxv_x + \frac{1}{2}\sigma^2 x^2 v_{xx} = rv , \quad (t, x) \in [0, T] \times [0, B)$$

with boundary and final conditions given by

$$\begin{cases} v(t, 0) = 0 , & 0 \le t \le T , \\ v(t, B) = 0 , & 0 \le t \le T , \\ v(T, x) = (x - K)^+ , & 0 \le x \le B . \end{cases}$$

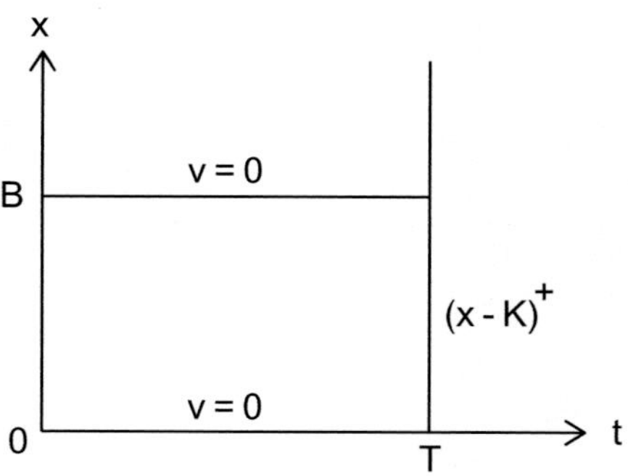

Fig. 18.4 Boundary and final conditions for an up-and-out call barrier option

(See Fig. 18.4.) Its solution is given by

$$\begin{aligned}v(t,x) &= x\left\{N\left(\delta_+(\tau,\frac{x}{K})\right) - N\left(\delta_+(\tau,\frac{x}{B})\right)\right\}\\&- e^{-r\tau}K\left\{N\left(\delta_-(\tau,\frac{x}{K})\right) - N\left(\delta_-(\tau,\frac{x}{B})\right)\right\}\\&- B\left(\frac{x}{B}\right)^{-2r/\sigma^2}\left\{N\left(\delta_+(\tau,\frac{B^2}{Kx})\right) - N\left(\delta_+(\tau,\frac{B}{x})\right)\right\}\\&+ e^{-r\tau}K\left(\frac{x}{B}\right)^{-2r/\sigma^2+1}\left\{N\left(\delta_-(\tau,\frac{B^2}{Kx})\right) - N\left(\delta_-(\tau,\frac{B}{x})\right)\right\}\end{aligned}$$

where $\tau = T - t$,

$$\delta_\pm(\tau,u) = \frac{1}{\sigma\sqrt{\tau}}\left\{\log u + (r \pm \frac{1}{2}\sigma^2)\tau\right\} ,$$

and $N(\cdot)$ *denotes the cumulative distribution function of the standard normal distribution.*

Proof Note that

$$\begin{aligned}&d(e^{-rt}v(t,S_t))\\&= e^{-rt}\left\{-rv\mathrm{d}t + v_t\mathrm{d}t + v_x\mathrm{d}S + \frac{1}{2}v_{xx}(\mathrm{d}S)^2\right\}\\&= e^{-rt}\left\{-rv + v_t + rSv_x + \frac{1}{2}\sigma^2S^2v_{xx}\right\}\mathrm{d}t + e^{-rt}\sigma Sv_x\mathrm{d}\widetilde{W} .\end{aligned}$$

The drift term should vanish for $0 \le t \le \tau$. Since (t, S_t) is an arbitrary point in $\{(t,x) : 0 \le t \le T, 0 \le x \le B\}$ before the up-and-out option strikes out, we obtain the Black–Scholes–Merton equation in the rectangle. □

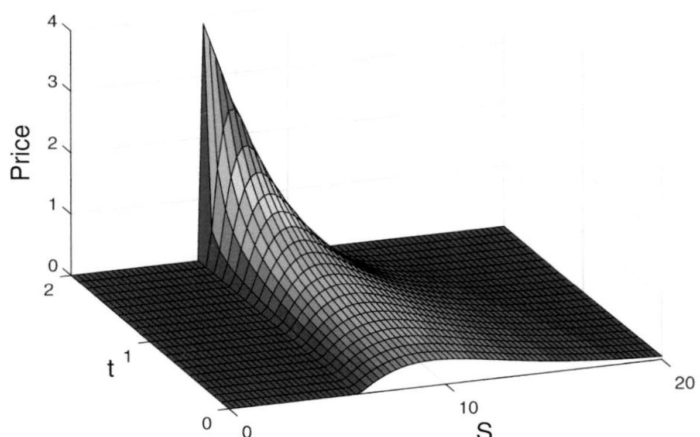

Fig. 18.5 Price of a down-and-out put option

Now we find the formula for the price of a down-and-out put barrier option. See Fig. 18.5 where we observe that the option price is zero if the asset price falls below the given barrier.

Theorem 18.3 (Down-and-Out Put) *The price of a down-and-out put option $P_{\text{d-o}}$ with strike price K with a barrier L, $L < K$, and expiry date T is given by*

$$\begin{aligned} P_{\text{d-o}} &= K\mathrm{e}^{-rT}N(-d_2^K) - S_0N(-d_1^K) \\ &\quad - K\mathrm{e}^{-rT}N(-d_2^L) + S_0N(-d_1^L) \\ &\quad - K\mathrm{e}^{-rT}\left(\frac{L}{S_0}\right)^{\frac{2r}{\sigma^2}-1}(N(d_4) - N(d_6)) \\ &\quad + S_0\left(\frac{L}{S_0}\right)^{\frac{2r}{\sigma^2}+1}(N(d_3) - N(d_5)) \end{aligned}$$

where

$$\begin{aligned} d_{1,2}^K &= \frac{\log\frac{S_0}{K} + (r \pm \frac{1}{2}\sigma^2)T}{\sigma\sqrt{T}}, \\ d_{1,2}^L &= \frac{\log\frac{S_0}{L} + (r \pm \frac{1}{2}\sigma^2)T}{\sigma\sqrt{T}}, \\ d_{3,4} &= \frac{\log\frac{L}{S_0} + (r \pm \frac{1}{2}\sigma^2)T}{\sigma\sqrt{T}}, \\ d_{5,6} &= \frac{\log\frac{L^2}{S_0K} + (r \pm \frac{1}{2}\sigma^2)T}{\sigma\sqrt{T}}. \end{aligned}$$

Proof Note that

$$
\begin{aligned}
P_{\text{d-o}} &= \mathrm{e}^{-rT}\mathbb{E}[(K-S_T)\mathbf{1}_{\{L<S_T<K\}}\mathbf{1}_{\{\min_{0\le t\le T} S_t>L\}}] \\
&= \mathrm{e}^{-rT}\mathbb{E}[(K-S_T)\mathbf{1}_{\{L<S_T<K\}}] \\
&- \mathrm{e}^{-rT}\mathbb{E}[(K-S_T)\mathbf{1}_{\{L<S_T<K\}}\mathbf{1}_{\{\min_{0\le t\le T} S_t\le L\}}] \\
&= \mathrm{e}^{-rT}\mathbb{E}[(K-S_T)\mathbf{1}_{\{S_T<K\}}] \\
&- \mathrm{e}^{-rT}\mathbb{E}[(K-S_T)\mathbf{1}_{\{S_T<L\}}] \\
&- \mathrm{e}^{-rT}\mathbb{E}[K\mathbf{1}_{\{L<S_T<K\}}\mathbf{1}_{\{\min_{0\le t\le T} S_t\le L\}}] \\
&+ \mathrm{e}^{-rT}\mathbb{E}[S_T\mathbf{1}_{\{L<S_T<K\}}\mathbf{1}_{\{\min_{0\le t\le T} S_t\le L\}}] \\
&= I_1 - I_2 - I_3 + I_4 \,.
\end{aligned}
$$

Observe that I_1 is the price of a European put option, and hence

$$
I_1 = K\mathrm{e}^{-rT}N(-d_2^K) - S_0N(-d_1^K) \,.
$$

Similarly, we see that

$$
\begin{aligned}
I_2 &= \mathrm{e}^{-rT}\mathbb{E}[(K-L+L-S_T)\mathbf{1}_{\{S_T<L\}}] \\
&= \mathrm{e}^{-rT}(K-L)\mathbb{P}(\{S_T<L\}) \\
&+ L\mathrm{e}^{-rT}N(-d_2^L) \\
&- S_0N(-d_1^L) \,.
\end{aligned}
$$

Let us compute $\mathbb{P}(\{S_T<L\})$. If W_t denotes the Brownian motion driving the price process, then we have

$$
\begin{aligned}
\mathbb{P}(\{S_T<L\}) &= \mathbb{P}\left(\left\{W_1 < \frac{\log\frac{L}{S_0} - (r-\frac{1}{2}\sigma^2)T}{\sigma\sqrt{T}}\right\}\right) \\
&= \mathbb{P}\left(\left\{-W_1 > \frac{\log\frac{S_0}{L} + (r-\frac{1}{2}\sigma^2)T}{\sigma\sqrt{T}}\right\}\right) \\
&= 1 - N(d_2^L) \\
&= N(-d_2^L)
\end{aligned}
$$

since $-W_1$ is a standard normal variable. Hence

$$
\begin{aligned}
I_2 &= \mathrm{e}^{-rT}(K-L)N(-d_2^L) + L\mathrm{e}^{-rT}N(-d_2^L) - S_0N(-d_1^L) \\
&= K\mathrm{e}^{-rT}N(-d_2^L) - S_0N(-d_1^L) \,.
\end{aligned}
$$

Now put

$$\lambda = \frac{1}{\sigma}\log\frac{L}{S_0}\,,$$

$$\kappa = \frac{1}{\sigma}\log\frac{K}{S_0}\,,$$

$$\theta = \frac{r}{\sigma} - \frac{1}{2}\sigma\,.$$

Since

$$\overline{W}_t = -W_t$$

is also a Brownian motion, we define a Brownian motion with drift by

$$\widehat{W}_t = \overline{W}_t - \theta t$$

and define

$$\widehat{M}_T = \max_{0\le t\le T}\widehat{W}_t\,.$$

Then

$$L < S_T < K$$

if and only if

$$-\kappa < \widehat{W}_T < -\lambda\,.$$

Also note that the following three statements are equivalent:

$$\min_{0\le t\le T} S_t \le L\,,$$

$$\min_{0\le t\le T}(W_t + \theta t) \le \lambda\,,$$

and

$$\max_{0\le t\le T}\widehat{W}_t \ge -\lambda\,.$$

Hence

$$\begin{aligned}&\mathbb{E}[\mathbf{1}_{\{L<S_T<K\}}\mathbf{1}_{\{\min_{0\le t\le T} S_t\le L\}}]\\&= \mathbb{P}\{-\kappa < \widehat{W}_T < -\lambda\,,\,\widehat{M}_T \ge -\lambda\}\end{aligned}$$

$$= \int_{-\kappa}^{-\lambda} \int_{-\lambda}^{\infty} \frac{2(2m-w)}{T\sqrt{2\pi T}} e^{-\theta w - \frac{1}{2}\theta^2 T - \frac{1}{2T}(2m-w)^2} \, dm \, dw$$

$$= \int_{-\kappa}^{-\lambda} \frac{1}{\sqrt{2\pi T}} e^{-\theta w - \frac{1}{2}\theta^2 T - \frac{1}{2T}(-2\lambda - w)^2} \, dw$$

$$= e^{2\theta\lambda} \int_{-\kappa}^{-\lambda} \frac{1}{\sqrt{2\pi T}} e^{-\frac{1}{2T}(w+2\lambda+\theta T)^2} \, dw$$

$$= e^{2\theta\lambda} \int_{z(-\kappa)}^{z(-\lambda)} \frac{1}{\sqrt{2\pi}} e^{-\frac{1}{2}z^2} \, dz$$

$$= e^{2\theta\lambda} \left\{ N\left(\frac{\lambda + \theta T}{\sqrt{T}}\right) - N\left(\frac{-\kappa + 2\lambda + \theta T}{\sqrt{T}}\right) \right\}$$

where we applied Theorem 9.1 in the second equality, and we used the identity

$$-\theta w - \frac{1}{2}\theta^2 T - \frac{1}{2T}(w + 2\lambda)^2 = -\frac{1}{2T}(w + 2\lambda + \theta T)^2 + 2\theta\lambda$$

in the fourth equality, and we took the substitution

$$z(w) = \frac{w + 2\lambda + \theta T}{\sqrt{T}}$$

in the fifth. Note that

$$e^{2\theta\lambda} = \exp\left(2\left(\frac{r}{\sigma} - \frac{1}{2}\sigma\right)\frac{1}{\sigma}\log\frac{L}{S_0}\right) = \left(\frac{L}{S_0}\right)^{\frac{2r}{\sigma^2}-1} ,$$

$$\frac{\lambda + \theta T}{\sqrt{T}} = \frac{\log\frac{L}{S_0} + (r - \frac{1}{2}\sigma^2)T}{\sigma\sqrt{T}} = d_4 ,$$

and

$$\frac{-\kappa + 2\lambda + \theta T}{\sqrt{T}} = \frac{-\log\frac{K}{S_0} + 2\log\frac{L}{S_0} + (r - \frac{1}{2}\sigma^2)T}{\sigma\sqrt{T}} = d_6 .$$

Hence

$$I_3 = K e^{-rT} \left(\frac{L}{S_0}\right)^{\frac{2r}{\sigma^2}-1} \{N(d_4) - N(d_6)\} .$$

Finally, since

$$S_T = S_0 e^{\sigma\theta T - \sigma\overline{W}_T} = S_0 e^{-\sigma(-\theta T + \overline{W}_T)} = S_0 e^{-\sigma\widehat{W}_T} ,$$

we have

$$
\begin{aligned}
&\mathbb{E}[S_T \mathbf{1}_{\{L<S_T<K\}} \mathbf{1}_{\{\min_{0\le t\le T} S_t \le L\}}] \\
&= \int_{-\kappa}^{-\lambda} \int_{-\lambda}^{\infty} S_0 e^{-\sigma w} \frac{2(2m-w)}{T\sqrt{2\pi T}} e^{-\theta w - \frac{1}{2}\theta^2 T - \frac{1}{2T}(2m-w)^2}\, dm\, dw \\
&= S_0 \int_{-\kappa}^{-\lambda} \frac{1}{\sqrt{2\pi T}} e^{-(\sigma+\theta)w - \frac{1}{2}\theta^2 T - \frac{1}{2T}(w+2\lambda)^2}\, dw \\
&= S_0 \int_{-\kappa}^{-\lambda} \frac{1}{\sqrt{2\pi T}} e^{-\frac{1}{2T}(w+2\lambda+(\sigma+\theta)T)^2 + 2\lambda(\sigma+\theta) + \frac{1}{2}\sigma^2 T + \sigma\theta T}\, dw \\
&= S_0 e^{2\lambda(\sigma+\theta) + \frac{1}{2}\sigma^2 T + \sigma\theta T} \int_{z(-\kappa)}^{z(-\lambda)} \frac{1}{\sqrt{2\pi}} e^{-\frac{1}{2}z^2}\, dz \\
&= S_0 e^{2\lambda(\sigma+\theta) + \frac{1}{2}\sigma^2 T + \sigma\theta T} \\
&\quad \times \left\{ N\left(\frac{\lambda + (\sigma+\theta)T}{\sqrt{T}} \right) + N\left(\frac{-\kappa + 2\lambda + (\sigma+\theta)T}{\sqrt{T}} \right) \right\}
\end{aligned}
$$

where we used the substitution

$$
z(w) = \frac{w + 2\lambda + (\sigma+\theta)T}{\sqrt{T}}
$$

in the fourth equality. Note that

$$
\begin{aligned}
&2\lambda(\sigma+\theta) + \frac{1}{2}\sigma^2 T + \sigma\theta T \\
&= 2\left(\frac{r}{\sigma^2} + \frac{1}{2}\right) \log \frac{L}{S_0} + rT \\
&= \log\left(\left(\frac{L}{S_0}\right)^{2r/\sigma^2+1} e^{rT} \right)
\end{aligned}
$$

and that

$$
\frac{\lambda + (\sigma+\theta)T}{\sqrt{T}} = \frac{\log \frac{L}{S_0} + (r + \frac{1}{2}\sigma^2)T}{\sigma\sqrt{T}} = d_3 ,
$$

$$
\frac{-\kappa + 2\lambda + (\sigma+\theta)T}{\sqrt{T}} = \frac{\log \frac{L^2}{S_0 K} + (r + \frac{1}{2}\sigma^2)T}{\sigma\sqrt{T}} = d_5 ,
$$

$$
I_4 = S_0 \left(\frac{L}{S_0}\right)^{2r/\sigma^2+1} (N(d_3) - N(d_5)) ,
$$

which completes the proof. □

18.3 Computer Experiments

Simulation 18.1 (Stopped Asset Price Process)
We plot five sample paths of the asset price process, some of which are knocked out by an upper barrier before T. See Fig. 18.2.

```
M = 5;  % number of sample paths
T = 1.5;
mu = 0.1;
sigma = 0.3;
a = 0;
b = 11;
S0 = 9;
N = 300;
dt = T/N;
t = 0:dt:T;
S = zeros(M,N+1);
S(:,1) = S0;
dW = sqrt(dt)*randn(M,N);
stop = (N+1)*ones(M,1);

for i = 1:M;
    for j = 1:N;
    S(i,j+1) = S(i,j) + mu*S(i,j)*dt + sigma*S(i,j)*dW(i,j);
      if ( (S(i,j+1) >= b) )
        S(i,j+1) = b;
        stop(i) = j+1;
        break;
    end
    end
end

for i = 1:M
    plot(0:dt:(stop(i)-1)*dt,S(i,1:stop(i)),'color',hsv2rgb([1-i/M 1 1]));
    hold on
end
```

Simulation 18.2 (Down-and-Out Put)
The surface defined by the Black–Scholes–Merton formula for a down-and-out put is given in Fig. 18.5.

```
K = 10;
L = 6; % a lower barrier
r = 0.05;
sigma = 0.3;
M= 75; % number of points on the asset axis including endpoints
N= 40; % number of points on the time axis including endpoints
T = 2;
dt = T/N;
S0 = 10;
S_max = 20;
dS = S_max/M;
S_range=0:dS:S_max; % Divide the interval into M subintervals.
```

```
t_range=0:dt:T;      % Divide the interval into N subintervals.
Put_do = zeros(M+1,N+1);
index_barrier = ceil(L/dS)

for i=1:M+1
    S = S_range(i);
    for j=1:N
    tau = T - t_range(j);
    d1K = (log(S/K) + (r+0.5*sigma^2)*tau)/sigma/sqrt(tau);
    d2K = (log(S/K) + (r-0.5*sigma^2)*tau)/sigma/sqrt(tau);
    d1L = (log(S/L) + (r+0.5*sigma^2)*tau)/sigma/sqrt(tau);
    d2L = (log(S/L) + (r-0.5*sigma^2)*tau)/sigma/sqrt(tau);
    d3 = (log(L/S) + (r+0.5*sigma^2)*tau)/sigma/sqrt(tau);
    d4 = (log(L/S) + (r-0.5*sigma^2)*tau)/sigma/sqrt(tau);
    d5 = (log(L^2/S/K) + (r+0.5*sigma^2)*tau)/sigma/sqrt(tau);
    d6 = (log(L^2/S/K) + (r-0.5*sigma^2)*tau)/sigma/sqrt(tau);

    I1(i,j)= K*exp(-r*tau)*normcdf(-d2K) - S*normcdf(-d1K);
    I2(i,j)= K*exp(-r*tau)*normcdf(-d2L) - S*normcdf(-d1L);
    I3(i,j)= K*exp(-r*tau)*(L/S)^(2*r/sigma^2-1)*(normcdf(d4)
             -normcdf(d6));
    I4(i,j)= S*(L/S)^(2*r/sigma^2+1)*(normcdf(d3)-normcdf(d5));

    Put_do(i,j) = I1(i,j) - I2(i,j) -I3(i,j) + I4(i,j);
    end
end
for i=1:index_barrier
    Put_do(i,:) =0;
end
for i=1:M+1
    Put_do(i,N+1) = max(K-S_range(i),0)*heaviside
      (S_range(i) - L);
end

[t_grid,S_grid]=meshgrid(t_range,S_range);
surf(S_grid,t_grid,Put_do);
```

Chapter 19
American Options

An option that can be exercised early, i.e., before or on the expiry date, is called an *American* option while a *European* option can be exercised only on the expiry date. Such an early exercise feature makes the pricing of American options harder. In the last section we introduce a very practical algorithm called the least squares Monte Carlo method, which is based on regression to estimate the continuation values from simulated sample paths.

19.1 American Call Options

Since an American option can be exercised at any time, it must be at least as valuable as an otherwise identical European option. For American call options, however, the right to early-exercise is worthless. In other words, the right on early exercise does not affect the price of an American call option.

Theorem 19.1 *The prices of an American call option and a European call option are equal if their expiry date and exercise price are equal.*

Proof The holder of an American call option would try to maximize his profit using various strategies. The first strategy is to early exercise the option at time $t < T$. If $S_t > K$, then the holder will gain profit $S_t - K$ at time t. This strategy is no better than the following second strategy: At the time t in the first strategy by short selling the holder receives S_t, and at expiry date T the holder has two choices: (i) either the holder exercises the option and buys a share of the stock at the exercise price K, or (ii) buys a share of the stock in the market at the price S_T. By the second strategy the holder of the option can buy a stock at expiry date paying at most K, and return the stock which was borrowed for short selling.

Since the value of K at time T is not greater than the value of K at time t, the second strategy is better than or as good as the first one. Note that the second strategy

G.H. Choe, *Stochastic Analysis for Finance with Simulations*, Universitext,
DOI 10.1007/978-3-319-25589-7_19

Table 19.1 Viewpoint of the holder of an American call option

Time	0	t	T
Without short sell	Option	Cash $S_t - K$	$e^{r(T-t)}(S_t - K)$
With short sell (I)	Option	Cash S_t and (-1) share of stock	$e^{r(T-t)}S_t - K$
With short sell (II)	Option	Cash S_t and (-1) share of stock	$e^{r(T-t)}S_t - S_T$

can be used for a European option. Hence the prices of an American call option and a European call option are equal if the expiry date and exercise price are identical. Consult Table 19.1. □

Remark 19.1 Consider European call options C_1 and C_2 on a stock with the same exercise price and different expiry dates T_1 and T_2, $T_1 < T_2$, respectively. Consider an American call option A_2 with expiry date T_2. It has the same value as C_2 by Theorem 19.1. However, A_2 can be exercised any time before T_2 including at time T_1. Hence it is worth as least as much as C_1. Thus C_2 is valuable at least as much as C_1. This implies that the European call option price is a monotonically increasing function of time to expiry.

19.2 American Put Options

Theorem 19.2 *Let $V(S,t)$ denote the price of an American put option with expiry date T and exercise price K. Then for each $t \in [0,T]$ there exists a number $S_t^* \in (0,\infty)$ such that for $0 \le S \le S_t^*$ and $0 \le t \le T$, we have*

$$V(S,t) = K - S \ \textit{and}\ \frac{\partial V}{\partial t} + \frac{1}{2}\sigma^2 S^2 \frac{\partial^2 V}{\partial S^2} + rS\frac{\partial V}{\partial S} < rV \,,$$

and for $S_t^* < S$ and $0 \le t \le T$,

$$V(S,t) > \max\{K - S, 0\} \ \textit{and}\ \frac{\partial V}{\partial t} + \frac{1}{2}\sigma^2 S^2 \frac{\partial^2 V}{\partial S^2} + rS\frac{\partial V}{\partial S} = rV \,.$$

The boundary condition at $S = S_t^*$ is that the option price is continuously differentiable with respect to S, is continuous in t, and

$$V(S_t^*, t) = \max\{K - S_t^*, 0\} \,, \quad \frac{\partial V}{\partial S}(S_t^*, t) = -1 \,.$$

Proof As in Sect. 15.1 the increment of V is given by

$$\delta V = \left(\frac{\partial V}{\partial t} + \frac{1}{2}\sigma^2 S^2 \frac{\partial^2 V}{\partial S^2}\right)\delta t + \frac{\partial V}{\partial S}\,\delta S \,.$$

Now from the viewpoint of an option seller we construct a self-financing and risk-free portfolio Π by

$$\Pi(S,t) = -V(S,t) + D(S,t) + \Delta(S,t)S \,. \tag{5}$$

In other words, Π consists of one option sold, and bank deposit D, and Δ units of underlying asset S. We do not know how much S will change as time passes by from t to $t+\delta t$. Thus we fix the hedge ratio Δ from t to $t+\delta t$, and suppose that S changes to $S+\delta S$ and Π changes to $\Pi+\delta\Pi$, respectively. The bank deposit D increases by the amount $\delta D = rD\delta t$. In summary, we have

$$\begin{aligned}\delta\Pi &= -\delta V + rD\delta t + \Delta\delta S \\ &= \left(-\frac{\partial V}{\partial t} - \frac{1}{2}\sigma^2 S^2\frac{\partial^2 V}{\partial S^2} + rD\right)\delta t + \left(-\frac{\partial V}{\partial S} + \Delta\right)\delta S \,.\end{aligned}$$

Take $\Delta = \frac{\partial V}{\partial S}$. Then the δS term disappears, and there is no risk in $\delta\Pi$ since there is no risky asset price movement. However, there remains the risk that the holder of an American option may early exercise the option and cause financial loss to the seller, and $\delta\Pi$ should be greater than the risk-free profit to compensate for the risk. Hence we have the inequality $\delta\Pi > r\,\Pi\,\delta t$, and hence

$$-\frac{\partial V}{\partial t} - \frac{1}{2}\sigma^2 S^2\frac{\partial^2 V}{\partial S^2} + rD \geq r\left(-V + D + \frac{\partial V}{\partial S}S\right) \,.$$

After canceling rD from both sides, we obtain the desired inequality. For the details of the proof, consult [75, 94]. □

The *optimal exercise boundary* for an American put option is a curve given by $\{(t, S_t^*) : 0 \leq t \leq T\}$. See Fig. 19.1. In the domain given by $\{(t,S) : S > S_t^*\}$ it is better to wait until the asset price goes down further without exercising early. As the

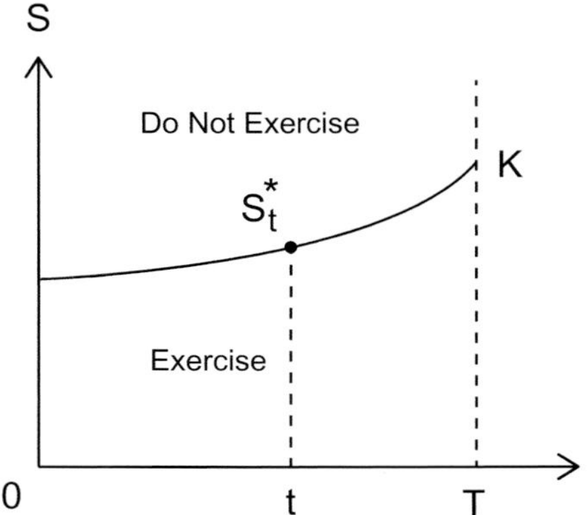

Fig. 19.1 Optimal exercise boundary for an American put option

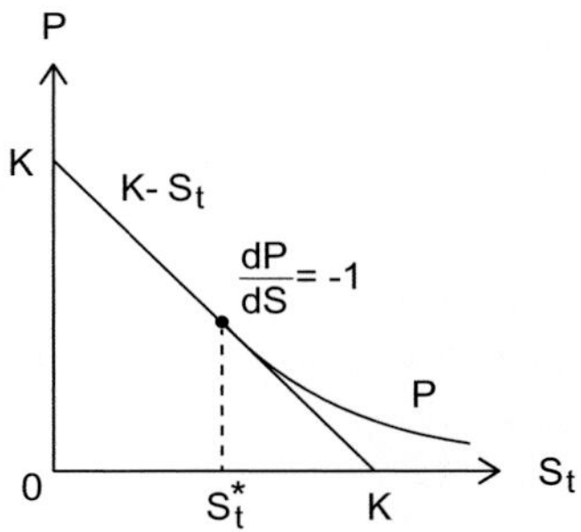

Fig. 19.2 The price P of an American put option when the asset price S_t takes the boundary price S_t^* at time t

expiry date approaches, we have smaller possibility that the asset price moves down further and the future payoff will increase. Hence we have to exercise the option at a higher price than before to expect the same level of payoff. Thus the function $t \mapsto S_t^*$ is monotonically increasing.

In the domain $\{(t, S) : S > S_t^*\}$ the inequality in Theorem 19.2 becomes an equality. The reason is that in the management of the portfolio Π, as in the case of a European option, we do not exercise the American put option before expiry date, and hence we have $\mathrm{d}\Pi = r\,\Pi\,\mathrm{d}t$. In this case, we give a boundary condition $\frac{\partial P}{\partial S} = -1$ along the optimal exercise boundary, and solve the Black–Scholes–Merton partial differential equation. (See Fig. 19.2.) For such a partial differential equation the boundary of the domain is not explicitly given, thus it is called a free boundary problem.

If $S_t < S_t^*$ at time t, i.e., it is better to early exercise the option, then the price $P(S_t, t)$ of the American put option at time t is given by a line segment $P = K - S_t$ with slope -1 as in Fig. 19.2. If $S_t \geq S_t^*$, then $P(S_t, t) > K - S_t$ as shown in the graph.

For $0 \leq t \leq T$ the price of an American put option at time t with $S_t = x$ is given by

$$P(x, t) = \sup_{t \leq \tau \leq T} \mathbb{E}\left[\mathrm{e}^{-r(\tau - t)}(K - S_\tau)^+ \middle| S_t = x \right]$$

where τ denotes a stopping time. Continuous time models are theoretically complicated and the explicit pricing formula is not yet known. For more information, consult [8, 47]. The binomial tree method is simple to implement in pricing American put options. At every step we compare the price of the corresponding European option if the option is not exercised and the profit if it is exercised, and choose the more profitable action. See [17] for a modified algorithm.

We say that an option is *perpetual* if the time to expiry is infinite. In the following result, note that $L < K$ and $V(S) > 0$ if $\sigma > 0$.

Theorem 19.3 (Perpetual American Put) *The price of a perpetual American put option is given by*

$$V(S) = \begin{cases} (K-L)\left(\dfrac{L}{S}\right)^{2r/\sigma^2}, & L < S < \infty\,, \\ K - S\,, & 0 \leq S \leq L\,, \end{cases}$$

where

$$L = \frac{2rK}{2r+\sigma^2}\,.$$

Proof Since the time to expiry is always infinite, the option price V does not depend on time t, and is a function of S only and the exercise boundary is of the form $S_t^* = L$ for some constant L for all $t > 0$. The Black–Scholes–Merton partial differential equation becomes an ordinary differential equation

$$\frac{1}{2}\sigma^2 S^2 \frac{\partial^2 V}{\mathrm{d}S^2} + rS\frac{\mathrm{d}V}{\mathrm{d}S} - rV = 0\,, \quad 0 < S < \infty\,.$$

We look for a general solution of the form

$$V(S) = c_1 S^{p_1} + c_2 S^{p_2}$$

for some constants c_1, c_2, p_1, p_2. Now use the boundary conditions $V(L) = K - L$, $\lim_{S\downarrow L} \frac{\mathrm{d}V}{\mathrm{d}S} = -1$ and $\lim_{S\to\infty} V(S) = 0$. □

19.3 The Least Squares Method of Longstaff and Schwartz

We introduce a practical and simple approach to pricing options with early exercise features, proposed by Longstaff and Schwartz, that combines the Monte Carlo method with a simple least squares regression. See [60]. A *Bermudan option* is an option that can be exercised only at a set of discrete times $0 < t_1 < \cdots < t_n = T$. By taking $\delta t_i = t_{i+1} - t_i$, $1 \leq i \leq n$, and $\max_{0\leq i\leq n-1} \delta t_i \to 0$ as $n \to \infty$, we can approximate an American option with Bermudan options. If an American option is in the money at the final exercise date T, the optimal strategy for the holder of the American option is to exercise it. Before the expiry date, however, it is optimal to exercise if the immediate exercise value is more valuable than the expected cash flow from continuation. Using the cross-sectional data in the simulated asset paths, we obtain the conditional expectation function which best represents the given data. This is done by regressing the realized cash flows from continuation on a set of basis functions, which allows us to find the optimal exercise rule for the option. The least squares Monte Carlo method is also applicable in path-dependent and multifactor situations.

Example 19.1 (American Put Option) Let $S_0 = 10.0$ and $r = 0.05$. Compute the price of an American put option with exercise price $K = 11.0$ and $T = 3$. Let us assume that the put option is exercisable at times $t = 1, 2, 3$, where $t = 3$ is the expiry date of the option. The discount rate is equal to $e^{-r\delta t} = 0.9512$. For the sake of simplicity, we explain the algorithm using eight sample paths for the price of the underlying asset. The sample paths are generated under the risk-neutral probability and are shown in Table 19.2.

To maximize the value of the option we find the stopping rule at each point along each sample path S_t, $t = 1, 2, 3$. The algorithm is recursive, and we proceed backward in time. Conditional on not exercising the option before $t = 3$, the cash flows $Y_3 = \max\{S_3 - K, 0\}$ received by the option holder at $t = 3$ are given by the last column of Table 19.2.

If the option is in the money at $t = 2$, the holder must decide whether to exercise the option immediately or continue until the final expiration date at $t = 3$. From the asset prices in Table 19.2, there are only four paths, the first, the fourth, the sixth and the seventh, for which the option is in the money at $t = 2$. Let X denote the stock prices at $t = 2$ for these four paths, and $Y = e^{-r\delta t} Y_3$ denote the corresponding discounted cash flows received at $t = 3$ if the put is not exercised at $t = 2$. By using only in-the-money paths we fit the approximating polynomial in the region where exercise is meaningful and reduce the computing time. See Table 19.3.

Table 19.2 Asset prices S_t at $t = 1, 2, 3$ and cash flow Y_3 at $t = 3$

Path	$t = 0$	$t = 1$	$t = 2$	$t = 3$	Y_3
1	10.00	10.60	10.10	11.70	0.00
2	10.00	11.30	13.70	14.20	0.00
3	10.00	10.10	11.50	10.80	0.20
4	10.00	11.80	10.90	10.20	0.80
5	10.00	10.40	11.60	12.10	0.00
6	10.00	10.70	10.20	10.50	0.50
7	10.00	10.80	9.50	10.40	0.60
8	10.00	9.50	11.70	12.60	0.00

Table 19.3 Regression at time 2

Path	$X = S_2$	$Y = e^{-r\delta t} Y_3$
1	10.10	0.0000
2	–	–
3	–	–
4	10.90	0.7610
5	–	–
6	10.20	0.4756
7	9.50	0.5707
8	–	–

At $t = 2$ we do not know the future asset price S_3 and the corresponding option value at $t = 3$. Thus we fit a quadratic polynomial to the data using the least squares method to estimate the expected cash flow from continuing the option's life conditional on S_2. The resulting conditional expectation function is $p_2(x) = \mathbb{E}[Y|X = x] = 0.8561x^2 - 17.3051x + 87.6840$. (We may employ basis functions defined by orthogonal polynomials such as Laguerre polynomials. See Example C.6 in Appendix C.)

With this conditional expectation function, we can now decide whether to exercise or continue by comparing the value of immediate exercise at $t = 2$ with the value from continuation given in Table 19.4. When the asset path is not in the money at $t = 2$ the option is continued.

The value of immediate exercise is equal to $K - S_2$ for the in-the-money paths, while the value from continuation is given by $p_2(S_2)$. If $S_2 - K > p_2(S_2)$ we exercise, otherwise we continue. We conclude that it is optimal to exercise the option at $t = 2$ for the first, the sixth and the seventh paths, and obtain Table 19.5 that displays the cash flows received by the option holder conditional on not exercising prior to time 2. When the option is exercised at $t = 2$, the cash flow in the final column becomes zero since there are no further cash flows once the option is exercised.

Moving backward in time recursively, we now check whether the option should be exercised at $t = 1$. From Table 19.2, there are six asset paths for which the option is in the money at $t = 1$. For these sample paths, we again define Y as the discounted value of the subsequent option cash flows. In defining Y at $t = 1$, we use actual realized cash flows along each sample path. As before, we choose only

Table 19.4 Optimal early exercise decision at time 2

Path	$S_2 - K$	$p_2(S_2)$	Decision
1	0.90	0.2365	Exercise
2	–	–	–
3	–	–	–
4	0.10	0.7755	Continuation
5	–	–	–
6	0.80	0.2439	Exercise
7	1.50	0.5514	Exercise
8	–	–	–

Table 19.5 Cash flow at time 2 conditional on not exercising before time 2

Path	$t = 1$	$t = 2$	$t = 3$
1	–	0.90	0.00
2	–	0.00	0.00
3	–	0.00	0.20
4	–	0.00	0.80
5	–	0.00	0.00
6	–	0.80	0.00
7	–	1.50	0.00
8	–	0.00	0.00

the paths satisfying $S_1 < K$, and take $X = S_1$. The vectors X and Y are given by the nondashed elements in Table 19.6. Since the option can only be exercised once, future cash flows occur at either $t = 2$ or $t = 3$, but not both. Cash flows received at $t = 2$ are discounted back one period to $t = 1$, and any cash flows received at $t = 3$ are discounted back two periods to $t = 1$. For example, discounted cash flows are $0.9 \times e^{-r\delta t} = 0.8561$ for the first path, and $0.2 \times e^{-2r\delta t} = 0.1810$ for the third.

The best fit quadratic polynomial at $t = 1$ by the least squares method is given by $p_1(x) = \mathbb{E}[Y|X = x] = 1.7138x^2 - 33.8220x + 166.6906$. Substituting the values of X, we obtain estimated continuation values, and we compare them with immediate exercise values at $t = 1$. See Table 19.7. Comparing two columns we see that exercise at $t = 1$ is optimal for the third, the fifth and the eighth paths.

Having identified the exercise strategy at $t = 1, 2, 3$, the stopping rule can now be given by Table 19.8, where the symbol '1' represents exercise dates at which the option is exercised, and '0' no exercise.

Following this stopping rule, we exercise the option at the exercise dates where there is a one in Table 19.8. This leads to the option cash flow presented in Table 19.9, and the option can now be valued by discounting each cash flow back to $t = 0$, and averaging over all paths. Applying this procedure results in a value of 0.8047 for the American put. See Simulation 19.3. For a more practical and efficient code with a large number of sample paths, see Simulation 19.4.

Table 19.6 Regression at time 1

Path	$X = S_1$	Y
1	10.60	0.8561
2	–	–
3	10.10	0.1810
4	–	–
5	10.40	0.0000
6	10.70	0.7610
7	10.80	1.4268
8	9.50	0.0000

Table 19.7 Optimal early exercise decision at time 1

Path	$S_1 - K$	$p_1(S_1)$	Decision
1	0.40	0.7167	Continuation
2	–	–	–
3	0.90	−0.0685	Exercise
4	–	–	–
5	0.60	0.3066	Exercise
6	0.30	0.9698	Continuation
7	0.20	1.2549	Continuation
8	1.50	0.0454	Exercise

Table 19.8 Stopping rule

Path	$t = 1$	$t = 2$	$t = 3$
1	0	1	0
2	0	0	0
3	1	0	0
4	0	0	1
5	1	0	0
6	0	1	0
7	0	1	0
8	1	0	0

Table 19.9 Complete option cash flow

Path	$t = 1$	$t = 2$	$t = 3$
1	0.00	0.90	0.00
2	0.00	0.00	0.00
3	0.90	0.00	0.00
4	0.00	0.00	0.80
5	0.60	0.00	0.00
6	0.00	0.80	0.00
7	0.00	1.60	0.00
8	1.50	0.00	0.00

19.4 Computer Experiments

Simulation 19.1 (American Call Option: The Binomial Tree Method)

We use the binomial tree method to price an American option. Observe that the prices of a European call and an American call are equal with the same parameter values.

```
S0 = 100;
K = 110;
T = 1;
r = 0.05;
sigma = 0.3;
M = 1000; % number of time steps
dt = T/M;
u = exp(sigma*sqrt(dt) + (r-0.5*sigma^2)*dt);
d = exp(-sigma*sqrt(dt) + (r-0.5*sigma^2)*dt);
p = 1/2;

Call_Am = max(S0*u.^([M:-1:0]').*d.^([0:1:M]')- K,0);
for i = M:-1:1
    S = S0*u.^([i-1:-1:0]').*d.^([0:1:i-1]');
    Call_Am = max(max(S-K,0), exp(-r*dt)*(p*C_Am(1:i)+(1-p)*C_Am(2:i+1)));
end
Call_Am

d1 = (log(S0/K)+(r+0.5*sigma^2)*T)/(sigma*sqrt(T));
```

```
d2 = d1-sigma*sqrt(T);
Call_Eu = S0*normcdf(d1) - K*exp(-r*T)*normcdf(d2)
```

The output is

```
Call_Am = 10.0212
Call_Eu = 10.0201
```

Simulation 19.2 (American Put Option: The Binomial Tree Method)
We use the binomial tree method to price an American option. Observe that the price of an American put option is higher than that of a European put when we take the same set of parameter values for both.

```
S0 = 100;
K = 110;
T = 1;
r = 0.05;
sigma = 0.3;
M = 1000; % number of time steps
dt = T/M;
u = exp(sigma*sqrt(dt) + (r-0.5*sigma^2)*dt);
d = exp(-sigma*sqrt(dt) + (r-0.5*sigma^2)*dt);
p = 1/2;

Put_Am = max(K - S0*u.^([M:-1:0]').*d.^([0:1:M]'),0);
for i = M:-1:1
    S = S0*u.^([i-1:-1:0]').*d.^([0:1:i-1]');
    Put_Am =max(max(K-S,0),exp(-r*dt)*(p*Put_Am(1:i)+(1-p)*Put_Am(2:i+1)));
end
Put_Am

d1 = (log(S0/K)+(r+0.5*sigma^2)*T)/(sigma*sqrt(T));
d2 = d1 - sigma*sqrt(T);
Put_Eu = K*exp(-r*T)*normcdf(-d2) - S0*normcdf(-d1)
```

The output is

```
Put_Am = 15.6189
Put_Eu = 14.6553
```

See also the result in Simulation 19.4.

Simulation 19.3 (Longstaff–Schwartz Method: A Step-by-Step Guide)
Here is a detailed explanation of Example 19.1.

```
K = 11.0;
T = 3;
r = 0.05;
L = 3; % number of time intervals
dt = T/L;
M = 8; % number of asset paths
```

Define the asset price paths given in Table 19.2.

```
S = [10.0  10.6  10.1  11.7;
     10.0  11.3  13.7  14.2;
     10.0  10.1  11.5  10.8;
     10.0  11.8  10.9  10.2;
     10.0  10.4  11.6  12.1;
     10.0  10.7  10.2  10.5;
     10.0  10.8   9.5  10.4;
     10.0   9.5  11.7  12.6]
```

Find cash flows at $t = 3$.

```
CF = zeros(M,L); % cash flow
CF(:,3) = max(K-S(:,4),0)
```

Compute the discount rate e^{-rdt}.

```
Discount_rate = exp(-r*dt);
```

Find the interpolating polynomial of degree 2 at time 2. The regression uses only the in-the-money asset paths at time 2 listed in Table 19.3.

```
j = 0; % index for the number of in-the-money at t=2 asset paths
for i=1:M % index for asset paths
         if S(i,3) < K % in-the-money at t=2
             j = j+1;
             index_ITM2(j) = i;
             X2(j) = S(i,3); % Asset price at t=2
             Y2(j) = CF(i,3)*exp(-r*dt); % discounted cash flow at t=2
         end
end
```

Print the number of the asset price paths in-the-money at time 2.

```
num_ITM2 = j
```

Print the indices corresponding to in-the-money asset paths at time 2.

```
index_ITM2(:)
```

Asset price X_2 and discounted cash flow Y_2 at time 2.

```
[X2(:),Y2(:)]
```

Find the interpolating polynomial $p_2(x) = \mathbb{E}[Y|X = x]$ of degree 2 based on the data (X_2, Y_2). Its coefficients are given in descending order.

```
p2 = polyfit(X2,Y2,2)
```

Find the optimal early exercise decision at time 2. See Table 19.4.

```
Exercise2 = max(K - X2(:),0);
Continuation2 = polyval(p2,X2(:));
[Exercise2,Continuation2]
```

Cash flow at time 2.

```
for j = 1:num_ITM2
        if Exercise2(j) >= Continuation2(j) % Early exercise
            CF(index_ITM2(j),2)=Exercise2(j); % cash flow on early exercise
            CF(index_ITM2(j),3)=0;
        else
            CF(index_ITM2(j),2)=0; % t=2 value on continuation
        end
end
sprintf('Cash flow at t=2')
CF
```

See Table 19.5.
Now we consider the regression at time 1. To find the regression polynomial we use only the asset paths in-the-money at time 1 listed in Table 19.6.

```
j = 0; % index for the number of in-the-money at t=2 asset paths
for i=1:M % index for asset paths
        if S(i,2) < K % in-the-money at t=1
            j = j+1;
            index_ITM1(j) = i;
            X1(j) = S(i,2); % Asset price at t=1
            Y1(j) = CF(i,2)*exp(-r*dt)+CF(i,3)*exp(-2*r*dt);
        end
end
```

In the above, in computing the discounted cash flow Y_1 at $t = 1$, only one term in the sum $CF(i,2)\mathrm{e}^{-rdt} + CF(i,3)\mathrm{e}^{-2rdt}$ is nonzero.
Print the number of the asset paths that are in-the-money at time 1.

```
num_ITM1 = j
```

Print the indices corresponding to in-the-money asset paths at time 1.

```
index_ITM1(:)
```

Asset price X_1 and discounted cash flow Y_1 at time 1.

```
[X1(:),Y1(:)]
```

Find the regression polynomial $p_1(x) = \mathbb{E}[Y|X = x]$ of degree 2 based on (X_1, Y_1). Its coefficients are given in descending order.

```
p1 = polyfit(X1,Y1,2)
```

Find the optimal early exercise decision at time 1. See Table 19.7.

```
Exercise1 = max(K - X1(:),0);
Continuation1 = polyval(p1,X1(:));
[Exercise1,Continuation1]

% Cash flow at time 1
for j = 1:num_ITM1
        if Exercise1(j) > Continuation1(j) % Early exercise
            CF(index_ITM1(j),1)=Exercise1(j); % cash flow on early exercise
            CF(index_ITM1(j),2)=0;
            CF(index_ITM1(j),3)=0;
```

```
        else
            CF(index_ITM1(j),1)=0;
        end
end
```

Find the option cash flow given in Table 19.9. Note that there is only one positive value along each asset price path as seen in Table 19.8.

```
CF
```

Compute the price of the American put option by taking the average of discounted future cash flows. We first compute the present value along each asset price path.

```
PV = exp(-r*dt)*CF(:,1) + exp(-r*2*dt)*CF(:,2) + exp(-r*3*dt)*CF(:,3);
mean(PV)
```

Simulation 19.4 (Longstaff–Schwartz Method)

We use the Longstaff–Schwartz method to price an American put option. The following is a modification of a code written by D.M. Lee.

```
S0 = 100;
K = 110;
T = 1;
r = 0.05;
sigma = 0.3;
L = 100; % number of time intervals
dt = T/L;
M = 10^4; % number of asset paths
Y = zeros(M,L);
S = S0*ones(M,L+1); % asset paths
    for k = 2:L+1
        S(:,k)=S(:,k-1).*exp((r-0.5*sigma^2)*dt+sigma*sqrt(dt)*randn(M,1));
    end
% Find payoff Y at expiry.
   for i=1:M
           Y(i,L) = max(K - S(i,L+1),0);
   end
% Find payoff Y at nodes for each time index.
for k = L+1:-1:3
    j = 0;
    for i=1:M
        if S(i,k-1) < K % in-the-money condition
            j = j+1;
            S1(j) = S(i,k-1); % in-the-money asset price
            Y1(j) = exp(-r*dt)*Y(i,k-1); % discounted cash flow
        end
    end
    p = polyfit(S1,Y1,2);
    for i = 1:M
        if K - S(i,k-1) > polyval(p,S(i,k-1)) % early exercise condition
            Y(i,k-2) = max(K - S(i,k-1),0);
        else
            Y(i,k-2) = exp(-r*dt)*Y(i,k-1);
        end
    end
```

```
end
American_put_LS = exp(-r*dt)*mean(Y(:,1))
```

The output is

```
American_put_LS = 15.6282
```

Compare it with the European put price given by the Black–Scholes–Merton formula.

```
d1 = (log(S0/K) + (r + 0.5*sigma^2)*T)/(sigma*sqrt(T));
d2 = d1 - sigma*sqrt(T);
N1 = normcdf(d1);
N2 = normcdf(d2);
European_put_BSM = S0*N1 - K*exp(-r*T)*N2 + K*exp(-r*T) - S0
```

The output is

```
European_put_BSM =  14.6553
```

Compare the American put price 15.6282 obtained by the Longstaff–Schwartz method in the above with the value obtained by the binomial tree method in Simulation 19.2.

Part VII
Portfolio Management

Chapter 20
The Capital Asset Pricing Model

How can we measure the performance of mutual funds and their investment risk? What is the use of a market index such as S&P 500? The portfolio theory can provide us with the answers. This chapter presents the Capital Asset Pricing Model (CAPM), which deals with an efficient portfolio management. For a historical introduction see [4].

To maximize the return of a portfolio consisting of more than one asset under a given risk level, we find the optimal weight $(w_1, \ldots, w_n)$ where w_i is the weight for the ith underlying asset. In this chapter the covariance matrix is denoted by C. All vectors are column vectors unless otherwise stated. The transpose of a matrix M, including row vectors and column vectors, is denoted by M^t. For example, if $\mathbf{v} = (v_1, \ldots, v_n)$ is a row vector then $\mathbf{v}^t$ is a column vector. Throughout the chapter except when explicitly mentioned we assume that all the assets under consideration are risky.

20.1 Return Rate and the Covariance Matrix

In this chapter we consider a single-period model for asset price movements. Let S_t denote the price of an asset S at time points $t = 0$ and $t = T$. If S_T is constant, then S is said to be risk-free. Otherwise, it is called a risky asset. See Fig. 20.1.

Definition 20.1 (Return) The *return* R_S of an asset S from time 0 to $T > 0$ is defined by

$$R_S = \frac{S_T - S_0}{S_0} \ .$$

G.H. Choe, *Stochastic Analysis for Finance with Simulations*, Universitext,
DOI 10.1007/978-3-319-25589-7_20

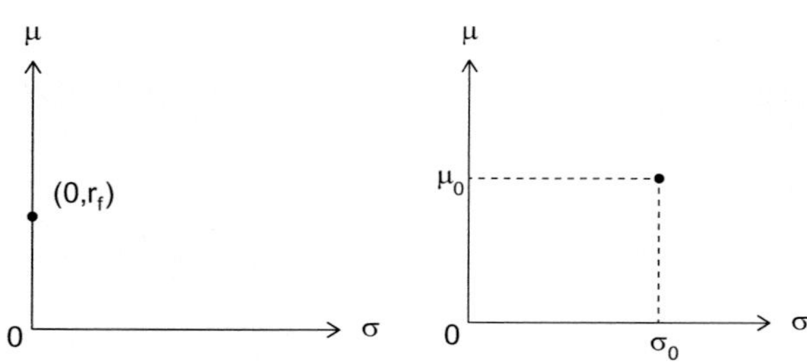

Fig. 20.1 A risk-free asset (*left*) and a risky asset (*right*)

Given several assets $S^1, \ldots, S^n$, we consider a portfolio

$$V = x_1 S^1 + \cdots + x_n S^n \ ,$$

where x_j is a real number. The case when $x_j < 0$ means the short-selling x_j units of the asset S^j.

Let R_j denote the return of asset S^j from time 0 to time $T > 0$. Then its *expected return* and *variance* are given by

$$\mu_j = \mathbb{E}[R_j] \quad \text{and} \quad \sigma_j^2 = \mathbb{E}[(R_j - \mu_j)^2] \ ,$$

respectively. By definition, the variance of the return of a *risk-free* asset is zero, while a *risky* asset has positive variance. Unless stated otherwise, all assets in this chapter are assumed to be risky.

Theorem 20.1 (Return of a Portfolio) *The return R_V of a portfolio*

$$V = x_1 S^1 + \cdots + x_n S^n$$

from time 0 to time $T > 0$ is given by

$$R_V = w_1 R_1 + \cdots + w_n R_n$$

where the weight w_j satisfies

$$w_j = \frac{x_j S_0^j}{V_0}$$

and

$$w_1 + \cdots + w_n = 1 \ .$$

Proof The values of V at time 0 and time T are $V_0 = x_1 S_0^1 + \cdots + x_n S_0^n$ and $V_T = x_1 S_T^1 + \cdots + x_n S_T^n$, respectively, and hence

$$\begin{aligned} R_V &= \frac{V_T - V_0}{V_0} \\ &= \frac{x_1(S_T^1 - S_0^1) + \cdots + x_1(S_T^n - S_0^n)}{V_0} \\ &= \frac{x_1 S_0^1}{V_0}\frac{S_T^1 - S_0^1}{S_0^1} + \cdots + \frac{x_n S_0^n}{V_0}\frac{S_T^n - S_0^n}{S_0^n}\,, \end{aligned}$$

which proves the statement. □

Definition 20.2 (Attainable Portfolio) The portfolio in the statement of Theorem 20.1 is called an attainable portfolio.

Definition 20.3 (Covariance) Define the covariance between the returns R_i and R_j by

$$C_{ij} = \mathrm{Cov}(R_i, R_j) = \mathbb{E}[(R_i - \mu_i)(R_j - \mu_j)]$$

and the covariance matrix by $C = \left[\, C_{ij} \,\right]$. Note that $C_{ii} = \mathrm{Var}(R_i) = \sigma_i^2$.

Remark 20.1 The covariance between a risky asset and a risk-free asset is 0. Thus if we consider a risk-free asset in defining the covariance matrix then a column and a row of a covariance matrix are zero vectors. In this chapter we assume that every asset is risky and the covariance matrix is invertible.

We define the expectation of a vector composed of random variables $\{X_i\}_{i=1}^n$ or the expectation of a matrix composed of random variables $\{Y_{ij}\}_{i,j=1}^n$ as the vector or the matrix given by the expectations of each component, i.e., $\mathbb{E}\left[\,(X_1, \ldots, X_n)\,\right] = (\mathbb{E}[X_1], \ldots, \mathbb{E}[X_n])$ and $\mathbb{E}\left[\,[Y_{ij}]_{ij}\,\right] = \left[\,\mathbb{E}[Y_{ij}]\,\right]_{ij}$. Using μ_i, $\boldsymbol{\mu}$ and μ respectively, we denote the expected return of the ith risky asset, the vector defined by n expected returns, and the expected return of a portfolio consisting of n risky assets. If we let $\mathbf{R} = (R_1, \ldots, R_n)^t$ and $\boldsymbol{\mu} = (\mu_1, \ldots, \mu_n)^t$, then

$$C = \mathbb{E}[(\mathbf{R} - \boldsymbol{\mu})(\mathbf{R} - \boldsymbol{\mu})^t]\,.$$

Thus C is a product of an $n \times 1$ matrix and a $1 \times n$ matrix, and hence C is an $n \times n$ matrix, and its (i, j)th component is $\mathbb{E}[(R_i - \mu_i)(R_j - \mu_j)]$.

Lemma 20.1 (Positive Semidefiniteness) *The covariance matrix C is symmetric and positive semidefinite. Hence its eigenvalues are nonnegative.*

Proof For $\mathbf{v} = (v_1, \ldots, v_n)^t$ we have

$$\mathbf{v}^t C\mathbf{v} = \mathbf{v}^t \mathbb{E}[(\mathbf{R} - \boldsymbol{\mu})(\mathbf{R} - \boldsymbol{\mu})^t]\mathbf{v} = \mathbb{E}[\mathbf{v}^t(\mathbf{R} - \boldsymbol{\mu})(\mathbf{R} - \boldsymbol{\mu})^t\mathbf{v}] = \mathbb{E}[u^2] \geq 0$$

where $u = \mathbf{v}^t(\mathbf{R} - \boldsymbol{\mu})$. □

Theorem 20.2 (Return of a Portfolio) *Consider the return of a portfolio V given by*

$$R_V = w_1 R_1 + \cdots + w_n R_n = \mathbf{w}^t \mathbf{R}$$

where $\mathbf{w} = (w_1, \ldots, w_n)^t$. *Its expected return* μ_V *and variance* σ_V^2 *are given by*

$$\mu_V = w_1\mu_1 + \cdots + w_n\mu_n = \mathbf{w}^t\boldsymbol{\mu} \ ,$$

$$\sigma_V^2 = \sum_{i=1}^{n}\sum_{j=1}^{n} w_i w_j c_{ij} = \mathbf{w}^t C\mathbf{w} \ .$$

Proof Use the fact that

$$\mu_V = \mathbb{E}\left[\sum_{i=1}^{n} w_i R_i\right] = \sum_{i=1}^{n} w_i \mu_i$$

and

$$\sigma_V^2 = \mathrm{Cov}\left(\sum_{i=1}^{n} w_i R_i, \sum_{i=1}^{n} w_i R_i\right) = \sum_{i=1}^{n}\sum_{j=1}^{n} w_i w_j C_{ij} \ .$$

□

20.2 Portfolios of Two Assets and More

Consider a portfolio consisting of two risky assets with their expected returns μ_1, μ_2 and their variances σ_1^2, σ_2^2, respectively.

Definition 20.4 (Correlation) The correlation coefficient ρ_{12} between the returns of two risky assets is given by

$$\rho_{12} = \frac{\mathrm{Cov}(R_1, R_2)}{\sigma_1\,\sigma_2} = \frac{C_{12}}{\sqrt{C_{11}C_{22}}} \ .$$

In this case, if we let w_1, w_2 be the weights then the expected return of the portfolio is equal to

$$\mu = w_1\mu_1 + w_2\mu_2$$

and its variance is given by

$$\begin{aligned}\sigma^2 &= C_{11}w_1^2 + C_{22}w_2^2 + 2C_{12}w_1w_2 \\ &= \sigma_1^2 w_1^2 + \sigma_2^2 w_2^2 + 2\rho_{12}\sigma_1\sigma_2 w_1 w_2\end{aligned}$$

where C_{ij} is the covariance between the returns. Consult also Definition 4.14.

If $\rho_{12} = 1$ then $\sigma^2 = (\sigma_1 w_1 + \sigma_2 w_2)^2$ and $\sigma = |\sigma_1 w_1 + \sigma_2 w_2|$. Hence the trajectory of the points

$$(\sigma, \mu) = (|\sigma_1 w_1 + \sigma_2 w_2|, w_1\mu_1 + w_2\mu_2)$$

on the (σ, μ)-plane is obtained by folding symmetrically along $\sigma = 0$ the left part of the straight line $w_1(\sigma_1, \mu_1) + w_2(\sigma_2, \mu_2)$. See the left graph in Fig. 20.2 where we choose $\boldsymbol{\mu} = \begin{bmatrix} 0.3 \\ 0.4 \end{bmatrix}$ and $C = \begin{bmatrix} 0.2 & 0.3 \\ 0.3 & 0.45 \end{bmatrix}$. The points on the thick line segment $0 \le w_1 \le 1, 0 \le w_2 \le 1$ represent portfolios without short selling.

If $\rho_{12} = -1$ then $\sigma^2 = (\sigma_1 w_1 - \sigma_2 w_2)^2$ and $\sigma = |\sigma_1 w_1 - \sigma_2 w_2|$. Hence the trajectory of the points

$$(\sigma, \mu) = (|\sigma_1 w_1 - \sigma_2 w_2|, w_1\mu_1 + w_2\mu_2)$$

on the (σ, μ)-plane is obtained by folding symmetrically along $\sigma = 0$ the left part of the straight line $w_1(\sigma_1, \mu_1) + w_2(-\sigma_2, \mu_2)$. See the right graph in Fig. 20.2 where we

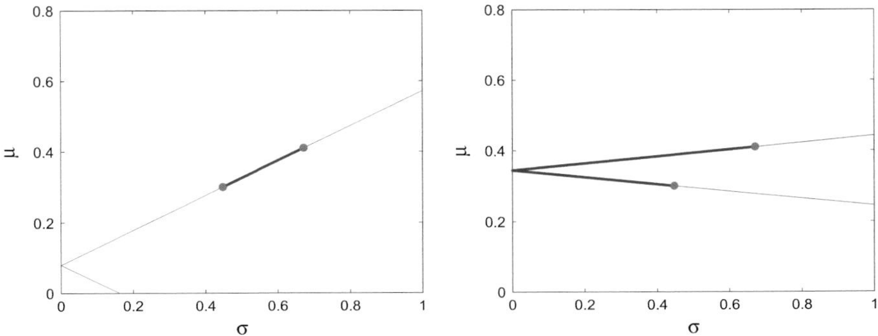

Fig. 20.2 A line representing portfolios for $\rho_{12} = 1$ (*left*) and a line for $\rho_{12} = -1$ (*right*)

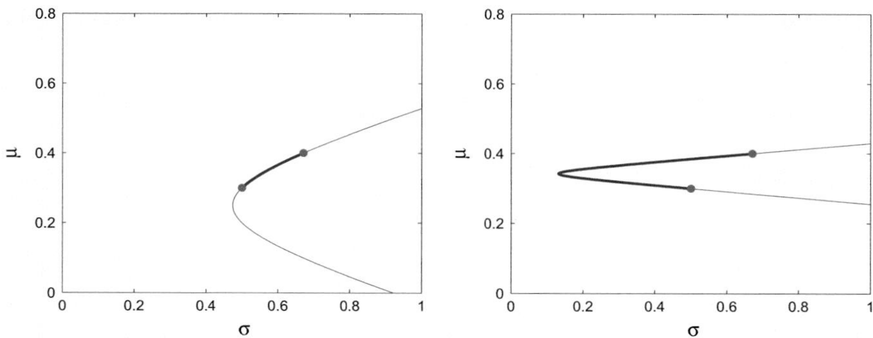

Fig. 20.3 A curve representing portfolios of two risky assets for $0 < \rho_{12} < 1$ (*left*) and a curve for $-1 < \rho_{12} < 0$ (*right*)

choose $\boldsymbol{\mu} = \begin{bmatrix} 0.3 \\ 0.4 \end{bmatrix}$ and $C = \begin{bmatrix} 0.2 & -0.3 \\ -0.3 & 0.45 \end{bmatrix}$. The points on the thick line segment $0 \le w_1 \le 1, 0 \le w_2 \le 1$ represent portfolios without short selling.

For $-1 < \rho_{12} < 1$ the portfolios are represented by the curves in Fig. 20.3, which do not intersect the μ-axis. The portfolios of two assets with $\rho_{12} \approx 1$ are represented by the curves in the left, and the portfolios with $\rho_{12} \approx -1$ in the right. We take $\boldsymbol{\mu} = \begin{bmatrix} 0.3 \\ 0.4 \end{bmatrix}$, $C = \begin{bmatrix} 0.25 & 0.3 \\ 0.3 & 0.45 \end{bmatrix}$ for $\rho_{12} \approx 0.8944$ and $C = \begin{bmatrix} 0.25 & -0.3 \\ -0.3 & 0.45 \end{bmatrix}$ for $\rho_{12} \approx -0.8944$. The points on the thick line segment $0 \le w_1 \le 1, 0 \le w_2 \le 1$ represent portfolios without short selling.

If the second asset is risk-free, i.e., the return R_2 is equal to r_f, then $\sigma_2 = 0$, and hence

$$(\sigma, \mu) = (|w_1|\sigma_1, w_1\mu_1 + w_2 r_f) = w_1(\sigma_1, \mu_1) + w_2(0, r_f)$$

for $w_1 \ge 0$ and

$$(\sigma, \mu) = (|w_1|\sigma_1, w_1\mu_1 + w_2 r_f) = w_1(-\sigma_1, \mu_1) + w_2(0, r_f)$$

for $w_1 < 0$. See Fig. 20.4 where $r_f = 0.3$ and $\sigma_1 = \sqrt{0.3} \approx 0.5477$. The points on the thick line segment for $0 \le w_1 \le 1, 0 \le w_2 \le 1$ represent portfolios without short selling.

Now we consider portfolios consisting of three or more assets, which are assumed to be risky. For simplicity of notation we only consider portfolios of three assets. Similar conclusions can be obtained for portfolios consisting of more assets. Let the points (σ_1, μ_1), (σ_2, μ_2) and (σ_3, μ_3) correspond to the given three assets,

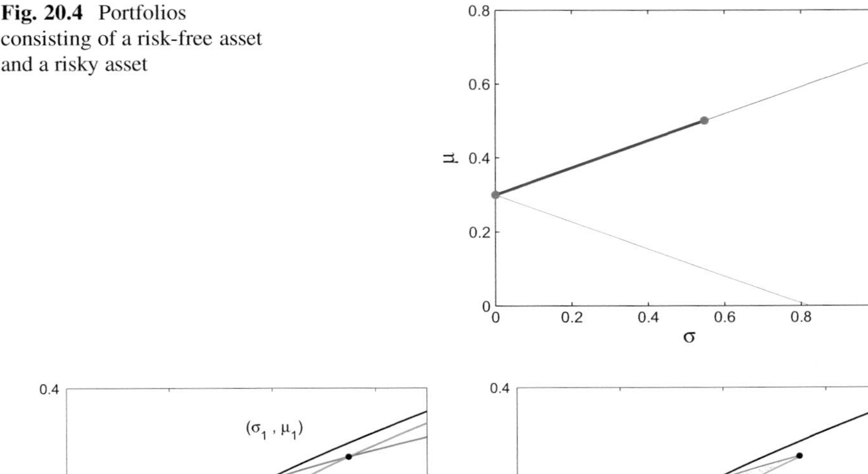

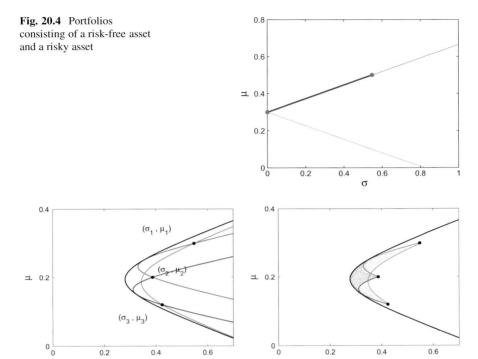

Fig. 20.4 Portfolios consisting of a risk-free asset and a risky asset

Fig. 20.5 The set of attainable portfolios consisting of three risky assets (*left*) and the image of the triangle $w_1 + w_2 + w_3 = 1, w_1, w_2, w_3 \geq 0$ (*right*)

respectively. We choose any two of them and apply the method in the previous section to plot curves representing the portfolios defined by the chosen two assets. Now we pick one portfolio from each of three curves, and form a new portfolio curve. Such a curve is a set of portfolios defined by a weighted sum of the original three portfolios represented by (σ_1, μ_1), (σ_2, μ_2) and (σ_3, μ_3). The set M of all such points is the right side of the curve in the left plot in Fig. 20.5, which is called the *Markowitz bullet*. In other words, if we let

$$W = \{(w_1, w_2, w_3) : w_1 + w_2 + w_3 = 1\}$$

and take $\mathbf{w}, \mathbf{w}' \in W$ and $c + c' = 1$, then $c\mathbf{w} + c'\mathbf{w}' \in W$. Thus, if $P, P' \in M$ then the new curve determined by P and P' is included in M.

On the right in Fig. 20.5 we have the image of the triangle $\{(w_1, w_2, w_3) : w_1 + w_2 + w_3 = 1, w_1, w_2, w_3 \geq 0\}$, which is the set of weights of three assets in a portfolio constructed without using short selling. We can see that the triangle is mapped into the first quadrant of the (σ, μ)-plane, being folded.

20.3 An Application of the Lagrange Multiplier Method

Now we solve the problem of finding the maximum or the minimum of a matrix function using the Lagrange multiplier method. From now on we write

$$\mathbf{1} = (1, \ldots, 1)^t$$

and define an $n \times 2$ matrix

$$M = [\mathbf{1}, \boldsymbol{\mu}] \tag{20.1}$$

where column vectors are given by $\mathbf{1}$ and $\boldsymbol{\mu} = (\mu_1, \ldots, \mu_n)^t$. In the following arguments to eliminate trivial cases we assume that $\boldsymbol{\mu} \neq \alpha\mathbf{1}$ for any arbitrary constant α. If we had $\boldsymbol{\mu} = \alpha\mathbf{1}$ for some α, then all the risky assets would have the same expected return with different risk levels.

Theorem 20.3 (Extremum of a Quadratic Form) *Let A be an invertible $n \times n$ symmetric matrix. A necessary condition that the function $\mathbf{x}^t A\mathbf{x}$ has its extremum at $\mathbf{x}$ under the constraint $\mathbf{1}^t\mathbf{x} - 1 = 0$ is*

$$\mathbf{x} = \frac{1}{\mathbf{1}^t A^{-1}\mathbf{1}} A^{-1}\mathbf{1} .$$

Proof To use Lagrange multiplier method, we put

$$f(\mathbf{x}) = \frac{1}{2}\mathbf{x}^t A\mathbf{x} ,$$

$$g(\mathbf{x}) = \mathbf{1}^t\mathbf{x} - 1 ,$$

and let

$$F(\mathbf{x}, \lambda) = f(\mathbf{x}) - \lambda g(\mathbf{x}) .$$

By Lemma A.2 we find the partial derivatives of F with respect to x_i, and set them equal to 0, and write the equations in a vector notation, and finally obtain

$$\mathbf{0} = \left(\frac{\partial F}{\partial x_1}, \ldots, \frac{\partial F}{\partial x_n}\right) = \nabla F(\mathbf{x}, \lambda) = \nabla f(\mathbf{x}) - \lambda \nabla g(\mathbf{x}) = \mathbf{x}^t A - \lambda\mathbf{1}^t .$$

By taking transposes we have $\mathbf{0} = A^t\mathbf{x} - \lambda\mathbf{1} = A\mathbf{x} - \lambda\mathbf{1}$ and $\mathbf{x} = \lambda A^{-1}\mathbf{1}$. Substituting the result in $\mathbf{1}^t\mathbf{x} = 1$, we obtain $\lambda = \dfrac{1}{\mathbf{1}^t A^{-1}\mathbf{1}}$. □

In the above result the symmetry condition can be relaxed. If we let $B = \frac{1}{2}(A + A^t)$, then B is symmetric and $\mathbf{x}^t A\mathbf{x} = \mathbf{x}^t B\mathbf{x}$. In applications, A is given by the covariance matrix, which is symmetric, and the quadratic form is the variance of the return of a portfolio of n assets. See the next section.

Theorem 20.4 (Extremum of a Quadratic Form) *Let A be an $n \times n$ symmetric matrix. Assume that A^{-1} and $(M^tA^{-1}M)^{-1}$ exist. If $\mathbf{x}^tA\mathbf{x}$ has its minimum at the point $\mathbf{x}$ under the constraints*

$$\begin{cases} \mathbf{1}^t\mathbf{x} - 1 = 0 , \\ \boldsymbol{\mu}^t\mathbf{x} - \mu_0 = 0 \end{cases}$$

for some constant μ_0, then x satisfies

$$\mathbf{x} = \lambda_1 A^{-1}\mathbf{1} + \lambda_2 A^{-1}\boldsymbol{\mu}$$

where λ_1, λ_2 satisfy the conditions

$$\begin{cases} \lambda_1 \mathbf{1}^t A^{-1}\mathbf{1} + \lambda_2 \mathbf{1}^t A^{-1}\boldsymbol{\mu} = 1 , \\ \lambda_1 \boldsymbol{\mu}^t A^{-1}\mathbf{1} + \lambda_2 \boldsymbol{\mu}^t A^{-1}\boldsymbol{\mu} = \mu_0 . \end{cases}$$

Proof To apply the Lagrange multiplier method we put

$$\begin{aligned} f(\mathbf{x}) &= \frac{1}{2}\mathbf{x}^tA\mathbf{x} , \\ g_1(\mathbf{x}) &= \mathbf{1}^t\mathbf{x} - 1 , \\ g_2(\mathbf{x}) &= \boldsymbol{\mu}^t\mathbf{x} - \mu_0 \end{aligned}$$

and let

$$F(\mathbf{x}, \lambda_1, \lambda_2) = f(\mathbf{x}) - \lambda_1 g_1(\mathbf{x}) - \lambda_2 g_2(\mathbf{x}) .$$

First, we compute the partial derivatives of F by Lemma A.2 and set them equal to 0, and finally obtain

$$\left(\frac{\partial F}{\partial x_1}, \ldots, \frac{\partial F}{\partial x_n} \right) = \nabla f - \lambda_1 \nabla g_1 - \lambda_2 \nabla g_2 = \mathbf{x}^tA - \lambda_1 \mathbf{1}^t - \lambda_2 \boldsymbol{\mu}^t = \mathbf{0} .$$

Hence $\mathbf{x}^tA = \lambda_1\mathbf{1}^t + \lambda_2\boldsymbol{\mu}^t$. By taking the transpose matrices, we obtain $A\mathbf{x} = \lambda_1\mathbf{1} + \lambda_2\boldsymbol{\mu}$ and

$$\mathbf{x} = \lambda_1 A^{-1}\mathbf{1} + \lambda_2 A^{-1}\boldsymbol{\mu} = A^{-1}M \begin{bmatrix} \lambda_1 \\ \lambda_2 \end{bmatrix} . \qquad (*)$$

We rewrite the constraints $\begin{cases} \mathbf{1}^t\mathbf{x} = 1 \\ \boldsymbol{\mu}^t\mathbf{x} = \mu_0 \end{cases}$ as $M^t\mathbf{x} = \begin{bmatrix} 1 \\ \mu_0 \end{bmatrix}$. Now using (∗), we convert the previous equation as

$$M^tA^{-1}M\begin{bmatrix} \lambda_1 \\ \lambda_2 \end{bmatrix} = \begin{bmatrix} 1 \\ \mu_0 \end{bmatrix} .$$

□

Remark 20.2 The conclusion of the above theorem can be rewritten as

$$\mathbf{x} = A^{-1}M\begin{bmatrix} \lambda_1 \\ \lambda_2 \end{bmatrix}$$

and

$$M^tA^{-1}M\begin{bmatrix} \lambda_1 \\ \lambda_2 \end{bmatrix} = \begin{bmatrix} 1 \\ \mu_0 \end{bmatrix} .$$

Hence

$$\mathbf{x} = A^{-1}M(M^tA^{-1}M)^{-1}\begin{bmatrix} 1 \\ \mu_0 \end{bmatrix} .$$

20.4 Minimum Variance Line

In this section we assume that the covariance matrix C is invertible, and that A is given by a covariance matrix in Theorem 20.4. Variance represents the level of risk in our model. Our objective in optimal portfolio management is to find the maximum of expected return when variance is given, or to find the minimum of variance when the expected return is given. See Fig. 20.6.

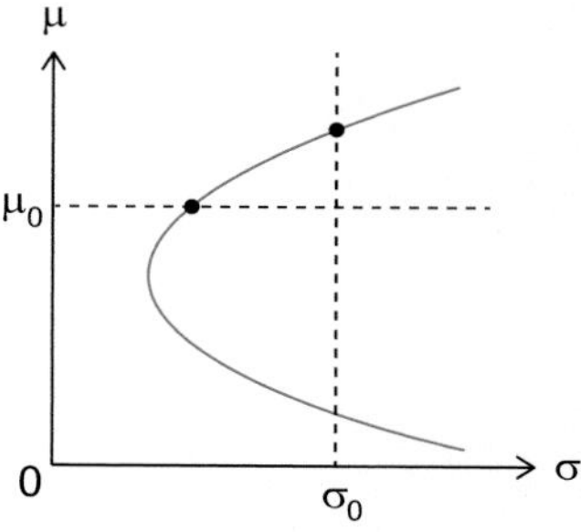

Fig. 20.6 Two optimization problems: Find the maximum of expected return when σ_0 is given, or find the minimum of variance when μ_0 is given.

Remark 20.3 The assumption that C is invertible is a plausible one. For, if C were not invertible, then there would exist $\mathbf{w} = (w_1, \ldots, w_n) \neq \mathbf{0}$ such that $C\mathbf{w} = \mathbf{0}$. Now consider a portfolio

$$V = w_1 S^1 + \cdots + w_n S^n .$$

Then its return R_V is given by

$$R_V = w_1 R_1 + \cdots + w_n R_n$$

and

$$\mathrm{Var}(R_V) = \sum_{ij} w_i w_j \mathrm{Cov}(R_i, R_j) = (C\mathbf{w}, \mathbf{w}) = 0 .$$

This implies that R_V is constant, which is nothing but $\sum_i w_i \mu_i$. Since

$$R_V = \sum_i w_i \frac{S_T^i - S_0^i}{S_0^i} = \sum_i \frac{w_i}{S_0^i} S_T^i - 1 ,$$

we have

$$\sum_i \frac{w_i}{S_0^i} S_T^i - 1 - \sum_i w_i \mu_i = 0 ,$$

which is a nontrivial linear combination of n random variables S_T^i representing risky assets and one risk-free asset. (The constant $-1-\sum_i w_i \mu_i$ is regarded as a coefficient $-1-\sum_i w_i \mu_i$ times the random variable representing a bond that pays 1 at time T.) Hence we would have linear dependence among random variables S_T^i, $1 \leq i \leq n$ and a bond given by a constant function 1. This is not a plausible assumption.

Definition 20.5 Since the covariance matrix C is assumed to be invertible, it has nonzero real eigenvalues. Furthermore, eigenvalues are positive, and we have $\mathbf{x}^t C \mathbf{x} > 0$ for every $\mathbf{x} \neq \mathbf{0}$, $\mathbf{x} \in \mathbb{R}^n$. The same property holds with C^{-1}. Define a bilinear form $(\cdot, \cdot)_C$ by

$$(\mathbf{x}, \mathbf{y})_C = \mathbf{y}^t C^{-1} \mathbf{x} ,$$

then $(\cdot, \cdot)_C$ is a positive definite inner product on $\mathbb{R}^n$.

Lemma 20.2 *Let $M = [\mathbf{1}, \boldsymbol{\mu}\,]$ be the $n \times 2$ matrix defined by (20.1). Then $M^t C^{-1} M$ is invertible, and* $\det(M^t C^{-1} M) > 0$.

Proof Note that

$$M^tC^{-1}M = \begin{bmatrix} \mathbf{1}^tC^{-1}\mathbf{1} & \mathbf{1}^tC^{-1}\boldsymbol{\mu} \\ \boldsymbol{\mu}^tC^{-1}\mathbf{1} & \boldsymbol{\mu}^tC^{-1}\boldsymbol{\mu} \end{bmatrix} .$$

By the Cauchy–Schwarz inequality we obtain

$$(\mathbf{x},\mathbf{y})_C{}^2 \le (\mathbf{x},\mathbf{x})_C\ (\mathbf{y},\mathbf{y})_C$$

i.e.,

$$|\mathbf{y}^tC^{-1}\mathbf{x}|^2 \le \mathbf{x}^tC^{-1}\mathbf{x}\ \mathbf{y}^tC^{-1}\mathbf{y}$$

and again we obtain

$$\mathbf{x}^tC^{-1}\mathbf{x}\ \mathbf{y}^tC^{-1}\mathbf{y} - \mathbf{y}^tC^{-1}\mathbf{x}\ \mathbf{x}^tC^{-1}\mathbf{y} \ge 0 .$$

(Check the necessary and sufficient condition given in Theorem B.1 for which equality holds.) Substituting $\mathbf{x} = \mathbf{1}$ and $\mathbf{y} = \boldsymbol{\mu}$, we obtain

$$\det(M^tC^{-1}M) = \mathbf{1}^tC^{-1}\mathbf{1}\ \boldsymbol{\mu}^tC^{-1}\boldsymbol{\mu} - \mathbf{1}^tC^{-1}\boldsymbol{\mu}\ \boldsymbol{\mu}^tC^{-1}\mathbf{1} \ge 0 .$$

Since $\boldsymbol{\mu}$ is not a scalar multiples of $\mathbf{1}$, the inequality is strict. □

If we define a 2×2 invertible matrix L by

$$L = (M^tC^{-1}M)^{-1} ,$$

then $\det L > 0$, $L^t = L$, and

$$L^{-1} = M^tC^{-1}M .$$

Note that

$$L = (\det L) \begin{bmatrix} \boldsymbol{\mu}^tC^{-1}\boldsymbol{\mu} & -\mathbf{1}^tC^{-1}\boldsymbol{\mu} \\ -\boldsymbol{\mu}^tC^{-1}\mathbf{1} & \mathbf{1}^tC^{-1}\mathbf{1} \end{bmatrix} .$$

When the expected return is equal to μ_0, the weight vector $\mathbf{w}$ that minimizes the variance is given by

$$\mathbf{w} = C^{-1}ML \begin{bmatrix} 1 \\ \mu_0 \end{bmatrix}$$

by Remark 20.2, and the corresponding variance satisfies

$$\begin{aligned}
\mathbf{w}^t C\mathbf{w} &= [1, \mu_0]\, L^t M^t C^{-1} C C^{-1} M L \begin{bmatrix} 1 \\ \mu_0 \end{bmatrix} \\
&= [1, \mu_0]\, L M^t C^{-1} M L \begin{bmatrix} 1 \\ \mu_0 \end{bmatrix} \\
&= [1, \mu_0]\, L \begin{bmatrix} 1 \\ \mu_0 \end{bmatrix} \\
&= L_{11} + 2\, L_{12}\, \mu_0 + L_{22}\, \mu_0^2
\end{aligned}$$

where $L = [L_{ij}]$. Since the covariance matrix is assumed to be invertible, Lemma 20.1 implies $\mathbf{w}^t C\mathbf{w} > 0$ for $\mathbf{w} \neq \mathbf{0}$, and hence $L_{22} > 0$.

Remark 20.4 Note that $L_{22} = (\det L)\mathbf{1}^t C^{-1}\mathbf{1}$.

A point on the (σ, μ)-plane will represent the standard deviation

$$\sigma = \sqrt{\mathbf{w}^t C\mathbf{w}}$$

and the expected return

$$\mu = \mathbf{w}^t \boldsymbol{\mu}$$

of the return of a portfolio consisting of risky assets with a weight vector $\mathbf{w}$ satisfying $\mathbf{1}^t\mathbf{w} = 1$. (In this book μ_i, $\boldsymbol{\mu}$, μ will denote respectively the expected return of the ith risky asset, ordered n-tuples of n expected returns, and the expected return of a portfolio consisting of n risky assets.

The set of the points on the (σ, μ)-plane corresponding to all the weight vectors is of the form

$$\{\ (\sigma, \mu) : \sigma \geq \sqrt{L_{11} + 2\, L_{12}\, \mu + L_{22}\, \mu^2}\ \}\ .$$

The boundary of the domain is given by

$$L_{11} + 2\, L_{12}\, \mu + L_{22}\, \mu^2 = \sigma^2\ .$$

Since $L_{22} > 0$, the above equation represents a hyperbola

$$\sigma_0^2 + L_{22}(\mu - \mu_0)^2 = \sigma^2\ .$$

The asymptotes are given by

$$\mu = \pm \frac{1}{\sqrt{L_{22}}}\sigma + \mu_0$$

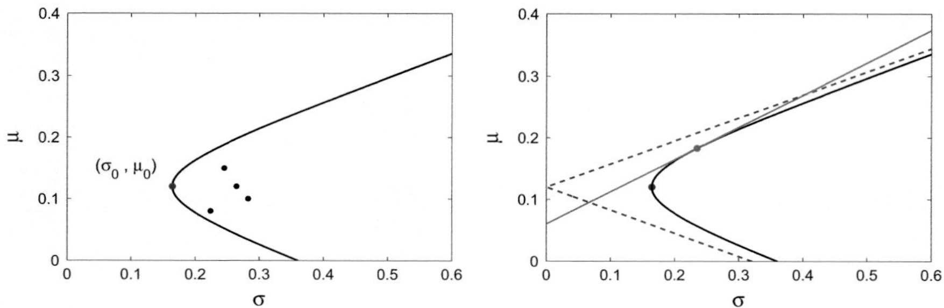

Fig. 20.7 The minimum variance line for risky assets and the point (σ_0, μ_0) representing the portfolio with the minimum variance among all portfolios (*left*), and asymptotes and a tangent line (*right*)

where

$$\mu_0 = -\frac{L_{12}}{L_{22}} = \frac{\mathbf{1}^t C^{-1}\boldsymbol{\mu}}{\mathbf{1}^t C^{-1}\mathbf{1}} ,$$

$$\sigma_0^2 = L_{11} - \frac{L_{12}^2}{L_{22}} = \frac{\det L}{L_{22}} = \frac{1}{\mathbf{1}^t C^{-1}\mathbf{1}} .$$

We used the result in Remark 20.4 in finding σ_0. Since the two asymptotes intersect at $(0, \mu_0)$ on the μ-axis, a tangent line with positive slope to the minimum variance line should have the μ-intercept $(0, r)$ for some $r < \mu_0$. See Fig. 20.7.

Definition 20.6 (Minimum Variance Line) The curve thus obtained is called the *minimum variance line*. A point on the minimum variance line represents the portfolio whose variance is the smallest among portfolios with the same expected return.

Among the points on the minimum variance line the point (σ_0, μ_0) represents the investment with the minimal risk. Its weight vector $\mathbf{w}_0$ is given by

$$\begin{aligned}
\mathbf{w}_0 &= C^{-1}ML \begin{bmatrix} 1 \\ \mu_0 \end{bmatrix} \\
&= C^{-1}ML \begin{bmatrix} 1 \\ \dfrac{\mathbf{1}^t C^{-1}\boldsymbol{\mu}}{\mathbf{1}^t C^{-1}\mathbf{1}} \end{bmatrix} \\
&= \frac{1}{\mathbf{1}^t C^{-1}\mathbf{1}} C^{-1}ML \begin{bmatrix} \mathbf{1}^t C^{-1}\mathbf{1} \\ \mathbf{1}^t C^{-1}\boldsymbol{\mu} \end{bmatrix} .
\end{aligned}$$

From the definition of L we have

$$\begin{bmatrix} \mathbf{1}^t C^{-1} \mathbf{1} \\ \mathbf{1}^t C^{-1} \boldsymbol{\mu} \end{bmatrix} = \begin{bmatrix} (L^{-1})_{11} \\ (L^{-1})_{21} \end{bmatrix} = \text{the first column of } L^{-1}$$

and

$$L \begin{bmatrix} \mathbf{1}^t C^{-1} \mathbf{1} \\ \mathbf{1}^t C^{-1} \boldsymbol{\mu} \end{bmatrix} = \begin{bmatrix} 1 \\ 0 \end{bmatrix} .$$

Hence

$$\mathbf{w}_0 = \frac{1}{\mathbf{1}^t C^{-1} \mathbf{1}} C^{-1} M \begin{bmatrix} 1 \\ 0 \end{bmatrix} = \frac{1}{\mathbf{1}^t C^{-1} \mathbf{1}} C^{-1} \mathbf{1} .$$

Note that we can apply Theorem 20.3 to obtain the same result.

20.5 The Efficient Frontier

Definition 20.7 (Dominating Portfolio) We are given two assets (or portfolios) with expected returns μ_1, μ_2, and standard deviations σ_1, σ_2, respectively. If $\mu_1 \geq \mu_2$ and $\sigma_1 \leq \sigma_2$ as in Fig. 20.8, we say that the first asset *dominates* the second asset.

Definition 20.8 (Efficient Frontier) A portfolio is said to be *efficient* when there is no other portfolio that dominates it. The set of all efficient portfolios is called the *efficient frontier*. In Fig. 20.9 the minimum variance line is given as the boundary of the set of attainable portfolios constructed by several assets, and the thick upper part is the efficient frontier.

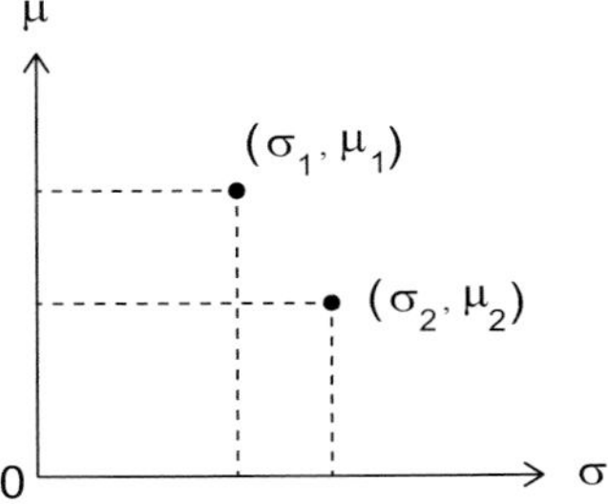

Fig. 20.8 The first asset dominates the second asset

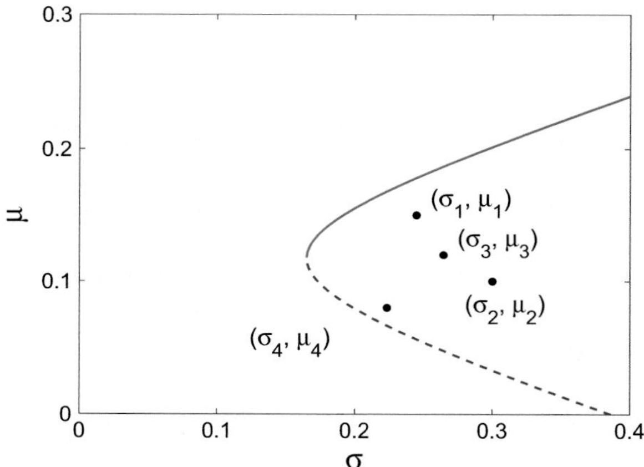

Fig. 20.9 The efficient frontier

Remark 20.5

(i) When the first asset dominates the second one, it is better to invest in the first one since the first asset has lower risk but higher expected return. Even when one asset dominates the other asset, we can combine two assets and construct a portfolio reducing risk as in Fig. 20.8.

(ii) For an efficient investment we should construct an efficient portfolio, and set the level of risk, determined by σ, according to the preference of the investor. To gain high return, an investor must accept high risk.

(iii) The efficient portfolio has the maximum expected return among the portfolios with the same standard deviation. Thus efficient frontier is a subset of the minimum variance line.

Lemma 20.3 *Let X_i, $1 \le i \le n$, be random variables. Then the covariance of two portfolios $\sum_i a_i X_i$ and $\sum_j b_j X_j$ is given by*

$$\mathrm{Cov}\left(\sum_{i=1}^{n} a_i X_i, \sum_{j=1}^{n} b_j X_j\right) = \sum_{i=1}^{n}\sum_{j=1}^{n} a_i b_j \mathrm{Cov}(X_i, X_j) \,.$$

In vector notation

$$\mathrm{Cov}(\mathbf{a}^t\mathbf{X}, \mathbf{b}^t\mathbf{X}) = \mathbf{a}^t C\mathbf{b} = \mathbf{b}^t C\mathbf{a}$$

where $\mathbf{a} = (a_1, \ldots, a_n)^t$, $\mathbf{b} = (b_1, \ldots, b_n)$, $\mathbf{X} = (X_1, \ldots, X_n)^t$ *and* $C = [\mathrm{Cov}(X_i, X_j)]_{ij}$.

Theorem 20.5 *All the efficient portfolios have the same covariance as the minimum variance portfolio represented by the point (σ_0, μ_0).*

Proof Let a point (σ, μ) represent a given efficient portfolio. Then

$$\sigma^2 - L_{22}\,\mu^2 - 2\,L_{12}\,\mu - L_{11} = 0$$

and the weight vector is given by

$$\mathbf{w} = C^{-1}ML\begin{bmatrix} 1 \\ \mu \end{bmatrix}.$$

In this case we have $\mathbf{w}^t\mathbf{R} = [1, \mu]LM^tC^{-1}\mathbf{R}$ and by Lemma 20.3 we obtain

$$\begin{aligned}
\mathrm{Cov}(\mathbf{w}_0^t\mathbf{R}, \mathbf{w}^t\mathbf{R}) &= \mathrm{Cov}(\mathbf{w}_0^t\mathbf{R}, [1, \mu]LM^tC^{-1}\mathbf{R}) \\
&= [1, \mu]LM^tC^{-1}C\mathbf{w}_0 \\
&= [1, \mu]LM^t\mathbf{w}_0 \\
&= [1, \mu]L\begin{bmatrix} \mathbf{1}^t \\ \boldsymbol{\mu}^t \end{bmatrix}\left(\frac{1}{\mathbf{1}^tC^{-1}\mathbf{1}}C^{-1}\mathbf{1}\right) \\
&= \frac{1}{\mathbf{1}^tC^{-1}\mathbf{1}}[1, \mu]L\begin{bmatrix} \mathbf{1}^tC^{-1}\mathbf{1} \\ \boldsymbol{\mu}^tC^{-1}\mathbf{1} \end{bmatrix} \\
&= \frac{1}{\mathbf{1}^tC^{-1}\mathbf{1}}[1, \mu]\begin{bmatrix} 1 \\ 0 \end{bmatrix} \\
&= \frac{1}{\mathbf{1}^tC^{-1}\mathbf{1}},
\end{aligned}$$

where we used

$$L^{-1} = M^tC^{-1}M = \begin{bmatrix} \mathbf{1}^tC^{-1}\mathbf{1} & \boldsymbol{\mu}^tC^{-1}\mathbf{1} \\ \boldsymbol{\mu}^tC^{-1}\mathbf{1} & \boldsymbol{\mu}^tC^{-1}\boldsymbol{\mu} \end{bmatrix}$$

in the fifth equality. □

20.6 The Market Portfolio

Now we consider a portfolio that contains a risk-free asset such as a bank deposit or bond. If we let $r_f > 0$ denote the risk-free interest rate, then such a risk-free asset is represented by a point $(0, r_f)$ on the (σ, μ)-plane. Consider an investment in which we invest into a risk-free asset by the ratio b, and into a portfolio π of risky assets by the ratio $1-b$. Assume that the return of the risky portfolio has its standard deviation and average given by (σ_π, μ_π) and that there exists a weight vector $\mathbf{w}$ such that the return is given by $\mathbf{w}^t\mathbf{R}$. Here we do not exclude the case that $b < 0$, which means

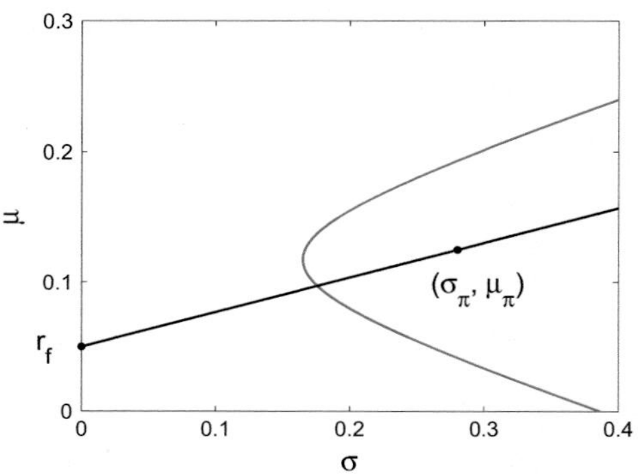

Fig. 20.10 A portfolio with a risk-free asset

that borrowed cash is invested more in the risky portfolio. The given risky portfolio is represented by the point (σ_π, μ_π) in Fig. 20.10.

Such a combination of investments has expected return

$$\mathbb{E}[b\, r_f + (1-b)\mathbf{w}^t\mathbf{R}] = b\, r_f + (1-b)\mu_\pi$$

and standard deviation

$$\sqrt{\mathrm{Var}(b\, r_f + (1-b)\mathbf{w}^t\mathbf{R})} = \sqrt{\mathrm{Var}((1-b)\mathbf{w}^t\mathbf{R})} = |1-b|\sigma_\pi \ ,$$

hence the corresponding point on the (σ, μ)-plane lies on the straight line

$$((1-b)\sigma_\pi, b\, r_f + (1-b)\mu_\pi) = b(0, r_f) + (1-b)(\sigma_\pi, \mu_\pi)$$

for $b \leq 1$. In other words, it is on the line passing through the points $(0, r_f)$ and (σ_π, μ_π).

In Fig. 20.10 the thick line segment corresponds to the range $0 \leq b \leq 1$, and the thin segment to $b < 0$. The line segment corresponding to $b > 1$ is obtained by reflecting the line in Fig. 20.10 using $\mu = r_f$ as the axis of symmetry. A portfolio represented by a point on the line segment has lower expected return than the case $b \leq 1$ but with the same risk, thus it is not regarded as a possible investment strategy.

When a risk-free asset is included, a necessary condition for an efficient investment is that the portfolio represented by (σ_π, μ_π) must be located on the efficient frontier for the portfolio without a risk-free asset. Furthermore, (σ_π, μ_π) should be the tangent point (σ_M, μ_M) of the tangent line passing through $(0, r_f)$ and tangent to the Markowitz bullet. Therefore, when a risk-free asset is included, the efficient frontier is the tangent line to the Markowitz bullet passing through $(0, r_f)$ as in Fig. 20.11.

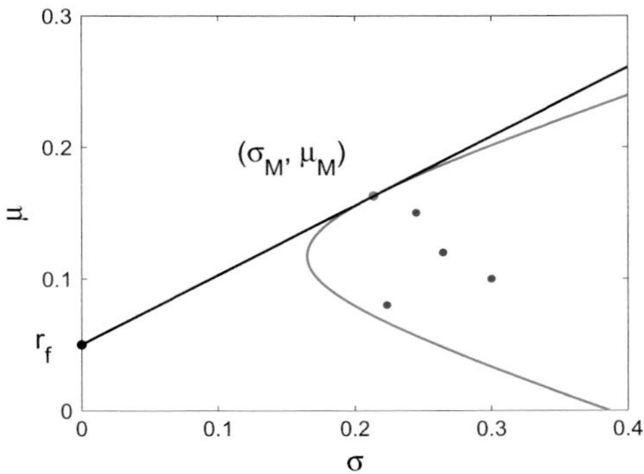

Fig. 20.11 The capital market line and the market portfolio

Definition 20.9 (Market Portfolio) The tangent line in the above is called the *capital market line*, and its equation is given by

$$\mu = r_f + \frac{\mu_M - r_f}{\sigma_M}\sigma \ .$$

The portfolio represented by (σ_M, μ_M) is called the *market portfolio* because it is employed by all the market participants for efficient investment. A stock price index is often used as a proxy for the market portfolio.

Definition 20.10 (Market Price of Risk) If a portfolio lying on the capital market line has risk σ, then

$$\frac{\mu_M - r_f}{\sigma_M}\sigma$$

is called the *risk premium* or *market price of risk*. It is a compensation for the investment risk. For more details consult [19].

Theorem 20.6 *The market portfolio satisfies the condition*

$$\gamma C\mathbf{w} = \boldsymbol{\mu} - r_f\mathbf{1}$$

where $\mathbf{w}$ *is the weight of the portfolio,* $\gamma = \dfrac{\mu_M - r_f}{\sigma_M^2}$*,* C *the covariance matrix,* $\boldsymbol{\mu} = (\mu_1, \ldots, \mu_n)$ *the expected return vector, and* $\mathbf{1} = (1, \ldots, 1)$.

Proof Among the lines passing through $(0, r_f)$ and a point in the interior of the efficient frontier, the capital market line has the maximal slope. See Figs. 20.10 and 20.11. If the market portfolio is represented by the point $(\sigma_M, \mu_M) = (\sqrt{\mathbf{w}^t C\mathbf{w}}, \boldsymbol{\mu}^t\mathbf{w})$ on the (σ, μ)-plane, then its slope is given by

$$\frac{\boldsymbol{\mu}^t\mathbf{w} - r_f}{\sqrt{\mathbf{w}^t C\mathbf{w}}} \ .$$

To find the maximum under the constraint $\mathbf{1}^t\mathbf{w} = 1$, we apply the Lagrange multiplier method. Let

$$F(\mathbf{w}, \lambda) = \frac{\boldsymbol{\mu}^t\mathbf{w} - r_f}{\sqrt{\mathbf{w}^t C\mathbf{w}}} - \lambda(\mathbf{1}^t\mathbf{w} - 1)$$

and take partial derivatives. Then

$$\frac{\partial F}{\partial w_i} = \frac{\mu_i \sqrt{\mathbf{w}^t C\mathbf{w}} - (\boldsymbol{\mu}^t\mathbf{w} - r_f)\dfrac{\sum_{j=1}^n C_{ij}w_j}{\sqrt{\mathbf{w}^t C\mathbf{w}}}}{\mathbf{w}^t C\mathbf{w}} - \lambda = 0$$

$$\frac{\partial F}{\partial \lambda} = \mathbf{1}^t\mathbf{w} - 1 = 0\ .$$

The first equation becomes

$$\mu_i\, \sigma_M^2 - (\mu_M - r_f)\sum_{j=1}^n C_{ij}w_j = \lambda\, \sigma_M^3\ ,$$

which is equal to

$$\boldsymbol{\mu} - \frac{\mu_M - r_f}{\sigma_M^2}\, C\mathbf{w} = \lambda\, \sigma_M \mathbf{1} \qquad (*)$$

in a vector form. Now we take the inner product with $\mathbf{w}$ and obtain

$$\mu_M - \frac{\mu_M - r_f}{\sigma_M^2}\sigma_M^2 = \lambda\, \sigma_M\ .$$

Hence we have

$$\lambda = \frac{r_f}{\sigma_M}\ ,$$

which yields

$$\mathbf{w} = \frac{\sigma_M^2}{\mu_M - r_f}\, C^{-1}(\boldsymbol{\mu} - r_f\mathbf{1})$$

in combination with $(*)$. Therefore we have $\gamma C\mathbf{w} = \boldsymbol{\mu} - r_f\mathbf{1}$. □

Remark 20.6

(i) As explained in Sect. 20.4, we need the condition $r_f < \mu_0$ for the existence of a tangent line with positive slope.

(ii) If $r_f \geq \mu_0$, the best strategy is to invest everything in the risk-free asset. If risky assets are independent and $\sigma_i = 1$, then we have

$$\mu_0 = \frac{\mathbf{1}^t C^{-1} \boldsymbol{\mu}}{\mathbf{1}^t C^{-1} \mathbf{1}} = \frac{\mu_1 + \cdots + \mu_n}{n}$$

since $C = \operatorname{diag}(1, \ldots, 1)$ is the identity matrix. Thus, if

$$r_f \geq \frac{\mu_1 + \cdots + \mu_n}{n}$$

then it is efficient to invest everything in the risk-free asset. In other words, when r_f is sufficiently high, bank deposit is better than stock.

(iii) For Figs. 20.7, 20.9, 20.11 we take

$$C = \begin{bmatrix} 0.06 & 0.05 & 0.02 & 0.01 \\ 0.05 & 0.08 & 0.05 & 0.01 \\ 0.02 & 0.05 & 0.07 & 0.01 \\ 0.01 & 0.01 & 0.01 & 0.05 \end{bmatrix} \quad \text{and} \quad \boldsymbol{\mu} = \begin{bmatrix} 0.15 \\ 0.1 \\ 0.12 \\ 0.08 \end{bmatrix} .$$

20.7 The Beta Coefficient

Definition 20.11 (Beta Coefficient) The *beta coefficient* β_V for a portfolio or an individual asset V is defined by

$$\beta_V = \frac{\operatorname{Cov}(R_V, R_M)}{\sigma_M^2} .$$

The beta coefficient measures the sensitivity of a given stock or portfolio price with respect to the stock price index. When we have a bull market, it is better to have a stock with large beta since the change in its price is larger than in other stocks in general. However, with a bear market a stock with smaller beta is better. Sectors with large betas are electronics, finance, medical equipments, and sectors with small betas are chemicals, medicine and food. Consumer-oriented sectors which are not affected during economic downturn have small betas.

Theorem 20.7 *For an arbitrary portfolio V with expected return μ_V, its beta coefficient β_V is given by*

$$\beta_V = \frac{\mu_V - r_f}{\mu_M - r_f} .$$

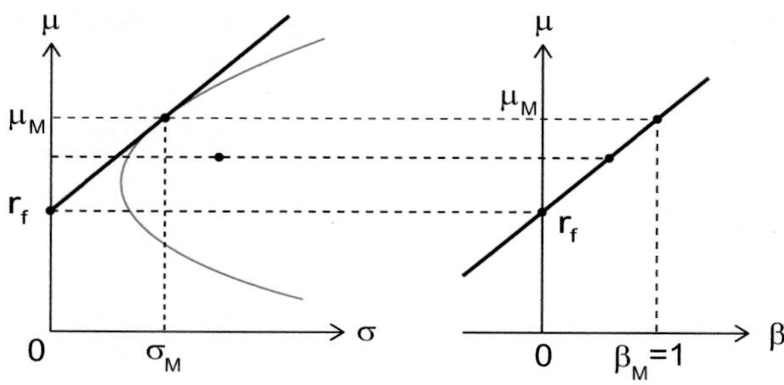

Fig. 20.12 The relation between the beta coefficient and expected return

Proof By Definition 20.11 we have

$$\beta_V = \frac{\mathbf{w}_V^t C \mathbf{w}_M}{\sigma_M^2} .$$

Theorem 20.6 implies that

$$\beta_V = \frac{1}{\mu_M - r_f} \mathbf{w}_V^t (\boldsymbol{\mu} - r_f \mathbf{1}) = \frac{1}{\mu_M - r_f} (\mathbf{w}_V^t \boldsymbol{\mu} - r_f \mathbf{w}_V^t \mathbf{1}) .$$

Now we use $\mathbf{w}_V^t \boldsymbol{\mu} = \mu_V$ and $\mathbf{w}_V^t \mathbf{1} = 1$. □

In Fig. 20.12 a portfolio V is represented by a point (β_V, μ_V) on the straight line

$$\mu = r_f + (\mu_M - r_f)\beta ,$$

called the *security market line* in the (β, μ)-plane. The term

$$(\mu_M - r_f)\beta$$

is regarded as the risk premium.

Remark 20.7 Let $\mathbf{w} = (w_1, \ldots, w_n)$ be the weight vector of the market portfolio. Then $w_i(\mu_i - r_f)$ is the portion of the contribution of the ith asset in the whole return of the portfolio.

20.8 Computer Experiments

Simulation 20.1 (Portfolios of Two Assets)
We take a covariance matrix

$$C = \begin{bmatrix} 0.2 & -0.3 \\ -0.3 & 0.45 \end{bmatrix}$$

which gives the correlation coefficient $\rho_{12} = -1$. See Fig. 20.2. For more experiments find a positive semidefinite symmetric matrix for C.

```
mu = [0.3; 0.41] ; % returns of assets
% Define a covariance matrix.
C= [ 0.2  -0.3 ;  -0.3  0.45 ] ;
eig(C) % Check the eigenvalues.
rho12 = C(1,2)/sqrt(C(1,1)*C(2,2)) % -1

N = 1000; % number of plotted points
mu_V = zeros(1,N);
sigma_V = zeros(1,N);
dt=1/N;
for i=1:N
w = [i*dt; 1-i*dt]; % weight
mu_V(i) = w' * mu;
sigma_V(i) = sqrt(w'*C*w);
end;
plot(sigma_V,mu_V, '.') ;
hold on

for i=N+1:2*N
w = [i*dt; 1-i*dt];
mu_V(i) = w' * mu;
sigma_V(i) = sqrt(w'*C*w);
end;
plot(sigma_V,mu_V, '.') ;
hold on

for i=N+1:2*N
w = [1-i*dt; i*dt];
mu_V(i) = w' * mu;
sigma_V(i) = sqrt(w'*C*w);
end;
plot(sigma_V,mu_V, '.') ;
hold on

plot(sqrt(C(1,1)),mu(1),'.') ;
plot(sqrt(C(2,2)),mu(2),'.') ;
```

Simulation 20.2 (Portfolios of Three Aassets)

We find the image of all attainable portfolios of three risky assets without short selling. See the right plot in Fig. 20.5.

```
mu_vec = [0.3; 0.2; 0.12] ; % returns of three risky assets
% covariance matrix
C = [ 0.3  0.02 0.01 ;
      0.02 0.15 0.03 ;
      0.01 0.03 0.18 ];

N = 1000; % number of plotted points
for i=1:N
w1 = rand;
w2 = 1-w1;
```

```
weight = [ w1; w2; 0];
mu12(i) = weight' * mu_vec;
sigma12(i) = sqrt(weight' * C * weight);
end ;

for i=1:N
w1 = rand;
w3 = 1-w1;
weight = [ w1; 0; w3];
mu13(i) = weight' * mu_vec;
sigma13(i) = sqrt(weight' * C * weight);
end ;

for i=1:N
w2 = rand;
w3 = 1-w2;
weight = [0; w2; w3];
mu23(i) = weight' * mu_vec;
sigma23(i) = sqrt(weight' * C * weight);
end ;

L = 5000;
for i=1:L
w2 = rand;
w1 = (1-w2)*rand;
weight = [ w1; w2; 1-w1-w2 ];
mu(i) = weight' * mu_vec ;
sigma(i) = sqrt(weight' * C * weight);
end ;

plot(sigma,mu,'.') ;
hold on
plot(sigma12,mu12,'.') ;
hold on
plot(sigma13,mu13,'.') ;
hold on
plot(sigma23,mu23,'.') ;
```

Simulation 20.3 (Capital Market Line)

Based on the data for four risky assets in Remark 20.6(iii), we plot the minimum variance portfolio, the efficient frontier, the minimum variance line, the market portfolio, and the capital market line.

```
mu = [ 0.15; 0.1; 0.12; 0.08]; % returns of assets
% the covariance matrix.
C = [ 0.06, 0.05, 0.02, 0.01 ;
      0.05, 0.08, 0.05, 0.01 ;
      0.02, 0.05, 0.07, 0.01 ;
      0.01, 0.01, 0.01, 0.05 ] ;

one_vec = ones(4,1); % a vector of 1's
inv_C = inv(C); % inverse of the covariance matrix
A1 = one_vec' * inv_C * one_vec; % 1'*C^(-1)*1
A2 = one_vec' * inv_C * mu; % 1'*C^(-1)*mu = mu'*C^(-1)*1
```

```
A3 = mu' * inv_C * mu; % mu'*C^(-1)*mu
L = inv([A1, A2; A2, A3]);

% minimum variance portfolio
mu_0 = - L(1,2)/L(2,2)
sigma_0 = sqrt(1/A1);

N = 200; % number of plotted points
max_sigma = 0.8
%max_mu = 1/sqrt(L(2,2)) * max_sigma + mu_0
max_mu = 0.5;
r = 0.06; % To have a tangent line we take r < mu_0.

% asymptotes
mu_asym = linspace(0,max_mu,N);
sigma_asym = zeros(1,N);
for i=1:N
    sigma_asym(i) = sqrt(L(2,2)*(mu_asym(i) - mu_0)^2);
end

% minimum variance line
mu_MVL = linspace(0,max_mu,N);
sigma_MVL = zeros(1,N);
for i=1:N
    sigma_MVL(i) = sqrt(L(2,2)*mu_MVL(i)^2 + 2*L(1,2)*mu_MVL(i) + L(1,1));
end

% Market Portfolio
mu_M = -(L(1,1)+L(1,2)*r)/(L(1,2)+L(2,2)*r);
sigma_M = sqrt(L(2,2)*mu_M^2 + 2*L(1,2)*mu_M + L(1,1));

% Capital Market Line
mu_CML = linspace(r,max_mu,N);
sigma_CML = zeros(1,N);
for i=1:N
    sigma_CML(i)=((mu_CML(i)-r)*sigma_M)/(mu_M-r);
end

plot(sigma_0, mu_0,'.'); % Minimum Variance Portfolio
hold on

plot(sigma_asym, mu_asym,'--'); % Asymptotes
hold on

plot(sigma_MVL, mu_MVL,'-'); % Minimum Variance Line
hold on

plot(sigma_M,mu_M,'.'); % Market Portfolio
hold on

plot(sigma_CML,mu_CML,'-'); % Capital Market Line
xlabel('\sigma') ;
ylabel('\mu');
```

Exercises

20.1 (Optimal Investment for Continuous Time) Consider the wealth consisting of a risk-free asset and a stock S_t. For more details consult [24]. Assume that S_t follows geometric Brownian motion $\mathrm{d}S_t = \alpha\, S_t\, \mathrm{d}t + \sigma\, S_t\, \mathrm{d}W_t$. Let X_t denote wealth, and let π_t be the amount of investment in S_t at t. We assume that X_t is self-financing. Since $X_t - \pi_t$ is the bank deposit, the increment in X_t is the sum of two increments: the first one from stock investment and the second from bank deposit. Since the number of shares of the stock is equal to π_t/S_t, we have $\mathrm{d}X_t = \frac{\pi_t}{S_t}\,\mathrm{d}S_t + (X_t - \pi_t)\, r\, \mathrm{d}t = \pi_t(\alpha \mathrm{d}t + \sigma \mathrm{d}W_t) + (X_t - \pi_t) r \mathrm{d}t$.

(i) Let $\widetilde{\pi}_t = \mathrm{e}^{-rt}\pi_t$ be a discounted portfolio. Show that the discounted wealth satisfies $\mathrm{d}\widetilde{X}_t = \widetilde{\pi}_t(\alpha - r)\,\mathrm{d}t + \widetilde{\pi}_t\,\sigma\,\mathrm{d}W_t$.

(ii) Define the market price of risk by $\theta = \frac{\alpha - r}{\sigma}$. To exclude trivial cases we assume that $\alpha > r$. Define $L_t = \mathrm{e}^{-\theta\, W_t - \frac{1}{2}\theta^2 t}$. Show that $\mathbb{E}[L_t^2] = \mathrm{e}^{\theta^2 t}$, and $\{L_t\widetilde{X}_t\}_{t\geq 0}$ is a martingale.

(iii) Under the condition that $\mathbb{E}[\widetilde{X}_T] = \mu$, the problem is to minimize the risk, i.e., to find the minimum of the variance of $\widetilde{X}$. Show that for such $\widetilde{X}_T$ we have $\widetilde{X}_T = C_1\mathbf{1} + C_2 L_T$ where

$$C_1 = \frac{X_0 - \mu\,\mathrm{e}^{\theta^2 T}}{1 - \mathrm{e}^{\theta^2 T}}\,, \qquad C_2 = \frac{\mu - X_0}{1 - \mathrm{e}^{\theta^2 T}}\,.$$

(iv) Show that $\widetilde{X}_t = C_1 + C_2 \mathrm{e}^{\theta^2 T}\mathrm{e}^{-\theta W_t - \frac{3}{2}\theta^2 t}$ in an optimal investment.

(v) Show that $\pi_t = -C_2 \frac{1}{\sigma}\mathrm{e}^{rt}\theta \mathrm{e}^{\theta^2 T}\mathrm{e}^{-\theta W_t - \frac{3}{2}\theta^2 t}$ in an optimal investment.

(vi) Show that $\sigma_{\widetilde{X}_T} = |\mu - X_0|/\sqrt{\mathrm{e}^{\theta^2 T} - 1}$ in the $(\sigma_{\widetilde{X}_T}, \mu)$-plane where μ represents $\mathbb{E}[\widetilde{X}_T]$.

Chapter 21
Dynamic Programming

We investigate continuous time models for combined problems of optimal portfolio selection and consumption. An optimal strategy maximizes a given utility, and the solution depends on what to choose as a utility function. Under the assumption that a part of wealth is consumed, we find an explicit solution for optimal consumption and investment under certain assumptions. Utility increases as more wealth is spent, however, less is reinvested and the capability for future consumption decreases, therefore the total utility over the whole period under consideration may decrease. See [9, 63, 64, 83] for more information.

In some of the previous chapters, we used the Lagrange multiplier method for optimization problems, however, in this chapter we employ the dynamic programming method to deal with stochastic control for optimization. The idea of dynamic programming is to optimize at every time step to achieve overall optimization.

21.1 The Hamilton–Jacobi–Bellman Equation

A classical method for finding minima or maxima is the calculus of variations, which is employed when the future event is deterministic. On the other hand, dynamic programming, which was introduced by the applied mathematician R. Bellman in the 1950s, results in an equation which is a modification of the Hamilton–Jacobi equation in classical mechanics.

Given an Itô stochastic process $\{X_t\}_{0\le t\le T}$, where X_t denotes wealth at time t, we consider an optimization problem defined by

$$\max_u \mathbb{E}\left[\int_0^T f(X_t, t, u_t)\,\mathrm{d}t + H(X_T, T)\right] .$$

G.H. Choe, *Stochastic Analysis for Finance with Simulations*, Universitext,
DOI 10.1007/978-3-319-25589-7_21

The Riemann integral represents the total utility from time $t = 0$ to time T. The case that $T = \infty$ is also possible. The term $H(X_T, T)$ represents the utility of the remaining wealth X_T by the time T. The expectation is the average over all possibilities.

In this problem the given constraint is given by the Itô process

$$\mathrm{d}X_t = \mu(X_t, t, u_t)\,\mathrm{d}t + \sigma(X_t, t, u_t)\,\mathrm{d}W_t \tag{21.1}$$

and the initial wealth $X_0 = x_0$. The stochastic process u_t is a control variable, in other words, optimal control is achieved by u_t based on the information X_t at time t. In general, the control u_t is of the form

$$u_t = g(X_t, t)$$

for some function $g(x, t)$. (When there is no danger of confusion we write u for simplicity. Even though $g(x, t)$ is a continuous function, it is not differentiable in general.)

In the above problem there might not exist a maximum over all possible choices for u, however the supremum always exists. In the problems presented in this chapter we assume that there exists a solution u for which the maximum exists. Thus we write maximum in place of supremum.

Instead of a rigorous derivation of the Hamilton–Jacobi–Bellman equation, which requires a lot of work, we present the idea in a heuristic manner. For arbitrary $0 \leq t \leq T$ define

$$J(X, t) = \max_u \mathbb{E}\left[\int_t^T f(X_s, s, u_s)\,\mathrm{d}s + H(X_T, T)\,\middle|\, X_t = X\right]$$

which is the sum of the total utility from time t to time T under the condition $X_t = X$ and the final remaining value at T given by $H(X_T, T)$. By definition,

$$J(X, T) = H(X, T)\ .$$

Now we derive a necessary condition under the assumption that the solution of a given optimization problem exists. Instead of presenting a rigorous proof we employ a heuristic method with the assumption that the given functions are sufficiently smooth as needed.

We assume that the solution of an optimization problem over the whole interval is optimal over each subinterval. Hence we have

$$J(X, t) = \max_u \mathbb{E}\left[\int_t^{t+\delta t} f(X_s, s, u_s)\,\mathrm{d}s + J(X + \delta X, t + \delta t)\,\middle|\, X_t = X\right] \tag{21.2}$$

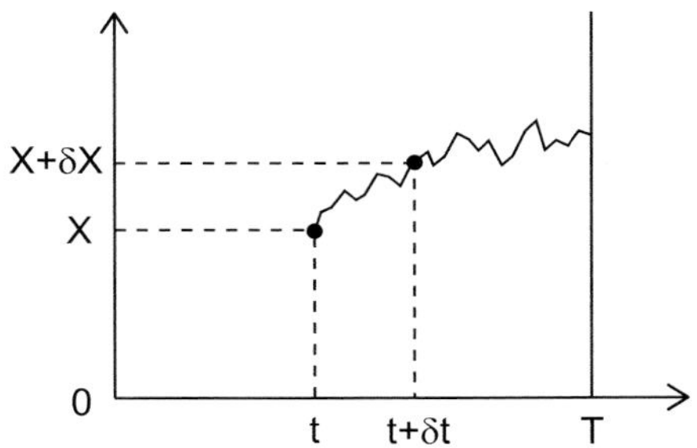

Fig. 21.1 Partition of time interval for total utility

where $X+\delta X$, under the condition that $X_t = X$ at time t, denotes $X_{t+\delta t}$ at time $t+\delta t$, which is given by (21.1). See Fig. 21.1. Here the integral is approximated by

$$\int_t^{t+\delta t} f(X_s, s, u_s)\,\mathrm{d}s \approx f(X_t, t, u_t)\,\delta t\ . \tag{21.3}$$

By the Itô formula we have

$$\begin{aligned}
&J(X+\delta X, t+\delta t)\\
&= J(X,t) + \frac{\partial J(X,t)}{\partial X}\delta X + \frac{\partial J(X,t)}{\partial t}\delta t + \frac{1}{2}\frac{\partial^2 J(X,t)}{\partial X^2}\sigma^2(X,t,u_t)\delta t + o(\delta t)\\
&= J(X,t) + \frac{\partial J(X,t)}{\partial X}\mu(X,t,u)\delta t + \frac{\partial J(X,t)}{\partial X}\sigma(X,t,u_t)\delta W\\
&\quad + \frac{\partial J(X,t)}{\partial t}\delta t + \frac{1}{2}\frac{\partial^2 J(X,t)}{\partial X^2}\sigma^2(X,t,u)\delta t + o(\delta t)\ .
\end{aligned} \tag{21.4}$$

Since the conditional expectation of the increment of Brownian motion is 0, we have

$$\mathbb{E}\left[\frac{\partial J}{\partial X}\sigma(X,t,u_t)\,\delta W\,\middle|\,X_t\right] = \mathbb{E}\left[\frac{\partial J}{\partial X}\sigma(X,t,u_t)\,\middle|\,X_t\right] \times \mathbb{E}\left[\delta W\middle|X_t\right] = 0\ . \tag{21.5}$$

Using (21.3), (21.4) and (21.5), we compute the right-hand side of (21.2), and obtain

$$\begin{aligned}
J(X,t) &= \max_u \mathbb{E}\Bigg[f(X,t,u_t)\,\delta t + J(X,t) + \frac{\partial J(X,t)}{\partial X}\mu(X,t,u_t)\,\delta t\\
&\qquad + \frac{\partial J(X,t)}{\partial X}\sigma(X,t,u_t)\,\delta W + \frac{\partial J(X,t)}{\partial t}\delta t\\
&\qquad + \frac{1}{2}\frac{\partial^2 J(X,t)}{\partial X^2}\sigma^2(X,t,u)\delta t + o(\delta t)\,\Bigg|\,X_t = X\Bigg]\\
&= \max_u \Big(f(X,t,u_t)\,\delta t + J(X,t) + \frac{\partial J(X,t)}{\partial X}\mu(X,t,u_t)\,\delta t\\
&\qquad + \frac{\partial J(X,t)}{\partial t}\delta t + \frac{1}{2}\frac{\partial^2 J(X,t)}{\partial X^2}\sigma^2(X,t,u_t)\delta t + o(\delta t)\Big)\ .
\end{aligned}$$

After subtracting $J(X, t)$ from both sides, then dividing by δt, and finally letting $\delta t \to 0$, we obtain the following conclusion.

Theorem 21.1 (Hamilton–Jacobi–Bellman Equation) *For $t \geq 0$ let X_t be a stochastic process given by*

$$\mathrm{d}X_t = \mu(X_t, t, u_t)\,\mathrm{d}t + \sigma(X_t, t, u_t)\,\mathrm{d}W_t \ .$$

To solve an optimization problem

$$\max_u \mathbb{E}\left[\int_0^T f(X_s, s, u_s)\,\mathrm{d}s + H(X_T, T)\middle| X_t = x_0\right]$$

define

$$J(X, t) = \max_u \mathbb{E}\left[\int_t^T f(X_s, s, u_s)\,\mathrm{d}s + H(X_T, T)\middle| X_t = X\right] .$$

Then J satisfies

$$\begin{aligned} &-\frac{\partial J}{\partial t}(X_t, t) \\ &= \max_u \left(f(X_t, t, u_t) + \frac{\partial J}{\partial x}(X_t, t)\,\mu(X_t, t, u_t) + \frac{1}{2}\frac{\partial^2 J}{\partial x^2}(X_t, t)\,\sigma^2(X_t, t, u_t) \right) \end{aligned}$$

with boundary condition

$$J(X_T, T) = H(X_T, T) \ .$$

Remark 21.1

(i) In general the necessary condition given in the above is a sufficient condition.
(ii) If an optimization problem is defined for the infinite horizon, i.e., $T = \infty$, then in general the condition $B = 0$ is given, and we find a solution satisfying $\lim_{t\to\infty} J(X, t) = 0$.

21.2 Portfolio Management for Optimal Consumption

Utility is a measure of satisfaction from the consumption of goods and services, and marginal utility is the amount of satisfaction from the consumption of an additional unit of goods and services. The marginal utility decreases as the quantity of consumed goods increases in general. Let $U(c)$ denote the utility function that measures the amount of utility where $c \geq 0$ is the consumption rate. That is, $U(c)\delta t$

denotes the total utility from the consumption of $c\,\delta t$ over a short time span δt. Since more consumption means more satisfaction in general, U is an increasing function.

By the law of diminishing marginal utility, U is a concave function. For example, for some constant $0 < \gamma < 1$ we have

$$U(c) = c^{\gamma}$$

or

$$U(c) = \log c \; .$$

The marginal utility is represented as U', and the law of diminishing marginal utility may be expressed as $U'' < 0$.

First, we assume that the price equation for a risky asset is given by a geometric Brownian motion $\mathrm{d}S_t = \alpha\, S_t\,\mathrm{d}t + \sigma\, S_t\,\mathrm{d}W_t$. (The symbol μ was already used to define X, and we use a new symbol α to define the asset price S.)

Let X_t denote the value of a portfolio at time t and let u_t be the ratio of investment into a risky asset. Suppose that $c_t\,\mathrm{d}t$ is consumed for a time period of length $\mathrm{d}t$. Assume that the portfolio is self-financing and there is no inflow of new investment from outside. To exclude the trivial case when one consumes as he/she wishes so that utility becomes very large, but the total wealth soon falls below zero, we impose a condition that $X_t \geq 0$. Hence the number of shares of a risky asset is equal to $u_t X_t/S_t$, and the amount invested in the risk-free asset is $(1 - u_t)X_t$. The increment of wealth after time length $\mathrm{d}t$ is equal to

$$\begin{aligned} \mathrm{d}X_t &= \frac{u_t X_t}{S_t}\mathrm{d}S_t + r(1-u_t)X_t\,\mathrm{d}t - c_t\,\mathrm{d}t \\ &= (\alpha u_t X_t + r(1-u_t)X_t - c_t)\,\mathrm{d}t + \sigma u_t X_t\,\mathrm{d}W_t \; . \end{aligned}$$

In the right-hand side of the first line $(u_t X_t/S_t)\mathrm{d}S_t$ represents the increment of the total stock price, and the second term $r\,(1 - u_t)X_t\,\mathrm{d}t$ is the interest on bank deposit. The term $-c_t\,\mathrm{d}t$ representing consumption has a negative coefficient since wealth decreases due to consumption.

Now it remains to find suitable u_t and $c_t \geq 0$ to maximize utility. For some constant $\rho > 0$ we consider $\mathrm{e}^{-\rho t}U(c_t)$ to discount the future value. In the following problem we take $B = 0$. Consult [63] or [66] for more general cases.

Theorem 21.2 (Power Utility) *For $0 < T < \infty$, we define the maximal utility over the period $0 \leq t \leq T$ by*

$$\max_{u,c} \mathbb{E}\left[\int_0^T \mathrm{e}^{-\rho t}U(c_t)\,\mathrm{d}t\right] .$$

If there exists $0 < \gamma < 1$ *such that* $U(c) = c^\gamma$, *then we have*

$$u_t = \frac{\alpha - r}{\sigma^2(1-\gamma)} = u^* \tag{21.6}$$

$$c_t = h(t)^{1/(\gamma-1)} X_t \tag{21.7}$$

where

$$\begin{aligned} A &= r\gamma + \frac{1}{2}\frac{\gamma}{1-\gamma}\frac{(\alpha-r)^2}{\sigma^2} - \rho \\ B &= 1-\gamma \\ h(t) &= \left\{\frac{B}{A}\left(e^{\frac{1}{1-\gamma}A(T-t)} - 1\right)\right\}^{1-\gamma} . \end{aligned}$$

The maximal utility over the period $0 \le t \le T$ *is given by*

$$\left\{\frac{B}{A}\left(e^{\frac{1}{1-\gamma}AT} - 1\right)\right\}^{1-\gamma} X_0^\gamma .$$

Proof For $0 \le t \le T$ define

$$J(x,t) = \max_{u,c} \mathbb{E}\left[\int_t^T e^{-\rho s} U(c_s)\, ds \,\middle|\, X_t = x\right] .$$

In the Hamilton–Jacobi–Bellman equation we take

$$f(t,c) = e^{-\rho t} c^\gamma$$

and need to maximize

$$\Gamma(u,c) \triangleq e^{-\rho t} c^\gamma + \{ux(\alpha - r) + (rx - c)\}\frac{\partial J}{\partial x} + \frac{1}{2}\sigma^2 u^2 x^2 \frac{\partial^2 J}{\partial x^2}$$

by Theorem 21.1. If a maximum is achieved at each fixed interior point (x,t), then we have

$$\begin{cases} 0 = \dfrac{\partial \Gamma}{\partial c} = e^{-\rho t}\gamma c^{\gamma-1} - \dfrac{\partial J}{\partial x} \\[2ex] 0 = \dfrac{\partial \Gamma}{\partial u} = x(\alpha - r)\dfrac{\partial J}{\partial x} + \sigma^2 u x^2 \dfrac{\partial^2 J}{\partial x^2} . \end{cases}$$

Hence c and u satisfy the conditions

$$\begin{cases} c^{\gamma-1} &= \frac{1}{\gamma} e^{\rho t} \dfrac{\partial J}{\partial x} \\ u &= \dfrac{\alpha - r}{\sigma^2} \left(-\dfrac{\partial J}{\partial x} \right) \left(x \dfrac{\partial^2 J}{\partial x^2} \right)^{-1} . \end{cases}$$

Now we look for J of the form

$$J(x,t) = e^{-\rho t} h(t) x^{\gamma} .$$

From the final condition $J(x,T) = 0$ we obtain the boundary condition

$$h(T) = 0 .$$

Note that

$$\begin{cases} \dfrac{\partial J}{\partial t} &= e^{-\rho t} h'(t) x^{\gamma} - \rho e^{-\rho t} h(t) x^{\gamma} \\ \dfrac{\partial J}{\partial x} &= \gamma e^{-\rho t} h(t) x^{\gamma-1} \\ \dfrac{\partial^2 J}{\partial x^2} &= \gamma(\gamma-1) e^{-\rho t} h(t) x^{\gamma-2} . \end{cases}$$

Hence

$$u = \frac{\alpha - r}{\sigma^2} \frac{(-1)\gamma e^{-\rho t} h(t) x^{\gamma-1}}{\gamma(\gamma-1) e^{-\rho t} h(t) x^{\gamma-1}} = \frac{1}{1-\gamma} \frac{\alpha - r}{\sigma^2} .$$

On the other hand,

$$c^{\gamma-1} = \frac{1}{\gamma} e^{\rho t} \gamma e^{-\rho t} h(t) x^{\gamma-1} = h(t) x^{\gamma-1} ,$$

and hence

$$c = h(t)^{1/(\gamma-1)} x .$$

To find h we substitute the preceding results in the Hamilton–Jacobi–Bellman equation and obtain

$$\begin{aligned}
0 &= \frac{\partial J}{\partial t} + \mathrm{e}^{-\rho t} c^{\gamma} + \{ux(\alpha - r) + (rx - c)\}\frac{\partial J}{\partial x} + \frac{1}{2}\sigma^2 u^2 x^2 \frac{\partial^2 J}{\partial x^2} \\
&= (\mathrm{e}^{-\rho t} h'(t) x^{\gamma} - \rho \mathrm{e}^{-\rho t} h(t) x^{\gamma}) + \mathrm{e}^{-\rho t} h(t)^{\frac{\gamma}{\gamma-1}} x^{\gamma} \\
&\quad + \left\{\frac{1}{1-\gamma}\frac{(\alpha - r)^2}{\sigma^2} x + (rx - h(t)^{\frac{1}{\gamma-1}} x)\right\} \gamma \mathrm{e}^{-\rho t} h(t) x^{\gamma-1} \\
&\quad + \frac{1}{2}\sigma^2 \frac{1}{(1-\gamma)^2}\frac{(\alpha - r)^2}{\sigma^4} x^2 \gamma(\gamma - 1)\mathrm{e}^{-\rho t} h(t) x^{\gamma-2} \\
&= (\mathrm{e}^{-\rho t} h'(t) x^{\gamma} - \rho \mathrm{e}^{-\rho t} h(t) x^{\gamma}) + \mathrm{e}^{-\rho t} h(t)^{\frac{\gamma}{\gamma-1}} x^{\gamma} \\
&\quad + \left\{\frac{1}{1-\gamma}\frac{(\alpha - r)^2}{\sigma^2} + r - h(t)^{\frac{1}{\gamma-1}}\right\} \gamma \mathrm{e}^{-\rho t} h(t) x^{\gamma} \\
&\quad + \frac{1}{2}\frac{-1}{(1-\gamma)}\frac{(\alpha - r)^2}{\sigma^2} \gamma \mathrm{e}^{-\rho t} h(t) x^{\gamma} .
\end{aligned}$$

After division by $\mathrm{e}^{-\rho t} x^{\gamma}$, the right-hand side becomes

$$h'(t) - \rho h(t) + h(t)^{\frac{\gamma}{\gamma-1}} + \left(r - h(t)^{\frac{1}{\gamma-1}}\right)\gamma h(t) + \frac{1}{2}\frac{1}{(1-\gamma)}\frac{(\alpha - r)^2}{\sigma^2}\gamma h(t) .$$

Hence we have

$$h'(t) + Ah(t) + Bh(t)^{-\frac{\gamma}{1-\gamma}} = 0$$

where the constants A and B are given by

$$\begin{aligned}
A &= r\gamma + \frac{1}{2}\frac{\gamma}{1-\gamma}\frac{(\alpha - r)^2}{\sigma^2} - \rho , \\
B &= 1 - \gamma .
\end{aligned}$$

To apply the result in Appendix C.2 we define a new variable

$$y(t) = h(t)^{\frac{1}{1-\gamma}} ,$$

then we obtain a linear differential equation

$$y'(t) + \frac{1}{1-\gamma} A y(t) + \frac{1}{1-\gamma} B = 0 .$$

Using an integrating factor

$$M(t) = \mathrm{e}^{\frac{1}{1-\gamma} A t},$$

we obtain

$$e^{-\frac{1}{1-\gamma}At}y(t) + \frac{B}{A} = C \,.$$

From the boundary condition $y(T) = 0$, we find

$$C = \frac{B}{A}e^{\frac{1}{1-\gamma}AT} \,,$$

and obtain

$$y(t) = -\frac{B}{A} + Ce^{-\frac{1}{1-\gamma}At} = \frac{B}{A}\left(e^{\frac{1}{1-\gamma}A(T-t)} - 1\right) \,.$$

□

Remark 21.2

(i) In Fig. 21.2, we plot ten sample paths of the optimal wealth process X_t, $X_0 = 100$, $0 \le t \le T = 30$, $\alpha = 0.1$, $r = 0.05$, $\sigma = 0.3$, $\gamma = 0.5$ and $\rho = 0.06$. In this case, $A = -0.0211$, $B = 0.5000$. We see that $X_t \to 0$ as $t \to T$, which agrees with common sense that there should be no remaining wealth after time T to maximize utility over the period $0 \le t \le T$.

(ii) An investor invests wealth into a risky asset with a constant rate u^*, which is proportional to $\alpha - r$ as can be seen in (21.6). For our choice of parameters, $u^* = 1.1111$.

(iii) As stock price can go up more on average, with larger values of α, we have to invest more in stocks as can be seen in (21.6).

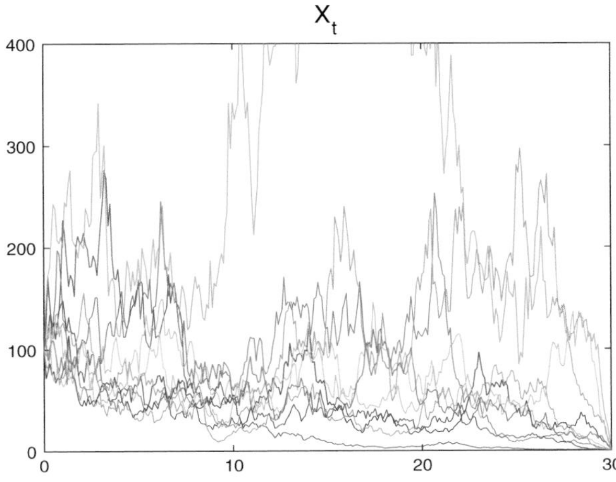

Fig. 21.2 Sample paths of optimal portfolio X_t with power utility for $0 \le t \le T$

(iv) For larger volatility σ, we should invest less in stocks as can be seen in (21.6).

(v) With higher interest rate r, less amount should be invested in stock while saving more in the bank account, as can be seen in (21.6).

(vi) An investor spends in proportion to the present wealth X_t. Even though the proportion coefficient $h(t)^{1/(\gamma-1)}$ in (21.7) increases to $+\infty$ as $t \to T$, the consumption rate

$$c_t = h(t)^{1/(\gamma-1)} X_t$$

can converge to a finite value since $X_t \to 0$ as $t \to T$. See Figs. 21.3, 21.4. (Consult Simulation 21.1.) The wealth process tends to vary widely due to the fact that there is no wage and the income of the investor is generated only by return on stock investment.

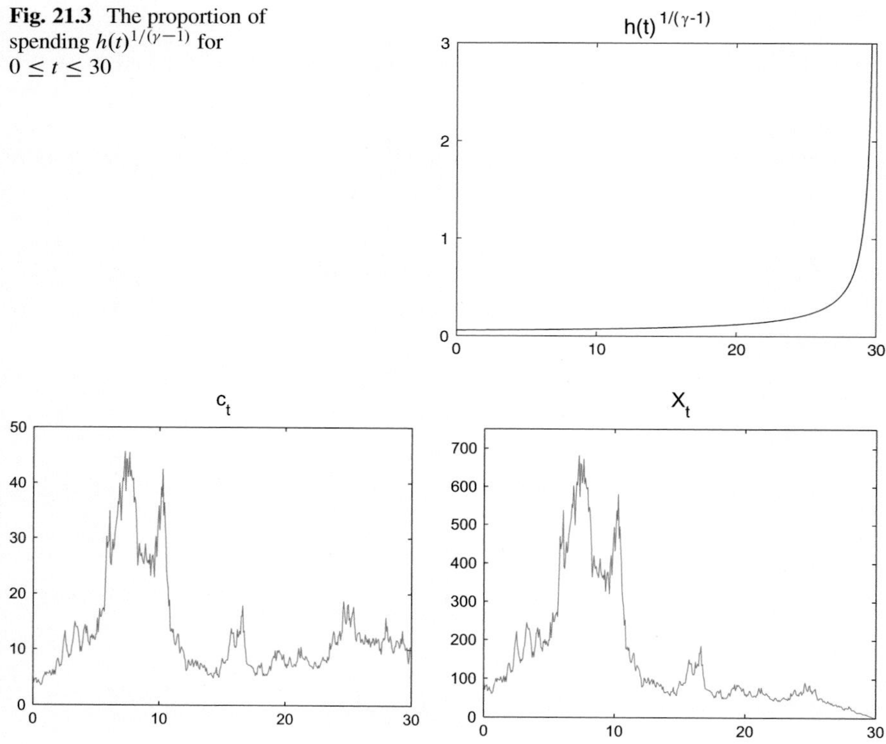

Fig. 21.3 The proportion of spending $h(t)^{1/(\gamma-1)}$ for $0 \le t \le 30$

Fig. 21.4 Optimal consumption and portfolio for $0 \le t \le 30$

Theorem 21.3 (Log Utility) *Over the period* $0 \le t < \infty$ *we define the maximum utility by*

$$\max_{u,c} \mathbb{E}\left[\int_0^\infty \mathrm{e}^{-\rho t} U(c_t)\,\mathrm{d}t\right] .$$

If $U(c) = \log c$, *then the optimal strategy is given by*

$$u_t = \frac{\alpha - r}{\sigma^2} , \quad c_t = \rho X_t .$$

The maximum utility over the period $0 \le t < \infty$ *is given by* $A \log X_0 + B$ *where*

$$A = \frac{1}{\rho} ,$$

$$B = \frac{1}{\rho}\left(\log \rho + \frac{r}{\rho} + \frac{(\alpha - r)^2}{2\rho\sigma^2} - 1\right) .$$

For the sample paths for the optimal portfolio X_t, see Fig. 21.5. Since it is impossible to do a simulation for the infinite time horizon problem, we take sufficiently large T for our simulation. In Simulation 21.2 we choose $T = 100$ which is justified by the fact that the estimate given by the Monte Carlo method with $T = 100$ is close to the closed-form solution.

Proof For $t \ge 0$ we define

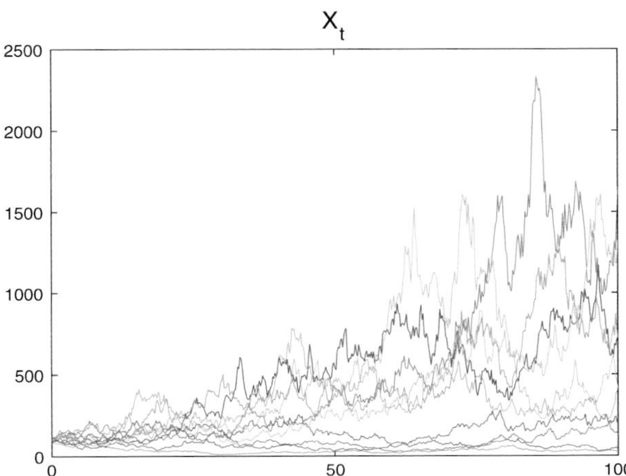

Fig. 21.5 Optimal portfolio X_t with log utility for $0 \le t < \infty$

$$J(x,t) = \max_{u,c} \mathbb{E}\left[\int_t^{\infty} \mathrm{e}^{-\rho s} U(c_s)\,\mathrm{d}s \middle| X_t = x\right] .$$

Take

$$f(t,c) = \mathrm{e}^{-\rho t} \log c$$

in the Hamilton–Jacobi–Bellman equation, and apply Theorem 21.1 to find the maximum of

$$\Gamma(u,c) \triangleq \mathrm{e}^{-\rho t} \log c + \{ux(\alpha - r) + rx - c\}\frac{\partial J}{\partial x} + \frac{1}{2}\sigma^2 u^2 x^2 \frac{\partial^2 J}{\partial x^2} .$$

Assuming the existence of a maximum for each fixed interior point (x, t), we obtain the following two equations:

$$\begin{cases} 0 = \dfrac{\partial \Gamma}{\partial c} = \mathrm{e}^{-\rho t}\dfrac{1}{c} - \dfrac{\partial J}{\partial x} \\[2ex] 0 = \dfrac{\partial \Gamma}{\partial u} = x(\alpha - r)\dfrac{\partial J}{\partial x} + \sigma^2 u x^2 \dfrac{\partial^2 J}{\partial x^2} . \end{cases}$$

Hence c and u satisfy the conditions

$$\begin{cases} c = \mathrm{e}^{-\rho t}\left(\dfrac{\partial J}{\partial x}\right)^{-1} \\[2ex] u = \dfrac{\alpha - r}{\sigma^2}\left(-\dfrac{\partial J}{\partial x}\right)\left(x\dfrac{\partial^2 J}{\partial x^2}\right)^{-1} . \end{cases}$$

Now we look for J of the form

$$J(x,t) = \mathrm{e}^{-\rho t}(A \log x + B) .$$

By taking partial derivatives, we obtain

$$\begin{cases} \dfrac{\partial J}{\partial t} &= -\rho \mathrm{e}^{-\rho t}(A \log x + B) \\[2ex] \dfrac{\partial J}{\partial x} &= \mathrm{e}^{-\rho t} A x^{-1} \\[2ex] \dfrac{\partial^2 J}{\partial x^2} &= -\mathrm{e}^{-\rho t} A x^{-2} . \end{cases}$$

Hence

$$\begin{cases} c = A^{-1}x \, , \\ u = \dfrac{\alpha - r}{\sigma^2} \left(-\mathrm{e}^{-\rho t} A x^{-1}\right) \left(-\mathrm{e}^{-\rho t} A x^{-1}\right)^{-1} = \dfrac{\alpha - r}{\sigma^2} \, . \end{cases}$$

To find constants A and B, we substitute the previous results in the Hamilton–Jacobi–Bellman equation and obtain

$$\begin{aligned} 0 &= \frac{\partial J}{\partial t} + \mathrm{e}^{-\rho t} \log c + \{ux(\alpha - r) + (rx - c)\} \frac{\partial J}{\partial x} + \frac{1}{2}\sigma^2 u^2 x^2 \frac{\partial^2 J}{\partial x^2} \\ &= -\rho\, \mathrm{e}^{-\rho t}(A \log x + B) + \mathrm{e}^{-\rho t}(-\log A + \log x) \\ &\quad + \left(\frac{(\alpha - r)^2}{\sigma^2} x + (rx - A^{-1}x) \right) \mathrm{e}^{-\rho t} A x^{-1} - \frac{1}{2}\sigma^2 \frac{(\alpha - r)^2}{\sigma^4} x^2 \mathrm{e}^{-\rho t} A x^{-2} \\ &= -\rho\, \mathrm{e}^{-\rho t}(A \log x + B) + \mathrm{e}^{-\rho t}(-\log A + \log x) \\ &\quad + \left(\frac{1}{2} \frac{(\alpha - r)^2}{\sigma^2} + (r - A^{-1}) \right) \mathrm{e}^{-\rho t} A \, . \end{aligned}$$

Dividing by $\mathrm{e}^{-\rho t}$ we obtain

$$0 = -\rho(A \log x + B) + (-\log A + \log x) + \left(\frac{1}{2} \frac{(\alpha - r)^2}{\sigma^2} + (r - A^{-1}) \right) A \, .$$

By comparing the coefficients of $\log x$, we obtain $A = \frac{1}{\rho}$. Thus

$$0 = -\rho B + \log \rho + \left(\frac{1}{2} \frac{(\alpha - r)^2}{\sigma^2} + (r - \rho) \right) \frac{1}{\rho} \, .$$

□

For a more general example of a utility function, see [48].

21.3 Computer Experiments

Simulation 21.1 (Power Utility)

We compare the Monte Carlo method with the closed-form solution for the utility maximization problem with $U(c) = \sqrt{c}$ in Theorem 21.2.

```
T = 30;
N = 500;
dt = T/N;
M = 10^5; % sample size for Monte Carlo
X0 = 100;  %initial portfolio
```

```
alpha = 0.1;
r = 0.05;
sigma = 0.3;
gamma = 0.5;
rho = 0.06;
u = (alpha-r)/(sigma^2)/(1-gamma) % optimal investment ratio
A = r*gamma + 0.5*(gamma/(1-gamma))*( (alpha-r)^2/sigma^2 ) - rho
B = 1 - gamma

h = zeros(1,N+1);
for i=1:N+1
    h(1,i) = ( (B/A)*(exp(A*(T-(i-1)*dt)/(1-gamma))-1) )^(1-gamma);
end
X = zeros(M,N+1);
c = zeros(M,N+1);
X(:,1) = X0;
c(:,1) = h(1,1)^(1/(gamma-1)) * X(:,1);

dW = sqrt(dt)*randn(M,N);
for j=1:N
    X(:,j+1)=X(:,j)+(alpha*u*X(:,j)+r*(1-u)*X(:,j)-c(:,j))*dt ...
                                + sigma*u*X(:,j).*dW(:,j);
    c(:,j+1)=h(j+1)^(1/(gamma-1)) * X(:,j+1);
end

integral = zeros(M,1);
for j=1:N
     integral(:) = integral(:) + exp(-rho*(j-1)*dt) * c(:,j).^gamma * dt;
end

plot(0:dt:T,h(1,1:N+1).^(1/(gamma-1)),'k');
plot(0:dt:T,c(1,1:N+1),'k');
plot(0:dt:T,X(1,1:N+1),'k');

max_utility_MC = mean(integral) % by Monte Carlo
max_utility_formula = (((B/A)*(exp(A*T/(1-gamma))-1))^(1-gamma))*(X0^gamma)
                                % by the closed-form formula
```

For the plots, see Figs. 21.3, 21.4. The Monte Carlo method and the closed-form formula give their results as follows:

```
max_utility_MC = 41.3021
max_utility_formula =  41.2441
```

Simulation 21.2 (Log Utility)

We compare the Monte Carlo method with the closed-form solution for the utility maximization problem with $U(c) = \log c$ in Theorem 21.3. Since it is impossible to do a simulation for an infinite horizon problem, we take $T = 100$. We take the same set of parameters as in Simulation 21.1. The estimate given by the Monte Carlo method agrees with the theoretical value with an acceptable error as one can see at the end of the paragraph. For the plot of the sample paths, see Fig. 21.5.

```
u = (alpha - r)/(sigma^2) % optimal investment ratio

A = 1/rho;
B = (log(rho) + r/rho + (alpha - r)^2/(2*rho*sigma^2) -1)/rho;

X = zeros(M,N+1);
c = zeros(M,N+1);
X0 = 100;            % initial portfolio
X(:,1) = X0;
c(:,1) = rho * X(1);

dW = sqrt(dt)*randn(M,N);
for j=1:N
    X(:,j+1)=X(:,j)+(alpha*u*X(:,j)+r*(1-u)*X(:,j)-c(:,j))*dt ...
                                +sigma*u*X(:,j).*dW(:,j);
    c(:,j+1)=rho*X(:,j+1);
end

integral = zeros(M,1);
for j=1:N
     integral(:) = integral(:) + exp(-rho*(j-1)*dt)*log(c(:,j))*dt;
end

max_utility_MC = mean(integral)       % by Monte Carlo
max_utility_formula =  A*log(X0) + B  % by the closed-form formula
```

The results of the Monte Carlo method and the closed-form formula are as follows:

```
max_utility_MC =  33.5208
max_utility_formula = 33.7443
```

Part VIII
Interest Rate Models

Chapter 22
Bond Pricing

In this chapter we derive a fundamental pricing equation for bond pricing under a general assumption on interest rate movements. For a comprehensive introduction to bonds, consult [100].

22.1 Periodic and Continuous Compounding

Definition 22.1 The *discount factor* $Z(t, T)$ is defined to be the rate of exchange between a given amount of money at time t and an amount of money at a later date T.

For example, if one hundred dollars at a future date T is worth ninety two dollars and fifty cents today ($t = 0$), then $Z(0, T) = 0.925$. *Compounded interest* means that there is interest on interest while *simple interest* means no interest on interest. A *compounding frequency* of interest accruals is the number of times per year when interest is paid. Throughout the book we consider only compounded interest rate. Suppose that we have the information that the interest rate per year is R without knowing the compounding method. Under the assumption of the periodic compounding let m be the compounding frequency, i.e., the interest is compounded m times per year. In T years the value of the investment of a principal A will be equal to $A[(1 + \frac{R}{m})^m]^T$. Since

$$A = Z(0, T) \times A\left[\left(1 + \frac{R}{m}\right)^m\right]^T ,$$

we have

$$Z(0, T) = \left(1 + \frac{R}{m}\right)^{-mT} .$$

G.H. Choe, *Stochastic Analysis for Finance with Simulations*, Universitext,
DOI 10.1007/978-3-319-25589-7_22

As $m \to \infty$ we compound more and more frequently, and obtain

$$\lim_{m\to\infty} A\left[\left(1+\frac{R}{m}\right)^m\right]^T = A\,\mathrm{e}^{RT}\,.$$

Since $(1+\frac{R}{m})^m$ is monotonically increasing and converges to e^R as $m \to \infty$ for $R > 0$, a depositor prefers continuous compounding to period compounding if the interest rate is positive. (See Exercise 22.1.) When \$A is invested at a constant continuously compounded rate R, then \$A grows to $\mathrm{e}^{RT} \times \$A$ at time T, or equivalently, \$A to be received at T is discounted to $\mathrm{e}^{-RT} \times \$A$ at $t = 0$, and hence the discount factor for continuous compounding is equal to $Z(0,T) = \mathrm{e}^{-RT}$.

Remark 22.1 (Periodic and Continuous Compounding) Let R_m denote the interest rate periodically compounding m times per year, e.g., $m = 2$ corresponds to semi-annual compounding,, and let R_c be the interest rate with continuous compounding. Suppose that we invest the initial amount A and the two compounding methods will produce the same amount at the end of n years. Since

$$A\,\mathrm{e}^{R_c n} = A\left[\left(1+\frac{R_m}{m}\right)^m\right]^n,$$

we have $\mathrm{e}^{R_c} = (1+\frac{R_m}{m})^m$. Hence $R_c = m\log(1+\frac{R_m}{m})$ and $R_m = m(\mathrm{e}^{R_c/m} - 1)$.

Remark 22.2 The discount factor $Z(t,T)$ is the time t price $P(t,T)$ of a zero coupon bond that pays \$1 with maturity T. In general, the curve $T \mapsto P(t,T)$ is assumed to be sufficiently smooth while $t \mapsto P(t,T)$ is regarded as stochastic.

22.2 Zero Coupon Interest Rates

Definition 22.2 Some bonds pay interest, called a *coupon*, periodically. They are called *coupon bonds*, or *fixed income securities*, because they pay a fixed amount of interest at regular intervals. If a bond does not pay any interest before maturity and provides a payoff only at time T, it is called a zero coupon bond. For a bond without coupon payments the interest and principal is realized at maturity. A *zero rate* (or *spot rate*), is the rate of interest earned on a zero coupon bond. See Definition 22.8(i) for the details.

Pricing a bond with coupon payments is not easy because a portion of interest is paid in the form of coupons before maturity. Thus we discount each coupon payment at the corresponding zero rate, which is obtained by the *bootstrap method* described as follows:

Consider the bonds traded in the bond market given in Table 22.1. For bonds with time to maturity equal to 1.5 years and 2 years, half of the annual coupon amount is paid every six months, i.e., compounding frequency is given by $m = 2$.

Table 22.1 Market data for bonds

Principal ($)	Time to maturity (years)	Annual coupon ($)	Bond price ($)	Compound. freq. (times per year)
100	0.25	0	98.31	4
100	0.50	0	96.54	2
100	1.00	0	93.14	1
100	1.50	6	98.10	2
100	2.00	10	104.62	2

Table 22.2 Continuously compounded zero rates determined by Table 22.1

Maturity (years)	Zero rate (%)
0.25	6.818
0.50	7.043
1.00	7.107
1.50	
2.00	

The first three bonds pay no coupons before maturity, and it is easy to calculate their zero rates.

(i) Let us consider the first bond which pays \$100, and we make the profit \$100 − \$98.31 = \$1.69, in three months. Note that $100 = A\left(1+\frac{R_m}{m}\right)^{mn}$ with $A = 98.31$, $m = 4$, $n = 0.25$. Hence $100 = 98.31\left(1+\frac{R_4}{4}\right)$ and $R_4 = \frac{4\times(100-98.31)}{98.31} = 6.876\%$. From the equation $A\,\mathrm{e}^{R_c n} = A\left(1+\frac{R_m}{m}\right)^{mn}$, we have $\mathrm{e}^{R_c\times 0.25} = 1+\frac{R_4}{4}$ and $R_c = 6.818\%$.

(ii) We do similarly for the next two bonds. Let us consider the second bond which pays \$100, and we make the profit \$100 − \$96.54 = \$3.46, in six months. Note that $100 = A\left(1+\frac{R_m}{m}\right)^{mn}$ with $A = 96.54$, $m = 2$, $n = 0.5$. Hence $100 = 96.54\left(1+\frac{R_2}{2}\right)$ and $R_2 = \frac{2\times(100-96.54)}{96.54} = 7.168\%$. From the equation $A\,\mathrm{e}^{R_c n} = A\left(1+\frac{R_m}{m}\right)^{mn}$, we have $\mathrm{e}^{R_c\times 0.5} = 1+\frac{R_2}{2}$ and $R_c = 7.043\%$.

(iii) Let us consider the third bond which pays \$100, and we make the profit \$100 − \$93.14 = \$6.86, in a year. Note that $100 = A\left(1+\frac{R_m}{m}\right)^{mn}$ with $A = 93.14$, $m = 1$, $n = 1$. Hence $100 = 93.14\left(1+\frac{R_1}{1}\right)$ and $R_1 = \frac{100-93.14}{93.14} = 7.365\%$. From the equation $A\,\mathrm{e}^{R_c n} = A\left(1+\frac{R_m}{m}\right)^{mn}$, we have $\mathrm{e}^{R_c\times 1} = 1+\frac{R_1}{1}$ and $R_c = 7.107\%$. Using the results from (i), (ii) and (iii) we can present a partially completed table as Table 22.2.

Now we consider the remaining cases.

(iv) Consider the fourth bond which has coupon payments of \$3 every six months. To calculate the cash price of a bond we discount each cash flow at the appropriate zero rate. The cash flow is given as follows:

→ 6 months: \$3
→→ 1 year: \$3
→→→ 1.5 years: \$103

Table 22.3 Continuously compounded zero rates determined by Table 22.1

Maturity (years)	Zero rate (%)
0.25	6.818
0.50	7.043
1.00	7.107
1.50	7.233
2.00	7.348

One arrow corresponds to half a year. Now we use the earlier results given in Table 22.2. Note that

$$98.10 = 3\,\mathrm{e}^{-0.07043\times 0.5} + 3\,\mathrm{e}^{-0.07107\times 1.0} + 103\,\mathrm{e}^{-R\times 1.5}$$

Hence $R = 0.07233$, and we can fill a blank space in the previous table.

(v) Consider the fifth bond which has coupon payments of \$5 every six months. To calculate the cash price of a bond we discount each cash flow at the appropriate zero rate. The cash flow is given as follows:

→ 6 months: \$5
→→ 1 year: \$5
→→→ 1.5 years: \$5
→→→→ 2.0 years: \$105

Now we use the previous results. Since

$$104.62 = 5\,\mathrm{e}^{-0.07043\times 0.5} + 5\,\mathrm{e}^{-0.07107\times 1.0} + 5\,\mathrm{e}^{-0.07233\times 1.5} + 105\,\mathrm{e}^{-R\times 2.0},$$

we have $R = 0.07348$, and finally we obtain Table 22.3.

22.3 Term Structure of Interest Rates

The *term structure of interest rates*, or *spot rate curve*, or *yield curve* at time t is the relation between the level of interest rates and their time to maturity. Usually, long-term interest rates are greater than short term interest rates, and hence the yield curve is upward moving.

There are several theories for the term structure of interest rates.

1. Expectations Theory: The interest rate on a long-term bonds will be equal to an average of the short-term interest rates that people expect to occur over the life of the long-term bond. This theory does not explain why the yield curve is upward moving in most cases since the short-term interest rate moves in either direction in general.

2. Segmented Market Theory: Short, medium and long maturity bonds are separate and segmented, and their interest rates are determined by their own supply and demand, and move independently of each other. This theory does not explain why the interest rates on bonds of different maturities tend to move together.
3. Liquidity Premium Theory: The interest rate on a long-term bond will equal an average of short-term interest rates expected to occur over the life of the long-term bonds plus a liquidity premium. Since investors tends to prefer short-term bonds to avoid the liquidity risk, for a long-term bond a positive liquidity premium must be added.

Example 22.1 (Management of Interest Rate) Suppose that the interest rates posted by a bank is given in Table 22.4. Depositors would choose to put their money in the bank only for one year because they want to have more financial flexibility. Otherwise longer term deposits would tie up the fund for a longer period of time. Now suppose that we want a mortgage. We would choose a five-year mortgage at 6% because it fixes the borrowing rate for the next five years and subjects us to less refinancing risk. Therefore, when the bank posts the rates shown in the above table, the majority of its customers would choose one-year deposits and five-year mortgages. This creates an asset/liability mismatch for the bank. There is no problem if interest rates fall. The bank will finance the five-year 6% loans with deposits that cost less than 3% in the future and net interest income will increase. However, if the rates rise, then the deposits that are financing these 6% loans will cost more than 3% in the future and net interest income will decline.

How can we ensure that the maturities of the assets on which interest is earned and the maturities of the liabilities on which interest is paid are matched? We can do it by increasing the five-year rate on both deposits and mortgages. For example, the bank can post new rates given in Table 22.5 where the five-year deposit rate is 4% and the five-year mortgage rate 7%. This would make five-year deposits relatively more attractive and one-year mortgages relatively more attractive. Some customers who chose one-year deposits when the rates were as in Table 22.4 will switch to five-year deposits, and some customers who chose five-year mortgages when the rates were as in Table 22.4 will choose one-year mortgages. This method may lead to the match of the maturities of assets and liabilities. If there is still an imbalance with depositors tending to choose a one-year maturity and borrowers a five-year maturity,

Table 22.4 Present interest rates

Maturity (years)	Deposit rate	Lending rate
1	3%	6%
5	3%	6%

Table 22.5 New interest rates

Maturity (years)	Deposit rate	Mortgage rate
1	3%	6%
5	4%	7%

five-year deposit and mortgage rates could be increased even further. Eventually the imbalance will disappear.

22.4 Forward Rates

Definition 22.3 The *forward rate* is the future zero rate implied by today's term structure of interest rates. See Definition 22.8(ii) for the details.

For example, consider two different strategies for bond investment. In the first strategy, we buy a bond with maturity equal to two years with interest rate $R_2 = 4.0\%$. In this case, an investment of \$1 will grow to $\$1 \times e^{R_2 \times 2}$ in two years. The second strategy is to buy a bond with maturity of one year which has interest rate $R_1 = 3.0\%$. Using the sum of the principal \$1 and the interest, which is equal to $\$1 \times e^{R_1 \times 1}$, we reinvest at time 1 year, and receive $\$1 \times e^{R_1 \times 1} e^{F_{12} \times (2-1)}$ where F_{12} is the forward rate that is predetermined today. To avoid arbitrage, two investment strategies should yield the same profit, i.e.,

$$e^{R_2 \times 2} = e^{R_1 \times 1} e^{F_{12} \times (2-1)} .$$

Hence $R_2 \times 2 = R_1 \times 1 + F_{12} \times (2-1)$. Therefore, we have $F_{12} = 5.0\%$. See Table 22.6.

Remark 22.3 Suppose that the zero rates for time periods T_1 and T_2 are R_1 and R_2 with both rates continuously compounded. The forward rate for the period between times T_1 and T_2 is

$$\frac{R_2 T_2 - R_1 T_1}{T_2 - T_1} .$$

A *forward rate agreement* (FRA) is an agreement that a certain rate, called the *forward rate*, will be exchanged for interest at the market rate to a notional amount during a certain future time period from T_1 to T_2. The agreement is so structured that neither party needs to make an upfront payment, i.e., the value of the contract is zero when the agreement is entered.

Example 22.2 Suppose that a company has agreed with a bank at time $t = 0$ that it will receive the forward rate 4% on the notional amount \$100 million for 3 months

Table 22.6 Calculation of forward rates

Year (n)	Zero rate for an n-year investment (% per annum)	Forward rate for the nth year (% per annum)
1	$R_1 = 3.0$	
2	$R_2 = 4.0$	$F_{12} = 5.0$

starting in $T_1 = 3$ years. That is, $T_2 = T_1 + \frac{3}{12} = 3.25$. Suppose that the interest rate turns out to be 4.5% (with quarterly compounding) at T_1. The company has to pay $\$125,000 = \$100\text{ million} \times 0.5\% \times \frac{3}{12}$ to the bank at time T_2, or equivalently, $\$123,630 = (1 + 0.045)^{-0.25} \times \$125,000$ at T_1.

22.5 Yield to Maturity

Consider a coupon bond that provides the holder with cash flows $c_1, \ldots, c_n$ on dates $0 < t_1 < \cdots < t_n = T$. (In most cases, c_i is a coupon payment of fixed amount c for $1 \le i \le n-1$, and c_n is a coupon payment plus the principal.) Let $Z(0, t_i)$ be the discount factors for each date t_i. Then the value B of the coupon bond is equal to

$$B = \sum_{i=1}^{n} c_i Z(0, t_i) . \tag{22.1}$$

Note that B is equal to the present value of the future cash flows.

Definition 22.4 The *yield to maturity*, or just YTM for short, is the single discount rate for all dates t_i that makes the present value of the future cash flows equal to the bond price. It is also called the *internal rate of return*.

(i) In the continuously compounding case, the yield y is defined by $Z(0, t_i) = \mathrm{e}^{-yt_i}$, i.e,

$$B = \sum_{i=1}^{n} c_i \mathrm{e}^{-yt_i} . \tag{22.2}$$

Since

$$\frac{\mathrm{d}}{\mathrm{d}y} B(y) = -\sum_{i=1}^{n} t_i c_i \mathrm{e}^{-yt_i} < 0 ,$$

there exists a unique value for y satisfying (22.2). Since

$$\frac{\mathrm{d}^2}{\mathrm{d}y^2} B(y) = \sum_{i=1}^{n} t_i^2 c_i \mathrm{e}^{-yt_i} > 0 ,$$

the graph of $B(y)$ is monotonically decreasing and convex.

(ii) In the periodic compounding case with the compounding frequency m, the yield is defined by $Z(0, t_i) = (1 + \frac{y}{m})^{-mt_i}$, i.e,

$$B = \sum_{i=1}^{n} c_i \left(1 + \frac{y}{m}\right)^{-mt_i} . \tag{22.3}$$

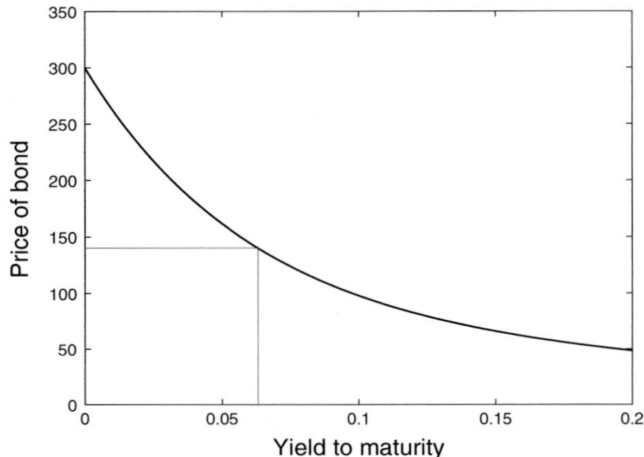

Fig. 22.1 Computation of the yield to maturity when the bond price is given

Note that the periodically compounded yield converges to the continuously compounded yield.

Example 22.3 Consider a coupon bond, with face value $P = \$100$, that pays $C = \$6$ every six months until maturity $T = 20$. Suppose that its present value is $B = \$140$. What is the yield to maturity? We have to solve

$$B = \sum_{k=1}^{2T} C\mathrm{e}^{-0.5ky} + P\mathrm{e}^{-Ty} = C\mathrm{e}^{-0.5y}\frac{1 - \mathrm{e}^{-Ty}}{1 - \mathrm{e}^{-0.5y}} + P\mathrm{e}^{-Ty}\,.$$

Using the Newton–Raphson method we find $y = 0.0632$. See Fig. 22.1 and Simulation 22.1. The present bond price $B = \$140$ is represented by the horizontal line, and the vertical line the yield to maturity.

22.6 Duration

The duration of a bond is a measure of how long on average the holder of the bond has to wait before receiving cash payment. A zero-coupon bond that matures in n years has a duration of n years. However, a coupon bond maturing in n years has a duration of less than n years, because the holder receives some of the payments prior to year n.

Definition 22.5 (Duration for Continuous Compounding) Consider a coupon bond that provides the holder with cash flows $c_1, \ldots, c_n$ on dates $0 < t_1 < \cdots < t_n = T$. Recall that the price B and the (continuously compounded) yield to maturity

y satisfy

$$B(y) = \sum_{i=1}^{n} c_i e^{-yt_i} .$$

The *duration* D of the bond is defined by

$$D = -\frac{B'(y)}{B(y)} = \frac{1}{B}\sum_{i=1}^{n} t_i c_i e^{-yt_i} .$$

Remark 22.4

(i) Duration was introduced by F. Macaulay in 1938.
(ii) We may rewrite the duration as

$$D = \sum_{i=1}^{n} \frac{c_i e^{-yt_i}}{B} t_i ,$$

which is the weighted average of the coupon payment dates t_i. Note that the sum of the weights equals 1 and that the ith weight is the proportion of the discounted cash flow at t_i.
(iii) When the bond yield increases, the bond price decreases.

Theorem 22.1 *Let δy be a small change in the continuously compounded yield to maturity, and let δB the corresponding change in the bond price. Then*

$$\frac{\delta B}{B} \approx -D\,\delta y .$$

Proof Note that $\delta B \approx B'(y)\,\delta y = \sum_{i=1}^{n}(-t_i)c_i e^{-yt_i}\,\delta y = -DB\,\delta y$. □

Example 22.4 Consider a three-year 8% coupon bond with a face value of \$100. Coupon payments of $\$100 \times \frac{0.08}{2} = \$4$ are made every six months. (There are six coupon payments before and on the maturity date $T = 3$.) Suppose that the yield to maturity is 7.5% per annum with continuous compounding, i.e., $y = 0.075$. We will show how to calculate the duration of the bond. First, the cash flow of the coupon bond is given in Table 22.7.

Then we fill the next columns for the present values of the cash flows, the weights, and the payment dates times the weights. The total of the last column in Table 22.8 equals the duration.

Definition 22.6 (Duration for Periodic Compounding) Let y be the yield to maturity for the periodic compounding case with a compounding frequency of m times per year. Then the *duration* D is defined by

$$D = \sum_{i} t_i \frac{c_i \left(1 + \frac{y}{m}\right)^{-mt_i}}{B}$$

Table 22.7 Calculation of duration: a bond with coupons

Time (years)	Cash flow ($)	Present value ($)	Weight	Time × Weight (years)
$t_1 = 0.5$	$c_1 = 4$			
$t_2 = 1.0$	$c_2 = 4$			
$t_3 = 1.5$	$c_3 = 4$			
$t_4 = 2.0$	$c_4 = 4$			
$t_5 = 2.5$	$c_5 = 4$			
$t_6 = 3.0$	$c_6 = 104$			
Total	124			

Table 22.8 Calculation of duration

Time (years)	Cash flow ($)	Present value ($)	Weight	Time × Weight (years)
0.5	4	3.853	0.038	0.019
1.0	4	3.711	0.037	0.037
1.5	4	3.574	0.035	0.053
2.0	4	3.443	0.034	0.068
2.5	4	3.316	0.033	0.082
3.0	104	83.046	0.823	2.468
Total	124	100.943	1.000	$D = 2.727$

and the *modified duration* D^* is defined by

$$D^* = \frac{D}{1 + \frac{y}{m}} \,.$$

Note that $D^* \to D$ as $m \to \infty$. Since

$$B'(y) = -\sum_i t_i c_i \left(1 + \frac{y}{m}\right)^{-mt_i - 1} ,$$

we have

$$\delta B \approx -\frac{BD}{1 + \frac{y}{m}} \delta y = -BD^* \delta y \,.$$

In other words,

$$\frac{B'(y)}{B(y)} \approx -D^* \delta y \,.$$

Definition 22.7 (Convexity)

(i) For a continuously compounded case, the *convexity* C of a bond with price B and yield to maturity y is defined by

$$C = \frac{B''(y)}{B(y)} = \frac{1}{B} \sum_i t_i^2 c_i \mathrm{e}^{-yt_i} .$$

(ii) For a periodically compounded case, the *convexity* is defined by

$$C^* = \frac{B''(y)}{B(y)} = \frac{1}{B} \sum_i t_i \left(t_i + \frac{1}{m}\right) c_i \left(1 + \frac{y}{m}\right)^{-mt_i} \frac{1}{(1 + \frac{y}{m})^2} .$$

Theorem 22.2

(i) For a continuously compounded case, we have

$$\frac{\delta B}{B} \approx -D\,\delta y + \frac{1}{2} C\,\delta y^2 .$$

(ii) For a periodically compounded case, we have

$$\frac{\delta B}{B} \approx -D^*\,\delta y + \frac{1}{2} C^*\,\delta y^2 .$$

Proof By the Taylor series expansion, $\delta B \approx B'(y)\,\delta y + \frac{1}{2} B''(y)\,\delta y^2$. Dividing both sides by B, we obtain the desired result. □

22.7 Definitions of Various Interest Rates

Let the present time be t, and let $P(t,T)$ and $P(t,S)$ denote the zero coupon bond prices with maturities T and S, $T < S$, respectively. Consider a bank that wants to lend money, say \$1, to a borrower for a future period $[T,S]$. Let the interest rate for lending be $f(t,T,S)$, which is agreed on at time t. The bank shorts $\frac{P(t,T)}{P(t,S)}$ units of the S-maturity zero coupon bond, and long one unit of the T-maturity zero coupon bond. (The price of $\frac{P(t,T)}{P(t,S)}$ units of the S-maturity zero coupon bond at time t equals $\frac{P(t,T)}{P(t,S)} \times P(t,S) = P(t,T)$, which is exactly equal to the cost of buying one unit of the T-maturity zero coupon bond. Hence there is no cash transfer.) At time T, the proceeds from the T-maturity bond are lent out for the period $[T,S]$ with the interest rate $f(t,T,S)$. At time S, the loan is paid back from the borrower and the short position of S-bond is recovered, i.e., the S-bond is returned by the bank, resulting in a net cash flow of

$$C = \$1 \times e^{f(t,T,S)(S-T)} - \$1 \times \frac{P(t,T)}{P(t,S)}$$

at time S. Since the bank did not need any money at time t, we have $C = 0$ by the no arbitrage principle. Thus

$$f(t,T,S) = \frac{1}{S-T} \log \frac{P(t,T)}{P(t,S)}$$

and

$$P(t,T)e^{-(S-T)f(t,T,S)} = P(t,S) .$$

Definition 22.8

(i) The *spot rate* $R(t,T)$ at t for maturity T is defined as the yield to maturity of the T-maturity bond, i.e., $P(t,T) = e^{-(T-t)R(t,T)}$. More precisely,

$$R(t,T) = -\frac{\log P(t,T)}{T-t} .$$

(ii) The *forward rate* $f(t,T,S)$ at t for $T < S$ is defined by

$$f(t,T,S) = \frac{1}{S-T} \log \frac{P(t,T)}{P(t,S)} .$$

Since $P(t,t) = 1$, taking $T = t$, we obtain

$$e^{-(S-t)f(t,t,S)} = P(t,S) ,$$

which in turn implies that

$$f(t,t,S) = R(t,S) .$$

(iii) The *instantaneous forward rate* $f(t,T)$, over the interval $[T, T + dt]$, is defined by

$$f(t,T) = \lim_{S \downarrow T} f(t,T,S) .$$

Note that

$$\begin{aligned} f(t,T) &= \lim_{S\downarrow T} \frac{1}{S-T} \log \frac{P(t,T)}{P(t,S)} \\ &= -\lim_{S\downarrow T} \frac{\log P(t,S) - \log P(t,T)}{S-T} \\ &= -\frac{\partial}{\partial T} \log P(t,T) \\ &= -\frac{1}{P(t,T)} \frac{\partial}{\partial T} P(t,T) \ . \end{aligned}$$

Hence

$$P(t,T) = \mathrm{e}^{-\int_t^T f(t,u)\mathrm{d}u}$$

and

$$R(t,T) = \frac{1}{T-t} \int_t^T f(t,u)\mathrm{d}u \ .$$

(iv) Define the *short rate* $r(t)$, over the interval $[t, t+\mathrm{d}t]$, by

$$r(t) = \lim_{T\downarrow t} R(t,T) = f(t,t) \ .$$

22.8 The Fundamental Equation for Bond Pricing

A zero coupon bond will pay the investor a riskless fixed amount at maturity, and it might be regarded as a riskless investment. However, if the interest rate r is stochastic, a bond is a risky asset since a change in the interest rate will cause a change in the present value of the pre-agreed payment at maturity. Especially when the investor wants to sell the bond before maturity, a sudden increase in the short-term interest rate may cause losses.

We assume that the short rate r_t satisfies a stochastic differential equation

$$\mathrm{d}r_t = \mu(r_t, t)\,\mathrm{d}t + \sigma(r_t, t)\,\mathrm{d}W_t$$

for some $\mu(r,t)$ and $\sigma(r,t)$. Note that in bond pricing there is no underlying asset with which to hedge the risk. Therefore we construct a portfolio in which one bond is hedged by another bond with different maturity in the derivation of the fundamental equation for bond pricing given in the following theorem due to O. Vasicek. The key concept is similar to the Delta-hedging idea in the derivation of the

Black–Scholes–Merton equation, and it is fundamental in pricing bonds and interest rate derivatives. See [99].

Theorem 22.3 (Fundamental Equation for Pricing) *Let $V(r,t)$ denote the price of a bond with payoff equal to 1 at maturity date T, or an interest rate derivative that pays $g(r_T)$ where r_T denotes the interest rate at T. Then it satisfies*

$$\frac{\partial V}{\partial t}+\frac{1}{2}\sigma^2\frac{\partial^2 V}{\partial r^2}+m(r,t)\frac{\partial V}{\partial r}=rV \tag{22.4}$$

for some function $m(r,t)$. The final condition is given by $V(r,T)=1$ for a bond, and $V(r,T)=g(r)$ for a derivative, respectively.

Proof Since a bond is an interest rate derivative with contingent claim given by $g(r_T)=1$ at maturity, we prove the theorem for general derivatives. We consider two arbitrarily chosen derivatives, and construct a portfolio in which one derivative is hedged by another derivative with different maturity. Let $V_i(r,t)$ denote the price at time t of the ith derivative with maturity T_i where $t\le\min\{T_1,T_2\}$. Set up a portfolio Π given by

$$\Pi=V_1-\Delta V_2$$

where Δ is the number of units of the second derivative that is shorted. Then the increment of Π after time increment $\mathrm{d}t$ is given by

$$\mathrm{d}\Pi=\frac{\partial V_1}{\partial t}\mathrm{d}t+\frac{\partial V_1}{\partial r}\mathrm{d}r+\frac{1}{2}\sigma^2\frac{\partial^2 V_1}{\partial r^2}\mathrm{d}t-\Delta\left(\frac{\partial V_2}{\partial t}\mathrm{d}t+\frac{\partial V_2}{\partial r}\mathrm{d}r+\frac{1}{2}\sigma^2\frac{\partial^2 V_2}{\partial r^2}\mathrm{d}t\right)$$

by Itô's lemma. To make the coefficient of the risky term $\mathrm{d}r$ equal to zero, eliminating randomness in the portfolio, we take

$$\Delta=\frac{\partial V_1}{\partial r}\Big/\frac{\partial V_2}{\partial r}\,. \tag{22.5}$$

Then

$$\mathrm{d}\Pi=\left\{\frac{\partial V_1}{\partial t}+\frac{1}{2}\sigma^2\frac{\partial^2 V_1}{\partial r^2}-\frac{\dfrac{\partial V_1}{\partial r}}{\dfrac{\partial V_2}{\partial r}}\left(\frac{\partial V_2}{\partial t}+\frac{1}{2}\sigma^2\frac{\partial^2 V_2}{\partial r^2}\right)\right\}\mathrm{d}t\,.$$

Since Π is risk-free, we have $\mathrm{d}\Pi=r\Pi\,\mathrm{d}t$, and hence

$$\frac{\partial V_1}{\partial t}+\frac{1}{2}\sigma^2\frac{\partial^2 V_1}{\partial r^2}-\frac{\dfrac{\partial V_1}{\partial r}}{\dfrac{\partial V_2}{\partial r}}\left(\frac{\partial V_2}{\partial t}+\frac{1}{2}\sigma^2\frac{\partial^2 V_2}{\partial r^2}\right)=r\left(V_1-\frac{\dfrac{\partial V_1}{\partial r}}{\dfrac{\partial V_2}{\partial r}}V_2\right)\,.$$

Hence

$$\frac{1}{\frac{\partial V_1}{\partial r}}\left(\frac{\partial V_1}{\partial t}+\frac{1}{2}\sigma^2\frac{\partial^2 V_1}{\partial r^2}\right)-\frac{1}{\frac{\partial V_2}{\partial r}}\left(\frac{\partial V_2}{\partial t}+\frac{1}{2}\sigma^2\frac{\partial^2 V_2}{\partial r^2}\right)=r\left(\frac{1}{\frac{\partial V_1}{\partial r}}V_1-\frac{1}{\frac{\partial V_2}{\partial r}}V_2\right).$$

By collecting V_1 terms on the left-hand side and V_2 terms on the right, we obtain

$$\frac{\dfrac{\partial V_1}{\partial t}+\dfrac{1}{2}\sigma^2\dfrac{\partial^2 V_1}{\partial r^2}-rV_1}{\dfrac{\partial V_1}{\partial r}}=\frac{\dfrac{\partial V_2}{\partial t}+\dfrac{1}{2}\sigma^2\dfrac{\partial^2 V_2}{\partial r^2}-rV_2}{\dfrac{\partial V_2}{\partial r}}.$$

The left-hand side depends only on V_1 not V_2, and the right-hand side depends only on V_2 not V_1, and hence both sides do not depend on the derivative price. Therefore the derivative price V satisfies

$$\frac{\dfrac{\partial V}{\partial t}+\dfrac{1}{2}\sigma^2\dfrac{\partial^2 V}{\partial r^2}-rV}{\dfrac{\partial V}{\partial r}}=-m(r,t)$$

for some function $m(r,t)$. The negative sign in front of $m(r,t)$ is for notational convenience in the final statement of the result. □

Example 22.5 If V_t denotes the value at time t of an interest rate option with maturity T and strike rate r_K, then its price satisfies the fundamental pricing equation given in Theorem 22.3 with the final condition $V(r,T)=N\times\max\{r-r_K,0\}$ for some nominal amount N.

Remark 22.5 As in the derivation of the Black–Scholes–Merton differential equation presented in Theorem 15.1, we can simulate the portfolio process $\Pi(r,t)=V_1(r,t)-\Delta V_2(r,t)$ where we take Δ given by (22.5) and the price processes of bond 1 and bond 2. The fundamental pricing equation depends on our choice of the interest rate model. For example, the Vasicek model assumes that $m(r,t)$ has the same form as the drift rate of the original interest rate process, although possibly with different parameter values. Consult Simulation 23.4 in Chap. 23.

Remark 22.6 When there is continuous coupon payment, we modify the model as

$$\frac{\partial V}{\partial t}+\frac{1}{2}\sigma^2\frac{\partial^2 V}{\partial r^2}+m(r,t)\frac{\partial V}{\partial r}-rV+K(r,t)=0$$

where $K(r,t)\,\mathrm{d}t$ is the amount received in a period of length $\mathrm{d}t$. For discrete coupon payment, we introduce the jump condition

$$V(r,t_c-)=V(r,t_c+)+K(r,t_c)$$

where t_c is the time of coupon payment and $K(r,t_c)$ is the amount of coupon.

Remark 22.7 (Risk-Neutral Measure) Let P_t be the price at time t of a bond with maturity T such that $P_T = 1$. Let $\mathrm{d}r_t = \mu(r_t, t)\mathrm{d}t + \sigma(r_t, t)\mathrm{d}W_t$ be a model of the movement of the physical interest rate where W_t is a $\mathbb{P}$-Brownian motion. Let $m(r, t)$ be the function given in Theorem 22.3. Put

$$\theta(r,t) = \frac{\mu(r,t) - m(r,t)}{\sigma(r,t)}$$

and $\theta_t = \theta(r_t, t)$, and we assume, throughout the rest of the chapter, that the following *Novikov condition* holds:

$$\mathbb{E}^{\mathbb{P}}\left[\exp\left(\frac{1}{2}\int_0^T \theta_t^2 \, \mathrm{d}t\right)\right] < \infty .$$

(Here we are assuming that $\sigma \neq 0$.) Define

$$\widetilde{W}_t = W_t + \int_0^t \theta_u \mathrm{d}u .$$

By Girsanov's theorem, there is a probability measure $\mathbb{Q}$ equivalent to $\mathbb{P}$, with the *Radon–Nikodym derivative*

$$\frac{\mathrm{d}\mathbb{Q}}{\mathrm{d}\mathbb{P}} = \exp\left(-\int_0^T \theta_t \, \mathrm{d}W_t - \frac{1}{2}\int_0^T \theta_t^2 \, \mathrm{d}t\right)$$

such that $\widetilde{W}_t$, $0 \leq t \leq T$, is a $\mathbb{Q}$-Brownian motion. The measure $\mathbb{Q}$ is called the *risk-neutral measure*. The interest rate now satisfies

$$\mathrm{d}r_t = \mu \mathrm{d}t + \sigma(\mathrm{d}\widetilde{W}_t - \theta \mathrm{d}t) = m \, \mathrm{d}t + \sigma \mathrm{d}\widetilde{W}_t ,$$

and it is called the *risk-neutral interest rate*.

Remark 22.8 (Market Price of Interest Rate Risk) Suppose that the coefficient $m(r, t)$ in Theorem 22.3 is of the form

$$-m(r,t) = \sigma(r,t)\theta(r,t) - \mu(r,t)$$

for some $\theta(r, t)$. (We just take $\theta = \frac{\mu - m}{\sigma}$.) Then (22.4) becomes

$$\frac{\partial V}{\partial t} + \frac{1}{2}\sigma^2 \frac{\partial^2 V}{\partial r^2} + (\mu - \sigma\theta)\frac{\partial V}{\partial r} - rV = 0 . \tag{22.6}$$

By Itô's lemma, we have

$$\begin{aligned}
\mathrm{d}V &= \left(\frac{\partial V}{\partial t} + \frac{1}{2}\sigma^2 \frac{\partial^2 V}{\partial r^2}\right)\mathrm{d}t + \frac{\partial V}{\partial r}\mathrm{d}r \\
&= \left(\frac{\partial V}{\partial t} + \frac{1}{2}\sigma^2 \frac{\partial^2 V}{\partial r^2}\right)\mathrm{d}t + \frac{\partial V}{\partial r}(\mu \mathrm{d}t + \sigma \mathrm{d}W) \\
&= \left(\frac{\partial V}{\partial t} + \frac{1}{2}\sigma^2 \frac{\partial^2 V}{\partial r^2} + \mu \frac{\partial V}{\partial r}\right)\mathrm{d}t + \sigma \frac{\partial V}{\partial r}\mathrm{d}W \\
&= \left(\sigma\theta \frac{\partial V}{\partial r} + rV\right)\mathrm{d}t + \sigma \frac{\partial V}{\partial r}\mathrm{d}W
\end{aligned}$$

where (22.6) is used for the last equality. Hence

$$\mathrm{d}V - rV\mathrm{d}t = \sigma \frac{\partial V}{\partial r}(\theta\, \mathrm{d}t + \mathrm{d}W) \tag{22.7}$$

and

$$\mathbb{E}^{\mathbb{P}}\left[\frac{\mathrm{d}V - rV\mathrm{d}t}{\sigma \frac{\partial V}{\partial r}}\right] = \theta\, \mathrm{d}t\,. \tag{22.8}$$

(Here we are assuming that $\sigma \frac{\partial V}{\partial r} \neq 0$.) The left-hand side of (22.8) is the risk-adjusted excess return above the risk-free rate, and hence $\theta = \dfrac{\mu - m}{\sigma}$ is called the *market price of interest rate risk*.

Let $m(r,t)$ be the function given in Theorem 22.3. Let θ, $\widetilde{W}_t$ and $\mathbb{Q}$ be defined as in Remark 22.7.

Theorem 22.4 (Discounted Feynman–Kac Theorem for Interest Rate Derivatives) *Consider an interest rate derivative with its payoff at maturity S given by X_S. Let V_t denote the price of the interest rate derivative at $0 \leq t \leq S$. Choose a numeraire given by the risk-free cash account process*

$$B_t = \mathrm{e}^{\int_0^t r_u \mathrm{d}u}$$

and consider the corresponding discounted derivative price process

$$\widetilde{V}_t = B_t^{-1} V_t = \mathrm{e}^{-\int_0^t r_u \mathrm{d}u} V_t\,.$$

Assume that at least one of following conditions holds:

(i) *There exists a random variable* $Y \geq 0$ *such that* $\mathbb{E}^{\mathbb{Q}}[Y] < \infty$ *and* $|\widetilde{V}_t| \leq Y$ *for every t.*

(ii) $\mathbb{E}^{\mathbb{Q}}\left[\int_0^S \left(B_t^{-1}\sigma\frac{\partial V}{\partial r}\right)^2 \mathrm{d}t\right] < \infty.$

Then

$$V_t = \mathbb{E}^{\mathbb{Q}}\left[\mathrm{e}^{-\int_t^S r_u \mathrm{d}u} X_S \middle| \mathcal{F}_t\right] , \quad 0 \leq t \leq S ,$$

where

$$\mathrm{d}r_t = m\,\mathrm{d}t + \sigma \mathrm{d}\widetilde{W}_t .$$

Proof Note that

$$\begin{aligned} \mathrm{d}\widetilde{V}_t &= \mathrm{e}^{-\int_0^t r_u \mathrm{d}u}\mathrm{d}V_t + \mathrm{e}^{-\int_0^t r_u \mathrm{d}u}(-r_t)V_t \mathrm{d}t \\ &= B_t^{-1}(\mathrm{d}V_t - r_t V_t \mathrm{d}t) \\ &= B_t^{-1}\sigma\frac{\partial V}{\partial r}\mathrm{d}\widetilde{W}_t , \qquad \text{(by Eq. (22.7))} \end{aligned}$$

and hence $\widetilde{V}_t$ is a local $\mathbb{Q}$-martingale. Since at least one of two given conditions holds, $\widetilde{V}_t$ is a $\mathbb{Q}$-martingale. (See Lemma 5.1 in [30] and Remark 7.2 in [49] for the proof.) Hence

$$\mathrm{e}^{-\int_0^t r_u \mathrm{d}u} V_t = \widetilde{V}_t = \mathbb{E}^{\mathbb{Q}}\left[\widetilde{V}_S \middle| \mathcal{F}_t\right] = \mathbb{E}^{\mathbb{Q}}\left[\mathrm{e}^{-\int_0^S r_u \mathrm{d}u} V_S \middle| \mathcal{F}_t\right] .$$

Since $\mathrm{e}^{-\int_0^t r_u \mathrm{d}u}$ is $\mathcal{F}_t$-measurable, we have

$$V_t = \mathbb{E}^{\mathbb{Q}}\left[\mathrm{e}^{\int_0^t r_u \mathrm{d}u}\mathrm{e}^{-\int_0^S r_u \mathrm{d}u} V_S \middle| \mathcal{F}_t\right] = \mathbb{E}^{\mathbb{Q}}\left[\mathrm{e}^{-\int_t^S r_u \mathrm{d}u} V_S \middle| \mathcal{F}_t\right]$$

for $0 \leq t \leq S$. Now use $V_S = X_S$. □

Corollary 22.1 (Discounted Feynman–Kac Theorem for Bonds) *The price* $P(t, S)$ *of a zero coupon bond with maturity S at time* $0 \leq t \leq S$ *is given by*

$$P(t, S) = \mathbb{E}^{\mathbb{Q}}\left[\mathrm{e}^{-\int_t^S r_u \mathrm{d}u} \middle| \mathcal{F}_t\right] .$$

Proof Note that the bond price $P(t, S)$ satisfies $P(S, S) = X_S = 1$. □

Example 22.6 Consider a forward contract in which a price K will be paid at time T in return for a repayment of \$1 at time S, $T < S$. Equivalently, K is paid at T in return for delivery at T of the S-bond that has a value $P(T, S)$ at T. Now the problem is to find how much the contract is worth at time $t < T$. Since the contract has value

$X = P(T, S) - K$ at T, we have

$$\begin{aligned}
V(t) &= \mathbb{E}^{\mathbb{Q}}[\mathrm{e}^{-\int_t^T r_u \mathrm{d}u}(P(T,S) - K)|\mathcal{F}_t] \\
&= \mathbb{E}^{\mathbb{Q}}[\mathrm{e}^{-\int_t^T r_u \mathrm{d}u}\, \mathbb{E}^{\mathbb{Q}}[\mathrm{e}^{-\int_T^S r_u \mathrm{d}u}|\mathcal{F}_T] \mid \mathcal{F}_t] - K\, \mathbb{E}^{\mathbb{Q}}[\mathrm{e}^{-\int_t^T r_u \mathrm{d}u}|\mathcal{F}_t] \\
&= \mathbb{E}^{\mathbb{Q}}[\mathbb{E}^{\mathbb{Q}}[\mathrm{e}^{-\int_t^T r_u \mathrm{d}u}\mathrm{e}^{-\int_T^S r_u \mathrm{d}u}|\mathcal{F}_T] \mid \mathcal{F}_t] - K\, \mathbb{E}^{\mathbb{Q}}[\mathrm{e}^{-\int_t^T r_u \mathrm{d}u}|\mathcal{F}_t] \\
&= \mathbb{E}^{\mathbb{Q}}[\mathrm{e}^{-\int_t^S r_u \mathrm{d}u}|\mathcal{F}_t] - K\, \mathbb{E}^{\mathbb{Q}}[\mathrm{e}^{-\int_t^T r_u \mathrm{d}u}|\mathcal{F}_t] \\
&= P(t,S) - KP(t,T)\ .
\end{aligned}$$

If K is chosen so that there is no need to exchange any money at time $t = 0$ for the forward contract, then $V(0) = 0$, and hence $K = \frac{P(0,S)}{P(0,T)}$.

Theorem 22.5 *As in the method for pricing an option on a stock given in Sect. 16.1, there exists a self-financing and replicating strategy for interest rate derivatives with maturity S.*

Proof Construct a portfolio composed of two assets: a bond P with maturity T, $S \leq T$, and the risk-free cash deposit B. Let $\mathbb{Q}$ be the risk-neutral measure constructed in Theorem 22.4. Assume that the bond pays \$1 at T and that the risk-free cash account satisfies $\mathrm{d}B_t = r_t B_t \mathrm{d}t$, $B_0 = 1$, for the interest rate process r_t. Then $B_t = \mathrm{e}^{\int_0^t r_u \mathrm{d}u}$. Let X_S denote the payoff at S of the interest rate derivative under consideration. Define a $\mathbb{Q}$-martingale

$$M_t = \mathbb{E}^{\mathbb{Q}}[B_S^{-1} X_S|\mathcal{F}_t]$$

for $0 \leq t \leq S$. Let $\widetilde{P}_t = B_t^{-1} P_t$ be the discounted bond price. Then, by Eq. (22.7),

$$\begin{aligned}
\mathrm{d}\widetilde{P}_t &= -r_t B_t^{-1} P_t \mathrm{d}t + B_t^{-1} \mathrm{d}P_t \\
&= B_t^{-1}(\mathrm{d}P_t - r_t P_t \mathrm{d}t) \\
&= B_t^{-1} \sigma \frac{\partial P}{\partial r} \mathrm{d}\widetilde{W}_t\ .
\end{aligned}$$

Since $\widetilde{P}_t$ and M_t are both $\mathbb{Q}$-martingales, by the Martingale Representation Theorem, there exists a predictable process ϕ_t such that

$$M_t = M_0 + \int_0^t \phi_u \mathrm{d}\widetilde{P}_u\ .$$

Define

$$\psi_t = M_t - \phi_t \widetilde{P}_t$$

and consider a portfolio V defined by

$$V_t = \phi_t P_t + \psi_t B_t \ .$$

Since

$$V_t = B_t(\phi_t \widetilde{P}_t + \psi_t) = B_t M_t \ ,$$

we have

$$\begin{aligned}
\mathrm{d}V_t &= B_t \mathrm{d}M_t + M_t \mathrm{d}B_t \\
&= B_t \phi_t \mathrm{d}\widetilde{P}_t + M_t \mathrm{d}B_t \\
&= B_t \phi_t (-r_t B_t^{-1} P_t \mathrm{d}t + B_t^{-1} \mathrm{d}P_t) + (\psi_t + \phi_t B_t^{-1} P_t) \mathrm{d}B_t \\
&= \phi_t (-r_t P_t \mathrm{d}t + \mathrm{d}P_t) + (\psi_t + \phi_t B_t^{-1} P_t) \mathrm{d}B_t \\
&= \phi_t \mathrm{d}P_t + \psi_t \mathrm{d}B_t + \phi_t B_t^{-1} P_t (-r_t B_t \mathrm{d}t + \mathrm{d}B_t) \\
&= \phi_t \mathrm{d}P_t + \psi_t \mathrm{d}B_t \ .
\end{aligned}$$

Hence V_t is self-financing. Finally, V_t replicates the contingent claim at time S since $V_S = B_S M_S = B_S \mathbb{E}^{\mathbb{Q}}[B_S^{-1} X_S | \mathcal{F}_S] = B_S B_S^{-1} X_S = X_S$. □

Example 22.7 If V itself is a bond with maturity S, and $X_S = V_S = 1$, then holding V is a self-financing and replicating strategy. In this case, $\phi_t = 1$ and $\psi_t = 0$.

22.9 Computer Experiments

Simulation 22.1 (Yield to Maturity)

Using the Newton–Raphson method we compute the yield to maturity when the present price of a coupon bond is given.

```
B = 140;    % current bond price
C = 5;      % Coupon is paid every six months.
T = 20;     % time to maturity of bond
P = 100;    % face value
r_max = 0.2;
L = 500;
dr = r_max/L;
r = 0:dr:r_max; % range of interest rates

PV = C*exp(-0.5*r(:)).*(1-exp(-T*r(:)))./(1-exp(-0.5*r(:)))+P*exp(-r(:)*T);
PV(1) = C*2*T + P;

plot(r,PV)
xlabel('Interest rate')
ylabel('Price of bond')
```

For the graph of bond price as a function of interest rate, see Fig. 22.1. Now we find the yield to maturity by Newton's method.

```
syms y % Create a symbolic variable y.
F = C*exp(-0.5*y)*(1-exp(-T*y))/(1-exp(-0.5*y)) + P*exp(-y*T) - B;
F1 = diff(F,y);
n = 7;
Y = zeros(n+1,1);
Y(1) = 0.01;
for i=1:n
    Y(i+1) = Y(i) - subs(F,Y(i)) / subs(F1,Y(i));
end
YTM = Y(n+1) % Yield to maturity
```

Or, we may use the following command to compute the yield to maturity.

```
solve(C*exp(-y/2)*(1-exp(-T*y))/(1-exp(-y/2))+P*exp(-y*T)==B,y,'Real',true)
```

Simulation 22.2 (Discounted Feynman–Kac Theorem)

We compute the present ($t = 0$) price of a zero coupon bond with maturity $T = 30$ years using Corollary 22.1 in the Vasicek model $\mathrm{d}r_t = \alpha(\bar{r} - r_t)\,\mathrm{d}t + \sigma\,\mathrm{d}W_t$ given in Sect. 23.2. The closed form formula for bond prices in the following code is from Theorem 23.3.

```
T= 30;
N = 1000;
dt = T/N;
t = 0:dt:T;
M = 10^5;    % the number of samples for the Monte Carlo method
r = zeros(M,1);
alpha = 0.4;
r_bar = 0.05;
sigma = 0.015;
r0 = 0.03;
r(:) = r0;
Integral = zeros(M,1);
for i = 1:N
    dW = sqrt(dt)*randn(M,1);
    r(:) = r(:) + alpha*(r_bar - r(:))*dt + sigma*dW(:);
    Integral(:) = Integral(:) + r(:)*dt;
end

B = 1/alpha*(1-exp(-alpha*T));
A = (B-T)*(r_bar - sigma^2/2/alpha^2)- sigma^2*B^2/4/alpha;
Vasicek_formula = exp(A-B*r0)

MC_bond_price = mean(exp(-Integral(:)))
sample_std = std(exp(-Integral(:)))/sqrt(M)
```

The output is given by

```
Vasicek_formula =
                         0.2389
MC_bond_price =
                         0.2390
sample_std =
                        1.4652e-04
```

Note that the theoretical formula under the Vasicek model (Theorem 23.3) and the Monte Carlo method produce the prices very close to each other. If we employ the Cox–Ingersoll–Ross model in Sect. 23.3 given by $dr_t = \alpha(\bar{r} - r_t)\,dt + \sigma\sqrt{r_t}\,dW_t$ with Matlab code

```
r(:) = max(0,r(:) + alpha*(r_bar - r(:))*dt + sigma*sqrt(r(:)).*dW(:));
```

using the same parameters as in the Vasicek model except for $\sigma = 0.015/\sqrt{\bar{r}} = 0.0671$, we obtain the bond price 0.2387, which is close to the value computed by the Vasicek model.

Simulation 22.3 (Distribution of the Discount Factor)

To plot the distributions of $\int_0^T r_t dt$ and the discount factor $\exp(-\int_0^T r_t dt)$ considered in Simulation 22.2 under the Vasicek model, we generate 10^6 sample paths. (We observe that all the sample values for the integral $\int_0^T r_t dt$ are positive even though a sample path r_t, $r_0 = 0.03$, can be negative for some $0 < t \leq T$.) The number of sample values belonging to the jth subinterval is given by `bin(j)` in the following code.

```
y = Integral(:);
histogram(y,50,'Normalization','pdf')

title('Distribution of \int_0^T r_t dt');
ylabel('Probability');
```

For the plots, see Fig. 22.2. Observe that the distributions of $\int_0^T r_t dt$ and $\exp(-\int_0^T r_t dt)$ are approximately normal and lognormal, respectively.

If the Cox–Ingersoll–Ross model in Sect. 23.3 is employed in bond pricing with the same parameters as in the Vasicek model except for $\sigma = 0.015/\sqrt{\bar{r}} = 0.0671$ to match the volatility level, we have the distributions given in Fig. 22.3. Observe that the distribution of $\int_0^T r_t dt$ is skewed to the right and $\exp(-\int_0^T r_t dt)$ is more symmetric than in the Vasicek model case. This is due to the fact that in the CIR model the interest rate is always nonnegative, not symmetrically distributed.

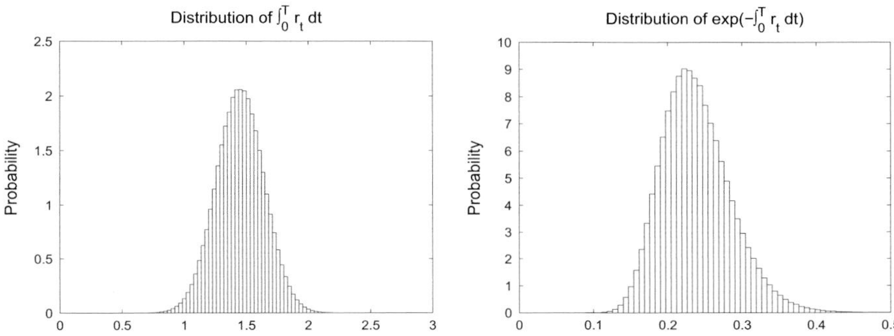

Fig. 22.2 Probability distributions of $\int_0^T r_t \mathrm{d}t$ and $\exp(-\int_0^T r_t \mathrm{d}t)$ in the Vasicek model

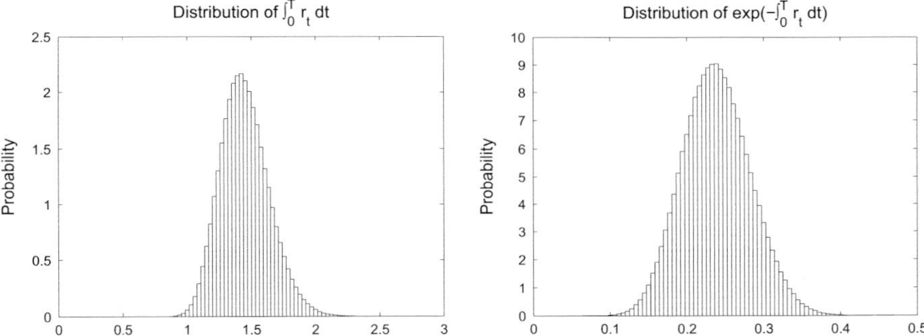

Fig. 22.3 Probability distributions of $\int_0^T r_t \mathrm{d}t$ and $\exp(-\int_0^T r_t \mathrm{d}t)$ in the CIR model

Exercises

22.1 Prove that $(1 + \frac{R}{m})^m$ is monotonically increasing and converges to e^R as $m \to \infty$.

22.2 Let $\Pi = V_1 - \Delta V_2$ as in the proof of Theorem 22.3 and assume that Π is hedged by choosing Δ as in (22.5). Show that the following holds:

$$\frac{1}{\Pi}\frac{\partial \Pi}{\partial t} + \frac{1}{2}\sigma^2 \frac{1}{\Pi}\frac{\partial^2 \Pi}{\partial r^2} = r \, .$$

22.3 Under the assumption that the bond price at t is given by $P(t, S) = \mathbb{E}^{\mathbb{Q}}[\mathrm{e}^{-\int_t^S r_u \mathrm{d}u} | \mathcal{F}_t]$, derive the fundamental equation for bond pricing given in Theorem 22.3.

Chapter 23
Interest Rate Models

While the assumption that the interest rate is constant produces reasonable estimates in option pricing, the same assumption would produce less reliable results in pricing bonds and interest rate derivatives. One of the reasons is that the bonds usually have much longer maturity. We investigate interest rate models expressed in terms of stochastic differential equations. In this chapter we introduce models for the short rate r_t, in particular. Recall the definition of the short rate given in Definition 22.8. If not stated otherwise, an interest rate means a short rate in this chapter.

23.1 Short Rate Models

Definition 23.1 (Affine Term Structure) Let $P(t, T)$ denote the price at time t of a bond with maturity T. Assume that $P(T, T) = 1$. If the bond price is given by a formula of the form

$$P(r, t, T) = \mathrm{e}^{A(t,T)-B(t,T)\,r}$$

for some sufficiently smooth deterministic functions $A(t, T)$ and $B(t, T)$ where r denotes the interest rate, then the model is said to have an *affine term structure*. Note that in the exponent the coefficient for r is $-B(t, T)$ so that we have $B(t, T) > 0$ in agreement with the fact that

$$\frac{\partial}{\partial r}P(r, t, T) = -B(t, T)P(r, t, T) < 0\ .$$

Theorem 23.1 *Consider a risk-neutral interest rate process r_t such that*

$$\mathrm{d}r_t = m(r_t, t)\,\mathrm{d}t + \sigma(r_t, t)\,\mathrm{d}\widetilde{W}_t$$

G.H. Choe, *Stochastic Analysis for Finance with Simulations*, Universitext,
DOI 10.1007/978-3-319-25589-7_23

where $\widetilde{W}_t$ is a $\mathbb{Q}$-Brownian motion. Assume that the interest rate model has an affine term structure. Suppose that its diffusion and drift terms are of the form

$$m(r,t) = a(t) + b(t)\,r$$

and

$$\sigma^2(r,t) = c(t) + d(t)\,r$$

for some deterministic functions a, b, c and d. (In other words, m and σ^2 are affine functions in r.) Then A and B given in Definition 23.1 satisfy the system of ordinary differential equations

$$\begin{cases} \dfrac{\partial}{\partial t}A(t,T) - a(t)B(t,T) + \dfrac{1}{2}c(t)B(t,T)^2 = 0\,, \quad A(T,T) = 0\,, \\ \dfrac{\partial}{\partial t}B(t,T) + b(t)B(t,T) - \dfrac{1}{2}d(t)B(t,T)^2 = -1\,, \quad B(T,T) = 0\,. \end{cases} \tag{23.1}$$

Proof Recall that the bond price $P(t,T)$ satisfies the fundamental equation in Theorem 22.3. By substituting $P(t,T) = \mathrm{e}^{A(t,T)-B(t,T)r}$ in place of V in the equation, we obtain the ordinary differential equations in (23.1). The boundary conditions follow since

$$1 = P(T,T) = \mathrm{e}^{A(T,T)-B(T,T)r}$$

for every r. □

Remark 23.1 The second equation in (23.1) for $B(t,T)$ is called a Riccati equation. (See Appendix C.2.) Once we solve for B, then we insert it into the first equation, and obtain A.

23.2 The Vasicek Model

Consider the Vasicek model [99] for an interest rate given by

$$\mathrm{d}r_t = \alpha(\bar{r} - r_t)\,\mathrm{d}t + \sigma\,\mathrm{d}\widetilde{W}_t$$

where r_t is the interest rate at t, and α, $\bar{r}$ and σ are positive constants. (For $\bar{r} = 0$ we obtain the Ornstein–Uhlenbeck process in Example 12.1.) This model has the mean reversion property that is observed in the financial market. That is, if the interest rate r_t is higher than $\bar{r}$, called the long-term mean, then $\alpha(\bar{r} - r_t)$, the coefficient of $\mathrm{d}t$, becomes negative, and the interest rate tends to decrease, and if the level of

r_t is lower than $\bar{r}$ then $\alpha(\bar{r} - r_t)$ becomes positive and the interest rate tends to go up. The coefficient α is called the reversion rate. In the long run, the interest rate is distributed around $\bar{r}$. Note that the interest rate can become negative in the Vasicek model.

Theorem 23.2 *The interest rate r_t in the Vasicek model has the normal distribution, and for $0 \leq s \leq t$ we have*

$$r_t = \bar{r} + (r_s - \bar{r})\,\mathrm{e}^{-\alpha(t-s)} + \sigma \int_s^t \mathrm{e}^{-\alpha(t-u)} \mathrm{d}\widetilde{W}_u \,.$$

Hence the conditional mean and variance are given by

$$\mathbb{E}^{\mathbb{Q}}[r_t|\mathcal{F}_s] = \bar{r} + (r_s - \bar{r})\,\mathrm{e}^{-\alpha(t-s)} \,,$$

$$\mathrm{Var}^{\mathbb{Q}}(r_t|\mathcal{F}_s) = \frac{\sigma^2}{2\alpha}(1 - \mathrm{e}^{-2\alpha(t-s)}) \,.$$

The limiting probability distribution of the interest rate r_t converges to a normal distribution with mean $\bar{r}$ and variance $\frac{\sigma^2}{2\alpha}$ as $t \to +\infty$.

Proof Recall that an Itô integral of a deterministic function is normally distributed by Theorem 10.3. Since r_t in the above is a sum of a deterministic part $\bar{r} + (r_s - \bar{r})\,\mathrm{e}^{-\alpha(t-s)}$ and an Itô integral of a deterministic function, it is also normally distributed. Note that

$$\begin{aligned}
\mathbb{E}^{\mathbb{Q}}[r_t|\mathcal{F}_s] &= \mathbb{E}^{\mathbb{Q}}\left[\bar{r} + (r_s - \bar{r})\mathrm{e}^{-\alpha(t-s)} + \sigma \int_s^t \mathrm{e}^{-\alpha(t-u)} \mathrm{d}\widetilde{W}_u \middle| \mathcal{F}_s\right] \\
&= \bar{r} + (r_s - \bar{r})\mathrm{e}^{-\alpha(t-s)} + \sigma\, \mathbb{E}^{\mathbb{Q}}\left[\int_s^t \mathrm{e}^{-\alpha(t-u)} \mathrm{d}\widetilde{W}_u \middle| \mathcal{F}_s\right] \\
&= \bar{r} + (r_s - \bar{r})\mathrm{e}^{-\alpha(t-s)} + \sigma\, \mathbb{E}^{\mathbb{Q}}\left[\int_s^t \mathrm{e}^{-\alpha(t-u)} \mathrm{d}\widetilde{W}_u\right] \\
&= \bar{r} + (r_s - \bar{r})\mathrm{e}^{-\alpha(t-s)} + \sigma \times 0
\end{aligned}$$

where the third equality is from the independence property of conditional expectation in Theorem 5.2. Note that

$$\begin{aligned}
\mathrm{Var}^{\mathbb{Q}}(r_t|\mathcal{F}_s) &= \mathbb{E}^{\mathbb{Q}}[(r_t - \mathbb{E}^{\mathbb{Q}}[r_t|\mathcal{F}_s])^2|\mathcal{F}_s] \\
&= \mathbb{E}^{\mathbb{Q}}\left[\left(\sigma \int_s^t \mathrm{e}^{-\alpha(t-u)} \mathrm{d}\widetilde{W}_u\right)^2 \middle| \mathcal{F}_s\right] \\
&= \sigma^2 \int_s^t \mathrm{e}^{-2\alpha(t-u)} \mathrm{d}u \quad \text{(by Itô identity)} \\
&= \frac{\sigma^2}{2\alpha}(1 - \mathrm{e}^{-2\alpha(t-s)}) \,.
\end{aligned}$$

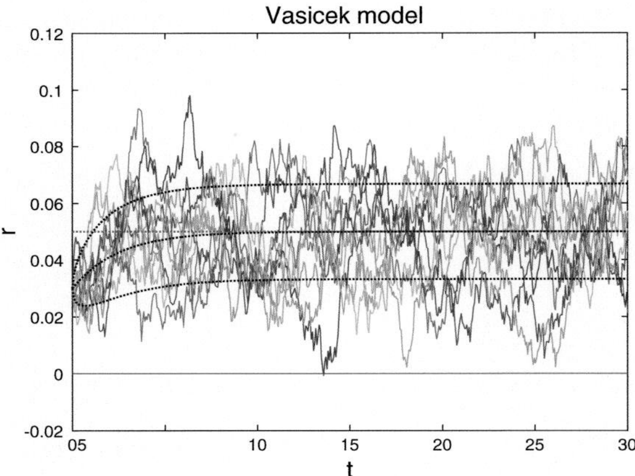

Fig. 23.1 Sample paths of the interest rate in the Vasicek model

To find the limiting distribution, we take the limit as $t \to +\infty$ and note that the limit of normal distributions is also normal. □

In Fig. 23.1 are plotted ten sample paths of the interest rate obtained from Simulation 23.1 where we take $\alpha = 0.4$, $\bar{r} = 0.05$, $\sigma = 0.015$, $T = 30$ and $r_0 = 0.3$. Note that r_t can take negative values. The solid line is the graph of the expectation of r_t that converges to the level $\bar{r}$, and the two dotted lines above and below the solid line represent the range of r_t within one standard deviation from the mean.

Theorem 23.3 *The bond price under the Vasicek model is given by*

$$P(t,T) = \mathrm{e}^{A(T-t)-B(T-t)\,r_t}$$

where

$$B(T-t) = \frac{1}{\alpha}(1-\mathrm{e}^{-\alpha(T-t)}) ,$$

$$A(T-t) = (B(T-t)-(T-t))\left(\bar{r}-\frac{\sigma^2}{2\alpha^2}\right) - \frac{\sigma^2}{4\alpha}B(T-t)^2 .$$

Proof The equations in (23.1) now become

$$\begin{cases} \dfrac{\partial}{\partial t}A(t,T) - \bar{r}\alpha B(t,T) + \dfrac{\sigma^2}{2}B(t,T)^2 = 0 , & A(T,T) = 0 , \\ \dfrac{\partial}{\partial t}B(t,T) - \alpha B(t,T) + 1 = 0 , & B(T,T) = 0 . \end{cases} \tag{23.2}$$

From the second equation in (23.2) we obtain

$$B(t,T) = \frac{1}{\alpha}(1 - e^{-\alpha(T-t)}) .$$

By substituting the result in the first equation, we can express $\frac{\partial}{\partial t}A(t,T)$ in terms of constant and exponential functions of t, which can be integrated easily. □

Remark 23.2 (Calibration) How can we estimate parameters α, $\bar{r}$ and σ in the interest rate model? First, using the historical data, we estimate σ. Then, using N bond prices beginning at a fixed time, say $t = 0$, with maturities $T_1, \ldots, T_N$, we find α and $\bar{r}$ minimizing the sum

$$\sum_{i=1}^{N} \left(P^{\text{Vasicek}}(0,T_i) - P^{\text{data}}(0,T_i)\right)^2$$

where $P^{\text{Vasicek}}(0,T_i)$ is the price by the Vasicek model and $P^{\text{data}}(0,T_i)$ is the market price.

Remark 23.3 (Term Structure of Interest Rates in the Vasicek Model) Note that the value of the zero coupon bond at time t depends only on time to maturity $\tau = T - t$ and the spot rate is given by

$$R(t,t+\tau) = -\frac{\log P(r_t, t+\tau)}{\tau} = -\frac{A(\tau)}{\tau} + \frac{B(\tau)}{\tau} r_t .$$

(See Definition 22.8.) Thus for fixed τ we have

$$\mathrm{d}R(t,t+\tau) = \frac{B(\tau)}{\tau}\,\mathrm{d}r_t = \frac{B(\tau)}{\tau}\gamma(\bar{r} - r_t)\,\mathrm{d}t + \frac{B(\tau)}{\tau}\,\sigma \mathrm{d}\widetilde{W}_t .$$

Note that the random component is given by $\frac{B(\tau)}{\tau}\sigma \mathrm{d}\widetilde{W}_t$ and that

$$\mathrm{Var}(\mathrm{d}R(t,t+\tau)) = \left(\frac{B(\tau)}{\tau}\right)^2 \sigma^2 \mathrm{d}t = \left(\frac{B(\tau)}{\tau}\right)^2 \mathrm{Var}(\mathrm{d}r_t) .$$

Since

$$0 < \frac{B(\tau)}{\tau} = \frac{1 - e^{-\alpha\tau}}{\alpha\tau} < 1 ,$$

financial shocks to short-term interest rates, $\mathrm{d}r_t$, have a milder effect on longer-term bonds, $\mathrm{d}R(t,\tau)$. Since $\frac{B(\tau)}{\tau}$ is a monotonically decreasing function of τ, which converges to 0 as $\tau \to \infty$ as shown in Fig. 23.2, we observe that the volatility $\frac{B(\tau)}{\tau}\sigma$ of $\mathrm{d}R(t,t+\tau)$ for long-term spot rates is smaller than the volatility of short-term yields.

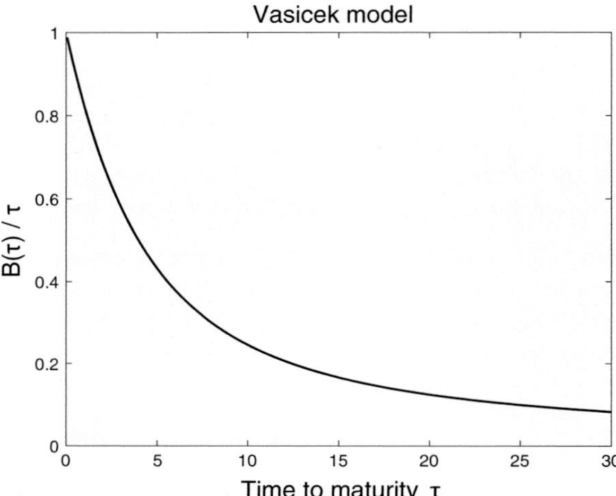

Fig. 23.2 $B(\tau)/\tau$ as a monotonically decreasing function of τ in the Vasicek model

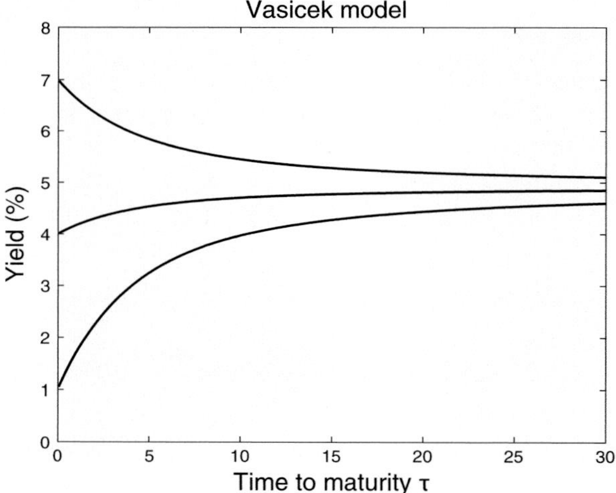

Fig. 23.3 Term structure of interest rates in the Vasicek model

As seen in Fig. 23.3 where three curves of $R(0, \tau)$ corresponding to $r_0 = 0.01, 0.04, 0.07$ are given, longer-term rates move less than short-term rates. This does not necessarily imply that the volatility of long-term bond prices is smaller than the volatility of short-term bond prices since the duration of a bond should be multiplied in computing bond price. See Theorem 22.1.

Figure 23.4 shows the surface given by the bond price under the Vasicek model with $\alpha = 0.4$, $\bar{r} = 0.05$, $\sigma = 0.015$, $T = 10$ and $0 \leq r \leq 0.5$.

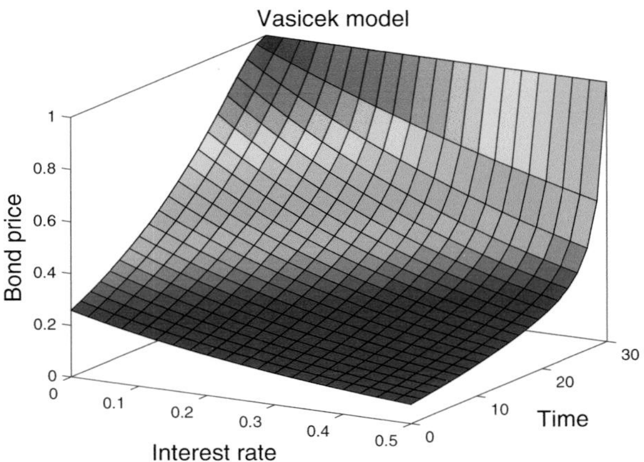

Fig. 23.4 Bond price by the Vasicek model

23.3 The Cox–Ingersoll–Ross Model

Consider the Cox–Ingersoll–Ross model [22], or the CIR model for short, given by

$$\mathrm{d}r_t = \alpha(\bar{r} - r_t)\,\mathrm{d}t + \sigma\sqrt{r_t}\,\mathrm{d}\widetilde{W}_t\ .$$

There are two major differences from the Vasicek model. First, due to $\sigma\sqrt{r_t}$, which is the coefficient of $\mathrm{d}W_t$, volatility itself has randomness. Second, the interest rate is always nonnegative. If r_t becomes zero, then the coefficient of $\mathrm{d}W_t$ is also zero and the coefficient of $\mathrm{d}t$ is positive, and interest rate becomes positive again.

In Fig. 23.5 are presented sample paths of r_t, where the solid line is the graph of the expectation of r_t that converges to the level $\bar{r}$, and the two dotted lines above and below the solid line represent the range of r_t within one standard deviation from the mean. See Simulation 23.2.

Theorem 23.4 *The interest rate r_t in the CIR model has the noncentral chi-squared distribution. More precisely, the probability density function of the interest rate r at time t, with $r = r_0$ at time $t = 0$, is given by*

$$f(r) = c_t\,\chi^2(c_t r, d, \lambda_t)$$

where the values of the degrees of freedom d, the non-centrality parameter λ_t, and normalizing constant c_t are given by

$$d = \frac{4\alpha\bar{r}}{\sigma^2}\ ,$$

$$\lambda_t = c_t r_0 \mathrm{e}^{-\alpha t}\ ,$$

$$c_t = \frac{4\alpha}{\sigma^2(1 - \mathrm{e}^{-\alpha t})}\ .$$

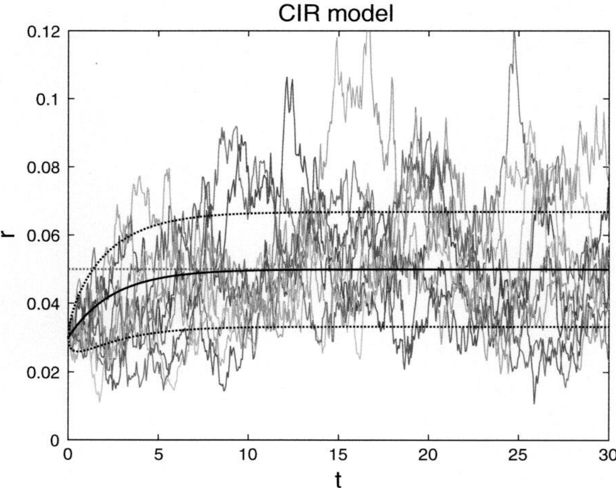

Fig. 23.5 Sample paths of the interest rate in the CIR model

For $s \le t$

$$\mathbb{E}[r_t|r_s] = r_s\,\mathrm{e}^{-\alpha(t-s)} + \bar{r}\left(1 - \mathrm{e}^{-\alpha(t-s)}\right)$$

and

$$\mathrm{Var}(r_t|r_s) = r_s\,\frac{\sigma^2}{\alpha}\left(\mathrm{e}^{-\alpha(t-s)} - \mathrm{e}^{-2\alpha(t-s)}\right) + \bar{r}\,\frac{\sigma^2}{2\alpha}\left(1 - \mathrm{e}^{-\alpha(t-s)}\right)^2 .$$

Note that $\lim_{t\to\infty}\mathbb{E}[r_t] = \bar{r}$ *and* $\lim_{t\to\infty}\mathrm{Var}(r_t) = \bar{r}\dfrac{\sigma^2}{2\alpha}$. *The limiting distribution itself is given by* $\dfrac{4\alpha}{\sigma^2}\chi^2\left(\dfrac{4\alpha}{\sigma^2}r, \dfrac{4\alpha}{\sigma^2}\bar{r}, 0\right)$.

Proof For the part concerning the noncentral chi-squared distribution, see [18]. For the part for the mean and variance, consult [14]. □

Remark 23.4 Let

$$\tau = \inf\{t > 0 : r_t \le 0\}$$

be the first hitting time that the interest rate r_t becomes zero or possibly negative. (By convention, $\inf\emptyset = +\infty$.) Note that $\{\tau = +\infty\}$ is the event that X never hits zero. If $2\alpha\bar{r} \ge \sigma^2$ then

$$\Pr(\tau = +\infty) = 1 ,$$

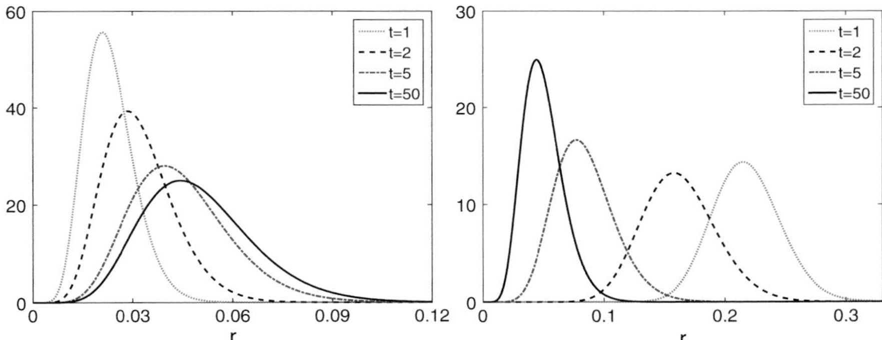

Fig. 23.6 Probability distribution of the interest rate r_t, $t = 1, 2, 5, 50$, in the CIR model: $r_0 = 0.01$ (*left*) and $r_0 = 0.3$ (*right*)

i.e., $r_t > 0$ for every $t > 0$ with probability 1. If $2\alpha\bar{r} < \sigma^2$, then

$$\Pr(\tau < +\infty) = 1 ,$$

i.e., the interest rate becomes zero or negative in finite time. For the proof, see [18].

For detailed information on the noncentral chi-squared distribution see Example 4.16. In Fig. 23.6, obtained from Simulation 23.3, the interest rate r_t is nonnegative. The probability distributions for r_t are plotted for $t = 1, 2, 5, 50$ with the initial condition $r_0 = 1\%$ in the left panel, and $r_0 = 30\%$ in the right panel. In both cases the distribution converges to a limiting distribution around $\bar{r} = 0.05$.

Theorem 23.5 *The bond price under the CIR model is given by*

$$P(r, t, T) = \mathrm{e}^{A(T-t)-B(T-t)\,r}$$

where

$$\gamma = \sqrt{\alpha^2 + 2\sigma^2} ,$$

$$A(\tau) = \frac{2\alpha\bar{r}}{\sigma^2} \log\left(\frac{2\gamma \mathrm{e}^{(\alpha+\gamma)\tau/2}}{(\alpha+\gamma)(\mathrm{e}^{\gamma\tau}-1)+2\gamma}\right),$$

$$B(\tau) = \frac{2(\mathrm{e}^{\gamma\tau}-1)}{(\alpha+\gamma)(\mathrm{e}^{\gamma\tau}-1)+2\gamma} .$$

Proof The equations in (23.1) now become

$$\begin{cases} \dfrac{\partial}{\partial t}A(t,T) = a(t)B(t,T) + \dfrac{1}{2}c(t)B(t,T)^2 , & A(T,T) = 0 , \\ \dfrac{\partial}{\partial t}B(t,T) = \alpha B(t,T) + \dfrac{\sigma^2}{2}B(t,T)^2 - 1 , & B(T,T) = 0 . \end{cases} \tag{23.3}$$

The second equation in (23.3), called the Riccati equation, has a solution given in the statement even though it is nonlinear. By substituting it in the first equation, we can express $\frac{\partial}{\partial t}A(t,T)$ in terms of constant and exponential functions of t, which can be integrated easily. □

Figure 23.7 shows the curves of $R(0,\tau)$ corresponding to $r_0 = 0.01, 0.04, 0.07$, and Fig. 23.8 shows the surface given by the bond price with $\alpha = 0.4$, $\bar{r} = 0.05$, $\sigma = 0.0671$, $T = 10$ and $0 \le r \le 0.5$ under the CIR model.

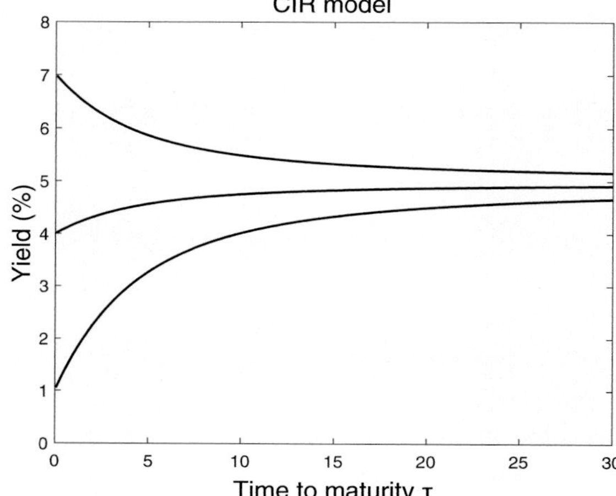

Fig. 23.7 Term structure of interest rates in the CIR model

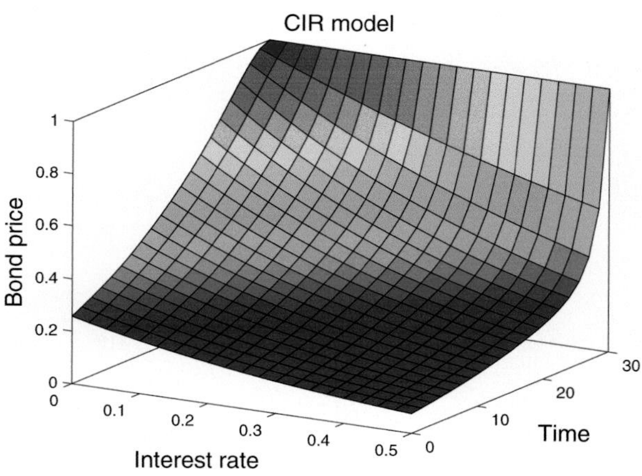

Fig. 23.8 Bond price by the CIR model

23.4 The Ho–Lee Model

Early term structure models such as the Vasicek, and Cox–Ingersoll–Ross models were not arbitrage-free models. Consider the Ho–Lee model [40] given by

$$\mathrm{d}r_t = \theta(t)\,\mathrm{d}t + \sigma\,\mathrm{d}\widetilde{W}_t$$

where $\widetilde{W}_t$ is a $\mathbb{Q}$-Brownian motion, and $\theta(t)$ and the constant σ are calibrated by the observed market data. The Ho–Lee model was the first arbitrage-free model.

Theorem 23.6 *The bond price under the Ho–Lee model is given by*

$$P(t,T) = \mathrm{e}^{A(t,T)-(T-t)\,r}$$

where

$$A(t,T) = \frac{\sigma^2}{6}\,(T-t)^3 - \int_t^T \theta(s)\,(T-s)\,\mathrm{d}s\,.$$

Proof In the Ho–Lee model, since $a(t) = \theta(t)$, $b(t) = 0$, $c(t) = \sigma^2$ and $d(t) = 0$ in (23.1), we have

$$\begin{cases} \dfrac{\partial}{\partial t}A(t,T) = \theta(t)B(t,T) - \dfrac{1}{2}\sigma^2 B(t,T)^2\,, & A(T,T) = 0\,, \\ \dfrac{\partial}{\partial t}B(t,T) = -1\,, & B(T,T) = 0\,. \end{cases}$$

Hence $B(t,T) = T - t$ and

$$\frac{\partial}{\partial t}A(t,T) = \theta(t)\,(T-t) - \frac{1}{2}\sigma^2 (T-t)^2\,.$$

Thus

$$\begin{aligned} A(t,T) &= A(t,T) - A(T,T) \\ &= -\int_t^T \frac{\partial A(s,T)}{\partial s}\,\mathrm{d}s \\ &= -\int_t^T \theta(s)\,(T-s)\,\mathrm{d}s + \frac{1}{2}\sigma^2 \int_t^T (T-s)^2 \mathrm{d}s\,, \end{aligned}$$

which completes the proof. □

Remark 23.5 Recall that the instantaneous forward rate is given by

$$\begin{aligned} f(0,T) &= -\frac{\partial}{\partial T}\log P(0,T) \\ &= -\frac{\partial}{\partial T}(A(0,T) - Tr_0) \\ &= -\frac{\partial A(0,T)}{\partial T} + r_0 \\ &= -\frac{\sigma^2}{2}T^2 + \int_0^T \theta(s)\,\mathrm{d}s\,. \end{aligned}$$

(See Definition 22.8.) We assume that at $t = 0$ for some T_1, say $T_1 = 30$ years, we have the information of the bond prices $P(0,T)$ at $t = 0$ for every maturity $0 \le T \le T_1$ from the market.[1] Hence $f(0,T)$, $0 \le T \le T_1$, is also known at $t = 0$, and used as input data into the formula. Since

$$f(0,T) = -\frac{\partial A(0,T)}{\partial T} + r_0 = -\frac{\sigma^2}{2}T^2 + \int_0^T \theta(s)\mathrm{d}s\,,$$

we have

$$\theta(T) = \frac{\partial f(0,T)}{\partial T} + \sigma^2 T\,.$$

Remark 23.6 (Drawbacks of the Ho–Lee Model) The interest rate process in the Ho–Lee model can grow to $\pm\infty$ as $T \to +\infty$, which is not realistic. Furthermore, the volatility of the changes in long-term interest rates, which is measured by the variance of $\mathrm{d}R(t, t+\tau)$, is equal to the volatility of the changes in the short-term interest rate since

$$R(t, t+\tau) = \frac{-A(\tau) + \tau r_t}{\tau} = -\frac{A(\tau)}{\tau} + r_t\,.$$

That is, for a fixed τ we have

$$\mathrm{Var}(\mathrm{d}R(t, t+\tau)) = \mathrm{Var}(\mathrm{d}r_t) = \sigma^2 \mathrm{d}t\,.$$

This is not in agreement with the market data which shows gradual decline as a function of τ.

---

[1] In practice, since $P(0,T)$ is given only for finitely many discrete values of T in real data, we apply some interpolation method by sufficiently smooth curves. Thus we may assume that it is differentiable in T as many times as needed.

23.5 The Hull–White Model

Consider the Hull–White model [42] given by

$$dr_t = (\theta(t) - \gamma\, r_t)\, dt + \sigma\, d\widetilde{W}_t$$

where $\widetilde{W}_t$ is a $\mathbb{Q}$-Brownian motion, and the deterministic function $\theta(t)$ and the constant σ are calibrated by the market data. More precisely, the function $\theta(t)$ is chosen first to match exactly the term structure of interest rates, and after that we choose σ and γ to fit the term structure of spot rate volatilities.

Theorem 23.7 *The bond pricing formula in the Hull–White model is given by*

$$P(r, t; T) = e^{A(t;T)-B(t;T)\, r}$$

where

$$B(t; T) = \frac{1}{\gamma}(1 - e^{-\gamma(T-t)})$$

and

$$A(t; T) = -\int_t^T B(s; T)\, \theta(s)\, ds + \frac{\sigma^2}{2\gamma^2}\left(T - t + \frac{1 - e^{-2\gamma(T-t)}}{2\gamma} - 2B(t; T)\right) .$$

Remark 23.7

(i) It can be shown that

$$\theta(t) = \frac{\partial f(0, t)}{\partial t} + \gamma f(0, t) + \frac{\sigma^2}{2\gamma}(1 - e^{-2\gamma t}) .$$

(ii) Since

$$R(t, t + \tau) = -\frac{A(\tau)}{\tau} + \frac{1}{\gamma}\frac{1 - e^{-\gamma\tau}}{\tau} r_t ,$$

we have

$$dR(t, t + \tau) = \frac{1}{\gamma}\frac{1 - e^{-\gamma\tau}}{\tau} dr_t .$$

Let

$$f(\tau) = \frac{1}{\gamma}\frac{1 - e^{-\gamma\tau}}{\tau} , \quad \tau > 0 .$$

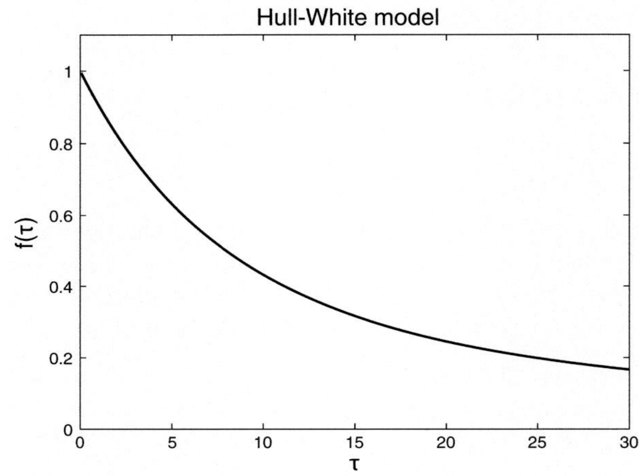

Fig. 23.9 Volatility decreases as time to maturity τ increases in the Hull–White model

Then

$$\mathrm{Var}(\mathrm{d}R(t, t+\tau)) = f(\tau)^2 \mathrm{Var}(\mathrm{d}r_t) = f(\tau)^2 \sigma^2 \mathrm{d}t ,$$

which is decreasing as τ increases since

$$\begin{aligned} \frac{\partial f}{\partial \tau}(\tau) &= \frac{1}{\gamma} \frac{\gamma \mathrm{e}^{-\gamma\tau}\tau - (1 - \mathrm{e}^{-\gamma\tau})}{\tau^2} \\ &= \frac{\mathrm{e}^{-\gamma\tau}(\gamma\tau + 1) - 1}{\gamma\tau^2} \\ &= \frac{(\gamma\tau + 1) - \mathrm{e}^{\gamma\tau}}{\gamma\tau^2 \mathrm{e}^{\gamma\tau}} < 0 \end{aligned}$$

in agreement with the real market term structure. See Fig. 23.9 for the plot of the graph of $f(\tau)$, $0 < \tau \leq 30$.

23.6 Computer Experiments

Simulation 23.1 (Vasicek Model)

We generate $M = 10$ sample paths of the interest rate in the Vasicek model. See Fig. 23.1.

```
T= 30;
N = 500;
dt = T/N;
t = 0:dt:T;
M = 10;
```

```
r = zeros(M,N);
alpha = 0.4;
r_bar = 0.05;
sigma = 0.015;
r0 = 0.03;
r(:,1) = r0 ;
dW = sqrt(dt)*randn(M,N);
for i = 1:N
    r(:,i+1) = r(:,i) + a*(r_bar - r(:,i))*dt + sigma*dW(:,i);
end
for j = 1:M
    plot(t,r(j,:),'color',hsv2rgb([1-j/M 1 1]));
    hold on
end
t = 0:dt:T;
plot(t,0,'k')
plot(t,r_bar,'k:')
% mean
plot(t,r_bar+(r0-r_bar)*exp(-alpha*t))

% standard deviation
stdev = sqrt(sigma^2/(2*alpha)*(1-exp(-2*alpha*t)));
plot(t,r_bar+(r0-r_bar)*exp(-alpha*t) + stdev,'k-.');
plot(t,r_bar+(r0-r_bar)*exp(-alpha*t) - stdev,'k-.');
```

Simulation 23.2 (Cox–Ingersoll–Ross Model)

We generate $M = 10$ sample paths of the interest rate in the CIR model. Except for $\sigma = 0.015/\sqrt{\bar{r}} = 0.0671$ to match the volatility level in the Vasicek model we take the same set of parameter values for T, N, M, α, $\bar{r}$ and r_0. See Fig. 23.5.

```
T= 30;
N = 500;
dt = T/N;
t = 0:dt:T;
M = 10;
r = zeros(M,N);
alpha = 0.4;
r_bar = 0.05;
sigma = 0.015/sqrt(r_bar)
r0 = 0.03;
r(:,1) = r0 ;
dW = sqrt(dt)*randn(M,N);

% Avoid the event that the discretized interest rate becomes negative!
for i = 1:N
    r(:,i+1) = max(0,r(:,i) + alpha*(r_bar-r(:,i))*dt...
                                  + sigma*sqrt(r(:,i)).*dW(:,i));
end

for j = 1:M
    plot(t,r(j,:),'color',hsv2rgb([1-j/M 1 1]));
    hold on
end
t = 0:dt:T;
```

```
plot(t,0,'k--')
plot(t,r_bar,'b')

% expectation
Exp_r = r0*exp(-alpha*t)+r_bar*(1-exp(-alpha*t));
plot(t,Exp_r)

% standard deviation
stdev = sqrt(r0*sigma^2/alpha*(exp(-alpha*t)-exp(-2*alpha*t))...
                          + r_bar*sigma^2/(2*alpha)*(1-exp(-alpha*t)).^2);
plot(t,Exp_r + stdev,'k-.');
plot(t,Exp_r - stdev,'k-.');
```

Simulation 23.3 (CIR Model: Probability Distribution)

We plot the probability density functions of r_t in the Cox–Ingersoll–Ross model for several values of t as shown in Fig. 23.6. We start with $r_0 = 1\%$ and observe that the limiting distribution of r_t for $t = 50$ is distributed around $\bar{r} = 5\%$.

```
alpha = 0.4;
r_bar = 0.05;
r0 = 0.01;
sigma = 0.015;
r_max = 0.07;
N = 500;
dr = r_max/N;

t0 = 1;
c_t = 4*alpha/sigma^2/(1-exp(-alpha*t0));
d = 4*alpha*r_bar/sigma^2;
lambda_t = c_t*r0*exp(-alpha*t0);
r = 0:dr:r_max;
y0=c_t*ncx2pdf(c_t*r,d,lambda_t);  % noncentral chi-squared distribution
plot(r,y0)
hold on;
```

We plot the same type of graphs for $t_1 = 2$, $t_2 = 5$ and $t_3 = 50$ in the same panel. See the left panel in Fig. 23.6.

Simulation 23.4 (The Fundamental Equation for Pricing)

We simulate the Delta hedging used in the derivation of the fundamental equation for pricing interest rate derivatives in Chap. 22. (See the proof of Theorem 22.3.) In the following discrete time version that is used for numerical simulation we check the self-financing condition on the risk-free cash account C_t. Consider a portfolio Π_t given by

$$\Pi_t = V_{1,t} - \Delta_t V_{2,t} + C_t \ . \tag{23.4}$$

Hence the self-financing condition implies that

$$\begin{aligned}\delta\Pi_t &= \delta V_{1,t} - \Delta_t \delta V_{2,t} + r_t C_t \delta t \\ &= (V_{1,t+\delta t} - V_{1,t}) - \Delta_t (V_{2,t+\delta t} - V_{2,t}) + r_t C_t \delta t \ .\end{aligned} \tag{23.5}$$

After discrete time rebalancing, we have

$$\Pi_{t+\delta t} = V_{1,t+\delta t} - \Delta_{t+\delta t} V_{2,t+\delta t} + C_{t+\delta t} \ . \tag{23.6}$$

If we take

$$\Delta = \frac{\partial V_1}{\partial r} \bigg/ \frac{\partial V_2}{\partial r}$$

in (23.5) as in the proof of Theorem 22.3, then $\delta\Pi_t$ in (23.5) becomes risk-free and Π_t behaves as if it is a risk-free bank account and satisfies

$$\delta\Pi_t = r_t \Pi_t \delta t \ . \tag{23.7}$$

The simulation result for (23.7) is represented by the vector `Pi_bank` in the MATLAB code given below. The last panel in Fig. 23.10 displays the risk-free investment

$$\Pi_t = \Pi_0 \exp\left(\int_0^t r_u \mathrm{d}u \right) , \quad 0 \le t \le T ,$$

with continuous compounding interest r_t.

On the other hand, (23.4),(23.5) and (23.6) together imply that

$$\begin{aligned} &(V_{1,t+\delta t} - \Delta_{t+\delta t} V_{2,t+\delta t} + C_{t+\delta t}) - (V_{1,t} - \Delta_t V_{2,t} + C_t) \\ &= (V_{1,t+\delta t} - V_{1,t}) - \Delta_t (V_{2,t+\delta t} - V_{2,t}) + r_t C_t \delta t \ , \end{aligned} \tag{23.8}$$

which is simplified to

$$- \Delta_{t+\delta t} V_{2,t+\delta t} + C_{t+\delta t} - C_t = -\Delta_t V_{2,t+\delta t} + r_t C_t \delta t \ . \tag{23.9}$$

Thus

$$C_{t+\delta t} = (1 + r_t \delta t) C_t + (\Delta_{t+\delta t} - \Delta_t) V_{2,t+\delta t} \ . \tag{23.10}$$

The simulation result for C_t is given in the fifth panel in Fig. 23.10, and the corresponding portfolio `Pi_hedging` is given in the last panel. Observe that the two graphs are almost identical.

In the following simulation for the interest rate r_t we employ the Vasicek model, and consider two bonds with maturity dates $T_1 = 3$ and $T_2 = 4$.

```
T1 = 3; T2 = 4;
N = 200; dt = T1/N; t = 0:dt:T1;
alpha = 0.35;
r_bar = 0.05;
sig = 0.015; % sigma
r0 = 0.03;
```

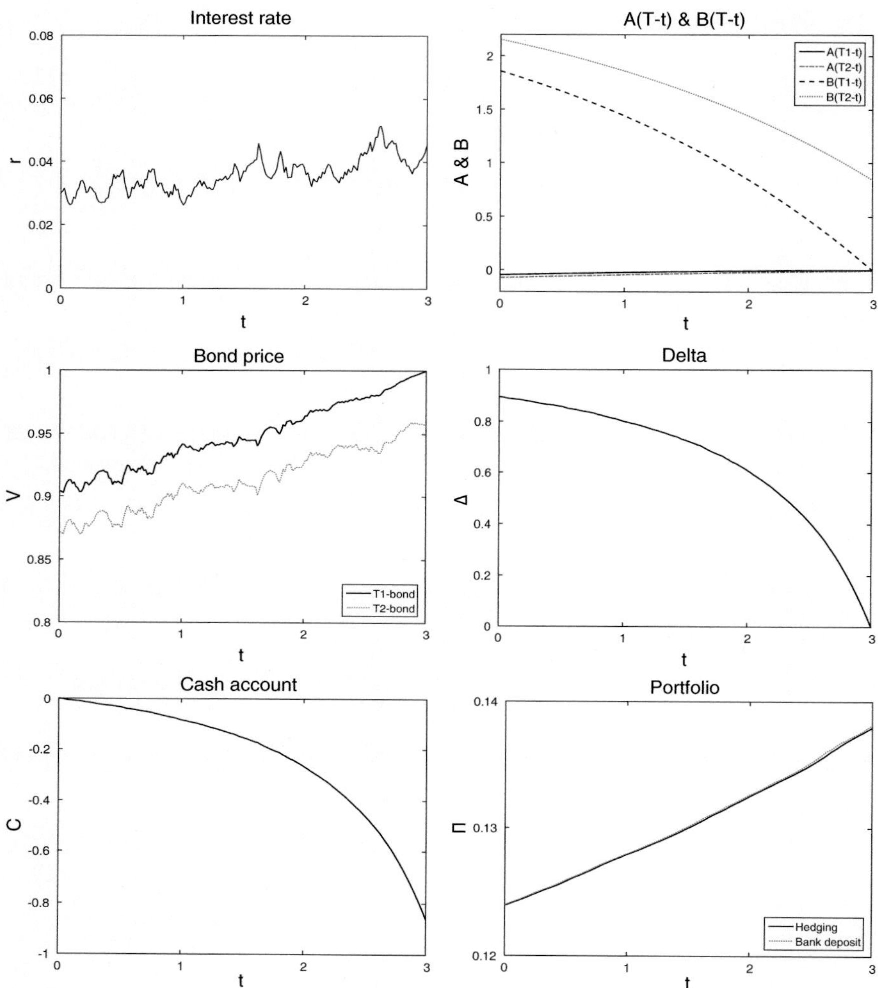

Fig. 23.10 Delta hedging in the derivation of the fundamental equation for pricing interest rate derivatives

```
r(1) = r0;
dW = sqrt(dt)*randn(1,N);
for i = 1:N
    r(i+1) = r(i) + alpha*(r_bar - r(i))*dt + sig*dW(i);
end

figure(1)
plot(t,r);
xlabel('t');
ylabel('r');
```

```
title('Interest rate');

figure(2)
V1 = zeros(N+1,1); % Bond 1 price by Vasicek
Y2 = zeros(N+1,1); % Bond 2 price by Vasicek
% Compute B_k(T_i-t), A_k(T_k-t) and V_k(t) for k=1,2
for i = 1:N+1
B1(i) = 1/alpha*(1-exp(-alpha*(T1-t(i))));
A1(i) = (B1(i)-(T1-t(i)))*(r_bar-sig^2/2/alpha^2)-sig^2*B1(i)^2/4/alpha;
V1(i) = exp(A1(i)-B1(i)*r(i));
end
for i = 1:N+1
B2(i) = 1/alpha*(1-exp(-alpha*(T2-t(i))));
A2(i) = (B2(i)-(T2-t(i)))*(r_bar-sig^2/2/alpha^2)-sig^2*B2(i)^2/4/alpha;
V2(i) = exp(A2(i)-B2(i)*r(i));
end
plot(t,A1,'-k',t,A2,'--r',t,B1,'-.k',t,B2,'r)
legend('A(T1-t)','A(T2-t)','B(T1-t)','B(T2-t)');
xlabel('t');
ylabel('A & B');
title('A(T-t) & B(T-t)');

figure(3)
plot(t,V1,'-k',t,V2,':r')
legend('T1-bond','T2-bond');
xlabel('t');
ylabel('V');
title('Bond price');

figure(4)
Delta = zeros(1,N+1); % hedge ratio
for i = 1:N+1
Delta(i) = B1(i)*V1(i)/(B2(i)*V2(i));
end
plot(t,Delta)
xlabel('t');
ylabel('\Delta');
title('Delta');


figure(5)
% Cash account C with continuously compounding interest
% Choose any amount for C(1).
C = zeros(1,N+1);
% self-financing condition
for i = 1:N
C(i+1) = (1+r(i)*dt)*C(i) + (Delta(i+1)-Delta(i))*V2(i+1);
end
plot(t,C,'Color','k')
xlabel('t');
ylabel('C');
title('Cash account');

figure(6)
```

```
Pi_hedging = zeros(1,N+1); % portfolio of V1 and V2 and Cash
for i = 1:N+1
Pi_hedging(i) = V1(i) - Delta(i)*V2(i) + C(i);
end

Pi_bank = zeros(1,N+1); % risk-free bank deposit
Pi_bank(1) = Pi_hedging(1);
cum_rate = zeros(1,N+1);
for i = 1:N
cum_rate(i+1) = cum_rate(i) + r(i);
Pi_bank(i+1) = Pi_hedging(1)*exp(cum_rate(i)*dt);
end

plot(t,Pi_bank,'k',t,Pi_hedging,':r')
xlabel('t');
ylabel('\Pi');
title('Portfolio');
legend('Hedging','Bank deposit');
```

Exercises

23.1 (Merton Model) Suppose that the interest rate r_t is given by a stochastic differential equation $\mathrm{d}r = \mu\,\mathrm{d}t + \sigma\,\mathrm{d}W$ for some constants μ and σ and that the market price of risk λ is constant.

(i) Show that the price of a zero-coupon bond is given by

$$P(r,t,T) = \exp\left(-r(T-t) - \frac{1}{2}(\mu - \lambda\sigma)(T-t)^2 + \frac{1}{6}\sigma^2(T-t)^3\right).$$

(ii) Show that the yield to maturity of a zero coupon bond in Definition 22.8 is given by

$$R(t,T) = r + \frac{1}{2}(\mu - \lambda\sigma)(T-t) - \frac{1}{6}\sigma^2(T-t)^2 .$$

Does this conclusion agree with common sense?

23.2 (Vasicek Model) For the Vasicek model prove the following facts:

(i) $\mathrm{Cov}(r_t, r_s) = \mathrm{e}^{-\alpha(t+s)}\left(\sigma_0 + \sigma^2\dfrac{\mathrm{e}^{2\alpha s} - 1}{2\alpha}\right).$

(ii) $\mathbb{E}[\int_s^t r_u \mathrm{d}u \,|\, \mathcal{F}_s] = \bar{r}(t-s) + (r_s - \bar{r})\dfrac{1 - \mathrm{e}^{-\alpha(t-s)}}{\alpha}.$

(iii) Derive the limiting probability density of the interest rate directly from the Kolmogorov equation.

23.3 (CIR Model) For the CIR model, derive the limiting pdf of the interest rate directly from the Kolmogorov equation. More precisely, show that the pdf $p(x)$ is

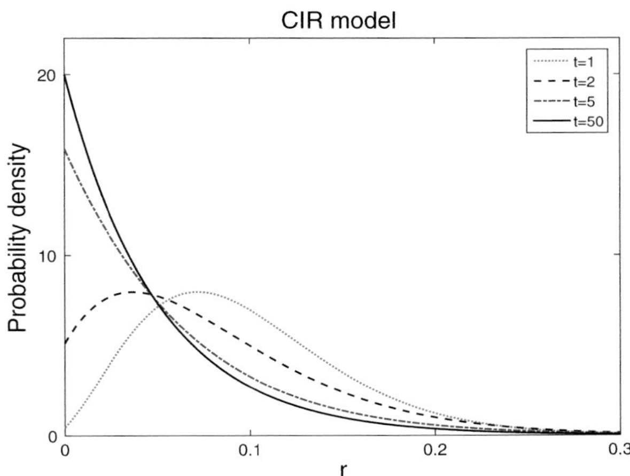

Fig. 23.11 Limit of the pdf of the interest rate r_t in the CIR model with $2\alpha\bar{r} = \sigma^2$

given by

$$p(x) = Cx^{(2\alpha\bar{r})/\sigma^2 - 1} \mathrm{e}^{-(2\alpha/\sigma^2)x} , \quad x \geq 0 ,$$

where $C > 0$ is the normalizing constant satisfying

$$\int_{-\infty}^{\infty} p(x)\mathrm{d}x = 1 .$$

To ensure that

$$\lim_{x \to 0+} p(x) = 0 ,$$

we impose the condition that $2\alpha\bar{r} > \sigma^2$. What happens if $2\alpha\bar{r} = \sigma^2$? (Hint: See Fig. 23.11.)

23.4 (Ho–Lee Model) Show that, in the Ho–Lee Model, for every fixed time to maturity $\tau = T - t$ we have $\mathrm{Var}(\mathrm{d}R(t, \tau)) = \mathrm{Var}(\mathrm{d}r_t) = \sigma^2 \mathrm{d}t$.

Chapter 24
Numeraires

A *numeraire* is a reference asset against which all other assets are evaluated. For example, the concept of time value of money is equivalent to discounting assets using the risk-free bond as a numeraire. Sometimes, a suitable choice of a numeraire makes the computation of option prices easier.

24.1 Change of Numeraire for a Binomial Tree Model

As an illustration of numeraire change we present a discrete time model. Consider a one period binomial tree model for an asset price movement from time $t = 0$ to $t = T$. The values of all the assets including the risk-free bank deposit B_t, the underlying asset (for a given option) S_t, and the replicating portfolio V_t are determined by the ratio to a numeraire, which is the reference asset in our computation of option price. Assume that $B_0 = 1$ and $B_T = e^{rT}$ where r is the risk-free interest rate for the time period of length T. Also assume that S_T can have two values uS_0 and dS_0 where $d < e^{rT} < u$ depending on up and down states at T. See Fig. 24.1. As usual, we take a probability space $\Omega = \{\mathrm{u}, \mathrm{d}\}$ where u and d represent up and down states respectively, and choose a filtration $\mathcal{F}_0 = \{\emptyset, \Omega\}$, $\mathcal{F}_T = \{\emptyset, \Omega, \{\mathrm{u}\}, \{\mathrm{d}\}\}$.

Now we choose the underlying asset S_t as a numeraire. Then the discounted assets corresponding to B_t and S_t are denoted by $\widetilde{B}_t = \frac{B_t}{S_t}$ and $\widetilde{S}_t = \frac{S_t}{S_t} = 1$, respectively. For example, $\widetilde{B}_0 = \frac{1}{S_0}$. See Fig. 24.2.

Note that $\widetilde{S}_t$ is a martingale with respect to any probability measure $\mathbb{Q} = (q_u, q_d)$, i.e.,

$$\mathbb{E}^{\mathbb{Q}}[\widetilde{S}_T|\mathcal{F}_0] = \widetilde{S}_0$$

since $\widetilde{S}_T = \widetilde{S}_0 = 1$. (See the diagram on the right in Fig. 24.2.)

G.H. Choe, *Stochastic Analysis for Finance with Simulations*, Universitext,
DOI 10.1007/978-3-319-25589-7_24

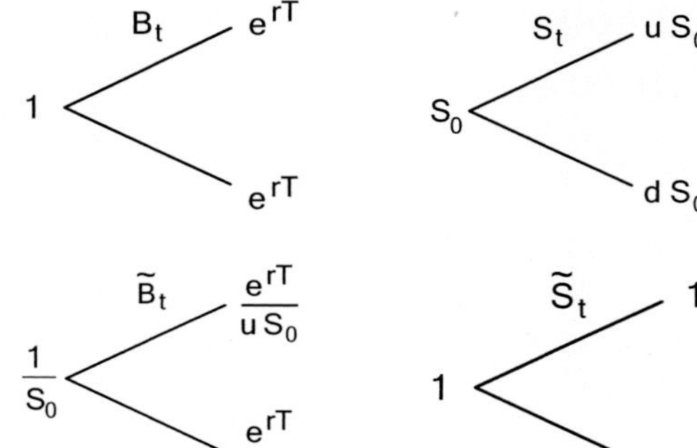

Fig. 24.1 The bank deposit and a risky asset price

Fig. 24.2 The discounted bank deposit and a discounted risky asset price when the risky asset is a numeraire

Lemma 24.1 *There exists a probability measure* $\mathbb{Q} = (q_u, q_d)$ *for which the following statements hold true:*

(i) The discounted bank deposit $\widetilde{B}_t = \frac{B_t}{S_t}$ *in terms of the numeraire* S_t *is a martingale, i.e.,*

$$\mathbb{E}^{\mathbb{Q}}[\widetilde{B}_T | \mathcal{F}_0] = \widetilde{B}_0 \, .$$

(ii) Consider a portfolio $V_t = c_1 S_t + c_2 B_t$. *Then* $\widetilde{V}_t = \frac{V_t}{S_t}$ *is a* $\mathbb{Q}$*-martingale.*

Proof (i) It suffices to show

$$q_u \left(\frac{\mathrm{e}^{rT}}{uS_0} \right) + q_d \left(\frac{\mathrm{e}^{rT}}{dS_0} \right) = \frac{1}{S_0} \tag{24.1}$$

together with

$$q_u + q_d = 1 \tag{24.2}$$

for some $q_u > 0$ and $q_d > 0$. Note that the system of linear equations defined by (24.1) and (24.2) is equivalent to

$$\begin{cases} dq_u + uq_d = ud\mathrm{e}^{-rT} \\ \ \ q_u + q_d \ \ = 1 \end{cases}$$

which has a solution

$$q_u = \frac{u(1 - d\mathrm{e}^{-rT})}{u - d} \, , \qquad q_d = \frac{d(u\mathrm{e}^{-rT} - 1)}{u - d} \, .$$

By the no arbitrage principle, we have $dS_0 < e^{rT} S_0 < uS_0$, and hence $q_u > 0$, $q_d > 0$.

(ii) By the linearity of conditional expectation, we have

$$\begin{aligned}\mathbb{E}^{\mathbb{Q}}[\widetilde{V}_T|\mathcal{F}_0] &= \mathbb{E}^{\mathbb{Q}}[c_1\widetilde{S}_T + c_2\widetilde{B}_T|\mathcal{F}_0] \\ &= c_1\mathbb{E}^{\mathbb{Q}}[\widetilde{S}_T|\mathcal{F}_0] + c_2\mathbb{E}^{\mathbb{Q}}[\widetilde{B}_T|\mathcal{F}_0] \\ &= c_1\widetilde{S}_0 + c_2\widetilde{B}_0 ,\end{aligned}$$

and hence $\mathbb{E}^{\mathbb{Q}}[\widetilde{V}_T|\mathcal{F}_0] = \widetilde{V}_0$. □

Now we apply the preceding idea to compute the option value at time 0.

Theorem 24.1 *With the underlying asset as a numeriare, we have the same option pricing formula given in Sect. 14.2 as follows:*

$$V_0 = e^{-rT}\left(\frac{e^{rT} - d}{u-d}V^u + \frac{u - e^{rT}}{u-d}V^d\right) .$$

Proof By Lemma 24.1 we have

$$\frac{V_0}{S_0} = \mathbb{E}^{\mathbb{Q}}\left[\frac{V_T}{S_T}\right] = q_u\frac{V^u}{uS_0} + q_d\frac{V^d}{dS_0} ,$$

and hence

$$V_0 = q_u\frac{1}{u}V^u + q_d\frac{1}{d}V^d = \frac{1 - de^{-rT}}{u-d}V^u + \frac{ue^{-rT} - 1}{u-d}V^d .$$

□

24.2 Change of Numeraire for Continuous Time

Using the stock price S_t as a numeraire in the continuous time model, we derive the formula for the price of a European call option. Let $B_t = e^{rt}$ be the risk-free asset price at time t, and take

$$\widetilde{B}_t = \frac{B_t}{S_t} .$$

We will find an equivalent probability measure $\mathbb{Q}$ for which $\widetilde{B}_t$ is a $\mathbb{Q}$-martingale. By Ito's lemma,

$$\mathrm{d}\left(\frac{1}{S_t}\right) = \left(-\frac{1}{S_t^2}\right)\mathrm{d}S_t + \frac{1}{2}\frac{2}{S_t^3}(\mathrm{d}S_t)^2 = -\frac{1}{S_t}\mu\mathrm{d}t - \frac{1}{S_t}\sigma\mathrm{d}W_t + \frac{1}{S_t}\sigma^2\mathrm{d}t\,.$$

Hence

$$S_t\,\mathrm{d}\left(\frac{1}{S_t}\right) = (\sigma^2-\mu)\mathrm{d}t - \sigma\mathrm{d}W_t\,. \tag{24.3}$$

Then

$$\begin{aligned}
\mathrm{d}\widetilde{B}_t &= \mathrm{d}\Big(\mathrm{e}^{rt}\frac{1}{S_t}\Big) \\
&= r\mathrm{e}^{rt}\frac{1}{S_t}\mathrm{d}t + \mathrm{e}^{rt}\frac{1}{S_t}\big((\sigma^2-\mu)\mathrm{d}t - \sigma\mathrm{d}W_t\big) \\
&= \frac{\mathrm{e}^{rt}}{S_t}\big((r+\sigma^2-\mu)\mathrm{d}t - \sigma\mathrm{d}W_t\big) \\
&= -\sigma\widetilde{B}_t\left(\frac{r+\sigma^2-\mu}{-\sigma}\mathrm{d}t + \mathrm{d}W_t\right)
\end{aligned}$$

where we used

$$\left[\mathrm{d}B_t, \mathrm{d}\left(\frac{1}{S_t}\right)\right]_t = 0\,.$$

Put

$$\theta = \frac{r+\sigma^2-\mu}{-\sigma} = \frac{\mu-r}{\sigma} - \sigma$$

and

$$X_t = W_t + \theta t\,.$$

Then

$$\mathrm{d}S_t = \mu S_t\mathrm{d}t + \sigma S_t\mathrm{d}W_t = (r+\sigma^2)S_t\mathrm{d}t + \sigma S_t\mathrm{d}X_t\,. \tag{24.4}$$

Let $\mathbb{Q}$ be a probability measure defined by

$$\mathrm{d}\mathbb{Q} = \exp\left(-\frac{1}{2}\theta^2T - \theta W_T\right)\mathrm{d}\mathbb{P}\,.$$

Then X_t is a $\mathbb{Q}$-Brownian motion by Girsanov's theorem, and $\widetilde{B}_t$ is a $\mathbb{Q}$-martingale since

$$d\widetilde{B}_t = -\sigma \widetilde{B}_t \, dX_t \ . \tag{24.5}$$

Define $\widetilde{V}_t$ and V_t by

$$\widetilde{V}_t = \frac{V_t}{S_t} = \mathbb{E}^{\mathbb{Q}}\left[\frac{C_T}{S_T} \middle| \mathcal{F}_t \right]$$

where C_T is the contingent claim on expiry date T. Then

$$V_T = S_T \mathbb{E}^{\mathbb{Q}}\left[\frac{C_T}{S_T} \middle| \mathcal{F}_T \right] = S_T \frac{C_T}{S_T} = C_T \ .$$

Now, to prove that V_t is the option price, it remains to show that V_t is self-financing. Since $\widetilde{V}_t$ is a $\mathbb{Q}$-martingale, by the Martingale Representation Theorem and (24.5) there exists a process α_t such that

$$\widetilde{V}_t = \widetilde{V}_0 + \int_0^t \alpha_u dX_u = \widetilde{V}_0 + \int_0^t \alpha_u \left(-\frac{1}{\sigma \widetilde{B}_u} \right) d\widetilde{B}_u \ .$$

Put

$$\phi_t = \alpha_t \left(-\frac{1}{\sigma \widetilde{B}_t} \right) \ .$$

Then

$$\widetilde{V}_t = \widetilde{V}_0 + \int_0^t \phi_u d\widetilde{B}_u \ ,$$

i.e.,

$$d\widetilde{V}_t = \phi_t \, d\widetilde{B}_t \ . \tag{24.6}$$

Now let

$$\psi_t = \widetilde{V}_t - \phi_t \widetilde{B}_t \ .$$

Then

$$\widetilde{V}_t = \phi_t \widetilde{B}_t + \psi_t \ ,$$

and

$$V_t = \widetilde{V}_t S_t = \phi_t B_t + \psi_t S_t \,. \tag{24.7}$$

Now let

$$\mathrm{d}V_t = a_t\,\mathrm{d}t + b_t\,\mathrm{d}W_t \tag{24.8}$$

for some a_t and b_t. Then (24.3) implies that the covariation of V_t and $1/S_t$ is given by

$$\left[\mathrm{d}V_t, \mathrm{d}\Big(\frac{1}{S_t}\Big)\right]_t = -\frac{\sigma b_t}{S_t}\mathrm{d}t \,. \tag{24.9}$$

From (24.6), we have

$$\frac{1}{S_t}\mathrm{d}V_t + V_t\,\mathrm{d}\left(\frac{1}{S_t}\right) + \left[\mathrm{d}V_t, \mathrm{d}\left(\frac{1}{S_t}\right)\right]_t = \phi_t\left\{\frac{1}{S_t}\mathrm{d}B_t + B_t\,\mathrm{d}\left(\frac{1}{S_t}\right)\right\} \,.$$

Hence (24.3) and (24.9) imply that

$$\mathrm{d}V_t + V_t\left((\sigma^2-\mu)\mathrm{d}t - \sigma\mathrm{d}W_t\right) - \sigma b_t\mathrm{d}t = \phi_t\left\{\mathrm{d}B_t + B_t\left((\sigma^2-\mu)\mathrm{d}t - \sigma\mathrm{d}W_t\right)\right\} \,.$$

Then (24.7) implies that

$$\mathrm{d}V_t = \phi_t\mathrm{d}B_t - \psi_t S_t\left((\sigma^2-\mu)\mathrm{d}t - \sigma\mathrm{d}W_t\right) + \sigma b_t\mathrm{d}t \,, \tag{24.10}$$

Thus, taking covariances with $\mathrm{d}(1/S_t)$ of both sides of (24.10), we have

$$b_t\left(-\frac{\sigma}{S_t}\right)\mathrm{d}t = (-\psi S_t)(-\sigma)\left(-\frac{\sigma}{S_t}\right)\mathrm{d}t \,.$$

Hence

$$b_t = \sigma\psi_t S_t \,. \tag{24.11}$$

Substituting (24.11) in (24.10), we obtain

$$\begin{aligned} \mathrm{d}V_t &= \phi_t\mathrm{d}B_t - \psi_t S_t\left((\sigma^2-\mu)\mathrm{d}t - \sigma\mathrm{d}W_t\right) + \sigma^2\psi_t S_t\mathrm{d}t \\ &= \phi_t\mathrm{d}B_t - \psi_t S_t\left(-\mu\mathrm{d}t - \sigma\mathrm{d}W_t\right) \\ &= \phi_t\mathrm{d}B_t + \psi_t\mathrm{d}S_t \,. \end{aligned} \tag{24.12}$$

Now (24.8) shows that V_t is self-financing.

Next, since $\widetilde{V}_t$ is a $\mathbb{Q}$-martingale, we have

$$\frac{V_0}{S_0} = \widetilde{V}_0 = \mathbb{E}^{\mathbb{Q}}[\widetilde{V}_T] = \mathbb{E}^{\mathbb{Q}}\left[\frac{V_T}{S_T}\right] = \mathbb{E}^{\mathbb{Q}}\left[\frac{C_T}{S_T}\right] .$$

For a European call option with strike price K and expiry date T, we have

$$C_T = (S_T - K)\mathbf{1}_{\{S_T \geq K\}} ,$$

and hence

$$\begin{aligned} V_0 &= S_0\mathbb{E}^{\mathbb{Q}}\left[\frac{(S_T - K)\mathbf{1}_{\{S_T \geq K\}}}{S_T}\right] \\ &= S_0\mathbb{E}^{\mathbb{Q}}\left[\mathbf{1}_{\{S_T \geq K\}}\right] - S_0 K\mathbb{E}^{\mathbb{Q}}\left[\frac{\mathbf{1}_{\{S_T \geq K\}}}{S_T}\right]. \end{aligned} \tag{24.13}$$

Note that

$$\begin{aligned} &\mathbb{E}^{\mathbb{Q}}[\mathbf{1}_{\{S_T \geq K\}}] \\ &= \mathbb{Q}(\{S_T \geq K\}) \\ &= \mathbb{Q}(\{S_0 \mathrm{e}^{(r+\frac{1}{2}\sigma^2)T+\sigma X_T} \geq K\}) && \text{(by (24.4))} \\ &= \mathbb{Q}\left(\left\{X_T \geq \frac{\log\frac{K}{S_0} - (r+\frac{1}{2}\sigma^2)T}{\sigma}\right\}\right) \\ &= \mathbb{Q}\left(\left\{X_1 \geq \frac{\log\frac{K}{S_0} - (r+\frac{1}{2}\sigma^2)T}{\sigma\sqrt{T}}\right\}\right) && \text{(since } X_T, \sqrt{T}X_1 \sim N(0,T)) \\ &= \mathbb{Q}\left(\left\{X_1 \leq \frac{\log\frac{S_0}{K} + (r+\frac{1}{2}\sigma^2)T}{\sigma\sqrt{T}}\right\}\right) && \text{(since } X_1 \sim N(0,1)) \\ &= N(d_1) , \end{aligned} \tag{24.14}$$

and that

$$\begin{aligned} &S_0\mathbb{E}^{\mathbb{Q}}\left[\frac{\mathbf{1}_{\{S_T \geq K\}}}{S_T}\right] \\ &= \mathbb{E}^{\mathbb{P}}\left[\mathbf{1}_{\{S_T \geq K\}} \exp\left(-(\mu - \frac{1}{2}\sigma^2)T - \sigma W_T\right) \exp\left(-\frac{1}{2}\theta^2 T - \theta\, W_T\right)\right]. \end{aligned}$$

Now put

$$\widetilde{\theta} = \sigma + \theta = \frac{\mu - r}{\sigma} .$$

Then

$$\mu - \frac{1}{2}\sigma^2 + \frac{1}{2}\theta^2 = r + \frac{1}{2}\widetilde{\theta}^2 .$$

Now define another equivalent probability measure $\mathbb{Q}'$ by

$$\frac{d\mathbb{Q}'}{d\mathbb{P}} = \exp\left(-\frac{1}{2}\widetilde{\theta}^2 T - \widetilde{\theta}\, W_T\right) .$$

Let

$$Y_t = W_t + \widetilde{\theta} t .$$

Then Y_t is a $\mathbb{Q}'$-Brownian motion, and $\widetilde{S}_t = e^{-rt}S_t = \dfrac{S_t}{B_t}$ is a $\mathbb{Q}'$-martingale. Then

$$\begin{aligned} S_0 K \mathbb{E}^{\mathbb{Q}}\left[\frac{\mathbf{1}_{\{S_T \geq K\}}}{S_T}\right] &= K e^{-rT} \mathbb{E}^{\mathbb{P}}\left[\mathbf{1}_{\{S_T \geq K\}} \exp\left(-\frac{1}{2}\widetilde{\theta}^2 T - \widetilde{\theta}\, W_T\right)\right] \\ &= K e^{-rT} \mathbb{E}^{\mathbb{Q}'}\left[\mathbf{1}_{\{S_T \geq K\}}\right] \\ &= K e^{-rT} \mathbb{Q}'(\{S_T \geq K\}) \\ &= K e^{-rT} \mathbb{Q}'(\{S_0 e^{(r-\frac{1}{2}\sigma^2)T+\sigma Y_T} \geq K\}) \\ &= K e^{-rT} N(d_2) . \end{aligned} \tag{24.15}$$

From (24.13)–(24.15) we may rewrite the Black–Scholes–Merton formula as

$$V_0 = S_0 \mathbb{Q}(\{S_T \geq K\}) - K e^{-rT} \mathbb{Q}'(\{S_T \geq K\}) .$$

In Table 24.1 we list three equivalent probability measures $\mathbb{P}$, $\mathbb{Q}$ and $\mathbb{Q}'$ introduced in the preceding discussion, the corresponding Brownian motions W_t, X_t and Y_t, respectively; and the martingales $\dfrac{B_t}{S_t}$ and $\dfrac{S_t}{B_t}$ with respect to $\mathbb{Q}$ and $\mathbb{Q}'$, respectively.

Remark 24.1 (i) An asset-or-nothing European call option with strike price K and expiry date T has payoff equal to $S_T \mathbf{1}_{\{S_T > K\}}$. Its price V_0^{asset} at time $t = 0$ satisfies

$$\frac{V_0^{\text{asset}}}{S_0} = \mathbb{E}^{\mathbb{Q}}\left[\frac{S_T \mathbf{1}_{\{S_T \geq K\}}}{S_T}\right] = \mathbb{Q}(\{S_T \geq K\}) .$$

Hence $V_0^{\text{asset}} = S_0 N(d_1)$. See also Exercise 17.2.

Table 24.1 Equivalent measures, Brownian motions and martingales

Probability	Brownian motion	Geometric Brownian motion	Martinagle
$\mathbb{P}$	W_t	$\mathrm{d}S_t = \mu S_t\mathrm{d}t + \sigma S_t\mathrm{d}W_t$	
$\mathbb{Q}$	$X_t = W_t + \theta t$	$\mathrm{d}S_t = (r + \sigma^2)S_t\mathrm{d}t + \sigma S_t\mathrm{d}X_t$	B_t/S_t
$\mathbb{Q}'$	$Y_t = W_t + \widetilde{\theta} t$	$\mathrm{d}S_t = rS_t\mathrm{d}t + \sigma S_t\mathrm{d}Y_t$	S_t/B_t

(ii) A cash-or-nothing European call option with strike price K and expiry date T has payoff equal to $\mathbf{1}_{\{S_T\geq K\}}$. Its price V_0^{cash} at time $t = 0$ satisfies

$$\frac{V_0^{\text{cash}}}{B_0} = \mathbb{E}^{\mathbb{Q}'}\left[\frac{\mathbf{1}_{\{S_T\geq K\}}}{B_T}\right] = \mathrm{e}^{-rT}\mathbb{Q}'(\{S_T \geq K\}) \ .$$

Hence $V_0^{\text{cash}} = \mathrm{e}^{-rT}N(d_2)$ as shown in Theorem 17.2.

24.3 Numeraires for Pricing of Interest Rate Derivatives

Let $P(r, t; T)$ be the price of the (zero coupon) bond with maturity T, used as a numeraire to discount an interest rate derivative $V(r, t; T)$, and let

$$\widetilde{V}(r, t; T) = \frac{V(r, t; T)}{P(r, t; T)} \ .$$

Once it is understood that the maturity dates for P and V are both equal to T, and that the time t has the range $0 \leq t \leq T$, we write $P(r, t)$, $V(r, t)$, $\widetilde{V}(r, t)$ instead of $P(r, t; T)$, $V(r, t; T)$, $\widetilde{V}(r, t; T)$, respectively, when there is no danger of confusion. In the following, as in the case for V, we have the fundamental equation for pricing $\widetilde{V}$, too.

Theorem 24.2 (Fundamental Equation for Pricing Discounted Interest Rate Derivatives) *Let*

$$\mathrm{d}r_t = m(r_t, t)\,\mathrm{d}t + \sigma(r_t, t)\,\mathrm{d}W_t$$

describe the interest rate process where W_t is a $\mathbb{Q}$-Brownian motion for a risk-neutral probability measure $\mathbb{Q}$. Then a discounted interest rate derivative $\widetilde{V}$ satisfies

$$\frac{\partial \widetilde{V}}{\partial t} + (m(r, t) + \sigma_P(r, t)\sigma(r, t))\,\frac{\partial \widetilde{V}}{\partial r} + \frac{1}{2}\sigma(r, t)^2\frac{\partial^2 \widetilde{V}}{\partial r^2} = 0 \qquad (24.16)$$

where

$$\sigma_P(r,t) = \frac{1}{P(r,t)} \frac{\partial P(r,t)}{\partial r} \sigma(r,t) \ .$$

Proof Since $V = P\widetilde{V}$, we have

$$\begin{cases} \dfrac{\partial V}{\partial t} = \dfrac{\partial P}{\partial t}\widetilde{V} + P\dfrac{\partial \widetilde{V}}{\partial t} \\ \dfrac{\partial V}{\partial r} = \dfrac{\partial P}{\partial r}\widetilde{V} + P\dfrac{\partial \widetilde{V}}{\partial r} \\ \dfrac{\partial^2 V}{\partial r^2} = \dfrac{\partial^2 P}{\partial r^2}\widetilde{V} + \dfrac{\partial P}{\partial r}\dfrac{\partial \widetilde{V}}{\partial r} + \dfrac{\partial P}{\partial r}\dfrac{\partial \widetilde{V}}{\partial r} + P\dfrac{\partial^2 \widetilde{V}}{\partial r^2} \ . \end{cases}$$

Since Theorem 22.3 holds for P, we have

$$\frac{\partial P}{\partial t} + m(r,t)\frac{\partial P}{\partial r} + \frac{1}{2}\sigma(r,t)^2\frac{\partial^2 P}{\partial r^2} = rP \ ,$$

and hence the following equation, obtained from Theorem 22.3,

$$\left(\frac{\partial P}{\partial t}\widetilde{V} + P\frac{\partial \widetilde{V}}{\partial t}\right) + m(r,t)\left(\frac{\partial P}{\partial r}\widetilde{V} + P\frac{\partial \widetilde{V}}{\partial r}\right) + \frac{1}{2}\sigma(r,t)^2\left(\frac{\partial^2 P}{\partial r^2}\widetilde{V} + 2\frac{\partial P}{\partial r}\frac{\partial \widetilde{V}}{\partial r} + P\frac{\partial^2 \widetilde{V}}{\partial r^2}\right) = rV$$

is simplified to

$$rP\widetilde{V} + P\left(\frac{\partial \widetilde{V}}{\partial t} + m(r,t)\frac{\partial \widetilde{V}}{\partial r} + \frac{1}{2}\sigma(r,t)^2\frac{\partial^2 \widetilde{V}}{\partial r^2}\right) + \sigma(r,t)^2\frac{\partial P}{\partial r}\frac{\partial \widetilde{V}}{\partial r} = rV \ ,$$

which is further reduced to

$$P\left(\frac{\partial \widetilde{V}}{\partial t} + m(r,t)\frac{\partial \widetilde{V}}{\partial r} + \frac{1}{2}\sigma(r,t)^2\frac{\partial^2 \widetilde{V}}{\partial r^2}\right) + \sigma(r,t)^2\frac{\partial P}{\partial r}\frac{\partial \widetilde{V}}{\partial r} = 0$$

since $rP\widetilde{V} = rV$. After dividing by P, we obtain

$$\frac{\partial \widetilde{V}}{\partial t} + m(r,t)\frac{\partial \widetilde{V}}{\partial r} + \frac{1}{2}\sigma(r,t)^2\frac{\partial^2 \widetilde{V}}{\partial r^2} + \sigma(r,t)^2\frac{1}{P}\frac{\partial P}{\partial r}\frac{\partial \widetilde{V}}{\partial r} = 0 \ .$$

Finally, we take $\sigma_P = \sigma\dfrac{1}{P}\dfrac{\partial P}{\partial r}$. □

Remark 24.2 The fundamental equation (24.16) for $\widetilde{V}$ in Theorem 24.2 is different from the result (22.4) for V in Theorem 22.3 in two ways: First, the left-hand side is zero, and the second, $\dfrac{\partial \widetilde{V}}{\partial r}$ has a coefficient that has one more term $\sigma_P(r,t)\sigma(r,t)$.

Definition 24.1 Put

$$\theta = \frac{m - (m + \sigma_P \sigma)}{\sigma} = -\sigma_P \ ,$$

and define an equivalent probability measure $\mathbb{Q}^T$, called the *forward risk-neutral* or *T-forward risk-neutral measure*, by

$$\mathbb{E}^{\mathbb{Q}}\left[\frac{\mathrm{d}\mathbb{Q}^T}{\mathrm{d}\mathbb{Q}}\middle|\mathcal{F}_t\right] = \exp\left(-\frac{1}{2}\int_0^t \theta_s^2 \mathrm{d}s - \int_0^t \theta_s \mathrm{d}W_s\right)$$

where $\{\mathcal{F}_t\}$ is the filtration defined by the information up to t.

Let

$$\widetilde{W}_t = W_t + \int_0^t \theta_s \mathrm{d}s \ .$$

Then, by Theorem 8.2, $\widetilde{W}_t$ is a $\mathbb{Q}^T$-Brownian motion and

$$\mathrm{d}r = m\,\mathrm{d}t + \sigma(\mathrm{d}\widetilde{W} - \theta\,\mathrm{d}t) = (m + \sigma_P\sigma)\mathrm{d}t + \sigma \mathrm{d}\widetilde{W} \ .$$

Theorem 24.3 *If $\widetilde{V}(r,T) = g_T$ is the payoff of a discounted interest rate derivative at maturity T, then*

$$\widetilde{V}(r,t) = \mathbb{E}^{\mathbb{Q}^T}[g_T \mid \mathcal{F}_t]$$

where the expectation is taken over the interest rate given by

$$\mathrm{d}r_t = (m(r_t,t) + \sigma_P(r,t)\sigma(r,t))\ \mathrm{d}t + \sigma(r_t,t)\,\mathrm{d}\widetilde{W}_t \ .$$

In other words,

$$V(r,t;T) = P(r,t;T)\,\mathbb{E}^{\mathbb{Q}^T}[g_T \mid \mathcal{F}_t] \ .$$

Proof Use Theorems 13.1 and 24.2. □

Remark 24.3 (i) The bond price $P(r,t;T)$ is known in the market at time t.
(ii) Recall that Theorem 22.4 states that

$$\begin{aligned} V(r,t;T) &= \mathbb{E}^{\mathbb{Q}}[\mathrm{e}^{-\int_t^T r_u \mathrm{d}u} g_T | \mathcal{F}_t] \\ &\neq \mathbb{E}^{\mathbb{Q}}[\mathrm{e}^{-\int_t^T r_u \mathrm{d}u} | \mathcal{F}_t]\, \mathbb{E}^{\mathbb{Q}}[g_T | \mathcal{F}_t] \\ &= P(r,t;T)\, \mathbb{E}^{\mathbb{Q}}[g_T | \mathcal{F}_t] \end{aligned}$$

since $\mathrm{e}^{-\int_t^T r_u \mathrm{d}u}$ and g_T are not necessarily uncorrelated with respect to $\mathbb{Q}$.

Part IX
Computational Methods

Chapter 25
Numerical Estimation of Volatility

Volatility is the most important parameter in the geometric Brownian motion model for asset price movement for option pricing. All other parameters such as asset price, strike price, time to expiry and the risk-free interest rate can be observed in the financial market. Volatility is used instead of option price for trading in the financial market. To estimate the volatility we employ various techniques, some of which are introduced in this chapter.

25.1 Historical Volatility

Historical volatility is an estimate of volatility based on the past data, usually collected over several years up to the current date. Assume that the asset price follows geometric Brownian motion

$$S(t) = S_t = S_0 \exp\left((\mu - \frac{1}{2}\sigma^2)t + \sigma W_t \right) .$$

Since

$$\log \frac{S_t}{S_u} = (\mu - \frac{1}{2}\sigma^2)(t-u) + \sigma(W_t - W_u)$$

for $0 \leq u < t$, we observe that $\log \frac{S_t}{S_u}$ follows the normal distribution with mean $(\mu - \frac{1}{2}\sigma^2)(t-u)$ and variance $\sigma^2(t-u)$.

G.H. Choe, *Stochastic Analysis for Finance with Simulations*, Universitext,
DOI 10.1007/978-3-319-25589-7_25

Suppose that historical asset price data $S(t_i)$ is available at equally spaced time values $t_i = i\,\delta t$, $1 \le i \le n$, and that $t_n = n \times \delta t$ is the present time. Define

$$L_i = \log \frac{S(t_i)}{S(t_{i-1})} .$$

Then L_i, $1 \le i \le n$, are independent and normally distributed with mean $(\mu - \frac{1}{2}\sigma^2)\delta t$ and variance $\sigma^2 \delta t$. Hence we may write

$$L_i = (\mu - \frac{1}{2}\sigma^2)\,\delta t + \sigma \sqrt{\delta t}\, Z_i \,,$$

where Z_i are independent standard normal variables. Define

$$A_m = \frac{1}{m} \sum_{j=1}^{m} L_{n-j+1} .$$

Assume that n is sufficiently large, and the moving window size m is small so that $n \gg m$. Since

$$\begin{aligned} A_m &= \frac{1}{m} \sum_{j=1}^{m} \big(\log S(t_{n-j+1}) - \log S(t_{n-j})\big) \\ &= \frac{1}{m} \log \frac{S(t_n)}{S(t_{n-m})} , \end{aligned}$$

we have

$$\begin{aligned} A_m &= \frac{1}{m} \Big\{ m(\mu - \frac{1}{2}\sigma^2)\delta t + \sigma (W(t_n) - W(t_{n-m})) \Big\} \\ &\sim N\big((\mu - \frac{1}{2}\sigma^2)\delta t, \frac{\sigma^2}{m}\delta t\big) \end{aligned}$$

where the symbol $\sim$ denotes the probability distribution. Since both the average and the variance of A_m are close to 0, we take

$$A_m = 0 ,$$

i.e., we may assume that the average of L_i is 0, and use

$$\frac{1}{m-1} \sum_{k=1}^{m} (L_{n-k+1} - 0)^2$$

to estimate $\sigma^2\delta t$. (We divide the sum by $m-1$ instead of m to obtain an unbiased estimate.) In other words, to estimate σ we use

$$\sqrt{\frac{1}{\delta t}\frac{1}{m-1}\sum_{k=1}^{m} L_{n-k+1}^2}\ .$$

As mentioned in Theorem 15.2 the dimension of σ is time$^{-1/2}$. If we want to find volatility per year using the daily data, we choose $\delta t = \frac{1}{N}$ where N is the number of trading days in a year, usually $N = 252$, and use the daily record of asset price movement, and estimate the yearly volatility. For the monthly volatility we take $N = 12$.

25.2 Implied Volatility

As inputs in the Black–Scholes–Merton equation we need parameters such as asset price, exercise price, time to expiry, risk-free interest rate and volatility, among which the volatility is not observable. However, we can find it when all other parameters and the option price are given. More precisely, as shown in Sect. 15.3,

$$\text{vega} = \frac{\partial V}{\partial \sigma} = S\sqrt{T-t}\,N'(d_1) > 0\ ,$$

and hence option price V is an increasing function of volatility σ while S_t, K, $T-t$ and r are fixed. Since we can observe S_t, K, $T-t$, r and V in the financial market, the remaining parameter σ is uniquely determined, which is called the *implied volatility*. Historical volatility is based on the past historical data while the implied volatility looks into the future movement of the asset price.

If we plot the implied volatility against exercise price, the middle of the graph is low and both ends are relatively high. Such a phenomena is called a *volatility smile*.

To find the implied volatility we employ numerical methods. First, construct an equation of the form $f(\sigma) = 0$ in terms of an unknown variable σ, and use either the bisection method or the Newton method.

25.2.1 The Bisection Method

In the bisection method we choose a sufficiently wide interval, say $[a, b]$. If $f(a) < 0$ and $f(b) > 0$ then the Intermediate Value Theorem implies that there exists an x^*, $a < x^* < b$, such that $f(x^*) = 0$. Note that if $f(\frac{a+b}{2}) < 0$ then $\frac{a+b}{2} < x^* < b$, and if $f(\frac{a+b}{2}) > 0$ then $a < x^* < \frac{a+b}{2}$ by the Intermediate Value Theorem again. In this way, we can reduce the width of the possible range of x^* by half. Applying the

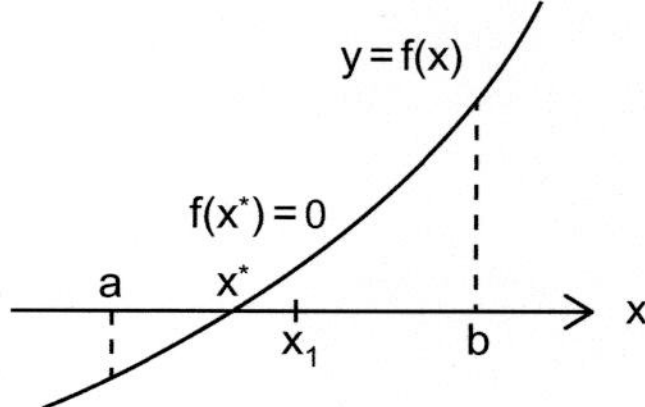

Fig. 25.1 Solution of $f(x^*) = 0$ by the bisection method

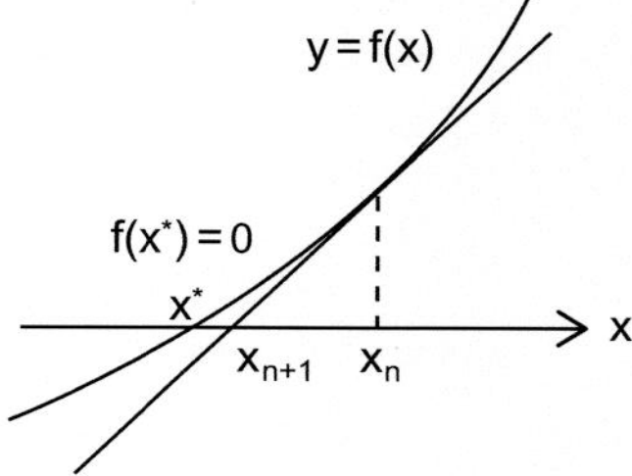

Fig. 25.2 Solution of $f(x^*) = 0$ by the Newton–Raphson method

method n times, we obtain an interval of length $(b-a)/2^n$ in which x^* exists, i.e.,

$$|x^* - x_n| \leq (b-a)\frac{1}{2^n} .$$

See Fig. 25.1.

25.2.2 *The Newton–Raphson Method*

In the Newton–Raphson method, or the Newton method, we take a tangent line to the curve $y = f(x)$ under the assumption that f is differentiable and $f' \neq 0$. At $x = x_n$ the equation for the tangent line is given by

$$y = f'(x_n)(x - x_n) + f(x_n)$$

and the x-intercept is denoted by x_{n+1}. See Fig. 25.2. It is known that $\{x_n\}_{n=1}^{\infty}$ converges to c such that $f(c) = 0$. We apply the iterative algorithm

$$x_{n+1} = x_n - \frac{f(x_n)}{f'(x_n)}$$

repeatedly.

The convergence speed of the Newton–Raphson method is much faster than that of the bisection method as can be seen in the following result.

Theorem 25.1 *If $f(x^*) = 0$ and $f'(x^*) \neq 0$, there exists a constant $A > 0$ such that for every $n \geq 1$ we have*

$$|x_{n+1} - x^*| \leq A\,|x_n - x^*|^2 .$$

Proof Since $f(x^*) = 0$, we have

$$\begin{aligned} x_{n+1} - x^* &= x_n - \frac{f(x_n) - f(x^*)}{f'(x_n)} - x^* \\ &= x_n - \frac{f'(x_n)(x_n - x^*) + O(|x_n - x^*|^2)}{f'(x_n)} - x^* \\ &= x_n - (x_n - x^*) + O(|x_n - x^*|^2) - x^* \\ &= O(|x_n - x^*|^2) \end{aligned}$$

where $O(h^k)$ denotes a quantity satisfying $\limsup_{h\to 0} \frac{O(h^k)}{h^k} \leq C$ for some constant $0 < C < \infty$.

Or, if we want to use the Taylor expansion, let $\delta_n = x^* - x_n$. Then

$$0 = f(x^*) = f(x_n + \delta_n) = f(x_n) + f'(x_n)\delta_n + O(\delta_n^2) ,$$

and hence

$$\delta_n = -\frac{f(x_n)}{f'(x_n)} + O(\delta_n^2) .$$

Therefore

$$x^* = x_n + \delta_n = x_n - \frac{f(x_n)}{f'(x_n)} + O(\delta_n^2) = x_{n+1} + O(\delta_n^2) ,$$

which shows that $\delta_{n+1} = x^* - x_{n+1} = O(\delta_n^2)$. □

Figure 25.3 shows the results of two methods for finding zeros of the equation $e^{-x^2} - \sin x = 0$.

Remark 25.1 (Nonexistence of a Solution) If $f(x) = x^2 + 1$, then there exists no solution for $f(x) = 0$ and the above algorithm generates a nonconvergent sequence. For a more theoretical analysis, consult Chap. 2 on invariant measures in [21].

Theorem 25.2 *Let $C(\sigma)$ denote the European call option price, which is regarded as a function of volatility σ while T, t, r, K, S_t are given. Then $C(\sigma)$ is a strictly monotonically increasing function of σ on $(0,\infty)$, and there exists a σ_0 such that*

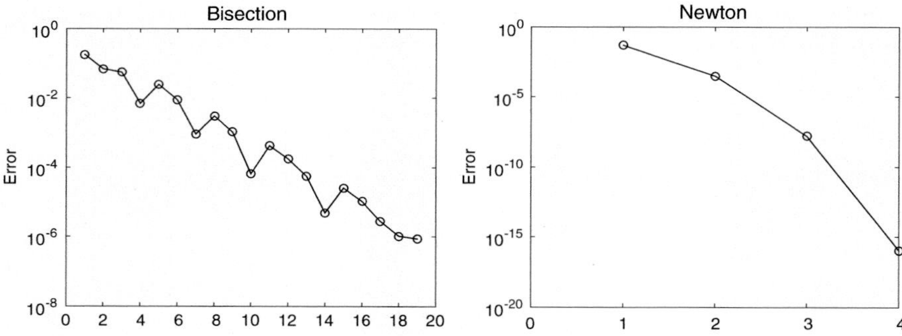

Fig. 25.3 Comparison of efficiency of the bisection method and the Newton method

$C(\sigma)$ is convex on $(0, \sigma_0)$ and concave on (σ_0, ∞). In fact, if $C''(\sigma_0) = 0$, then

$$\sigma_0 = \sqrt{2\frac{|\log(S_t/K) + r(T-t)|}{T-t}} .$$

Proof Note that

$$\frac{\partial C}{\partial \sigma} = \text{vega} = S\sqrt{T-t}\,N'(d_1) > 0 .$$

Also note that

$$\frac{\partial^2 C}{\partial \sigma^2} = -\frac{S\sqrt{T-t}}{\sqrt{2\pi}} \exp\left(-\frac{1}{2}d_1^2\right) d_1 \frac{\partial d_1}{\partial \sigma} = -\frac{\partial C}{\partial \sigma} d_1 \frac{\partial d_1}{\partial \sigma} .$$

Since

$$\frac{\partial d_1}{\partial \sigma} = -\frac{\log(S/K) + r(T-t)}{\sigma^2\sqrt{T-t}} + \frac{1}{2}\sqrt{T-t} = -\frac{d_2}{\sigma} ,$$

we have

$$\frac{\partial^2 C}{\partial \sigma^2} = \frac{\partial C}{\partial \sigma}\frac{d_1 d_2}{\sigma} .$$

Hence either $d_1(\sigma_1) = 0$ or $d_2(\sigma_2) = 0$ for some σ_1, σ_2. If $d_1(\sigma_1) = 0$ then

$$\sigma_1^2 = 2\left(-\frac{1}{T-t}\log\frac{S}{K} - r\right) \geq 0 ,$$

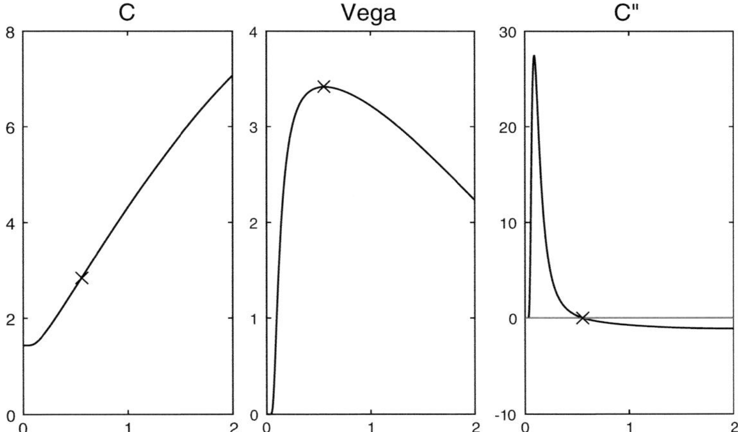

Fig. 25.4 The call price $C(\sigma)$ and its first and second order derivatives for $0 \le \sigma \le 2$

and if $d_2(\sigma_2) = 0$ then

$$\sigma_2^2 = 2\left(\frac{1}{T-t}\log\frac{S}{K} + r\right) \ge 0\,.$$

If both cases occur, then $\log\frac{S}{K} + r(T-t) = 0$, and we would have $d_{1,2} = \pm\frac{1}{2}\sigma\sqrt{T-t}$. Hence $\sigma_1 = \sigma_2 = 0$. □

In Fig. 25.4 we take $T = 1$, $t = 0$, $r = 5\%$, $K = 9$, $S_0 = 10$ and plot the graphs of $C(\sigma)$, $C'(\sigma)$, $C''(\sigma)$. The points marked by 'x' represent the inflection point of $C(\sigma)$, the maximum point of $C'(\sigma)$, and the point where $C''(\sigma) = 0$, respectively.

The following results shows the monotone convergence of the estimated value in computing the implied volatility by the Newton–Raphson algorithm.

Theorem 25.3 *Let*

$$\sigma_{n+1} = \sigma_n - \frac{F(\sigma_n)}{F'(\sigma_n)}$$

where $F(\sigma) = C(\sigma) - C^*$ *where* C^* *denotes the market value of the European call option. Let* σ^* *be a solution of* $F(\sigma^*) = 0$. *Then* $F'(\sigma)$ *takes its maximum at some point* $\sigma = \widehat{\sigma}$, *and the error* $\sigma_n - \sigma^*$ *decreases to 0 monotonically as n increases.*

Proof Note that

$$F'(\sigma) = \frac{\partial C}{\partial \sigma} = \text{vega} = S\sqrt{T-t}\,N'(d_1) > 0\,.$$

Since $F(\sigma^*) = 0$, we have

$$\sigma_{n+1} - \sigma^* = \sigma_n - \frac{F(\sigma_n)}{F'(\sigma_n)} - \sigma^* = \sigma_n - \sigma^* - \frac{(\sigma_n - \sigma^*)F'(\xi_n)}{F'(\sigma_n)}$$

for some ξ_n between σ^* and σ_n by the Mean Value Theorem. Hence

$$\frac{\sigma_{n+1} - \sigma^*}{\sigma_n - \sigma^*} = 1 - \frac{F'(\xi_n)}{F'(\sigma_n)} .$$

Since $F'(\sigma) > 0$ and $F'(\sigma)$ takes its maximum at $\widehat{\sigma}$, if we start from $\sigma_0 = \widehat{\sigma}$ in the Newton–Raphson algorithm, we have

$$0 < \frac{\sigma_1 - \sigma^*}{\sigma_0 - \sigma^*} < 1 ,$$

which implies that the error in σ_1 is smaller than the error in σ_0. Note that they have the same sign. To proceed we suppose that $\widehat{\sigma} < \sigma^*$. Then $\sigma_0 < \sigma_1 < \sigma^*$. Hence $0 < F'(\xi_1) < F'(\sigma_1)$, and

$$0 < \frac{\sigma_2 - \sigma^*}{\sigma_1 - \sigma^*} < 1 .$$

Continuing the same argument, we have

$$0 < \frac{\sigma_{n+1} - \sigma^*}{\sigma_n - \sigma^*} < 1$$

for every $n \geq 1$. Thus the error decreases monotonically as n increases. The same result can be obtained under the assumption that $\widehat{\sigma} > \sigma^*$. □

Remark 25.2 (Volatility Smile) In theory, the implied volatility should be equal for any choice of S and K, but in practice the graph of implied volatility versus moneyness $\frac{S}{K}$ is convex. Such a pattern is called a *volatility smile.* Options did not show a volatility smile before the stock market crash in 1987, but the investors afterwards began to reassess the probabilities of rare events such as financial disasters, and caused higher evaluations of out-of-the-money options. This reflects the fact that the standard Black–Scholes–Merton model assumes log-normal distributions of underlying asset returns and that normal distributions have small tail probabilities. Sometimes, the implied volatility curve is concave, which is called a *volatility frown.*

25.3 Computer Experiments

Simulation 25.1 (Bisection Method and Newton Method)

We compare the bisection method and Newton method. For the output see Fig. 25.3.

```
fun = @(x) exp(-x^2) - sin(x); % function
x0 = 1.5; % initial point
exact_sol= fzero(fun,x0)
% Bisection
a = 0.0;
b = 2.0;
k = 1;
k_max = 20;
F = @(x) exp(-x^2) - sin(x);
if ((F(a)*F(b)) > 0)
        disp('ERROR: F(a) and F(b) must have different signs')
end
xmid = (a + b)/2;
while (k < k_max)
    if ((F(a)*F(xmid)) < 0)
        b = xmid;
    else
        a = xmid;
    end
    xmid = (a + b)/2;
    bis_err(k) = abs(xmid - exact_sol);
    k = k+1;
end
subplot(1,2,1)
semilogy(bis_err,'-o')
title('Bisection')

% Newton
x0 = 1;
x = x0;
increment = 1;
k = 1;
while (k < k_max)
      Fval =  exp(-x^2) - sin(x) ;
      Fprime =  -2*x*exp(-x^2) - cos(x);
      increment = Fval/Fprime;
      x = x - increment;
      newt_err(k) = abs(x - exact_sol);
      k = k+1;
end
subplot(1,2,2)
semilogy(newt_err,'-o')
title('Newton')
```

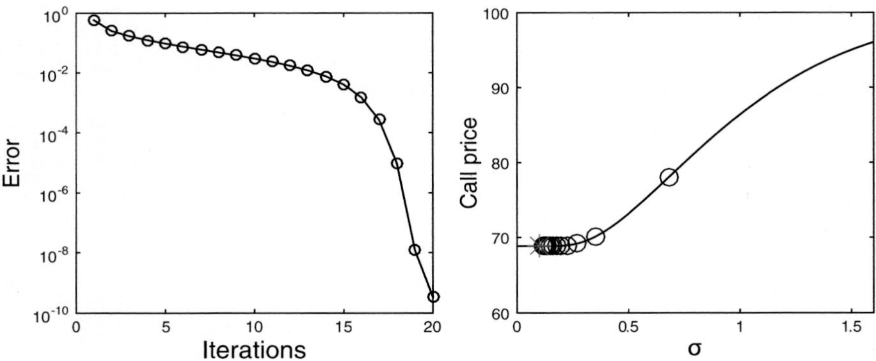

Fig. 25.5 Convergence of σ_n to σ^* as $n \to \infty$

Simulation 25.2 (Implied Volatility)

We compute the implied volatility. For the output see Fig. 25.5.

```
S0 = 100;
K =  40; % A deep in-the-money case chosen for the best looking plot.
r = 0.05;
T = 5;
sigma_market = 0.1; % a given condition
n = 20; %number of iterations

sigma0 = sqrt(2*abs((log(S0/K)+r*T)/T))

% definitions of functions
d1 = @(x) (log(S0/K)+(r+x.^2/2)*T)./(x*sqrt(T));
d2 = @(x) (log(S0/K)+(r-x.^2/2)*T)./(x*sqrt(T));
C = @(x) S0*normcdf(d1(x))-K*exp(-r*(T))*normcdf(d2(x));
vega = @(x) S0*sqrt(T)*normpdf(d1(x));

C_market=C(sigma_market); % observed in the market
F = @(x) C(x)-C_market;

sigma = zeros(1,n);
sigma(1)=sigma0;
for i=1:1:n-1
    sigma(i+1)=sigma(i) - F(sigma(i))/vega(sigma(i));
end

figure(1);
error=zeros(1,n);
for i=1:n;
    error(i)=abs(sigma_market-sigma(i));
end
semilogy(error,'o-');
hold off
```

```
figure(2);
x=0:0.01:2;
plot(x,C(x));
hold on
plot(sigma,C(sigma));
hold on
plot(sigma_market,C_market,'r*');
```

Exercises

25.1 Let $\delta t = \frac{T}{L}$ and $t_i = i \times \delta t$. Consider the geometric Brownian motion for the asset price S

$$\frac{S(t_{i+1}) - S(t_i)}{S(t_i)} = \mu\,\delta t + \sigma\sqrt{\delta t}\;Y_i$$

where $\mu, \sigma > 0$ are constant and $Y_0, Y_1, Y_2, \ldots$ are independent standard normal variables. Find

$$\lim_{\delta t \to 0+} \frac{1}{\delta t}\mathbb{E}\left[\left(\frac{S(t_{i+1}) - S(t_i)}{S(t_i)}\right)^2\right] .$$

Chapter 26
Time Series

A time series is a sequence of collected data over a time period. Theoretically, we usually assume that a given sequence is infinitely long into the future and sometimes also into the past, and it is regarded as an observed random sample realized from a sequence of random variables. Time series models are used to analyze historical data and to forecast future movement of market variables. Since financial data is collected at discrete time points, sometimes it is appropriate to use recursive difference equations rather than differential equations which are defined for continuous time. Throughout the chapter we consider only the discrete time models. A process is stationary if all of its statistical properties are invariant in time, and a process is weakly stationary if its mean, variance and covariance are invariant in time. We introduce time series models and study their applications in finance, especially in forecasting volatility. For an introduction to time series, consult [27, 59, 97].

26.1 The Cobweb Model

Before we formally start the discussion of time series, we present a basic example from economics to illustrate the application of time series models, and the criterion for its convergence.

Example 26.1 (Cobweb Model) Here is a simple application of the theory of difference equations in economics. We assume that the following relations hold at time t among demand d_t, supply s_t and market price p_t for some $a, b, \gamma > 0, \beta > 0$:

$$\begin{cases} d_t = a - \gamma p_t \\ s_t = b + \beta p_t^* + \varepsilon_t \\ s_t = d_t \end{cases} \tag{26.1}$$

G.H. Choe, *Stochastic Analysis for Finance with Simulations*, Universitext,
DOI 10.1007/978-3-319-25589-7_26

where p_t^* is the expected market price at time t, i.e.,

$$p_t^* = \mathbb{E}[p_t|\mathcal{F}_{t-1}]$$

and ε_t is a sequence of independent random shocks with zero mean. The first two assumptions in the model (26.1) are based on an economic common sense that demand is inversely proportional to present price and that supply is proportional to expected price. The third assumption in (26.1) implies that supply and demand are in a dynamical equilibrium in an ideal market.

In the left panel in Fig. 26.1 we start at time t from the economic state represented by the point A_1. If the price is high at p_t initially, then the demand shrinks to d_t and the price goes down to p_{t+1}, and the demand rises to d_{t+1}. In the right panel in Fig. 26.1, if the price is high then the supply is high and the price eventually drops. Combining both panels in Fig. 26.1, we obtain the diagram in Fig. 26.2 where the horizontal axis represents the quantity for supply and demand. As time progresses, the economic state moves from A_1 to A_2, then to A_3, and A_4.

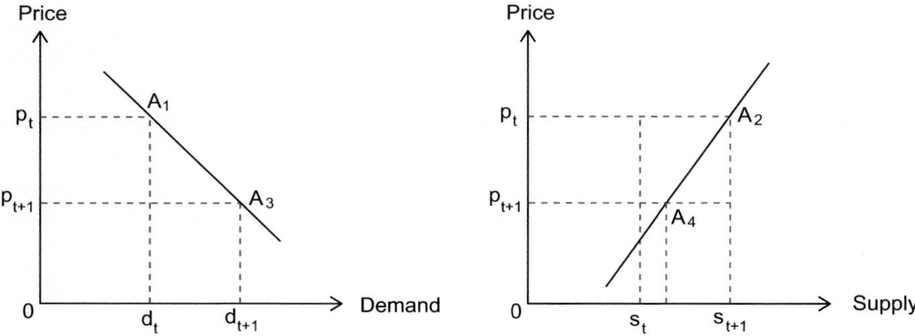

Fig. 26.1 Relations between demand and price (*left*), and supply and price (*right*)

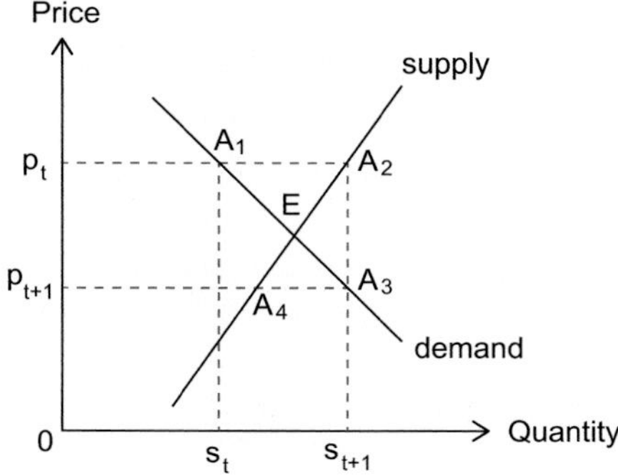

Fig. 26.2 The cobweb model for supply and demand

To simplify the model and derive a difference equation we suppose the following holds:

$$p_t^* = p_{t-1} \ .$$

Then the second condition in (26.1) becomes

$$s_t = b + \beta\, p_{t-1} + \varepsilon_t \ .$$

Hence the third equation in (26.1) yields

$$b + \beta\, p_{t-1} + \varepsilon_t = a - \gamma p_t \ ,$$

which produces a difference equation

$$p_t = \frac{a-b}{\gamma} - \frac{\beta}{\gamma} p_{t-1} - \frac{1}{\gamma}\varepsilon_t \ . \tag{26.2}$$

Its general solution is given by

$$\begin{aligned} p_t &= \frac{a-b}{\gamma}\sum_{i=0}^{t-1}\left(-\frac{\beta}{\gamma}\right)^i + \left(-\frac{\beta}{\gamma}\right)^t p_0 - \frac{1}{\gamma}\sum_{i=0}^{t-1}\left(-\frac{\beta}{\gamma}\right)^i \varepsilon_{t-i} \\ &= \frac{a-b}{\gamma}\frac{1-\left(-\frac{\beta}{\gamma}\right)^t}{1-\left(-\frac{\beta}{\gamma}\right)} + \left(-\frac{\beta}{\gamma}\right)^t p_0 - \frac{1}{\gamma}\sum_{i=0}^{t-1}\left(-\frac{\beta}{\gamma}\right)^i \varepsilon_{t-i} \\ &= \frac{a-b}{\gamma+\beta} + \left(-\frac{\beta}{\gamma}\right)^t \left(p_0 - \frac{a-b}{\gamma+\beta}\right) - \frac{1}{\gamma}\sum_{i=0}^{t-1}\left(-\frac{\beta}{\gamma}\right)^i \varepsilon_{t-i} \ . \end{aligned}$$

The cumulative error or shock has mean 0 and variance

$$\frac{1}{\gamma^2}\sum_{i=0}^{t-1}\left(\frac{\beta}{\gamma}\right)^{2i} \mathrm{Var}(\varepsilon_1) = \frac{1}{\gamma^2}\frac{1-\left(\frac{\beta}{\gamma}\right)^{2t}}{1-\left(\frac{\beta}{\gamma}\right)^2}\mathrm{Var}(\varepsilon_1) \ .$$

Thus p_t converges to $\dfrac{a-b}{\gamma+\beta}$ as $t \to \infty$ if $\frac{\beta}{\gamma} < 1$ as shown on the left in Fig. 26.3, and p_t diverges if $\frac{\beta}{\gamma} > 1$ as on the right.

Remark 26.1

(i) The cobweb model given here is multi-dimensional, however, it can be converted into a one-dimensional time series as given in (26.2).

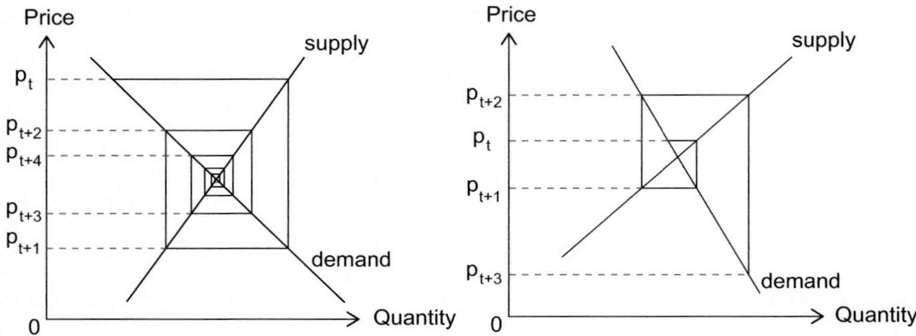

Fig. 26.3 Cobweb models: convergent (*left*) and divergent (*right*) cases

(ii) Cobweb models are also studied in nonlinear dynamical systems theory where the system is deterministic. The evolution of the system may become unpredictable due to nonlinearity. Consult [21] for more information.

26.2 The Spectral Theory of Time Series

We give a brief mathematical background of time series models.

Definition 26.1 (Stationary Time Series) A time series is a stochastic process $\{x_t\}_{t=1}^{\infty}$ or $\{x_t\}_{t=-\infty}^{\infty}$ indexed by time t. It is *weakly stationary* if $\mathbb{E}[x_t]$ and $\mathrm{Cov}(x_t, x_{t+k})$ do not depend on t, i.e., they are invariant in time. Then, $\mathbb{E}[x_t]$ is called the *mean* and $\gamma_k = \mathrm{Cov}(x_t, x_{t+k})$ is called the *autocovariance function* of the time series as a function of k. The time difference k is called *lag*. Note that $\gamma_0 = \mathrm{Var}(x_t)$ for every t, and

$$\gamma_{-k} = \mathrm{Cov}(x_t, x_{t-k}) = \mathrm{Cov}(x_{t+k}, x_{(t-k)+k}) = \mathrm{Cov}(x_{t+k}, x_t) = \gamma_k \ .$$

A time series $\{x_t\}_{t=1}^{\infty}$ is *strongly stationary* or just *stationary* if the joint probability distribution of $(x_{t+k_1}, \ldots, x_{t+k_n})$ does not depend on $t \geq 1$ for any $n \geq 1$ and $0 \leq k_1 < \cdots < k_n$. Clearly, a stationary times series is weakly stationary.

For a one-sided infinite sequence $\{\gamma_k\}_{k\geq 0}$, we define $\gamma_{-k} = \gamma_k$ for $k \geq 1$, and obtain a two-sided infinite sequence, i.e., γ_k is defined for every $k \in \mathbb{Z}$. Thus we always assume that a given sequence is two-sided and $\gamma_{-k} = \gamma_k$.

Definition 26.2 (Autocorrelation) The autocorrelation function (ACF) for a weakly stationary time series, as a function of k, is defined by

$$\rho_k = \frac{\mathrm{Cov}(x_t, x_{t+k})}{\sqrt{\mathrm{Var}(x_t)\mathrm{Var}(x_{t+k})}} = \frac{\gamma_k}{\sqrt{\gamma_0\gamma_0}} = \frac{\gamma_k}{\gamma_0} \ .$$

Definition 26.3 (Ergodicity) For a stationary process $\{x_t\}$, let $\Omega = \prod_t \mathbb{R}_t$ where $\mathbb{R}_t = \mathbb{R}^1$ for every t, and define $\mathbb{P}$ on Ω by taking the time invariant probability of the given time series, i.e.,

$$\mathbb{P}(C) = \Pr\{(x_1, \ldots, x_k) \in C\}$$

for a cylinder subset $C = I_1 \times \cdots \times I_k$ for $k \geq 1$ where $I_1, \ldots, I_k$ are arbitrary intervals. Then, by the Kolmogorov Extension Theorem, $\mathbb{P}$ can be extended to a probability measure defined on the σ-algebra $\mathcal{F}$ generated by all the cylinder subsets. The given time series is said to be *ergodic* if any shift-invariant (in other words, stationary) subset $C \in \mathcal{F}$ satisfies $\mathbb{P}(C) = 0$ or 1. We usually assume that the time series under consideration is ergodic. An ergodic stationary time series is an example of ergodic measure preserving shift transformations. For more information, see [21, 86].

Theorem 26.1 (Birkhoff Ergodic Theorem) *For an ergodic stationary time series* $\{x_t\}$, *we have*

$$\lim_{n\to\infty} \frac{1}{n} \sum_{t=1}^{n} f(x_t, \ldots, x_{t+k-1}, \ldots) = \mathbb{E}[f(x_1, \ldots, x_k, \ldots)] \tag{26.3}$$

where the expectation on the right-hand side exists.

Remark 26.2

(i) The Birkhoff Ergodic Theorem states that *time mean* equals *space mean*, which enables us to estimate mean, variance, covariance, etc. of the given time series by taking time averages of a single sample series. See Example 26.2.
(ii) By taking $f = \mathbf{1}_C$ for $C \in \mathcal{F}$, we can show that (26.3) is equivalent to the ergodicity condition. Thus (26.3) is usually employed as the definition of ergodicity in the study of time series.

Example 26.2 If a time series $\{x_t\}$ is stationary and ergodic with $\mathbb{E}[|x_t|] < \infty$, then

$$\lim_{n\to\infty} \frac{1}{n} \sum_{t=1}^{n} x_t = \mathbb{E}[x_1]$$

almost surely where $\mathbb{E}[x_1] = \mathbb{E}[x_2] = \cdots$. To see why, choose $f = x_1$.

Definition 26.4 (Positive Definite Sequence) A sequence $\{\gamma_k\}_{k=-\infty}^{\infty}$ is *positive definite* if

$$\sum_{t=-\infty}^{\infty} \sum_{s=-\infty}^{\infty} c_t \overline{c_s} \gamma_{t-s} \geq 0$$

for any sequence $\{c_k\}_{k=-\infty}^{\infty}$, $c_k \in \mathbb{C}$, such that $c_k = 0$ for $|k| \geq K$ for some $K > 0$.

Theorem 26.2 *The autocovariance function $\{\gamma_k\}$ of a weakly stationary time series is positive definite.*

Proof For any sequence $\{c_k\}_{k=-\infty}^{\infty}$, $c_k \in \mathbb{R}$, such that $\sum_{k=-\infty}^{\infty} |c_k|^2 < \infty$, we have

$$
\begin{aligned}
&\sum_{t=-\infty}^{\infty} \sum_{s=-\infty}^{\infty} c_t \overline{c_s} \gamma_{t-s} \\
&= \sum_t \sum_s c_t \overline{c_s} \mathrm{Cov}(x_s, x_{s+(t-s)}) \\
&= \sum_t \sum_s c_t \overline{c_s} \mathrm{Cov}(x_s, x_t) \\
&= \sum_t \sum_s c_t \overline{c_s} \mathbb{E}[(x_s - \mathbb{E}[x_s])(x_t - \mathbb{E}[x_t])] \\
&= \mathbb{E}\Big[\sum_s \overline{c_s}(x_s - \mathbb{E}[x_s]) \sum_t c_t (x_t - \mathbb{E}[x_t]) \Big] \\
&= \mathbb{E}\Big[\overline{\sum_s c_s(x_s - \mathbb{E}[x_s])} \sum_t c_t (x_t - \mathbb{E}[x_t]) \Big] \\
&= \mathbb{E}\Big[\Big| \sum_t c_t (x_t - \mathbb{E}[x_t]) \Big|^2 \Big] \geq 0 \,.
\end{aligned}
$$

□

Example 26.3 (Fourier Coefficients of a Measure) Suppose that

$$
a_k = \int_0^1 \mathrm{e}^{-2\pi \mathrm{i} k\theta} \mathrm{d}\nu(\theta)
$$

for some (positive) measure ν on $[0, 1]$. Then

$$
\begin{aligned}
\sum_{t=-\infty}^{\infty} \sum_{s=-\infty}^{\infty} c_t \overline{c_s} a_{t-s} &= \sum_{t=-\infty}^{\infty} \sum_{s=-\infty}^{\infty} c_t \overline{c_s} \int_0^1 \mathrm{e}^{-2\pi \mathrm{i}(t-s)\theta} \mathrm{d}\nu \\
&= \int_0^1 \sum_{t=-\infty}^{\infty} c_t \mathrm{e}^{-2\pi \mathrm{i} t\theta} \overline{\sum_{s=-\infty}^{\infty} c_s \mathrm{e}^{-2\pi \mathrm{i} s\theta}} \mathrm{d}\nu \\
&= \int_0^1 \Big| \sum_{t=-\infty}^{\infty} c_t \mathrm{e}^{-2\pi \mathrm{i} t\theta} \Big|^2 \mathrm{d}\nu \geq 0
\end{aligned}
$$

for $c_k \in \mathbb{C}$ such that $c_k = 0$ for $|k| \geq K$ for some $K > 0$. Hence $\{a_k\}$ is positive definite. Assume further that a_k is real. In this case, since

$$a_{-k} = \int_0^1 e^{-2\pi i(-k)\theta} d\nu = \overline{a_k} = a_k ,$$

we have $a_{-k} = a_k$.

The following theorem states that Example 26.3 is the only example of a positive definite sequence.

Theorem 26.3 (Herglotz) *A positive definite sequence* $\{a_k\}$ *is a Fourier transform of a (positive and finite) measure on* $[0, 1]$, *that is, there exists a positive measure* ν *on* $[0, 1]$ *such that*

$$a_k = \int_0^1 e^{-2\pi ik\theta} d\nu(\theta) , \quad k \in \mathbb{Z} .$$

Proof See [37, 46]. □

Corollary 26.1 *For the autocovariance function* $\{\gamma_k\}$ *of a weakly stationary time series* $\{x_t\}$, *there exists a positive measure* ν *on* $[0, 1]$ *such that*

$$\gamma_k = \int_0^1 e^{-2\pi ik\theta} d\nu(\theta) , \quad k \in \mathbb{Z} . \tag{26.4}$$

Proof Now apply Theorems 26.2 and 26.3. □

Definition 26.5 (Spectral Density Function) If the measure ν in (26.4) in Corollary 26.1 is absolutely continuous with respect to Lebesgue measure, then the function $f(x)$ such that $d\nu = f(x)\,dx$ is called the *spectral density function*. If $f(x)$ is m-times differentiable and $f^{(m)}(x)$ is integrable, then $\gamma_k = o(|k|^{-m})$ as $|k| \to \infty$. Consult [46] for its proof.

The following fact due to H. Wold is a theoretical foundation of time series models. For the proof, see [53]. For more information consult [32, 59, 79, 101].

Theorem 26.4 (Wold Decomposition) *A weakly stationary time series* $\{x_t\}$ *with its spectral density function can be written as*

$$x_t = \mu + \sum_{j=0}^{\infty} \psi_j \, \varepsilon_{t-j} \tag{26.5}$$

where $\psi_0 = 1$, *and* ε_t *are uncorrelated random variables with* $\mathbb{E}[\varepsilon_t] = 0$ *and* $\mathrm{Var}(\varepsilon_t) = \sigma^2$. *The random variables* ε_t *are called innovations.*

For an example when the spectrum is discrete, see Exercise 26.2.

26.3 Autoregressive and Moving Average Models

Now we introduce an abstract model for time series.

Definition 26.6 (MA Model) The right-hand side of (26.5) converges in the L^2-sense, and it can be approximated by a finite sum given by

$$x_t = \mu + \varepsilon_t + \psi_1 \varepsilon_{t-1} + \cdots + \psi_q \varepsilon_{t-q} ,$$

which is called a ***m**oving **a**verage model of order q*, or MA(q).

Definition 26.7 (Backshift Operator) Let $\{x_t\}$ be a time series and let B denote the *backward shift operator*, or *backshift operator*, defined by $Bx_t = x_{t-1}$. (A rigorous interpretation of this commonly used notation would be the following: Let $\mathcal{X}$ be the collection of all sample time series $\mathbf{x} = \{x_t\}$. Here we regard $\mathbf{x} = \{x_t\}$ as a bi-infinite sequence with its t-th component x_t . Define $B : \mathcal{X} \to \mathcal{X}$ by $(B\mathbf{x})_t = \mathbf{x}_{t-1}$. Thus the sequence $B\mathbf{x}$ is obtained by shifting $\mathbf{x}$ to the right by one time step.) Note that a constant time series $\boldsymbol{\mu} = (\ldots, \mu, \mu, \mu, \ldots)$ is invariant under B, i.e., $B\mu = \mu$.

Iterating the backshift operator B, we obtain $B^2 x_t = B(Bx_t) = x_{t-2}, \ldots, B^j x_t = x_{t-j}$. Hence we can rewrite (26.5) as

$$x_t = \mu + \psi(B)\varepsilon_t \tag{26.6}$$

where

$$\psi(B) = 1 + \psi_1 B + \psi_2 B^2 + \cdots .$$

Now we formally rewrite (26.6) as

$$\varepsilon_t = \frac{1}{\psi(B)}(x_t - \mu) = \phi(B)(x_t - \mu) = \phi(B)x_t - \widetilde{\mu} \tag{26.7}$$

for some constant $\widetilde{\mu}$ where the formal power series

$$\phi(B) = 1 - \phi_1 B - \phi_2 B^2 - \cdots \tag{26.8}$$

is defined by

$$\phi(B) = \frac{1}{\psi(B)} = \frac{1}{1 + \psi_1 B + \psi_2 B^2 + \cdots} .$$

By truncating high order terms of the power series $\phi(B)$ given in (26.8), we obtain a polynomial

$$1 - \phi_1 B - \cdots - \phi_p B^p ,$$

which approximately solves (26.7). Thus we have a time series model

$$\begin{aligned}\varepsilon_t &= (1 - \phi_1 B - \cdots - \phi_p B^p)x_t - \widetilde{\mu} \\ &= x_t - \phi_1 x_{t-1} - \cdots - \phi_p x_{t-p} - \widetilde{\mu} \ . \end{aligned} \tag{26.9}$$

Definition 26.8 (AR Model) The equation (26.9) may be rewritten as

$$x_t = \widetilde{\mu} + \phi_1 x_{t-1} + \cdots + \phi_p x_{t-p} + \varepsilon_t \ , \tag{26.10}$$

which is called an ***autoregressive model of order*** p, or AR(p).

Remark 26.3 (Invertibility Condition) It can be shown that the formal power series (26.8) converges if $\psi(z) \neq 0$ for complex numbers z such that $|z| \leq 1$, in other words, the zeros of $\psi(z)$ are outside the unit circle.

Definition 26.9 (ARMA Model) We combine the AR(p) and MA(q) models and construct an ARMA(p,q) model of the form

$$x_t = \mu + \phi_1 x_{t-1} + \cdots + \phi_p x_{t-p} + \varepsilon_t + \psi_1 \varepsilon_{t-a} + \cdots + \psi_q \varepsilon_{t-q} \ ,$$

which is an approximation of the formal power series by a formal rational function of B. Small values of p and q are sufficient in many applications.

Example 26.4 Suppose that x_t satisfies the AR(1) model

$$x_t = a_0 + a_1 x_{t-1} + \varepsilon_t$$

where ε_t denotes a random shock at t. We assume that ε_t are uncorrelated and identically distributed with zero mean. Then we have

$$\begin{aligned} x_t &= a_0 + a_1(a_0 + a_1 x_{t-2} + \varepsilon_{t-1}) + \varepsilon_t \\ &= a_0 + a_1 a_0 + a_1^2 x_{t-2} + a_1 \varepsilon_{t-1} + \varepsilon_t \\ &= (a_0 + a_1 a_0) + a_1^2(a_0 + a_1 x_{t-3} + \varepsilon_{t-2}) + (\varepsilon_t + a_1 \varepsilon_{t-1}) \\ &\ \ \vdots \\ &= a_0 \sum_{i=0}^{t-1} a_1^i + a_1^t x_0 + \sum_{i=0}^{t-1} a_1^i \varepsilon_{t-i} \ . \end{aligned}$$

Example 26.5 Consider the AR(2) model

$$x_t = a_0 + a_1 x_{t-1} + a_2 x_{t-2} + \varepsilon_t \ .$$

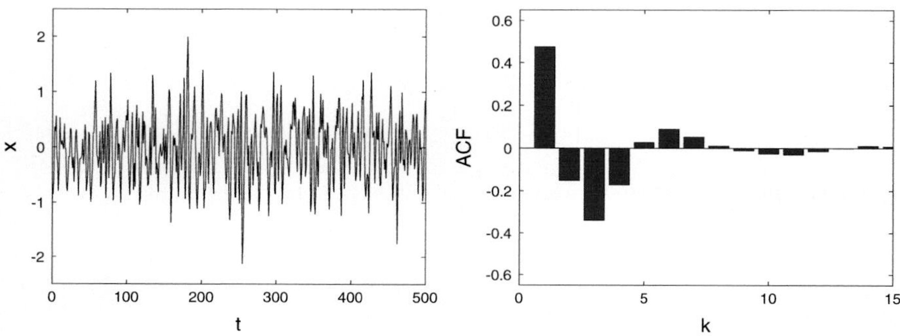

Fig. 26.4 The AR(2) model: time series x_t (*left*) and autocorrelation function (*right*)

It is necessary to have the condition that the zeros of $1 - a_1 z - a_2 z^2$ are outside the unit disk. In Fig. 26.4 a time series x_t, $1 \le t \le 500$, and the autocorrelation function γ_k/γ_0, $1 \le k \le 15$, are plotted in the left and right panels, respectively, for the AR(2) model. See Simulation 26.1.

26.4 Time Series Models for Volatility

First, we introduce some basic ideas for the forecasting of future volatility movement. Let S_n be the asset price at the end of the nth day. Let

$$u_n = \log \frac{S_n}{S_{n-1}} = \log \left(\frac{S_n - S_{n-1}}{S_{n-1}} + 1 \right)$$

denote the log return of the day n, and S_0 be the asset price at the start of the day 1. Since $\log(1 + x) \approx x$ for $x \approx 0$, we may use the definition

$$u_n = \frac{S_n - S_{n-1}}{S_{n-1}} \ .$$

Let $\mathcal{F}_n$ be the σ-algebra defined by the market information including asset prices up to day n. Then, by definition, S_n is $\mathcal{F}_n$-measurable and so is u_n. Let m be the length of the observation window for the estimation of u_n, and we use the time series data $u_{n-m}, u_{n-m+1}, \ldots, u_{n-1}$, and define $\bar{u}$ by

$$\bar{u} = \frac{1}{m} \sum_{i=1}^{m} u_{n-i} \ , \qquad (26.11)$$

which is $\mathcal{F}_{n-1}$-measurable. Define the sample variance

$$\sigma_n^2 = \frac{1}{m-1}\sum_{i=1}^{m}(u_{n-i} - \bar{u})^2$$

which is also $\mathcal{F}_{n-1}$-measurable. Since the average of u_i is practically regarded as zero and m is large, we take $\bar{u} = 0$, and may use

$$\sigma_n^2 = \frac{1}{m}\sum_{i=1}^{m} u_{n-i}^2 \tag{26.12}$$

to forecast the daily volatility for the day n. More generally, we may use a weighted average

$$\sigma_n^2 = \sum_{i=1}^{m} \alpha_i u_{n-i}^2\,, \quad \alpha_i \geq 0\,, \quad \sum_{i=1}^{m} \alpha_i = 1\,. \tag{26.13}$$

I. The EWMA Model

The **e**xponentially **w**eighted **m**oving **a**verage model, EWMA for short, is defined as follows: For a constant $0 < \lambda < 1$, take

$$\alpha_i = \frac{1-\lambda}{\lambda}\lambda^i$$

in (26.13). Define

$$\sigma_n^2 = \frac{1-\lambda}{\lambda}\sum_{i=1}^{\infty} \lambda^i u_{n-i}^2\,.$$

Since

$$\sigma_{n-1}^2 = \frac{1-\lambda}{\lambda}\sum_{i=1}^{\infty} \lambda^i u_{n-1-i}^2 = \frac{1-\lambda}{\lambda^2}\sum_{i=2}^{\infty} \lambda^i u_{n-i}^2\,,$$

we have

$$\sigma_n^2 = \frac{1-\lambda}{\lambda}(\lambda u_{n-1}^2 + \sum_{i=2}^{\infty} \lambda^i u_{n-i}^2) = (1-\lambda)u_{n-1}^2 + \lambda\sigma_{n-1}^2\,.$$

Thus

$$\sigma_n^2 = (1-\lambda)u_{n-1}^2 + \lambda\sigma_{n-1}^2\,. \tag{26.14}$$

II. The ARCH Model

The economist Robert Engel [28] introduced the ARCH(m) model defined by

$$\sigma_n^2 = \gamma \bar{V} + \sum_{i=1}^{m} \alpha_i u_{n-i}^2$$

and

$$u_n = \sigma_n \varepsilon_n , \quad \varepsilon_n \sim N(0, 1)$$

where ε_n are independent, and

$$\gamma > 0 , \quad \alpha_i \geq 0 , \quad \gamma + \sum_{i=1}^{m} \alpha_i = 1 .$$

It is a special case of (26.13). The acronym ARCH stands for **auto**regressive **c**onditional **h**eteroskedasticity. (The term 'heteroskedasticity' means variability of variance.) ARCH models are employed in modeling financial time series that exhibit time-varying volatility. There are many extensions of ARCH models.

Let $\mathcal{F}_n$ be the σ-algebra generated by the information up to time $n-1$. Then σ_n is $\mathcal{F}_{n-1}$-measurable, and

$$\mathbb{E}[u_n] = \mathbb{E}[\mathbb{E}[u_n|\mathcal{F}_{n-1}]] = \mathbb{E}[\mathbb{E}[\sigma_n \varepsilon_n|\mathcal{F}_{n-1}]] = \mathbb{E}[\sigma_n \mathbb{E}[\varepsilon_n|\mathcal{F}_{n-1}]] = 0$$

and

$$\mathbb{E}[u_n^2|\mathcal{F}_{n-1}] = \mathbb{E}[\sigma_n^2 \varepsilon_n^2|\mathcal{F}_{n-1}] = \sigma_n^2 \mathbb{E}[\varepsilon_n^2|\mathcal{F}_{n-1}] = \sigma_n^2 \mathbb{E}[\varepsilon_n^2] = \sigma_n^2 ,$$

thus the conditional variance is stochastic.

III. The GARCH Model

Tim Bollerslev [10] introduced the GARCH(p,q) model defined by

$$\sigma_n^2 = \alpha_0 + \sum_{i=1}^{p} \alpha_i u_{n-i}^2 + \sum_{j=1}^{q} \beta_j \sigma_{n-j}^2 \tag{26.15}$$

and

$$u_n = \sigma_n \varepsilon_n , \quad \varepsilon_n \sim N(0, 1)$$

where ε_n are independent and

$$\alpha_0 > 0 , \quad \alpha_i \geq 0 , \quad \beta_j \geq 0 , \quad \sum_{i=1}^{p} \alpha_i + \sum_{j=1}^{q} \beta_j < 1 .$$

Remark 26.4 GARCH stands for **g**eneralized ARCH. Note that the EWMA model is a special case of the GARCH(1,1) model with $\gamma = 0$, and that the ARCH(1) model is a special case of the GARCH(1,1) model with $\beta = 0$.

Let us focus on the GARCH(1,1) model defined by

$$\sigma_n^2 = \gamma \bar{V} + \alpha u_{n-1}^2 + \beta \sigma_{n-1}^2 \tag{26.16}$$

where

$$\gamma + \alpha + \beta = 1 .$$

Note that u_n is $\mathcal{F}_{n-1}$-measurable, and $u_n = \sigma_n \varepsilon_n$. Hence

$$\mathbb{E}[u_n^2|\mathcal{F}_{n-1}] = \mathbb{E}[\sigma_n^2 \varepsilon_n^2|\mathcal{F}_{n-1}] = \sigma_n^2\, \mathbb{E}[\varepsilon_n^2|\mathcal{F}_{n-1}] = \sigma_n^2\, \mathbb{E}[\varepsilon_n^2] = \sigma_n^2 .$$

Since, for $j > 0$,

$$\sigma_{n+j}^2 = \gamma \bar{V} + \alpha u_{n+j-1}^2 + \beta \sigma_{n+j-1}^2 ,$$

we have

$$\begin{aligned}
\mathbb{E}[\sigma_{n+j}^2|\mathcal{F}_{n+j-2}] &= \gamma \bar{V} + \alpha\, \mathbb{E}[u_{n+j-1}^2|\mathcal{F}_{n+j-2}] + \beta\, \mathbb{E}[\sigma_{n+j-1}^2|\mathcal{F}_{n+j-2}] \\
&= \gamma \bar{V} + \alpha \sigma_{n+j-1}^2 + \beta \sigma_{n+j-1}^2 \\
&= \gamma \bar{V} + (\alpha + \beta) \sigma_{n+j-1}^2 .
\end{aligned}$$

Hence

$$\mathbb{E}[\sigma_{n+j}^2 - \bar{V}|\mathcal{F}_{n+j-2}] = (\alpha + \beta)(\sigma_{n+j-1}^2 - \bar{V}) .$$

By the tower property of conditional expectation,

$$\begin{aligned}
\mathbb{E}[\sigma_{n+j}^2 - \bar{V}|\mathcal{F}_{n+j-3}] &= \mathbb{E}[\mathbb{E}[\sigma_{n+j}^2 - \bar{V}|\mathcal{F}_{n+j-2}]|\mathcal{F}_{n+j-3}] \\
&= \mathbb{E}[(\alpha + \beta)(\sigma_{n+j-1}^2 - \bar{V})|\mathcal{F}_{n+j-3}] \\
&= (\alpha + \beta)\, \mathbb{E}[\sigma_{n+j-1}^2 - \bar{V}|\mathcal{F}_{n+j-3}] \\
&= (\alpha + \beta)^2 (\sigma_{n+j-2}^2 - \bar{V}) .
\end{aligned}$$

By repeating the procedure, we have

$$\mathbb{E}[\sigma_{n+j}^2 - \bar{V}|\mathcal{F}_{n-1}] = (\alpha + \beta)^j (\sigma_n^2 - \bar{V}) . \tag{26.17}$$

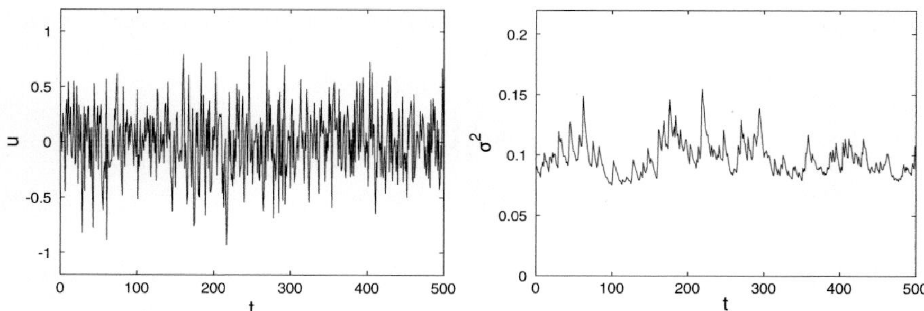

Fig. 26.5 The GARCH(1,1) model: time series for u_t (*left*) and σ_t^2 (*right*)

Hence we can forecast the volatility on day $n + j$ at the start of the day n, or equivalently, at the end of day $n-1$. Note that $\mathbb{E}[\sigma_{n+j}^2|\mathcal{F}_n]$ converges to $\bar{V}$ as $j \to \infty$.

Remark 26.5 Consider the GARCH(1,1) model.

(i) (Volatility clustering) A large value of u_{n-1}^2 or σ_{n-1}^2 generates a large value of σ_n^2, which implies that a large value of u_{n-1}^2 tends to be followed by a large value of u_n^2, which is called *volatility clustering* in financial time series. See Fig. 26.5 where we take $a_0 = 0.01$, $\alpha = 0.05$ and $\beta = 0.85$ for the simulation.

(ii) (Fat tail) It is known that the kurtosis κ satisfies

$$\kappa = \frac{\mathbb{E}[u_n^4]}{(\mathbb{E}[u_n^2])^2} = \frac{3\{1-(\alpha+\beta)^2\}}{1-(\alpha+\beta)^2-2\alpha^2} > 3$$

when the denominator is positive, which implies that the tail distribution of a GARCH(1,1) process is heavier than a normal distribution for which the kurtosis is equal to 3. For a simulation consult Simulation 26.2.

26.5 Computer Experiments

Simulation 26.1 (AR(2) Model)

We consider the AR(2) model and compute the autocorrelation function. See Fig. 26.4.

```
N = 10000 ; % the length of the time series
x = zeros(N,1);
x(1) = 1;
x(2) = 1;
a0 = 0;
a1 = 0.7;
a2 = -0.5;
```

```
% The polynomial a2*z^2 + a1*z + a0 is represented as poly = [-a2 -a1 1]
% The roots of the polynomial are returned in a column vector.
r = roots(poly)
abs(r)
sigma2 = 0.2; % variance for epsilon
for t=1:N
    epsilon(t) = randn * sqrt(sigma2) ;
end
for t=3:N
    x(t)= a0 + a1*x(t-1) + a2*x(t-2) + epsilon(t);
end

% Discard the first 100 samples to remove transient samples.
x(1:100) = [];
epsilon(1:100) = [];
mu = mean(x);
var_x = var(x); % variance of x
N = N - 100;
% autocorrelation
L = 15 ; % L denotes lag.
rho = zeros(1,L);

for k=1:L
    for t=1:N-k
    rho(k) = rho(k) + (x(t)-mu)*(x(t+k)-mu);
    end
rho(k) = rho(k)/(N-k)/var_x;
end

figure(1)
plot(x(1:500));
xlabel('t');
ylabel('x');

figure(2)
bar(rho)
xlabel('k');
ylabel('ACF');
```

Simulation 26.2 (GARCH(1,1) Model)

We consider the GARCH(1,1) model and observe the volatility clustering. See Fig. 26.5. In the following code, the command `mean(X > a)` produces the average of the indicator function of the event $\{X > a\}$, i.e., the probability of $\{X > a\}$.

```
sigma2(1) = 0.1;
a0 = 0.01;
alpha = 0.05;
beta = 0.85;
T = 100000; % Tail estimation requires large T
for t = 1:1:T
    u(t) = sqrt(sigma2(t))*randn;
    sigma2(t+1) = a0 + alpha*u(t)^2 + beta*sigma2(t);
end
u(1:100) = []; % Discard the first 100 samples to remove transient samples
```

```
sigma2(1:100) = [];
figure(1)
plot(u(1:300));
xlabel('u');
figure(2)
plot(sigma2(1:300));
xlabel('\sigma^2');
u = u./std(u); % Normalize u
for k = 1:1:4
    P(1,k) = 1 - normcdf(k); % Tail probability of the normal distribution
    P(2,k) = mean(u > k); % Estimate the tail probability of u
end
format long
P
```

In the following output we print only a few significant digits to save space. We observe that u has heavier tail than the standard normal distribution.

```
P =
    0.15865525    0.02275013    0.00134989    0.00003167
    0.15813813    0.02307307    0.00176176    0.00004004
```

Exercises

26.1 Show that if the measure $d\nu$ in (26.4) is the Dirac delta measure at a single point $\theta_0 = \frac{1}{2}$, then $\gamma_k = e^{-2\pi i k\theta_0} = (-1)^k$ and $d\nu = dF$ where $F(\theta) = 0$ for $\theta < \frac{1}{2}$, and $F(\theta) = 1$ for $\theta \geq \frac{1}{2}$.

26.2 (ACF with discrete spectrum) Define $x_t = \sum_{i=1}^{L} a_i \sin(2\pi\theta_i t + U_i)$ where a_i are constant, $U_i \sim U(0,1)$ are independent random variables, and $0 < \theta_1 < \theta_2 < \cdots < \theta_L < 1$. Show $\mathbb{E}[x_t] = 0$. Compute $\gamma_k = \mathbb{E}[x_t x_{t+k}]$. Find a monotonically increasing step function $F(\theta)$ such that $\gamma_k = \int_0^1 e^{2\pi i\theta k} dF(\theta)$ where the given integral is the Riemann–Stieltjes integral.

26.3 (ACF with absolute continuous spectrum) Consider $x_t = \phi x_{t-1} + \varepsilon_t$ where $|\phi| < 1$, $\{\varepsilon_t\}$ are independent and $\varepsilon_t \sim N(0,1)$. Find the spectral distribution function.

26.4 Find $\mathbb{E}[x_{t+1}|x_t, x_{t-1}, \ldots, x_{t-p+1}]$ in the AR(p) model.

26.5 Let x_t be the MA(m) process with equal weights $\frac{1}{m+1}$ at all lags defined by

$$x_t = \sum_{q=0}^{m} \frac{1}{m+1} \varepsilon_{t-q}$$

where ε_t are uncorrelated random variables with $\mathbb{E}[\varepsilon_t] = 0$ and $\text{Var}(\varepsilon_t) = \sigma^2$. Prove that the ACF of the process is given by

$$\rho(k) = \begin{cases} \frac{m+1-k}{m+1}\,, & 0 \le k \le m\,, \\ 0\,, & k > m\,. \end{cases}$$

26.6 Define an estimator for the yearly volatility σ^* from (26.12).

26.7 In the geometric Brownian motion, find the distribution of $\bar{u}$ defined by (26.11).

26.8 Note that the EWMA model is a special case of the GARCH model. Show that $\mathbb{E}[\sigma_{n+j}^2|\mathcal{F}_{n-1}] = \sigma_n^2$ in the EWMA model.

Chapter 27
Random Numbers

The Monte Carlo method was invented by John von Neumann and Stanislaw Ulam. It is based on probabilistic ideas and can solve many problems at least approximately. The method is powerful in option pricing which will be investigated in Chap. 28. We need random numbers to apply the method. For efficiency, we need a good random number generator.

The most widely used algorithm is given by a linear congruential generator $x_{n+1} = ax_n + b \pmod{M}$ for some natural numbers a, b, M, x_0. To obtain a sufficiently long sequence of random numbers without repetition, we choose a very large value for M. Even for such a seemingly simple algorithm it is very difficult to find suitable constants a, b, M, x_0 either by theoretical methods or computational methods. Most users of random numbers choose well-known generators and corresponding parameters.

In practice most random numbers are generated by deterministic algorithms, and they are called pseudorandom numbers. For a comprehensive reference on random number generation, consult [51]. For elementary introductions, see [7, 76]. Consult also [21].

27.1 What Is a Monte Carlo Method?

Suppose that we want to integrate a function of many variables, i.e., the dimension of the domain of the integrand is high, we apply Monte Carlo integration to estimate an integral. For $1 \leq k \leq s$ let $I_k = [0, 1]$, and consider a function $f(x_1, \ldots, x_s)$ on the s-dimensional cube $Q = \prod_{k=1}^{s} I_k \subset \mathbb{R}^s$. If we want to numerically integrate f on Q by Riemann integral, then we partition each interval I_k into n subintervals, and obtain n^s small cubes. Then choose a point q_i, $1 \leq i \leq n^s$, from each subcube, and evaluate $f(q_i)$ then compute their average. This is not practical since there are n^s numbers whose average would require a long time to compute even for small numbers such as $s = 10$ and $n = 10$. To avoid

G.H. Choe, *Stochastic Analysis for Finance with Simulations*, Universitext,
DOI 10.1007/978-3-319-25589-7_27

such a difficulty we use Monte Carlo method. First, choose a reasonably large number m that is considerably small compared with n^s. Then select m samples from $\{q_i : 1 \leq i \leq n^s\}$ and find the average of f at those m points. By the Central Limit Theorem we see that the error of the method is proportional to $\frac{1}{\sqrt{m}}$.

Here is another example of a Monte Carlo method. Suppose that we want to estimate $\pi = 3.14\ldots$. As shown in Fig. 27.1 we choose n uniformly distributed random points from the unit square, and count the number of points inside the circle. If there are k points, as $n \to \infty$, the relative frequency $\frac{k}{n}$ converges to $\frac{\pi}{4}$. Such a probabilistic or statistical method is called a Monte Carlo method, which was named after Monte Carlo which is famous for casinos.

As the third example, we consider the estimation of the gamma function defined by

$$\Gamma(p) = \int_0^\infty x^{p-1} \mathrm{e}^{-x}\,\mathrm{d}x\,.$$

Recall that $\Gamma(p) = (p-1)!$ for a natural number p. To estimate $\Gamma(p)$, we use an exponentially distributed random variable X with the density function $h(x) = \mathrm{e}^{-x}$, $x \geq 0$, and find $\mathbb{E}[X^{p-1}]$. See Fig. 27.2 where $\Gamma(6) = 5! = 120$ is estimated using 2^n random numbers in the Monte Carlo method.

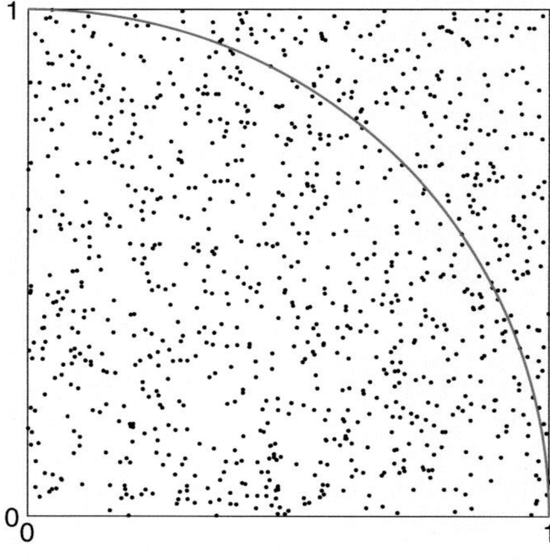

Fig. 27.1 Estimation of π by using random numbers

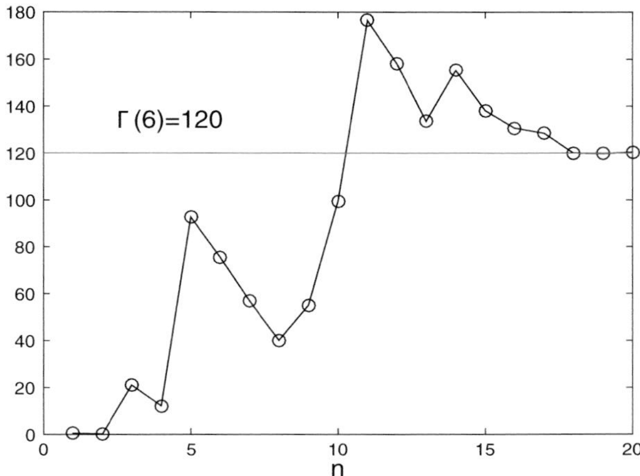

Fig. 27.2 Estimation of $\Gamma(p)$, $p = 6$, by the Monte Carlo method

27.2 Uniformly Distributed Random Numbers

Here is how to generate a sequence of real numbers with uniform distribution in the unit interval $[0, 1]$. First, choose a very large integer $M > 0$ and generate integers $0 \le x_n < M$, then divide them by M thus obtaining x_n/M. If M is very large, those fractions can approximate well any real numbers. Thus we may concentrate on the generation of random numbers satisfying $0 \le x_n < M$.

A *linear congruential generator* is an algorithm defined by

$$x_{n+1} = ax_n + b \pmod{M} ,$$

which is denoted by $LCG(M, a, b, x_0)$. It is important to choose suitable constants x_0,a,b and M.

Fact 27.1 *The period of $LCG(M, a, b, x_0)$ is equal to M if and only if the following conditions hold:*

(i) b and M are relatively prime,
(ii) if M is divisible by a prime p, then $a - 1$ is also divisible by p,
(iii) if M is divisible by 4, then $a - 1$ is also divisible by 4.

For a proof, see [51]. In the 1960s IBM developed a linear congruential generator called Randu, whose quality turned out to be less than expected. It is not used now except when we need to gauge the performance of a newly introduced generator against it. ANSI (American National Standards Institute) C and Microsoft C include linear congruential generators in their C libraries.

For a prime number p an inversive congruential generator is defined by

$$x_{n+1} = ax_n^{-1} + b \pmod{p}$$

Table 27.1 Pseudorandom number generators

Name	Algorithm	Period
Randu	LCG(2^{31}, 65539, 0, 1)	2^{29}
ANSI	LCG(2^{31}, 1103515245, 12345, 141421356)	2^{31}
Microsoft	LCG(2^{31}, 214013, 2531011, 141421356)	2^{31}
ICG	ICG(2^{31} - 1, 1, 1, 0)	$2^{31} - 1$
Ran0	LCG(2^{31}-1, 16807, 0, 3141)	$2^{31} - 2$
MT19937	A matrix linear recurrence over a finite field F_2	$2^{19937} - 1$

and denoted by $ICG(p, a, b, x_0)$, where x^{-1} is the inverse of x modulo p multiplication. In this case, since p is prime, the set $\mathbb{Z}_p$ is a field where division is possible. For example, $\mathbb{Z}_7 = \{0, 1, 2, 3, 4, 5, 6\}$ and all nonzero elements have inverse elements given by $1^{-1} = 1, 2^{-1} = 4, 3^{-1} = 5, 4^{-1} = 2, 5^{-1} = 3, 6^{-1} = 6$.

Table 27.1 lists various random number generators. All the generators except Randu have been in use until recently. Mersenne twister (MT19937) is the most widely used generator now. It is employed by MATLAB and many other computer software packages. Its name is derived from the fact that its period length is a Mersenne prime. For more information consult [51, 62, 78].

27.3 Testing Random Number Generators

To check the efficiency of a given pseudorandom number generator we apply not only theoretical number theory but also test it statistically. One such methods is the following: For a sequence $\{x_n\}$, $0 \leq x_n < M$, generated by a pseudorandom number generator we define a point in the three-dimensional cube $[0, M] \times [0, M] \times [0, M]$ by $(x_{3i+1}, x_{3i+2}, x_{3i+3})$ for $i \geq 0$ and check whether they reveal a lattice structure. The more visible the lattice structure is, the less random the generator is.

A lattice structure determined by a bad linear congruential generator $LCG(M, a, b, x_0)$ is presented in Fig. 27.3 where there are a small number of planes determined by the points generated by a linear congruential generator with $M = 2^{30}$, $a = 65539$, $b = 0$ and $x_0 = 1$. The points are on fifteen planes inside a cube, and such a lattice structure is a weakness of linear congruential generators. In practice, to make sure that thus generated n-tuples are more evenly distributed, we need to use generators which can produce sufficiently many $(n-1)$-dimensional hyperplanes inside an n-dimensional cube.

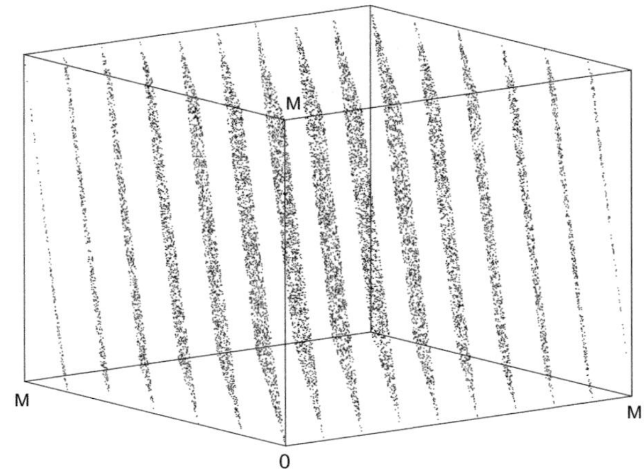

Fig. 27.3 Plane structure determined by three-dimensional points obtained from a linear congruential generator

27.4 Normally Distributed Random Numbers

Example 27.1 Here is a convenient way to generate random numbers with approximate standard normal distribution: Take independent and uniformly distributed random numbers $u_1, u_2, \ldots, u_N$ and let

$$X = \sqrt{\frac{12}{N}} \left(\sum_{k=1}^{N} u_k - \frac{N}{2} \right) ,$$

which has mean 0 and variance 1. By the Central Limit Theorem the sum X is *approximately normally distributed* for sufficiently large N. For $N = 12$ the factor $\sqrt{12/N}$ conveniently becomes 1. Let $U_1, \ldots, U_{12}$ be independent and uniformly distributed in $[0, 1]$. Then $\mathbb{E}[U_i] = \frac{1}{2}$ and $\text{Var}[U_i] = \int_0^1 (x - \frac{1}{2})^2 \, dx = \frac{1}{12}$. Hence

$$Y = U_1 + \cdots + U_{12} - 6$$

satisfies $\mathbb{E}[Y] = 0$ and $\text{Var}(Y) = \text{Var}(U_1) + \cdots + \text{Var}(U_{12}) = 1$. To find the skewness $\mathbb{E}\left[(Y - \mathbb{E}[Y])^3\right]$ we note that

$$(Y - \mathbb{E}[Y])^3 = Y^3 = \left(\sum_{i=1}^{12} (U_i - \frac{1}{2}) \right)^3 = \sum_{i,j,k} (U_i - \frac{1}{2})(U_j - \frac{1}{2})(U_k - \frac{1}{2}) .$$

If i, j, k are all distinct, we have

$$\mathbb{E}\big[(U_i - \frac{1}{2})(U_j - \frac{1}{2})(U_k - \frac{1}{2})\big] = \mathbb{E}\big[U_i - \frac{1}{2}\big] \, \mathbb{E}\big[U_j - \frac{1}{2}\big] \, \mathbb{E}\big[U_k - \frac{1}{2}\big] = 0 .$$

If $i \neq j = k$, then

$$\mathbb{E}[(U_i - \frac{1}{2})(U_j - \frac{1}{2})(U_k - \frac{1}{2})] = \mathbb{E}[U_i - \frac{1}{2}]\,\mathbb{E}[(U_j - \frac{1}{2})^2] = 0 \times \frac{1}{12} = 0 \,.$$

Finally, if $i = j = k$ then

$$\mathbb{E}[(U_i - \frac{1}{2})^3] = \int_0^1 (u - \frac{1}{2})^3 \mathrm{d}u = 0 \,.$$

Hence the skewness is equal to 0. As for the kurtosis, we have $\mathbb{E}[Y^4] = \frac{29}{10}$, which is very close to the theoretical value 3 of the kurtosis of the standard normal distribution. (See Exercise 4.12.) Therefore we may use Y as a crude approximation of a standard normal random variable in a numerical simulation. See Fig. 27.4 and Simulation 27.4.

Now we consider a rigorous method for generating random numbers with standard normal distribution. The Box–Muller algorithm [11] converts a pair of independent uniformly distributed random numbers into a pair of normally distributed random numbers.

Theorem 27.2 (Box–Muller) *For two independent random variables U_1 and U_2 uniformly distributed in $(0, 1)$, define*

$$(\rho, \theta) = (\sqrt{-2 \log U_1}, 2\pi U_2) \,,$$

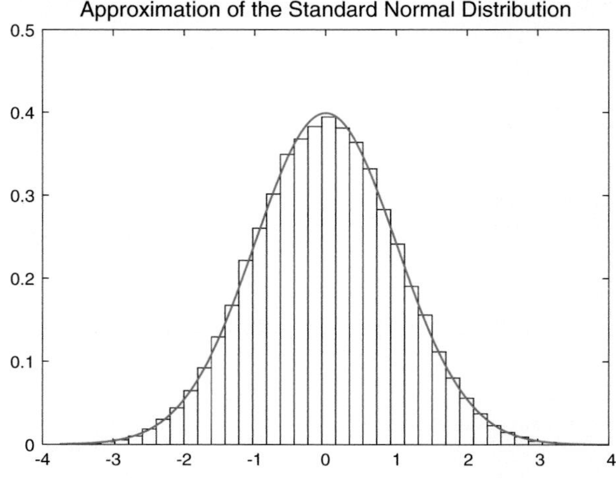

Fig. 27.4 The histogram for the distribution of $U_1 + \cdots + U_{12} - 6$ with the pdf of the standard normal distribution

and let

$$(Z_1, Z_2) = (\rho\cos\theta, \rho\sin\theta)\ .$$

Then Z_1 and Z_2 are independent, and have the standard normal distribution.

Proof Define a one-to-one mapping $\Phi : [0, 1) \times [0, 1) \to \mathbb{R}^2$ by

$$\Phi(u_1, u_2) = (\sqrt{-2\log u_1}\cos 2\pi u_2, \sqrt{-2\log u_1}\sin 2\pi u_2) = (z_1, z_2)\ .$$

Note that $z_1^2 + z_2^2 = -2\log u_1$. Since the Jacobian of Φ is given by

$$J_\Phi = \det\begin{bmatrix} -\dfrac{1}{\sqrt{-2\log u_1}}\dfrac{\cos 2\pi u_2}{u_1} & \sqrt{-2\log u_1}(-2\pi\sin 2\pi u_2) \\ -\dfrac{1}{\sqrt{-2\log u_1}}\dfrac{\sin 2\pi u_2}{u_1} & \sqrt{-2\log u_1}(-2\pi\cos 2\pi u_2) \end{bmatrix} = -\frac{2\pi}{u_1}\ ,$$

we have, for $A \subset [0, 1) \times [0, 1)$,

$$\begin{aligned} \iint_A f_{U_1,U_2}(u_1, u_2)\mathrm{d}u_1\mathrm{d}u_2 &= \Pr((U_1, U_2) \in A) \\ &= \Pr((Z_1, Z_2) \in \Phi(A)) \\ &= \iint_{\Phi(A)} f_{Z_1,Z_2}(z_1, z_2)\mathrm{d}z_1\mathrm{d}z_2 \\ &= \iint_A f_{Z_1,Z_2}(\Phi(u_1, u_2))|J_\Phi|\mathrm{d}u_1\mathrm{d}u_2 \\ &= \iint_A f_{Z_1,Z_2}(\Phi(u_1, u_2))\frac{2\pi}{u_1}\mathrm{d}u_1\mathrm{d}u_2\ . \end{aligned}$$

Since the joint probability density is given by

$$f_{U_1,U_2}(u_1, u_2) = 1\ , \quad (u_1, u_2) \in [0, 1) \times [0, 1)\ ,$$

we have

$$f_{Z_1,Z_2}(z_1, z_2) = \frac{u_1}{2\pi} = \frac{1}{2\pi}\mathrm{e}^{-\frac{1}{2}(z_1^2+z_2^2)}\ .$$

Since

$$f_{Z_1,Z_2}(z_1, z_2) = \frac{1}{\sqrt{2\pi}}\mathrm{e}^{-\frac{1}{2}z_1^2}\frac{1}{\sqrt{2\pi}}\mathrm{e}^{-\frac{1}{2}z_2^2} = f_{Z_1}(z_1)f_{Z_2}(z_2)\ ,$$

we conclude that Z_1 and Z_2 are independent. □

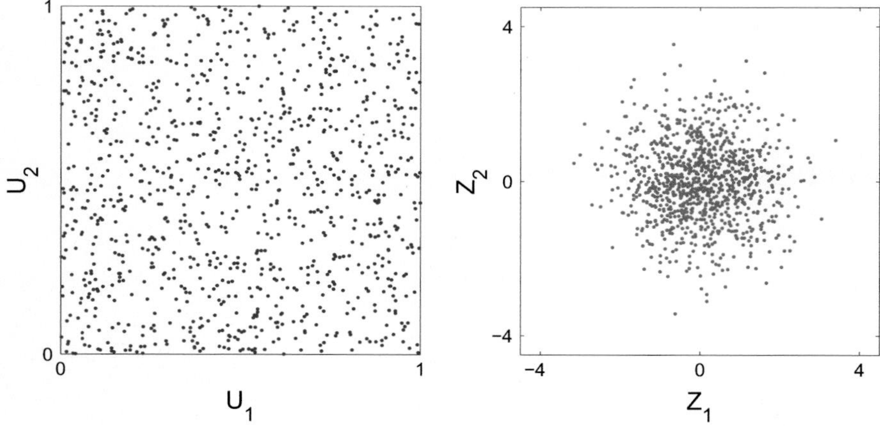

Fig. 27.5 Generation of normally distributed random points by the Box–Muller algorithm

In Fig. 27.5 are plotted 10^3 uniformly distributed random points (left) and the corresponding normally distributed points generated by the Box–Muller algorithm (right). Consult Simulation 27.5.

The following algorithm, called Marsaglia's polar method, is a modification of the Box–Muller method, and does not use trigonometric functions, which can reduce computation time even though some of the generated points by the pair (U_1, U_2) are wasted without being used.

Theorem 27.3 (Marsaglia) *For independent random variables U_1 and U_2 uniformly distributed in $(0, 1)$, define $V_1 = 2U_1 - 1$ and $V_2 = 2U_2 - 1$. Let $R = V_1^2 + V_2^2$ and select the points (V_1, V_2) such that $0 < R < 1$ and define*

$$(Z_1, Z_2) = \left(V_1 \sqrt{-2 \frac{\log R}{R}}, V_2 \sqrt{-2 \frac{\log R}{R}} \right) .$$

Then Z_1 and Z_2 are independent and have the standard normal distribution.

Proof Note that the accepted points (V_1, V_2) are uniformly distributed inside the unit disk D with the uniform probability density $\frac{1}{\pi}$. Define a mapping $\Psi : D \to [0, 1] \times [0, 1]$ by $\Psi(V_1, V_2) = (R, \Theta)$ where $R = V_1^2 + V_2^2$ and $\tan(2\pi\Theta) = \frac{V_2}{V_1}$, $\Theta \in [0, 1)$. Since, for $0 \le \alpha \le 1$,

$$\Pr(R \le \alpha) = \Pr(V_1^2 + V_2^2 \le \alpha) = \Pr(\sqrt{V_1^2 + V_2^2} \le \sqrt{\alpha}) = \pi(\sqrt{\alpha})^2 \frac{1}{\pi} = \alpha ,$$

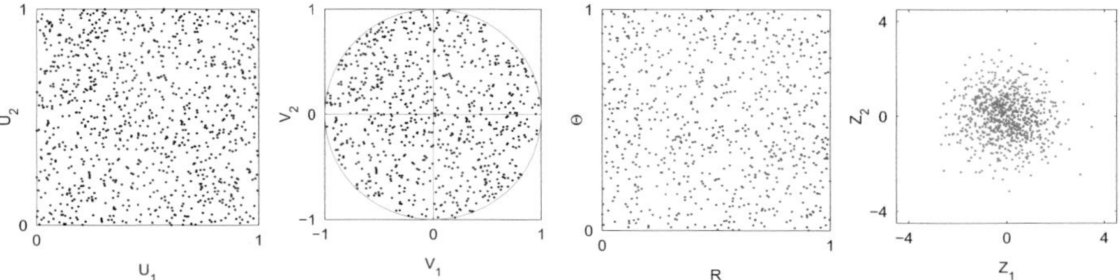

Fig. 27.6 Generation of normally distributed random points by the Marsaglia algorithm

we see that (R, Θ) is uniformly distributed in the unit square. Note that R and Θ are independent since

$$\begin{aligned}
&\Pr(\alpha_1 \le R \le \alpha_2,\ \beta_1 \le \Theta \le \beta_2) \\
&= \Pr(\sqrt{\alpha_1} \le \sqrt{V_1^2 + V_2^2} \le \sqrt{\alpha_2},\ \beta_1 \le \Theta \le \beta_2) \\
&= \left(\frac{1}{\pi}(\sqrt{\alpha_2})^2 - \frac{1}{\pi}(\sqrt{\alpha_1})^2\right)(\beta_2 - \beta_1) \\
&= \Pr(\alpha_1 \le R \le \alpha_2)\Pr(\beta_1 \le \Theta \le \beta_2)\ ,
\end{aligned}$$

and apply the Box–Muller method to obtain the standard normal variables Z_1 and Z_2. Finally we note that $\cos(2\pi\Theta) = V_1/\sqrt{V_1^2 + V_2^2}$ and $\sin(2\pi\Theta) = V_2/\sqrt{V_1^2 + V_2^2}$. □

In Fig. 27.6 we plot 10^3 uniformly distributed random points, the accepted points inside the unit disk, their images under ψ, and the normally distributed points generated by the Marsaglia algorithm (from left to right). Consult Simulation 27.6.

Remark 27.1 In practice, if memory size does not matter, we can obtain normally distributed random numbers very quickly using MATLAB, which can generate normally distributed random numbers as fast as it can generate uniformly distributed numbers. Almost instantly it can generate millions of random numbers on a desktop computer. For a practical algorithm employed by MATLAB, consult [70].

27.5 Computer Experiments

Simulation 27.1 (Estimation of π)

We plot the lattice structure in the unit cube determined by ordered triples of random numbers. See Fig. 27.1.

```
a = rand(1000,1);
b = rand(1000,1);
```

```
plot(a,b,'.');
hold on

t=0:pi/300:pi/2;
plot(cos(t),sin(t))
axis equal

N=10^6;
xx=rand(N,1);
yy=rand(N,1);
frequency = 0;
for i = 1:N
    if xx(i)^2+yy(i)^2 < 1
        count = 1;
    else
        count = 0;
    end
    frequency = frequency + count;
end
frequency/N*4   % This produces a number close to pi = 3.1415...
```

Simulation 27.2 (Estimation of the Gamma Function)
We test the formula $\Gamma(6) = 5!$ using exponentially distributed random numbers. See Fig. 27.2.

```
p = 6;
N = 20;
num = 2^N;
% (p-1)th power of exponentially distributed random numbers
A = exprnd(1,num,1) .^ (p-1);

for i = 1:N
    gamma(i) = mean(A(1:2^i));
end

x = 0:0.02:N;
plot(gamma,'o-');
hold on
plot(x,factorial(p-1));
title('\Gamma(6)=120');
xlabel('n');
```

Simulation 27.3 (Lattice Structure of Ordered Triples)
We plot the lattice structure in the unit cube determined by ordered triples of uniformly distributed random numbers. See Fig. 27.3.

```
a=65539;
b=0;
M=2^30;
x0=1;
N=3000;
x=zeros(N);
y=zeros(N);
z=zeros(N);
```

```
x(1)=x0;
y(1)=mod(a*x(1)+b,M);
z(1)=mod(a*y(1)+b,M);
for i=2:3*N
    x(i)=mod(a*z(i-1)+b,M);
    y(i)=mod(a*x(i)+b,M);
    z(i)=mod(a*y(i)+b,M);
end;
x=x./M;
y=y./M;
z=z./M;
plot3(x,y,z,'.');
```

Simulation 27.4 (Sum of Twelve Uniform Variables)

We show that the probability distribution of the sum of twelve uniform variables is close to the standard normal distribution. See Example 27.1.

```
SampleSize=100000;
U12=random('unif',0,1,12, SampleSize);
Y=sum(U12)-6;
mu = mean(Y)
sigma2 = mean(Y.^2)
skew = mean(Y.^3)
kurto = mean(Y.^4)

Number_of_bins = 40;
N = Number_of_bins - 1;

width = (max(Y) - min(Y))/N;

minimum_value = min(Y) - width/2;
maximum_value = max(Y) + width/2;

bin = zeros(1, Number_of_bins);
for i=1:SampleSize
j_bin=ceil((Y(i) - minimum_value)/width);
bin(j_bin)=bin(j_bin)+1;
end

x = minimum_value:0.01:maximum_value;
y = 1/sqrt(2*pi)*exp(-x.^2/2);

mid_points = minimum_value + width/2:width:maximum_value;
bar(mid_points,bin/SampleSize/width, 1, 'w')
hold on

plot(x,y)
```

The output is

```
mu = 0.0038
sigma2 = 1.0026
skew = 0.0069
kurto = 2.9085
```

For a plot of the histogram see Fig. 27.4.

Simulation 27.5 (Box–Muller Method)

We generate random numbers with the standard normal distribution using the Box–Muller algorithm.

```
N = 10000;
U1 = rand(N,1);
U2 = rand(N,1);

r = sqrt(-2*(log(U1)));
theta = 2*pi*U2;
Z1 = r.*cos(theta);
Z2 = r.*sin(theta);

plot(U1,U2,'.')
plot(Z1,Z2,'.')
```

For the plot see Fig. 27.5.

Simulation 27.6 (Marsaglia's Polar Method)

We generate random numbers with the standard normal distribution using Marsaglia's polar method. Note that the MATLAB command `atan` for $\arctan(x)$, $-\infty < x < \infty$, has its values in the range $[-\frac{\pi}{2}, \frac{\pi}{2}]$.

```
N = 10000;
index = zeros(N,1);
U1 = rand(N,1);
U2 = rand(N,1);
V1 = 2*U1 - 1;
V2 = 2*U2 - 1;

k = 0;
for i = 1:N
R = V1(i)^2 + V2(i)^2;
if (0 < R && R < 1)
k = k+1;
index(k)=i;
C = sqrt(-2*log(R)/R);
W1(k) = V1(i);
W2(k) = V2(i);
Z1(k) = V1(i)*C;
Z2(k) = V2(i)*C;
end
end

K = k
V1 = W1(1:K);
V2 = W2(1:K);

R = V1.^2 + V2.^2;

for k=1:K
    if V1(k) >= 0 && V2(k) >= 0
```

```
Theta(k) = (1/(2*pi))*atan(V2(k)/V1(k));
    elseif V1(k) >= 0 && V2(k) < 0
Theta(k) = (1/(2*pi))*atan(V2(k)/V1(k)) + 1;
    else
Theta(k) = (1/(2*pi))*atan(V2(k)/V1(k)) + 1/2;
    end
end

plot(U1,U2,'k.','markersize',1);
plot(V1,V2,'k.','markersize',1);
plot(cos(2*pi*t),sin(2*pi*t),'r-','LineWidth',1);
plot(R,Theta,'k.','markersize',1);
plot(Z1,Z2,'k.','markersize',1);
```

For the plot see Fig. 27.6.

Exercises

27.1 Show that if a random variable U is uniformly distributed in $(0, 1)$ then $Y = -\log U$ has exponential distribution.

27.2 Fix a constant x_0. Explain how to generate random numbers with the restricted normal distribution with the probability density function given by

$$f(x) = \frac{1}{1 - N(x_0)} \mathbf{1}_{[x_0,+\infty)}(x) \frac{1}{\sqrt{2\pi}} \mathrm{e}^{-x^2/2}$$

where

$$N(x) = \int_{-\infty}^{x} \frac{1}{\sqrt{2\pi}} \mathrm{e}^{-z^2/2} \, \mathrm{d}z \, .$$

(For an application see Exercise 28.7.)

Chapter 28
The Monte Carlo Method for Option Pricing

Option price is expressed as an expectation of a random variable representing a payoff. Thus we generate sufficiently many asset price paths using random number generators, and evaluate the average of the payoff. In this chapter we introduce efficient ways to apply the Monte Carlo method. The key idea is variance reduction, which increases the precision of estimates for a given sample size by reducing the sample variance in the application of the Central Limit Theorem. The smaller the variance is, the narrower the confidence interval becomes, for a given confidence level and a fixed sample size.

28.1 The Antithetic Variate Method

Here is a basic example of the antithetic variate method for reducing variance. Given a function $f(x)$ defined on the unit interval $[0, 1]$, choose a random variable U uniformly distributed in $[0, 1]$. Note that $1 - U$ is also uniformly distributed. To find the expectation $\mathbb{E}[f(U)]$ we compute

$$\frac{\mathbb{E}[f(U)] + \mathbb{E}[f(1-U)]}{2} = \mathbb{E}\left[\frac{f(U) + f(1-U)}{2}\right] .$$

In this case, $1 - U$ is called an *antithetic variate*. This method is effective especially when f is monotone.

Example 28.1 Note that $\int_0^1 e^x dx = e - 1$. If we want to find the integral by the Monte Carlo method, we estimate $\mathbb{E}[e^U]$ where U is uniformly distributed in $[0, 1]$. Let

$$X = \frac{e^U + e^{1-U}}{2} ,$$

G.H. Choe, *Stochastic Analysis for Finance with Simulations*, Universitext,
DOI 10.1007/978-3-319-25589-7_28

and find $\mathbb{E}[X]$ to estimate $\mathbb{E}[e^U]$. Note that

$$\begin{aligned}\mathrm{Var}(X) &= \frac{\mathrm{Var}(e^U) + \mathrm{Var}(e^{1-U}) + 2\,\mathrm{Cov}(e^U, e^{1-U})}{4} \\ &= \frac{\mathrm{Var}(e^U) + \mathrm{Cov}(e^U, e^{1-U})}{2} .\end{aligned}$$

Note that

$$\mathbb{E}[e^U e^{1-U}] = \int_0^1 e^x e^{1-x} dx = e \approx 2.71828$$

and

$$\mathrm{Cov}(e^U, e^{1-U}) = \mathbb{E}[e^U e^{1-U}] - \mathbb{E}[e^U]\,\mathbb{E}[e^{1-U}] = e - (e-1)^2 \approx -0.2342 .$$

Since

$$\mathrm{Var}(e^U) = \mathbb{E}[e^{2U}] - (\mathbb{E}[e^U])^2 = \frac{e^2-1}{2} - (e-1)^2 \approx 0.2420 ,$$

we have

$$\mathrm{Var}(X) \approx 0.0039 .$$

Hence the variance of X is greatly reduced in comparison with the variance of e^U. (See also Exercise 28.2 for a similar example.)

Theorem 28.1 *For any monotonically increasing (or decreasing) functions f and g, and for any random variable X, we have*

$$\mathbb{E}[f(X)g(X)] \geq \mathbb{E}[f(X)]\,\mathbb{E}[g(X)] ,$$

or, equivalently,

$$\mathrm{Cov}(f(X), g(X)) \geq 0 .$$

Proof First, consider the case when f and g are monotonically increasing. Note that

$$(f(x) - f(y))(g(x) - g(y)) \geq 0$$

for every x, y since two differences are either both nonnegative ($x \geq y$) or both non-positive ($x \leq y$). Thus, for any random variables X, Y, we have

$$(f(X) - f(Y))(g(X) - g(Y)) \geq 0 ,$$

and hence

$$\mathbb{E}[(f(X) - f(Y))(g(X) - g(Y))] \geq 0 .$$

Hence

$$\mathbb{E}[f(X)g(X)] + \mathbb{E}[f(Y)g(Y)] \geq \mathbb{E}[f(X)g(Y)] + \mathbb{E}[f(Y)g(X)] .$$

If X and Y are independent and identically distributed, then

$$\mathbb{E}[f(X)g(X)] = \mathbb{E}[f(Y)g(Y)]$$

and

$$\mathbb{E}[f(X)g(Y)] = \mathbb{E}[f(X)]\,\mathbb{E}[g(Y)] = \mathbb{E}[f(X)]\,\mathbb{E}[g(X)] ,$$
$$\mathbb{E}[f(Y)g(X)] = \mathbb{E}[f(Y)]\,\mathbb{E}[g(X)] = \mathbb{E}[f(X)]\,\mathbb{E}[g(X)] .$$

Hence

$$2\,\mathbb{E}[f(X)g(X)] \geq 2\,\mathbb{E}[f(X)]\,\mathbb{E}[g(X)] .$$

Therefore,

$$\mathrm{Cov}(f(X), g(X)) = \mathbb{E}[f(X)g(X)] - \mathbb{E}[f(X)]\,\mathbb{E}[g(X)] \geq 0 .$$

When f and g are monotonically decreasing, then we simply replace f and g by $-f$ and $-g$, respectively, and note that $-f$ and $-g$ are monotonically increasing and that $\mathrm{Cov}(f(X), g(X)) = \mathrm{Cov}(-f(X), -g(X)) \geq 0$. □

Corollary 28.1 *Let $f(x)$ be a monotone function and X be any random variable. Then we have*

(i) $\mathrm{Cov}(f(X), f(1-X)) \leq 0$,
(ii) $\mathrm{Cov}(f(X), f(-X)) \leq 0$.

Proof

(i) First, consider the case when f is monotonically increasing. Define $g(x) = -f(1-x)$. Then $g(x)$ is also increasing. Hence

$$\mathrm{Cov}(f(X), f(1-X)) = \mathrm{Cov}(f(X), -g(X)) = -\mathrm{Cov}(f(X), g(X)) \leq 0 .$$

If $f(x)$ is decreasing, then consider instead $-f(x)$ which is increasing, and note that $\mathrm{Cov}(f(X), f(1-X)) = \mathrm{Cov}(-f(X), -f(1-X)) \leq 0$.

(ii) The proof is similar to that of (i). □

Remark 28.1

(i) If $f(x)$ is monotone and U is uniformly distributed in $[0, 1]$, then $\mathrm{Cov}(f(U), f(1-U)) \leq 0$.
(ii) If $f(x)$ is monotone and Z is a standard normal variable, then $\mathrm{Cov}(f(Z), f(-Z)) \leq 0$.

Example 28.2 (Option Pricing) Consider a European call option with expiry date T and payoff $C(S_T) = \max\{S_T - K, 0\}$ where S_T is the asset price at T. If Z denotes a standard normal variable, we consider Z and $-Z$ to apply the antithetic variate method. Since $S_T = S_0 e^{(r-\frac{1}{2}\sigma^2)T+\sigma\sqrt{T}Z}$ and $\widetilde{S}_T = S_0 e^{(r-\frac{1}{2}\sigma^2)T-\sigma\sqrt{T}Z}$ are correlated, two payoffs $C(S_T)$ and $C(\widetilde{S}_T)$ are also correlated. Since

$$\mathbb{E}^{\mathbb{Q}}[C(S_T)] = \mathbb{E}^{\mathbb{Q}}[C(\widetilde{S}_T)] ,$$

we have

$$\mathbb{E}^{\mathbb{Q}}[C(S_T)] = \mathbb{E}^{\mathbb{Q}}\left[\frac{C(S_0 e^{(r-\frac{1}{2}\sigma^2)T+\sigma\sqrt{T}Z}) + C(S_0 e^{(r-\frac{1}{2}\sigma^2)T-\sigma\sqrt{T}Z})}{2}\right] .$$

In Simulation 28.1 the ratio between the variances for the standard Monte Carlo and the antithetic variate method is approximately equal to 2.7080.

When we price a path dependent option, we need to generate sample paths of a geometric Brownian motion by the discretized algorithm

$$S(t_{i+1}) = S(t_i) + \mu S(t_i)\delta t + \sigma S(t_i)\sqrt{\delta t}Z_i$$

where Z_i are independent standard normal variables. The antithetic path $\widetilde{S}_t$ is given by the algorithm

$$\widetilde{S}(t_{i+1}) = \widetilde{S}(t_i) + \mu\widetilde{S}(t_i)\delta t - \sigma\widetilde{S}(t_i)\sqrt{\delta t}Z_i .$$

The antithetic variate method employs both S_t and $\widetilde{S}_t$ as a pair. Since S_T and $\widetilde{S}_T$ are correlated, $f(S_T)$ and $f(\widetilde{S}_T)$ are also correlated for any function f. In Fig. 28.1 we take $\mu = 0.1$ and $\sigma = 0.3$.

The efficiency of the antithetic variate method in pricing European options depends on the parameter values as shown in Fig. 28.2. We take $T = 1$, $S_0 = 10$, $r = 0.05$, $\sigma = 0.3$, and choose various values of K. Payoffs C_1 are plotted by solid lines, and the corresponding antithetic variates C_2 by dotted lines as functions of z, the values of the standard normal distribution. The averages $(C_1 + C_2)/2$ are also plotted together. In Fig. 28.2 the reduction rate for the cases corresponding to $K = 8, 10, 12$ are approximately equal to 86%, 70%, 59%, respectively. Consult Simulation 28.2.

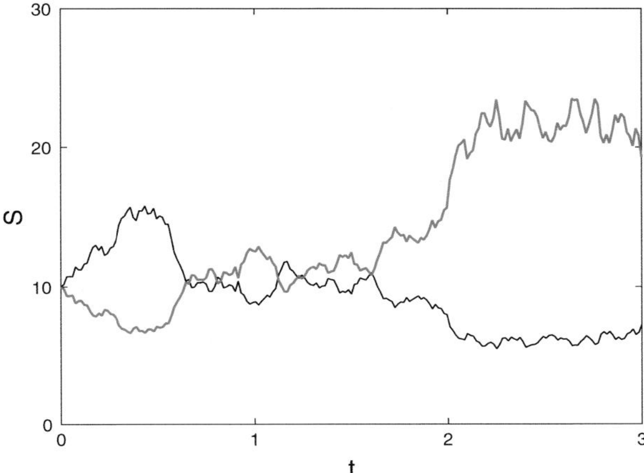

Fig. 28.1 A pair of sample paths of geometric Brown motion obtained by antithetic variate method

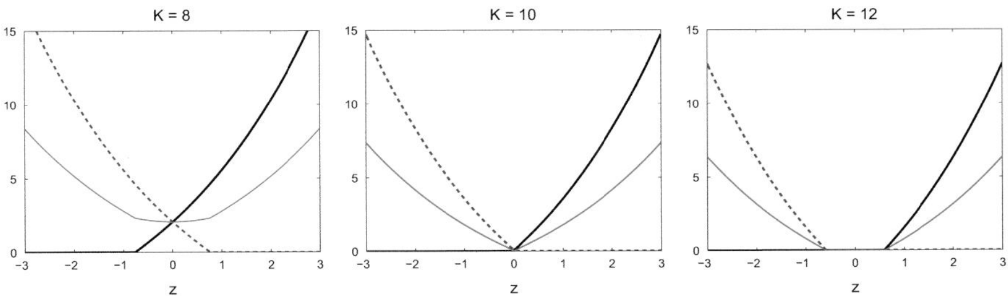

Fig. 28.2 Payoffs of European call options and the corresponding antithetic variates for $S_0 = 10$ and $K = 8, 10, 12$

28.2 The Control Variate Method

The control variate method for variance reduction in estimating $\mathbb{E}[X]$ employs another random variable Y whose properties are well-known. For example, the mean $\mathbb{E}[Y]$ is known. The random variable Y is called a *control variate*.

Suppose that $\mathbb{E}[Y]$ is already known. Define

$$\widetilde{X} = X + c\,(Y - \mathbb{E}[Y])$$

for a real constant c. Then

$$\mathbb{E}[\widetilde{X}] = \mathbb{E}[X] \,.$$

We try to find an optimal value $\widetilde{c}$ of c for which

$$\mathrm{Var}(\widetilde{X}) = \mathrm{Var}(X) + c^2\mathrm{Var}(Y) + 2c\,\mathrm{Cov}(X, Y) \tag{28.1}$$

becomes minimal. Differentiating with respect to c, we have

$$\widetilde{c} = -\frac{\mathrm{Cov}(X, Y)}{\mathrm{Var}(Y)}\,, \tag{28.2}$$

and by choosing $c = \widetilde{c}$ in (28.1) we obtain

$$\begin{aligned}\min_c \mathrm{Var}(\widetilde{X}) &= \mathrm{Var}(X) - \frac{\mathrm{Cov}(X, Y)^2}{\mathrm{Var}(Y)}\\ &= \mathrm{Var}(X)\left(1 - \frac{\mathrm{Cov}(X, Y)^2}{\mathrm{Var}(X)\mathrm{Var}(Y)}\right)\\ &= \mathrm{Var}(X)(1 - \rho(X, Y)^2)\\ &\le \mathrm{Var}(X)\,.\end{aligned}$$

Thus the maximal rate of variance reduction rate is equal to

$$\frac{\mathrm{Var}(X) - \min_c \mathrm{Var}(\widetilde{X})}{\mathrm{Var}(X)} = \rho(X, Y)^2\,.$$

Hence the more X and Y are correlated, the better the variance reduction gets.

Example 28.3 Let U be uniformly distributed in the interval $(0, 1)$ and

$$X = \mathrm{e}^U\,.$$

We want to estimate $\mathbb{E}[X]$. Take a control variate $Y = U$. Note that $\mathbb{E}[Y] = \int_0^1 y\,\mathrm{d}y = \frac{1}{2}$. Define

$$\widetilde{X} = \mathrm{e}^U - \widetilde{c}\ (U - \frac{1}{2})$$

for the optimal value $\widetilde{c}$ given by (28.2). Note that

$$\begin{aligned}\mathrm{Cov}(\mathrm{e}^U, U) &= \mathbb{E}[\mathrm{e}^U U] - \mathbb{E}[\mathrm{e}^U]\,\mathbb{E}[U]\\ &= \int_0^1 \mathrm{e}^x x\,\mathrm{d}x - (\mathrm{e} - 1)\frac{1}{2}\end{aligned}$$

$$= \mathrm{e} - (\mathrm{e}-1) - (\mathrm{e}-1)\frac{1}{2}$$
$$= \frac{3}{2} - \frac{1}{2}\mathrm{e} \approx 0.140859$$

and

$$\widetilde{c} = -\frac{\mathrm{Cov}(\mathrm{e}^U, U)}{\mathrm{Var}(U)} = -\frac{\frac{3}{2} - \frac{1}{2}\mathrm{e}}{\frac{1}{12}} = 6(\mathrm{e}-3) \approx -1.690309\,.$$

Recall that

$$\mathrm{Var}(\mathrm{e}^U) = \mathbb{E}[\mathrm{e}^{2U}] - \mathbb{E}[\mathrm{e}^U]^2 = \frac{\mathrm{e}^2-1}{2} - (\mathrm{e}-1)^2 \approx 0.242036\,.$$

Hence the variance reduction rate is approximately equal to

$$\rho(\mathrm{e}^U, U)^2 = \frac{\mathrm{Cov}(\mathrm{e}^U, U)^2}{\mathrm{Var}(\mathrm{e}^U)\mathrm{Var}(U)} \approx \frac{0.140859^2}{0.242036 \times \frac{1}{12}} \approx 0.9837\,,$$

and the variance is greatly reduced by 98.37%. See also Exercise 28.6.

Example 28.4 (European Call Option) To estimate $\mathbb{E}[X]$ for

$$X = (S_T - K)^+\,,$$

we choose the underlying asset itself as a control variate, i.e., we take $Y = S_T$. Recall that $\mathbb{E}[Y] = S_0\mathrm{e}^{rT}$ and $\mathrm{Var}(Y) = S_0^2\mathrm{e}^{2rT}(\mathrm{e}^{\sigma^2 T} - 1)$. To compute the covariance of X and Y we may use the Monte Carlo method since we do not need the precise value for $\widetilde{c}$ in (28.2). For exotic options with complicated payoffs, sometimes it might be convenient to take the payoffs of standard European calls and puts as control variates.

Example 28.5 (Asian Option with Arithmetic Average) We use the control variate method to compute the price of an Asian option with arithmetic average, denoted by V_{arith}, using the price of the corresponding Asian option with geometric average, V_{geo}, as a control variate. The formula for the price of an Asian option with geometric average, $V_{\text{geo formula}}$, is given in Theorem 18.1. Since the expectation of V_{geo} is equal to $V_{\text{geo formula}}$, the expectation of

$$\widetilde{V} = V_{\text{arith}} + c(V_{\text{geo}} - V_{\text{geo formula}})$$

is equal to the expectation of V_{arith}, however, with a suitable choice of c the variance of $\widetilde{V}$ is reduced in comparison to the variance of V_{arith} which is computed by the

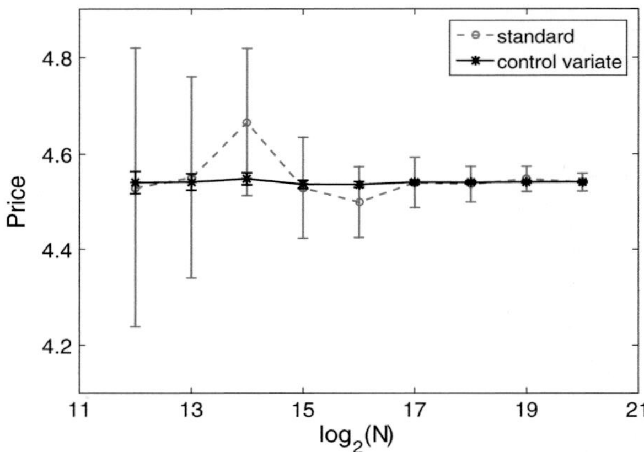

Fig. 28.3 Comparison of the standard Monte Carlo method and the control variate method for computing Asian option price

standard Monte Carlo method. The optimal value for c is

$$\widetilde{c} = -\frac{\mathrm{Cov}(V_{\mathrm{arith}}, V_{\mathrm{geo}})}{\mathrm{Var}(V_{\mathrm{geo}})} .$$

For our simulation we simply choose $\widetilde{c} = 1$, which is not the best choice, however, we have a reasonable reduction in variance. See Fig. 28.3 where the confidence intervals are presented for each sample size $N = 2^j$, $12 \le j \le 20$. Consult Simulation 28.4.

28.3 The Importance Sampling Method

The importance sampling method for Monte Carlo integration modifies a given probability density before computing the expectation. The idea is to assign more weight to a region where the given random variable has more importance. Consider an integral

$$\mathbb{E}^{\mathbb{P}}[X] = \int_{\Omega} X \,\mathrm{d}\mathbb{P}$$

with respect to a probability measure $\mathbb{P}$. Let $\mathbb{Q}$ be an equivalent probability measure with its Radon–Nikodym derivative with respect to $\mathbb{P}$ denoted by $\frac{\mathrm{d}\mathbb{Q}}{\mathrm{d}\mathbb{P}}$. Recall that the Radon–Nikodym derivative of $\mathbb{P}$ with respect to $\mathbb{Q}$, denoted by $\frac{\mathrm{d}\mathbb{P}}{\mathrm{d}\mathbb{Q}}$, satisfies

$$\frac{\mathrm{d}\mathbb{P}}{\mathrm{d}\mathbb{Q}} = \frac{1}{\frac{\mathrm{d}\mathbb{Q}}{\mathrm{d}\mathbb{P}}}$$

and

$$\int_\Omega X\,\mathrm{d}\mathbb{P} = \int_\Omega X\frac{\mathrm{d}\mathbb{P}}{\mathrm{d}\mathbb{Q}}\,\mathrm{d}\mathbb{Q}\,.$$

Example 28.6 Let $\mathbb{P}$ denote the Lebesgue measure on the unit interval $(0, 1)$. Let U be uniformly distributed in $(0, 1)$, and take $X = U^5$ and consider $\mathbb{E}[U^5]$. Note that $\mathbb{E}^{\mathbb{P}}[X] = \frac{1}{6}$ and

$$\mathrm{Var}^{\mathbb{P}}(X) = \mathbb{E}^{\mathbb{P}}[U^{10}] - (\mathbb{E}^{\mathbb{P}}[U^5])^2 = \frac{1}{11} - \frac{1}{36} = \frac{25}{396} \approx 0.0631\,.$$

Now we use the importance sampling method, putting more weight near the value 1. For example, we define a probability measure $\mathbb{Q}$ by

$$\mathbb{Q}(A) = \int_A 2x\,\mathrm{d}x$$

for $A \subset (0, 1)$, i.e., $\frac{\mathrm{d}\mathbb{Q}}{\mathrm{d}\mathbb{P}} = 2x$. Note that $\frac{\mathrm{d}\mathbb{P}}{\mathrm{d}\mathbb{Q}} = \frac{1}{2x}$ and

$$\mathbb{E}^{\mathbb{P}}[U^5] = \mathbb{E}^{\mathbb{Q}}\left[U^5\frac{\mathrm{d}\mathbb{P}}{\mathrm{d}\mathbb{Q}}\right] = \mathbb{E}^{\mathbb{Q}}\left[\frac{1}{2}U^4\right]\,.$$

Also note that

$$\begin{aligned}\mathrm{Var}^{\mathbb{Q}}\left(\frac{1}{2}U^4\right) &= \mathbb{E}^{\mathbb{Q}}\left[\frac{1}{4}U^8\right] - \left(\mathbb{E}^{\mathbb{Q}}\left[\frac{1}{2}U^4\right]\right)^2 \\ &= \frac{1}{4}\left\{\int_0^1 2u^9\,\mathrm{d}u - \left(\int_0^1 2u^5\,\mathrm{d}u\right)^2\right\} \\ &= \frac{1}{4}\left(\frac{1}{5} - \frac{1}{9}\right) = \frac{1}{45} \approx 0.0222\,.\end{aligned}$$

Thus the variance is reduced by approximately $\frac{0.0631-0.0222}{0.0631} \approx 64.8\%$.

The transformation rule in Theorem 4.6 corresponding to the pdf $f_Y(y) = 2y$ is $Y = \sqrt{U}$. (See Example 4.14.) Hence $\mathbb{E}^{\mathbb{P}}[U^5] = \mathbb{E}^{\mathbb{Q}}\left[\frac{1}{2}U^4\right] = \mathbb{E}^{\mathbb{Q}}\left[\frac{1}{2}Y^2\right]$ and $\mathrm{Var}^{\mathbb{Q}}\left(\frac{1}{2}U^4\right) = \mathrm{Var}^{\mathbb{Q}}\left(\frac{1}{2}Y^2\right)$. See Exercise 28.8 for a related problem, and for the numerical experiment see Simulation 28.5.

Example 28.7 (Deep Out-of-the-Money Asian Option) Consider an option whose payoff becomes zero if the underlying asset price falls below an agreed level K at expiry or before expiry. In computing the price of such an option using the standard Monte Carlo method, too many sample asset paths are wasted without contributing in option valuation, especially when K is very high compared to S_0 so that the option is deep out-of-the-money.

We move upward the asset paths as time increases by replacing a given drift coefficient r by a larger value r_1 in simulating geometric Brownian motion so that more sample paths thus generated would have nonzero contribution in evaluating the average of payoff. Girsanov's theorem allows us to do this as long as we multiply by the Radon–Nikodym derivative. We adopt this idea in Simulation 28.6, and compute the price of an Asian call option with arithmetic average where $S_0 = 100$, $K = 150$ and $r = 0.05$.

Here is a theoretical background. Let

$$C_T(S_{t_1}, \ldots, S_{t_L}) = \max\left\{\frac{1}{L}(S_{t_1} + \cdots + S_{t_L}) - K, 0\right\}$$

be the payoff of an Asian option with arithmetic average. The standard Monte Carlo method for option pricing computes

$$\mathrm{e}^{-rT}\mathbb{E}^{\mathbb{P}}[C_T(S_{t_1}, \ldots, S_{t_L})]$$

using the sample paths of the geometric Brownian motion

$$\mathrm{d}S_t = rS_t\mathrm{d}t + \sigma S_t\mathrm{d}W_t$$

where $\mathbb{P}$ denotes the martingale measure and W_t is a $\mathbb{P}$-Brownian motion. Since S_0 is too low compared to K, future values of S_t, $0 < t \leq T$, will tend to stay low so that the arithmetic average will also be far below K in most cases, and most sample values for Monte Carlo simulation will be zero.

Now choose a constant $r_1 > r$ and let

$$\theta = \frac{r - r_1}{\sigma} .$$

By Girsanov's theorem there exists an equivalent probability measure $\mathbb{Q}$ with Radon–Nikodym derivative $\frac{\mathrm{d}\mathbb{Q}}{\mathrm{d}\mathbb{P}}$ satisfying

$$\mathbb{E}\left[\frac{\mathrm{d}\mathbb{Q}}{\mathrm{d}\mathbb{P}}\middle|\mathcal{F}_t\right] = \mathrm{e}^{-\frac{1}{2}\theta^2 t - \theta W_t}$$

and $X_t = W_t + \theta t$ is a $\mathbb{Q}$-Brownian motion. Note that

$$\frac{\mathrm{d}\mathbb{P}}{\mathrm{d}\mathbb{Q}} = \mathrm{e}^{\frac{1}{2}\theta^2 T + \theta W_T} = \mathrm{e}^{\frac{1}{2}\theta^2 T + \theta(X_T - \theta T)} = \mathrm{e}^{-\frac{1}{2}\theta^2 T + \theta X_T}$$

and

$$\mathrm{d}S_t = r_1 S_t\mathrm{d}t + \sigma S_t\mathrm{d}X_t .$$

Hence the option price is equal to

$$\begin{aligned}
e^{-rT}\mathbb{E}^{\mathbb{P}}[C_T] &= e^{-rT}\mathbb{E}^{\mathbb{Q}}\left[C_T \frac{d\mathbb{P}}{d\mathbb{Q}}\right] \\
&= e^{-rT}\mathbb{E}^{\mathbb{Q}}\left[C_T\, e^{\frac{1}{2}\theta^2 T + \theta W_T}\right] \\
&= e^{-rT}\mathbb{E}^{\mathbb{Q}}\left[C_T\, e^{-\frac{1}{2}\theta^2 T + \theta X_T}\right].
\end{aligned}$$

Note that the same idea can be applied for the computation of European options. For more theoretical details, consult Chap. 16.

28.4 Computer Experiments

Simulation 28.1 (Antithetic Variate Method)

We use the antithetic variate method to compute the price of a European call option, and compare the results with the Black–Scholes–Merton formula. The price of the vanilla call option according to the Black–Scholes–Merton formula is approximately equal to `call_vanilla` = 10.0201, while the estimated values by the standard Monte Carlo method and by the antithetic variate method are `a` = 9.9271 and `a_anti` = 10.0823, respectively. The corresponding variances are `b` = 379.5477 and `b_anti` = 140.1554 with their ratio given by 2.7080.

```
S0 = 100;
K = 110;
r = 0.05;
sigma = 0.3;
T = 1;
d1 = (log(S0/K) + (r + 0.5*sigma^2)*T)/(sigma*sqrt(T));
d2 = (log(S0/K) + (r - 0.5*sigma^2)*T)/(sigma*sqrt(T));
call_vanilla = S0*normcdf(d1) - K*exp(-r*T)*normcdf(d2)

N = 1000;
dt = T/N;
M = 10^4;
S = ones(1,N+1);
S2 = ones(1,N+1);
V = zeros(M,1);
V2 = zeros(M,1);
V_anti = zeros(M,1);
for i = 1:M
    S(1,1) = S0;
    S2(1,1) = S0;
    dW = sqrt(dt)*randn(1,N);
    for j = 1:N
    S(1,j+1) = S(1,j)*exp((r-0.5*sigma^2)*dt + sigma*dW(1,j));
    S2(1,j+1) = S2(1,j)*exp((r-0.5*sigma^2)*dt - sigma*dW(1,j));
```

```
      end
         V(i) = exp(-r*T)*max(S(1,N+1)-K,0);
         V2(i) = exp(-r*T)*max(S2(1,N+1)-K,0);
      V_anti(i) = (V(i)+V2(i))/2;
end
a = mean(V)
b = var(V)
a_anti = mean(V_anti)
b_anti = var(V_anti)
ratio = b/b_anti
```

Simulation 28.2 (Reduction Rate of Variance)

We estimate the reduction rate of variance in the antithetic variate method for various choices of parameters. The option price is given by taking the average of $e^{-rT}(C_1 + C_2)/2$ with respect to the standard normal density. If the graph of $(C_1 + C_2)/2$ is close to being flat for z belonging to some significant range where substantial probability is concentrated, say $-2 < z < 2$, then the variance itself is small.

```
S0 = 10;
K = 8;    %Choose other values for K.
r = 0.05;
sigma = 0.3;
T = 1;
Z = randn(10^7,1);
C1 = max(S0*exp((r-0.5*sigma^2)*T +sigma*sqrt(T)*Z)-K,0);
C2 = max(S0*exp((r-0.5*sigma^2)*T -sigma*sqrt(T)*Z)-K,0);
C = (C1+C2)/2;
Variance_classical = var(C1)
Variance_anti = var(C)
reduction_rate = 1- Variance_anti / Variance_classical
```

Simulation 28.3 (Antithetic Variate Method for a Barrier Option)

We use the antithetic variate method to compute the price of a down-and-out put barrier option taking the same set of parameters in Simulation 28.1. The price of the down-and-out put barrier option according to the formula is approximately equal to P_do $= 10.6332$, while the estimated values by the standard Monte Carlo method and by the antithetic variate method are a $= 10.6506$ and a_anti $= 10.6381$, respectively. The corresponding variances are b $= 174.3541$ and b_anti $=$ 36.8697 with their ratio given by 4.7289.

```
L = 60; % a lower barrier

d1K = (log(S0/K) + (r + 0.5*sigma^2)*T)/(sigma*sqrt(T));
d2K = (log(S0/K) + (r - 0.5*sigma^2)*T)/(sigma*sqrt(T));
d1L = (log(S0/L) + (r + 0.5*sigma^2)*T)/(sigma*sqrt(T));
d2L = (log(S0/L) + (r - 0.5*sigma^2)*T)/(sigma*sqrt(T));
d3 = (log(L/S0) + (r + 0.5*sigma^2)*T)/(sigma*sqrt(T));
d4 = (log(L/S0) + (r - 0.5*sigma^2)*T)/(sigma*sqrt(T));
d5 = (log(L^2/S0/K) + (r + 0.5*sigma^2)*T)/(sigma*sqrt(T));
d6 = (log(L^2/S0/K) + (r - 0.5*sigma^2)*T)/(sigma*sqrt(T));
```

```
put_vanilla = K*exp(-r*T)*normcdf(-d2K) - S0*normcdf(-d1K);
P2 = - K*exp(-r*T)*normcdf(-d2L) + S0*normcdf(-d1L) ;
P3 = - K*exp(-r*T)*(L/S0)^(2*r/sigma^2-1)*(normcdf(d4)-normcdf(d6));
P4 = S0*(L/S0)^(2*r/sigma^2+1)*(normcdf(d3)-normcdf(d5));
P_do = put_vanilla + P2 + P3 + P4

N = 10000;
dt = T/N;
M = 10^4;
V1 = zeros(M,1);
V2 = zeros(M,1);
V_anti = zeros(M,1);
S1 = ones(1,N+1);
S2 = ones(1,N+1);
for i=1:M
    S1(1,1) = S0;
    S2(1,1) = S0;
    dW = sqrt(dt)*randn(1,N);
    for j = 1:N
    S1(1,j+1) = S1(1,j)*exp((r-0.5*sigma^2)*dt + sigma*dW(1,j));
    S2(1,j+1) = S2(1,j)*exp((r-0.5*sigma^2)*dt - sigma*dW(1,j));
    end
S1_min = min(S1(1,:));
S2_min = min(S2(1,:));
if S1_min > L
       V1(i) = exp(-r*T)*max(K - S1(1,N+1),0);
else
        V1(i) = 0;
end
if S2_min > L
        V2(i) = exp(-r*T)*max(K - S2(1,N+1),0);
else
        V2(i)=0;
end
    V_anti(i) = (V1(i)+V2(i))/2;
end
```

Simulation 28.4 (Control Variate Method)

We use the control variate method to compute the price of an Asian option with arithmetic average using the price of the corresponding Asian option with geometric average as a control variate. The formula for an Asian option with geometric average is known, and given by `V_geo_formula` in the program. The expectation of `V_control = V - V_geo + V_geo_formula` is equal to that of the price `V` of an Asian option with arithmetic average, but the variance is reduced. See Fig. 28.3 for the output.

```
S0 = 100;
K = 110;
r = 0.05;
sigma = 0.3;
T = 1;
L = 12; % number of observations
dt = T/L;
```

```
sigma_bar = sqrt( sigma^2*(L+1)*(2*L+1)/(6*L^2));
mu_bar = 1/2*sigma_bar^2 + (r-1/2*sigma^2)*(L+1) / (2*L);

d1 = (log(S0/K) + (mu_bar+1/2*sigma_bar^2)*T)/(sigma_bar*sqrt(T));
d2 = (log(S0/K) + (mu_bar-1/2*sigma_bar^2)*T)/(sigma_bar*sqrt(T));

V_geo_formula = S0*exp((mu_bar -r)*T)*normcdf(d1) -K*exp(-r*T)*normcdf(d2);

J = 20;
ave = zeros(J,1);
ave_control = zeros(J,1);
error = zeros(J,1);
error_control = zeros(J,1);
ratio = ones(J,1);

S = ones(2^J,L);
dW = sqrt(dt)*randn(2^J,L);
for i=1:2^J
    S(i,1) = S0*exp((r-1/2*sigma^2)*dt +sigma*dW(i,1)); %asset price at T_1
    for j=2:L
        S(i,j) = S(i,j-1) *exp((r-1/2*sigma^2)*dt+ sigma*dW(i,j));
    end
end

J1 = 12;
for n=J1:J
    N = 2^n;
    V_arith = exp(-r*T) * max( mean(S(1:N,:),2) - K , 0);
    ave(n) = mean(V_arith);
    var_V_arith = var(V_arith);
    error(n) = 1.96*sqrt(var_V_arith)/sqrt(N);
    V_geo = exp(-r*T) * max( exp(mean(log(S(1:N,:)),2)) - K , 0);
    V = V_arith - V_geo + V_geo_formula;
    ave_control(n) = mean(V);
    var_control = var(V);
    error_control(n) = 1.96*sqrt(var_control)/sqrt(N);
    ratio(n) = var_V_arith/var_control;
end

errorbar(J1:J, ave(J1:J), error(J1:J), 'ro--')
hold on
errorbar(J1:J, ave_control(J1:J), error_control(J1:J),'k*-','linewidth',2)
legend('standard', 'control variate');
xlabel('log_2(N)');
ylabel('Price');
```

Simulation 28.5 (Importance Sampling Method)

The following is for the experiment of Example 28.6.

```
M=10^6;
U = rand(M,1);

Ave1 = mean(U.^5)
```

```
Var1 = var(U.^5)

% Importance sampling method is applied now.
Ave2 = mean(0.5*U.^2)
Var2 = var(0.5*U.^2)
```

Compare the output!

```
Ave1 =  0.1667
Var1 =  0.0630
Ave2 =  0.1668
Var2 =  0.0222
```

Simulation 28.6 (Importance Sampling Method)

We shift the average of the asset prices using the Girsanov theorem which allows us to replace r by r_1 in generating sample paths of geometric Brownian motion. This example computes the price of an Asian option with arithmetic average when the option is in deep out-of-the-money. (Consult Example 28.7.) We monitor the asset price $L = 12$ times until the expiry date $T = 1$. The price obtained by the standard Monte Carlo method is `price` $= 0.1928$ with variance 3.9979 while the price obtained by the importance sampling method is `price1` $= 0.1956$ and the variance `variance1` $= 0.2363$. The variance is reduced by `ratio` $= 16.9185$. This method is comparable to the control variate method in the efficiency.

```
S0 = 100;
K = 150;
r = 0.05;
sigma = 0.3;
T = 1;
L = 12; % number of measurements
dt = T/L;

r1 = r + 0.5;
theta = (r - r1)/sigma;

M = 10^6;
dW = sqrt(dt)*randn(M,L);
W = sum(dW,2);

RN = exp(-0.5*theta^2*T + theta*W); % Radon-Nikodym derivative dQ/dQ1

S = zeros(M,L);
for i=1:M
    S(i,1) = S0*exp((r-1/2 *sigma^2)*dt + sigma*dW(i,1));
    for j=2:L
        S(i,j) = S(i,j-1) * exp((r-1/2*sigma^2)*dt+ sigma*dW(i,j));
    end
end

S1 = zeros(M,L);
for i=1:M
    S1(i,1) = S0*exp((r1-1/2 *sigma^2)*dt + sigma*dW(i,1));
    for j=2:L
```

```
            S1(i,j) = S1(i,j-1) * exp((r1-1/2*sigma^2)*dt+ sigma*dW(i,j));
        end
    end

    V = exp(-r*T) * max( mean(S(1:M,:),2) - K, 0);
    price = mean(V)
    variance = var(V)

    V1 = exp(-r*T) * max( mean(S1(1:M,:),2) - K, 0);
    price1 = mean(V1.*RN)
    variance1 = var(V1.*RN)

    ratio = variance / variance1
```

Exercises

28.1 Let U denote a uniformly distributed variable in $(0, 1)$. Let F be a cumulative distribution function of a random variable X. Assume that F^{-1} exists. Show that $F^{-1}(U)$ and $F^{-1}(1-U)$ are identically distributed, but negatively correlated.

28.2

(i) Prove that $\int_0^1 \mathrm{e}^{\sqrt{x}}\mathrm{d}x = 2$ by direct computation.
(ii) Estimate $\mathbb{E}[\mathrm{e}^{\sqrt{U}}]$ using the antithetic variate method to find the above integral where U is uniformly distributed in $[0, 1]$.

28.3

(i) If $f(x)$ is monotone on $[a, b]$, how can we apply the antithetic variate method for $\mathbb{E}[f(V)]$ where the random variable V is uniformly distributed in $[a, b]$. (Hint: If U is uniformly distributed in $[0, 1]$, then both $a + (b-a)U$ and $b + (a-b)U$ are uniformly distributed in $[a, b]$.)
(ii) If $0 = a_1 < b_1 = a_2 < b_2 = a_3 < \cdots < b_n = 1$, and f is monotone on each subinterval $[a_i, b_i]$, explain how to apply the antithetic variate method for $\mathbb{E}[f(U)]$ where U is uniformly distributed in $[0, 1]$. (Hint: Use the antithetic variate method on each subinterval.)

28.4 Let f be a symmetric function, i.e., $f(-x) = f(x)$, and let Z denote the standard normal variable. Is the antithetic variate method for the estimation of $\mathbb{E}[f(Z)]$ more efficient than the classical Monte Carlo method without variance reduction?

28.5

(i) Plot the payoffs of straddles as functions of z, the values of the standard normal distribution, for $T = 1$, $r = 0.05$, $\sigma = 0.3$, $S_0 = 10$ and $K = 8, 10, 12$.
(ii) Discuss the efficiency of the antithetic variate method for the pricing of a straddle.

28.6 Estimate $\mathbb{E}[e^{\sqrt{U}}]$ using the control variate method for $\int_0^1 e^{\sqrt{x}}dx = 2$ where U is uniformly distributed in $[0, 1]$. (For the exact evaluation of the integral, consult Exercise 28.2(i).)

28.7 (Continuation of Exercise 27.2) Suppose that we know how to generate random numbers with the restricted normal distribution with the pdf given by

$$f(x) = \frac{1}{1 - N(x_0)} \mathbf{1}_{[x_0,+\infty)}(x) \frac{1}{\sqrt{2\pi}} e^{-x^2/2}$$

where

$$N(x) = \int_{-\infty}^{x} \frac{1}{\sqrt{2\pi}} e^{-z^2/2} \, dz \, .$$

Explain how to use such random numbers to improve the standard Monte Carlo method for the pricing of a European call option whose payoff is zero for $S_T \leq K$.

28.8 In Example 28.6 take $\frac{d\mathbb{Q}}{d\mathbb{P}} = 5x^4$ and check the reduction of variance in this case.

28.9 Consider a deep out-of-the-money European call option with a short time to expiry date T. Many sample paths of the asset price S_t fall into the region where $S_T \leq K$ and produce zero values for the payoff at T. Thus these samples are wasted without contributing much to the evaluation of the expectation in the standard Monte Carlo method. To overcome such a problem, we increase the mean and variance of the asset price used in the simulation so that more sample values of S_T exceed K than before. Explain how to achieve the goal without relying on the Girsanov's theorem.

Chapter 29
Numerical Solution of the Black–Scholes–Merton Equation

The price of a European call option is given by the Black–Scholes–Merton partial differential equation with the payoff function $(x - K)^+$ as the final condition. However, for a more general option with an arbitrary payoff function there is no simple formula for option price, and we have to resort to numerical methods studied in this Chapter. For further information, the reader is referred to [38, 92, 98].

29.1 Difference Operators

Let $y(x)$ be a function and let y_m denote $y(mh)$ for some fixed small h. If not stated otherwise, we assume that $h > 0$ and that functions are evaluated at $x = mh$ for $m \in \mathbb{Z}$. The first and higher order ordinary and partial derivatives of y are approximated by Taylor series of finite order, which are again approximated by difference operators in more than one way. The most elementary example of difference is given by $y_{m+1} - y_m$ which approximates $hy'(mh)$. Some of the most frequently used difference operators are listed in Table 29.1 with corresponding Taylor series approximations where y and its derivatives are evaluated at $x = mh$.

Proof The Taylor series approximations given in Table 29.1 can be proved from the following observations: First, note that

$$y_{m+1} = y(mh + h) = y_m + hy' + \frac{1}{2}h^2y'' + \frac{1}{6}h^3y''' + \frac{1}{24}h^4y^{(iv)} + \cdots \tag{29.1}$$

where the derivatives are evaluated at $x = mh$. Now we compute $y_{m+1} - y_m$. Replacing h by $-h$ in (29.1), we have

$$y_{m-1} = y(mh + (-h)) = y_m - hy' + \frac{1}{2}h^2y'' - \frac{1}{6}h^3y''' + \frac{1}{24}h^4y^{(iv)} + \cdots$$

G.H. Choe, *Stochastic Analysis for Finance with Simulations*, Universitext,
DOI 10.1007/978-3-319-25589-7_29

Table 29.1 The first order difference operators and their Taylor series approximations

Operator	Symbol	Definition	Taylor Series
Forward	Δ	$y_{m+1} - y_m$	$hy' + \frac{1}{2}h^2y'' + \frac{1}{6}h^3y''' + \frac{1}{24}h^4y^{(iv)} + \cdots$
Backward	∇	$y_m - y_{m-1}$	$hy' - \frac{1}{2}h^2y'' + \frac{1}{6}h^3y''' - \frac{1}{24}h^4y^{(iv)} + \cdots$
Half Central	δ	$y_{m+\frac{1}{2}} - y_{m-\frac{1}{2}}$	$hy' + \frac{1}{24}h^3y''' + \cdots$
Full Central	Δ_0	$\frac{1}{2}(y_{m+1} - y_{m-1})$	$hy' + \frac{1}{6}h^3y''' + \cdots$

Table 29.2 The second order central difference operator and its Taylor approximation

Operator	Symbol	Definition	Taylor Series
Second order central	δ^2	$y_{m+1} - 2y_m + y_{m-1}$	$h^2y'' + \frac{1}{12}h^4y^{(iv)} + \cdots$

and we obtain $y_m - y_{m-1}$. Similarly, by choosing increment size $\frac{1}{2}h$, we have

$$y_{m+\frac{1}{2}} = y(mh + \frac{1}{2}h) = y_m + \frac{1}{2}hy' + \frac{1}{8}h^2y'' + \frac{1}{48}h^3y''' + \cdots$$

$$y_{m-\frac{1}{2}} = y(mh - \frac{1}{2}h) = y_m - \frac{1}{2}hy' + \frac{1}{8}h^2y'' - \frac{1}{48}h^3y''' + \cdots$$

Hence

$$\delta y = y_{m+\frac{1}{2}} - y_{m-\frac{1}{2}} = hy' + \frac{1}{24}h^3y''' + \cdots .$$

□

Proof For the 2nd order central difference in Table 29.2, we note that

$$\begin{aligned}\delta^2 y &= (y_{m+1} - y_m) - (y_m - y_{m-1}) \\ &= (hy' + \frac{1}{2}h^2y'' + \frac{1}{6}h^3y''' + \frac{1}{24}h^4y^{(iv)} + \cdots) \\ &\quad -(hy' - \frac{1}{2}h^2y'' + \frac{1}{6}h^3y''' - \frac{1}{24}h^4y^{(iv)} + \cdots) \\ &= h^2y'' + \frac{1}{12}h^4y^{(iv)} + \cdots .\end{aligned}$$

□

29.2 Grid and Finite Difference Methods

Consider a heat equation (or a diffusion equation)

$$\frac{\partial u}{\partial t} = \frac{\partial^2 u}{\partial x^2}\,, \qquad 0 \le x \le L\,, \quad t \ge 0$$

with an initial condition $u(x, 0) = g(x)$ and boundary conditions

$$u(0, t) = a(t)\,, \qquad u(L, t) = b(t)\,.$$

Example 29.1 Consider the heat heat equation given above. Take $L = \pi$, $g(x) = \sin x$, $a(t) = b(t) = 0$. Then $u(x, t) = \mathrm{e}^{-t} \sin x$.

Let N_x and N_t denote the numbers of subintervals in the partitions of the intervals $[0, L]$ and $[0, T]$, respectively. Put

$$h = \frac{L}{N_x}\,, \qquad k = \frac{T}{N_t}$$

and consider a grid (or a mesh) given by

$$\{(jh, ik) : 0 \le j \le N_x\,,\ 0 \le i \le N_t\}$$

and use the symbol U_j^i to denote the value of the finite difference solution at the grid point (jh, ik). Note that U_j^i approximates $u(jh, ik)$. See Fig. 29.1.

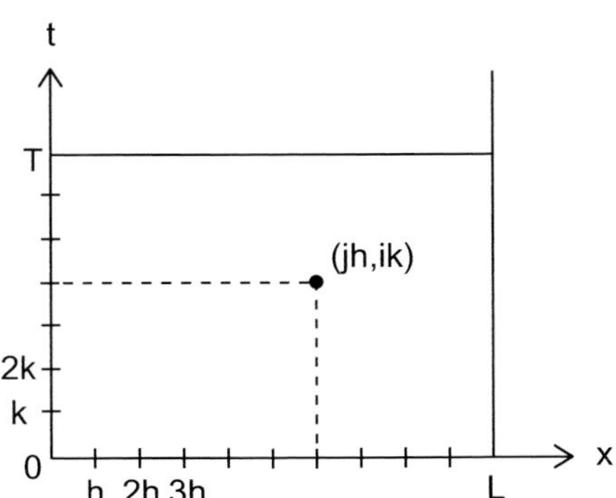

Fig. 29.1 A grid for a finite difference method

29.2.1 Explicit Method

We may approximate the given heat equation using the discretizations

$$\frac{\partial}{\partial t} \longrightarrow \frac{1}{k}\Delta_t \, ,$$
$$\frac{\partial^2}{\partial x^2} \longrightarrow \frac{1}{h^2}\delta_x^2 \, .$$

Then we obtain a finite difference equation

$$\frac{1}{k}\Delta_t U_j^i - \frac{1}{h^2}\delta_x^2 U_j^i = 0 \, ,$$

in other words,

$$\frac{1}{k}(U_j^{i+1} - U_j^i) - \frac{1}{h^2}(U_{j+1}^i - 2U_j^i + U_{j-1}^i) = 0 \, .$$

Hence we have a convex linear combination of U_{j+1}^i, U_j^i, U_{j-1}^i given by

$$U_j^{i+1} = \nu U_{j+1}^i + (1 - 2\nu)U_j^i + \nu U_{j-1}^i \tag{29.2}$$

where

$$\nu = \frac{k}{h^2}$$

is the *mesh ratio*. The $(j + 1)$st value is computed explicitly by the jth values on the right-hand side. Thus it is called an *explicit method*. The algorithm takes the average of current states to obtain the next stage in time, and is called FTCS, an acronym for forward difference in time and central difference in space (Fig. 29.2).

To check the stability of the FTCS algorithm, consider the heat equation defined on $0 \le x \le 2$, $0 \le t \le 1$ where the initial condition is given by the point mass at $x = 1$. (See the left and middle plots in Fig. 29.3 and the first part of Simulation 29.1.

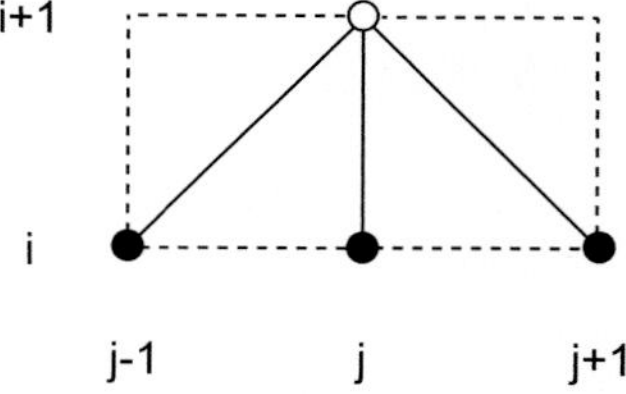

Fig. 29.2 FTCS: Taking average on the grid

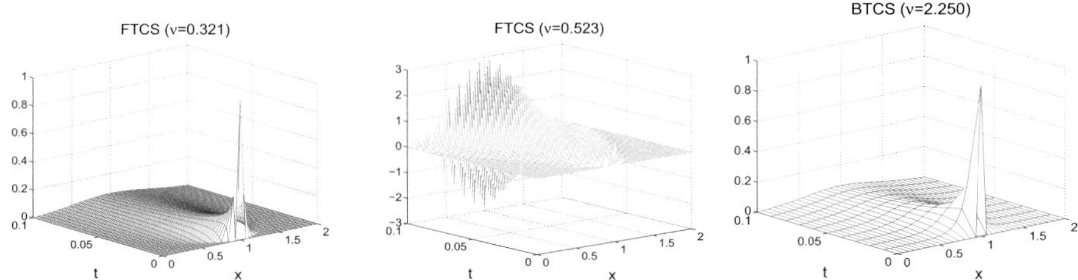

Fig. 29.3 Comparison of FTCS and BTCS for the heat equation with initial condition given by the Dirac delta functional: FTCS with $\nu < \frac{1}{2}$, by $\nu > \frac{1}{2}$, and BTCS with $\nu > \frac{1}{2}$ (from left to right)

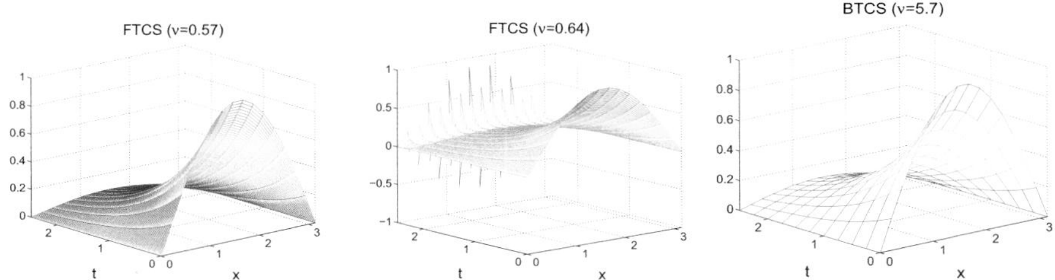

Fig. 29.4 Comparison of FTCS and BTCS for the heat equation with $u(x, 0) = \sin x$: FTCS with $\nu \approx \frac{1}{2}$ and $\nu > \frac{1}{2}$, and BTCS with $\nu > \frac{1}{2}$ (from left to right)

Table 29.3 Numerical values obtained by FTCS for $\nu = 4$

1.0	0	−1792	5728	−10864	13345	−10864	5728	−1792	0
0.75	0	64	−336	780	−1015	780	−336	64	0
0.5	0	0	16	−56	81	−56	16	0	0
0.25	0	0	0	4	−7	4	0	0	0
0.0	0	0	0	0	1	0	0	0	0
t \ x	0.0	0.25	0.5	0.75	1.0	1.25	1.5	1.75	2.0

See also Fig. 29.4.) Take $N_x = 8$ and $N_t = 4$, then $h = k = \frac{1}{4}$, $\nu = 4$ and (29.2) becomes

$$U_j^{i+1} = 4U_{j+1}^i - 7U_j^i + 4U_{j-1}^i \tag{29.3}$$

which produce numerical values given in Table 29.3. Note that the magnitude of numerical values are literally exploding contrary to physical intuition and that the signs of nonzero numerical values are alternating as shown in Table 29.4.

To find out why the algorithm is unstable, suppose that the initial condition is given by a point mass at $x = 1$ and $\nu > \frac{1}{2}$ in (29.2). If the signs of U_j^i are given by

$$\dots, \; U_{j-1}^i \geq 0 \;, \; U_j^i \leq 0 \;, \; U_{j+1}^i \geq 0 \;, \; U_{j+2}^i \leq 0 \;, \; \dots$$

Table 29.4 Signs of numerical values obtained by FTCS for $\nu = 4$

1.0	0	−	+	−	+	−	+	−	0
0.75	0	+	−	+	−	+	−	+	0
0.5	0	0	+	−	+	−	+	0	0
0.25	0	0	0	+	−	+	0	0	0
0.0	0	0	0	0	+	0	0	0	0
t \ x	0.0	0.25	0.5	0.75	1.0	1.25	1.5	1.75	2.0

then

$$U_j^{i+1} = \nu \times \text{positive} + (1-2\nu) \times \text{negative} + \nu \times \text{positive} \geq 0 \,,$$
$$U_{j+1}^{i+1} = \nu \times \text{negative} + (1-2\nu) \times \text{positive} + \nu \times \text{negative} \leq 0 \,,$$

and so on. Hence we have an alternating pattern again at $(j+1)$st time step. Let

$$\mathbf{U}^i = \left(U_0^i, \ldots, U_{N_x}^i\right)$$

and

$$||(a_0, \ldots, a_n)||_1 = \sum_{j=0}^{n} |a_j| \,.$$

Then

$$\begin{aligned} ||\mathbf{U}^{i+1}||_1 &= \sum_{j=0}^{N_x} |\nu U_{j+1}^i + (1-2\nu)U_j^i + \nu U_{j-1}^i| \\ &= \sum_j |\nu U_{j+1}^i| + |(1-2\nu)U_j^i| + |\nu U_{j-1}^i| \\ &= \nu \sum_j |U_j^i| + (2\nu - 1) \sum_j |U_j^i| + \nu \sum_j |U_j^i| \\ &= (4\nu - 1) \sum_j |U_j^i| \\ &= (4\nu - 1)^{i+1} \, ||\mathbf{U}^0||_1 \,, \end{aligned}$$

which shows that the norm of $\mathbf{U}^{i+1}$ increases exponentially as i increases since $4\nu - 1 > 1$. Therefore it is necessary to have the condition $\nu \leq \frac{1}{2}$ to have the stability of the numerical scheme given by FTCS.

To view the iteration scheme from the viewpoint of matrix computation, define an $(N_x - 1) \times (N_x - 1)$ tridiagonal matrix A by

$$A = \begin{bmatrix} 1-2\nu & \nu & 0 & \cdots & \cdots & 0 \\ \nu & 1-2\nu & \nu & 0 & & \vdots \\ 0 & \ddots & \ddots & \ddots & \ddots & \vdots \\ \vdots & \ddots & \ddots & \ddots & \ddots & 0 \\ \vdots & & 0 & \nu & 1-2\nu & \nu \\ 0 & \cdots & \cdots & 0 & \nu & 1-2\nu \end{bmatrix}$$

and $(N_x - 1)$-dimensional vectors $\mathbf{U}^0$, $\mathbf{U}^i$, $1 \leq i \leq N_t$, and $\mathbf{p}^i$ by

$$\mathbf{U}^0 = \begin{bmatrix} g(h) \\ g(2h) \\ \vdots \\ \vdots \\ g((N_x-1)h) \end{bmatrix}, \quad \mathbf{U}^i = \begin{bmatrix} U_1^i \\ U_2^i \\ \vdots \\ \vdots \\ U_{N_x-1}^i \end{bmatrix}, \quad \text{and} \quad \mathbf{r}^i = \begin{bmatrix} \nu\, a(ik) \\ 0 \\ \vdots \\ \vdots \\ 0 \\ \nu\, b(ik) \end{bmatrix}.$$

Then

$$\mathbf{U}^{i+1} = A\mathbf{U}^i + \mathbf{r}^i\,, \quad 0 \leq i \leq N_t - 1\,.$$

Remark 29.1 (Discretization Error) The error of a numerical scheme arising from discretization at the grid point (jh, ik) is called *local accuracy*. The local accuracy ε_j^i for FTCS can be estimated as follows: Let u_j^i denote the exact solution $u(jh, ik)$. Then, under the assumption that the derivatives are bounded if needed, we have

$$\begin{aligned} \varepsilon_j^i &= \frac{1}{k}\Delta_t u_j^i - \frac{1}{h^2}\delta_x^2 u_j^i \\ &= \left\{ \frac{\partial u}{\partial t} + \frac{1}{2}k\frac{\partial^2 u}{\partial t^2} + O(k^2) \right\} - \left\{ \frac{\partial^2 u}{\partial x^2} + \frac{1}{12}h^2\frac{\partial^4 u}{\partial x^4} + O(h^4) \right\} \\ &= \frac{1}{2}k\frac{\partial^2 u}{\partial t^2} - \frac{1}{12}h^2\frac{\partial^4 u}{\partial x^4} + O(k^2) + O(h^4) \\ &= O(k) + O(h^2) \end{aligned}$$

where we used the Taylor series expansions given in Tables 29.1 and 29.2.

29.2.2 Implicit Method

For the discretization of the heat equation

$$\frac{\partial u}{\partial t} - \frac{\partial^2 u}{\partial x^2} = 0$$

we use the backward difference in time and the 2nd order central difference in space as follows:

$$\frac{1}{k}\nabla_t U_j^i - \frac{1}{h^2}\delta_x^2 U_j^i = 0 \, .$$

Then we have

$$\frac{U_j^i - U_j^{i-1}}{k} - \frac{U_{j+1}^i - 2U_j^i + U_{j-1}^i}{h^2} = 0 \, .$$

Now, for the sake of notational convenience, we replace i by $i + 1$, and obtain

$$\frac{U_j^{i+1} - U_j^i}{k} - \frac{U_{j+1}^{i+1} - 2U_j^{i+1} + U_{j-1}^{i+1}}{h^2} = 0 \, .$$

Hence

$$U_j^{i+1} = U_j^i + \frac{k}{h^2}(U_{j+1}^{i+1} - 2U_j^{i+1} + U_{j-1}^{i+1}) \, ,$$

which is equivalent to

$$(1 + 2\nu)\, U_j^{i+1} = U_j^i + \nu U_{j+1}^{i+1} + \nu U_{j-1}^{i+1}$$

where

$$\nu = \frac{k}{h^2} \, .$$

The algorithm is called BTCS, an acronym for backward difference in time and central difference in space. Note that U_j^{i+1} is a weighted average of U_j^i, U_{j+1}^{i+1}, U_{j-1}^{i+1}, and the algorithm is called an *implicit method* (Fig. 29.5).

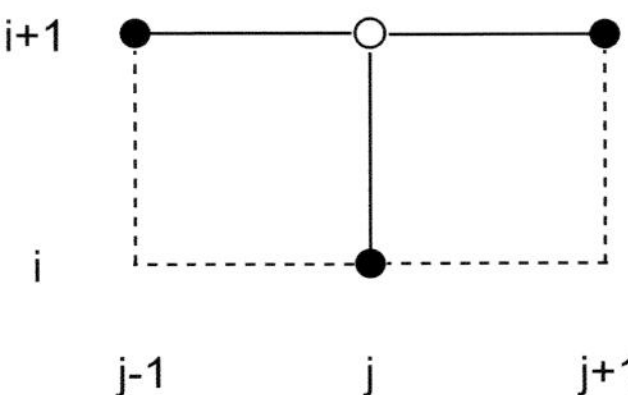

Fig. 29.5 BTCS: Taking average on the grid

Define a $(N_x - 1) \times (N_x - 1)$ tridiagonal matrix B by

$$B = \begin{bmatrix} 1+2\nu & -\nu & 0 & \cdots & \cdots & 0 \\ -\nu & 1+2\nu & -\nu & 0 & & \vdots \\ 0 & \ddots & \ddots & \ddots & \ddots & \vdots \\ \vdots & \ddots & \ddots & \ddots & \ddots & 0 \\ \vdots & & 0 & -\nu & 1+2\nu & -\nu \\ 0 & \cdots & \cdots & 0 & -\nu & 1+2\nu \end{bmatrix}$$

and $(N_x - 1)$-dimensional vectors $\mathbf{s}^i$ by

$$\mathbf{s}^i = \begin{bmatrix} \nu\, a((i+1)k) \\ 0 \\ \vdots \\ \vdots \\ 0 \\ \nu\, b((i+1)k) \end{bmatrix} .$$

Then

$$B\mathbf{U}^{i+1} = \mathbf{U}^i + \mathbf{s}^i ,$$

or equivalently,

$$\mathbf{U}^{i+1} = B^{-1}(\mathbf{U}^i + \mathbf{s}^i) .$$

Remark 29.2 The local accuracy for BTCS is given by

$$\begin{aligned} \varepsilon_j^i &= \frac{1}{k}\nabla_t u_j^i - \frac{1}{h^2}\delta_x^2 u_j^i \\ &= \left\{ \frac{\partial u}{\partial t} - \frac{1}{2}k\frac{\partial^2 u}{\partial t^2} + O(k^2) \right\} - \left\{ \frac{\partial^2 u}{\partial x^2} + \frac{1}{12}h^2\frac{\partial^4 u}{\partial x^4} + O(h^4) \right\} \end{aligned}$$

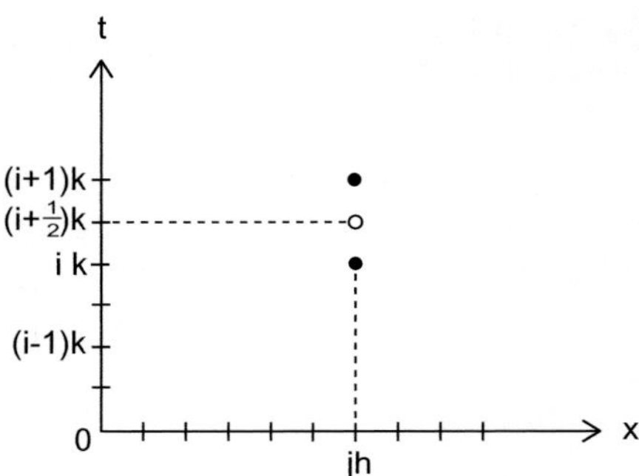

Fig. 29.6 Time average on the grid

$$
\begin{aligned}
&= -\frac{1}{2}k\frac{\partial^2 u}{\partial t^2} - \frac{1}{12}h^2\frac{\partial^4 u}{\partial x^4} + O(k^2) + O(h^4) \\
&= O(k) + O(h^2) \ .
\end{aligned}
$$

29.2.3 Crank–Nicolson Method

We consider the intermediate time level at $(i+\frac{1}{2})k$ and the corresponding discretization of the heat equation given by

$$\frac{1}{k}\delta_t U_j^{i+\frac{1}{2}} - \frac{1}{h^2}\delta_x^2 U_j^{i+\frac{1}{2}} = 0 \ .$$

(See Fig. 29.6). Taking the time average of the right hand side, we obtain a new relation

$$\frac{1}{k}\delta_t U_j^{i+\frac{1}{2}} - \frac{1}{h^2}\delta_x^2 \mu_t U_j^{i+\frac{1}{2}} = 0 \ ,$$

where μ_t denotes the averaging operation by half distance in time, in other words,

$$\frac{1}{k}\delta_t U_j^{i+\frac{1}{2}} - \frac{1}{h^2}\delta_x^2 \frac{U_j^{i+1} + U_j^i}{2} = 0$$

and

$$\frac{1}{k}\delta_t U_j^{i+\frac{1}{2}} - \frac{1}{h^2}\frac{(U_{j+1}^{i+1} - 2U_j^{i+1} + U_{j-1}^{i+1}) + (U_{j+1}^i - 2U_j^i + U_{j-1}^i)}{2} = 0 \ .$$

Finally, we obtain the Crank–Nicolson scheme given by

$$U_j^{i+1} - U_j^i - \frac{1}{2}\nu(U_{j+1}^{i+1} - 2U_j^{i+1} + U_{j-1}^{i+1} + U_{j+1}^i - 2U_j^i + U_{j-1}^i) = 0 \ .$$

29.3 Numerical Methods for the Black–Scholes–Merton Equation

Example 29.2 (Explicit Method) We use the FTCS algorithm to compute the numerical solution of the Black–Scholes–Merton equation for the price of a European put option with strike price $K = 5$ and expiry date $T = 1$. We take $\sigma = 0.3$ and $r = 0.05$ in our computation. It is easier to plot the price surface of a put option than a call option since for large values of S the price of a European put option is close to zero, and hence we can choose an upper bound $S_{\max}$ for S using a relatively small value. In this example, we choose $S_{\max} = 10$. See Fig. 29.7 and Simulation 29.2.

Example 29.3 (Implicit Method) We use the BTCS algorithm to compute the numerical solution of the Black–Scholes–Merton equation for the price of a binary put option. See Fig. 29.8 and Simulation 29.3.

Example 29.4 (Crank–Nicolson Method) We use the Crank–Nicolson algorithm to find the numerical solution of the Black–Scholes–Merton equation for the price of a down-and-out put option. See Fig. 29.9 and Simulation 29.4.

29.4 Stability

Theorem 29.1 (Lax Equivalence Theorem) *A numerical scheme for a finite difference method converges to a true solution if and only if its local accuracy tends to zero as $k, h \to 0$, and it satisfies a certain stability condition.*

One such stability condition is the von Neumann stability condition.

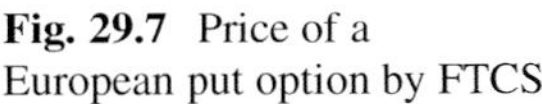
Fig. 29.7 Price of a European put option by FTCS

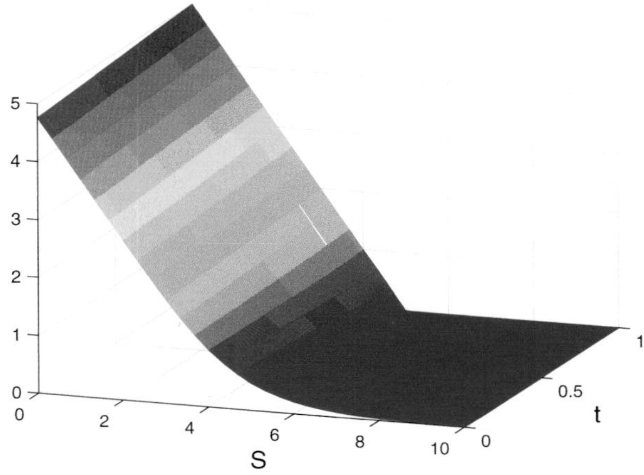

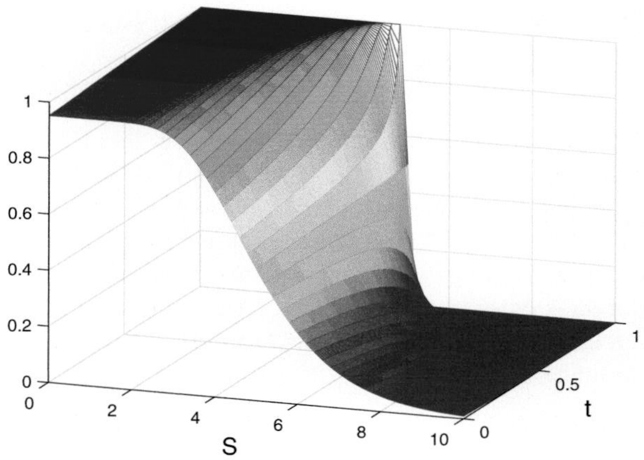

Fig. 29.8 Price of a binary put option by BTCS

Definition 29.1 A finite difference method is stable in the sense of von Neumann if, disregarding initial conditions and boundary conditions, under the substitution $U_i^j = \xi^i e^{\sqrt{-1}\beta jh}$ we have $|\xi| \leq 1$ for every $\beta h \in [-\pi, \pi]$.

Example 29.5 In FTCS for the diffusion equation we have

$$U_j^{i+1} = \nu U_{j+1}^i + (1-2\nu)U_j^i + \nu U_{j-1}^i ,$$

and hence

$$\xi^{i+1} e^{\sqrt{-1}\beta jh} = \nu \xi^i e^{\sqrt{-1}\beta jh} e^{\sqrt{-1}\beta h} + (1-2\nu)\xi^i e^{\sqrt{-1}\beta jh} + \nu \xi^i e^{\sqrt{-1}\beta jh} e^{-\sqrt{-1}\beta h}.$$

Thus

$$\begin{aligned}
\xi &= \nu e^{\sqrt{-1}\beta h} + (1-2\nu) + \nu e^{-\sqrt{-1}\beta h} \\
&= 1 + \nu(e^{\sqrt{-1}\beta h/2} - e^{-\sqrt{-1}\beta h/2})^2 \\
&= 1 + \nu(2\sqrt{-1}\sin(\frac{1}{2}\beta h)^2 \\
&= 1 - 4\nu \sin^2(\frac{1}{2}\beta h) .
\end{aligned}$$

The condition that $|\xi| \leq 1$ is equivalent to

$$-1 \leq 1 - 4\nu \sin^2(\frac{1}{2}\beta h) \leq 1 ,$$

which is again equivalent to

$$0 \leq \nu \sin^2(\frac{1}{2}\beta h) \leq \frac{1}{2} .$$

The condition is satisfied for any β and h if $0 \le \nu \le \frac{1}{2}$. It can be shown that BTCS is unconditionally stable, i.e., the von Neumann stability condition is satisfied for every $\nu > 0$.

Remark 29.3 The binomial tree method for option pricing can be shown to be a special case of the BTCS algorithm under the condition $h^2 = \sigma^2 k$ for the Black–Scholes–Merton equation after taking the change of variables. (See Exercise 29.3.)

29.5 Computer Experiments

Simulation 29.1 (FTCS and BTCS for the Heat Equation)

We compare numerical solutions of the heat equation defined on $0 \le x \le 2$, $0 \le t \le 0.1$ obtained by FTCS and BTCS for various choices of N_x and N_t where the initial condition is given by the Dirac functional at $x = 1$. (See Fig. 29.3.) For the second experiment with $u(x, 0) = \sin x$, replace `U(Nx/2+1,1)=1` by `U(2:Nx,1)=sin(dx:dx:(Nx-1)*dx)`. (See Fig. 29.4.)

```
T = 0.1;
Nx = 30; % Choose an even integer.
L = 2;
x = linspace(0,L,Nx+1);
% L = pi;
dx = L / Nx;

%%% FTCS (nu < 1/2)
Nt = 70;
t = linspace(0,T,Nt+1);
dt = T / Nt;
nu = dt / dx^2
A = (1-2*nu)*eye(Nx-1,Nx-1) + nu*diag(ones(Nx-2,1),1)
                          ... + nu*diag(ones(Nx-2,1),-1);
U = zeros(Nx+1,Nt+1);
U(Nx/2+1,1)=1;
for i=1:Nt
    U(2:Nx,i+1) = A * U(2:Nx,i);
end
figure(1);
mesh(0:dx:L,0:dt:T,U');

%%% FTCS (nu > 1/2)
Nt = 43;
dt = T / Nt;
nu = dt / dx^2
A = (1-2*nu)*eye(Nx-1,Nx-1) + nu*diag(ones(Nx-2,1),1)
                          ... + nu*diag(ones(Nx-2,1),-1);
U = zeros(Nx+1,Nt+1);
U(Nx/2+1,1)=1;
for i=1:Nt
    U(2:Nx,i+1) = A * U(2:Nx,i);
```

```
end
figure(2);
mesh(0:dx:L,0:dt:T,U');

%%% BTCS (nu > 1/2)
Nt = 10;
dt = T / Nt;
nu = dt / dx^2
B = (1+2*nu)*eye(Nx-1,Nx-1) - nu*diag(ones(Nx-2,1),1)
                        ... - nu*diag(ones(Nx-2,1),-1);
U = zeros(Nx+1,Nt+1);
U(Nx/2+1,1)=1;
for i=1:Nt
    U(2:Nx,i+1) = B\U(2:Nx,i);
end
figure(3);
mesh(0:dx:L,0:dt:T,U');
```

Simulation 29.2 (FTCS for a Put)

We compute the price of a European put option by the FTCS method and plot the graph. See Example 29.2 and Fig. 29.7.

```
K = 5;  % strike price
sigma = 0.3;
r = 0.05;
T = 1;
S_max = 10;
M = 20;  % number of partitions of asset price interval
N = 30;  % number of partitions of time interval
dt = T/N;
dS = S_max/M;

V=zeros(M+1,N+1);
V(:,N+1) = max(K-(0:dS:S_max)',0);
V(1,:) = K*exp(-r*dt*(N-[0:N]));
V(M+1,:) = 0;

a = 0.5*dt*(sigma^2*(0:M).^2 - r.*(0:M));
b = 1 - dt*(sigma^2*(0:M).^2 + r);
c = 0.5*dt*(sigma^2*(0:M).^2 + r.*(0:M));

for i=N:-1:1  % backward computation
    for j=2:M
        V(j,i) = a(j)*V(j-1,i+1) + b(j)*V(j,i+1)+c(j)*V(j+1,i+1);
    end
end

mesh((0:dS:S_max),(0:dt:T),V')
```

Simulation 29.3 (BTCS for a Binary Put)

We compute the price of a binary put option by the BTCS method and plot the graph. See Example 29.3 and Fig. 29.8.

```
K = 5;  % strike price
sigma = 0.3;
r = 0.05;
T = 1;
S_max = 20;
M = 25;  % number of partitions of asset price interval
N = 30;  % number of partitions of time interval
dt = T/N;
dS = S_max/M;

V = zeros(M+1,N+1);
V(:,N+1) = heaviside(K-(0:dS:S_max)');
V(1,:) = exp(-r*dt*(N-[0:N]));

a =  0.5*dt*(-sigma^2*(0:M).^2 + r*(0:M));
b =  1 + dt*(sigma^2*(0:M).^2 + r);
c = -0.5*dt*(sigma^2.*(0:M).^2 + r*(0:M));

TriDiag = diag(a(3:M),-1) + diag(b(2:M)) + diag(c(2:M-1),1);
B = zeros(M-1,1);
for i = N:-1:1  % backward computation
    B(1) = -a(2)*V(1,i);
    V(2:M,i) = TriDiag \ (V(2:M,i+1) + B);
end

mesh((0:dS:S_max),(0:dt:T),V')
```

Simulation 29.4 (Crank–Nicolson for a Down-and-Out Put)

We compute the price of a down-and-out put option by the Crank–Nicolson method and plot the graph. See Example 29.4 and Fig. 29.9.

```
K = 4;  % strike price
sigma = 0.3;
r = 0.05;
T = 1;
```

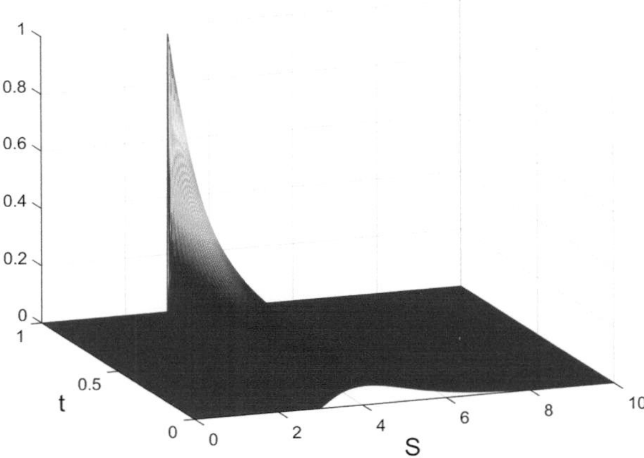

Fig. 29.9 Price of a down-and-out put option by the Crank–Nicolson method

```
S_max = 10;
L = 3;   % the lower barrier
M = 50;   % number of partitions of asset price interval
N = 30;   % number of partitions of time interval
dt = T/N;
dS = (S_max - L)/M; % We need dS < K-L.

V = zeros(M+1,N+1);
V(:,N+1) = max(K-(L:dS:S_max)',0);
V(1,:) = 0;   % boundary condition at S=0
V(M+1,:) = 0;   % boundary condition at S=large

M1 = L/dS;
M2 = S_max/dS;

a =  0.25*dt*(sigma^2*(M1:1:M2).^2 - r*(L/dS:1:S_max/dS));
b =  -0.5*dt*(sigma^2*(M1:1:M2).^2 + r);
c =  0.25*dt*(sigma^2*(M1:1:M2).^2 + r*(L/dS:1:S_max/dS));

A = -diag(a(3:M),-1) + diag(1-b(2:M)) - diag(c(2:M-1),1);
B =  diag(a(3:M),-1) + diag(1+b(2:M)) + diag(c(2:M-1),1);

for i=N:-1:1  % backward computation
    V(2:M,i) = A \ B*V(2:M,i+1);
end

J = floor(L/dS);
U = zeros(J, N+1);
W = [U;V];

mesh((L-J*dS:dS:S_max), (0:dt:T),W')
```

Exercises

29.1 Check the formulas given in Tables 29.1 and 29.2 for functions $f(x) = x^3$ and $f(x) = x^4$.

29.2 Show that the Crank–Nicolson method has local accuracy $O(k^2) + O(h^2)$ and that it is unconditionally stable.

29.3 Take $x = \log S$ and $v = e^{-rt}V$ in the Black–Scholes–Merton equation in Theorem 15.1, and obtain

$$\frac{\partial v}{\partial t} + \frac{1}{2}\sigma^2 \frac{\partial^2 v}{\partial x^2} + (r - \frac{1}{2}\sigma^2)\frac{\partial v}{\partial x} = 0 .$$

Choose h and k in BTCS under the condition that $h^2 = \sigma^2 k$, and check that the resulting numerical algorithm is the binomial tree method in Sect. 14.3.

Chapter 30
Numerical Solution of Stochastic Differential Equations

Stochastic differential equations (SDEs) including the geometric Brownian motion are widely used in natural sciences and engineering. In finance they are used to model movements of risky asset prices and interest rates. The solutions of SDEs are of a different character compared with the solutions of classical ordinary and partial differential equations in the sense that the solutions of SDEs are stochastic processes. Thus it is a nontrivial matter to measure the efficiency of a given algorithm for finding numerical solutions. In this chapter we introduce two methods for numerically solving stochastic differential equations. For more details consult [50, 92].

30.1 Discretization of Stochastic Differential Equations

Given an SDE

$$\mathrm{d}X_t = a(t, X_t)\,\mathrm{d}t + b(t, X_t)\,\mathrm{d}W_t\ , \quad X_0 = x_0$$

defined on a time interval $[t_0, T]$, we consider its corresponding time discretization

$$Y_{n+1} = Y_n + a(t_n, Y_n)\,\Delta_n + b(t_n, Y_n)\,\Delta W_n\ , \quad n = 0, 1, \cdots, N-1$$

where $t_0 < t_1 < \cdots < t_N = T$, $\Delta_n = t_{n+1} - t_n$, $\Delta W_n = W_{t_{n+1}} - W_{t_n}$, and study iterative algorithms to find numerical solutions.

To plot a sample path for Y_t on an interval $t \in [t_0, T]$ we plot a piecewise linear function defined by

$$Y_t = Y_{t_n} + \frac{t - t_n}{t_{n+1} - t_n}(Y_{t_{n+1}} - Y_{t_n})\ , \quad t_n \le t \le t_{n+1}\ ,$$

which reflects the nondifferentiability of a sample path of X_t.

G.H. Choe, *Stochastic Analysis for Finance with Simulations*, Universitext,
DOI 10.1007/978-3-319-25589-7_30

Definition 30.1 (Strong Convergence) Let $\delta = \max_n |t_{n+1} - t_n|$. Suppose that an SDE for X_t has a discrete solution Y_n such that there exists $\delta_0 > 0$, $\gamma > 0$ and $K > 0$ for which

$$\mathbb{E}[|X_T - Y_N|] \leq K\delta^\gamma \quad \text{for every } 0 < \delta < \delta_0 \, .$$

Then we say that Y converges to X in the *strong* sense and call γ the order of strong convergence.

Definition 30.2 (Weak Convergence) Let $\delta = \max_n |t_{n+1} - t_n|$. Suppose that an SDE for X_t has a discrete solution Y_n such that there exists $\delta_0 > 0$, $\beta > 0$ and $K > 0$ for which

$$|\mathbb{E}[g(X_T)] - \mathbb{E}[g(Y_N)]| \leq K\delta^\beta \quad \text{for every } 0 < \delta < \delta_0$$

for an arbitrary nice function g such as a polynomial or a piecewise linear function. Then we say that Y converges to X in the *weak sense* and call β the order of weak convergence.

Remark 30.1 (i) If we take $g(x) = x$ and $g(x) = (x - \mathbb{E}[X_T])^2$ in the previous definition of weak convergence, we can obtain the average and variance of X_T, respectively.

(ii) Let $\phi(x)$ be a convex function. Then, by Jensen's inequality we have $\mathbb{E}[\phi(X)] \geq \phi(\mathbb{E}[X])$. If we take $\phi(x) = |x|$, then we obtain

$$\mathbb{E}[|X_T - Y_N|] \geq |\mathbb{E}[X_T - Y_N]| = |\mathbb{E}[X_T] - \mathbb{E}[Y_N]| \, .$$

(iii) In many applications we need not find the values of X_t for all $0 \leq t \leq T$. For example, to compute the price of a European option where C_T denotes the payoff function at maturity T it suffices to know $C_T(X_T)$. That is, it is enough to consider the weak convergence speed.

(iv) Since the root mean square of ΔW_n is not δ but $\delta^{1/2}$, the discrete approximate solution of an SDE has a smaller order of convergence than the discrete approximate solution of an ordinary differential equation, in general.

(v) Consider a computer simulation for strong convergence where $t_0 = 0$. We take the time step $\delta = \frac{T}{N}$, and obtain a discrete solution Y and its values at T, denoted by Y_T^j, $1 \leq j \leq J$, and compute

$$\varepsilon(\delta) = \frac{1}{J} \sum_{j=1}^{J} |X_T^j - Y_N^j|$$

and finally plot the graph of $-\log \varepsilon(\delta)$ against $-\log \delta$ for the values $\delta = 2^{-3}, 2^{-4}, 2^{-5}$, and so on.

From now on we treat only the case when the functions a and b do not depend on t.

30.2 Stochastic Taylor Series

In this section we present the Taylor series expansion of a stochastic process given by an SDE $\mathrm{d}X_t = a(t, X_t)\mathrm{d}t + b(t, X_t)\mathrm{d}W_t$. If $a(t, x) = a(t)$ and $b(t, x) = 0$, then we have the usual Taylor series expansion. In the following discussion, for the sake of notational simplicity, we consider the case when $a(t, x)$ and $b(t, x)$ are functions of x only.

30.2.1 Taylor Series for an Ordinary Differential Equation

Given a sufficiently smooth function $a(x) : \mathbb{R} \to \mathbb{R}$, we consider a one-dimensional autonomous ordinary differential equation $\frac{\mathrm{d}}{\mathrm{d}t}X_t = a(X_t)$ on the time interval $[t_0, T]$ which has a solution X_t with an initial data X_{t_0}. We may rewrite the equation as

$$X_t = X_{t_0} + \int_{t_0}^{t} a(X_s)\,\mathrm{d}s\ . \qquad (30.1)$$

Given a C^1 function $f : \mathbb{R} \to \mathbb{R}$, we have

$$\frac{\mathrm{d}}{\mathrm{d}t}(f(X_t)) = f'(X_t)\frac{\mathrm{d}}{\mathrm{d}t}X_t = a(X_t)\frac{\partial}{\partial x}f(X_t)$$

by the chain rule. If we let L be a differential operator defined by

$$L = a(x)\frac{\partial}{\partial x}\ ,$$

then

$$\frac{\mathrm{d}}{\mathrm{d}t}(f(X_t)) = Lf(X_t)\ .$$

Equivalently,

$$f(X_t) = f(X_{t_0}) + \int_{t_0}^{t} Lf(X_s)\,\mathrm{d}s\ . \qquad (30.2)$$

If we substitute $f(x) = a(x)$ in (30.2), then we obtain

$$a(X_t) = a(X_{t_0}) + \int_{t_0}^{t} L\,a(X_s)\,\mathrm{d}s\,. \tag{30.3}$$

Substituting (30.3) back into (30.1), we have

$$\begin{aligned} X_t &= X_{t_0} + \int_{t_0}^{t} \left(a(X_{t_0}) + \int_{t_0}^{s} La(X_z)\mathrm{d}z \right) \mathrm{d}s \\ &= X_{t_0} + a(X_{t_0})\,(t - t_0) + \int_{t_0}^{t} \int_{t_0}^{s} La(X_z)\,\mathrm{d}z\,\mathrm{d}s\,. \end{aligned} \tag{30.4}$$

Similarly, if we substitute $f = La$ in (30.2) then we obtain

$$La(X_t) = La(X_{t_0}) + \int_{t_0}^{t} L^2 a(X_u)\,\mathrm{d}u\,, \tag{30.5}$$

and, by substituting (30.5) back into (30.4), we obtain

$$\begin{aligned} X_t &= X_{t_0} + a(X_{t_0}) \int_{t_0}^{t} \mathrm{d}s + \int_{t_0}^{t} \int_{t_0}^{s} \left(La(X_{t_0}) + \int_{t_0}^{z} L^2 a(X_u)\mathrm{d}u \right) \mathrm{d}z\,\mathrm{d}s \\ &= X_{t_0} + a(X_{t_0})\,(t - t_0) + \frac{1}{2} La(X_{t_0})\,(t - t_0)^2 + R(t_0; t)\,, \end{aligned} \tag{30.6}$$

where

$$R(t_0; t) = \int_{t_0}^{t} \int_{t_0}^{s} \int_{t_0}^{z} L^2 a(X_u)\,\mathrm{d}u\,\mathrm{d}z\,\mathrm{d}s\,.$$

The idea is to keep on substituting the nth order approximation of X_t into the original equation (30.1) to obtain the $(n + 1)$-st order approximation.

30.2.2 Taylor Series for a Stochastic Differential Equation

Now we consider the Taylor series expansion for an SDE

$$\mathrm{d}X_t = a(X_t)\,\mathrm{d}t + b(X_t)\,\mathrm{d}W_t\,. \tag{30.7}$$

By Itô's lemma, we have

$$\begin{aligned} f(X_t) &= f(X_{t_0}) + \int_{t_0}^{t} \left(a(X_s) \frac{\partial f}{\partial x}(X_s) + \frac{1}{2} b^2(X_s) \frac{\partial^2 f}{\partial x^2}(X_s) \right) \mathrm{d}s \\ &\quad + \int_{t_0}^{t} b(X_s) \frac{\partial f}{\partial x}(X_s)\, \mathrm{d}W_s \\ &= f(X_{t_0}) + \int_{t_0}^{t} L^0 f(X_s)\, \mathrm{d}s + \int_{t_0}^{t} L^1 f(X_s)\, \mathrm{d}W_s \end{aligned} \tag{30.8}$$

where

$$L^0 = a\frac{\partial}{\partial x} + \frac{1}{2} b^2 \frac{\partial^2}{\partial x^2} , \quad L^1 = b\frac{\partial}{\partial x} .$$

If we take $f(x) = x$ in (30.8), we have

$$L^0 f = a , \quad L^1 f = b$$

and recover the original equation (30.7) rewritten in the integral form as

$$X_t = X_{t_0} + \int_{t_0}^{t} a(X_s)\, \mathrm{d}s + \int_{t_0}^{t} b(X_s)\, \mathrm{d}W_s . \tag{30.9}$$

We take

$$f = a , \quad f = b$$

in (30.8) and obtain

$$a(X_t) = a(X_{t_0}) + \int_{t_0}^{t} L^0 a(X_s)\, \mathrm{d}s + \int_{t_0}^{t} L^1 a(X_s)\, \mathrm{d}W_s \tag{30.10}$$

and

$$b(X_t) = b(X_{t_0}) + \int_{t_0}^{t} L^0 b(X_s)\, \mathrm{d}s + \int_{t_0}^{t} L^1 b(X_s)\, \mathrm{d}W_s . \tag{30.11}$$

We substitute (30.10) and (30.11) into (30.9), and obtain

$$\begin{aligned} X_t &= X_{t_0} + \int_{t_0}^{t} \left(a(X_{t_0}) + \int_{t_0}^{s} L^0 a(X_u)\, \mathrm{d}u + \int_{t_0}^{s} L^1 a(X_u)\, \mathrm{d}W_u \right) \mathrm{d}s \\ &\quad + \int_{t_0}^{t} \left(b(X_{t_0}) + \int_{t_0}^{s} L^0 b(X_u)\, \mathrm{d}u + \int_{t_0}^{s} L^1 b(X_u)\, \mathrm{d}W_u \right) \mathrm{d}W_s \\ &= X_{t_0} + a(X_{t_0})\,(t - t_0) + b(X_{t_0})\,(W_t - W_{t_0}) + R(t_0; t) \end{aligned} \tag{30.12}$$

where

$$
\begin{aligned}
R(t_0;t) = &\int_{t_0}^{t}\int_{t_0}^{s} L^0 a(X_u)\,\mathrm{d}u\,\mathrm{d}s + \int_{t_0}^{t}\int_{t_0}^{s} L^1 a(X_u)\,\mathrm{d}W_u\,\mathrm{d}s \\
&+ \int_{t_0}^{t}\int_{t_0}^{s} L^0 b(X_u)\,\mathrm{d}u\,\mathrm{d}W_s + \int_{t_0}^{t}\int_{t_0}^{s} L^1 b(X_u)\,\mathrm{d}W_u\,\mathrm{d}W_s \,.
\end{aligned}
$$

From (30.8) we have

$$
\begin{aligned}
f(X_t) = f(X_{t_0}) + L^0 f(X_{t_0}) &\int_{t_0}^{t} \mathrm{d}s + L^1 f(X_{t_0}) \int_{t_0}^{t} \mathrm{d}W_s \\
&+ c(X_{t_0}) \int_{t_0}^{t}\int_{t_0}^{s_2} \mathrm{d}W_{s_1}\mathrm{d}W_{s_2} + R(f, t_0; t)
\end{aligned}
$$

where

$$
c(x) = b(x)\left\{b(x)f''(x) + b'(x)f'(x)\right\} \,. \tag{30.13}
$$

Note that

$$
\begin{aligned}
\int_{t_0}^{t}\int_{t_0}^{s_2} \mathrm{d}W_{s_1}\mathrm{d}W_{s_2} &= \int_{t_0}^{t} (W_{s_2} - W_{t_0})\,\mathrm{d}W_{s_2} \\
&= \int_{t_0}^{t} W_{s_2}\,\mathrm{d}W_{s_2} - W_{t_0}\int_{t_0}^{t} \mathrm{d}W_{s_2} \\
&= \frac{1}{2}\{W_t^2 - W_{t_0}^2 - (t - t_0)\} - W_{t_0}(W_t - W_{t_0}) \\
&= \frac{1}{2}\{(W_t - W_{t_0})^2 - (t - t_0)\} \,.
\end{aligned}
$$

Hence (30.13) becomes

$$
\begin{aligned}
f(X_t) = f(X_{t_0}) &+ L^0 f(X_{t_0})(t - t_0) + L^1 f(X_{t_0})(W_t - W_{t_0}) \\
&+ \frac{1}{2} c(X_{t_0})\{(W_t - W_{t_0})^2 - (t - t_0)\} + R(f, t_0; t) \,.
\end{aligned} \tag{30.14}
$$

If we take $f(x) = x$ in (30.14), then

$$
\begin{aligned}
X_t = X_{t_0} &+ a(X_{t_0})\,(t - t_0) + b(X_{t_0})\,(W_t - W_{t_0}) \\
&+ \frac{1}{2} b(X_{t_0}) b'(X_{t_0})\{(W_t - W_{t_0})^2 - (t - t_0)\} + R(f, t_0; t).
\end{aligned} \tag{30.15}
$$

30.3 The Euler Scheme

Consider an SDE

$$dX_t = a(X_t)\,dt + b(X_t)\,dW_t\ , \quad X_0 = x_0$$

defined on the time interval $[t_0, T]$. Here, for the sake of notational simplicity, we consider the case when $a(t,x)$ and $b(t,x)$ are functions of x only. The *Euler scheme* is a numerical method based on the approximation given by (30.12), after truncation of the remainder term $R(t_0; t)$, to find a numerical solution of

$$Y_{n+1} = Y_n + a(Y_n)\Delta_n + b(Y_n)\Delta W_n\ , \quad n = 0, 1, \ldots, N-1$$

at $0 = t_0 < t_1 < \cdots < t_N = T$, where $Y_0 = x_0$, $\Delta_n = t_{n+1} - t_n$, $\Delta W_n = W_{t_{n+1}} - W_{t_n}$. The increment ΔW_n has normal distribution with average 0 and variance Δ_n. The increments ΔW_n are independent of each other and obtained by random number generators in computer simulations.

If a and b are bounded and Lipschitz continuous, then the Euler scheme has strong order $\gamma = 0.5$. On the other hand, the strong order of a discretized numerical solution of an ordinary differential equation is equal to 1. The weak order of the Euler scheme is equal to $\beta = 1$.

In Fig. 30.1 is plotted the speed of numerical approximation by the Euler scheme for geometric Brownian motion

$$dS_t = \mu\, S_t\, dt + \sigma\, S_t\, dW_t$$

with $\mu = 0.5$, $\sigma = 0.6$. The points $(n, -\log \varepsilon_n)$, $4 \leq n \leq 10$, are plotted. In this case, since we have a closed form solution, we can compare the numerical solution obtained by the Euler scheme with the theoretical solution, where we take time step $\delta = 2^{-n}$, $4 \leq n \leq 10$, and sample size 10^4.

In the case of strong convergence the error satisfies $\varepsilon \approx K\delta^{\gamma}$, and hence we have

$$-\log_2 \varepsilon \approx -\log_2 K + \gamma\, n\ .$$

Thus the slope of the regression line is approximately equal to γ if we plot $-\log_2 \varepsilon$ for each n. In the case of weak convergence the slope is approximately equal to β. In Fig. 30.1 we observe that the slope in the first graph is close to 0.5 and in the second graph the slope is approximately equal to 1, and thus the speed of convergence to zero is exponential.

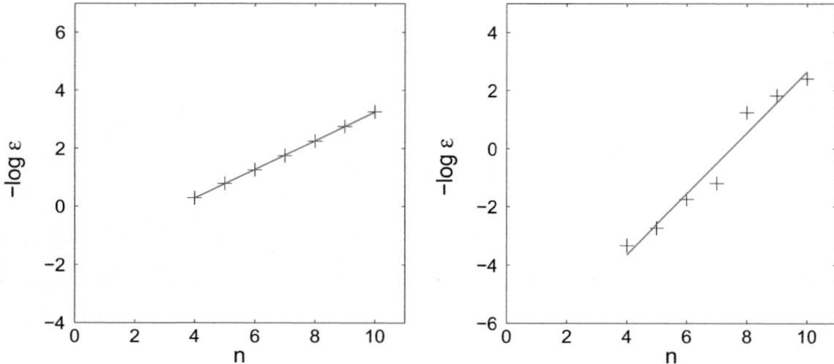

Fig. 30.1 The Euler Scheme: speeds of strong convergence (*left*) and weak convergence (*right*)

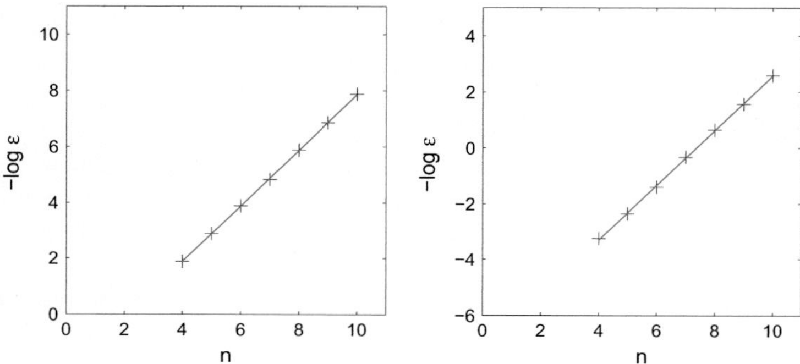

Fig. 30.2 The Milstein Scheme: speeds of strong convergence (*left*) and weak convergence (*right*)

30.4 The Milstein Scheme

The *Milstein scheme* is a numerical method based on the approximation given by (30.15), after truncation of the remainder term $R(f, t_0; t)$, to find a numerical solution of

$$Y_{n+1} = Y_n + a(Y_n)\Delta_n + b(Y_n)\Delta W_n + \frac{1}{2}b(Y_n)\,b'(Y_n)\left\{(\Delta W_n)^2 - \Delta_n\right\}$$

for $n = 0, 1, \ldots, N-1$ at $0 = t_0 < t_1 < \cdots < t_N = T$, where $b'(x)$ denotes the derivative of $b(x)$ with respect to x. It was named after Grigori N. Milstein [69].

If $\mathbb{E}[(X_0)^2] < \infty$ and if a and b are twice continuously differentiable and the second order derivatives are Lipschitz continuous, then the order of strong convergence of the Milstein scheme is $\gamma = 1.0$. The order of weak convergence is also $\beta = 1.0$.

Figure 30.2 displays numerical results from the Milstein scheme for geometric Brownian motion with $\mu = 0.5$, $\sigma = 0.6$. The sample size is 10^4. Observe that the slopes are approximately equal to 1 in both cases.

30.5 Computer Experiments

Simulation 30.1 (Euler Scheme: Weak Convergence)
We test the speed of weak convergence of the Euler scheme for solving the geometric Brownian motion when we take $g(x) = (x - \mathbb{E}[X_T])^2$.

```
T = 1;
N = [2^4,2^5,2^6,2^7,2^8,2^9,2^10];
J = 10^4;
mu = 0.5;
sigma = 0.6;
X_0 = 10;

X_T=zeros(1,J);
Y_N=zeros(1,J);
for n=1:length(N)
    dt = T/N(n);
    for j=1:J
        W(1) = 0;
        Y(1) = X_0;
        for i=1:N(n)
            dW = sqrt(dt)*randn;
            W(i+1) = W(i) + dW;
            Y(i+1) = Y(i) + mu*Y(i)*dt + sigma*Y(i)*dW;
        end
        Y_N(j) = Y(N(n)+1);
        X_T(j) = X_0*exp((mu - 0.5*sigma^2)*T + sigma*W(N(n)+1));
    %epsilon(n) = abs(mean(X_T) - mean(Y_T));
    epsilon(n) = abs(var(X_T) - var(Y_N));
    end
end
line_fit = polyfit(log(N)/log(2),-log(epsilon)/log(2),1)

plot(log(N)/log(2),-log(epsilon)/log(2),'+')
hold on
plot(log(N)/log(2),line_fit(1)*log(N)/log(2) + line_fit(2), ':')
```

Simulation 30.2 (Milstein Scheme: Strong Convergence)
We test the speed of strong convergence of the Milstein scheme for solving the geometric Brownian motion.

```
T = 1;
N = [2^4,2^5,2^6,2^7,2^8,2^9,2^10];
J = 10^4;
Error=zeros(1,J);
mu = 0.5;
sigma = 0.6;

X_0 = 10;
for n=1:length(N)
    dt = T/N(n);
    t = [0:dt:T];
        for j=1:J
```

```
            W(1) = 0;
            Y(1) = X_0;
            for i=1:N(n)
            dW = sqrt(dt)*randn;
            W(i+1) = W(i)+dW;
            Y(i+1) =Y(i)+mu*Y(i)*dt+sigma*Y(i)*dW+sigma^2/2*Y(i)*(dW^2-dt);
            end
         X_T = X_0*exp((mu - 0.5*sigma^2)*T + sigma*W(N(n)+1));
         Error(j) = abs(X_T - Y(N(n)+1));
        end
    epsilon(n) = mean(Error);
end
line_fit = polyfit(log(N)/log(2),-log(epsilon)/log(2),1)

plot(log(N)/log(2),-log(epsilon)/log(2),'+')
hold on
plot(log(N)/log(2),line_fit(1)*log(N)/log(2) + line_fit(2), ':')
```

Exercises

30.1 Let $\delta t = \frac{T}{L}$ and $t_i = i\,\delta t$. Consider the geometric Brownian motion

$$\frac{S(t_{i+1}) - S(t_i)}{S(t_i)} = \mu\,\delta t + \sigma\sqrt{\delta t}\,Y_i$$

where μ, σ are positive constants and $Y_0, Y_1, Y_2, \ldots$ are independent standard normal variables.

(i) What is the distribution of $\log\left(\dfrac{S(t)}{S_0}\right)$? Justify your answer.

(ii) Find

$$\lim_{\delta t \to 0+} \frac{1}{\delta t}\,\mathbb{E}\left[\left(\frac{S(t_{i+1}) - S(t_i)}{S(t_i)}\right)^2\right].$$

30.2 Compare the exact solution obtained in Problem 12.3 for the SDE

$$\mathrm{d}X_t = \mathrm{d}t + 2\sqrt{X_t}\,\mathrm{d}W_t$$

with a numerical solution obtained by the Milstein scheme.

30.3 Compare the exact solution obtained in Problem 12.4 for the SDE

$$\mathrm{d}X_t = -X_t(2\log X_t + 1)\mathrm{d}t + 2X_t\sqrt{-\log X_t}\,\mathrm{d}W_t$$

with a numerical solution obtained by the Milstein scheme.

Appendix A
Basic Analysis

In this chapter we introduce the definitions, notations and facts for sets, functions and metric spaces.

A.1 Sets and Functions

We denote the sets of the natural numbers, integers, rational numbers, real numbers, complex numbers by $\mathbb{N}$, $\mathbb{Z}$, $\mathbb{Q}$, $\mathbb{R}$, $\mathbb{C}$, respectively.[1] The *difference* of two sets A and B is defined by $A \setminus B = \{x : x \in A,\ x \notin B\}$, and their *symmetric difference* is defined by $A \triangle B = (A \cup B) \setminus (A \cap B)$. If $A \subset X$, then A^c denotes the complement of A, i.e., $A^c = X \setminus A$.

Given two sets X and Y, a function $f : X \to Y$ (or a map or a mapping) from X to Y is a rule that assigns for every $x \in X$ a unique element $f(x) \in Y$. The sets X and Y are called the domain and range (or codomain) of f, respectively. For each $x \in X$ the element $f(x)$ is called the image of x under f. Sometimes the range of f means the subset $\{f(x) : x \in X\}$.

If $\{f(x) : x \in X\} = Y$, then f is called an onto function. If $f(x_1) \neq f(x_2)$ whenever $x_1 \neq x_2$, then f is called a one-to-one function. If $f : X \to Y$ is onto and one-to-one, it is called a bijection or a one-to-one correspondence, and in this case there exists an inverse function f^{-1}.

Even when the inverse function of $f : X \to Y$ does not exist, the inverse image of $E \subset Y$ under f is defined by $f^{-1}(E) = \{x : f(x) \in E\}$. The operation of taking inverse images satisfies $f^{-1}(E \cup F) = f^{-1}(E) \cup f^{-1}(F)$, $f^{-1}(E \cap F) = f^{-1}(E) \cap f^{-1}(F)$ and $f^{-1}(Y \setminus E) = X \setminus f^{-1}(E)$.

[1]The symbol $\mathbb{Z}$ is from the German word 'Zahl' for number, and $\mathbb{Q}$ from the Italian word 'quoziente' for quotient.

G.H. Choe, *Stochastic Analysis for Finance with Simulations*, Universitext,
DOI 10.1007/978-3-319-25589-7

The number of elements of a set A is called the cardinality of A, and denoted by card$A$. If $f : X \to Y$ is a one-to-one correspondence, then X and Y have the same cardinality. If X has finitely many elements, or X and $\mathbb{N}$ have the same cardinality, then X is called a countable set. Examples of countable sets are $\mathbb{N}$, $\mathbb{Z}$, $\mathbb{Q}$, etc. A countable union of countable sets is also countable, more precisely, if A_i is a countable set for each $i = 1, 2, \ldots$ then $\bigcup_{i=1}^{\infty} A_i$ is also a countable set. If X is not countable, X is called an uncountable set. Examples of uncountable sets are $\mathbb{R}$ and $\mathbb{C}$. If we add uncountably many positive numbers, the sum is always infinite, and that is why we consider only a countable sum of positive numbers and use the notation $\sum$.

For a subset $A \subset X$, the characteristic function (or the indicator function) of A, denoted by $\mathbf{1}_A$, is defined by

$$\mathbf{1}_A(x) = \begin{cases} 1 \, , \, x \in A \, , \\ 0 \, , \, x \notin A \, . \end{cases}$$

A.2 Metric Spaces

Definition A.1 A metric (or distance) on a set X is a function $d : X \times X \to \mathbb{R}$ satisfying the following conditions:

(i) $d(x, y) \geq 0$ for every $x, y \in X$. $d(x, y) = 0$ holds only for $x = y$.
(ii) Symmetry holds, i.e., $d(x, y) = d(y, x)$ for every $x, y \in X$.
(iii) The triangle inequality holds, i.e., $d(x, z) \leq d(x, y) + d(y, z)$ for every $x, y, z \in X$.

A set X on which a metric d is defined is called a metric space and denoted by (X, d).

Example A.1 For any two points $x = (x_1, \ldots, x_n)$, $y = (y_1, \ldots, y_n)$ in the Euclidean space $\mathbb{R}^n$ define $d(x, y) = \left(\sum_{i=1}^{n} (x_i - y_i)^2\right)^{1/2}$, then $(\mathbb{R}^n, d)$ is a metric space. It is the standard metric on the Euclidean spaces.

Definition A.2 A norm on a vector space V is a function $|| \cdot || : V \to \mathbb{R}$ satisfying the following conditions:

(i) For every $v \in V$, $||v|| \geq 0$, and $||v|| = 0$ holds only for $v = 0$.
(ii) For a scalar c and a vector $v \in V$ we have $||cv|| = |c| \, ||v||$.
(iii) The triangle inequality $||v_1 + v_2|| \leq ||v_1|| + ||v_2||$ holds.

A vector space X equipped with a norm $|| \cdot ||$ is denoted by $(V, || \cdot ||)$ and called a normed space.

Fact A.1 *On a normed space a metric can be defined by* $d(x, y) = ||x - y||$. *Hence a normed space is a metric space.*

Example A.2 For $1 \leq p < \infty$ and $\mathbf{x} \in \mathbb{R}^n$ define $||\mathbf{x}||_p = \left(\sum_{i=1}^{n} |x_i|^p\right)^{1/p}$, and for $p = \infty$ define $||\mathbf{x}||_\infty = \max_{1 \leq i \leq n} |x_i|$, then $||\cdot||_p$ is a norm for every $1 \leq p \leq \infty$.

Given a sequence $\{x_n\}_{n=1}^{\infty}$ of points in a metric space (X, d), if the sequence of real numbers $\{d(x_n, x)\}_{n=1}^{\infty}$ converges to 0 then we say that the sequence $\{x_n\}_{n=1}^{\infty}$ *converges* to the limit x in X and write $x = \lim_{n\to\infty} x_n$ or $x_n \to x$. More precisely, x_n converges to x if for any arbitrary $\varepsilon > 0$ there exists an $N \geq 1$ such that $d(x, x_n) < \varepsilon$ for any $n \geq N$.

Given a sequence $x_1, x_2, x_3, \ldots$ of points of a metric space (X, d) if for any arbitrary $\epsilon > 0$ there exists an N such that $d(x_m, x_n) < \varepsilon$ for any $m, n \geq N$, then $\{x_n\}$ is called a Cauchy sequence.

A convergent sequence is a Cauchy sequence. A Cauchy sequence in a space X need not converge to a point in X. For example, a sequence $1, 1.4, 1.41, 1.414, \ldots$ in $X = \mathbb{Q}$ does not have its limit $\sqrt{2}$ in X. If any arbitrary Cauchy sequence converges in a metric space X, then it is called a complete space. For example, $\mathbb{R}^n$ is a complete space.

A.3 Continuous Functions

Throughout this section a function $f : X \to Y$ is given where (X, d_X) and (Y, d_Y) are metric spaces.

Definition A.3 (Continuity) If a function f satisfies $\lim_{n\to\infty} f(x_n) = f(x)$ as $\lim_{n\to\infty} x_n = x$, then f is said to be continuous at x. If this property holds for every $x \in X$, then we say that f is said to be continuous on X.

Fact A.2 *The following statements are equivalent:*

(i) f is continuous at $x \in X$.
(ii) For any $\varepsilon > 0$ there exists a $\delta > 0$ such that $d_X(x, x') < \delta$ implies $d_Y(f(x), f(x')) < \varepsilon$. The constant δ depends on x and ε, i.e., $\delta = \delta(\varepsilon, x)$.

Definition A.4 (Uniform Continuity) If for any $\varepsilon > 0$ there exists a $\delta > 0$ such that $d_X(x, x') < \delta$ implies $d_Y(f(x), f(x')) < \varepsilon$, and if δ does not depend on x but only on ε, then f is said to be uniformly continuous on X.

Example A.3

(i) Let $X = [0, 1]$ and $X_0 = (0, 1]$. The function $f(x) = \frac{1}{x}$ on X_0 is continuous but not uniformly continuous. It cannot be extended to X. This fact is consistent with Theorem A.11.
(ii) If $f : \mathbb{R}^1 \to \mathbb{R}^1$ is Lipschitz continuous, i.e., there exists a constant $0 \leq M < \infty$ such that $|f(x) - f(y)| < M|x - y|$ for every x, y, then f is uniformly continuous.
(iii) If $f : \mathbb{R}^1 \to \mathbb{R}^1$ is differentiable and if $|f'(x)| \leq M < \infty$, then by the Mean Value Theorem f satisfies $|f(x) - f(y)| < |f'(z)||x - y| \leq M|x - y|$, and hence f is Lipschitz continuous and uniformly continuous.

Now we consider the metric space consisting of continuous functions.

Example A.4 Let $C([0,1],\mathbb{R})$ and $C([0,1],\mathbb{C})$ be the sets of all continuous functions f defined on $[0,1]$ with their values in $\mathbb{R}$ and $\mathbb{C}$, respectively. They are vector spaces with the norm defined by $||f|| = \max_{x\in[0,1]} |f(x)|$, called the uniform norm. We can define a metric d by

$$d(f,g) = \max_{x\in[0,1]} |f(x) - g(x)|$$

for two continuous functions f, g, and hence $C([0,1],\mathbb{R})$ and $C([0,1],\mathbb{C})$ are metric spaces. They are complete metric spaces.

Definition A.5 (Open Ball) On a metric space (X,d), for $r > 0$ the subset $B_r(x_0) = \{x \in X : d(x_0,x) < r)$ is called the *open ball* of radius r centered at x_0. Let $U \subset X$. If for any point x_0 of U there exists an $r = r(x_0) > 0$ such that $B_r(x_0) \subset U$, then U is said to be *open*. A set is *closed* if its complement is open. Trivial examples of open subsets are $\emptyset$ and X.

Example A.5

(i) If $X = \mathbb{R}^1$, then open intervals (a,b), $(-\infty,b)$ and (a,∞) are open sets.
(ii) If X is a normed space with a norm $||\cdot||$, then the open unit ball $B_1(\mathbf{0}) = \{x \in X : ||x|| < 1\}$ is open.

If a subset $K \subset X$ is covered by subsets $\{U_\lambda : \lambda \in \Lambda\}$, i.e., $K \subset \bigcup_\lambda U_\lambda$, then $\{U_\lambda : \lambda \in \Lambda\}$ is called a cover of K. If there exists a $\Lambda_0 \subset \Lambda$ such that $K \subset \bigcup_{\lambda\in\Lambda_0} U_\lambda$, then we call $\{U_\lambda : \lambda \in \Lambda_0\}$ a subcover of K, or a subcover of $\{U_\lambda : \lambda \in \Lambda\}$. If every U_λ is an open set, then $\{U_\lambda : \lambda \in \Lambda\}$ is called an open cover. If Λ_0 is a finite set, then $\{U_\lambda : \lambda \in \Lambda\}$ is called a finite subcover.

Definition A.6 (Compact Set) If, for any open cover of a set K, there exists a finite subcover, then K is said to be *compact*.

Fact A.3 (Heine–Borel Theorem) *A subset of* $\mathbb{R}^n$ *is compact if and only if it is closed and bounded.*

Fact A.4 (Bolzano–Weierstrass Theorem) *Every bounded sequence in* $\mathbb{R}^n$ *has a convergent subsequence.*

Example A.6 A metric space (X,d) is given.

(i) The empty set $\emptyset$ and a finite set are compact.
(ii) For $X = \mathbb{R}^1$, a closed and bounded interval $[a,b]$ is compact.
(iii) For $X = \mathbb{R}^1$, closed and unbounded intervals such as $\mathbb{R}^1$, $(-\infty,b]$ and $[a,\infty)$ are not compact.

Fact A.5 *Given two metric spaces (X, d_X), (Y, d_Y) and a continuous function $f : X \to Y$. If X is a compact set, then the following holds:*

(i) $f(X)$ is a compact subset of Y.
(ii) For $Y = \mathbb{R}^1$ f assumes its maximum and minimum on X, i.e., there exist x_1, $x_2 \in K$ such that

$$f(x_1) = \max_{x \in K} f(x) , \quad f(x_2) = \min_{x \in K} f(x) .$$

(iii) f is uniformly continuous.

Definition A.7 (Dense Subset) A subset X_0 of a metric space X is *dense* in X if for every $x \in X$ and for every $r > 0$ the open ball $B_r(x)$ satisfies $B_r(x) \cap X_0 \neq \emptyset$. Equivalently, for every $x \in X$ there exists a sequence $x_n \in X_0$ that converges to x.

Example A.7

(i) The set of integers $\mathbb{Z}$ is not dense in $\mathbb{R}$.
(ii) The set of rational numbers $\mathbb{Q}$ is dense in the set of real numbers $\mathbb{R}$.
(iii) The set of irrational numbers $\mathbb{R} \setminus \mathbb{Q}$ is dense in $\mathbb{R}$.
(iv) The open interval (a, b) is dense in the interval $[a, b]$.

Fact A.6 (Weierstrass) *A real-valued trigonometric function of period 1 is a function of the form*

$$\sum_{n=1}^{N} a_n \cos(2\pi nx) + \sum_{n=1}^{N} b_n \sin(2\pi nx) , \; a_n, b_n \in \mathbb{R}, \; N \in \mathbb{N} .$$

The set of all such functions is dense in $(C[0, 1], \mathbb{R})$.

Fact A.7 (Stone–Weierstrass) *A complex-valued trigonometric function of period 1 is a function of the form*

$$\sum_{n=-N}^{N} c_n \mathrm{e}^{2\pi \mathrm{i} nx}, \; c_n \in \mathbb{C}, \; N \in \mathbb{N}$$

where $\mathrm{i}^2 = -1$. *The set of all such functions is dense in* $(C[0, 1], \mathbb{C})$.

A.4 Bounded Linear Transformations

We are given two normed spaces $(X, ||\cdot||_X)$, $(Y, ||\cdot||_Y)$. When there is no danger of confusion we write $||\cdot||$ to denote both $||\cdot||_X$ and $||\cdot||_Y$. If a map $T : X \to Y$ satisfies $T(x_1 + x_2) = T(x_1) + T(x_2)$ for $x_1, x_2 \in X$ and $T(cx) = cf(x)$ for $x \in X$

and scalar c, then T is called a linear transformation. For a linear transformation T we usually write Tx instead of $T(x)$.

Define a *norm* $||T||$ of a linear transformation $T : X \to Y$ by

$$||T|| = \sup_{x \neq 0} \frac{||Tx||}{||x||} = \sup_{||x||=1} ||Tx|| \, .$$

The transformation T is called a bounded transformation if $||T|| < \infty$, which is equivalent to the condition that there exists an $M < \infty$ such that

$$||Tx|| \leq M||x||$$

for every x. In fact, $||T||$ is the infimum of such constants M.

Example A.8 For $X = \mathbb{R}^n$, $Y = \mathbb{R}^m$ and an $m \times n$ matrix A, the map $T : X \to Y$ defined by $T\mathbf{x} = A\mathbf{x}$ is linear.

(i) Define a norm on $\mathbb{R}^k$ by

$$||\mathbf{v}||_1 = \sum_{i=1}^{k} |v_i| \, , \quad \mathbf{v} = (v_1, \ldots, v_k) \, .$$

Then the norm of $T : X \to Y$ is given by

$$||T|| = \max_{1 \leq j \leq n} \sum_{i=1}^{m} |A_{ij}| \, .$$

(ii) Define a norm on $\mathbb{R}^k$ by

$$||\mathbf{v}||_\infty = \max_{1 \leq i \leq k} |v_i| \, , \quad \mathbf{v} = (v_1, \ldots, v_k) \, .$$

Then the norm of $T : X \to Y$ is given by

$$||T|| = \max_{1 \leq i \leq m} \sum_{j=1}^{n} |A_{ij}| \, .$$

(iii) Let $m = n$ and define a norm on $\mathbb{R}^n$ by

$$||\mathbf{v}||_2 = \left(\sum_{i=1}^{n} |v_i|^2 \right)^{1/2} \, , \quad \mathbf{v} = (v_1, \ldots, v_n) \, .$$

Let λ_i denote the eigenvalues of the symmetric matrix A^tA or the singular values of A. Then the norm of $T : \mathbb{R}^n \to \mathbb{R}^n$ is given by

$$||T|| = \max_{1 \le i \le n} \sqrt{\lambda_i} \ .$$

Fact A.8 *Any two norms* $||\cdot||_1$ *and* $||\cdot||_2$ *on a finite dimensional vector space* X *are equivalent, i.e., there exist constants* $0 < A \le B$ *such that* $A||x||_1 \le ||x||_2 \le B||x||_1$ *for every* $x \in X$. *In this case, the metrics induced by the all the norms are equivalent since* $A||x - y||_1 \le ||x - y||_2 \le B||x - y||_1$ *for every* $x, y \in X$, *and a convergent sequence with respect to one norm is also convergence with respect to the other norm.*

Theorem A.9 *Let* $(X, ||\cdot||_X)$ *and* $(Y, ||\cdot||_Y)$ *be normed spaces, and let* $f : X \to Y$ *be a linear mapping. Then the following statements are equivalent:*

(i) f *is bounded.*
(ii) f *is continuous.*
(iii) f *is uniformly continuous.*

Proof (i) $\Rightarrow$ (iii) If there exists an $M < \infty$ such that $||f(x)|| \le M||x||$ for every x, then by the linearity we have $||f(x_1) - f(x_2)|| \le M||x_1 - x_2||$ for $x_1, x_2 \in X$. For every $\epsilon > 0$ choose $\delta = \epsilon/M$, then $||x_1 - x_2|| < \delta$ implies $||f(x_1) - f(x_2)|| < \epsilon$. In other words, f is uniformly continuous.

(iii) $\Rightarrow$ (ii) This is obvious by definition.

(ii) $\Rightarrow$ (i) If $||f(x)|| \le M||x||$ does not hold for all x, then there exists a sequence $x_n \in X$ such that $||x_n|| = 1$ and $||f(x_n)|| > n$. Define a new sequence $z_n = \frac{1}{n}x_n$. Then $||f(z_n)|| > 1$. Hence $z_n \to 0$, but $f(z_n)$ does not converge to $0 \in Y$. In other words, f is not continuous, which is a contradiction. □

Definition A.8 Let (X, d) be a metric space. A mapping $f : X \to X$ is called a *contraction* if there is a constant $0 < \alpha < 1$ such that for every $x, y \in X$ we have $d(f(x), f(y)) \le \alpha\, d(x, y)$.

Theorem A.10 (Banach Fixed Point Theorem) *If* f *is a contraction defined on a complete metric space* X, *then* f *has a unique fixed point* $x^* \in X$, *i.e.,* $f(x^*) = x^*$. *Furthermore, for any starting point* $x_0 \in X$ *we have the convergence of* $x_m = f^m(x_0)$ *to* x^* *and the error bound is given by*

$$d(x_m, x^*) \le \frac{\alpha^m}{1-\alpha}\, d(x_0, x_1) \ ,$$

which shows that the speed of convergence to a limit is exponential.

Proof Let $x_n = f^n(x_0)$, $n \ge 1$. Then there exists $0 < \alpha < 1$ such that

$$d(x_{n+1}, x_n) = d(f(x_n), f(x_{n-1})) \le \alpha\, d(x_n, x_{n-1}) \le \cdots \le \alpha^n d(x_1, x_0) \ .$$

Hence, for $n > m$,

$$\begin{aligned} d(x_m, x_n) &\le d(x_m, x_{m+1}) + d(x_{m+1}, x_{m+2}) + \cdots + d(x_{n-1}, x_n) \\ &\le (\alpha^m + \alpha^{m-1} + \cdots + \alpha^{n-1})\, d(x_0, x_1) \\ &= \alpha^m \frac{1-\alpha^{n-m}}{1-\alpha}\, d(x_0, x_1) \\ &\le \frac{\alpha^m}{1-\alpha}\, d(x_0, x_1)\ . \qquad (*) \end{aligned}$$

Since $d(x_n, x_m) \to 0$ as $n, m \to +\infty$, the sequence $x_0, x_1, x_2, \ldots$ is Cauchy, and hence it converges to some point x^* since X is complete. Since

$$d(x^*, f(x^*)) \le d(x^*, x_n) + d(x_n, f(x^*)) \le d(x^*, x_n) + \alpha\, d(x_{n-1}, x^*)\ ,$$

we can make $d(x^*, f(x^*))$ arbitrarily small by choosing large n. Hence it must be zero, and $f(x^*) = x^*$. To prove the uniqueness, take x, y such that $f(x) = x$ and $f(y) = y$. Then $d(x, y) = d(f(x), f(y)) \le \alpha\, d(x, y)$, which is true only when $d(x, y) = 0$. To find an error bound, let $n \to +\infty$ in the inequality $(*)$. □

For an application of the Fixed Point Theorem in numerical linear algebra, see Sect. B.7.

A.5 Extension of a Function

We are given two sets X, Y and a function $f : X_0 \to Y$ where X_0 is a subset of X. If there exists a function $F : X \to Y$ such that $F(x) = f(x)$ for every $x \in X_0$, then F is called an *extension* of f to X. In this case f is called the *restriction* of F to X_0.

Now we consider the extension of a continuous function defined on a dense subset to the whole space.

Lemma A.1 *Let (X, d_X) and (Y, d_Y) be metric spaces. If $f : X \to Y$ is a uniformly continuous function, then f maps a Cauchy sequence $\{x_n\}$ to a Cauchy sequence $\{f(x_n)\}$.*

Proof For every $\epsilon > 0$ the uniform discontinuity of f implies that there exists a $\delta > 0$ such that $d_X(z, z') < \delta$ implies $d_Y(f(z), f(z')) < \epsilon$. If $\{x_n\}$ is a Cauchy sequence, then there exists a sufficiently large N such that $d_X(x_m, x_n) < \delta$ for $m, n \ge N$. Hence $d_Y(f(x_m), f(x_n)) < \epsilon$. □

Theorem A.11 *Let (X, d_X) and (Y, d_Y) be metric spaces, and let $X_0 \subset X$ be a dense subspace of X. Suppose that Y is a complete space. If $f : X_0 \to Y$ is uniformly continuous, then f can be extended to X as a continuous and linear map.*

Proof First, we will define $f(x)$ for any $x \in X$. Take a sequence $x_n \in X_0$ converging to x. By Lemma A.1 $f(x_n)$ is a Cauchy sequence. Since Y is a complete space, $f(x_n)$ converges to a limit, which is defined to be $f(x)$.

To show that the preceding definition is well-defined and independent of any particular sequence $x_n \in X_0$, take another sequence $x'_n \in X_0$ and show $\lim_{n\to\infty} f(x_n) = \lim_{n\to\infty} f(x'_n)$. Arranging elements of two sequences alternatingly we obtain a new sequence $x_1, x'_1, x_2, x'_2, x_3, x'_3, \ldots$, which also converges to x, hence is a Cauchy sequence. By Lemma A.1, $f(x_1), f(x'_1), f(x_2), f(x'_2), f(x_3), f(x'_3), \ldots$ is also a Cauchy sequence. Since Y is a complete space, there exists a unique limit, which is the common limit of two subsequences $f(x_1), f(x_2), \ldots$ and $f(x'_1), f(x'_2), \ldots$, and hence they have the same limit. □

Corollary A.1 *Let $(X, ||\cdot||_X)$ and $(Y, ||\cdot||_Y)$ be normed spaces, and let $X_0 \subset X$ be a dense vector subspace of X. Assume that Y is a complete space. If a linear transformation $f : X_0 \to Y$ is continuous, then f is uniformly continuous, and hence by Theorem A.11 f can be extended to X as a continuous and linear transformation.*

A.6 Differentiation of a Function

Definition A.9 (Differentiation) If a function $f : \mathbb{R}^n \to \mathbb{R}^m$ satisfies the condition that for every point $\mathbf{x}$ there exists an $m \times n$ matrix $Df(\mathbf{x})$ such that

$$\lim_{||h||\to 0} \frac{||f(\mathbf{x}+\mathbf{h}) - f(\mathbf{x}) - Df(\mathbf{x})\mathbf{h}||}{||h||} = 0 ,$$

we say that f is differentiable and call the matrix $Df(\mathbf{x})$, or the associated linear transformation, the *derivative* of f.

Remark A.1 The derivative of a differentiable function $f = (f_1, \ldots, f_m) : \mathbb{R}^n \to \mathbb{R}^m$ is defined by

$$Df(\mathbf{x}) = \left[\frac{\partial f_i}{\partial x_j}\right]_{ij} .$$

For $m = 1$ we have

$$Df(\mathbf{x}) = \nabla f(\mathbf{x}) = \left(\frac{\partial f}{\partial x_1}, \ldots, \frac{\partial f}{\partial x_n}\right) ,$$

where

$$f(\mathbf{x}+\mathbf{h}) - f(\mathbf{x}) \approx \nabla f(\mathbf{x})\mathbf{h}$$

and the right-hand side is a product of a $1 \times n$ matrix and an $n \times 1$ matrix.

Now we consider the partial derivatives of a function defined by a matrix.

Lemma A.2

(i) For a column vector $\mathbf{a} \in \mathbb{R}^n$ *define* $g : \mathbb{R}^n \to \mathbb{R}^1$ *by* $g(\mathbf{x}) = \mathbf{a} \cdot \mathbf{x} = \mathbf{a}^t\mathbf{x}$. *Then* $\nabla g(\mathbf{x}) = \mathbf{a}^t$.

(ii) For an $n \times n$ *matrix A define* $f : \mathbb{R}^n \to \mathbb{R}^1$ *by* $f(\mathbf{x}) = \mathbf{x}^t A\mathbf{x}$. *Then* $\nabla f(\mathbf{x}) = \mathbf{x}^t(A + A^t)$.

Proof

(i) The proof is a direct application of the definition. Intuitively, if $\mathbf{h} \approx \mathbf{0}$ then $g(\mathbf{x} + \mathbf{h}) - g(\mathbf{x}) \approx \nabla g(\mathbf{x})\mathbf{h}$, and hence

$$\mathbf{a}^t(\mathbf{x} + \mathbf{h}) - \mathbf{a}^t\mathbf{x} = \mathbf{a}^t\mathbf{h} \approx \nabla g(\mathbf{x})\mathbf{h} .$$

Thus $\nabla g(\mathbf{x}) = \mathbf{a}^t$.

(ii) As in the first part, for $\mathbf{h} \approx \mathbf{0}$ we have

$$\begin{aligned} f(\mathbf{x} + \mathbf{h}) - f(\mathbf{x}) &= (\mathbf{x} + \mathbf{h})^t A(\mathbf{x} + \mathbf{h}) - \mathbf{x}^t A\mathbf{x} \\ &= \mathbf{x}^t A\mathbf{x} + \mathbf{x}^t A\mathbf{h} + \mathbf{h}^t A\mathbf{x} + \mathbf{h}^t A\mathbf{h} - \mathbf{x}^t A\mathbf{x} \\ &= \mathbf{x}^t A\mathbf{h} + \mathbf{h}^t A\mathbf{x} + \mathbf{h}^t A\mathbf{h} \\ &\approx \mathbf{x}^t A\mathbf{h} + \mathbf{h}^t A\mathbf{x} \\ &= \mathbf{x}^t A\mathbf{h} + (A\mathbf{x})^t\mathbf{h} \\ &= \mathbf{x}^t A\mathbf{h} + \mathbf{x}^t A^t\mathbf{h} \\ &= \mathbf{x}^t(A + A^t)\mathbf{h} \end{aligned}$$

and $\nabla f(\mathbf{x}) = \mathbf{x}^t(A + A^t)$. □

Definition A.10 (Variation) For a partition $a = t_0 < t_1 < \cdots < t_n = b$ and a function $f : [a, b] \to \mathbb{R}$, put $\delta f_i = f(t_{i+1}) - f(t_i)$. Define the variation of f by

$$V_a^b(f) = \sup \sum_{i=0}^{n-1} |\delta f_i|$$

where the supremum is taken over all finite partitions of $[a, b]$.

Appendix B
Linear Algebra

Linear algebra deals with operations of vectors and matrices, and it is widely used in many areas because it is an essential tool in solving a system of linear equations. The goal of this chapter is to introduce some basic concepts and terminology in linear algebra.

B.1 Vectors

Theorem B.1 (Cauchy–Schwarz Inequality) *Let V be a vector space with an inner product $(\cdot, \cdot)$. Then we have $(v, w)^2 \leq (v, v)(w, w)$ for $v, w \in V$. Equality holds if and only if one of the vectors v, w is a constant multiple of the other one.*

Proof First, consider the case when the scalar field is real. If $v = 0$ or $w = 0$, then the inequality clearly holds. It suffices to prove the theorem for the case $v \neq 0$ and $w \neq 0$. Since

$$f(t) = (tv + w, tv + w) = t^2(v, v) + 2(v, w)t + (w, w) \geq 0$$

for every real t, $f(t)$ has nonnegative minimum

$$f(t_0) = \frac{(v, v)(w, w) - (v, w)^2}{(v, v)}$$

at $t_0 = -\dfrac{(v, w)}{(v, v)}$. The quadratic function $f(t)$ is equal to 0 if and only if there exists a real number t such that $tv + w$ is the zero vector, and otherwise $f(t) > 0$. □

G.H. Choe, *Stochastic Analysis for Finance with Simulations*, Universitext,
DOI 10.1007/978-3-319-25589-7

Given a linear map $T : V \to V$, if there exists a nonzero vector v such that $Tv = \lambda v$ for some scalar λ, then v is called a characteristic vector or an eigenvector, and λ is called a characteristic value or an eigenvalue.

B.2 Matrices

Let V and W be two vector spaces with bases $\{v_1, \ldots, v_m\}$ and $\{w_1, \ldots, w_n\}$, respectively. A convenient way of expressing a linear transformation $T : V \to W$ is to compute $T(v_j) = \sum_i a_{ij} w_i$. Then the nm scalars a_{ij} have all the information on T. The following basic facts regarding matrix multiplication are frequently used in Part VII.

Fact B.2

(i) *Put* $\mathbf{1} = (1, \ldots, 1)^t \in \mathbb{R}^n$. *Then* $\mathbf{1}^t\mathbf{1} = n$, *and* $\mathbf{1}\mathbf{1}^t$ *is an* $n \times n$ *matrix all of whose* n^2 *components are* 1. *Note that* $\mathbf{1}^t\mathbf{x} = x_1 + \cdots + x_n$.

(ii) *Let* L *and* M *be matrices of sizes* $m \times n$ *and* $n \times \ell$, *respectively. Then*

$$LM = \left[LM^1 \cdots LM^\ell \right]$$

where M^j *denotes the* j*th column of* M. *Note that* LM^j *is an* $m \times 1$ *matrix, i.e., an* m*-dimensional column vector, since it is the product of matrices of sizes* $m \times n$ *and* $n \times 1$, *respectively.*

(iii) *Let* M *and* L *be matrices of sizes* $n \times \ell$ *and* $\ell \times k$. *Then*

$$ML = \begin{bmatrix} M_1 L \\ \vdots \\ M_n L \end{bmatrix}$$

where M_i *is the* i*th row of* M. *Note that* $M_i L$ *is a* k*-dimensional row vector since it is a product of matrices of sizes* $1 \times \ell$ *and* $\ell \times k$.

If a basis $\mathcal{B}$ of V consists of eigenvectors $v_1, \ldots, v_n$ of T, then the matrix $A = [a_{ij}]$ of T with respect to $\mathcal{B}$ satisfies $a_{ij} = 0$ for $i \neq j$ and $T(v_i) = a_{ii} v_i$ for every i. Thus A is a diagonal matrix. One of the goals of linear algebra is to take a suitable basis so that the corresponding matrix of T is close to a diagonal matrix. The *trace* of A is defined by $\sum_{i=1}^n A_{ii}$ and is equal to the sum of eigenvalues.

The *column rank* of a matrix A is the dimension of the column space of A, and the *row rank* of A is the dimension of the row space of A. The column rank and the row rank are equal, and $\text{rank}(A) \leq \min(m, n)$. If $\text{rank}(A) = \min\{m, n\}$, then A is said to have full rank.

The linear transformation $T(\mathbf{x}) = A\mathbf{x}$ is injective if and only if $\text{rank}(A) = n$, i.e., A has full column rank. The linear transformation $T(\mathbf{x}) = A\mathbf{x}$ is surjective if and only if $\text{rank}(A) = m$, i.e., A has full row rank. If $m = n$, then A is invertible if and only if $\text{rank}(A) = n$.

If B is an $n \times k$ matrix, then $\text{rank}(AB) \le \min\{\text{rank}\, A, \text{rank}\, B\}$. If B is an $n \times k$ matrix of rank n, then $\text{rank}(AB) = \text{rank}(A)$. If C is an $\ell \times m$ matrix of rank m, then $\text{rank}(CA) = \text{rank}(A)$.

B.3 The Method of Least Squares

Let V be a vector space with an inner product $(\cdot, \cdot)$. The case when V is infinite dimensional is not excluded in the following argument. Consider the problem of approximating a vector $v \in V$ by a subspace V_0. We introduce the least squares method, which finds the minimum

$$\min_{w \in V_0} ||v - w||^2 = (v - w, v - w) .$$

We consider the case when V_0 is finite-dimensional. There are two methods. For the first method we choose $V_0 = \text{span}\{w_1, \ldots, w_k\}$ and define

$$f(c_1, \ldots, c_k) = ||v - (c_1 w_1 + \cdots + c_k w_k)||^2 ,$$

then take partial derivatives with respect to c_j, and set them equal to 0.

The second method is based on the fact that the distance minimizing vector is the orthogonal projection of v onto V_0, and it solves the system of equations $(w_j, v - (c_1 w_1 + \cdots + c_k w_k)) = 0$.

For $\Omega = \{\omega_1, \ldots, \omega_n\}$ with the normalized counting measure $\mathbb{P}$, i.e., $\mathbb{P}(\{\omega_i\}) = \frac{1}{n}$ for $1 \le i \le n$, we may regard $L^2(\Omega)$ as $\mathbb{R}^n$. Table B.1 lists the corresponding concepts in Lebesgue integral theory and linear algebra. For example, the constant function $1 = \mathbf{1}_\Omega$ is identified with the vector $(1, \ldots, 1)$.

For $V = L^2(\Omega)$ let V_0 be a two-dimensional subspace of V spanned by the constant function $1 = \mathbf{1}_\Omega$ and a nonconstant function X, i.e., $V_0 = \{a1 + bX : a, b \in \mathbb{R}\}$. Fix $Y \in V$. If a and b solve the minimization problem

$$\min_{a,b} ||Y - (a1 + bX)||_{L^2} .$$

Table B.1 Comparison of Lebesgue integral on a finite space and linear algebra

Lebesgue Integral	Linear Algebra
A random variable X	A vector $\mathbf{v}$
The constant function 1	The vector $(1, \ldots, 1)$
$\int_\Omega XY \mathrm{d}\mathbb{P} = \mathbb{E}[XY]$	$\frac{1}{n}\mathbf{v} \cdot \mathbf{w}$
$\int_\Omega X^2 \mathrm{d}\mathbb{P} = \mathbb{E}[X^2] = \lVert X\rVert_{L^2}^2$	$\frac{1}{n}\lVert\mathbf{v}\rVert^2$
$L^2(\Omega, \mathbb{P})$	$\mathbb{R}^n$

To solve for a and b, either we set

$$f(a,b) = ||Y - (a1 + bX)||_{L^2}^2 = (Y - (a1 + bX), Y - (a1 + bX))$$

and take the partial derivatives with respect to a and b, set the derivatives equal to 0, or we use the orthogonality relations

$$\begin{cases} (1, Y - (a1 + bX)) = 0 \\ (X, Y - (a1 + bX)) = 0 , \end{cases}$$

or, equivalently

$$\begin{bmatrix} 1 & \mathbb{E}[X] \\ \mathbb{E}[X] & \mathbb{E}[X^2] \end{bmatrix} \begin{bmatrix} a \\ b \end{bmatrix} = \begin{bmatrix} \mathbb{E}[Y] \\ \mathbb{E}[XY] \end{bmatrix} . \tag{B.1}$$

Since 1 and X are linearly independent, the determinant $\mathbb{E}[X^2] - \mathbb{E}[X]^2$ is nonzero by the Cauchy–Schwarz inequality. (In fact, it is positive.) Hence the solution of (B.1) is given by $b = \frac{\mathrm{Cov}(X,Y)}{\mathrm{Var}(X)}$ and $a = \mathbb{E}[Y] - b\,\mathbb{E}[X]$.

Example B.1 (Linear Regression) If we try to approximately express data $\{(x_1, y_1), \ldots, (x_n, y_n)\}$ for $n \geq 2$ using a line of best fit $Y = a + bX$, we need to solve the least squares problem defined by

$$\min_{a,b} \sum_{i=1}^{n} |y_i - (a + bx_i)|^2 .$$

If we put $\mathbf{x} = (x_1, \ldots, x_n)$, $\mathbf{y} = (y_1, \ldots, y_n)$, $\mathbf{1} = (1, \ldots, 1)$, then the above problem is converted into

$$\min_{a,b} ||\mathbf{y} - (a\mathbf{1} + b\mathbf{x})||^2 ,$$

which has a solution

$$b = \frac{\sum_{i=1}^{n} x_i y_i - \sum_{i=1}^{n} x_i \sum_{i=1}^{n} y_i}{\sum_{i=1}^{n} x_i^2 - (\sum_{i=1}^{n} x_i)^2}$$

by the preceding computation. The hidden assumption in this problem is the condition that $\mathbf{1}$ and $\mathbf{x}$ are linearly independent, which is equivalent to the condition that not all x_i are equal. See Fig. B.1.

Example B.2 (Quadratic Regression) Given the points $(x_1, y_1), \ldots, (x_n, y_n)$, the objective is to find (a, b, c) for which $\sum_{i=1}^{n} |(ax_i^2 + bx_i + c) - y_i|^2$ is minimized. Let $\mathbf{v}_1 = (x_1^2, \ldots, x_n^2)^t$, $\mathbf{v}_2 = (x_1, \ldots, x_n)^t$, $\mathbf{v}_3 = (1, \ldots, 1)^t$ and $\mathbf{y} = (y_1, \ldots, y_n)^t$. Assume that the vectors are linearly independent. Then the given problem is

Fig. B.1 The line of best fit and the least squares method

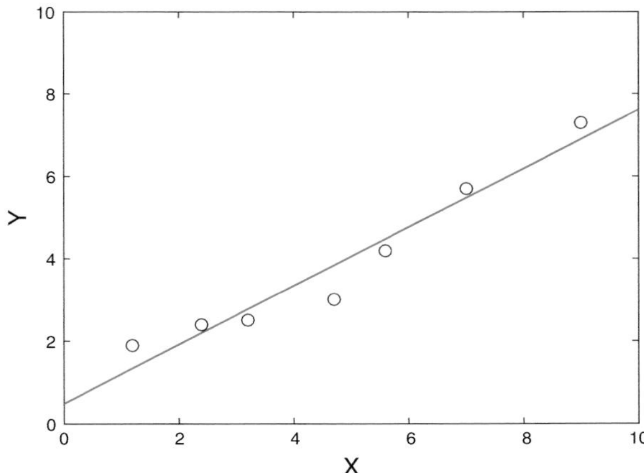

equivalent to the minimization of $||\mathbf{y}-(a\mathbf{v}_1+b\mathbf{v}_2+c\mathbf{v}_3)||$. Since $\mathbf{y}-(a\mathbf{v}_1+b\mathbf{v}_2+c\mathbf{v}_3)$ is orthogonal to $\mathbf{v}_1$, $\mathbf{v}_2$ and $\mathbf{v}_3$, we have

$$\begin{cases} a\,\mathbf{v}_1\cdot\mathbf{v}_1 + b\,\mathbf{v}_1\cdot\mathbf{v}_2 + c\,\mathbf{v}_1\cdot\mathbf{v}_3 = \mathbf{y}\cdot\mathbf{v}_1 \\ a\,\mathbf{v}_2\cdot\mathbf{v}_1 + b\,\mathbf{v}_2\cdot\mathbf{v}_2 + c\,\mathbf{v}_2\cdot\mathbf{v}_3 = \mathbf{y}\cdot\mathbf{v}_2 \\ a\,\mathbf{v}_3\cdot\mathbf{v}_1 + b\,\mathbf{v}_3\cdot\mathbf{v}_2 + c\,\mathbf{v}_3\cdot\mathbf{v}_3 = \mathbf{y}\cdot\mathbf{v}_3 \end{cases}$$

which is equivalent to

$$A^tA\begin{bmatrix} a \\ b \\ c \end{bmatrix} = \begin{bmatrix} \mathbf{y}\cdot\mathbf{v}_1 \\ \mathbf{y}\cdot\mathbf{v}_2 \\ \mathbf{y}\cdot\mathbf{v}_3 \end{bmatrix} \tag{B.2}$$

where the columns of A are given by $\mathbf{v}_1$, $\mathbf{v}_2$ and $\mathbf{v}_3$. Since A^tA is symmetric, its eigenvalues are real. If $A^tA\mathbf{v} = \lambda\mathbf{v}$ for some $\lambda \in \mathbb{R}$, $\mathbf{v} \neq \mathbf{0}$, then

$$||A\mathbf{v}||^2 = (A\mathbf{v}, A\mathbf{v}) = (A^tA\mathbf{v}, \mathbf{v}) = (\lambda\mathbf{v}, \mathbf{v}) = \lambda||\mathbf{v}||^2 ,$$

and hence $\lambda > 0$. Otherwise, $A\mathbf{v} = \mathbf{0}$ and A would have rank less than 3. Since A^tA has positive eigenvalues, it is invertible, and the above matrix equation (B.2) has a unique solution. For a numerical simulation see Fig. F.1.

B.4 Symmetric Matrices

Theorem B.3 *Let A be a symmetric matrix, then the following statements are equivalent:*

(i) *A is positive definite.*
(ii) *There exists a symmetric positive definite matrix B such that $A = B^2$.*
(iii) *There exists an invertible matrix B such that $A = B^t B$.*

Proof To prove that the statement (i) implies the statement (ii), recall that A is orthogonally diagonalizable. Hence $P^t AP = D$ where P is orthogonal and D is diagonal. Note that the diagonal entries of D are eigenvalues of A, which are positive. Hence we have $D = D_0^2$ where D_0 is the diagonal matrix whose entries are positive square roots of eigenvalues of A. Thus

$$A = PD_0^2 P^t = PD_0 D_0 P^t = PD_0 P^t PD_0 P^t = (PD_0 P^t)(PD_0 P^t) = B^2$$

where $B = PD_0 P^t$.

To prove (ii) $\Rightarrow$ (iii), note that $A = BB = B^t B$.

To prove (iii) $\Rightarrow$ (i), note that $\mathbf{x}^t A\mathbf{x} = \mathbf{x}^t B^t B\mathbf{x} = (B\mathbf{x})^t (B\mathbf{x}) = ||B\mathbf{x}||^2 > 0$, $\mathbf{x} \neq \mathbf{0}$. □

Theorem B.4 *Let A be a nonnegative definite matrix. Then there exists a lower triangular matrix L having nonnegative diagonal elements such that $A = LL^t$. Furthermore, if A is positive definite, then the matrix L is unique and has positive diagonal elements.*

Proof See p.147, [91]. □

Definition B.1 If A is symmetric and positive definite, then there exists a unique lower triangular matrix L with positive diagonal entries such that $A = LL^t$. This is called the *Cholesky decomposition* of A.

Example B.3 Here is how to transform a random vector so that the resulting random vector is normalized, i.e., its mean is zero and the covariance matrix is the identity matrix. Suppose that an $m \times 1$ random vector $\mathbf{X}$ has a mean vector $\boldsymbol{\mu} = \mathbb{E}[\mathbf{X}]$ and a positive-definite covariance matrix Σ. Choose a matrix L satisfying $\Sigma = LL^t$, and put $\mathbf{Z} = L^{-1}(\mathbf{X} - \boldsymbol{\mu})$. Then

$$\mathbb{E}[\mathbf{Z}] = \mathbb{E}[L^{-1}(\mathbf{X} - \boldsymbol{\mu})] = L^{-1}\mathbb{E}[\mathbf{X} - \boldsymbol{\mu}] = L^{-1}(\boldsymbol{\mu} - \boldsymbol{\mu}) = \mathbf{0}$$

and

$$\begin{aligned} \mathrm{Var}(\mathbf{Z}) &= \mathrm{Var}(L^{-1}\mathbf{X} - L^{-1}\boldsymbol{\mu}) \\ &= \mathrm{Var}(L^{-1}\mathbf{X}) \\ &= \mathbb{E}[L^{-1}\mathbf{X}(L^{-1}\mathbf{X})^t] \end{aligned}$$

$$
\begin{aligned}
&= \mathbb{E}[L^{-1}\mathbf{X}\mathbf{X}^t(L^{-1})^t] \\
&= L^{-1}\mathbb{E}[\mathbf{X}\mathbf{X}^t](L^{-1})^t \\
&= L^{-1}\mathrm{Var}(\mathbf{X})(L^{-1})^t \\
&= L^{-1}\Sigma(L^{-1})^t = I_m \; .
\end{aligned}
$$

B.5 Principal Component Analysis (PCA)

Let A be a nonnegative-definite $n \times n$ matrix, i.e., $A\mathbf{x} \cdot \mathbf{x} \geq 0$ for $\mathbf{x} \in \mathbb{R}^n$. Assume further that A is symmetric. (For example, A is a covariance matrix or correlation matrix.) Let $\lambda_1 \geq \lambda_2 \geq \cdots \geq \lambda_n \geq 0$ be the eigenvalues of A with corresponding eigenvectors $\mathbf{v}_1, \ldots, \mathbf{v}_n$ of norm 1. (All vectors are column vectors.) Recall that they are pairwise orthogonal. It can be shown that

$$\lambda_1 = A\mathbf{v}_1 \cdot \mathbf{v}_1 = \max_{||\mathbf{v}||=1} A\mathbf{v} \cdot \mathbf{v} \; .$$

Next, it can be proved that

$$\lambda_2 = A\mathbf{v}_2 \cdot \mathbf{v}_2 = \max_{\substack{||\mathbf{v}||=1 \\ \mathbf{v}\cdot\mathbf{v}_1=0}} A\mathbf{v} \cdot \mathbf{v} \; .$$

Proceeding inductively, we have

$$\lambda_{k+1} = A\mathbf{v}_2 \cdot \mathbf{v}_2 = \max_{\substack{||\mathbf{v}||=1 \\ \mathbf{v}\cdot\mathbf{v}_i=0,\ 1\leq i\leq k}} A\mathbf{v} \cdot \mathbf{v} \; .$$

(If A is a covariance matrix of a random vector $\mathbf{X} = (X_1, \ldots, X_n)$, then $\lambda_i = \mathrm{Var}(\mathbf{v}_i \cdot \mathbf{X})$.) The projection of $\mathbf{x}$ in the direction of $\mathbf{v}_i$ is called the ith *principal component* of $\mathbf{x}$. Note that we have the following decompositions:

$$
\begin{aligned}
I_n &= \mathbf{v}_1\mathbf{v}_1^t + \cdots + \mathbf{v}_n\mathbf{v}_n^t \\
A &= \lambda_1\mathbf{v}_1\mathbf{v}_1^t + \cdots + \lambda_n\mathbf{v}_n\mathbf{v}_n^t
\end{aligned}
$$

where I_n is the $n \times n$ identity matrix. The principal component analysis seeks to find some small k for which $(\lambda_1 + \cdots + \lambda_k)/(\lambda_1 + \cdots + \lambda_n)$ is close to 1 so that the approximation $\lambda_1\mathbf{v}_1\mathbf{v}_1^t + \cdots + \lambda_k\mathbf{v}_k\mathbf{v}_k^t$ is sufficiently close to A.

B.6 Tridiagonal Matrices

Lemma B.1 *Consider a uniform tridiagonal matrix*

$$B = \begin{bmatrix} a & b & 0 & \cdots & \cdots & 0 \\ b & a & b & \ddots & & \vdots \\ 0 & b & a & \ddots & \ddots & \vdots \\ \vdots & \ddots & \ddots & \ddots & b & 0 \\ \vdots & & \ddots & b & a & b \\ 0 & \cdots & \cdots & 0 & b & a \end{bmatrix}$$

Then B has the eigenvalues $\lambda_k = a + 2b\cos\frac{k\pi}{n+1}$, $1 \le k \le n$.

Proof First, we consider a uniform tridiagonal matrix

$$A = \begin{bmatrix} 0 & 1 & 0 & \cdots & \cdots & 0 \\ 1 & 0 & 1 & \ddots & & \vdots \\ 0 & 1 & 0 & \ddots & \ddots & \vdots \\ \vdots & \ddots & \ddots & \ddots & 1 & 0 \\ \vdots & & \ddots & 1 & 0 & 1 \\ 0 & \cdots & \cdots & 0 & 1 & 0 \end{bmatrix}$$

We will show that the eigenvalues of A are given by $\lambda_k = 2\cos\frac{k\pi}{n+1}$, $1 \le k \le n$. Let $\mathbf{x} = (x_1, \ldots, x_n)^t$ be an eigenvector of A with corresponding eigenvalue λ. Then, after defining $x_0 = x_{n+1} = 0$, we have the difference equation

$$x_{j-1} + x_{j+1} = \lambda x_j \, , \quad 1 \le j \le n \, .$$

Its characteristic polynomial is given by

$$p(r) = r^2 - \lambda r + 1 \, .$$

Suppose that $\lambda = 2$. Then $p(r) = 0$ has a double root $r = 1$, and hence $x_j = c_1 r^j + c_2 j r^j = c_1 + c_2 j$, $0 \le j \le n+1$, for some constants c_1, c_2. Since $x_0 = x_{n+1} = 0$, we have $c_1 = c_2 = 0$ and $\mathbf{x} = \mathbf{0}$, which contradicts the fact that $\mathbf{x} \ne \mathbf{0}$. Hence $\lambda \ne 2$. Now suppose that $\lambda = -2$. Then $p(r) = 0$ has a double root $r = -1$, and hence $x_j = c_1(-1)^j + c_2 j(-1)^j = (-1)^j(c_1 + c_2 j)$, $0 \le j \le n+1$, which would lead to a contradiction, too. Thus $\lambda \ne 2$.

Since $\lambda^2 \ne 4$, the discriminant of $p(r) = 0$ is equal to $\lambda^2 - 4 \ne 0$, and there are two distinct (not necessarily real) roots λ_1, λ_2 of $p(r) = 0$. Then the general solution is of the form $x_j = c_1 r_1^j + c_2 r_2^j$ for some constants c_1, c_2. Since $x_0 = 0$, we have $c_1 + c_2 = 0$, and $x_j = c_1(r_1^j - r_2^j)$. Note that $c_1 \neq 0$ since $\mathbf{x} \neq \mathbf{0}$. Since $x_{n+1} = 0$, we have $r_1^{n+1} - r_2^{n+1} = 0$, and hence $\left(\frac{r_1}{r_2}\right)^{n+1} = 1$. Since $r_1 r_2 = 1$, we have $\frac{1}{r_2} = r_1$ and $r_1^{2(n+1)} = 1$. Hence $r_1, r_2 = \exp(\pm 2\pi \mathrm{i} \frac{k}{2(n+1)})$, $1 \leq k \leq n$. Thus $\lambda = r_1 + r_2 = 2\cos(2\pi \frac{k}{2(n+1)})$.

Finally, let us consider a general case. It suffices to consider the case when $b \neq 0$. Since $A = \frac{1}{b}(B - aI)$, we have

$$\det(\lambda I - A) = \det\left(\lambda I - \frac{1}{b}(B - aI)\right) = \frac{1}{b^n} \det((b\lambda + a)I - B) .$$

Now we apply the previous result for A. □

Definition B.2 Let $A = [a_{ij}]$ be an $n \times n$ matrix where a_{ij} are complex numbers, $1 \leq i, j \leq n$. It is said to be strictly diagonally dominant (by rows) if $|a_{ii}| > \sum_{j=1, j\neq i}^{n} |a_{ij}|$ for every $1 \leq i \leq n$.

Fact B.5 (Gershgorin Circle Theorem) *Let $A = [a_{ij}]$ be an $n \times n$ matrix where a_{ij} are complex numbers, $1 \leq i, j \leq n$. Let $r_i = \sum_{j\neq i} |a_{ij}|$ and let $D(a_{ii}, r_i)$ denote the disk of radius r_i centered at a_{ii} in the complex plane. Then the eigenvalues of A lie inside $\bigcup_{i=1}^{n} D(a_{ii}, r_i)$.*

Proof Take an eigenvalue λ of A with a corresponding eigenvector $\mathbf{v} = (v_1, \ldots, v_n)$. Choose $i^* \in \{1, \ldots, n\}$ satisfying $|v_{i^*}| = \max_j |v_j|$. Note that $|v_{i^*}| > 0$. Since $A\mathbf{v} = \lambda\mathbf{v}$, the ith row satisfies $\sum_j a_{ij} v_j = \lambda v_i$ for every i. Hence $\sum_{j\neq i} a_{ij} v_j = \lambda v_i - a_{ii} v_i$. Thus

$$|\lambda - a_{i^* i^*}| = \left| \frac{\sum_{j\neq i^*} a_{i^* j} v_j}{v_{i^*}} \right| \leq \sum_{j\neq i^*} |a_{i^* j}| = r_{i^*} .$$

□

Remark B.1 Put

$$B = \begin{bmatrix} 1+2\nu & -\nu & 0 & \cdots & \cdots & 0 \\ -\nu & 1+2\nu & -\nu & \ddots & & \vdots \\ 0 & -\nu & 1+2\nu & \ddots & \ddots & \vdots \\ \vdots & \ddots & \ddots & \ddots & -\nu & 0 \\ \vdots & & \ddots & -\nu & 1+2\nu & -\nu \\ 0 & \cdots & \cdots & 0 & -\nu & 1+2\nu \end{bmatrix}$$

(i) Since B is strictly diagonally dominant, it is invertible by Gershgorin's theorem for $|\nu| \le \frac{1}{2}$.

(ii) Lemma B.1 implies that B has eigenvalues

$$\lambda_k = 1 + 2\nu - 2\nu \cos \frac{k\pi}{n+1}, \quad 1 \le k \le n .$$

For $|\nu| \le \frac{1}{2}$, we have $\lambda_k > 0$ and B is invertible.

(iii) Invertibility alone can proved as follows: Let B_n be the $n \times n$ tridiagonal matrix given as above. Let $C_n = \frac{1}{\nu} B_n$. For the sake of notational simplicity we put $c_n = \det C_n$. Then $\det B_n = \nu^n c_n$ and

$$c_n = \left(\frac{1}{\nu} + 2\right) c_{n-1} - c_{n-2} .$$

It suffices to show that $c_n > 0$, $n \ge 2$. By mathematical induction we show that $c_n > 0$ and $c_n > c_{n-1}$ for $n \ge 2$. For $n = 1$, $c_1 = \frac{1}{\nu} + 2 > 0$. For $n = 2$

$$c_2 = \left(\frac{1}{\nu} + 2\right)^2 - 1 = \left(\frac{1}{\nu}\right)^2 + 4\frac{1}{\nu} + 3 > 0$$

and

$$c_2 - c_1 = \left(\frac{1}{\nu}\right)^2 + 3\frac{1}{\nu} + 1 > 0 .$$

Now we assume that $c_k > 0$ and $c_k > c_{k-1}$ for $2 \le k < n$. Then

$$c_n = \left(\frac{1}{\nu} + 2\right) c_{n-1} - c_{n-2} = \left(\frac{1}{\nu} + 1\right) c_{n-1} + (c_{n-1} - c_{n-2}) > 0$$

and

$$c_n - c_{n-1} = \frac{1}{\nu} c_{n-1} + (c_{n-1} - c_{n-1}) > 0 .$$

B.7 Convergence of Iterative Algorithms

Given an $n \times n$ matrix A and $\mathbf{b} \in \mathbb{R}^n$, consider the equation

$$A\mathbf{x} = \mathbf{b} . \tag{B.3}$$

Take C and D such that $A = D + C$. Then (B.3) becomes

$$\mathbf{x} = -D^{-1}C\mathbf{x} + D^{-1}\mathbf{b} . \tag{B.4}$$

Let $B = -D^{-1}C = I - D^{-1}A$ and $\mathbf{d} = D^{-1}\mathbf{b}$, and define $T : \mathbb{R}^n \to \mathbb{R}^n$ by

$$T(\mathbf{x}) = B\mathbf{x} + \mathbf{d} .$$

Then (B.4) is equivalent to

$$T(\mathbf{x}) = \mathbf{x} \tag{B.5}$$

and Theorem A.10 implies that if T is a contraction then (B.5) has a unique solution, which is the limit of $T^n(\mathbf{x}_0)$ for any $\mathbf{x}_0$. Since $||T(\mathbf{x}) - T(\mathbf{y})|| = ||B(\mathbf{x} - \mathbf{y})|| \leq ||B||\,||\mathbf{x}-\mathbf{y}||$, we choose a norm so that $||B|| < 1$ to have a contraction. For example, if we choose a norm $||\mathbf{x}||_1 = \sum_{i=1}^{n} |x_i|$, then

$$||T|| = ||B||_1 = \max_{1\leq j\leq n} \sum_{i=1}^{n} |B_{ij}| ,$$

and if we choose a norm $||\mathbf{x}||_\infty = \max_{1\leq i\leq n} |x_i|$, then

$$||T|| = ||B||_\infty = \max_{1\leq i\leq n} \sum_{j=1}^{n} |B_{ij}| .$$

Example B.4 (Jacobi Algorithm) If we choose $D = \mathrm{diag}(a_{11}, \ldots, a_{nn})$ with $a_{ii} \neq 0$, then a sufficient condition for the convergence is

$$||I - D^{-1}A||_\infty = \max_{1\leq i\leq n} \sum_{j\neq 1}^{n} \left|\frac{a_{ij}}{a_{ii}}\right| < 1 ,$$

which is equivalent to $|a_{ii}| > \sum_{j=1}^{n} |a_{ij}|$ for every i. The iteration is given by

$$x_i^{(k+1)} = \frac{b_i - \sum_{j\neq i} a_{ij}x_j^{(k)}}{a_{ii}} , \quad 1 \leq i \leq n .$$

For more information, consult [12, 55].

Appendix C
Ordinary Differential Equations

An ordinary differential equation is an equation defined by derivatives of a function of one variable. For a general introduction to ordinary differential equations, consult [39, 96].

Lemma C.1 (Gronwall's Inequality) *Let $g(t) : [0,a] \to \mathbb{R}$ be continuous and nonnegative. Suppose that there exist constants $C \geq 0$, $K \geq 0$ such that*

$$g(t) \leq C + K \int_0^t g(s)\,\mathrm{d}s\ ,$$

for every $t \in [0,a]$. Then $g(t) \leq C\mathrm{e}^{Kt}$ for every $t \in [0,a]$.

Proof First, suppose that $C > 0$. Define $G(t)$ by

$$G(t) = C + K \int_0^t g(s)\,\mathrm{d}s\ .$$

Then $G(t) \geq g(t)$ and $G(t) \geq C > 0$. Since $\frac{\mathrm{d}}{\mathrm{d}t}G(t) = Kg(t) \leq KG(t)$, we have $\frac{\mathrm{d}}{\mathrm{d}t}(\log G(t)) \leq K$, and $\log G(t) \leq \log G(0) + Kt$. Since $G(0) = c$, we conclude that $g(t) \leq G(t) \leq G(0)\mathrm{e}^{Kt} = C\mathrm{e}^{Kt}$. □

C.1 Linear Differential Equations with Constant Coefficients

A differential operator L is said to be linear if it satisfies $L(c_1 f_1 + c_2 f_2) = c_1 L(f_1) + c_2 L(f_2)$ and $L(cf) = cL(f)$ for arbitrary constants c_1, c_2 and c. To solve a linear ordinary differential equation

$$a_n \frac{\partial^n f}{\partial x^n} + \cdots + a_1 \frac{\partial f}{\partial x} + a_0 f = g(x) \tag{C.1}$$

G.H. Choe, *Stochastic Analysis for Finance with Simulations*, Universitext,
DOI 10.1007/978-3-319-25589-7

where $a_n, \dots, a_1, a_0$ are constants, we first solve the homogeneous equation

$$a_n \frac{\partial^n f}{\partial x^n} + \cdots + a_1 \frac{\partial f}{\partial x} + a_0 f = 0 \tag{C.2}$$

and find a homogeneous solution f_H, then take the sum with a particular solution f_P of (C.1). To find a general solution of (C.2) we solve a polynomial equation

$$a_n \lambda^n + \cdots + a_1 \lambda + a_0 = 0 \, . \tag{C.3}$$

(i) If there exist distinct real solutions $\lambda_1, \dots, \lambda_n$ for (C.3), then a general solution is of the form $c_1 e^{\lambda_1 x} + \cdots + c_n e^{\lambda_n x}$.

(ii) If some of them are double roots, say $\lambda_1 = \lambda_2$, then the corresponding solution is of the form

$$c_1 e^{\lambda_1 x} + c_2 x e^{\lambda_1 x} + c_3 e^{\lambda_3 x} + \cdots + c_n e^{\lambda_n x} \, .$$

If some of them are triple roots, say $\lambda_1 = \lambda_2 = \lambda_3$, then the corresponding solution is of the form

$$c_1 e^{\lambda_1 x} + c_2 x e^{\lambda_1 x} + c_3 x^2 e^{\lambda_3 x} + \cdots + c_n e^{\lambda_n x}$$

and so on.

(iii) If there are complex roots $\lambda = a + b\sqrt{-1}, b \neq 0$, then there exist solutions of the form $e^x \cos(bx)$ and $e^x \sin(bx)$. A general solution is a linear combination of such functions. The set of general solutions of (C.1) is an n-dimensional vector space V, and the set of solutions of (C.2) is a parallel translate of the form $f_P + V$.

Example C.1 To solve a differential equation

$$f'' - 2af' + a^2 f = (D - a)^2 f = 0$$

where $D = \frac{d}{dx}$, we solve $(t - a)^2 = 0$ and find a double root $t = a$. Put $g(x) = e^{-ax} f(x)$. Then $D^2 g = 0$, and hence $g(x) = c_1 + c_2 x$. Thus $f(x) = c_1 e^{-x} + c_2 x e^{-x}$. The same idea can be used for a triple root as in a differential equation $(D-1)^3 f = 0$, and so on.

Example C.2 To solve a differential equation

$$((D - \alpha)^2 + \beta^2) f = 0$$

where α, β are real constants, we put $g(x) = e^{-\alpha x} f(x)$. Then $D^2 g = -\beta^2 g$, $g(x) = c_1 \cos \beta x + c_2 \sin \beta x$, and finally $f(x) = c_1 e^{\alpha x} \cos \beta x + c_2 e^{\alpha x} \sin \beta x$.

Example C.3 To solve an ordinary differential equation

$$f'' + f = 1$$

we first solve the homogeneous equation $f'' + f = 0$ and find the two-dimensional vector space $V = \{c_1 \sin x + c_2 \cos x : c_1, c_2 \text{ constant}\}$ consisting of homogeneous solutions and translate the space by $f_P = 1$, and find the desired solution set $V + f_P = \{c_1 \sin x + c_2 \cos x + 1 : c_1, c_2 \text{ constant}\}$.

C.2 Linear Differential Equations with Nonconstant Coefficients

The linear equation

$$\frac{\mathrm{d}y}{\mathrm{d}x} + P(x)y = Q(x)$$

can be solved by multiplying by an integrating factor $\mathrm{e}^{\int P(x)\mathrm{d}x}$. Then we have

$$\frac{\mathrm{d}}{\mathrm{d}x}(\mathrm{e}^{\int P(x)\mathrm{d}x}y) = Q(x)\mathrm{e}^{\int P(x)\mathrm{d}x} ,$$

which produces the solution

$$y = \mathrm{e}^{-\int P(x)\mathrm{d}x}\left(\int Q(x)\mathrm{e}^{\int P(x)\mathrm{d}x}\mathrm{d}x + C\right) .$$

Example C.4 An ordinary differential equation need not have a global solution. For example, consider $y' = y^2$, which has a trivial solution $y = 0$ and a nontrivial solution of the form $y = \dfrac{1}{-x + C}$ where C is a constant.

Example C.5 The initial value problem $y' = 2\sqrt{xy}$, $y(0) = 1$, has a solution $y = (1 + \frac{2}{3}x^{\frac{3}{2}})^2$.

Example C.6 Consider *Laguerre's equation* defined by

$$xy'' + (1 - x)y' + py = 0$$

where p is a constant. It is known that the only solutions bounded near the origin are constant multiples of $F(-p, 1, x)$ where $F(a, c, x)$ denotes the confluent hypergeometric function defined by

$$F(a,c,x) = 1 + \sum_{n=1}^{\infty} \frac{a(a+1)\cdots(a+n-1)}{n!c(c+1)\cdots(c+n-1)} x^n .$$

The solutions are polynomials if p is a nonnegative integer. The functions

$$L_n(x) = F(-n, 1, x) , \quad n \geq 0 ,$$

are called *Laguerre polynomials.* They form an orthonormal basis in the Hilbert space $L^2([0,\infty), e^{-x}dx)$ where the weight of the measure is given by the density e^{-x}, $x \geq 0$. Or, we may consider the Hilbert space $L^2([0,\infty), dx)$ and take the following definition of Laguerre polynomials:

$$L_n(x) = e^{-x/2} \frac{1}{n!} e^x \frac{d^n}{dx} (x^n e^{-x}) , \quad n \geq 1 .$$

For example, $L_0(x) = e^{-x/2}$, $L_1(x) = e^{-x/2}(1-x)$, $L_2(x) = e^{-x/2}(1 - 2x + \frac{x^2}{2})$. Consult [96] for more details.

C.3 Nonlinear Differential Equations

Definition C.1 An ordinary differential equation of the form

$$\frac{dy}{dx} + P(x)y = Q(x)y^n \tag{C.4}$$

is called a Bernoulli equation.

For $n \neq 0, 1$, multiplying both sides of (C.4) by $(1-n)y^{-n}$, we obtain

$$(1-n)y^{-n}\frac{dy}{dx} + (1-n)P(x)y^{1-n} = (1-n)Q(x) .$$

By the change of variable $z = y^{1-n}$, we have

$$\frac{dz}{dx} = (1-n)y^{-n}\frac{dy}{dx} .$$

Hence the nonlinear equation given by (C.4) is converted into the linear equation

$$\frac{dz}{dx} + (1-n)P(x)z = (1-n)Q(x) .$$

Definition C.2 An ordinary differential equation of the form

$$\frac{\mathrm{d}y}{\mathrm{d}x} = p(x) + q(x)y + r(x)y^2 \tag{C.5}$$

is called a Riccati equation.

There is no closed-form solution for (C.5) and we have to resort to numerical solutions, in general. However, if we can find a particular solution $y_1(x)$, then the general solution is of the form $y(x) = y_1(x) + z(x)$ where $z(x)$ is the general solution of

$$\frac{\mathrm{d}z}{\mathrm{d}x} - \{q(x) + 2r(x)y_1(x)\}\, z(x) = r(x)z(x)^2 \ ,$$

which is the Bernoulli equation with $n = 2$.

C.4 Ordinary Differential Equations Defined by Vector Fields

A function $\mathbf{F} : \mathbb{R}^n \to \mathbb{R}^n$ is called a vector field. If a curve $\boldsymbol{\phi} : (a, b) \to \mathbb{R}^n$ satisfies the differential equation

$$\frac{\mathrm{d}\boldsymbol{\phi}}{\mathrm{d}t} = \mathbf{F}(\boldsymbol{\phi}(t)) \ ,$$

then $\boldsymbol{\phi}$ is called an integral curve of the vector field $\mathbf{F}$.

Definition C.3 (Exponential of a Matrix) Let A denote an $n \times n$ matrix. Define e^A by

$$\exp(A) = \mathrm{e}^A = \sum_{k=0}^{\infty} \frac{1}{k!} A^k$$

where the infinite sum is defined by the convergence of finite sums of matrices in the n^2-dimensional Euclidean space. To show that the sum converges, first we regard $A = (A_{ij})_{i,j=1}^n$ as a point in the Euclidean space $\mathbb{R}^{n^2}$ with the norm $||\cdot||$ defined by

$$||A|| = \max_{ij} |A_{ij}| \ .$$

Recall that $\mathbb{R}^{n^2}$ is a complete metric space. Then we use the fact that the partial sums $\sum_{n=1}^{N} \frac{1}{n!} A^n$ form a Cauchy sequence in $\mathbb{R}^{n^2}$, and converges to a limit as $N \to \infty$.

Theorem C.1 (Differential Equation Defined by a Matrix) *Let* $\mathbf{x} : \mathbb{R}^1 \to \mathbb{R}^n$ *denote a differentiable curve. Consider an ordinary differential equation*

$$\frac{\mathrm{d}}{\mathrm{d}t}\mathbf{x}(t) = A\,\mathbf{x}(t)\ , \quad \mathbf{x}(0) = \mathbf{x}_0$$

defined by an $n \times n$ *matrix* A. *Then its solution is given by*

$$\mathbf{x}(t) = \mathrm{e}^{tA}\mathbf{x}_0\ .$$

C.5 Infinitesimal Generators for Vector Fields

Example C.7 Let A be an $n \times n$ matrix. Put $g_t = \mathrm{e}^{tA}$ for $t \in \mathbb{R}$. Then $\{g_t\}_{t\in\mathbb{R}}$ is a one-parameter group of linear transformations acting on $\mathbb{R}^n$. Note that

$$\mathcal{A}\mathbf{v} = \lim_{t\to 0} \frac{1}{t}(\mathrm{e}^{tA}\mathbf{v} - \mathbf{v}) = A\mathbf{v}$$

and that $\mathcal{A}$ is the linear transformation given by multiplication by A.

Example C.8 (Ordinary Differential Equation) Consider a system of linear differential equations given by

$$\frac{\mathrm{d}}{\mathrm{d}t}\mathbf{x}(t) = A\mathbf{x}(t)\ , \quad \mathbf{x}(0) = \mathbf{x}_0\ .$$

Then clearly $\mathbf{x}(t) = g_t\mathbf{x}_0 = \mathrm{e}^{tA}\mathbf{x}_0$ and

$$\left.\frac{\mathrm{d}}{\mathrm{d}t}\right|_{t=0}\mathbf{x}(t) = \lim_{t\to 0}\frac{1}{t}(\mathrm{e}^{tA}\mathbf{x}_0 - \mathbf{x}_0) = Ax_0\ .$$

Hence $\mathcal{A}$ is multiplication by A as expected. For example, consider $x'' = x$. Then put $y = x'$. Then we have a system of equations $x' = y$, $y' = x$. Let $\mathbf{x} = \begin{bmatrix} x \\ y \end{bmatrix}$ and $A = \begin{bmatrix} 0 & 1 \\ 1 & 0 \end{bmatrix}$. Then $\frac{\mathrm{d}}{\mathrm{d}t}\mathbf{x} = A\mathbf{x}$. Its solution is given by $\mathbf{x}(t) = \mathrm{e}^{tA}\mathbf{x}_0$.

Example C.9 (Exponential of Differential Operators) Consider differential operators which act on the set of sufficiently smooth functions defined on the real line.

(i) The exponential of differentiation is translation, i.e.,

$$\exp\left(t\frac{\mathrm{d}}{\mathrm{d}x}\right)\phi(x) = \phi(x+t)$$

in a suitable sense.

(ii) The exponential of Laplacian solves the diffusion equation since

$$\exp\left(t\frac{\mathrm{d}^2}{\mathrm{d}x^2}\right)\phi(x) = (\phi * G_t)(x)$$

in a suitable sense where G_t is the heat kernel and $*$ denotes convolution. For more information, consult [55].

Appendix D
Diffusion Equations

D.1 Examples of Partial Differential Equations

An equation consisting of partial derivatives of a function of several variables is called a partial differential equation. When one of the variables represents time, it is usually denoted by t. The following are typical examples of second order linear partial differential equations with constant coefficients.

Example D.1 The partial differential equation

$$\frac{\partial^2 f}{\partial x_1^2} + \cdots + \frac{\partial^2 f}{\partial x_n^2} = 0$$

is called the *Laplace equation*, and its solutions are called harmonic functions. For the existence and uniqueness of the solution, we impose a boundary condition along the boundary of the domain.

Example D.2 The partial differential equation

$$\frac{\partial^2 f}{\partial t^2} - \left(\frac{\partial^2 f}{\partial x_1^2} + \cdots + \frac{\partial^2 f}{\partial x_n^2} \right) = 0$$

is called the *wave equation*, and describes propagation of waves in n-dimensional space. For the existence and uniqueness of the solution, we need an initial condition, and in some cases a boundary condition.

Example D.3 The partial differential equation

$$\frac{\partial f}{\partial t} = a^2 \frac{\partial^2 f}{\partial x^2}$$

G.H. Choe, *Stochastic Analysis for Finance with Simulations*, Universitext,
DOI 10.1007/978-3-319-25589-7

is called the *heat equation* or *diffusion equation*, and represents dissipation of heat or chemicals. The Black–Scholes–Merton partial differential equation can be converted to the heat equation through changes of variables.

D.2 The Fourier Transform

Define the Fourier transform $F(s)$ of a function $f(t)$ by

$$F(s) = \mathscr{F}[f(t)](s) = \frac{1}{2\pi}\int_{-\infty}^{\infty} \mathrm{e}^{-\mathrm{i}st} f(t)\,\mathrm{d}t\ .$$

Note that $\mathscr{F}$ is a linear transformation.

Example D.4 (Solution of Heat Equation) Using the Fourier transformation we solve the heat equation given by

$$\frac{\partial f}{\partial t} = a^2 \frac{\partial^2 f}{\partial x^2}\ , \quad -\infty < x < \infty\ , \quad t > 0 \tag{D.1}$$

together with the initial condition

$$f(x,0) = g(x)\ , \quad -\infty < x < \infty$$

and the boundary condition

$$f(x,t) \to 0\ , \quad \frac{\partial f}{\partial x}(x,t) \to 0 \quad \text{as } x \to \pm\infty\ , \quad t > 0\ .$$

Apply the Fourier transform to both sides of (D.1), and derive and solve the resulting equation in terms of $F(s)$, and apply the inverse Fourier transform. Then we have

$$f(x,t) = \frac{1}{2a\sqrt{\pi t}}\int_{-\infty}^{\infty} g(s)\exp\left(-\frac{(x-s)^2}{4a^2 t}\right)\mathrm{d}s\ .$$

If $g(s)$ is given by the Dirac delta functional $\delta_0(s)$ concentrated at 0, then

$$f(x,t) = \frac{1}{2a\sqrt{\pi t}}\exp\left(-\frac{x^2}{4a^2 t}\right)\ .$$

D.3 The Laplace Transform

Define the Laplace transform $F(s)$ of a function $f(t)$, $t \geq 0$, by

$$F(s) = \mathscr{L}[f(t)](s) = \int_0^\infty e^{-st} f(t)\, dt\ .$$

For some functions the Laplace transform is not defined. For example, for $s > 0$,

$$\mathscr{L}\left[\frac{1}{t}\right] = \int_0^1 \frac{e^{-st}}{t} dt + \int_1^\infty \frac{e^{-st}}{t} dt > \int_\varepsilon^1 \frac{e^{-s}}{t} dt \to +\infty$$

as $\varepsilon \downarrow 0$. If $f(t)$ is piecewise continuous on $t \geq 0$ and satisfies the condition that $|f(t)| \leq M e^{at}$, $a \leq 0$, then $\mathscr{L}[f(t)](s)$ exists for all $s > a$. It is easy to see that $\mathscr{L}$ is a linear transformation.

Fact D.1 *The following basic facts are known:*

(i) $\mathscr{L}[f(at)] = \frac{1}{a} F\left(\frac{s}{a}\right)$, $a \neq 0$.
(ii) $\mathscr{L}[t^\alpha] = \frac{\Gamma(\alpha+1)}{s^{\alpha+1}}$, $\alpha \geq 0$. *In particular,* $\mathscr{L}[1] = \frac{1}{s}$, $s > 0$.
(iii) $\mathscr{L}[e^{at}] = \frac{1}{s-a}$, $s > a$.
(iv) $\mathscr{L}[f'(t)] = s\mathscr{L}[f] - f(0)$.
(v) $\mathscr{L}[f''(t)] = s^2\mathscr{L}[f] - sf(0) - f'(0)$.
(vi) $\mathscr{L}[tf(t)] = -F'(s)$.
(vii) $\mathscr{L}[\int_0^t f(u) du] = \frac{1}{s}\mathscr{L}[f(t)]$.
(viii) $\mathscr{L}\left[\frac{f(t)}{t}\right] = \int_s^\infty F(v) dv$.
(ix) $\mathscr{L}[e^{at} f(t)] = F(s-a)$.
(x) $\mathscr{L}[f(t-a)\, H(t-a)] = e^{-as}\mathscr{L}[f]$, $a > 0$, *where* $H(x)$ *is the Heaviside function defined by* $H(x) = 1$, $x \geq 0$, *and* $H(x) = 0$, $x < 0$.
(xi) $\mathscr{L}[\delta(t-a)] = e^{-as}$, $a > 0$.

Example D.5

$$\mathscr{L}\left[t^{-\frac{1}{2}}\right] = \frac{\sqrt{\pi}}{\sqrt{s}}\ .$$

To see why, note that

$$\begin{aligned}
\mathscr{L}\left[t^{-\frac{1}{2}}\right] &= \int_0^\infty e^{-st} t^{-\frac{1}{2}}\, dt \qquad (\text{Take } st = u.)\\
&= s^{-\frac{1}{2}} \int_0^\infty e^{-u} u^{-\frac{1}{2}} du \qquad (\text{Take } u = z^2.)\\
&= 2s^{-\frac{1}{2}} \int_0^\infty e^{-z^2} dz = 2s^{-\frac{1}{2}} \frac{\sqrt{\pi}}{2}\ .
\end{aligned}$$

Fact D.2 *The following facts are known:*

(i) Define the convolution by

$$(f * g)(t) = \int_0^t f(t-u)\, g(u)\, \mathrm{d}u \,.$$

Then

$$\mathscr{L}[f * g] = \mathscr{L}[f]\,\mathscr{L}[g] \,.$$

(ii) The inverse of the Laplace transform is given by

$$\mathscr{L}^{-1}[F(s)] = f(t) = \frac{1}{2\pi \mathrm{i}} \int_{c-\mathrm{i}\infty}^{c+\mathrm{i}\infty} F(s)\, \mathrm{e}^{st} \mathrm{d}s \,.$$

Definition D.1 For $-\infty < x < \infty$ define the *error function* by

$$\mathrm{erf}(x) = \frac{2}{\sqrt{\pi}} \int_0^x \mathrm{e}^{-u^2} \mathrm{d}u \,.$$

Note that $\lim_{x\to\infty} \mathrm{erf}(x) = 1$. Let $N(\cdot)$ denote the cumulative distribution function of the standard normal distribution. Then

$$\frac{1 + \mathrm{erf}(x)}{2} = \int_{-\infty}^{\sqrt{2}x} \frac{1}{\sqrt{2\pi}} \mathrm{e}^{-u^2/2} \mathrm{d}u = N(\sqrt{2}x) \,.$$

Fact D.3 *The error function satisfies*

$$\mathscr{L}\big[\mathrm{erf}(\sqrt{t})\,\big] = \frac{1}{s\sqrt{s+1}} \,.$$

To see why, note that

$$\mathrm{erf}(\sqrt{t}) = \frac{2}{\sqrt{\pi}} \int_0^{\sqrt{t}} \mathrm{e}^{-x^2} \mathrm{d}x = \frac{1}{\sqrt{\pi}} \int_0^t u^{-\frac{1}{2}} \mathrm{e}^{-u} \mathrm{d}u \,.$$

Hence

$$\mathscr{L}\big[\mathrm{erf}(\sqrt{t})\,\big] = \frac{1}{\sqrt{\pi}} \mathscr{L}\left[\int_0^t u^{-\frac{1}{2}} \mathrm{e}^{-u} \mathrm{d}u\right] = \frac{1}{\sqrt{\pi}} \frac{1}{s} \mathscr{L}\left[t^{-\frac{1}{2}} \mathrm{e}^{-t}\right] .$$

Now use $\mathscr{L}[t^{-\frac{1}{2}} \mathrm{e}^{-t}] = \frac{\sqrt{\pi}}{\sqrt{s+1}}$.

Fact D.4 *For $C \geq 0$ the error function satisfies*

$$\mathscr{L}\left[\operatorname{erf}\left(\frac{C}{\sqrt{t}}\right)\right] = \frac{1}{s}(1 - \mathrm{e}^{-2C\sqrt{s}}) .$$

Example D.6 (Diffusion Equation) Using the Laplace transformation we solve

$$\frac{\partial f}{\partial t} = a^2 \frac{\partial^2 f}{\partial x^2} \tag{D.2}$$

for $a > 0$. The initial condition is given by the Dirac delta function $f(x, 0) = \delta_0(x)$. For $t > 0, f(x, t)$ represents a probability density and satisfies $\lim_{x\to\pm\infty} f(x, t) = 0$ and $\lim_{x\to\pm\infty} \frac{\partial f}{\partial x}(x, t) = 0$. This problem is not easy to solve because the initial condition is given by a measure. We make it smoother in the following way: First, put

$$u(y, t) = \int_{-\infty}^{y} f(x, t)\,\mathrm{d}x .$$

From (D.2) we obtain

$$\int_{-\infty}^{y} \frac{\partial f}{\partial t}(x, t)\,\mathrm{d}x = a^2 \int_{-\infty}^{y} \frac{\partial^2 f}{\partial x^2}(x, t)\,\mathrm{d}x .$$

Then the left- and the right-hand sides are respectively equal to $\frac{\partial u}{\partial t}(x, t)$ and

$$a^2 \int_{-\infty}^{y} \frac{\partial^2 f}{\partial x^2}(x, t)\,\mathrm{d}x = a^2 \left(\frac{\partial f}{\partial x}(y, t) - \frac{\partial f}{\partial x}(-\infty, t)\right) = a^2 \frac{\partial f}{\partial x}(y, t) .$$

Hence we obtain

$$\frac{\partial u}{\partial t} = a^2 \frac{\partial^2 u}{\partial x^2} \tag{D.3}$$

with $u(x, 0) = H(x)$ where $H(x)$ is the Heaviside function. Recall that from distribution theory, established by Laurent Schwartz, we have $H'(x) = \delta_0(x)$.

Let $U(x, s)$ denote the Laplace transform of $u(x, t)$ with respect to t. Since

$$\mathscr{L}\left[\frac{\partial u}{\partial t}\right](x, s) = s\mathscr{L}[u](x, s) - u(x, 0) ,$$

by taking the Laplace transforms of the both sides of (D.3), we obtain

$$a^2 \frac{\partial^2 U}{\partial x^2} - s\,U = -H(x) . \tag{D.4}$$

To solve it, we first try to find a solution of the corresponding homogeneous equation, which is a second order linear differential equation, and there exists two linearly independent solutions. Since $s > 0$, the solution is of the form $c_1\,\mathrm{e}^{(\sqrt{s}/a)x} + c_2\,\mathrm{e}^{-(\sqrt{s}/a)x}$. Now, for $x > 0$, (D.4) has a particular solution $U_P(x, s) = \frac{1}{s}$, and hence (D.4) has a general solution of the form

$$U(x, s) = c_1\,\mathrm{e}^{(\sqrt{s}/a)x} + c_2\,\mathrm{e}^{-(\sqrt{s}/a)x} + \frac{1}{s}\,.$$

Since $\lim_{x\to\infty} u(x, t) = 1$, we have $\lim_{x\to\infty} U(x, s) = \frac{1}{s}$. Hence $c_1 = 0$ and (D.4) has a general solution of the form

$$U(x, s) = c_2\,\mathrm{e}^{-(\sqrt{s}/a)x} + \frac{1}{s}\,.$$

There still remains an unused condition. The initial data is symmetric and the differential equation is also symmetric so that the solution, which is a probability density function $f(x, t)$, is also symmetric with respect to $x = 0$. Hence $u(0, t) = \frac{1}{2}$ and hence $U(0, s) = \frac{1}{2s}$. Since

$$c_2\mathrm{e}^{-(\sqrt{s}/a)\times 0} + \frac{1}{s} = \frac{1}{2s}\,,$$

we have $c_2 = -\frac{1}{2s}$. A general solution of (D.4) is of the form

$$U(x, s) = -\frac{1}{2s}\mathrm{e}^{-(\sqrt{s}/a)x} + \frac{1}{s}\,. \tag{D.5}$$

Finally, taking the inverse Laplace transform of (D.5), we obtain

$$\begin{aligned}
u(x, t) &= \mathscr{L}^{-1}\left(-\frac{1}{2s}\mathrm{e}^{-(\sqrt{s}/a)x} + \frac{1}{s}\right) \\
&= \frac{1}{2}\mathscr{L}^{-1}\left(\frac{1}{s}(1 - \mathrm{e}^{-(\sqrt{s}/a)x})\right) + \mathscr{L}^{-1}\left(\frac{1}{2s}\right) \\
&= \frac{1}{2}\mathrm{erf}\left(\frac{x}{2a\sqrt{t}}\right) + \frac{1}{2} \\
&= N\left(\frac{x}{a\sqrt{2t}}\right) \\
&= \frac{1}{\sqrt{2\pi}}\int_{-\infty}^{x/a\sqrt{2t}} \exp\left(-\frac{y^2}{2}\right)\,\mathrm{d}y\,.
\end{aligned}$$

Therefore

$$f(x,t) = \frac{\partial u}{\partial x}(x,t) = \frac{1}{2a\sqrt{\pi t}} \exp\left(-\frac{x^2}{4a^2 t}\right) .$$

D.4 The Boundary Value Problem for Diffusion Equations

Consider the diffusion equation $u_t = u_{xx}$ with an initial condition (IC) and/or a boundary condition (BC) on a domain $0 \le x \le L$, $0 \le t \le T$ with

$$\begin{cases} IC & u(x,0) = g(x) , \\ BC & u(0,t) = a(t) , \quad u(L,t) = b(t) . \end{cases}$$

We do not exclude the cases $L = \infty$ and/or $T = \infty$ (Fig. D.1).

Example D.7 $g(x) = \sin(\frac{\pi}{L}x)$, $a = b = 0$. Then $u(x,t) = \mathrm{e}^{-t}\sin(x)$.

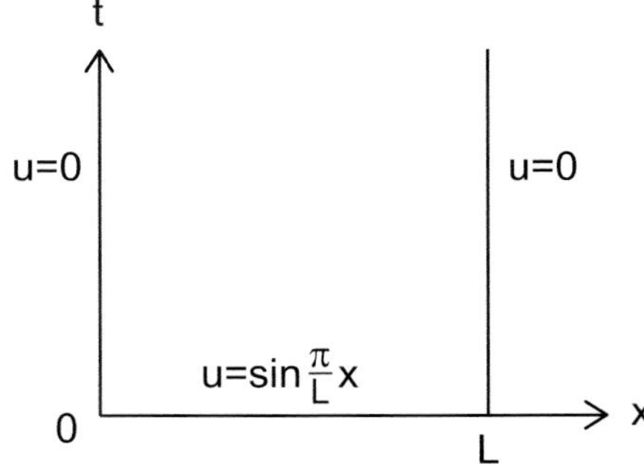

Fig. D.1 A domain and a boundary condition for a diffusion equation

Appendix E
Entropy

E.1 What Is Entropy?

Claude Elwood Shannon created *information theory* by publishing a paper on data transmission in 1948, laying the foundation of digital communication technology. One of his contributions is on data compression, where entropy is the amount of information content or randomness contained in a digital signal. He adopted the term 'entropy' from thermodynamics in physics to measure information content after he met John von Neumann. According to legend, von Neumann said, "There are two reasons why the term entropy should be used. First, no one knows what entropy really is, and there will be no one who will challenge the new theory. Second, a fashionable word such as entropy will attract many people to a new theory." The mathematical definition of Shannon's entropy has the same form as the entropy used in thermodynamics, and they share the same conceptual root in the sense that both measure the amount of randomness.

Shannon's entropy can be explained in the following way: Consider an infinitely long binary sequence. Shannon showed that the maximal possible rate of data compression is equal to the entropy. For example, if 0 and 1 appear independently with probability $\frac{1}{2}$ each, then the entropy is $\log 2$. The most widely used lossless data compression algorithm is the Ziv–Lempel algorithm. For more information consult [21].

Definition E.1 (Discrete Entropy) Let $(p_1, \ldots, p_n)$, $p_i \geq 0$, be a discrete probability distribution. Its entropy is defined by

$$H = \sum_{i=1}^{n} p_i \log \frac{1}{p_i}$$

G.H. Choe, *Stochastic Analysis for Finance with Simulations*, Universitext,
DOI 10.1007/978-3-319-25589-7

where the natural logarithm is employed and $0 \log 0 = 0$ by convention. (Any logarithmic base may be used in the definition.) Note that the discrete entropy is nonnegative.

Example E.1 Consider a coin tossing problem where the probability of obtaining heads is p, and the probability of obtaining tails is $q = 1 - p$. Then the entropy is equal to $-p \log p - q \log q$. Note that the entropy of the probability distribution $(p_1, \ldots, p_n)$ is bounded by $\log n$. For the proof see Example 3.7.

Definition E.2 (Continuous Entropy) For a continuous distribution the probability density function $\rho(x)$, the entropy is defined by

$$H(\rho) = -\int_{-\infty}^{\infty} \rho(x) \log \rho(x) \, dx .$$

Note that the continuous entropy can be negative.

Example E.2 The uniform distribution in $[a, b]$ has density function

$$\rho(x) = \begin{cases} \frac{1}{b-a}, & x \in [a, b] \\ 0, & x \notin [a, b] \end{cases}$$

and its entropy is equal to $-\int_a^b \frac{1}{b-a} \log \frac{1}{b-a} \, dx = \log(b - a)$.

Example E.3 Consider a normal distribution with the pdf given by

$$\rho(x) = \frac{1}{\sqrt{2\pi\sigma^2}} \exp\left(-\frac{(x-\mu)^2}{2\sigma^2}\right) .$$

Then $H(\rho) = \frac{1}{2}\log(2\pi\sigma^2) + \frac{1}{2}$.

E.2 The Maximum Entropy Principle

When the full information is not available for a probability distribution for a given problem, we attempt to guess the probability distribution under a constraint determined by partial information. Since entropy is regarded as the amount of randomness, any additional information on the probability distribution would reduce entropy, thus the entropy maximizing distribution already contains all the available information, which is called the Maximum Entropy Principle. In other words, the desired distribution is the one with maximum entropy. For an application in finance, see [16].

Example E.4 (Discrete Entropy) We are given a small cube. When rolled, the cube comes to rest showing on its upper surface one to six spots with probabilities equal to $p_i > 0$, $p_1 + \cdots + p_6 = 1$, which are unknown to us. If it is known that the expectation of the number of spots on the cube is equal to 4, i.e., $\sum_{i=1}^{6} i \times p_i = 4$, what is a reasonable way to estimate p_i? To solve the problem we employ the Maximum Entropy Principle. Since the sum of probabilities is equal to 1, we have $g_1(p_1, \ldots, p_6) = \sum_{i=1}^{6} p_i - 1 = 0$, and the given condition is written as $g_2(p_1, \ldots, p_6) = \sum_{i=1}^{6} i \times p_i - 4 = 0$. Now we apply the Lagrange multiplier method to the following function F defined by

$$F(p_1, \ldots, p_6, \lambda_1, \lambda_2) = \sum_i p_i \log \frac{1}{p_i} + \lambda_1 g_1(p_1, \ldots, p_6) + \lambda_2 g_2(p_1, \ldots, p_6)$$

and find $p_1, \ldots, p_6$. Then the distribution with maximal entropy is given by $p_i = Cr^i$, $1 \le i \le 6$, for some C and r. For more details consult [33].

For a continuous entropy, we use calculus of variations to find ρ. We assume that there exists a unique pdf $\rho_{\max}$ such that $H(\rho_{\max}) = H_{\max}$ is maximal among all possible values $H(\rho)$ where ρ is a pdf satisfying given conditions. Choose a pdf ϕ and define

$$\rho_\varepsilon(x) = \frac{\rho_{\max}(x) + \varepsilon\phi(x)}{1 + \varepsilon} \ . \tag{E.1}$$

We assume that $\rho_\varepsilon \ge 0$ for some sufficiently small $\eta > 0$, $-\eta < \varepsilon < \eta$, so that ρ_ε is a pdf. Furthermore, we choose ϕ in such a way that ρ_ε satisfies the given conditions. Consider a function $\varepsilon \mapsto H(\rho_\varepsilon)$, $-\eta < \varepsilon < \eta$, which takes the maximum at $\varepsilon = 0$. We derive a condition on $\rho_{\max}$ from $\frac{\mathrm{d}}{\mathrm{d}\varepsilon}\Big|_{\varepsilon=0} H(\rho_\varepsilon) = 0$ and show $\frac{\mathrm{d}^2}{\mathrm{d}\varepsilon^2}\Big|_{\varepsilon=0} H(\rho_\varepsilon) < 0$ to check whether $H(\rho_{\max}) = H_{\max}$ is maximal. Note that

$$\frac{\mathrm{d}}{\mathrm{d}\varepsilon} H(\rho_\varepsilon) = -\int_{-\infty}^{\infty} \left\{ \left(\frac{\mathrm{d}}{\mathrm{d}\varepsilon}\rho_\varepsilon \right) \log \rho_\epsilon + \frac{\mathrm{d}}{\mathrm{d}\varepsilon}\rho_\varepsilon \right\} \mathrm{d}x \ ,$$

$$\frac{\mathrm{d}^2}{\mathrm{d}\varepsilon^2} H(\rho_\varepsilon) = -\int_{-\infty}^{\infty} \left\{ \left(\frac{\mathrm{d}^2}{\mathrm{d}\varepsilon^2}\rho_\varepsilon \right) \log \rho_\epsilon + \left(\frac{\mathrm{d}}{\mathrm{d}\varepsilon}\rho_\varepsilon \right)^2 \frac{1}{\rho_\varepsilon} + \frac{\mathrm{d}^2}{\mathrm{d}\varepsilon^2}\rho_\varepsilon \right\} \mathrm{d}x \ .$$

Since $\frac{\mathrm{d}}{\mathrm{d}\varepsilon}\rho_\varepsilon\Big|_{\varepsilon=0} = \phi(x) - \rho_{\max}(x)$, we have

$$\frac{\mathrm{d}}{\mathrm{d}\varepsilon} H(\rho_\varepsilon)\Big|_{\varepsilon=0} = -\int \{(\phi - \rho_{\max}) \log \rho_{\max} + (\phi - \rho_{\max})\} \, \mathrm{d}x \ .$$

Since $\frac{\mathrm{d}}{\mathrm{d}\varepsilon} H(\rho_\varepsilon)\big|_{\varepsilon=0} = 0$, and since

$$\int (\phi - \rho_{\max}) \, \mathrm{d}x = 1 - 1 = 0 \ , \tag{E.2}$$

we have

$$\int (\phi - \rho_{\max}) \log \rho_{\max} \, dx = 0 \,. \tag{E.3}$$

Since $\frac{d^2}{d\varepsilon^2}\rho_\varepsilon\Big|_{\varepsilon=0} = -2(\phi(x) - \rho_{\max}(x))$, (E.2) and (E.3) together imply

$$\frac{d^2}{d\varepsilon^2}H(\rho_\varepsilon)\Big|_{\varepsilon=0} = -\int_{-\infty}^{\infty} \frac{(\phi - \rho_{\max})^2}{\rho_{\max}} dx < 0 \,.$$

Theorem E.1 (Uniform Distribution) *Among the continuous probability distributions taking values in $[a, b]$, the uniform distribution has the maximal entropy.*

Proof Take s and t such that $a \le s < t \le b$. Choose a pdf $\phi(x)$ which is almost concentrated at s and t with probabilities p and $1 - p$, respectively. Since we will take the limit as ϕ converges to a discrete probability measure concentrated at s and t, we assume that $\phi(x)$ itself is such a discrete probability measure. By (E.3), we have

$$H_{\max} = -p \log \rho_{\max}(s) - (1 - p) \log \rho_{\max}(t) = -p \log \frac{\rho_{\max}(s)}{\rho_{\max}(t)} - \log \rho_{\max}(t)$$

for $0 < p < 1$. Hence $\log \frac{\rho_{\max}(s)}{\rho_{\max}(t)} = 0$ and $H_{\max} = -\log \rho_{\max}(t)$. Hence $\rho_{\max}(s) = \rho_{\max}(t)$ and $\rho_{\max}(t) = e^{-H_{\max}}$. Thus $\rho_{\max}(x) = e^{-H_{\max}}$. Since $\int_a^b \rho_{\max}(x)dx = 1$, we have $\rho_{\max}(x) = \frac{1}{b-a}$ and $H_{\max} = \log(b - a)$. □

Theorem E.2 (Exponential Distribution) *Among all the continuous probability distributions taking values in the interval $[0, \infty)$ with average $1/\lambda$, the exponential distribution has the maximal entropy.*

Proof Take s and t such that $0 \le s < \frac{1}{\lambda} < t < \infty$. Choose a pdf $\phi(x)$ which is almost concentrated at s and t with probabilities p and $1-p$, respectively. We assume that $\phi(x)$ itself is a discrete probability measure concentrated at s and t. Choose p in such a way that the average of ϕ satisfies $sp + t(1-p) = \frac{1}{\lambda}$, i.e., $p = (t - \frac{1}{\lambda})/(t - s)$, and define ρ_ε as in (E.1). By (E.3), we have $H_{\max} = p(\log \rho_{\max}(t) - \log \rho_{\max}(s)) - \log \rho_{\max}(t)$. Hence

$$\frac{\log \rho_{\max}(t) - (-H_{\max})}{t - \frac{1}{\lambda}} = \frac{\log \rho_{\max}(t) - \log \rho_{\max}(s)}{t - s} \,. \tag{E.4}$$

As s and t both converge to $\frac{1}{\lambda}$, the right-hand side of (E.4) converges to the limit $\rho'_{\max}(\frac{1}{\lambda})/\rho_{\max}(\frac{1}{\lambda})$. Hence $\rho_{\max}(\frac{1}{\lambda}) = e^{-H_{\max}}$ since the left-hand side of (E.4) also has to converge to a limit. Thus (E.4) is rewritten as

$$\frac{\log \rho_{\max}(t) - \log \rho_{\max}(\frac{1}{\lambda})}{t - \frac{1}{\lambda}} = \frac{\log \rho_{\max}(t) - \log \rho_{\max}(s)}{t - s} \,. \tag{E.5}$$

Note that the left- and right-hand sides of (E.5) are the average slopes of $y = \log \rho_{\max}(x)$ over the intervals $[s, t]$ and $[\frac{1}{\lambda}, t]$. Since they are equal, $y = \log \rho_{\max}(x)$ is a straight line. Therefore, $\rho_{\max}(x) = C\mathrm{e}^{-ax}$ for some $C > 0$ and $a > 0$. Since $\int_0^\infty C\mathrm{e}^{-ax}\mathrm{d}x = 1$, we have $C = a$. Since $\frac{1}{\lambda} = \int_0^\infty xa\mathrm{e}^{-ax}\mathrm{d}x = \frac{1}{a}$, we have $a = \lambda$, $\rho_{\max}(x) = \lambda\mathrm{e}^{-\lambda x}$ and $H_{\max} = -\log \rho_{\max}(\frac{1}{\lambda}) = \log \lambda + 1$. □

Theorem E.3 (Normal Distribution) *Among the continuous probability distributions taking values in $(-\infty, \infty)$ with average μ and variance σ^2, the normal distribution $N(\mu, \sigma^2)$ has the maximal entropy.*

Proof For notational simplicity we assume that $\mu = 0$ and $\sigma = 1$. Let $\rho_{\max}$ be the pdf such that $H(\rho_{\max}) = H_{\max}$ is maximal among all possible values $H(\rho)$ where ρ is a pdf satisfying

$$\int_{-\infty}^{\infty} x\rho(x)\mathrm{d}x = 0 \tag{E.6}$$

and

$$\int_{-\infty}^{\infty} x^2\rho(x)\mathrm{d}x = 1 \,. \tag{E.7}$$

First, note that $\rho_{\max}(-x) = \rho_{\max}(x)$. If not, let $\widetilde{\rho}(x) = \rho_{\max}(-x)$. Then

$$H(\widetilde{\rho}) = -\int_{-\infty}^{+\infty} \rho_{\max}(-x) \log \rho_{\max}(-x)\,\mathrm{d}x = \int_{+\infty}^{-\infty} \rho_{\max}(y) \log \rho_{\max}(y)\,\mathrm{d}y$$

and hence $H(\widetilde{\rho}) = H_{\max}$ and $\rho_{\max} = \widetilde{\rho}$ by the uniqueness.

Based on the fact that $\rho_{\max}$ is an even function, we take a symmetric set $J = [-b, -a] \cup [a, b]$, $0 < a < b$, such that $\rho_{\max}(x) \geq C$ for $x \in J$ for some $C > 0$. Choose $\phi(x) \geq 0$ such that

$$\{x : \phi(x) > 0\} \subset [a + \delta, b - \delta] \cup [-(b - \delta), -(a + \delta)]$$

for some small $\delta > 0$ such that $a + \delta < b - \delta$ with the properties

$$\int_{-\infty}^{\infty} \phi(x)\mathrm{d}x = 1 \,, \tag{E.8}$$

$$\int_{-\infty}^{\infty} x\phi(x)\mathrm{d}x = 0 \tag{E.9}$$

and

$$\int_{-\infty}^{\infty} x^2\phi(x)\mathrm{d}x = 1\ . \tag{E.10}$$

Define

$$\rho_\varepsilon(x) = \frac{\rho_{\max}(x) + \varepsilon\phi(x)}{1+\varepsilon}\ .$$

Then $\int_{-\infty}^{\infty} \rho_\varepsilon \mathrm{d}x = 1$ and $\rho_\varepsilon \geq 0$ for some $\eta > 0$, $-\eta < \varepsilon < \eta$, and ρ_ε satisfies (E.6) and (E.6). Define a mapping $\varepsilon \mapsto H(\rho_\varepsilon)$ and derive (E.2) and (E.3).

To find a condition on $\rho_{\max}(x)$ on $|x| \geq 1$, fix $a \geq 1$, and a pdf ϕ which is almost concentrated at $x = \pm a$ with probability $0 < p < \frac{1}{2}$ each, and at 0 with probability $1-2p$. Since we take the limit as ϕ converges to a discrete probability concentrated at $\pm a$ and 0, we assume that ϕ itself is such a discrete probability. Then p and a satisfy $a^2p + (-a)^2p + 0^2(1-2p) = 1$ by (E.10). Hence $2p = \frac{1}{a^2}$, and (E.3) implies $-H_{\max} = \int \phi \log \rho_{\max} \mathrm{d}x = \frac{1}{a^2} \log \rho_{\max}(a) + (1 - \frac{1}{a^2}) \log \rho_{\max}(0)$, and hence

$$\rho_{\max}(x) = \rho_{\max}(0)\mathrm{e}^{-(H_{\max} + \log \rho_{\max}(0))x^2}\ , \quad |x| \geq 1\ . \tag{E.11}$$

To find a condition on $\rho_{\max}(x)$ for $|x| \leq 1$, fix $a \geq 1$ and take $\phi(x)$ which is almost concentrated at $x = \pm a, \pm\frac{1}{a}$ with probabilities p, p, $\frac{1}{2} - p$, $\frac{1}{2} - p$. From (E.10), $a^2p + \frac{1}{a^2}(\frac{1}{2} - p) + (-a)^2p + \frac{1}{(-a)^2}(\frac{1}{2} - p) = 1$. Hence $a^2p + \frac{1}{a^2}(\frac{1}{2} - p) = \frac{1}{2}$, $p = \frac{1}{2}\frac{1}{a^2+1}$. Since (E.3) implies $-H_{\max} = 2\left\{\log \rho_{\max}(a) \times p + \log \rho_{\max}\left(\frac{1}{a}\right) \times (\frac{1}{2} - p)\right\}$, we have, for $|a| \geq 1$,

$$\begin{aligned}
&\log \rho_{\max}\left(\frac{1}{a}\right) \\
&= \frac{1}{\frac{1}{2} - p}\left(-\frac{1}{2}H_{\max} - p \log \rho_{\max}(a)\right) \\
&= \frac{2(a^2+1)}{a^2}\left[-\frac{1}{2}H_{\max} - \frac{1}{2}\frac{1}{a^2+1}\left\{\log \rho_{\max}(0) - (H_{\max} + \log \rho_{\max}(0))a^2\right\}\right] \\
&= \log \rho(0) - \left(\frac{1}{a^2}H_{\max} + \log \rho_{\max}(0)\right)\ ,
\end{aligned}$$

and hence, by taking $x = \frac{1}{a}$, we conclude that (E.11) holds for $|x| \leq 1$, too.

Since (E.8) and (E.10) hold for $\rho(x)$, we have

$$\int_{-\infty}^{\infty} \rho(0)\mathrm{e}^{-(H_{\max} + \log \rho(0))x^2}\mathrm{d}x = 1 \tag{E.12}$$

and

$$\int_{-\infty}^{\infty} x^2 \rho(0) e^{-(H_{\max} + \log \rho(0))x^2} dx = 1 . \tag{E.13}$$

Substituting $y = \sqrt{2}\sqrt{H_{\max} + \log \rho_{\max}(0)}x$ in (E.12), we obtain

$$\rho_{\max}(0) = \frac{\sqrt{H_{\max} + \log \rho_{\max}(0)}}{\sqrt{\pi}} . \tag{E.14}$$

Hence (E.13) becomes $\int_{-\infty}^{\infty} x^2 \rho_{\max}(0) \exp(-\pi \rho_{\max}(0)^2 x^2) \, dx = 1$. Taking $z = \sqrt{2\pi}\rho(0)x$, we have $2\pi\rho_{\max}(0)^2 = 1$, $\rho_{\max}(0) = \frac{1}{\sqrt{2\pi}}$, and $H_{\max} = \frac{1}{2} + \frac{1}{2}\log(2\pi)$ by (E.14). Thus $\rho_{\max}(x) = \frac{1}{\sqrt{2\pi}} e^{-\frac{1}{2}x^2}$. □

Consider a risky asset S. If the average of the return $\frac{\delta S}{S}$ over time δt is given by $\mu\delta t$ and the variance by $\sigma^2\delta t$, then by applying the maximal entropy principle we assume that $\frac{\delta S}{S}$ has the normal distribution with average $\mu\delta t$ and variance $\sigma^2\delta t$. Thus the maximal entropy principle gives another plausible explanation for hypothesis on the geometric Brownian motion of S.

Appendix F
MATLAB Programming

We present a brief introduction to the software MATLAB,[1] providing just enough to understand and modify the programs given in this book. For a more comprehensive introduction to various examples of MATLAB programs, see [70].

F.1 How to Start

The name of a MATLAB file ends with the extension '.m'. The command `clear all` removes all stored variables from the current workspace, freeing up system memory. The command `clc` clears the area on the screen called the Command Window, and the output is displayed in the same starting position on the screen. The command `clf` clears the current figure window. In the beginning of a MATLAB file it is convenient to include a line consisting of three commands `clear all; clc; clf;`. To save space the above three commands are usually not shown in MATLAB codes presented in this book.

If a command or a name of a variable ends without a semicolon (;) then the output is printed in the Command Window. If there is no need to see the output, just type a semicolon at the end of the line, then MATLAB will perform the command, but will not show the result on the screen. The following code lists a few basic mathematical operations.

```
1+4
9-2
5*6
2^7
```

[1]MATLAB® is a registered trademark of The MathWorks, Inc., and was developed to deal with mathematical problems mostly in applied mathematics. It stands for MATrix LABoratory, and as the name suggests, it is efficient in doing numerical computations involving vectors and matrices.

G.H. Choe, *Stochastic Analysis for Finance with Simulations*, Universitext,
DOI 10.1007/978-3-319-25589-7

produces

```
ans =
          5
ans =
          7
ans =
         30
ans =
        128
```

The following code

```
p=1/3
format long
p
a=exp(-1.2)
b=log(a)
get(0,'format') % Check the format for display of decimal numbers.
format short
b
```

produces

```
p =
         0.3333
p =
        0.333333333333333
a =
        0.301194211912202
b =
      -1.200000000000000

ans =
long

b =
        -1.2000
```

For trigonometric functions

```
pi
sin(pi)
cos(pi/2)
tan(pi/4)
```

produces

```
ans =
         3.1416
ans =
        1.2246e-16
ans =
        6.1232e-17
ans =
         1.0000
```

F.2 Random Numbers

To generate 2×5 matrices A, B, C consisting of random numbers with the uniform distribution in $(0, 1)$, with the standard normal distribution, and with the uniform distribution in the set of integers $-1 \leq n \leq 4$, respectively, we use

```
A = rand(2,5)
B = randn(2,5)
C = randi([-1,3],2,5)
```

produces

```
A =
     0.4363    0.0261    0.4306    0.7624    0.6800
     0.1739    0.9547    0.9616    0.0073    0.7060
B =
     0.0831   -0.5279   -0.8499    0.7253   -0.3864
     0.1578    0.7231   -0.7964    1.6865   -0.5051
C =
      2     3     0     1    -1
      3     1     3     3     1
```

F.3 Vectors and Matrices

MATLAB can process a given group of objects more efficiently if they are expressed as vectors or matrices.

```
u = [1 2 3 4]
v = transpose(u)
w = zeros(4,1)
v + w
3*v
A = ones(3,4)
A*v
B = [1 2 3; 0 1 4; 2 3 1; -1 0 7]
C = A*B
```

produces

```
u =
          1     2     3     4
v =
          1
          2
          3
          4
w =
          0
          0
          0
          0
```

```
ans =
         1
         2
         3
         4
ans =
         3
         6
         9
        12
A =
         1     1     1     1
         1     1     1     1
         1     1     1     1
ans =
        10
        10
        10
B =
         1     2     3
         0     1     4
         2     3     1
        -1     0     7
C =
         2     6    15
         2     6    15
         2     6    15
```

There are two ways to compute the inverse of a matrix.

```
D = rand(3,3)
det(D) % Check the determinant to see whether the matrix is invertible.
I = diag(ones(3,1)) % the identity matrix
I / D
D \ I
inv(D)
```

produces

```
D =
        0.8842    0.3990    0.7360
        0.0943    0.0474    0.7947
        0.9300    0.3424    0.5449
ans =
        0.0480
I =
         1     0     0
         0     1     0
         0     0     1
ans =
       -5.1301    0.7197    5.8793
       14.3272   -4.2228  -13.1923
       -0.2460    1.4249    0.0894
ans =
       -5.1301    0.7197    5.8793
       14.3272   -4.2228  -13.1923
```

```
        -0.2460    1.4249    0.0894
ans =
        -5.1301    0.7197    5.8793
        14.3272   -4.2228  -13.1923
        -0.2460    1.4249    0.0894
```

It is better to use `E/D` instead of `E*inv(D)` when we solve a linear system of equations $XD = E$, and use `D\E` instead of `inv(D)*E` when we solve a linear system of equations $DX = E$.

```
E = rand(3,3)
det(E)

E / D
E*inv(D)

D \ E
inv(D)*E
```

produces

```
E =
         0.6862    0.3037    0.7202
         0.8936    0.0462    0.7218
         0.0548    0.1955    0.8778
ans =
        -0.1712
ans =
         0.6531    0.2377    0.0929
        -4.1002    1.4765    4.7091
         2.3036    0.4647   -2.1782
ans =
         0.6531    0.2377    0.0929
        -4.1002    1.4765    4.7091
         2.3036    0.4647   -2.1782
ans =
        -2.5551   -0.3753    1.9857
         5.3352    1.5768   -4.3100
         1.1094    0.0086    0.9297
ans =
        -2.5551   -0.3753    1.9857
         5.3352    1.5768   -4.3100
         1.1094    0.0086    0.9297
```

The following defines a tridiagonal matrix.

```
a = -1;
b = 2;
c = 3;
m = 6;
T = diag(a*ones(m,1)) + diag(b*ones(m-1,1),1) + diag(c*ones(m-1,1),-1)
```

produces

```
T =
    -1     2     0     0     0     0
     3    -1     2     0     0     0
     0     3    -1     2     0     0
     0     0     3    -1     2     0
     0     0     0     3    -1     2
     0     0     0     0     3    -1
```

F.4 Tridiagonal Matrices

The following produces a tridiagonal matrix. Consult Lemma B.1 for the interpretation of the output.

```
a = -1;
b = 2;
c = 2;
n = 6;
%The following produces a tridiagonal matrix.
T = diag(a*ones(n,1)) + diag(b*ones(n-1,1),1) + diag(c*ones(n-1,1),-1)

rank(T) % rank
det(T) % determinant
eig(T) % eigenvalues

lambda(1:n,1) = a + 2*b*cos([1:n]*pi/(n+1))
```

produces

```
T =
        -1     2     0     0     0     0
         2    -1     2     0     0     0
         0     2    -1     2     0     0
         0     0     2    -1     2     0
         0     0     0     2    -1     2
         0     0     0     0     2    -1
ans =
         6
ans =
        13
ans =
       -4.6039
       -3.4940
       -1.8901
       -0.1099
        1.4940
        2.6039
lambda =
        2.6039
        1.4940
```

```
-0.1099
-1.8901
-3.4940
-4.6039
```

In the last sentence of the code, `lambda(1:n,1)` and `cos([1:n]*pi/(n+1))` are regarded as vectors.

F.5 Loops for Iterative Algorithms

As an example of an iterative algorithm, we present a method of testing the uniform distribution of `M` pseudorandom numbers generated by the command `rand`. First, we divide the unit interval into `N` subintervals of equal length, $[0, \frac{1}{N}), [\frac{1}{N}, \frac{2}{N}), \ldots, [\frac{N-1}{N}, 1)$, and count how many times the pseudorandom numbers belong to each subinterval.

```
M = 500;
N = 5;
x = rand(M,1);
width = 1/N;
bin = zeros(1,N);  % initial values for each bin

for i = 1:M
j = ceil(x(i)/width);
bin(j) = bin(j) + 1;
end

bin
```

produces

```
bin =
        95    104    111     95     95
```

Note that `ceil(x(i)/width)` will find to which bin the ith random number `x(i)` belongs, where `ceil(x)` for 'ceiling' is the smallest integer greater than or equal to `x`.

For the second example of an iterative algorithm, consider the Newton method for computing $\sqrt{5}$. Take $f(x) = x^2 - 5$.

```
format long
x0 = 2.0; % Choose a suitable starting point.
f =  x0^2 - 5;
fprime =  2*x0;
x = x0 - f/fprime

while abs(x - x0) > 10^(-10)
    x0 = x;
    f =  x^2 - 5;
    fprime =  2*x;
```

```
    x = x - f/fprime
end
```

produces

```
x =
        2.250000000000000
x =
        2.236111111111111
x =
        2.236067977915804
x =
        2.236067977499790
x =
        2.236067977499790
```

F.6 How to Plot Graphs

As an example, let use plot the standard normal density function $y = 1/(\sqrt{2\pi})\exp(-x^2/2)$ on the interval $-5 \le x \le 5$. First, take 1001 equally spaced points in the interval where the function is evaluated.

```
x = -5:0.01:5;
```

Without the semicolon (;) at the end of the sentence, there would be a display of $1001 = \frac{5-(-5)}{0.01} + 1$ points starting from -5 to 5 with equal increment 0.01 because x is regarded as a *vector* by MATLAB, i.e., x $= -5.0000, -4.9900, -4.9800, -4.9700, -4.9600$, and so on. To evaluate the function we use

```
y = 1/(sqrt(2*pi))*exp(-x.^2/2);
```

where

```
x.^2
```

means that x^2 is computed for each entry x of the *vector* x.

```
plot(x,y,'Color','b','LineWidth',1.5)
```

produces a graph where b means the color blue. Other abbreviations for available colors are k for black, r for red, g for green, and y for yellow. The line width can be changed to 1 or 2, or any positive number.
For a specification of the range of the plot, we use

```
set(gca,'xlim',[-5 5],'ylim',[0 0.5],'xtick',[-4:1:4],'ytick',[0:0.1:0.5]);
```

where gca means the current axes for plotting a graph, and xlim and ylim specify the ranges for the axes, and xtick and ytick specify the locations where tick marks are drawn.

F.7 Curve Fitting

For a set of two-dimensional data we find the best fitting polynomial in the least squares sense. The command `polyfit(X,Y,n)` finds the polynomial of degree n that best fits the data. In the following we find the best fitting quadratic polynomial $Y = aX^2 + bX + c$ (Fig. F.1).

```
x = 0:0.1:10;
X = [1.2 2.4 3.2 4.7 5.6 7.0 9.0];
Y = [1.9 2.4 2.5 3.0 4.2 5.7 7.3];
p = polyfit(X,Y,2)
f = polyval(p,x);
plot(x,f)
hold on
plot(X,Y,'o')
```

The coefficients a, b, c are given below:

```
p =
    0.0631     0.0722     1.7048
```

For a linear fit see Fig. B.1.

F.8 How to Define a Function

A simple method of defining a function such as $f(x) = \frac{\sin x}{x}$ is to use the command `inline` as follows:

```
f = inline('sin(x)/x')
```

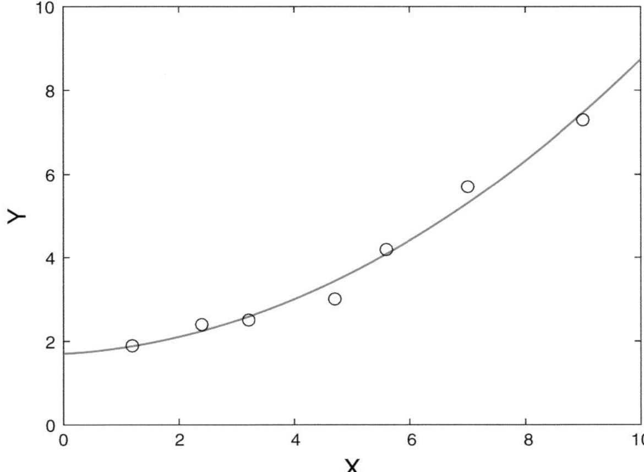

Fig. F.1 Polynomial fitting of data

```
x = 0.1;
f(x)
```

which produces

```
f =
      Inline function:
      f(x) = sin(x)/x
ans =
    0.9983
```

For a complicated definition such as the Black–Scholes–Merton formula we define a separate MATLAB file and call it from the inside of another file. Here is an example. Define a function `BlackScholes()` of five variables by

$$\texttt{BlackScholes}(S, K, T, r, \sigma) = SN(d_1) - Ke^{-rT}N(d_2)$$

where

$$d_{1,2} = \frac{\log \frac{S}{K} + (r \pm \frac{1}{2}\sigma^2)T}{\sigma\sqrt{T}} .$$

(We may choose any function name instead of `BlackScholes`, however, an informative name will be helpful to avoid confusion with other functions. The function name itself does not matter. Only the name of the file containing the definition is called from another file.) Save the definition of `BlackScholes()` as a separate MATLAB file with a title such as BSM.m as follows:

```
function [Call_price, Put_price] = BlackScholes(S,K,T,r,sigma)
% an example of a function file
d1 = (log(S/K)+(r + 0.5*sigma^2)*T)/(sigma*sqrt(T));
d2 = d1 - sigma*sqrt(T);
N1 = 0.5*( 1+erf(d1/sqrt(2)) );
N2 = 0.5*( 1+erf(d2/sqrt(2)) );
Call_price = S*N1 - K*exp(-r*T)*N2;
Put_price = Call_price +K*exp(-r*T) - S;
```

Input variables, `S,K,T,r,sigma`, are written inside a pair of parentheses, and outputs, `Call_price, Put_price`, are written between a pair of brackets. From another MATLAB file we call the function file name `BSM` as follows:

```
S = 100;
K = 110;
T = 1;
r = 0.05;
sigma = 0.3;
[Call_price, Put_price] = BSM(S,K,T,r,sigma)
```

Then we obtain the following output in the command window:

```
Call_price =
               10.0201

Put_price =
               14.6553
```

F.9 Statistics

Compute the cumulative probability $\Pr(Z < a)$ where $Z \sim N(0, 1)$ using the command `normcdf(a)`.

```
normcdf(0)
normcdf(1)
normcdf(2)
normcdf(3)
```

produces

```
ans =
        0.5000
ans =
        0.8413
ans =
        0.9772
ans =
        0.9987
```

The command `mean(X)` computes the average of sample values of a random variable X while `mean(X > a)` gives the probability of the event $\{X > a\}$. That is, `mean(X)` $= \mathbb{E}[X]$ and `mean(X > a)` $= \Pr(X > a) = \mathbb{E}\left[\mathbf{1}_{\{X>a\}}\right]$.

```
X = randn(100000,1);
mean(X.^2)
1 - mean(X > 1)
1 - mean(X > 2)
1 - mean(X > 3)
```

produces

```
ans =
        1.0043
ans =
        0.8418
ans =
        0.9770
ans =
        0.9986
```

The command `norminv(u)` gives the inverse cumulative probability of the standard normal distribution. That is, $u = \Pr(Z \leq \texttt{norminv(u)})$, $0 \leq u \leq 1$.

```
norminv(0)
norminv(0.5)
norminv(0.99)
norminv(1)
normcdf(norminv(0.25))
```

produces

```
ans =
       -Inf
ans =
          0
ans =
         2.3263
ans =
        Inf
ans =
         0.2500
```

Note that `-Inf` and `Inf` represent $-\infty$ and $+\infty$, respectively.

Solutions for Selected Problems

Problems of Chap. 1

1.1 Let K be the closed cone given by all convex linear combinations of $\mathbf{d}_1, \ldots, \mathbf{d}_M$ with nonnegative coefficients. Then either $\mathbf{d}_0 \in K$ or $\mathbf{d}_0 \notin K$. If $\mathbf{d}_0 \in K$, then (i) has a solution. In this case, $\boldsymbol{\pi} \cdot \mathbf{d}_0 = \lambda_1 \boldsymbol{\pi} \cdot \mathbf{d}_1 + \cdots + \lambda_M \boldsymbol{\pi} \cdot \mathbf{d}_M$, and (ii) does not hold. If $\mathbf{d}_0 \notin K$, then there exists a hyperplane H in $\mathbb{R}^N$ that separates $\mathbf{d}_0$ and K. Choose a normal vector $\boldsymbol{\beta} \neq \mathbf{0}$ to H and pointing in the direction where K is located. (See Fig. 1.) Then $\boldsymbol{\pi}$ satisfies the second statement, but not the first statement. (*Remark.* Suppose that the market is arbitrage free. If we take $\mathbf{d}_0 = \mathbf{S}_0$, then case (ii) does not happen and case (i) holds. Thus $\mathbf{S}_0 = \lambda_1 \mathbf{d}_1 + \cdots + \lambda_M \mathbf{d}_M$ for some $\lambda_1, \ldots, \lambda_M \geq 0$.)

1.2

(i) Consider the equation $\begin{bmatrix} C^u \\ C^d \end{bmatrix} = D^t \begin{bmatrix} a \\ b \end{bmatrix}$. Unless $(S_T^2(\omega_u), S_T^2(\omega_d))$ is a constant multiple of $(S_T^1(\omega_u), S_T^1(\omega_d))$, there is a unique solution for every (C^u, C^d). That is, unless S^1 and S^2 are essentially the same asset with a different number of units, a unique solution exists and the market is complete.

(ii) The given system of equations becomes

$$\begin{cases} C^u = auS_0 + b(1+r)B_0 \\ C^d = adS_0 + b(1+r)B_0 \end{cases} .$$

Since $d \neq u$, the determinant of $D^t = \begin{bmatrix} uS_0 & b(1+r) \\ dS_0 & b(1+r) \end{bmatrix}$ is nonzero, and hence there exists a unique solution.

G.H. Choe, *Stochastic Analysis for Finance with Simulations*, Universitext,
DOI 10.1007/978-3-319-25589-7

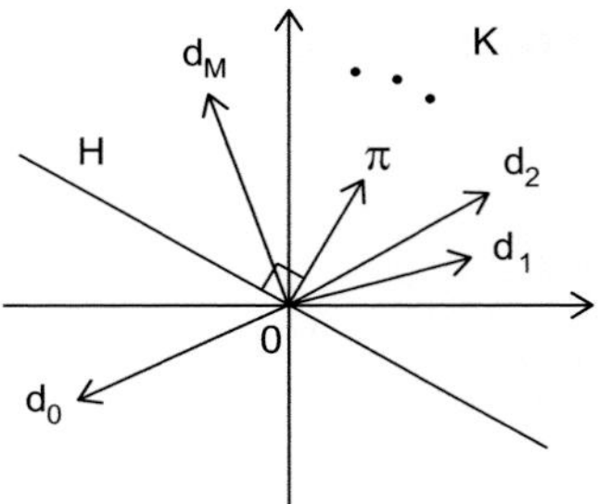

Fig. 1 The closed convex cone K in the proof of Farkas' lemma

Problems of Chap. 2

2.1 The payoff of the portfolio is given by $-S_T + (S_T - K)^+$ which can be written as a piecewise linear function $f(S_T)$ defined by $f(S_T) = -S_T$, $0 \le S_T \le K$, and $f(S_T) = -K$, $S_T \ge K$. The maximum loss is equal to $-K$.

2.2 A call option provides insurance if the holder of the call option has a short position in the asset on which the option is written. A put option provides insurance if the holder of the put option owns the asset on which the option is written.

2.3 A strangle is a sum of a put option with payoff $(K_1 - S_T)^+$ and a call option with payoff $(S_T - K_2)^+$.

Problems of Chap. 3

3.7 Since $A = \bigcup_{k=1}^{\infty} \left\{ x = \sum_{n=1}^{k} a_n 2^{-n} : a_n = 0, 1 \right\}$, A has countably many points, and its Lebesgue measure is 0.

3.8 Show that $\frac{1}{4} = 0.0202020202\ldots$ in the ternary expansion. If we consider an experiment with three outcomes represented by three symbols '0', '1', '2', each having probability $\frac{1}{3}$, then the unit interval is regarded as the set of every possible outcome represented by a number x in the ternary expansion $x = a_1 a_2 a_3 \ldots$ where $a_i \in \{0, 1, 2\}$. The set A consists of the results of obtaining either 0 or 2 in each trial, and hence the probability of A is $\frac{2}{3} \times \frac{2}{3} \times \frac{2}{3} \times \cdots = 0$.

3.9 Use the sequence constructed in the construction of the Lebesgue integral in Sect. 3.3.

3.13 If we take $\phi(x) = \frac{1}{x}$ in Jensen's inequality, then $\frac{1}{\int X \, d\mathbb{P}} \le \int \frac{1}{X} \, d\mathbb{P}$ and hence $\int X \, d\mathbb{P} \int \frac{1}{X} \, d\mathbb{P} \ge 1$. Since $Y \ge \frac{1}{X}$, we conclude that $\int X \, d\mathbb{P} \int Y \, d\mathbb{P} \ge 1$.

Problems of Chap. 4

4.1 Note that

$$\sum_{j=0}^{\infty} \mathbb{P}(\{X \geq j\}) = \sum_{j=0}^{\infty} \sum_{k=j}^{\infty} \mathbb{P}(\{k \leq X < k+1\}) .$$

4.2 Since $\phi(X) = 1$, we have $0 = \phi(\mathbb{E}[X]) < \mathbb{E}[\phi(X)] = 1$.

4.3 $F(x) = \begin{cases} 0, \ x < a \ , \\ 1, \ x \geq a \ . \end{cases}$

4.4 Substituting $t = \sqrt{x}$, we have

$$\int_0^1 \mathrm{e}^{\sqrt{x}} \mathrm{d}x = \int_0^1 \mathrm{e}^t 2t \, \mathrm{d}t = \left[\mathrm{e}^t 2t\right]_0^1 - \int_0^1 \mathrm{e}^t 2 \, \mathrm{d}t = 2\mathrm{e} - 2(\mathrm{e} - 1) = 2 .$$

4.5 Let $f(x)$ denote the pdf of the standard normal distribution. Since $f(-x) = f(x)$, we have $N(\alpha) + N(-\alpha) = \int_{-\infty}^{\alpha} f(x)\mathrm{d}x + \int_{-\infty}^{-\alpha} f(x)\mathrm{d}x = \int_{-\infty}^{\alpha} f(x)\mathrm{d}x + \int_{\alpha}^{\infty} f(x)\mathrm{d}x = \int_{-\infty}^{\infty} f(x)\mathrm{d}x = 1$.

4.6 Since $\Pr(F^{-1}(U) \leq x) = \Pr(U \leq F(x)) = F(x)$, we see that $F^{-1}(U)$ and X have the same distribution function F.

4.8

$$\begin{aligned}
\mathbb{E}[\mathrm{e}^X] &= \int_{-\infty}^{\infty} \mathrm{e}^x \frac{1}{\sqrt{2\pi}\sigma} \mathrm{e}^{-(x-\mu)^2/2\sigma^2} \mathrm{d}x \qquad (\text{Take } z = \frac{x-\mu}{\sigma}.) \\
&= \int_{-\infty}^{\infty} \mathrm{e}^{\sigma z + \mu} \frac{1}{\sqrt{2\pi}} \mathrm{e}^{-z^2/2} \mathrm{d}z \\
&= \int_{-\infty}^{\infty} \mathrm{e}^{\mu + \sigma^2/2} \frac{1}{\sqrt{2\pi}} \mathrm{e}^{-(z-\sigma)^2/2} \mathrm{d}z \\
&= \mathrm{e}^{\mu + \sigma^2/2} \int_{-\infty}^{\infty} \frac{1}{\sqrt{2\pi}} \mathrm{e}^{-z^2/2} \mathrm{d}z = \mathrm{e}^{\mu + \sigma^2/2} .
\end{aligned}$$

Since $2X$ has mean 2μ and variance $4\sigma^2$, we have $\mathbb{E}[\mathrm{e}^{2X}] = \mathrm{e}^{2\mu + 2\sigma^2}$. Thus

$$\begin{aligned}
\mathrm{Var}(\mathrm{e}^X) &= \mathbb{E}[(\mathrm{e}^X - \mathrm{e}^{\mu + \sigma^2/2})^2] \\
&= \mathbb{E}[\mathrm{e}^{2X} - 2\mathrm{e}^{\mu + \sigma^2/2} \mathrm{e}^X + \mathrm{e}^{2\mu + \sigma^2}] \\
&= \mathrm{e}^{2\mu + 2\sigma^2} - 2\mathrm{e}^{\mu + \sigma^2/2} \mathrm{e}^{\mu + \sigma^2/2} + \mathrm{e}^{2\mu + \sigma^2} \\
&= \mathrm{e}^{2\mu + \sigma^2} (\mathrm{e}^{\sigma^2} - 1) .
\end{aligned}$$

4.9 Since αZ is normal with mean 0 and variance α^2, by Exercise 4.8 we have

$$||e^{\alpha Z}||_p^p = \mathbb{E}[e^{\alpha p Z}] = e^{\alpha^2 p^2/2} < \infty\,.$$

4.10

$$\begin{aligned}
\mathbb{E}[e^{\alpha Z^2}] &= \int_{-\infty}^{\infty} \frac{1}{\sqrt{2\pi}} e^{\alpha x^2} e^{-x^2/2}\,dx \\
&= \int_{-\infty}^{\infty} \frac{1}{\sqrt{2\pi}} e^{-(1-2\alpha)x^2/2}\,dx \qquad (z = \sqrt{1-2\alpha}\,x.) \\
&= \frac{1}{\sqrt{1-2\alpha}} \int_{-\infty}^{\infty} \frac{1}{\sqrt{2\pi}} e^{-z^2/2}\,dz = \frac{1}{\sqrt{1-2\alpha}}\,.
\end{aligned}$$

4.12 Note that

$$\mathbb{E}[Y^4] = \mathbb{E}\Big[\sum_i \sum_j \sum_k \sum_\ell (U_i - \frac{1}{2})(U_j - \frac{1}{2})(U_k - \frac{1}{2})(U_\ell - \frac{1}{2})\Big]\,.$$

There are five cases: (i) all of i, j, k, ℓ are distinct, (ii) there exists only one equal pair, (iii) there are two equal pairs, (iv) one equal triple, and (v) all are equal. The cases (i),(ii),(iv) produce zero, while (iii),(v) yield $\frac{1}{144}\binom{12}{2}\binom{4}{2}\binom{2}{2} = \frac{11}{4}$ and $\frac{1}{80}\binom{12}{1}\binom{4}{4} = \frac{3}{20}$, respectively. Hence $\mathbb{E}[Y^4] = \frac{11}{4} + \frac{3}{20} = \frac{29}{10}$.

4.13 Since $h(x) = x^2$, we have $h^{-1}(y) = \pm\sqrt{y}$ for $0 \le y \le 1$ and

$$f_Y(y) = \frac{1}{|-2\sqrt{y}|} f_X(-\sqrt{y}) + \frac{1}{|2\sqrt{y}|} f_X(\sqrt{y}) = \frac{1}{2\sqrt{y}}\,, \qquad 0 < y \le 1\,.$$

4.14 Since X and Y are independent, the joint pdf is the product of f_X and f_Y, i.e., $f_{X,Y}(x, y) = \frac{1}{2\pi} e^{-(x^2+y^2)/2}$. Hence for $-\frac{\pi}{2} < \theta_1 < \theta_2 < \frac{\pi}{2}$,

$$\begin{aligned}
\Pr\left(\theta_1 < \tan^{-1}\frac{Y}{X} < \theta_2\right) &= \Pr\left(\tan\theta_1 < \frac{Y}{X} < \tan\theta_2\right) \\
&= \iint_D \frac{1}{2\pi} e^{-(x^2+y^2)/2}\,dxdy \\
&= \frac{1}{\pi} \int_0^{\infty} \int_{\theta_1}^{\theta_2} e^{-r^2/2} r\,drd\theta = \frac{\theta_2 - \theta_1}{\pi}
\end{aligned}$$

where $D = \{(x, y) : \theta_1 < \tan^{-1}\frac{y}{x} < \theta_2\}$. (Note that D has two pieces!)

4.15 By the chain rule and the invariance of probability, we have

$$\begin{aligned}\iint_Q f_{V,W}(\Phi(x,y))|J_\Phi(x,y)|\mathrm{d}x\mathrm{d}y &= \iint_{\Phi(Q)} f_{V,W}(v,w)\mathrm{d}v\mathrm{d}w \\ &= \iint_Q f_{X,Y}(x,y)\mathrm{d}x\mathrm{d}y\end{aligned}$$

for any square $Q \subset \mathbb{R}^2$. Thus $f_{V,W}(\Phi(x,y))|J_\Phi(x,y)| = f_{X,Y}(x,y)$.

4.17 Take $X \sim N(0,1)$ and $Y = X^2$. Then X and Y are not independent, and $\mathrm{Cov}(X,Y) = \mathbb{E}[XY] - \mathbb{E}[X]\mathbb{E}[Y] = \mathbb{E}[X^3] - \mathbb{E}[X]\mathbb{E}[X^2] = 0$.

4.18 Define $\begin{bmatrix} X_1 \\ X_2 \end{bmatrix} = \begin{bmatrix} a_{11} & a_{12} \\ a_{21} & a_{22} \end{bmatrix} \begin{bmatrix} Z_1 \\ Z_2 \end{bmatrix}$ where the vectors $\mathbf{v}_1 = (a_{11}, a_{12})$, $\mathbf{v}_2 = (a_{21}, a_{22})$ are of length 1 so that X_1, X_2 have variances equal to 1. A necessary and sufficient condition for $\mathbb{E}[X_1X_2] = \rho$ is that $\mathbf{v}_1 \cdot \mathbf{v}_2 = \rho$. Now we choose $\mathbf{v}_1 = (1,0)$ and find $\mathbf{v}_2$.

4.19

(i) Let $f_X(x)$ denote the pdf of X. Then

$$\begin{aligned}\int_0^\infty x^2 f_X(x)\mathrm{d}x &= \int_0^\infty \left(\int_0^x 2t\,\mathrm{d}t\right) f_X(x)\mathrm{d}x \\ &= \int_0^\infty \left(\int_0^\infty 2t\,\mathbf{1}_{[0,x]}(t)\,\mathrm{d}t\right) f_X(x)\mathrm{d}x \\ &= \int_0^\infty \int_0^\infty 2t\,\mathbf{1}_{[0,x]}(t) f_X(x)\mathrm{d}x\,\mathrm{d}t \quad \text{(by Fubini's theorem)} \\ &= \int_0^\infty \int_0^\infty 2t\,\mathbf{1}_{[t,\infty)}(x) f_X(x)\mathrm{d}x\,\mathrm{d}t \\ &= 2\int_0^\infty t\left(\int_0^\infty \mathbf{1}_{[t,\infty)}(x) f_X(x)\mathrm{d}x\right)\mathrm{d}t \\ &= 2\int_0^\infty t\,\mathbb{P}(X > t)\,\mathrm{d}t\,.\end{aligned}$$

The formula has a geometric interpretation: Consider the volume obtained by rotating the graph $y = 1 - F_X(x)$, $x \geq 0$, around the y-axis. The left-hand side is obtained by adding horizontal disks of radius x with thickness $|\mathrm{d}y| = f_X(x)\mathrm{d}x$, while the right-hand side is obtained by adding the volumes of infinitesimally thin vertical cylinders.

(ii) Note that

$$\begin{aligned}\sum_{n=0}^{\infty} n^2 \mathbb{P}(X=n) &= \sum_{n=1}^{\infty}\left(\sum_{k=1}^{n}(2k-1)\right)\mathbb{P}(X=n)\\ &= \sum_{k=1}^{\infty}(2k-1)\sum_{n=k}^{\infty}\mathbb{P}(X=n)\\ &= \sum_{k=2}^{\infty}(2k-1)\mathbb{P}(X\ge k)\ .\end{aligned}$$

4.21 Since the joint pdf of X and Y is equal to $f_X(x)f_Y(y)$, we have

$$\mathbb{P}(XY\le a) = \int_0^{\infty}\int_{-\infty}^{a/y} f_X(x)f_Y(y)\,\mathrm{d}x\mathrm{d}y + \int_{-\infty}^{0}\int_{a/y}^{\infty} f_X(x)f_Y(y)\,\mathrm{d}x\mathrm{d}y\ .$$

Now we differentiate the both sides with respect to a, and obtain

$$\begin{aligned}f_{XY}(a) &= \int_0^{\infty} f_X\left(\frac{a}{y}\right)\frac{1}{y}f_Y(y)\,\mathrm{d}y - \int_{-\infty}^{0} f_X\left(\frac{a}{y}\right)\frac{1}{y}f_Y(y)\,\mathrm{d}y\\ &= \int_0^{\infty} f_X\left(\frac{a}{y}\right)\frac{1}{|y|}f_Y(y)\,\mathrm{d}y + \int_{-\infty}^{0} f_X\left(\frac{a}{y}\right)\frac{1}{|y|}f_Y(y)\,\mathrm{d}y\\ &= \int_{-\infty}^{\infty} f_X\left(\frac{a}{y}\right)\frac{1}{|y|}f_Y(y)\,\mathrm{d}y\ .\end{aligned}$$

4.22 By Exercise 4.21, we have

$$f_{XY}(a) = \int_0^1 \frac{1}{y}f_X\left(\frac{a}{y}\right)\mathrm{d}y = \int_a^1 \frac{1}{y}\,\mathrm{d}y = -\log a$$

for $0\le a\le 1$, and $f_{XY}(a) = 0$ elsewhere.

4.23

(i) $\mathbb{E}[\mathbf{X}] = \boldsymbol{\mu} + P\sqrt{D}\,\mathbb{E}[\mathbf{Z}] = \boldsymbol{\mu}$ and $\mathrm{Var}(\mathbf{X}) = \mathrm{Var}(P\sqrt{D}\,\mathbf{Z}) = \mathbb{E}[P\sqrt{D}\,\mathbf{Z}(P\sqrt{D}\mathbf{Z})^t] = \mathbb{E}[P\sqrt{D}\,\mathbf{Z}\mathbf{Z}^t\sqrt{D}^tP^t] = P\sqrt{D}\,\mathbb{E}[\mathbf{Z}\mathbf{Z}^t]\sqrt{D}^tP^t = P\sqrt{D}\,\mathrm{Var}(\mathbf{Z})\sqrt{D}^tP^t = P\sqrt{D}\sqrt{D}^tP^t = PDP^t = \Sigma$. Thus $\mathbf{X}\sim N(\boldsymbol{\mu},\Sigma)$.

(ii) Use the fact that the columns of $P\sqrt{D}$ are $\sqrt{\lambda_1}\mathbf{v}_1,\dots,\sqrt{\lambda_n}\mathbf{v}_n$.

4.24

(i)

$$
\begin{aligned}
f_{Z_1^2+Z_2^2}(x) &= \int_{-\infty}^{\infty} f_{Z_1^2}(x-y) f_{Z_2^2}(y) \mathrm{d}y \\
&= \int_0^x f_{Z_1^2}(x-y) f_{Z_2^2}(y) \mathrm{d}y \\
&= \int_0^x \frac{1}{\sqrt{2\pi(x-y)}} \mathrm{e}^{-(x-y)/2} \frac{1}{\sqrt{2\pi y}} \mathrm{e}^{-y/2} \mathrm{d}y \\
&= \frac{1}{2\pi} \mathrm{e}^{-x/2} \int_0^x \frac{1}{\sqrt{y(x-y)}} \mathrm{d}y = \frac{1}{2} \mathrm{e}^{-x/2} .
\end{aligned}
$$

For the last equality we used the fact that

$$
\int_0^x \frac{1}{\sqrt{y(x-y)}} \mathrm{d}y = \int_0^1 \frac{1}{\sqrt{u(1-u)}} \mathrm{d}u = \pi. \quad \text{(Substitute } y = xu.\text{)}
$$

(ii) Use the convolution!

Problems of Chap. 5

5.2

(i) $\mathcal{G} = \{\emptyset, \{a,b\}, \{c,d\}, \Omega\}$.
(ii) For a Borel subset $A \subset \mathbb{R}^1$, we have $Z^{-1}(A) = \Omega \in \mathcal{G}$ if $1 \in A$. Otherwise, $Z^{-1}(A) = \emptyset \in \mathcal{G}$. Hence Z is $\mathcal{G}$-measurable.
(iii) Since $Y^{-1}(\{5\})$ is $\mathcal{G}$-measurable and contains a, $Y^{-1}(\{5\}) = \{a,b\}$ or $Y^{-1}(\{5\}) = \Omega$. In either case, $Y(b) = 5$.
(iv) Since $\mathcal{H}$ contains $X^{-1}(\{0\}) = \{a\}$, $X^{-1}(\{3\}) = \{b,c\}$, $X^{-1}(\{1\}) = \{d\}$, we have $\mathcal{H} = \sigma(X) = \{\emptyset, \{a\}, \{b,c\}, \{d\}, \{a,b,c\}, \{a,d\}, \{b,c,d\}, \Omega\}$.
(v) Note that W is $\mathcal{H}$-measurable since $W^{-1}(\{10\}) = \{a\} \in \mathcal{H}$, $W^{-1}(\{20\}) = \{b,c,d\} \in \mathcal{H}$. Hence $\mathbb{E}[W|X] = W$.

5.6 As $\delta y \to 0$, we have

$$
\begin{aligned}
\mathbb{E}[X|Y] &\approx \mathbb{E}[X | y \leq Y \leq y + \delta y] \\
&= \frac{\iint_{[0,1]\times[y,y+\delta y]} X \,\mathrm{d}\mathbb{P}}{\mathbb{P}([0,1] \times [y, y+\delta y])} \\
&= \frac{\iint_{[0,1]\times[y,y+\delta y]} x(x+y) \,\mathrm{d}x\mathrm{d}y}{\iint_{[0,1]\times[y,y+\delta y]} (x+y) \,\mathrm{d}x\mathrm{d}y}
\end{aligned}
$$

$$\approx \frac{\int_{[0,1]} x(x+y)\,\mathrm{d}x\,\delta y}{\int_{[0,1]} (x+y)\,\mathrm{d}x\,\delta y}$$

$$\to \frac{\int_{[0,1]} x(x+y)\,\mathrm{d}x}{\int_{[0,1]} (x+y)\,\mathrm{d}x} = \frac{\frac{1}{2}y + \frac{1}{3}}{y + \frac{1}{2}}$$

where the symbol '$\approx$' represents approximation and '$\to$' denotes convergence in a suitable sense, respectively. Hence $\mathbb{E}[X|Y] = \frac{3Y+2}{6Y+3}$.

5.7 First, we have

$$\begin{aligned}
\mathrm{Var}(X|Y) &= \mathbb{E}[(X - \mathbb{E}[X|Y])^2|Y] \\
&= \mathbb{E}[X^2|Y] - 2\mathbb{E}[X|Y]^2 + \mathbb{E}[X|Y]^2 \\
&= \mathbb{E}[X^2|Y] - \mathbb{E}[X|Y]^2 .
\end{aligned}$$

Hence

$$\mathbb{E}[\mathrm{Var}(X|Y)] = \mathbb{E}[\mathbb{E}[X^2|Y]] - \mathbb{E}[\mathbb{E}[X|Y]^2] = \mathbb{E}[X^2] - \mathbb{E}[\mathbb{E}[X|Y]^2] .$$

On the other hand,

$$\mathrm{Var}(\mathbb{E}[X|Y]) = \mathbb{E}[\mathbb{E}[X|Y]^2] - \mathbb{E}[\mathbb{E}[X|Y]]^2 = \mathbb{E}[\mathbb{E}[X|Y]^2] - \mathbb{E}[X]^2 .$$

Therefore, $\mathbb{E}[\mathrm{Var}(X|Y)] + \mathrm{Var}(\mathbb{E}[X|Y]) = \mathbb{E}[X^2] - \mathbb{E}[X]^2$.

5.8

(i) Note that for B such that $\mathbb{P}(B) > 0$, we have $\mathbb{P}(B|A) = \frac{\mathbb{P}(A|B)\,\mathbb{P}(B)}{\mathbb{P}(A)}$. Now use $\mathbb{P}(A) = \sum_{k=1}^{n} \mathbb{P}(A|B_k)\,\mathbb{P}(B_k)$.

(ii) Let A denote the event of being infected, and B represent the positive reaction by the test. The superscript c represents complement, e.g., A^c means the event of being healthy. We summarize all the given information as follows: $\mathbb{P}(A) = 0.001$, $\mathbb{P}(B|A) = 0.95$, $\mathbb{P}(B|A^c) = 0.01$. The given exercise problem is to find $\mathbb{P}(A|B)$. Since

$$\mathbb{P}(A \cap B) = \mathbb{P}(B|A)\,\mathbb{P}(A) = 0.95 \times 0.001 = 0.00095$$

and

$$\mathbb{P}(A^c \cap B) = \mathbb{P}(B|A^c)\,\mathbb{P}(A^c) = 0.01 \times 0.999 = 0.00999 ,$$

we have

$$\mathbb{P}(A|B) = \frac{\mathbb{P}(A \cap B)}{\mathbb{P}(B)} = \frac{0.00095}{0.01094} \approx 8.68\% .$$

In the following table the necessary probabilities are printed in bold (Table 1).

Table 1 Bayes' rule and probabilities

1	$\mathbb{P}(A) = \mathbf{0.00100}$	$\mathbb{P}(A^c) = \mathbf{0.99900}$
$\mathbb{P}(B) = \mathbf{0.01094}$	$\mathbb{P}(A \cap B) = \mathbf{0.00095}$	$\mathbb{P}(A^c \cap B) = \mathbf{0.00999}$
$\mathbb{P}(B^c) = 0.98906$	$\mathbb{P}(A \cap B^c) = 0.00005$	$\mathbb{P}(A^c \cap B^c) = 0.98901$

Problems of Chap. 6

6.1 We need to show that

$$P(X_u \le y \mid \sigma(X_s), \sigma(X_t)) = P(X_u \le y \mid \sigma(X_t)) \tag{F.1}$$

for a Markov process $\{X_t\}$. Recall that $P(A|\mathcal{G}) = \mathbb{E}[\mathbf{1}_A|\mathcal{G}]$ for a measurable subset A and a sub-σ-algebra $\mathcal{G}$. Hence (F.1) is equivalent to

$$\mathbb{E}[\mathbf{1}_{\{X_u \le y\}} \mid \sigma(X_s), \sigma(X_t)] = \mathbb{E}[\mathbf{1}_{\{X_u \le y\}} \mid \sigma(X_t)] \ . \tag{F.2}$$

The left-hand side is equal to $\mathbb{E}[\mathbb{E}[\mathbf{1}_{\{X_u \le y\}}|\mathcal{F}_t] \mid \sigma(X_s), \sigma(X_t)]$ by the tower property since $\sigma(X_s), \sigma(X_t) \subset \mathcal{F}_t$. Since

$$\mathbb{E}[\mathbf{1}_{\{X_u \le y\}}|\mathcal{F}_t] = \mathbb{E}[\mathbf{1}_{\{X_u \le y\}} \mid \sigma(X_t)]$$

by the Markov property, the left-hand side is equal to

$$\mathbb{E}[\mathbb{E}[\mathbf{1}_{\{X_u \le y\}}|\sigma(X_t)] \mid \sigma(X_s), \sigma(X_t)] = \mathbb{E}[\mathbf{1}_{\{X_u \le y\}}|\sigma(X_t)] \ .$$

6.2 If $\omega \in \Omega$ is of the form $\omega = a_1 \ldots a_n \ldots$ for $a_i \in \{0, 1\}$, then $S_n(\omega) = k-(n-k)$ where k is the number of 1's among $a_1, \ldots, a_n$. Note that

$$\begin{aligned}
&\mathbb{E}[S_{n+1}|\mathcal{F}_n](\omega) \\
&= \frac{1}{\mathbb{P}([a_1, \ldots, a_n])} \int_{[a_1,\ldots,a_n]} S_{n+1} \mathrm{d}\mathbb{P} \\
&= \frac{1}{\mathbb{P}([a_1, \ldots, a_n])} \left(\int_{[a_1,\ldots,a_n,0]} S_{n+1} \mathrm{d}\mathbb{P} + \int_{[a_1,\ldots,a_n,1]} S_{n+1} \mathrm{d}\mathbb{P} \right) \\
&= \frac{1}{p^k q^{n-k}} \left((k-(n-k)+1) p^k q^{n-k} p + (k-(n-k)-1) p^k q^{n-k} q \right) \\
&= (k-(n-k)+1)p + (k-(n-k)-1)q \\
&= k-(n-k)+p-q \\
&\ne S_n(\omega)
\end{aligned}$$

since $p \neq q$. More formally,

$$\begin{aligned}\mathbb{E}[S_{n+1}|\mathcal{F}_n] &= \mathbb{E}[S_n + Z_{n+1}|\mathcal{F}_n] \\ &= \mathbb{E}[S_n|\mathcal{F}_n] + \mathbb{E}[Z_{n+1}|\mathcal{F}_n] \\ &= S_n + \mathbb{E}[Z_{n+1}|\mathcal{F}_n] \\ &= S_n + \mathbb{E}[Z_{n+1}] \\ &= S_n + p - q \neq S_n \ .\end{aligned}$$

6.5 Use $\mathbb{E}[\,\mathbf{1}_{\{|X_t|>n\}}\,|X_t|\,] \leq \mathbb{E}[\,\mathbf{1}_{\{|Y|>n\}}\,|Y|\,]$.

6.6

(i) For $n \geq 2$, we have

$$\begin{aligned}\Pr(N > n) &= \Pr(\min\{k : U_1 \leq U_2 \leq \cdots \leq U_{k-1} > U_k\} > n) \\ &= \Pr(U_1 \leq U_2 \leq \cdots \leq U_n) \\ &= \int_0^1 \int_0^{u_n} \cdots \int_0^{u_2} \mathrm{d}u_1 \cdots \mathrm{d}u_{n-1}\mathrm{d}u_n \\ &= \int_0^1 \int_0^{u_n} \cdots \int_0^{u_3} u_2\,\mathrm{d}u_2 \cdots \mathrm{d}u_{n-1}\mathrm{d}u_n \\ &= \int_0^1 \int_0^{u_n} \cdots \int_0^{u_4} \frac{1}{2}u_3^2\,\mathrm{d}u_3 \cdots \mathrm{d}u_{n-1}\mathrm{d}u_n \\ &= \qquad \vdots \\ &= \int_0^1 \frac{1}{(n-1)!}\,u_n^{n-1}\mathrm{d}u_n = \frac{1}{n!} \ .\end{aligned}$$

This is intuitively clear since $\Pr(U_{i_1} \geq U_{i_2} \geq \cdots \geq U_{i_n})$ are all equal to $\frac{1}{(n-2)!}$ for any permutation $(i_1, i_2, \ldots, i_n)$ of $(1, 2, \ldots, n)$, and

$$\Pr(N > n) = \Pr(U_1 \geq U_2 \geq \cdots \geq U_n) = \frac{1}{n!} \ .$$

(ii) Since

$$\Pr(M = n) = \Pr(M > n-1) - \Pr(M > n) = \frac{1}{(n-1)!} - \frac{1}{n!} = \frac{1}{n!}(n-1)$$

for $n \geq 2$, we have

$$\mathbb{E}[M] = \sum_{n=2}^{\infty} n \Pr(M = n) = \sum_{n=2}^{\infty} n \frac{1}{n!}(n-1) = \sum_{n=2}^{\infty} \frac{1}{(n-2)!} = \sum_{n=0}^{\infty} \frac{1}{n!} = \mathrm{e} \, .$$

6.9 Hint: Since $\mathcal{G}_t \subset \mathcal{H}_t$, we have

$$\mathbb{E}[X_t|\mathcal{G}_s] = \mathbb{E}[\mathbb{E}[X_t|\mathcal{H}_s]|\mathcal{G}_s] = \mathbb{E}[X_s|\mathcal{G}_s] = X_s$$

for $s < t$.

6.10 Since $X_{n-1} = \mathbb{E}[X_n|\mathcal{F}_{n-1}] = X_n$ for every n, we have

$$X_n = X_{n-1} = \cdots = X_1 = X_0 = \mathbb{E}[X_1|\mathcal{F}_0] \, ,$$

and X_n is constant.

6.12 Note that $\{\lambda_B = n\} = \{X_n \in B\} \cap \left(\bigcap_{k>n}\{X_k \notin B\}\right)$, which is not necessarily $\mathcal{F}_n$-measurable. Hence the last hitting time λ_B is not a stopping time.

6.13 Note that $\{\tau_1 \vee \tau_2 = n\} = \{\tau_1 = n, \tau_2 \leq n\} \cup \{\tau_1 \leq n, \tau_2 = n\} \in \mathcal{F}_n$. Thus $\tau_1 \vee \tau_2$ is a stopping time.

Problems of Chap. 7

7.5 Let $f(t,x) = x^3 - 3tx$. Then $\mathrm{d}f(t, W_t) = (f_t + \frac{1}{2}f_{xx})\mathrm{d}t + f_x \mathrm{d}W_t = -3t\mathrm{d}W_t$, and hence $f(t, W_t) = \int_0^t (3W_s^2 - 3s)\mathrm{d}W_s$, which is a martingale.

7.8 $\mathbb{E}[X_t^2] = \mathbb{E}[W_t^2 + t^2 W_1^2 - 2tW_t W_1] = t + t^2 - 2t^2 = t - t^2$.

7.10

$$\begin{aligned}
\mathbb{E}\left[W_t \mathrm{e}^{\theta W_t}\right] &= \frac{1}{\sqrt{2\pi t}} \int_{-\infty}^{\infty} x\,\mathrm{e}^{\theta x} \mathrm{e}^{-\frac{x^2}{2t}}\,\mathrm{d}x \\
&= \frac{1}{\sqrt{2\pi t}} \int_{-\infty}^{\infty} x\,\mathrm{e}^{-\frac{(x-\theta t)^2 - \theta^2 t^2}{2t}}\,\mathrm{d}x \\
&= \mathrm{e}^{\frac{1}{2}\theta^2 t} \frac{1}{\sqrt{2\pi t}} \int_{-\infty}^{\infty} (y + \theta t)\,\mathrm{e}^{-\frac{y^2}{2t}}\,\mathrm{d}x \\
&= \mathrm{e}^{\frac{1}{2}\theta^2 t}(0 + \theta t)
\end{aligned}$$

7.11

$$\begin{aligned}\mathbb{E}\left[W_t^2 e^{\theta W_t}\right] &= \frac{1}{\sqrt{2\pi t}}\int_{-\infty}^{\infty} x^2\, e^{\theta x} e^{-\frac{x^2}{2t}}\, dx \\ &= \frac{1}{\sqrt{2\pi t}}\int_{-\infty}^{\infty} x^2\, e^{-\frac{(x-\theta t)^2-\theta^2 t^2}{2t}}\, dx \\ &= e^{\frac{1}{2}\theta^2 t}\frac{1}{\sqrt{2\pi t}}\int_{-\infty}^{\infty} (y+\theta t)^2\, e^{-\frac{y^2}{2t}}\, dy \\ &= e^{\frac{1}{2}\theta^2 t}(t + 0 + \theta^2 t^2)\,.\end{aligned}$$

7.12 We check the conditions in Definition 7.1. All the processes start at 0 and have mean 0.

(i) The increments are independent, and we have

$$\mathrm{Var}(X_t - X_s) = \alpha^2 \mathrm{Var}(W_{t/\alpha^2} - W_{s/\alpha^2}) = \alpha^2(t/\alpha^2 - s/\alpha^2) = t - s\,.$$

(ii) Take $\alpha = a^{-1/2}$ in Part (i).

(iii) The increments are independent, and we have $\mathrm{Var}(X_t - X_s) = \mathrm{Var}((W_{T+t} - W_T) - (W_{T+s} - W_T)) = \mathrm{Var}(W_{T+t} - W_{T+s}) = t - s$.

(iv) For $s < t \le T$ and $T < s < t$ we have increments $X_t - X_s$ given by $W_t - W_s$ or $-(W_t - W_s)$. For $s \le T < t$ we have $X_t - X_s = 2W_T - W_t - W_s = (W_T - W_t) + (W_T - W_s) = -(W_t - W_T) + (W_T - W_s)$. Hence for every partition $0 \le t_1 < t_2 \le \cdots \le t_{2n-1} < t_{2n}$ the increments $X_{t_2} - X_{t_1}, \ldots, X_{t_{2n}} - X_{t_{2n-1}}$ are independent. Furthermore,

$$\mathrm{Var}(X_t - X_s) = \mathrm{Var}(W_t - W_T) + \mathrm{Var}(W_T - W_s) = (t - T) + (T - s) = t - s\,.$$

7.13 We will find a function f such that $\mathbb{E}[W_s|W_t] = f(W_t)$. For such a function we have

$$\int_{\{W_t\in B\}} W_s\, d\mathbb{P} = \int_{\{W_t\in B\}} \mathbb{E}[W_s|W_t]\, d\mathbb{P} = \int_{\{W_t\in B\}} f(W_t)\, d\mathbb{P}$$

for every interval B. Note that

$$\begin{aligned}\int_{\{W_t\in B\}} W_s\, d\mathbb{P} &= \int W_s\, \mathbf{1}_{\{W_t\in B\}}\, d\mathbb{P} \\ &= \int_{-\infty}^{\infty}\int_B x\, p(s; 0, x)\, p(t-s; x, y)\, dy\, dx \\ &= \int_B \int_{-\infty}^{\infty} x\, p(s; 0, x)\, p(t-s; x, y)\, dx\, dy\end{aligned}$$

and

$$\int_{\{W_t\in B\}} f(W_t)\,\mathrm{d}\mathbb{P} = \int_B f(y)\,p(t;0,y)\,\mathrm{d}y\,.$$

Hence

$$\begin{aligned}
& f(y)p(t;0,y) \\
&= \int_{-\infty}^{\infty} x p(s;0,x)p(t-s;x,y)\,\mathrm{d}x \\
&= \int_{-\infty}^{\infty} x\frac{1}{\sqrt{2\pi s}}\exp\left(-\frac{x^2}{2s}\right)\frac{1}{\sqrt{2\pi(t-s)}}\exp\left(-\frac{(x-y)^2}{2(t-s)}\right)\mathrm{d}x \\
&= \frac{1}{\sqrt{2\pi}}\int_{-\infty}^{\infty} x\frac{1}{\sqrt{2\pi s(t-s)}}\exp\left(-\frac{(t-s)x^2+sx^2+sy^2-2sxy}{2s(t-s)}\right)\mathrm{d}x \\
&= \frac{1}{\sqrt{2\pi}}\int_{-\infty}^{\infty} x\frac{1}{\sqrt{2\pi s(t-s)}}\exp\left(-\frac{(\sqrt{t}x-\frac{s}{\sqrt{t}}y)^2}{2s(t-s)}\right)\exp\left(-\frac{y^2}{2t}\right)\mathrm{d}x \\
&= p(t;0,y)\int_{-\infty}^{\infty} xp\left(\frac{s(t-s)}{t};\frac{s}{t}y,x\right)\mathrm{d}x = p(t;0,y)\frac{s}{t}y\,.
\end{aligned}$$

Thus $f(y) = \dfrac{s}{t}y$.

7.15 Choose $u = \frac{1}{t}$ and $v = \frac{1}{s}$ for $0 < s < t$. Then $u < v$, and $\mathbb{E}[X_v|X_u] = X_u$. In other words, $\mathbb{E}[vW_{1/v}|uW_{1/u}] = uW_{1/u}$. Hence $\mathbb{E}[vW_s|uW_t] = uW_t$. Note that $\sigma(uW_t) = \sigma(W_t)$ and that $\mathbb{E}[vW_s|uW_t] = \mathbb{E}[vW_s|W_t]$.

7.17 Since $\mathbb{E}[\mathrm{e}^{at}\cos W_t|\mathcal{F}_s] = \mathrm{e}^{at}\mathbb{E}[\cos W_t|\mathcal{F}_s] = \mathrm{e}^{at}\mathrm{e}^{-(t-s)/2}\cos W_s$, we need $\mathrm{e}^{at}\mathrm{e}^{-(t-s)/2} = \mathrm{e}^{as}$, which would imply $(a-\frac{1}{2})t + \frac{1}{2}s = as$, and $a = \frac{1}{2}$.

Problems of Chap. 8

8.2

(i) Since $\mathbb{E}^{\mathbb{Q}}[W_t] = \mathbb{E}^{\mathbb{P}}[W_t\mathrm{e}^{-\frac{1}{2}\theta^2 t-\theta W_t}] = \mathrm{e}^{-\frac{1}{2}\theta^2 t}\mathbb{E}^{\mathbb{P}}[W_t\mathrm{e}^{-\theta W_t}]$, we have $\mathbb{E}^{\mathbb{Q}}[W_t] = \mathrm{e}^{-\frac{1}{2}\theta^2 t}(-\theta t)\mathrm{e}^{\frac{1}{2}\theta t} = -\theta t$.
(ii) Since $\mathbb{E}^{\mathbb{Q}}[W_t^2] = \mathbb{E}^{\mathbb{P}}[W_t^2\mathrm{e}^{-\frac{1}{2}\theta^2 t-\theta W_t}] = \mathrm{e}^{-\frac{1}{2}\theta^2 t}\mathbb{E}^{\mathbb{P}}[W_t^2\mathrm{e}^{-\theta W_t}]$, we have $\mathbb{E}^{\mathbb{Q}}[W_t^2] = \mathrm{e}^{-\frac{1}{2}\theta^2 t}(\theta^2 t^2 + t)\mathrm{e}^{\frac{1}{2}\theta t} = \theta^2 t^2 + t$.

8.3 $\mathbb{E}^{\mathbb{Q}}[\mathrm{e}^{\alpha X_t}] = \mathbb{E}^{\mathbb{P}}[\mathrm{e}^{\alpha X_t}L] = \mathbb{E}^{\mathbb{P}}[\mathbb{E}^{\mathbb{P}}[\mathrm{e}^{\alpha X_t}L|\mathcal{F}_t]] = \mathbb{E}^{\mathbb{P}}[\mathrm{e}^{\alpha X_t}\mathbb{E}^{\mathbb{P}}[L|\mathcal{F}_t]] = \mathbb{E}^{\mathbb{P}}[\mathrm{e}^{\alpha X_t}L_t] = \mathbb{E}[\mathrm{e}^{\alpha(W_t+\theta t)}\mathrm{e}^{-\frac{1}{2}\theta^2 t-\theta W_t}] = \mathrm{e}^{\alpha\theta t-\frac{1}{2}\theta^2 t}\mathbb{E}[\mathrm{e}^{(\alpha-\theta)W_t}]$, which is equal to $\mathrm{e}^{\alpha\theta t-\frac{1}{2}\theta^2 t}\mathrm{e}^{\frac{1}{2}(\alpha-\theta)^2 t} = \mathrm{e}^{\frac{1}{2}\alpha^2 t} = \mathbb{E}[\mathrm{e}^{\alpha W_t}]$.

8.4 Since $W_t - W_s$ and $\mathcal{F}_s$ are independent, we have

$$\begin{aligned}\mathbb{E}[(W_t - W_s)\,\mathrm{e}^{\theta(W_t - W_s)}|\mathcal{F}_s] &= \mathbb{E}[(W_t - W_s)\,\mathrm{e}^{\theta(W_t - W_s)}] \\ &= \mathbb{E}[W_{t-s}\,\mathrm{e}^{\theta W_{t-s}}] \\ &= \theta(t-s)\,\mathrm{e}^{\frac{1}{2}\theta^2(t-s)}\,.\end{aligned}$$

Hence

$$\begin{aligned}\mathbb{E}[W_t\,\mathrm{e}^{\theta W_t}|\mathcal{F}_s] - W_s\mathbb{E}[\mathrm{e}^{\theta W_t}|\mathcal{F}_s] &= \mathbb{E}[(W_t - W_s)\,\mathrm{e}^{\theta W_t}|\mathcal{F}_s] \\ &= \theta(t-s)\,\mathrm{e}^{\frac{1}{2}\theta^2(t-s)}\mathrm{e}^{\theta W_s}\,.\end{aligned}$$

8.5 Since $W_t - W_s$ and $\mathcal{F}_s$ are independent, we have

$$\begin{aligned}\mathbb{E}[(W_t - W_s)^2\,\mathrm{e}^{\theta(W_t - W_s)}|\mathcal{F}_s] &= \mathbb{E}[(W_t - W_s)^2\,\mathrm{e}^{\theta(W_t - W_s)}] \\ &= \mathbb{E}[W_{t-s}^2\,\mathrm{e}^{\theta W_{t-s}}] \\ &= ((t-s) + \theta^2(t-s)^2)\,\mathrm{e}^{\frac{1}{2}\theta^2(t-s)}\,.\end{aligned}$$

On the other hand,

$$\begin{aligned}&\mathrm{e}^{\theta W_s}\mathbb{E}[(W_t - W_s)^2\,\mathrm{e}^{\theta(W_t - W_s)}|\mathcal{F}_s] \\ &= \mathbb{E}[(W_t - W_s)^2\,\mathrm{e}^{\theta W_t}|\mathcal{F}_s] \\ &= \mathbb{E}[W_t^2\,\mathrm{e}^{\theta W_t}|\mathcal{F}_s] - 2W_s\,\mathbb{E}[W_t\,\mathrm{e}^{\theta W_t}|\mathcal{F}_s] + W_s^2\,\mathbb{E}[\mathrm{e}^{\theta W_t}|\mathcal{F}_s]\,.\end{aligned}$$

Hence

$$\begin{aligned}&\mathbb{E}[W_t^2\,\mathrm{e}^{\theta W_t}|\mathcal{F}_s] \\ &= \mathrm{e}^{\theta W_s}\mathbb{E}[(W_t - W_s)^2\mathrm{e}^{\theta(W_t - W_s)}|\mathcal{F}_s] + 2W_s\mathbb{E}[W_t\mathrm{e}^{\theta W_t}|\mathcal{F}_s] - W_s^2\mathbb{E}[\mathrm{e}^{\theta W_t}|\mathcal{F}_s] \\ &= \{(t-s) + \theta^2(t-s)^2 + W_s^2 + 2\theta(t-s)W_s\}\,\mathrm{e}^{\frac{1}{2}\theta^2(t-s)}\mathrm{e}^{\theta W_s}\,.\end{aligned}$$

8.6 Take $s < t$. From Exercise 8.4,

$$\mathbb{E}[W_t L_t|\mathcal{F}_s] = \mathrm{e}^{-\frac{1}{2}\theta^2 t}\mathbb{E}[W_t\mathrm{e}^{-\theta W_t}|\mathcal{F}_s] = (W_s - \theta(t-s))\,L_s\,.$$

From Exercise 8.5,

$$\mathbb{E}[W_t^2 L_t \mid \mathcal{F}_s] = \mathrm{e}^{-\frac{1}{2}\theta^2 t}\mathbb{E}[W_t^2\mathrm{e}^{-\theta W_t} \mid \mathcal{F}_s] = \{(t-s) + (W_s - \theta(t-s))^2\}\,L_s\,.$$

Hence we have

$$
\begin{aligned}
&\mathbb{E}^{\mathbb{Q}}[X_t^2 - t \mid \mathcal{F}_s] \\
&= \mathbb{E}[X_t^2 L_t L_s^{-1} \mid \mathcal{F}_s] - t \quad \text{(by Lemma 8.1)} \\
&= L_s^{-1}\, \mathbb{E}[(W_t + \theta t)^2 L_t \mid \mathcal{F}_s] - t \\
&= L_s^{-1} \left(\mathbb{E}[W_t^2 L_t \mid \mathcal{F}_s] + 2\theta t\, \mathbb{E}[W_t L_t \mid \mathcal{F}_s] + \theta^2 t^2 \mathbb{E}[L_t \mid \mathcal{F}_s] \right) - t \\
&= \{(t-s) + (W_s - \theta(t-s))^2\} + 2\theta t\{W_s - \theta(t-s)\} + \theta^2 t^2 - t \\
&= (W_s + \theta s)^2 - s \\
&= X_s^2 - s\,.
\end{aligned}
$$

8.7 Let

$$A = \left\{ \lim_{t\to\infty} \frac{W_t}{t} = 0 \right\}\,.$$

Then $\mathbb{P}(A) = 1$ by the Law of Large Numbers (Corollary 7.2). Note that

$$\lim_{t\to\infty} \frac{X_t}{t} = \lim_{t\to\infty} \frac{W_t + \theta t}{t} = \theta \neq 0$$

on A. Hence $\mathbb{Q}(A) = 0$ since

$$\mathbb{Q}\left(\left\{ \lim_{t\to\infty} \frac{X_t}{t} = 0 \right\} \right) = 1\,.$$

Thus $\mathbb{P}$ and $\mathbb{Q}$ are not equivalent.

8.8 To have a fair game the expected reward must be equal to \$0. Hence $pA - (1-p)B = 0$, which implies that $B = \frac{p}{1-p}A > A$. In other words, if we interpret the event of having heads and tails as upward and downward movements respectively, we need to give more weight to the downward movement than the upward movement to compensate the bias.

Problems of Chap. 10

10.4 Put $M_t = S_0 e^{-\frac{1}{2}\sigma^2 t + \sigma W_t}$. Then $M_t = M_0 + \int_0^t \sigma M_u \,dW_u$. Hence M_t is a martingale, and for every $t \geq 0$ we have $\mathbb{E}[M_t] = \mathbb{E}[M_0] = S_0$, which implies that $\mathbb{E}[S_0 e^{-\frac{1}{2}\sigma^2 t + \sigma W_t}] = S_0$, and $\mathbb{E}[S_t] = S_0 e^{\mu t}$.

10.8 Note that

$$
\begin{aligned}
&\mathrm{Cov}(X_t, X_{t+u})\\
&= \mathrm{Cov}\left(\int_0^t f(s)\mathrm{d}W_s, \int_0^{t+u} f(s)\mathrm{d}W_s\right)\\
&= \mathbb{E}\left[\left(\int_0^t f(s)\mathrm{d}W_s - 0\right)\left(\int_0^{t+u} f(s)\mathrm{d}W_s - 0\right)\right]\\
&= \mathbb{E}\left[\int_0^t f(s)\mathrm{d}W_s \left(\int_0^t f(s)\mathrm{d}W_s + \int_t^{t+u} f(s)\mathrm{d}W_s\right)\right]\\
&= \mathbb{E}\left[\int_0^t f(s)\mathrm{d}W_s \int_0^t f(s)\mathrm{d}W_s\right] + \mathbb{E}\left[\int_0^t f(s)\mathrm{d}W_s \int_t^{t+u} f(s)\mathrm{d}W_s\right]\\
&= \int_0^t \mathbb{E}\left[f(s)^2\right]\mathrm{d}s + \mathbb{E}\left[\int_0^t f(s)\mathrm{d}W_s\right]\mathbb{E}\left[\int_t^{t+u} f(s)\mathrm{d}W_s\right]\\
&= \int_0^t f(s)^2\mathrm{d}s + 0 \times 0
\end{aligned}
$$

where we used the fact that $\int_0^t f(s)\mathrm{d}W_s$ and $\int_t^{t+u} f(s)\mathrm{d}W_s$ are independent in the fifth equality.

10.9 By the result of Exercise 4.10 we have

$$\mathbb{E}\left[X_t^2\right] = \mathbb{E}\left[\mathrm{e}^{2W_t^2}\right] = \mathbb{E}\left[\mathrm{e}^{2tW_1^2}\right] = \frac{1}{\sqrt{1-4t}}, \quad 0 \le t < \frac{1}{4}.$$

Hence the identity in Theorem 10.1 does not hold.

10.10 Note that $\mathrm{Cov}(X, Y) = \mathbb{E}[XY] = \int_0^T t(T-t)\,\mathrm{d}t = \frac{1}{6}T^3$.

10.11 By the Martingale Representation Theorem, $f(t, W_t) = f(0, W_0) + \int_0^t \alpha_s \mathrm{d}W_s$. Now use the Itô formula.

10.12 The function $f(t,x)$ satisfies two differential equations $f_t + \frac{1}{2}f_{xx} = 0$ and $f_x = f$. From the second equation, we see that $f(t,x) = \mathrm{e}^x g(t)$ for some g. From the first equation we have $g' + \frac{1}{2}g = 0$, and hence $g(t) = g(0)\mathrm{e}^{-\frac{1}{2}t}$. Thus $X_t = X_0\mathrm{e}^{W_t - \frac{1}{2}t}$.

10.14

(i) $\int_0^t f(s, W_s)\mathrm{d}W_s$ is equal to X_t.
(ii) To prove that X_t is a martingale, either use the definition of a martingale or apply the Martingale Representation Theorem since X_t is an Itô integral.

10.17 Suppose that X_t is a martingale.

Solution 1. Since

$$\mathrm{d}X_t = \left\{ g'(t)\mathrm{e}^{W_t+g(t)} + \frac{1}{2}\mathrm{e}^{W_t+g(t)} \right\} \mathrm{d}t + \mathrm{e}^{W_t+g(t)}\mathrm{d}W_t \, ,$$

we have

$$g'(t)\mathrm{e}^{W_t+g(t)} + \frac{1}{2}\mathrm{e}^{W_t+g(t)} = 0$$

by the Martingale Representation Theorem. Hence $g(t) = -\frac{1}{2}t + C$ for some constant C.

Solution 2. Since $\mathbb{E}[X_t] = \mathrm{e}^{\frac{1}{2}t+g(t)}$, $t \geq 0$, is constant, we have $g(t) = -\frac{1}{2}t + C$.

10.20 $\mathrm{d}X_t = (2tW_t + 0)\mathrm{d}t + t^2\mathrm{d}W_t - 2tW_t\mathrm{d}t = t^2\mathrm{d}W_t$.

Problems of Chap. 11

11.2 If δ denotes the continuous dividend yield, then the modified model for the stock price movement is given by $\mathrm{d}S_t = (\mu - \delta)S_t\mathrm{d}t + \sigma S_t\mathrm{d}W_t$.

11.4

(i) Let $f(t,x) = \mathrm{e}^x$. Then $\mathrm{e}^{X_t} = f(t, X_t)$. Hence

$$\mathrm{d}(\mathrm{e}^{X_t}) = \mathrm{e}^{X_t}\mathrm{d}X_t + \frac{1}{2}\mathrm{e}^{X_t}(\mathrm{d}X_t)^2 = \mathrm{e}^{X_t}(a_t + \frac{1}{2}b_t^2)\mathrm{d}t + \mathrm{e}^{X_t}b_t\mathrm{d}W_t \, .$$

(ii) Take $X_t = -\int_0^t r_s\mathrm{d}s$. Then $\mathrm{d}X_t = -r_t\mathrm{d}t$, $a_t = -r_t$, $b_t = 0$. Since $Z_t = \mathrm{e}^{X_t}$, we have $\mathrm{d}Z_t = Z_t(-r_t)\mathrm{d}t$.

11.5 $X_t = (1 + W_t)^2$.

11.6 For $t = 0$, we have $\int_0^0 W_s\mathrm{d}W_s = f(0, W_0)$. Hence $f(0,0) = 0$. By Itô's lemma we have

$$\begin{cases} f_t + \frac{1}{2}f_{xx} = 0 \, , \\ f_x = x \, . \end{cases}$$

From the second equation we have $f(t,x) = \frac{1}{2}x^2 + \phi(t)$. Combining the result with the first equation, we obtain $\phi(t) = -\frac{1}{2} + C$ for some constant C. Thus $f(t,x) = \frac{1}{2}x^2 - \frac{1}{2}t + C$. Since $f(0,0) = 0$, we conclude that $\int_0^t W_s\mathrm{d}W_s = f(t, W_t) = \frac{1}{2}W_t^2 - \frac{1}{2}t$.

11.7

(i) Suppose that there exists $f(t,x)$ such that $\int_0^t W_s^2\,\mathrm{d}W_s = f(t, W_t)$. This assumption is wrong and here is why. Suppose that the assumption were right. Taking $t=0$, we find $0 = f(0,0)$. By Itô's lemma, we have $W_t\,\mathrm{d}W_t = (f_t + \frac{1}{2}f_{xx})\,\mathrm{d}t + f_x\,\mathrm{d}W_t$. Hence

$$\begin{cases} f_t + \frac{1}{2}f_{xx} = 0\,, \\ f_x \qquad\quad = x^2\,. \end{cases}$$

Hence $f = \frac{1}{3}x^3 + \phi(t)$ for some $\phi(t)$. Thus $\phi'(t) + x = 0$, which is impossible.

(ii) Now let us modify the assumption! Assume that there exists $f(t,x)$ and $g(t,x)$ such that $\int_0^t W_s^2 \mathrm{d}W_s = f(t,W_t) + \int_0^t g(s,W_s)\,\mathrm{d}s$. Then

$$\begin{cases} f_t + \frac{1}{2}f_{xx} + g = 0\,, \\ f_x \qquad\qquad\quad = x^2. \end{cases}$$

From the second equation we have $f(t,x) = \frac{1}{3}x^3 + \phi(t)$. Combining the result with the first equation, we obtain $\phi'(t) + x + g(t,x) = 0$, and hence

$$\int_0^t W_s^2 \mathrm{d}W_s = \frac{1}{3}W_t^3 + \phi(t) + \int_0^t \left(-\phi'(s) - W_s\right)\,\mathrm{d}s = \frac{1}{3}W_t^3 - \int_0^t W_s\,\mathrm{d}s\,.$$

11.8 Assume that $\int_0^t s\,\mathrm{d}W_s = f(t,W_t) + \int_0^t g(s,W_s)\,\mathrm{d}s$ for some f and g. By Itô's lemma, we have

$$\begin{cases} f_t + \frac{1}{2}f_{xx} + g = 0\,, \\ f_x \qquad\qquad\quad = t\,. \end{cases}$$

Hence $f(t,x) = tx + \phi(t)$ for some $\phi(t)$. Note that $0 = f(0,0) = \phi(0)$. Since $x + \phi'(t) + g = 0$, we have $g(t,x) = -\phi'(t) - x$. Thus

$$\int_0^t s\,\mathrm{d}W_s = tW_t + \phi(t) + \int_0^t (-\phi'(s) - W_s)\,\mathrm{d}s = tW_t - \int_0^t W_s\,\mathrm{d}s\,.$$

11.10 Assume that there exist f and g such that $\int_0^t \mathrm{e}^{W_s}\,\mathrm{d}W_s = f(t,W_t) + \int_0^t g(s,W_s)\,\mathrm{d}s$. By Itô's lemma, we have $\mathrm{e}^{W_t}\,\mathrm{d}W_t = (f_t + \frac{1}{2}f_{xx})\,\mathrm{d}t + f_x\,\mathrm{d}W_t + g\,\mathrm{d}t$. Hence

$$\begin{cases} f_t + \frac{1}{2}f_{xx} + g = 0\,, \\ f_x \qquad\qquad\quad = \mathrm{e}^x\,. \end{cases}$$

Hence $f(t, x) = e^x + \phi(t)$ for some $\phi(t)$. Note that $0 = f(0, 0) = 1 + \phi(0)$. Since $\phi'(t) + \frac{1}{2}e^x + g = 0$, we have $g(t, x) = -\phi'(t) - \frac{1}{2}e^x$. Thus

$$\int_0^t e^{W_s}\,dW_s = e^{W_t} + \phi(t) + \int_0^t (-\phi'(t) - \frac{1}{2}e^{W_s})\,ds = e^{W_t} - 1 - \frac{1}{2}\int_0^t e^{W_s}\,ds\,.$$

11.11

(i) Let $f(t, x) = x^k$, $k \geq 2$. Then

$$W_t^k = \int_0^t \frac{1}{2}k(k-1)W_s^{k-2}ds + \int_0^t kW_s^{k-1}dW_s\,.$$

Hence

$$a_k(t) = \mathbb{E}[W_t^k] = \int_0^t \frac{1}{2}k(k-1)\mathbb{E}[W_s^{k-2}]ds = \int_0^t \frac{1}{2}k(k-1)a_{k-2}(s)ds\,.$$

(ii) Since the pdf $f(x)$ of W_t is an even function, we have $\mathbb{E}[W_t^{2k+1}] = \int_{-\infty}^{\infty} x^{2k+1}f(x)dx = 0$. Or, we may use mathematical induction on the recursive relation $a_{2k+1}(t) = \frac{1}{2}k(k-1)\int_0^t a_{2k-1}(s)\,ds$ together with $a_1 = 0$. Similarly, by induction,

$$a_{2k}(t) = \frac{1}{2}(2k)(2k-1)\int_0^t \frac{(2k-2)!s^{k-1}}{2^{k-1}(k-1)!}ds = \frac{(2k)!t^k}{2^k k!}\,.$$

11.14 First, by taking expectations on both sides, we obtain $\alpha(t) = \alpha(0) + \int_0^t (a(s)\alpha(s) + b(s))ds$, and hence $\alpha'(t) = a(t)\alpha(t) + b(t)$. For $\beta(t)$, note that

$$X_t^2 = X_0^2 + \int_0^t \left[2X_s\{a(s)X_s + b(s)\} + \{c(s)X_s + d(s)\}^2\right]ds$$
$$+ \int_0^t [2X_s\{c(s)X_s + d(s)\}]\,dW_s$$

and take expectations on both sides, then take derivatives.

Problems of Chap. 12

12.2 For $a \geq 0$, let $X_t = 0$, $0 \leq t \leq a$, and $X_t = (W_t - a)^3$, $t > a$.

12.6 Take $\alpha_t = \gamma_t = 0$. Then $\mathrm{d}X_t = \beta_t X_t \mathrm{d}t + \delta_t X_t \mathrm{d}W_t$. Let Y_t be an Itô process defined by $\mathrm{d}Y_t = \beta_t \mathrm{d}t + \delta_t \mathrm{d}W_t$. Then $\mathrm{d}[Y, Y]_t = \delta_t^2 \mathrm{d}t$ and $\mathrm{d}X_t = X_t \mathrm{d}Y_t$, and hence

$$
\begin{aligned}
X_t &= X_0 \exp\left(Y_t - Y_0 - \frac{1}{2}[Y, Y]_t\right) \\
&= X_0 \exp\left(\int_0^t \beta_s \mathrm{d}s + \int_0^t \delta_s \mathrm{d}W_s - \frac{1}{2}\int_0^t \delta_s^2 \mathrm{d}s\right) \\
&= X_0 \exp\left(\int_0^t (\beta_s \mathrm{d}s - \frac{1}{2}\delta_s^2)\mathrm{d}s + \int_0^t \delta_s \mathrm{d}W_s\right) .
\end{aligned}
$$

For arbitrary α_t, γ_t, we look for a solution of the form $X_t = G_t H_t$ where

$$
\begin{aligned}
\mathrm{d}G_t &= \beta_t G_t \mathrm{d}t + \delta_t G_t \mathrm{d}W_t , \quad G_0 = 1 , \\
\mathrm{d}H_t &= a_t \mathrm{d}t + b_t \mathrm{d}W_t , \quad H_0 = X_0 .
\end{aligned}
$$

12.7 Assume that $X_t = f(t, W_t)$ for some $f(t, x)$. Then

$$
(f_t + \frac{1}{2}f_{xx})\mathrm{d}t + f_x \mathrm{d}W_t = f^3 \mathrm{d}t - f^2 \mathrm{d}W_t .
$$

Hence

$$
\begin{cases}
f_t + \frac{1}{2}f_{xx} = f^3 , \\
f_x = -f^2 .
\end{cases}
$$

From the second equation, we have $-\mathrm{d}f/f^2 = \mathrm{d}x$, $1/f = x + g(t)$, $f(t, x) = 1/(x + g(t))$. Substituting the result in the first equation, we obtain $g(t) = C$ for some constant C. Since $X_0 = 1$, we have $C = 1$. Thus $X_t = 1/(W_t + 1)$.

12.8 Note that

$$
\begin{aligned}
\mathrm{d}Y_t &= 2X_t \mathrm{d}X_t + \frac{1}{2} \times 2(\mathrm{d}X_t)^2 \\
&= 2X_t(-\alpha X_t \mathrm{d}t + \sigma \mathrm{d}W_t) + (-\alpha X_t \mathrm{d}t + \sigma \mathrm{d}W_t)^2 \\
&= (\sigma^2 - 2\alpha X_t^2)\mathrm{d}t + 2\sigma X_t \mathrm{d}W_t .
\end{aligned}
$$

Thus, by substituting (12.3), we obtain

$$
\mathrm{d}Y_t = (\sigma^2 - 2\alpha Y_t)\mathrm{d}t + 2\sigma \left(e^{-\alpha t}x_0 + e^{-\alpha t}\sigma \int_0^t e^{\alpha s}\mathrm{d}W_s\right)\mathrm{d}W_t .
$$

12.9 Let $X_t = f(t, W_t)$ for some $f(t, x)$. Then $\frac{1}{2}f_{xx} + f_t = \frac{3}{4}f^2$ and $f_x = -f^{3/2}$. From the second equation, $f(t, x) = 4/(c + x)^2$ for some constant C. Hence $X_t =$

$4/(C + W_t)^2$. Since $X_0 = 4/(C + 0)^2$, $C = \pm 2/\sqrt{X_0}$. Thus

$$X_t = 4/(\pm 2/\sqrt{X_0} + W_t)^2 \ .$$

Note that both solutions have the same distribution.

Problems of Chap. 15

15.5 (ii) The answer is $-S_t + Ke^{-r(T-t)}$ or 0. (iii) $Ke^{-r(T-t)}$.

15.7 Observe first

$$\begin{aligned}
\frac{\partial d_1}{\partial K} &= \frac{K}{S_0}\left(-\frac{S_0}{K^2}\right)\frac{1}{\sigma\sqrt{T}} = -\frac{1}{K\sigma\sqrt{T}} \ , \\
\frac{\partial d_2}{\partial K} &= \frac{\partial}{\partial K}(d_1 - \sigma\sqrt{T}) = \frac{\partial d_1}{\partial K} \ .
\end{aligned}$$

Hence

$$\begin{aligned}
&\frac{\partial C}{\partial K} \\
&= S_0 N'(d_1)\frac{\partial d_1}{\partial K} - e^{-rT}N(d_2) - Ke^{-rT}N'(d_2)\frac{\partial d_2}{\partial K} \\
&= -S_0\frac{1}{\sqrt{2\pi}}e^{-d_1^2/2}\frac{1}{K\sigma\sqrt{T}} - e^{-rT}N(d_2) + Ke^{-rT}\frac{1}{\sqrt{2\pi}}e^{-d_2^2/2}\frac{1}{K\sigma\sqrt{T}} \\
&= -S_0\frac{1}{K\sigma\sqrt{2\pi T}}e^{-d_1^2/2} - e^{-rT}N(d_2) + \frac{1}{\sigma\sqrt{2\pi T}}e^{-d_2^2/2-rT} \ .
\end{aligned}$$

Now we note that

$$\begin{aligned}
-\frac{d_2^2}{2} - rT &= -\frac{(d_1 - \sigma\sqrt{T})^2}{2} - rT \\
&= -\frac{d_1^2}{2} + \sigma d_1\sqrt{T} - \frac{\sigma^2 T}{2} - rT \\
&= -\frac{d_1^2}{2} + \log\frac{S_0}{K} \ .
\end{aligned}$$

Therefore

$$\begin{aligned}\frac{\partial C}{\partial K} &= -S_0 \frac{1}{\sigma K\sqrt{2\pi T}} \mathrm{e}^{-d_1^2/2} - \mathrm{e}^{-rT} N(d_2) + \frac{1}{\sigma\sqrt{2\pi T}} \mathrm{e}^{-d_1^2/2+\log(S_0/K)} \\ &= -\frac{S_0}{K} \frac{1}{\sigma\sqrt{2\pi T}} \mathrm{e}^{-d_1^2/2} - \mathrm{e}^{-rT} N(d_2) + \frac{S_0}{K} \frac{1}{\sigma\sqrt{2\pi T}} \mathrm{e}^{-d_1^2/2} \\ &= -\mathrm{e}^{-rT} N(d_2) \,.\end{aligned}$$

15.10

(i) Since $\Delta = \frac{\partial V}{\partial S} \approx \frac{\delta V}{\delta S}$, we have

$$\Omega \approx \frac{\frac{\delta V}{V}}{\frac{\delta S}{S}} = \frac{\delta V}{\delta S}\frac{S}{V} \approx \Delta \frac{S}{V} \,.$$

(ii) Since $V \leq SN(d_1)$ and $\Delta = N(d_1)$, we have $\Omega \geq 1$.

(iii) Note that a European put option loses its value if the asset price increases, i.e., $\delta V \leq 0$ if $\delta S > 0$. Hence $\Omega \leq 0$. Or, we may use the fact that the deltas of the call and put options with the same strike price, denoted by Δ_C and Δ_P respectively, satisfy the relation $\Delta_P = \Delta_C - 1 \leq 0$.

Problems of Chap. 16

16.1 The price V_0 at $t = 0$ is given by the expectation with respect to a martingale measure $\mathbb{Q}$, and $V_0 = \mathrm{e}^{-rT}\mathbb{E}^{\mathbb{Q}}[1] = \mathrm{e}^{-rT}$, which is nothing but the risk-free bond price. If μ is replaced by r in the expectation with respect to a physical measure $\mathbb{P}$, we would have the same answer since $\mathrm{e}^{-rT}\mathbb{E}^{\mathbb{P}}[1] = \mathrm{e}^{-rT}$.

Problems of Chap. 17

17.1 Using $C_T = S_T - F$, we find F such that $\mathbb{E}^{\mathbb{Q}}[\mathrm{e}^{-rT}(S_T - F)|\mathcal{F}_0] = 0$. Hence

$$F = \mathbb{E}^{\mathbb{Q}}[S_T] = \mathbb{E}^{\mathbb{Q}}[A\mathrm{e}^{(r-\frac{1}{2}\sigma^2)T+\sigma X_T}] = A\mathrm{e}^{rT}\mathbb{E}^{\mathbb{Q}}[\mathrm{e}^{-\frac{1}{2}\sigma^2 T+\sigma X_T}] = A\mathrm{e}^{rT} \,,$$

where we used the fact $\mathbb{E}^{\mathbb{Q}}[\mathrm{e}^{-\frac{1}{2}\sigma^2 T+\sigma X_T}] = \mathbb{E}^{\mathbb{Q}}[\mathrm{e}^{-\frac{1}{2}\sigma^2 0+\sigma X_0}] = 1$ since $\mathrm{e}^{-\frac{1}{2}\sigma^2 t+\sigma X_t}$ is a $\mathbb{Q}$-martingale. □

17.2 Since an asset-or-nothing call can be decomposed into the sum of a European call and a cash-or-nothing call as can be seen in the identity

$$\underbrace{S_T \mathbf{1}_{\{S_T>K\}}}_{\text{asset-or-nothing}} = \underbrace{(S_T - K)^+}_{\text{European call}} + \underbrace{K\mathbf{1}_{\{S_T>K\}}}_{\text{cash-or-nothing}} ,$$

the European call price is equal to the asset-or-nothing call price minus the cash-or-nothing call price.

Problems of Chap. 20

20.1

(i) By the Itô formula for a product, we have

$$\begin{aligned} \mathrm{d}(\mathrm{e}^{-rt}X_t) &= -r\mathrm{e}^{-rt}X_t\mathrm{d}t + \mathrm{e}^{-rt}[((\alpha - r)\pi_t + rX_t)\mathrm{d}t + \sigma\pi_t\mathrm{d}W_t] \\ &= \mathrm{e}^{-rt}[(\alpha - r)\pi_t\mathrm{d}t + \sigma\pi_t\mathrm{d}W_t] . \end{aligned}$$

(ii) Let $Y_t = \mathrm{e}^{-2\theta W_t - \frac{1}{2}(2\theta)^2 t}$. Since Y_t is nothing but L_t with 2θ in place of θ, it is a martingale, and hence $\mathbb{E}[\, Y_t] = \mathbb{E}[\, Y_0] = 1$, $t \geq 0$. Then $L_t^2 = \mathrm{e}^{-2\theta W_t - \theta^2 t} = Y_t\,\mathrm{e}^{\theta^2 t}$, and $\mathbb{E}[L_t^2] = \mathbb{E}[\mathrm{e}^{-2\theta W_t - \theta^2 t}] = \mathbb{E}[\, Y_t]\mathrm{e}^{\theta^2 t} = \mathrm{e}^{\theta^2 t}$, which proves statement (a). Note that $\mathrm{d}L\,\mathrm{d}\widetilde{X} = (-\theta L)(\widetilde{\pi}\sigma)\,\mathrm{d}t = -(\alpha - r)L\widetilde{\pi}\,\mathrm{d}t$. By the Itô formula for a product, we have

$$\begin{aligned} \mathrm{d}(L\widetilde{X}) &= \widetilde{X}\mathrm{d}L + L\mathrm{d}\widetilde{X} + \mathrm{d}L\mathrm{d}\widetilde{X} \\ &= -\theta L\widetilde{X}\mathrm{d}W + L(\widetilde{\pi}(\alpha - r)\mathrm{d}t + \widetilde{\pi}\sigma\mathrm{d}W) - (\alpha - r)L\widetilde{\pi}\,\mathrm{d}t \\ &= -\theta L\widetilde{X}\mathrm{d}W + L\widetilde{\pi}\sigma\mathrm{d}W = L(-\theta\widetilde{X} + \pi\sigma)\,\mathrm{d}W . \end{aligned}$$

Hence $L_t\widetilde{X}_t = X_0 + \int_0^t L_s(\pi_s\sigma - \theta\widetilde{X}_s)\mathrm{d}W_s$ and $L_t\widetilde{X}_t$ is a martingale, which proves (b). (Here we need some conditions on the integrand to define the Itô integral, and in this theorem we assume that such conditions are satisfied.)

(iii) Note that, by the properties of a martingale, $\mathbb{E}[L_t\widetilde{X}_t] = \mathbb{E}[L_0\widetilde{X}_0] = X_0$, $0 \leq t \leq T$. This condition is implicitly given in the problem. Consider $L^2(\Omega, \mathcal{F}_T)$ with the inner product defined by $(X, Y)_{L^2} = \mathbb{E}[XY]$ for $X, Y \in L^2$. Define $V = \{\phi : \mathbb{E}[\phi] = \mathbb{E}[L_T\phi] = 0\}$. Since $(1, \phi)_{L^2} = \mathbb{E}[\phi] = 0$ and $(L_T, \phi)_{L^2} = \mathbb{E}[L_T\phi] = 0$, the subspace V is spanned by the functions orthogonal to 1 and L_T. Using the Gram–Schmidt orthogonalization, we choose an orthonormal basis $\{\phi_i : i \geq 1\}$ for V. For $(b_1, \ldots, b_n) \in \mathbb{R}^n$ let $Y = \widetilde{X}_T + b_1\phi_1 + \cdots + b_n\phi_n$. Then $\mathbb{E}[Y] = \mathbb{E}[\widetilde{X}_T] = \mu$ and $\mathbb{E}[L_T Y] = \mathbb{E}[L_T\widetilde{X}_T] = X_0$. Define

$$f(b_1, \ldots, b_n) = \mathrm{Var}(\widetilde{X}_T + b_1\phi_1 + \cdots + b_n\phi_n) = \mathrm{Var}(Y) .$$

Then $f(b_1, \ldots, b_n) = \mathbb{E}[\widetilde{X}_T^2] + \sum_{i=1}^n b_i^2 + 2\sum_{i=1}^n b_i\, \mathbb{E}[\widetilde{X}_T\phi_i] - \mathbb{E}[\widetilde{X}_T]^2$. Since f takes its minimum $\mathrm{Var}(\widetilde{X}_T)$ at $(0, \ldots, 0)$, we have $\frac{\partial f}{\partial b_j}\Big|_{(0,\ldots,0)} = 0$, and hence $\mathbb{E}[\widetilde{X}_T\phi_j] = 0$, $1 \leq j \leq n$. Since the above holds for every n we have $\mathbb{E}[\widetilde{X}_T\phi] = 0$ for arbitrary $\phi \in V$. Hence $\widetilde{X}_T$ belongs to $V^\perp$. From part (ii) we obtain a system of linear equations

$$\mu = \mathbb{E}[\widetilde{X}_T] = \mathbb{E}[C_1 + C_2L_T] = C_1 + C_2$$
$$X_0 = \mathbb{E}[L_T\widetilde{X}_T] = \mathbb{E}[C_1L_T + C_2L_T^2] = C_1 + C_2\mathrm{e}^{\theta^2 T} ,$$

which has the desired solution.

(iv) Since $\{L_t\widetilde{X}_t\}_t$ is a martingale, using $\widetilde{X}_T = C_1\mathbf{1} + C_2L_T$ we obtain

$$\begin{aligned} L_t\widetilde{X}_t &= \mathbb{E}[L_T\widetilde{X}_T|\mathcal{F}_t] = \mathbb{E}[C_1L_T + C_2L_T^2|\mathcal{F}_t] \\ &= C_1\mathbb{E}[L_T|\mathcal{F}_t] + C_2\mathrm{e}^{\theta^2 T}\mathbb{E}\left[\mathrm{e}^{-2\theta W_T - \frac{1}{2}(2\theta)^2 T}\middle|\mathcal{F}_t\right] \\ &= C_1L_t + C_2\mathrm{e}^{\theta^2 T}\mathrm{e}^{-2\theta W_t - \frac{1}{2}(2\theta)^2 t} . \end{aligned}$$

To obtain the last equality we used the fact that L_t and $\mathrm{e}^{-2\theta W_t - \frac{1}{2}(2\theta)^2 t}$ are martingales. Now we divide both sides by L_t.

(v) Let $\widetilde{\pi}_t = \mathrm{e}^{-rt}\pi_t$ be the discounted efficient portfolio. By part (iv) the discounted wealth $\widetilde{X}_t$ obtained by the efficient investment satisfies

$$\mathrm{d}\widetilde{X}_t = C_2\mathrm{e}^{\theta^2 T}\mathrm{e}^{-\theta W_t - \frac{3}{2}\theta^2 t}(-\theta)\,(\theta\mathrm{d}t + \mathrm{d}W_t) .$$

By part (i), we have $\mathrm{d}\widetilde{X}_t = \widetilde{\pi}_t\sigma(\theta\mathrm{d}t + \mathrm{d}W_t)$. Now compare the coefficients to obtain $\widetilde{\pi}_t$.

(vi) By part (iii) we have

$$\begin{aligned} &(1 - \mathrm{e}^{\theta^2 T})^2\,(C_1^2 + 2C_1C_2 + C_2^2\,\mathrm{e}^{\theta^2 T} - \mu^2) \\ &= (X_0 - \mu\mathrm{e}^{\theta^2 T})^2 + 2(X_0 - \mu\mathrm{e}^{\theta^2 T})(\mu - X_0) + (\mu - X_0)^2\mathrm{e}^{\theta^2 T} - (1 - \mathrm{e}^{\theta^2 T})^2\mu^2 \\ &= -(X_0 - \mu)^2(1 - \mathrm{e}^{\theta^2 T}) . \end{aligned}$$

Hence $\sigma_{\widetilde{X}_T}^2 = \mathbb{E}[\widetilde{X}_T^2] - \mu^2 = C_1^2 + 2C_1C_2 + C_2^2\,\mathrm{e}^{\theta^2 T} - \mu^2 = \frac{(\mu - X_0)^2}{\mathrm{e}^{\theta^2 T} - 1}$.

Problems of Chap. 22

22.1 Assume that $R \geq 0$. Let $f(x) = (1 + \frac{R}{x})^x$, $x > 0$. To show that $f(x)$ is monotonically increasing, we prove that $g(x) = \log f(x)$ is monotonically increasing. Since

$$g'(x) = \log(1 + \frac{R}{x}) + x\frac{-\frac{R}{x^2}}{1 + \frac{R}{x}} = \log(1 + \frac{R}{x}) - \frac{R}{x + R} ,$$

we let $h(R) = \log(1 + \frac{R}{x}) - \frac{R}{x+R}$ for fixed $x > 0$, and observe that

$$h'(R) = \frac{\frac{1}{x}}{1 + \frac{R}{x}} - \frac{(x + R) - R}{(x + R)^2} = \frac{1}{x + R} - \frac{x}{(x + R)^2} = \frac{R}{(x + R)^2} > 0$$

for $R > 0$ and $h(0) = 0$. Since $h(R) > 0$ for $R > 0$, we have $g'(x) > 0$ for every $x > 0$, and $f(x)$ is monotonically increasing.

22.2 Since $\frac{\partial \Pi}{\partial r} = 0$ after Π is hedged by choosing Δ as given by (22.5), we obtain $\frac{1}{\Pi}\frac{\partial \Pi}{\partial t} + \frac{1}{2}\frac{1}{\Pi}\frac{\partial^2 \Pi}{\partial r^2}\sigma^2 = r$ from the fundamental equation for bond pricing.

Problems of Chap. 23

23.1 Recall that the pricing equation for P is given by

$$\frac{\partial P}{\partial t} + (\mu - \lambda\sigma)\frac{\partial P}{\partial r} + \frac{1}{2}\sigma^2\frac{\partial^2 P}{\partial r^2} = rP . \tag{F.3}$$

The boundary condition for a zero-coupon bond is given by

$$P(r, T, T) = 1 . \tag{F.4}$$

(i) We try to find a solution of the form

$$P(r, t, T) = A(T - t)\,\mathrm{e}^{-rB(T-t)} \tag{F.5}$$

for some functions A and B. Then

$$\frac{\partial P}{\partial r} = -A(T-t)B(T-t)\,\mathrm{e}^{-rB(T-t)}\,,$$

$$\frac{\partial^2 P}{\partial r^2} = A(T-t)B(T-t)^2\,\mathrm{e}^{-rB(T-t)}\,,$$

$$\frac{\partial P}{\partial t} = -A'(T-t)\mathrm{e}^{-rB(T-t)} + rA(T-t)B'(T-t)\,\mathrm{e}^{-rB(T-t)}\,.$$

After cancellation of common factors, the pricing equation becomes

$$-A' + rAB' - (\mu - \lambda\sigma)AB + \frac{1}{2}\sigma^2 AB^2 = rA\,,$$

which must hold for every r. Hence $-A' - (\mu - \lambda\sigma)AB + \frac{1}{2}\sigma^2 AB^2 = 0$ and

$$A(B'-1) = 0\,. \tag{F.6}$$

From (F.4) and (F.5) we have $A(0)\mathrm{e}^{-rB(0)} = 1$ for every r. Hence $A(0) = 1$ and $B(0) = 0$. Thus we have $B(T-t) = T-t$ from (F.6). Then A satisfies

$$\frac{A'(T-t)}{A(T-t)} = -(\mu-\lambda\sigma)(T-t) + \frac{1}{2}\sigma^2(T-t)^2$$

with $A(0) = 1$. Hence $A(T-t) = \exp(-\frac{1}{2}(\mu-\lambda\sigma)(T-t)^2 + \frac{1}{6}\sigma^2(T-t)^3)$.

(ii) The result is not realistic because the yield is negative for large T.

23.2

(i) Use $\mathrm{Cov}(r_t, r_s) = \sigma_0 \mathrm{e}^{-\alpha(t+s)} + \sigma^2 \int_0^s \mathrm{e}^{-\alpha(s-u)}\mathrm{e}^{-\alpha(t-u)}\mathrm{d}u$.
(ii) Use $\mathbb{E}[\int_s^t r_u \mathrm{d}u \,|\, \mathcal{F}_s] = \int_s^t \mathbb{E}[r_u|\mathcal{F}_s]\mathrm{d}u$.
(iii) To find the limiting distribution, we may take $t \to \infty$, or if we want to find it directly, then we take $\frac{\partial p}{\partial t} = 0$ in the Kolmogorov forward equation, and solve

$$0 = -\frac{\partial}{\partial x}(\alpha(\bar{r}-x)p) + \frac{\partial^2}{\partial x^2}\left(\frac{1}{2}\sigma^2 p\right)\,.$$

Hence

$$C = -\alpha(\bar{r}-x)p + \frac{\partial}{\partial x}\left(\frac{1}{2}\sigma^2 p\right)$$

for some constant C. If we assume that a solution satisfies the condition that $\lim_{x\to+\infty} xp(x) = 0$ and $\lim_{x\to+\infty} p'(x) = 0$, then $C = 0$. From

$$\frac{p'(x)}{p(x)} = \frac{\alpha(\bar{r} - x)}{\frac{1}{2}\sigma^2} ,$$

we obtain

$$p(x) = \frac{1}{\sqrt{\pi\frac{\sigma^2}{\alpha}}} \exp\left(-\frac{(x-\bar{r})^2}{\frac{\sigma^2}{\alpha}}\right) .$$

The coefficient is chosen so that $\int_{-\infty}^{\infty} p(x)\mathrm{d}x = 1$.

23.3 For the limiting probability density of the interest rate, we take $\frac{\partial p}{\partial t} = 0$ in the Kolmogorov forward equation, and solve

$$0 = -\frac{\partial}{\partial x}(\alpha(\bar{r} - x)p(x)) + \frac{\partial^2}{\partial x^2}\left(\frac{1}{2}\sigma^2 xp(x)\right) .$$

As in Exercise 23.2(iii), we solve

$$C_1 = -\alpha(\bar{r} - x)p(x) + \frac{\partial}{\partial x}\left(\frac{1}{2}\sigma^2 xp(x)\right) .$$

Hence

$$C_1 = -\alpha(\bar{r} - x)p(x) + \frac{1}{2}\sigma^2 p(x) + \frac{1}{2}\sigma^2 xp'(x) .$$

If we assume that $p(0) = 0$ and $\lim_{x\to 0} xp'(x) = 0$, then $C_1 = 0$. From

$$\frac{p'(x)}{p(x)} = \frac{\alpha\bar{r} - \frac{1}{2}\sigma^2 - \alpha x}{\frac{1}{2}\sigma^2 x} ,$$

we obtain

$$\log p(x) = \left(\frac{2\alpha\bar{r}}{\sigma^2} - 1\right)\log|x| - \frac{2\alpha}{\sigma^2}x + C_2$$

for x such that $p(x) > 0$. Thus $p(x) = Cx^{(2\alpha\bar{r})/\sigma^2 - 1}\mathrm{e}^{-(2\alpha/\sigma^2)x}$, $x \geq 0$. If $2\alpha\bar{r} = \sigma^2$, then the limit is an exponential function.

23.4 Let $\tau = T - t$ be fixed. For the sake of notational convenience, we shall write $R(t, t+\tau)$ and $P(t, t+\tau)$ in place of $R(t,T)$ and $P(t,T)$, respectively. From Definition 22.8(i),

$$R(t, t+\tau) = -\frac{\log P(t, t+\tau)}{\tau} = -\frac{A(t, t+\tau)}{\tau} + r_t \, .$$

Hence $\mathrm{d}R(t, t+\tau) = \mathrm{d}r_t$ and

$$\mathrm{Var}(\mathrm{d}R(t, t+\tau)) = \mathrm{Var}(\mathrm{d}r_t) = \mathrm{Var}(\sigma \mathrm{d}\widetilde{W}_t) = \sigma^2 \mathrm{d}t \, .$$

Problems of Chap. 26

26.1 $\gamma_k = \mathrm{e}^{-2\pi \mathrm{i} k \theta_0} = (-1)^k$, and $F(\theta) = 0$, $\theta < \frac{1}{2}$, and $F(\theta) = 1$, $\theta \geq \frac{1}{2}$.

26.2 This is an example of a discrete spectrum, i.e., the measure $\mathrm{d}\nu$ is discrete. First, note that

$$\begin{aligned}
\mathbb{E}[x_t] &= \int \cdots \int_{[0,1]^n} \sum_{i=1}^{L} a_i \sin(2\pi\theta_i t + u_i)\, \mathrm{d}u_1 \cdots \mathrm{d}u_n \\
&= \sum_{i=1}^{L} \int_0^1 a_i \sin(2\pi\theta_i t + u_i)\, \mathrm{d}u_i = 0 \, .
\end{aligned}$$

Now for the autocovariance, note that

$$\begin{aligned}
&\mathbb{E}[x_t x_{t+k}] \\
&= \int \cdots \int \sum_{i=1}^{L} a_i \sin(2\pi\theta_i t + u_i) \sum_{j=1}^{L} a_j \sin(2\pi\theta_j (t+k) + u_j)\, \mathrm{d}u_1 \cdots \mathrm{d}u_n \\
&= \int \cdots \int \sum_{i=1}^{L} a_i^2 \sin(2\pi\theta_i t + u_i) \sin(2\pi\theta_j (t+k) + u_i)\, \mathrm{d}u_1 \cdots \mathrm{d}u_n \\
&+ \int \cdots \int \sum_{i \neq j} a_i a_j \sin(2\pi\theta_i t + u_i) \sin(2\pi\theta_j (t+k) + u_j)\, \mathrm{d}u_1 \cdots \mathrm{d}u_n \\
&= \int_0^1 \sum_{i=1}^{L} a_i^2 \sin(2\pi\theta_i t + u_i) \sin(2\pi\theta_i (t+k) + u_i)\, \mathrm{d}u_i \\
&+ \int_0^1 \int_0^1 \sum_{i \neq j} a_i a_j \sin(2\pi\theta_i t + u_i) \sin(2\pi\theta_j (t+k) + u_j)\, \mathrm{d}u_i \mathrm{d}u_j \, .
\end{aligned}$$

Hence

$$\begin{aligned}\mathbb{E}[x_t x_{t+k}] &= \sum_{i=1}^{L} a_i^2 \int_0^1 \frac{\cos(2\pi\theta_i k) - \cos(2\pi\theta_i(2t+k) + 2u_i)}{2}\,\mathrm{d}u_i \\ &\quad + \sum_{i\neq j} a_i a_j \int_0^1 \sin(2\pi\theta_i t + u_i)\,\mathrm{d}u_i \int_0^1 \sin(2\pi\theta_j(t+k) + u_j)\,\mathrm{d}u_j \\ &= \sum_{i=1}^{L} \frac{1}{2} a_i^2 \cos(2\pi\theta_i k)\ .\end{aligned}$$

Thus $\gamma_0 = \frac{1}{2}\sum_{i=1}^{L} a_i^2$, $\gamma_k = \sum_{i=1}^{L} b_i \cos(2\pi\theta_i k)$, $k \neq 0$, for some b_i, and

$$\gamma_k = \int_0^1 \cos(2\pi\theta k)\mathrm{d}F(\theta) = \int_0^1 \mathrm{e}^{2\pi\mathrm{i}\theta k}\mathrm{d}F(\theta)$$

where $F(\theta)$ is a monotonically increasing step function with jumps of size b_i at $\theta = \theta_i$. Note that the discrete measure $\mathrm{d}F$ has point masses of size b_i at $\theta = \theta_i$ for $1 \leq i \leq L$.

26.3 Since

$$\begin{aligned}x_{t+k} &= \phi x_{t+k-1} + \varepsilon_{t+k} \\ &= \phi(\phi x_{t+k-2} + \varepsilon_{t+k-1}) + \varepsilon_{t+k} \\ &= \ \vdots \\ &= \phi^k x_t + \sum_{i=0}^{k-1} \phi^i \varepsilon_{t+k-i}\ ,\end{aligned}$$

we have $\gamma(k) = \gamma(0)\phi^{|k|}$ and

$$f(\theta) = \sum_{k=-\infty}^{\infty} \gamma(k)\mathrm{e}^{2\pi\mathrm{i}k\theta} = \gamma(0)\frac{1-\phi^2}{1-2\phi\cos 2\pi\theta + \phi^2} \geq 0\ ,$$

for $0 \leq \theta \leq 1$.

26.4 $\mathbb{E}[x_{t+1}|x_t, x_{t-1}, \ldots, x_{t-p+1}] = \phi_1 x_t + \cdots + \phi_p x_{t-p+1}$.

26.6 Since $(\sigma^*)^2\delta t = \frac{1}{m-1}\sum_{i=1}^{m} u_{n-i}^2$, define

$$\sigma^* = \left(\frac{1}{\delta t}\frac{1}{m-1}\sum_{i=1}^{m} u_{n-i}^2\right)^{1/2}.$$

26.7 Since $u_n = (\mu - \frac{1}{2}\sigma^2)\delta t + \sigma\sqrt{\delta t}\xi_n$ where $\xi_n \sim N(0,1)$ for $n \geq 1$, $\bar{u}$ is normally distributed with mean $(\mu - \frac{1}{2}\sigma^2)\delta t$ and variance $\sigma^2\delta t/m$.

26.8 Choose $p + q = 1$ and $\alpha + \beta = 1$.

Problems of Chap. 27

27.1 We find $\phi : [0,1] \to [0,\infty)$ such that $Y = \phi(U)$ has exponential distribution, i.e., its pdf is given by $f_Y(y) = \mathrm{e}^{-y}$, $y \geq 0$. Since

$$\mathrm{e}^{-\phi(x)} = \begin{cases} \dfrac{1}{\phi'(x)}\,, & 0 \leq x \leq 1\,, \\ 0\,, & \text{elsewhere}\,, \end{cases}$$

we have $\mathrm{e}^{-\phi(x)}\phi'(x) = 1$ on $[0,1]$, and $-\mathrm{e}^{-\phi(x)} = x + C$ for some constant C. If $\phi(0) = 0$, then $C = -1$, $\phi(x) = -\log(-x+1)$, and hence $Y = -\log(1-U)$. If $\phi(0) = \infty$, then $\phi'(x) < 0$, $\mathrm{e}^{-\phi(x)}\phi'(x) = -1$, $C = 0$, and hence $Y = -\log U$.

27.2 Let U denote a random variable uniformly distributed in $(0,1)$. Let $F(x)$ denote the cumulative distribution function of a random variable generating the random numbers with the given distribution. Then

$$F(x) = \frac{1}{1 - N(x_0)} \mathbf{1}_{[x_0,+\infty)}(x)\,(N(x) - N(x_0))\,.$$

Then $Y = F^{-1}(U)$ and X have the identical distribution. Here we regard F as a one-to-function defined on $[x_0,+\infty)$. (For the proof, see Exercise 4.6.) Since

$$U = F(Y) = \frac{N(Y) - N(x_0)}{1 - N(x_0)}\,,$$

we have $Y = N^{-1}((1 - N(x_0))\,U + N(x_0))$. Thus, we first generate uniformly distributed numbers u_i, $i \geq 1$, using a random number generator, and compute $N^{-1}((1 - N(x_0))\,u_i + N(x_0))$, $i \geq 1$.

Problems of Chap. 28

28.1 Since $1 - U$ is also uniformly distributed in $(0,1)$, we have

$$\Pr(F^{-1}(1-U) \leq x) = \Pr(1 - U \leq F(x)) = \Pr(U \leq F(x)) = F(x)\,.$$

Hence $F^{-1}(1-U)$ has the cumulative distribution function F. Since the cdf of $F^{-1}(U)$ is $F(x)$ by Exercise 4.6, we see that $F^{-1}(U)$ and $F^{-1}(1-U)$ have the same cdf. To show that the correlation is negative, use Corollary 28.1.

28.2

(i) Put $y=\sqrt{x}$. Then $y^2=x$, $2y\,\mathrm{d}y=\mathrm{d}x$. Hence $\int_0^1 \mathrm{e}^{\sqrt{x}}\mathrm{d}x=\int_0^1 \mathrm{e}^y 2y\,\mathrm{d}y$, which is equal to $\left[\mathrm{e}^y 2y\right]_0^1-\int_0^1 \mathrm{e}^y 2\,\mathrm{d}y=2\mathrm{e}-2(\mathrm{e}-1)=2$.

(ii) To estimate $\mathbb{E}[\mathrm{e}^{\sqrt{U}}]$ by the Monte Carlo method, we consider $X=\frac{1}{2}(\mathrm{e}^{\sqrt{U}}+\mathrm{e}^{\sqrt{1-U}})$ where $U\sim U(0,1)$. Note that

$$\begin{aligned}\mathrm{Var}(X)&=\frac{\mathrm{Var}(\mathrm{e}^{\sqrt{U}})+\mathrm{Var}(\mathrm{e}^{\sqrt{1-U}})+2\,\mathrm{Cov}(\mathrm{e}^{\sqrt{U}},\mathrm{e}^{\sqrt{1-U}})}{4}\\&=\frac{\mathrm{Var}(\mathrm{e}^{\sqrt{U}})+\mathrm{Cov}(\mathrm{e}^{\sqrt{U}},\mathrm{e}^{\sqrt{1-U}})}{2}\;.\end{aligned}$$

Note that

$$\mathbb{E}[\mathrm{e}^{\sqrt{U}}\mathrm{e}^{\sqrt{1-U}}]=\int_0^1 \mathrm{e}^{\sqrt{x}}\mathrm{e}^{\sqrt{1-x}}\mathrm{d}x\approx 3.8076$$

and

$$\mathrm{Cov}(\mathrm{e}^{\sqrt{U}},\mathrm{e}^{\sqrt{1-U}})=\mathbb{E}[\mathrm{e}^{\sqrt{U}}\mathrm{e}^{\sqrt{1-U}}]-\mathbb{E}[\mathrm{e}^{\sqrt{U}}]\,\mathbb{E}[\mathrm{e}^{\sqrt{1-U}}]\approx -0.1924\;.$$

By direct calculation, we have $\mathrm{Var}(\mathrm{e}^{\sqrt{U}})=\frac{\mathrm{e}^2-7}{2}\approx 0.1945$. Hence $\mathrm{Var}(X)\approx 0.001$, which is a great reduction in comparison with the variance of $\mathrm{e}^{\sqrt{U}}$.

28.3

(i) We estimate $\frac{b-a}{2}(\mathbb{E}[f(a+(b-a)U)]+\mathbb{E}[f(b+(a-b)U)])$. Note that $a+(b-a)U$ and $a+(b-a)(1-U)=b+(a-b)U$ are antithetic.

(ii) Since $\mathbb{E}[f(U)]=\int_0^1 f(x)\,\mathrm{d}x=\sum_{i=1}^n\int_{a_i}^{b_i} f(x)\,\mathrm{d}x$, we estimate

$$\sum_{i=1}^n \frac{b_i-a_i}{2}(\mathbb{E}[f(a_i+(b_i-a_i)U)]+\mathbb{E}[f(b_i+(a_i-b_i)U)])\;.$$

28.4 No. Due to the symmetry of the function under consideration, the antithetic variate method is equivalent to the standard Monte Carlo method.

28.5

(i) For the plots of the payoffs of straddles see Fig. 2.

(ii) In the second case where the payoff function of the straddle is close to being symmetric, the efficiency of the antithetic variate method is not good.

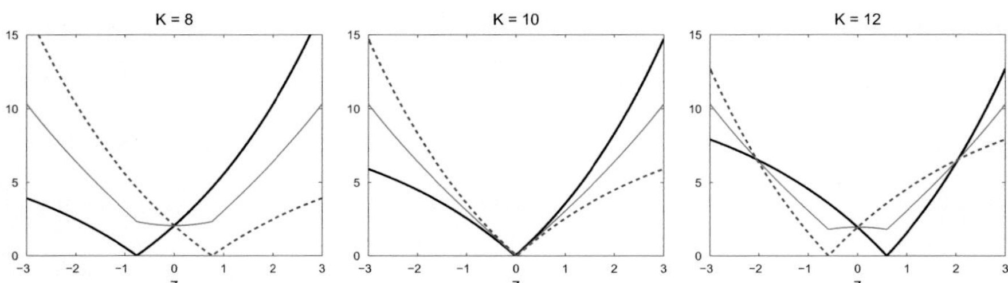

Fig. 2 Payoffs of straddles and the corresponding antithetic variates for $S_0 = 10$ and $K = 8, 10, 12$

28.6 Take $Y = \mathrm{e}^U$. Note that $\mathbb{E}[Y] = \int_0^1 \mathrm{e}^y \mathrm{d}y = \mathrm{e} - 1$. Using $Y = \mathrm{e}^U$ as a control variate, define $Z = \mathrm{e}^{\sqrt{U}} - (\mathrm{e}^U - (\mathrm{e} - 1))$.

28.7 The option price is equal to

$$(1 - N(x_0))\mathrm{e}^{-rT}\mathbb{E}[S_0\,\mathrm{e}^{\sigma\sqrt{T}Y+(r-\frac{1}{2}\sigma^2)T} - K]$$

where the random variable Y has the restricted normal distribution. The standard Monte Carlo method computes the average of payoff over the whole real line $(-\infty, \infty)$. Since the payoff is zero on the interval $(-\infty, x_0)$, the generated random numbers belonging to the interval are not used in actual computation and wasted. Therefore the evaluation based on the restricted normal distribution is more efficient than the standard Monte Carlo method. See (16.5) in Sect. 16.3 for the exact value of x_0.

28.8 Choose $g(x) = 5x^4$ and define a probability measure $\mathbb{Q}$ by $\frac{\mathrm{d}\mathbb{Q}}{\mathrm{d}\mathbb{P}} = 5x^4$. Note that $\mathbb{E}^{\mathbb{P}}[U^5] = \mathbb{E}^{\mathbb{Q}}\left[U^5 \frac{\mathrm{d}\mathbb{P}}{\mathrm{d}\mathbb{Q}}\right] = \mathbb{E}^{\mathbb{Q}}\left[\frac{1}{5}U\right]$. Also note that

$$\begin{aligned}
\mathrm{Var}^{\mathbb{Q}}\left(\frac{1}{5}U\right) &= \frac{1}{25}\left(\mathbb{E}^{\mathbb{Q}}\left[U^2\right] - \left(\mathbb{E}^{\mathbb{Q}}\left[U\right]\right)^2\right) \\
&= \frac{1}{25}\left\{\int_0^1 5x^6\,\mathrm{d}x - \left(\int_0^1 5x^5\,\mathrm{d}x\right)^2\right\} \\
&= \frac{1}{25}\left\{\frac{5}{7} - \frac{25}{36}\right\} = \frac{1}{35} - \frac{1}{36} = \frac{1}{1260} \approx 0.00079\,.
\end{aligned}$$

Hence the variance is reduced by $\frac{0.0631-0.00079}{0.0631} \approx 98.7\%$.

28.9 First, we sample $\widetilde{Z}$ from the normal distribution $N(\frac{m}{\sigma\sqrt{T}}, s^2)$ so that $\sigma\sqrt{T}\widetilde{Z} \sim N(m, \sigma^2 T s^2)$. Then the option price is given by

$$
\begin{aligned}
&\mathrm{e}^{-rT}\mathbb{E}^{\mathbb{Q}}[(S_T - K)^+] \\
&= \mathrm{e}^{-rT}\mathbb{E}^{\mathbb{Q}}\left[\left(S_0 \mathrm{e}^{(r-\frac{1}{2}\sigma^2)T+\sigma\sqrt{T}X} - K\right)^+\right] \quad \text{where } X \sim N(0,1) \quad \text{w.r.t. } \mathbb{Q} \\
&= \mathrm{e}^{-rT}\mathbb{E}^{\widetilde{\mathbb{Q}}}\left[\left(S_0 \mathrm{e}^{(r-\frac{1}{2}\sigma^2)T+\sigma\sqrt{T}X} - K\right)^+ \frac{\mathrm{d}\mathbb{Q}}{\mathrm{d}\widetilde{\mathbb{Q}}}\right], \ X \sim N(\frac{m}{\sigma\sqrt{T}}, s^2) \text{ w.r.t. } \widetilde{\mathbb{Q}}
\end{aligned}
$$

where

$$
\begin{aligned}
\frac{\mathrm{d}\mathbb{Q}}{\mathrm{d}\widetilde{\mathbb{Q}}} &= \frac{\frac{1}{\sqrt{2\pi}}\exp(-X^2/2)}{\frac{1}{\sqrt{2\pi s^2}}\exp\left(-(X - \frac{m}{\sigma\sqrt{T}})^2/(2s^2)\right)} \\
&= s\exp\left(-\frac{X^2}{2} + \frac{(X - \frac{m}{\sigma\sqrt{T}})^2}{2s^2}\right) \\
&= s\exp\left(-\frac{(\frac{m}{\sigma\sqrt{T}} + sZ)^2}{2} + \frac{Z^2}{2}\right)
\end{aligned}
$$

where $Z \sim N(0,1)$ with respect to $\widetilde{\mathbb{Q}}$. Thus the option price is equal to

$$
s\,\mathrm{e}^{-rT}\mathbb{E}\left[\left(S_0 \mathrm{e}^{(r-\frac{1}{2}\sigma^2)T+m+s\sigma\sqrt{T}Z} - K\right)^+ \exp\left(\frac{Z^2}{2} - \frac{(\frac{m}{\sigma\sqrt{T}} + sZ)^2}{2}\right)\right].
$$

Problems of Chap. 29

29.3 Take $x = \log S$. Then

$$
\frac{\partial V}{\partial S} = \frac{\partial V}{\partial x}\frac{\mathrm{d}x}{\mathrm{d}S} = \frac{\partial V}{\partial x}\frac{1}{S},
$$

$$
\frac{\partial^2 V}{\partial S^2} = \frac{\partial V}{\partial S}\left(\frac{\partial V}{\partial x}\frac{1}{S}\right) = \frac{\partial V}{\partial x}\left(\frac{\partial V}{\partial x}\frac{1}{S}\right)\frac{1}{S} = \left(\frac{\partial^2 V}{\partial x^2}\frac{1}{S} + \frac{\partial V}{\partial x}\left(-\frac{1}{S^2}\right)\mathrm{e}^x\right)\frac{1}{S},
$$

and hence

$$
S^2\frac{\partial^2 V}{\partial S^2} = \frac{\partial^2 V}{\partial x^2} - \frac{\partial V}{\partial x}.
$$

Thus the Black–Scholes–Merton equation is transformed into

$$\frac{\partial V}{\partial t} + \frac{1}{2}\sigma^2 \left(\frac{\partial^2 V}{\partial x^2} - \frac{\partial V}{\partial x} \right) + r\frac{\partial V}{\partial x} = rV \,.$$

Now let $u = \mathrm{e}^{-rt} V$. Then we have

$$\frac{\partial u}{\partial t} + \frac{1}{2}\sigma^2 \frac{\partial^2 u}{\partial x^2} + (r - \frac{1}{2}\sigma^2)\frac{\partial u}{\partial x} = 0 \,.$$

Applying BTCS we obtain

$$\frac{U_j^{i+1} - U_j^i}{k} + \frac{1}{2}\sigma^2 \frac{U_{j+1}^{i+1} - 2U_j^{i+1} + U_{j-1}^{i+1}}{h^2} + (r - \frac{1}{2}\sigma^2)\frac{U_{j+1}^{i+1} - U_{j-1}^{i+1}}{2h} = 0 \,.$$

Choose h and k satisfying $h^2 = \sigma^2 k$. Then

$$U_j^i = q\, U_{j+1}^{i+1} + (1-q)\, U_{j-1}^{i+1}$$

where

$$q = \frac{1}{2} + \frac{\sqrt{k}}{2\sigma}\left(r - \frac{1}{2}\sigma^2 \right) \,.$$

Since $U_j^i = \mathrm{e}^{-rik} V_j^i$ for every i, j, we have

$$V_j^i = \mathrm{e}^{-rk} \left(q\, V_{j+1}^{i+1} + (1-q)\, V_{j-1}^{i+1} \right) \,.$$

Glossary

Accrued Interest Interest on a bond that has accrued but has not yet been paid.

Agent Someone authorized to do business on behalf of a client.

American Option An option that can be exercised at any time before or on expiry.

Arbitrage An opportunity to make risk-free profit by buying and selling in different markets utilizing the disparity between market prices. When more broadly used, arbitrage means an opportunity to make a profit without risking any future loss.

Asian Option An option whose payoff at expiry depends on an average of the underlying asset prices over the life of the option.

Ask Price The price at which a dealer or market-maker offers to sell an asset. Offer price.

Asset A financial claim or a piece of property that has monetary value such as cash, shares and bonds.

At-the-Money Option An option for which the current market price of the underlying asset is equal to the strike price.

Barrier Option An option that becomes activated or ceases to exist when the underlying asset price hits the given barrier(s).

Basis Point 0.01%, that is, 0.0001.

Bermudan Option An option that can be exercised at a set of times unlike European options, which can be exercised only at the expiry and American options, which can be exercised any time. The name comes from the fact that Bermuda is geographically located between America and Europe.

Beta The percentage change in the price of a security for a 1% change in the market portfolio.

G.H. Choe, *Stochastic Analysis for Finance with Simulations*, Universitext,
DOI 10.1007/978-3-319-25589-7

Bid Price The price at which a dealer or market-maker buys an asset. The interest rate a dealer will pay to borrow funds in the money market.

Bid-Ask Spread The difference between the bid price and the ask price.

Binary Option An option whose payoff is either nothing or a certain amount of cash or shares of the underlying asset. Also called a digital option.

Binomial Tree A diagram for a possible movement of asset price over time in which asset price can move up or down in each time period.

Black–Scholes–Merton Equation The second order partial differential equation for the price of a European call option where two variables for partial differentiations represent time and asset price.

Black–Scholes–Merton Formula The formula for the price of a European call option given by the asset price, strike price, interest rate, time to expiry, and the volatility. The solution of the Black–Scholes–Merton equation.

Bond A debt security that promises to make payments periodically for a specified period of time.

Bund Bond issued by the Federal German government.

Broker An agent, either an individual or firm, who executes buy and sell orders submitted by an investor.

Call Option An financial contract that gives the holder of the option the right, but not the obligation, to buy the underlying asset at a specified price.

Cash Flows Cash payments to the holder of a security.

Collateral Cash or securities pledged to the lender to guarantee payment in the event that the borrower is not able to make payments. In the futures market a trader has to put up collateral called the initial margin against the possibility that he will lose money on the trade.

Common Stock A security that is a claim on the earnings and assets of a company with voting right. An ordinary share.

Convexity A measure of the curvature of the graph for the price of a bond plotted against the interest rate.

Consumption Spending by consumers on nondurable goods and services.

Contingent Claim A claim that can be made depending on whether one or more specified outcomes occur or do not occur in the future.

Counterparty The other party to a trade or contract.

Coupon Interest payment from a bond.

Coupon Bond A bond that periodically pays the owner fixed interest payments, called coupons, until the maturity date at which time a specified final amount, called the principal, is repaid.

Coupon Rate The amount of the yearly coupon payment expressed as a percentage of the face value of a bond.

Credit Risk The risk that arises from the possibility that the borrower might default.

Currency Paper money and coins.

Dealer A person who buys and sells securities, acting as principal in trading for its own account, as opposed to a broker who acts as an agent in executing orders on behalf of its clients. Most dealers act as brokers.

Default An event in which the issuer of a debt security is unable to make interest payments or pay off the amount owed when the security matures.

Delivery The act of the seller to supply the underlying asset to the buyer. Some derivatives contracts include the physical delivery of the underlying commodity or security.

Delta The rate of change in the price of a derivative due to the change in the underlying asset. It measures the sensitivity of the price of the derivative to the price of the underlying asset.

Delta Hedging Hedging a derivative position using the underlying asset whose amount is determined by delta.

Delta-Neutral An option is said to be delta-neutral if it is delta-hedged. It is protected against small changes in the price of the underlying asset.

Derivative A financial instrument whose value is determined or derived by the price of another asset.

Digital Option Another name for a binary option.

Discount Bond A bond that is bought at a price below its face value and whose face value is repaid at the maturity date. There is no interest payment, and hence it is also called a zero-coupon bond.

Discount Factor The rate of exchange between a given amount of money at time t and an amount of money at a later date T.

Diversification Investing in a portfolio of assets whose returns move in different ways so that the combined risk is lower than for the individual assets.

Dividend The payment per share made by a company to its shareholders. On ordinary shares the amount varies with the profitability of the firm. On preferred shares it is usually fixed.

Dow Jones Industrial Average (DJIA) US index based on the stock prices of 30 leading industrial companies traded on the NYSE. It is the simple average of share prices not weighted by market capitalization.

Down-and-In Option A knock-in option that pays the specified payoff only if the underlying asset price hits a barrier.

Down-and-Out Option A knock-out option that becomes worthless once the underlying asset price hits a barrier.

Drift The expected change per unit time period in an asset price.

Duration A measure of the value-weighted average life of a bond.

Equity Common stock. An equity holder is a part-owner of a company, and receives dividend payment if the company makes a profit.

Exchange Stock exchange. A marketplace in which securities are traded.

Exchange Rate The price of one currency expressed in terms of another currency.

European Option An option that can be exercised only on expiry date.

Exercise The purchase or sale of the underlying asset at the strike price specified in the option contract.

Exercise Price Another name for the strike price of an option. The amount that is paid for the underlying asset.

Exotic Option A derivative contract with a nonstandard characteristic.

Expiry The last date of a contract beyond which it becomes worthless if not exercised.

Face Value The principal or par value of a bond, which is repaid at maturity.

Fixed Income Security A security that pays a fixed amount on a regular basis until maturity. A generic term for bonds.

Forward Contract A financial contract on the price and quantity of an asset or commodity to sell or buy at a specified time in the future. An agreement between two parties which is not traded on an exchange.

Forward Price The price specified in advance in a forward contract for a given asset or commodity. The forward price makes the forward contract have zero value when the contract is written.

Futures Contract A financial contract similar to a forward contract that is traded publicly in a futures exchange. The buyer and the seller have to place a margin to avoid a default on the contract, and the futures contract is marked-to-market periodically.

Forward Rate Agreement (FRA) is an agreement that a forward rate will be exchanged for interest at the market rate to a notional amount during a certain future time period. The agreement is so structured that neither party needs to make an upfront payment.

Gamma The rate of change in delta of an option for a small change in the price of the underlying asset. It measures the sensitivity of delta to the price of the underlying asset.

Gilt A bond issued by the UK government. The bond was gilt-edged in older days.

Greeks The rates of changes, denoted by Greek letters except Vega, in the price of a financial derivative when the parameters such as the price of the underlying asset, time, interest rate and volatility change by one unit.

Hedge Ratio The ratio of the quantity of the hedge instrument to be bought or sold such as the futures contracts to the quantity of the asset to be hedged.

Hedge To protect oneself by eliminating the risk of loss in an investment.

Implied Volatility The volatility implied by the market price of an option. The price computed by applying the Black–Scholes–Merton formula using the implied volatility equals the market price.

Instrument A share, bond or some other tradable security.

Interest Rate The cost of borrowing money.

Interest Rate Risk The possibility of loss associated with changes in interest rates.

In-the-Money Option An option that would have positive value if exercised now. For example, if the asset price is above the strike price then a call option is in-the-money.

Intrinsic Value The value of an option if it can be exercised immediately.

Lambda Another name for vega. Note that Λ is an upside down V.

Lognormal Distribution A probability distribution of a random variable whose logarithm is normally distributed.

Long Position A position that can profit from an increase in the price of an asset under consideration.

Margin A deposit required for both buyers and sellers of a futures contract. It eliminates or reduces the level of loss in case that the buyer or seller fails to meet the obligations of the contract.

Margin Call The required amount to be added to the margin account by the owner of a margined position. This can be due to a loss on the position or an increase in the margin requirement.

Market Capitalization The total value of a company on the stock market, which is equal to the current share price times the number of shares issued.

Market Maker A trader in stocks, bonds, commodities or derivatives who offers to buy at the bid price (lower) and sell at the offer price (higher) at the same time, thereby making a market.

Mark-to-Market Updated evaluation of a portfolio or a position based on the current market prices.

Maturity Time to the expiration date of a debt instrument.

Negotiable Security A security that can be traded freely after it is issued.

No Arbitrage Principle The principle that one cannot create positive value out of nothing, i.e., there is no free lunch.

Numeraire The units in which a payoff is denominated.

Offer Price The price at which a trader offers to sell an asset. The interest rate a dealer asks for lending funds in the money market. Ask price.

Open Outcry A trading system in which traders gather in one physical location and convey their offers to sell and buy by gesturing and shouting.

Out-of-the-Money Option An option that would have zero value if it is exercised now. For example, if the asset price is below the strike price then a call option is out-of-the-money.

Over-the-Counter Transactions Trades and deals that occur without the involvement of a regulated exchange. OTC in abbreviation.

Par The face or nominal value of a bond, normally repaid at maturity.

Par Bond A bond that is trading at par value.

Payoff The value of a position usually at expiry or maturity.

Physical Measure The probability measure defined by the movement of the asset price in the real market.

Position An investor who buys (sells) securities has a long (short) position.

Practitioner A person who works in finance industry in contrast to an academician who teaches at a college.

Preference Shares Preference stock. Preference shares pay a fixed dividend and do not carry voting rights.

Premium In the options market the premium is the price of an option. Sometimes it means additional amount above a certain reference level.

Present Value Today's value of a cash flow in the future.

Put-Call Parity A relationship between the prices of European call and put options.

Put Option A contract giving the holder the right, but not the obligation, to sell the underlying asset at a specified price.

Rebalancing Resetting of the weights of assets in a portfolio. It involves periodically buying or selling assets in the portfolio to maintain the desired level of asset allocation. For example, consider a portfolio consisting of assets A and B. Suppose that the original target asset allocation was 70% for the asset A and 30% for the asset B. If the asset A performed well in a given period and the weight for the asset A is now 80%, then we sell some units (or shares) of the asset A and buy the asset B in order to maintain the original target allocation.

Replication When a financial derivative can be duplicated by constructing a portfolio of consisting of a suitable combination of the underlying asset and the risk-free asset, we say that the portfolio replicates the derivative.

Risk The uncertainty associated with the return on an investment.

Risk-Neutral Measure The probability measure for which the expected return on the asset is the risk-free interest rate.

Risk Premium The difference between the expected return on an asset and the risk-free interest rate. The spread between the interest rate on bonds with default risk and the risk-free interest rate.

Rho The rate of change in value of a derivative due to a change in the interest rate, usually a basis change.

Rollover To delay the payment of a debt. To roll a position over one expiry or delivery month to a later month.

S&P 500 Standard and Poor's 500. An index of five hundred shares traded on the New York Stock Exchange (NYSE).

Security A tradable financial instrument such as a stock and bond.

Self-Financing Portfolio A portfolio without additional investment or withdrawal to manage.

Sharpe Ratio The ratio of the risk premium to the standard deviation of the return.

Short Position A position that can profit from a decrease in the price of an asset under consideration.

Short Selling A transaction in which an investor borrows stock from a broker, sells it in the market, then buys it back and returns it to the lender. A profit will be made if the price has fallen.

Sovereign Bond A bond issued by a government.

Split In stock split one share is divided into multiple shares without raising new capital.

Spot Price The price of a security for immediate delivery. However, the actual settlement and delivery may take place a few days later.

Spot Rate Zero coupon rate.

Spread The difference between two prices and interest rates. For example, the difference between the bid and offer prices for a dealer, or the difference between lending and funding interest rates for a commercial bank.

Stock A security that is a claim on the earnings and assets of a company.

Stock Exchange An organized and regulated market for trading securities. Some stock exchanges use open outcry trading method, while others use telecommunication systems to connect dealers.

Stock Index An average of the prices of a collection of stocks. The average is either equally weighted, or weighted with weights proportional to market capitalization. The most commonly used quoted index in the US are the Dow Jones Industrial Average (DJIA) and the S&P 500.

Strike Price The same as the exercise price.

Straddle The purchase of a call and a put with the same strike price and time to expiry.

Strangle The purchase of a call with a higher strike and a put with a lower strike and the same time to expiry.

Systematic Risk Undiversifiable or market risk.

Systemic Risk The risk that an event such as a bank failure might have a domino effect on the whole financial system.

Tenor Time to maturity.

Term Structure of Interest Rates Behavior of interest rates for a range of maturities.

Theta The rate of change in the value of a derivative due to the passage of time.

Tick Size The minimum amount of asset price movement allowed in a price quotation.

Tick Value The value of one tick movement in the quoted price on the whole contract size.

Time Value The difference between an option's price and its intrinsic value.

Time Value of Money This expression refers to the fact that money to be received in the future is worth less than the same amount of money today if interest rate is positive.

Trader A person who buys and sells securities, either for herself, or on behalf of someone else.

Transaction Cost The money spent in exchanging financial assets or goods.

Treasury Bill A short-term debt security issued by a government.

Treasury Note US Treasury notes have maturities between 1 and 10 years when issued.

Treasury Bond US Treasury bonds have maturities between 10 and 30 years when issued.

Underlying The asset that is referred to in a derivative contract to define the payoff. Its price determines the price of the derivative.

Up-and-In Option A knock-in option that is activated only when the asset price exceeds the barrier.

Up-and-Out Option A knock-out option that becomes worthless when the asset price exceeds the barrier.

Vanilla Option A standard option such as European calls and European puts.

Vega The rate of change in the price of a derivative due to a change in volatility.

Volatility The standard deviation of the continuously compounded return on an asset. A key parameter in option pricing.

Volatility Skew The difference in implied volatility between in-the-money and out-of-the-money options, which makes the implied volatility graph skewed.

Volatility Smile The phenomenon that the graph of the implied volatility for a range of strike prices looks like a smile when both in-the-money and out-of-the-money options have higher values of volatility than at-the-money options.

Volatility Surface A three-dimensional graph in which volatility is plotted against strike price and time to expiry.

Warrant A long-dated option as a security that can be freely traded.

Writer The seller of an option.

Yield A generic term for the return on an investment, or more specifically, the yield to maturity of a bond.

Yield Curve A plot of the interest rates for a given type of bond against time to maturity. It is upward-sloping if the rates on shorter maturity bonds are lower than those on longer maturity bonds, and downward-sloping when short-term yields are higher.

Yield to Maturity The interest rate that equates the present value of the payments received from a bond in the future with its current market price.

Zero Coupon Bond A discount bond without coupon payments.

Zero Coupon Rate Also spot rate. The interest rate that is applicable to a specific future date.

References

1. M. Avellaneda and P. Laurence, *Quantitative Modeling of Derivative Securities: From Theory to Practice*, Chapman & Hall/CRC, Boca Raton, 1999.
2. J. Baz and G. Chacko, *Financial Derivatives: Pricing, Applications, and Mathematics*, Cambridge University Press, Cambridge, 2004.
3. M. Baxter and A. Rennie, *Financial Calculus: An Introduction to Derivative Pricing*, Cambridge University Press, Cambridge, 1996.
4. P.L. Bernstein, *Capital Ideas: The Improbable Origins of Modern Wall Street*, Free Press, New York, 1992.
5. P.L. Bernstein, *Against the Gods: The Remarkable Story of Risk*, John Wiley & Sons, New York, 1996.
6. F. Black and M. Scholes, *The Pricing of options and corporate liabilities*, Journal of Political Economy, **81** (1973), 637–654.
7. D.J. Bennett, *Randomness*, Havard University Press, Cambridge, 1998.
8. N.H. Bingham and R. Kiesel, *Risk-Neutral Valuation: Pricing and Hedging of Financial Derivatives, 2nd ed.*, Springer, London, 2004.
9. T. Bjork, *Arbitrage Theory in Continuous Time, 3rd ed.*, Oxford University Press, Oxford, 2009.
10. T. Bollerslev, *Generalized autoregressive conditional heteroskedasticity*, Journal of Econometrics, **31** (1986), 307–327.
11. G.E.P. Box and M.E. Muller, *A note on the generation of random normal deviates*, The Annals of Mathematical Statistics, **29** (1958), 610–611.
12. P. Brandimarte, *Numerical Methods in Finance*, John Wiley & Sons, New York, 2002.
13. P. Brémaud, *Fourier Analysis and Stochastic Processes*, Springer, Cham, 2014.
14. D. Brigo and F. Mercurio, *Interest Rate Models - Theory and Practice, 2nd ed.*, Springer-Verlag, Berlin, 2006.
15. Z. Brzeźniak and T. Zastawniak, *Basic Stochastic Processes*, Springer-Verlag, Berlin, 1999.
16. P.W. Buchen and M. Kelly, *The maximum entropy distribution of an asset inferred from option prices*, Journal of Financial and Quatitative Analysis, **31** (1996), 143–159.
17. S.J. Byun and I.J. Kim, *Optimal exercise boundary in a binomial option pricing model*, Journal of Financial Engineering, **3** (1994), 137–158.
18. A.J.G. Cairns, *Interest Rate Models: An Introduction*, Princeton University Press, Princeton, 2004.
19. M. Capinski and T. Zastawniak, *Mathematics for Finance – An Introduction to Financial Engineering*, Springer-Verlag, London, 2003.

G.H. Choe, *Stochastic Analysis for Finance with Simulations*, Universitext,
DOI 10.1007/978-3-319-25589-7

20. F. Chang, *Stochastic Optimaization in Continuous Time*, Cambridge University Press, Cambridge, 2004.
21. G. Choe, *Computational Ergodic Theory*, Springer, Heidelberg, 2005.
22. J. Cox, J. Ingersoll and S. Ross, *A theory of the term structure of interest rates*, Econometrica, **53** (1985), 385–408.
23. J.C. Cox, S.A. Ross and M. Rubinstein, *Option pricing: a simplified approach*, Journal of Financial Economics, **7** (1979), 229–263.
24. J. Cvitanić and F. Zapatero, *Introduction to the Economics and Mathematics of Financial Markets*, MIT Press, Cambridge, 2004.
25. M. Davis and A. Etheridge, *Louis Bachelier's Theory of Speculation: The Origins of Modern Finance*, Priceton University Press, Priceton, 2006.
26. J.L. Doob, *Stochastic Processes*, Wiley, New York, 1953.
27. W. Enders, *Applied Econometric Time Series, 3rd ed.*, John Wiley & Sons, Hoboken, 2004.
28. R. Engel, *Autoregressive conditional heteroskedasticity with estimates of the variance of UK inflation*, Econometrica, **50** (1982), 987–1008.
29. A. Etheridge, *A Course in Financial Calculus*, Cambridge University Press, Cambridge, 2002.
30. D. Filipović, *Term-Structure Models: A Graduate Course*, Springer, Berlin, 2009.
31. G.B. Folland, *Real Analysis: Modern Techniques and Their Applications, 2nd ed.*, John Wiley & Sons, New York, 1999.
32. J.D. Hamilton, *Time Series Analysis*, Princeton University Press, Princeton, 1994.
33. R.W. Hamming, *The Art of Probability*, Addison-Wesley, Redwood City, 1991.
34. J.M. Harrison and D. Kreps, *Martingales and arbitrage in multiperiod securities markets*, Journal of Economic Theory, **20** (1979), 381–408.
35. J.M. Harrison and S. Pliska, *Martingales and stochastic integrals in the theory of continuous trading*, Stochastic Processes and Their Applications, **11** (1981), 215–260.
36. E.G. Haug, *The Complete Guide to Option Pricing Formulas, 2nd ed.*, McGraw-Hill, New York, 2007.
37. H. Helson, *Harmonic Analysis*, Addison-Wesley, Reading, 1983; revised 2nd ed., Hindustan Book Agency and Helson Publishing Co., 1995.
38. D. Higham, *An Introduction to Financial Option Valuation: Mathematics, Stochastics and Computation*, Cambridge Unversity Press, Cambridge, 2004.
39. M.W. Hirsh and S. Smale, *Differential Equations, Dynamical Systems, and Linear Algebra*, Academic Press, San Diego, 1974.
40. T. Ho and S.B. Lee, *Term structure movements and pricing interest rate contingent claims*, Journal of Finance, **41** (1986), 1011–1029.
41. J. Hull, *Options, Futures, and Other Derivatives, 8th ed.*, Prentice Hall, Upper Saddle River, 2011.
42. J. Hull and A. White, *Pricing interest-rate-derivative securities*, The Review of Financial Studies, **3** (1990), 573–392.
43. P. James, *Option Theory*, John Wiley & Sons, Hoboken, 2003.
44. M. Joshi, *The Concepts and Practice of Mathematical Finance*, Cambridge University Press, Cambridge, 2003.
45. I. Karatzas and S.E. Shreve, *Brownian Motion and Stochastic Calculus, 2nd ed.*, Springer, New York, 1998.
46. Y. Katznelson, *An Introduction to Harmonic Analysis*, Dover, New York, 1976.
47. I.J. Kim, *The analytic valuation of American options*, The Review of Financial Studies, **3** (1990), 547–572.
48. T.S. Kim and E. Omberg, *Dynamic non-myopic portfolio behavior*, The Review of Financial Studies, **9** (1996), 141–161.
49. F.C. Klebaner, *Introduction to Stochastic Calculus with Applications, 2nd ed.*, Imperial College Press, London, 2005.
50. P.E. Kloeden and E. Platen, *Numerical Solution of Stochastic Differential Equations*, Springer-Verlag, Berlin, 1999.

51. D. Knuth, *The Art of Computer Programming, Vol. 2, 3rd ed.*, Addison-Wesley, Reading, 1997.
52. A.N. Kolmogorov, *Grundbegriffe der Wahrscheinlichkeitsrechnung*, Springer-Verlag, Berlin, 1933.
53. L.H. Koopmans, *The Spectral Analysis of Time Series*, Academic Press, New York, 1974.
54. R. Korn and E. Korn, *Option Pricing and Portfolio Optimization: Modern Methods of Financial Mathematics*, American Mathematical Society, Providence, 2001.
55. E. Kreyszig, *Introductory Functional Analysis with Applications*, John Wiley & Sons, New York, 1978.
56. M.P. Kritzman, *Puzzles of Finance: Six Practical Problems and their Remarkable Solutions*, John Wiley & Sons, New York, 2000.
57. S.S.S. Kumar, *Financial Derivatives*, New Delhi, Prentice Hall India, 2007.
58. H. Kuo, *Introduction to Stochastic Integration*, Springer, New York, 2006.
59. T.L. Lai and H. Xing, *Statistical Models and Methods for Financial markets*, Springer, New York, 2008.
60. F.A. Longstaff and E.S. Schwartz, *Valuing American options by simulation: a simple least-squares approach*, The Review of Financial Studies, **14** (2001), 113–147.
61. R.L. McDonald, *Derivatives Markets, 3rd ed.*, Pearson, 2013.
62. M. Matsumoto and T. Nishimura, *Mersenne twister: a 623-dimensionally equidistributed uniform pseudo-random number generator*, ACM Transactions on Modeling and Computer Simulation, **8** (1998), 3–30.
63. R.C. Merton, *Lifetime portfolio selection under uncertainty: the continuous-time case*, Review of Economics and Statistics, **51** (1969), 247–257.
64. R.C. Merton, *Optimum consumption and portfolio rules in a continuous-time model*, Journal of Economic Theory, **3** (1971), 373–413.
65. R.C. Merton, *Theory of rational option pricing*, Bell Journal of Economics and Management Science, **4** (1973), 141–183.
66. R.C. Merton, *Continuous-Time Finance*, Blackwell Publishing, Oxford, 1990.
67. P.A. Meyer, *A decomposition theorem for supermartingales*, Illinois Journal of Mathematics, **6** (1962), 193–205.
68. P.A. Meyer, *Decomposition of supermartingales: the uniqueness theorem*, Illinois Journal of Mathematics, **7** (1963), 1–17.
69. G.N. Milstein, *Approximate integration of stochastic differential equations*, Theory of Probability and Its Applications, **19** (1974), 557–562.
70. C.B. Moler, *Numerical Computing with Matlab*, Society for Industrial and Applied Mathematics, Philadelphia, 2004.
71. P. Mörters and Y. Peres, *Brownian Motion*, Cambridge University Press, Cambridge, 2010.
72. R.J. Muirhead, *Aspects of Multivariate Statistical Theory, 2nd ed.*, New York, Wiley, 2005.
73. E. Nelson, *Dynamical Theories of Brownian Motion, 2nd ed.*, Priceton University Press, Princeton, 2001.
74. B. Oksendal, *Stochastic Differential Equations: An Introduction with Applications, 6th ed.*, Springer, Berlin, 2003.
75. A. Pascucci, *PDE and Martingale Methods in Option Pricing*, Springer, Berlin, 2011.
76. I. Peterson, *The Jungle of Randomness: A Mathematical Safari*, John Wiley & Sons, New York, 1998.
77. S.R. Pliska, *Introduction to Mathematical Finance: Discrete Time Models*, Blackwell Publishers, Oxford, 1997.
78. S. Park and K. Miller, *Random number generators: good ones are hard to find*, Communications of the Association for Computing Machinery **31** (1988), 1192–1201.
79. M.B. Priestley, *Spectral Analysis and Time Series, Vol. 2: Multivariate Series, Prediction and Control*, Academic Press, London, 1981.
80. P.E. Protter, *Stochastic Integration and Differential Equations, 2nd ed.*, Springer, Berlin, 2005.
81. S.I. Resnick, *A Probability Path*, Birkhäuser, Boston, 1998.

82. A.J. Roberts, *Elementary Calculus of Financial Mathematics*, Society for Industrial and Applied Mathematics, Philadelphia, 2009.
83. L.C.G. Rogers, *Optimal Investment*, Springer, Berlin, 2013.
84. L.C.G. Rogers and D. Williams, *Diffusions, Markov Processes and Martingales, Vol. 1*, Cambridge University Press, Cambridge, 2000.
85. L.C.G. Rogers and D. Williams, *Diffusions, Markov Processes and Martingales, Vol. 2*, Cambridge University Press, Cambridge, 2000.
86. G.G. Roussas, *An Introduction to Measure-Theoretic Probability, 2nd Ed.*, Academic Press, Waltham, 2014.
87. H.L. Royden, *Real Analysis, 3rd ed.*, Macmillan, New York, 1989.
88. W. Rudin, *Real and Complex Analysis, 3rd ed.*, McGraw-Hill, New York, 1986.
89. D. Rupert, *Statistics and Finance: An Introduction*, Springer, New York, 2004.
90. R.L. Schilling, *Measures, Integrals and Martingales*, Cambridge University Press, Cambridge, 2005.
91. J.R. Schott, *Matrix Analysis for Statistics*, John Wiley & Sons, Hoboken, 2005.
92. R. Seydel, *Tools for Computational Finance, 3rd ed.*, Springer, Berlin, 2006.
93. S.E. Shreve, *Stochastic Calculus for Finance I: The Binomial Asset Pricing Model*, Springer, New York, 2004.
94. S.E. Shreve, *Stochastic Calculus for Finance II: Continuous Time Models*, Springer, New York, 2004.
95. R. Shukla and M. Tomas, *Black–Scholes option pricing formula*, retrieved from http://myweb.whitman.syr.edu/rkshukla/Research/BS.pdf
96. G.F. Simmons, *Differential Equations with Applications and Historical Notes, 2nd ed.*, McGraw-Hill, New York, 1991.
97. R.S. Tsay, *Analysis of Financial Time Series, 2nd ed.*, John Wiley & Sons, Hoboken, 2005.
98. Ö. Uğur, *An Introduction to Computational Finance*, Imperial College Press, London, 2009.
99. O.A. Vasicek, *An equilibrium characterization of the term structure*, Journal of Financial Economics **5** (1977), 177–188.
100. P. Veronesi, *Fixed Income Securiies: Valuation, Risk, and Risk Management*, John Wiley & Sons, Hoboken, 2010.
101. W. Wei, *Time Series Analysis: Univariate and Multivariate Methods, 2nd ed.*, Pearson, Boston, 2005.
102. U. Wiersema, *Brownian Motion Calculus*, John Wiley & Sons, Chichester, 2008.
103. D. Williams, *Probability with Martingales*, Cambridge University Press, Cambridge, 1991.
104. P. Wilmott, S. Howison and J. Dewynne, *The Mathematics of Financial Derivatives: A Student Introduction*, Cambridge University Press, Cambridge, 1995.
105. G. Woo, *The Mathematics of Natural Catastrophes*, Imperial College Press, London, 1999.

Index

G.H. Choe, *Stochastic Analysis for Finance with Simulations*, Universitext,
DOI 10.1007/978-3-319-25589-7

Printed in the United States
By Bookmasters